Polarization
Phenomena
in
Nuclear
Reactions

following the oral presentation of papers was not recorded, but speakers were asked to submit their comments in writing, and only such written comments are included in the Proceedings.

We wish to thank Prof. T. B. Clegg, Prof. S. E. Darden, and Dr. G. R. Plattner for their assistance in editing and proofreading. We are grateful to Mrs. Rosemary Bohannan for the excellent typing of the entire book.

H. H. Barschall
W. Haeberli

Madison, Wisconsin
October 1970

Just as we were preparing these Proceedings for the printer, we received the sad news of the death of our friend, teacher, and colleague Paul Huber.

Paul Huber was one of the pioneers in the study of Polarization Phenomena in Nuclear Reactions; he first observed polarized neutrons from nuclear reactions in 1953 and he guided the construction of one of the earliest sources of polarized ions in 1960. He was the organizer of the first symposium of this series and the guiding spirit behind the second and third symposia. He will be remembered and missed at future symposia.

H. H. Barschall
W. Haeberli

February 1971

The Third International Symposium on Polarization Phenomena in Nuclear Reactions took place from August 31 to September 4, 1970, at the University of Wisconsin in Madison. The first symposium in this series was held in 1960 in Basel, Switzerland, and the Proceedings were published as Supplement VI to *Helvetica Physica Acta* (Birkhäuser Verlag, Basel, 1961). The Proceedings of the second symposium, which was held in 1965 in Karlsruhe, Germany, were published as Supplement 12 to *Experientia* (Birkhäuser Verlag, Basel, 1966). Publication of the Proceedings of the Madison Conference is supported by the National Science Foundation and the University of Wisconsin. In the present volume, the Proceedings of the Basel symposium are referred to as "First Polarization Symp."; the Proceedings of the Karlsruhe symposium, as "Second Polarization Symp."; and the Proceedings of the Madison symposium, as "Third Polarization Symp."

The conference was supported by the International Union of Pure and Applied Physics, by the United States Atomic Energy Commission, by the National Science Foundation, and by the University of Wisconsin. The Organizing Committee consisted of:

H. H. Barschall, University of Wisconsin—Madison
S. E. Darden, University of Notre Dame
W. Haeberli, University of Wisconsin—Madison
P. Huber, University of Basel
G. R. Satchler, Oak Ridge National Laboratory
H. A. Weidenmüller, University of Heidelberg

The Organizing Committee decided to follow the pattern of the previous symposia both regarding the scope of topics and regarding invited papers and research contributions. The invited papers were intended to be surveys to cover advances since the Karlsruhe meeting, but not necessarily to include contributions to the Madison meeting. All participants had the privilege of presenting one brief paper orally. More than half the 175 contributions submitted were presented by one of the authors.

In the Proceedings the survey papers form the first part and are printed in the order in which they were given. The research contributions make up the second part and are arranged roughly in the same order as the subject matter of the survey papers, which is not necessarily the order in which they were presented orally. The discussion

Published 1971
The University of Wisconsin Press
Box 1379, Madison, Wisconsin 53701

The University of Wisconsin Press, Ltd.
27–29 Whitfield Street, London, W.1

First printing

Printed in the United States of America
Cushing-Malloy, Inc., Ann Arbor, Michigan

ISBN 0-299-05890-5; LC 71-143762

Polarization Phenomena in Nuclear Reactions

Proceedings of the

Third International Symposium

Madison, 1970

edited by **H. H. Barschall** and **W. Haeberli**

The University of Wisconsin Press

Madison

Contents

FORMAL DESCRIPTION AND THEORY

Contents

LIGHT NUCLEI

Contents

ELASTIC SCATTERING OF NUCLEONS

SCATTERING OF DEUTERONS AND TRITONS

DESIGN AND CALIBRATION OF ION SOURCES

Contents

The Madison Convention

I

Polarization effects involving spin-one particles should be described either by spherical tensor operators τ_{kq}, with normalization given by $\mathrm{Tr}\{\tau_{kq}\tau_{k'q'}^{\dagger}\} = 3\delta_{kk'}\delta_{qq'}$, or by Cartesian operators S_i, $(3/2)(S_i S_j + S_j S_i) - 2\delta_{ij}$ $(i = x, y, z)$. S_i denotes the usual spin-one angular momentum operators.

II

The state of spin orientation of an assembly of particles, referred to as *polarization*, should be denoted by the symbols t_{kq} (spherical) or p_i, p_{ij} (Cartesian). These quantities should be referred to a right-handed coordinate system in which the positive z-axis is along the direction of momentum of the particles, and the positive y-axis is along $\vec{k}_{in} \times \vec{k}_{out}$ for the nuclear reaction which the polarized particles initiate, or from which they emerge.

III

Terms used to describe the effect of initial polarization of a beam or target on the differential cross section for a nuclear reaction should include the modifiers *analyzing* or *efficiency*, and should be denoted by T_{kq} (spherical) or A_i, A_{ij} (Cartesian). These quantities should be referred to a right-handed coordinate system in which the positive z-axis is along the beam direction of the incident particles and the y-axis is along $\vec{k}_{in} \times \vec{k}_{out}$ for the reaction in question.

IV

In the expression for a nuclear reaction A(b, c)D an arrow placed over a symbol denotes a particle which is initially in a polarized state or whose state of polarization is measured.

□□□□□□□□□□□□

Editors' remarks: For several months before the symposium a number of participants discussed the need for conventions concerning notations and coordinate systems for describing polarization phenomena, especially for spin-one particles. During the symposium about twenty-five physicists met and formulated specific recommendations. The recommendations provide as much continuity as possible with current usage. They do not include transfer phenomena, but they are designed to facilitate generalization to transfer phenomena.

The recommendations of the working group were presented to the whole conference and were accepted without dissent. P. Huber proposed that these recommendations be called "The Madison Convention," and this proposal was adopted unanimously. The substance of the Madison Convention is given on the preceding page. The report of the working group which formulated the convention follows.

1. TENSORS

In discussing polarization effects involving spin-one particles, either spherical tensors (as given by Lakin [Phys. Rev. 98 (1955) 139]) or Cartesian tensors (as given by Goldfarb [Nucl. Phys. 7 (1958) 622]) should be used. The two sets of tensor operators are given below in terms of the spin-one angular momentum operators and the unit matrix.

$$1 = \begin{pmatrix} 1 & 0 & 0 \\ 0 & 1 & 0 \\ 0 & 0 & 1 \end{pmatrix}; \quad S_x = (1/\sqrt{2})\begin{pmatrix} 0 & 1 & 0 \\ 1 & 0 & 1 \\ 0 & 1 & 0 \end{pmatrix}; \quad S_y = (1/\sqrt{2})\begin{pmatrix} 0 & -i & 0 \\ i & 0 & -i \\ 0 & i & 0 \end{pmatrix}; \quad S_z = \begin{pmatrix} 1 & 0 & 0 \\ 0 & 0 & 0 \\ 0 & 0 & -1 \end{pmatrix}$$

Spherical

$$\tau_{00} = 1$$

$$\tau_{10} = (\sqrt{3/2})S_z$$

$$\tau_{1\pm1} = \mp(1/2)\sqrt{3}(S_x \pm iS_y)$$

$$\tau_{20} = (1/\sqrt{2})(3S_z^2 - 2)$$

$$\tau_{2\pm1} = \mp(1/2)\sqrt{3}[(S_x \pm iS_y)S_z + S_z(S_x \pm iS_y)]$$

$$\tau_{2\pm2} = (1/2)\sqrt{3}(S_x \pm iS_y)^2$$

In general, the τ_{kq} satisfy

$$\tau_{kq} = (-1)^q \tau_{k-q}^\dagger$$

Cartesian

$$\mathcal{P}_x = S_x$$

$$\mathcal{P}_y = S_y$$

$$\mathcal{P}_z = S_z$$

$$\mathcal{P}_{xx} = 3S_x^2 - 2$$

$$\mathcal{P}_{yy} = 3S_y^2 - 2$$

$$\mathcal{P}_{zz} = 3S_z^2 - 2$$

$$\mathcal{P}_{xy} = (3/2)(S_x S_y + S_y S_x)$$

$$\mathcal{P}_{xz} = (3/2)(S_x S_z + S_z S_x)$$

$$\mathcal{P}_{yz} = (3/2)(S_y S_z + S_z S_y)$$

In the corresponding spin-1/2 case one has

Spherical

Cartesian

$\tau_{00} = 1$ (2×2 unit matrix)

1, (2×2 unit matrix)

$\tau_{10} = \sigma_z$

$\sigma_x, \sigma_y, \sigma_z$

$\tau_{1\pm1} = \mp(1/\sqrt{2})(\sigma_x \pm i\sigma_y)$

2. NOMENCLATURE

   a. When describing the state of spin orientation of an assembly of particles (e.g., a beam or a target), one should include the word polarization.
   b. When describing the effect of initial polarization of a beam or target on the differential cross section for a nuclear reaction, one should include the word analyzing or efficiency, as, for example, in the terms analyzing power, analyzing tensor, or efficiency tensor.
   c. Terms such as polarization phenomena, polarization effects, may still be used in a general sense to refer to measurements or phenomena associated with the interaction or detection of polarized particles.

3. COORDINATE SYSTEMS

   The polarization of beams of particles should be referred to a right-handed coordinate system in which the positive z-axis is along the direction of momentum of the particles, and the positive y-axis is along $\vec{k}_{in} \times \vec{k}_{out}$ for the nuclear reaction which the polarized particles initiate, or from which they are emerging. In the latter case it should be explicitly stated whether $\vec{k}_{out}$ is in the c.m. or lab system.
   The analyzing power (efficiency tensors, etc.) should be referred to a right-handed coordinate system in which the positive z-axis is along the beam direction ($\vec{k}_{in}$) of the incident particles, and the y-axis is along $\vec{k}_{in} \times \vec{k}_{out}$ for the reaction in question.

4. NOTATION

   One should use different symbols in order to distinguish between polarization and analyzing quantities. The following notation is recommended:

	Polarization	Analyzing Power	
Spherical	t_{kq}	T_{kq}	$|q| \le k$
Cartesian	p_i, p_{ij}	A_i, A_{ij}	$i, j = x, y, z$

These quantities are then defined as the expectation values of the corresponding operators specified in sect. 1. For example, if the transition matrix is M, the polarization of particles emerging from a reaction initiated by unpolarized particles is given by the spherical tensors

$$t_{kq} \equiv \langle \tau_{kq} \rangle_{\text{final}} = \frac{\text{Tr}(MM^\dagger \tau_{kq})}{\text{Tr}(MM^\dagger)} .$$

The analyzing power of the reaction is given by the tensors

$$T_{kq} = \frac{\text{Tr}(M \tau_{kq} M^\dagger)}{\text{Tr}(MM^\dagger)} .$$

In Cartesian form, the quantities p and A are given by similar expressions with the τ operators replaced by the $\mathcal{P}$ operators of sect. 1.

In this notation, the differential cross section for a reaction initiated by a beam with tensor components t_{kq} is given by

$$\sigma = \sigma_0 (\sum_{k,q} t_{kq} T_{kq}^*),$$

where σ_0 is the cross section for unpolarized particles. With the choice of axes recommended in sect. 3,

$$T_{kq} = (-)^{k-q} T_{k-q} ,$$

if parity is conserved, so that $T_{10} = 0$, $T_{11} = T_{1-1} =$ pure imaginary and $T_{2q} = (-)^q T_{2-q} =$ pure real. Then the expression for the cross section may be written explicitly (for $k \le 2$) in terms of spherical tensors as

$$\sigma = \sigma_0 [1 + 2iT_{11} \, \text{Re}(it_{11}) + T_{20} t_{20} + 2T_{21} \, \text{Re}(t_{21}) + 2T_{22} \, \text{Re}(t_{22})].$$

Similarly, the cross section for a reaction initiated by polarized particles with spin-1 may be written in terms of the Cartesian tensors as

$$\sigma = \sigma_0 [1 + \frac{3}{2} p_y A_y + \frac{1}{2} p_{zz} A_{zz} + \frac{2}{3} p_{xz} A_{xz} + \frac{1}{6} (p_{xx} - p_{yy})(A_{xx} - A_{yy})].$$

5. USE OF ARROWS TO DENOTE POLARIZATION MEASUREMENTS

When reporting polarization phenomena in nuclear reactions, it is proposed that to the symbolic notation A(b,c)D be added an arrow placed over the symbols denoting those particles which are prepared in a polarized state or whose state of polarization is measured. For example, if the polarization is measured for a particle, c, emerging from a reaction between unpolarized particles A and b, the reaction is written $\vec{A}$(b,$\vec{c}$)D. The notations $\vec{A}$(b,c)D and A($\vec{b}$,$\vec{c}$)D denote polarized target and polarization transfer experiments, respectively.

APPENDIX

The spherical and Cartesian tensor moments are related as follows:

$$t_{10} = \sqrt{3/2}\, p_z$$

$$t_{1\pm1} = \mp(\sqrt{3}/2)(p_x \pm i\,p_y)$$

$$t_{20} = p_{zz}/\sqrt{2}$$

$$t_{2\pm1} = \mp(1/\sqrt{3})(p_{xz} \pm i\,p_{yz})$$

$$t_{2\pm2} = (1/2\sqrt{3})(p_{xx} - p_{yy} \pm 2i\,p_{xy})$$

$$p_x = (-1/\sqrt{3})(t_{11} - t_{1-1})$$

$$p_y = (i/\sqrt{3})(t_{11} + t_{1-1})$$

$$p_z = (\sqrt{2/3})\,t_{10}$$

$$p_{xx} = (\sqrt{3}/2)(t_{22} + t_{2-2}) - t_{20}/\sqrt{2}$$

$$p_{yy} = -(\sqrt{3}/2)(t_{22} + t_{2-2}) - t_{20}/\sqrt{2}$$

$$p_{zz} = \sqrt{2}\,t_{20}$$

$$p_{xy} = p_{yx} = -i(\sqrt{3}/2)(t_{22} - t_{2-2})$$

$$p_{xz} = p_{zx} = -(\sqrt{3}/2)(t_{21} - t_{2-1})$$

$$p_{yz} = p_{zy} = i(\sqrt{3}/2)(t_{21} + t_{2-1})$$

The same relations hold between the T_{kq} and the A_i, A_{ij}.

PARTICIPANTS IN THE DISCUSSIONS WHICH LED TO THE DRAFTING OF THE MADISON CONVENTION

G. R. Satchler, *Chairman*	W. Grüebler	G. R. Plattner	R. G. Seyler
	W. Haeberli	P. A. Quin	P. Shanley
J. Arvieux	G. M. Hale	J. Raynal	H. Sherif
S. D. Baker	P. W. Keaton, Jr.	N. Rohrig	J. E. Simmons
T. B. Clegg	E. J. Ludwig	S. Roman	M. Simonius
S. E. Darden	J. S. C. McKee	G. Roy	L. D. Tolsma
L. J. B. Goldfarb	G. G. Ohlsen	P. Schwandt	
J. A. R. Griffith	F. G. Perey	F. Seiler	

Polarization
Phenomena
in
Nuclear
Reactions

Introduction

P. HUBER, University of Basel, Switzerland

With particular pleasure I am opening this Third International Symposium on Polarization Phenomena. When I first visited the United States in 1951, the nuclear laboratory here at Madison was certainly familiar to me through its excellent scientific work, although by American standards it was not one of the larger centers. In the interim this level of excellence has not only continued but has been increased through a succession of further fundamental investigations. This symposium could scarcely be held at a more fitting place.

The year 1960 saw the opening of the first symposium in this field in Basel. Five years later the second symposium followed in Karlsruhe. Now a further five years have passed before this third symposium. No planning stands behind these five-year intervals. It is much more the expression of natural growth and the significance of the direction of research which brings about an essentially new impulse for the acquisition of fundamental knowledge. The development of new methods, the supply of experimental and theoretical results, and the resultant need to exchange and discuss all of these, grows sufficiently within five years so that a symposium becomes indispensable.

At the Basel symposium, the Basel group was able to introduce the first source of polarized particles, which was built, with some important differences and less ambitious goals, on the principle of the Erlangen group. Shortly thereafter the Harwell, Saclay, and Minnesota sources were also operational. The nuclear physics measurements reported at this first symposium, with the exception of the $^3H(\vec{d},n)^4He$ reaction, were undertaken exclusively with polarized particles from nuclear reactions. At the Karlsruhe symposium, this situation had fundamentally changed. A large number of conventional polarized-particle sources with improved properties were described. The work of Donnally and Sawyer proved to be especially important. They made possible the construction of excellent practical sources of polarized negative hydrogen or deuteron ions, in which corresponding metastable atoms are polarized and ionized by electron attachment. Appropriate methods for the production and ionization of metastable hydrogen atoms could be found. McKibben and Lawrence reported at that time on the construction of a corresponding source, the successful operation and quality of which has since been reported. It was already apparent in 1965 that important properties of nuclear interactions could be investigated with exceptional precision with the help of polarized particles from these ingenious sources.

The development just sketched has enormously intensified since the Karlsruhe symposium. A cursory glance through the program of the present symposium makes this obvious. Not only have a great number of new sources come into operation; much more impressive are the nuclear physics results, which demonstrate the effectiveness of the utilization of polarized particles and polarization-sensitive phenomena.

As well as the purely scientific information, about which we are all very curious and for which we have come to Madison, this symposium will also be concerned with the written notation describing reactions with polarized projectiles, polarized targets, and polarized outgoing particles. As the Basel convention has established a definition for the polarization direction for particles with spin 1/2, a Madison convention for the notation of reactions with polarized particles in the incoming and outgoing channels should now be established.

Now let me express, for all of us here, our sincere thanks to Professors Barschall and Haeberli and the Organizing Committee for taking upon themselves the hard work the organization of such a symposium brings. In particular, the local committee has had to excite many organizational and financial "quantum-jumps" of not too great a lifetime to get all the necessary services for a symposium ready. For this we are all especially thankful. The efforts thus made are not only to the personal advantage of each individual, but, more importantly, they make possible an international community of scientists interested in similar problems. They also create a real possibility for mutual human understanding over all political borders, which today represents an urgent need. With the knowledge that this symposium will also yield fundamental contributions in all cited fields, I open it.

The Origin of the Spin-Orbit Interaction

P. SIGNELL, Michigan State University, East Lansing, USA

The spin-orbit splitting has long been considered a characteristic feature of the shell model of the nucleus, and values for it have been produced in innumerable Hartree-Fock calculations. As an example, probably the most calculated is the j-splitting of the p levels of ^{16}O. Since the residual ^{16}O nucleus is spinless, the only way that the $P_{1/2}$ and $P_{3/2}$ levels can be split in an independent-particle model is for the effective potential to contain a spin-orbit term. Such a term couples the spin vector of a single particle or hole with its own orbital angular momentum vector to form a scalar interaction.

The origin of the effective shell model spin-orbit interaction has been pursued for many years. It was apparent quite early that a *two body* spin-orbit interaction would produce a first-order j-splitting of the many-body single-particle levels. The first attempts to produce such a two-body interaction from theory applied the methods used in single-photon exchange to the problem of single-pion exchange. The result [1] was strong central and tensor forces, but no spin-orbit force. Some of the theorists who were convinced of the soundness of the model spent considerable effort trying to derive the shell model spin-orbit splitting as a higher order effect of the two-body tensor force. These calculations invariably produced a splitting which was either of the wrong sign or of a magnitude much too small [1]. One of the last of these calculations [2] was presented at the London Conference in 1958, where over a fourth of the experimental P-wave splitting in ^{16}O was produced as a second-order effect of a two-body tensor force.

Just prior to that time, a number of polarization experiments had been carried out on the two-nucleon system. It soon became clear [3] that these experimental data not only required a strong two-body spin-orbit interaction, but required it to be of the right sign for the shell-model spin-orbit splitting! Since one-pion exchange had produced no such interaction, the theorists began looking for it in the simultaneous exchange of two pions between the two nucleons, with nucleon intermediate states (fig. 1). In other words, the shell model splitting had previously been considered to be due to a higher order nucleon-nucleon effect in the many-body system. Now it was assumed to be

due to a higher order pion–nucleon effect in the
two–body system. However, before this could
be confirmed or disproved, the elementary–particle
physicists entered the land of baryon and meson
resonances. Since these resonances had been virtually unforeseen by strong–interaction perturbation theory, the old methods of calculation were
discarded. Most theorists lost interest in trying
to work with the very complex nucleon–nucleon
interaction when they had their hands full with
the similar, but algebraically much simpler,
pion–nucleon and pion–pion systems.

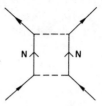

Fig. 1. Exchange of
two pions between
two nucleons.

   Thus the burden of delineating the two–nucleon interaction fell entirely upon the experimentalists and phenomenologists. This collaboration had perhaps its finest hour in 1962, with the publication of the
Hamada–Johnston [4] and Yale [5] potentials. For nuclear physics purposes, these potentials are still among the best energy–dependent
phase shift representations we have. Although their overall χ^2 fit to
the data may not be as good as some of the latest representations,
they reproduce correctly the crucial experimental properties of the
deuteron, while some late–model representations do not [6, 7].
   During the past eight years a number of other potentials have been
produced with features alternate to those of the Hamada–Johnston potential. Some of these gave slightly improved fits to the data, while
some gave much worse fits [8]. But during this time the reaction matrix technique was developed for adapting the hard–core potentials,
like the Hamada–Johnston one, for nuclear calculations. The basic
idea is that one should solve the equation

$$ G = V - V \frac{Q}{E} G $$

for Brueckner's reaction matrix G. The purpose is that, although the
matrix elements of V are infinite for hard–core potentials in the plane
wave or oscillator representations, G is well behaved. By careful
approximations to the two–nucleon potential V, to the Pauli operator
Q, and to the energy denominator e, procedures have been developed
for obtaining fairly accurate values for G without dealing explicitly
with the infinite matrix elements of V. The procedure followed by Kuo
and Brown [9] in their tabulation of Hamada–Johnston–potential reaction
matrix elements included use of the Moszkowski–Scott separation
method for even–parity states and the reference spectrum method for
odd–parity states, where the separation method runs afoul of monotonic
repulsion.
   Quite recently Lande and Svenne [10] have used the Kuo–Brown procedure to investigate which part of the Hamada–Johnston potential
produces the spin–orbit doublet splitting in ^{16}O and ^{40}Ca. Lande and

Svenne found the Hartree-Fock energy minimum, corresponding to a particular value of the oscillator parameter, and computed the physical results there. If a large enough oscillator basis had been used, of course, the energy curve would have been flat: with a complete set of states, any value of the oscillator parameter would do. However, when Lande and Svenne tried evaluating the spin-orbit splitting at a value which gave the *correct* nuclear radius, they got a very different answer for the spin-orbit splitting. For example, the D-state splitting in ^{40}Ca changed by a factor of six! Enlarging the basis also produced a rather different answer. The set of oscillator states used, then, was far from complete for this problem, and the variational solution should be the one used.

At the variational solution, Lande and Svenne found that the splittings produced by the individual parts of the Hamada-Johnston potential added up to a splitting very nearly equal to the total produced by all the parts acting together. Among the various parts, they found that by far the largest contribution was from the spin-orbit potential in the two-nucleon triplet P states, $^{3}P_{0}$, $^{3}P_{1}$, and $^{3}P_{2}$. In contrast to the two-body spin-orbit contribution, all others were found to be no more than about 20% of the total calculated splitting and were of the wrong sign. This last contradicts the 1958 London Conference report and other earlier papers. The total calculated P-splitting in ^{16}O and the D-splitting in ^{40}Ca were each about twice the experimental value. However, too low a value of the splitting was produced at the value of the oscillator parameter which gave the experimental radius, for whatever that is worth.

Although there have been detailed examinations of the various approximations used by Kuo and Brown, one might be left with the uneasy feeling that a statistical devil might have caused the errors to act in concert. This could be unfortunate when one is dealing with the spin-orbit splitting, which shows such sensitivity to the position of the variational minimum. Fortunately, Grillot and McManus [11] have solved the equation for the reaction matrix directly on the computer, finding precisely the matrix elements given by Kuo and Brown.[†] Thus it seems reasonably assured that the main part of the shell model spin-orbit splitting originates in the P-wave spin-orbit part of the Hamada-Johnston two-nucleon potential. Let us now turn to the experimental and then theoretical evidence regarding that part of the two-body potential.

The task of relating the two-body P-wave spin-orbit potential to the experimental data is greatly simplified by decomposition of the

---

[†]This contradicts the numbers displayed by D. Koltun, Phys. Rev. Lett. 19 (1967) 910. However, those numbers attributed by Koltun to Grillot and McManus were from a preliminary calculation which did not include the Pauli operator Q (private communication from D. Grillot).

experimental phase shifts into central, tensor, and spin-orbit parts [8].
These are defined by the equations

$$^3P_0: \quad \delta_0 = \Delta_C - 4\Delta_T - 2\Delta_{LS}$$

$$^3P_1: \quad \delta_1 = \Delta_C - 2\Delta_T - \Delta_{LS}$$

$$^3P_2: \quad \delta_2 = \Delta_C - 0.4\Delta_T + \Delta_{LS} ,$$

where the coefficients of the Δ-phases are the appropriate matrix elements of the central, tensor, and spin-orbit operators. The results [12] of single-energy analyses of the data using the Δ-phases are shown in fig. 2, although for clarity only the Δ_{LS} data points are shown. The central and tensor errors are comparable. If one used the Born approximation to relate these phases to the corresponding potentials, the spin-orbit potential would be stronger and of shorter range as compared to the tensor potential [8]. In fig. 3 the Δ_{LS} data points are shown again, along with the predictions of two of the most popular two-body potentials. The latter are obviously quite good as representations of the data.

The final link is provided by the work of Heller and Sher [13], who found that although the Born approximation is grossly violated for the usual strong, short-range spin-orbit potential V_{LS}, the value of Δ_{LS} appears to be almost completely determined by V_{LS}. In turn, V_{LS} is quite uncorrelated with Δ_C and Δ_T. We have completed the argument: the currently most reliable type of nuclear Brueckner-Hartree-Fock calculations say that by far the largest contributor to the many-body spin-orbit splitting is the P-wave two-body spin-orbit potential $V_{LS}^{(P)}$. In turn, $V_{LS}^{(P)}$ is correlated with the two-nucleon P-wave spin-orbit phase shift $\Delta_{LS}^{(P)}$ which is very accurately determined by polarization measurements in proton-proton scattering. One can still hope that the correct magnitudes for the nuclear splittings will emerge along with the correct radii when the proper central, tensor, and spin-orbit potentials are used in a Brueckner-Hartree-Fock calculation of sufficient accuracy.

We now come to our final topic. Where does the nucleon-nucleon spin-orbit interaction come from? By this we mean: what elementary-particle processes are chiefly responsible for the observed Δ_{LS}? As mentioned earlier, it was originally thought that the exchange of two pions, three pions, etc., between nucleon lines would produce it. Then the Δ_δ (N* or 33) resonance was discovered, but the possibility of double counting prevented one from trusting completely its inclusion as another intermediate-state elementary particle. Neither, however, could one pretend that the higher-mass baryons did not exist and just

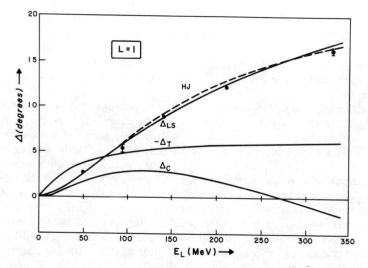

Fig. 2. The solid curves represent the experimental behavior of the central, tensor, and spin-orbit phase shifts
(see text) for the two-nucleon triplet P-waves. The actual
data points shown are for the spin-orbit phase only. An
error bar is not shown when it is the size of the central
circle or smaller.

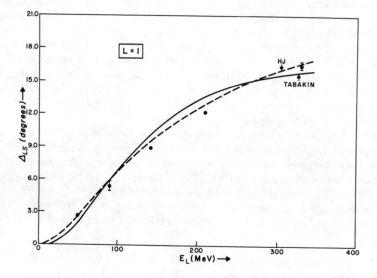

Fig. 3. Predictions of two potentials compared to the
phase data of fig. 2.

use old-fashioned perturbative field theory with nucleons alone. Discovery of the boson resonances and the existence of at least local duality provided the final blow. In the present state of elementary particle theory, then, there is no reason to believe that we are about to turn the corner on understanding the P-wave nucleon-nucleon spin-orbit interaction.

Nevertheless, particle theory is not completely helpless. Many hadronic interactions can be calculated with success, especially if they are sufficiently peripheral so that they only involve the exchange of relatively well-understood systems. In our case, the experimentally well-defined peripheral states with the same relevant quantum numbers as those we are interested in are the F-waves. That is, the F-wave Δ_{LS} comes from the long range part of the same interaction seen by the P-wave Δ_{LS}. The F-wave experimental Δ_{LS} values [12] are shown in fig. 4. Δ_{LS} is now much smaller than Δ_T and is very much smaller than Δ_C and Δ_T at low energies. This is what one would expect from its shorter range. Let us see what particle theory has to say about it.

The most peripheral contributor is one-pion exchange, but it produces no spin-orbit interaction. The next most peripheral contributor is two-pion exchange. The total energy of the two pions, which is called the "mass" of the exchanged system, varies continuously upwards from the rest mass energy of the two pions. Three-pion exchange begins at the rest mass energy of three pions, and so on. One strikes the first strong boson-resonance exchanges, the ρ, ω, and ϵ, near the threshold for the exchange of six pions.[†] This corresponds to a very short range indeed when compared to the two-pion threshold. The exchange of two particles is so difficult a calculation that few attempt it; three is virtually hopeless. Thus as one goes upward from the three-pion threshold, the calculated amplitude (which includes only two-pion exchange) should become a less and less accurate representation of the true amplitude. One might guess that it probably should not be trusted beyond about four-pion masses, and it is usually cut off at that point.

We now attempt to determine whether the F-wave Δ_{LS} is sufficiently peripheral to be accounted for solely by the low-mass (long-range) part of two-pion exchange. For F-waves, one finds that contributions with nucleon intermediate states alone dominate: those contributions in which one or both of the nucleons has become for a time a higher-mass baryon are much smaller. The F-wave Δ_{LS} data are shown in fig. 5, along with a typical calculated two-pion-exchange contribution

---

[†]That is, $m_{\rho,\omega,\epsilon} \simeq 6m_{\pi}$ with $m_{\pi} = 13$ MeV. The scalar, isoscalar ϵ is now believed to have a "mass" of perhaps 900 MeV but with a very large width. See, for example, D. Morgan and G. Shaw, Phys. Rev. D2 (1970) 520.

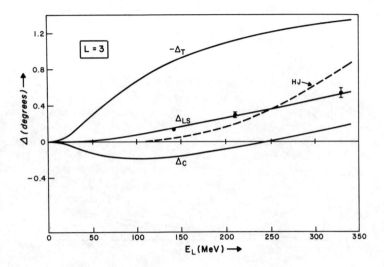

Fig. 4. Same as in fig. 2, but for the triplet F-waves.

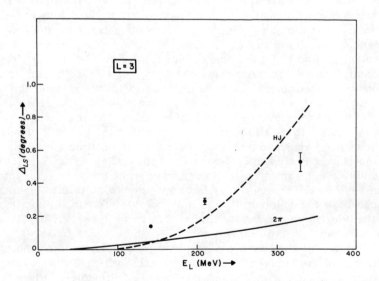

Fig. 5. The data of fig. 4, and the low-mass two-pion exchange contribution.

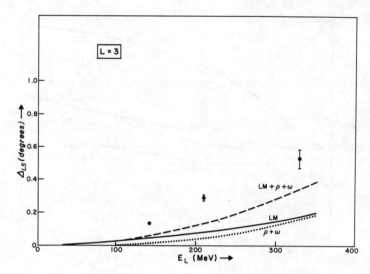

Fig. 6. The "2π" curve of fig. 5 is here labeled "LM."
The (high-mass) one-boson-exchange ρ and ω contributions are also shown.

curve. The latter is obviously inadequate. One can also go up to the
higher exchanged-mass region and estimate the contributions of the
exchanged boson resonances, using parameters derived from other
elementary-particle experiments. These contributions for the ρ and ω
resonances are shown in fig. 6, along with the previously discussed
low-mass (LM) part and the sum of these. The sum is still far from
sufficient, even at the quite moderate energy of 140 MeV. Nor do the
phenomenological potentials seem much better. For example, the
latest Bryan-Scott [14] one-boson-exchange phenomenological potential produces an F-wave Δ_{LS} which only falls along the dashed total
curve in fig. 6. This is in spite of the fact that the ρ and ω coupling
constants used by Bryan and Scott were factors of three and four larger
than the Lomon-Feshbach [15] ones used for fig. 6. In any case, we
obviously do not yet understand even the part of the spin-orbit interaction seen by the F-waves. Perhaps we need to go to the H-waves,
although their two-nucleon Δ_{LS} is only poorly determined experimentally.
If one only achieves understanding at that level, however, it will be
a long way down to the P-waves.

One concludes that we are a long way from understanding the origin
of that part of the nucleon-nucleon interaction which we believe produces the main part of the Brueckner-Hartree-Fock spin-orbit splitting.

I would like to thank David Grillot and Arthur Kerman for helpful
discussions, and Jonas Holdeman and Bruce VerWest for invaluable
assistance in calculations.

REFERENCES

[1] L. Eisenbud and E. P. Wigner, Nuclear Structure (Princeton University Press, 1958). See esp. p. 61.
[2] A. Arima and T. Terasawa, Nuclear Forces and the Few-Nucleon Problem, eds., T. Griffy and E. Power (Pergamon Press, New York, 1960).
[3] P. Signell and R. Marshak, Phys. Rev. 106 (1957) 832; J. Gammel and R. Thaler, Phys. Rev. 109 (1957) 1229.
[4] T. Hamada and I. D. Johnston, Nucl. Phys. 34 (1962) 382.
[5] K. E. Lassila et al., Phys. Rev. 126 (1962) 881.
[6] R. A. Arndt and L. D. Roper, Phys. Rev. D1 (1970) 129.
[7] P. Signell, Proc. 1969 Midwest Theory Conference (Physics Dept., University of Iowa, 1970), and Phys. Rev. C2 (1970) 1171.
[8] P. Signell, in Advances in Nuclear Physics, vol. 2, eds. M. Baranger and E. Vogt (Plenum Press, New York, 1969).
[9] T. T. S. Kuo and G. E. Brown, Nucl. Phys. 85 (1966) 40.
[10] A. Lande and J. P. Svenne, Nucl. Phys. A124 (1969) 241.
[11] D. Grillot and H. McManus, private communication.
[12] B. VerWest and P. Signell, to be published.
[13] L. Heller and M. S. Sher, Phys. Rev. 182 (1969) 1031.
[14] R. Bryan and B. L. Scott, Phys. Rev. 177 (1969) 1435.
[15] E. L. Lomon and H. Feshbach, Ann. Phys. (N.Y.) 48 (1968) 94.

DISCUSSION

Breit:

Present evidence does not *necessarily* indicate the presence of important physical effects that have not already been pointed out in the literature. It is certain that there are grievous technical difficulties in carrying out the calculations. The OBE theories have obvious limitations caused by neglect of wave-function distortions.

Polarization and Symmetry Properties

HANS A. WEIDENMÜLLER, University of Heidelberg and Max-Planck-Institut für Kernphysik, Heidelberg, Germany

1. INTRODUCTION

Symmetry properties of the Hamiltonian have always received special attention in the development of quantum physics. In the absence of a full dynamical theory, they provide a classification scheme for the states of a system, and in many cases allow the derivation of selection rules for transitions between the states. We know that the fundamental interactions can be grouped not only according to their strengths, but also according to the degree of their inherent symmetry violation. The strong interactions carry the highest symmetry, while the weak interactions break several symmetries completely. In the study of the symmetry properties of the fundamental interactions, nuclear polarization experiments have been of crucial importance. Let me just remind you of the famous Wu experiment in which the parity violation of the weak interactions was first found. Conversely, symmetry considerations play a central role in the setup of nuclear polarization experiments and in the interpretation of polarization phenomena observed in nuclear reactions. In accord with this two-fold role of symmetry in polarization experiments, this review mainly concerns itself with the following two questions: (i) Which symmetry properties of the fundamental interactions can be tested through nuclear polarization experiments? (ii) Given the fact that the fundamental interactions have certain symmetry properties, which restrictions does this impose upon the data in nuclear polarization experiments?

Many of the symmetries of the fundamental interactions cannot be studied through polarization experiments involving nuclei. This applies, for instance, to charge-conjugation (which always involves particles and antiparticles), to strangeness, to SU(3) symmetry, etc. It also applies to G-parity, although there exists some speculation [1] that beta-transitions in light nuclei indicate that the weak interactions do not conserve G-parity. Aside from the symmetries of the fundamental interactions, in nuclei we often deal with symmetries which do not hold for the full Hamiltonian, but which do hold approximately for a certain class of low-lying nuclear states. I have in mind symmetries like the SU(3) scheme in s-d shell nuclei, the j-j coupling scheme for nuclei in the $f_{7/2}$ shell, etc. The study of such approximate symmetries in nuclear polarization experiments is often hampered by the fact that the reaction mechanism may not conserve the symmetry in question. Reliable information on the symmetry properties of the nuclear states can therefore only be obtained if the reaction mechanism is well understood.

Two symmetry operations are of special importance for nuclear polarization experiments. These are the operations of parity (P) and time-reversal (T). They are discussed in some detail in this paper. A review of the present status of our understanding of these two symmetries in nuclear physics is given in sect. 2. Restrictions on polarization experiments and relations between observables which follow if P and T commute with the nuclear Hamiltonian are discussed in sect. 3. Sect. 4 concerns itself with another symmetry, i.e., invariance under rotations in isospin space and its consequences for nuclear polarization experiments.

2. PRESENT STATUS OF P AND T INVARIANCE

The status of these two symmetries in nuclear physics was recently reviewed by Henley [2]. Aside from the more recent experimental data, most of the material of the present section is taken from ref. [2], where additional references can be found. Also, a useful collection of pertinent data can be found in ref. [3].

2.1. *Parity*
There is now good evidence that the strong and electromagnetic interactions conserve parity, while the weak interaction does not. As a consequence, second-order weak processes, i.e., the virtual exchange of a lepton pair between two nucleons, give rise to an effective internucleon force, induced by the weak interaction, which violates parity. However, the violation of parity produced by such a force is about seven orders of magnitude too small to be detectable by present techniques. It is the purpose of the nuclear parity experiments to investigate whether the weak interaction can be understood as the interaction of a current with itself, i.e., whether it has the form

$$\sqrt{2}\, G\, J_\mu^+ J_\mu \tag{1}$$

where G is the coupling constant and J_μ the four vector of the beta-current, given by

$$J_\mu = \left(\psi_p \, \gamma_\mu \, \frac{1 + \gamma_5}{\sqrt{2}} \, \psi_n \right) + \left(\psi_\nu^{(e)} \, \gamma_\mu \, \frac{1 + \gamma_5}{\sqrt{2}} \, \psi_e \right) + \left(\psi_\nu^{(\mu)} \, \gamma_\mu \, \frac{1 + \gamma_5}{\sqrt{2}} \, \psi_\mu \right) + \cdots$$

$$(2)$$

(Actually, the hadronic part of J_μ should be written in the form of the Cabbibo theory; this is suppressed in what follows.) Evidence for the validity of the form (1) comes from the weak interactions involving leptons. If the form (1) is generally valid, it gives rise to a parity-violating force between nucleons which is of first order in G. The parity-violating effects of this force are of the order of one part in 10^{-7}. The experimental search for such effects thus aims at an understanding of the structure of the weak interaction. The validity of the form (1) is also of great interest in astrophysics.

The violation of invariance against the parity operation in nuclear physics has been tested in three different ways:

(i) By testing the validity of selection rules derived from parity.

(ii) By looking for a circular polarization of gamma rays emitted from nuclei.

(iii) By searching for an asymmetry of gamma rays emitted from polarized nuclei.

Of the experiments of class (i), most look for the alpha decay of the 8.88 MeV 2^- level in ^{16}O, the decay of which to the 0^+ ground state of ^{12}C is parity forbidden. So far, only an upper limit $\Gamma_\alpha \leq 1.1 \times 10^{-9}$ eV has been found [2] for the width of this level; recent unpublished results [4] yield $\Gamma_\alpha = (2 \pm 1) \times 10^{-10}$ eV and indicate that we are at the verge of finding violation of parity in these experiments.[†] Another possibility consists in looking for terms proportional to an odd power of $\cos \theta$ in a β-γ angular correlation function, where θ is the angle between electron and photon. These terms are also forbidden by parity conservation. Such an experiment was recently performed by Baker and Hamilton [5] who investigated the β-γ angular correlation in the beta decay of ^{203}Hg. The asymmetry coefficient A_1 multiplying the first Legendre polynomial P_1 was found to be $(6 \pm 8) \times 10^{-5}$. This result is consistent with the number given on the last line of table 1 below. The authors hope to improve the statistical accuracy of their method.

While experiments of class (i) have not yet given definite proof for parity violation in nuclei, the situation is different for experiments of class (ii). If parity is conserved, the circular polarization of

---

[†]At the conference, the author learned that the experiment by Wäffler et al. has now given positive evidence of parity violation. The value of Γ_α is $\Gamma_\alpha = (2.2 \pm 0.8) \times 10^{-10}$ eV (H. Wäffler et al., Phys. Rev. Lett., to be published).

gamma rays emitted from unoriented nuclei must vanish. If parity is not conserved, such a polarization arises from the interference between transition amplitudes of the same multipolarity, but different parity. In the experiments one chooses transitions where the parity-allowed transition amplitude is strongly hindered. One thereby hopes to improve the chances of detecting the interference term.

At the time of my last report on this subject [6] at the meeting in Karlsruhe in 1965, the situation was unclear. The experimental difficulties in detecting a circular polarization of order 10^{-5} are formidable. The number of counts required to reach the necessary statistical accuracy is tremendous. This situation was dramatically improved by the work of Lobashov and collaborators [7], who decided not to count individual photons but rather to measure the integrated flux of photons, thus avoiding pile-up problems. This method is now widely used. Many of the more recent results have been obtained in this way.

Table 1 gives a survey of the circular polarization measurements. Positive indications of parity violation have been obtained by one experimental group each for the nuclei ^{41}K, ^{175}Lu, and ^{180}Hf. The only nucleus which has been studied by several groups is ^{181}Ta. The situation is not completely clear. The results of four groups are essentially in agreement and give a small value around -0.6×10^{-5} for the polarization, while two groups (refs. [34,35]) find substantially larger

Table 1. Measurements of the circular polarization of nuclear gamma rays

Nucleus	Transition	Polarization $\times 10^5$	Refs.
^{41}K	$7/2^- \rightarrow 3/2^+$	$+1.9 \pm 0.3$	[21]
^{57}Fe	$9/2^- \rightarrow 7/2^+$	$+2 \pm 6$	[22]
^{175}Lu	$9/2^- \rightarrow 7/2^+$	$+4 \pm 1$	[23]
	$5/2^+ \rightarrow 7/2^+$	$+2 \pm 3$	[24]
^{180}Hf	$8^- \rightarrow 6^+$	$|P| < 200$	[25]
		-1400 ± 900	[26]
		-280 ± 45	[27]
^{181}Ta	$5/2^+ \rightarrow 7/2^+$	$+3 \pm 21$	[28]
		-0.60 ± 0.25	[29]
		-0.6 ± 0.1	[7,30]
		-0.67 ± 0.07	[31]
		-1 ± 4	[24]
		-0.39 ± 0.18	[32]
		-1.3 ± 0.7	[33]
		-9 ± 6	[34]
		-2.8 ± 0.6	[35]
^{203}Tl	$3/2^+ \rightarrow 1/2^+$	-2 ± 3	[24]

values. Still, all groups agree that the polarization differs from zero, and it is probably safe to say that parity violation in nuclei has been found.

The experiments of class (iii) are all of the same type. Polarized thermal neutrons are captured by ^{113}Cd to form the 9.05 MeV 1^+ resonance in ^{114}Cd. This resonance decays by gamma emission. A small branch leads to the ground state. One looks for a parity-forbidden asymmetry of the emitted 9 MeV photons with respect to the direction of polarization of the neutrons. The experiment, originally proposed by a group in Brookhaven, utilizes the high-level density near neutron threshold. Many 1^- resonances are expected to occur close to the 9.05 MeV 1^+ resonance, so that the parity non-conserving nuclear force can mix these states particularly well. In this way one expects a considerable enhancement of the asymmetry. The experiment was performed by two groups in recent years. The results are given in table 2. It is somewhat disconcerting to see that the most recent measurement of the asymmetry [41] has a 75% likelihood of being zero.

Table 2. Asymmetry of 9.05 MeV gammas after capture of polarized neutrons by ^{113}Cd

Energies accepted (MeV)	Uncorrected asymmetry $\times 10^4$	Refs.
8.3-9.3	1.2 ± 7.8	[36]
8.1-9.4	-3.7 ± 0.9	[37,38]
8.5-9.5	-3.5 ± 1.2	[39]
8.8-9.5	-2.5 ± 2.2	[40]
7.99-9.15	-0.6 ± 1.9	[41]

In summary we can say that there exists [8] now rather good evidence that the strong and electromagnetic interactions conserve parity, that the effects of parity violation in nuclei exist and are of the order of one part in 10^{-6} or 10^{-7}, and that these effects can be ascribed to the current-current form (1) of the weak interaction. Uncertainties exist at three states of the investigations: (i) in the experimental data, where closer agreement between the results of different groups and still smaller statistical errors in some of the data, although certainly most difficult to obtain, would be highly desirable; (ii) in the evaluation of the nuclear effects, where the estimates of enhancement factors, of parity-forbidden admixtures, and of parity-forbidden transition amplitudes hopefully can be improved; and (iii) in the understanding of the details of the basic effective parity-violating nucleon-nucleon interaction which emerges as a consequence of the current-current hypothesis. It is of particular interest to find out whether this effective interaction contains an isovector component or not, since such a component can only be caused by the strangeness-changing part of the weak interaction. Two experiments have been proposed to study this question. Both the circular polarization of retarded electromagnetic interactions in self-conjugate nuclei and the asymmetry of photons emitted after the capture of polarized thermal neutrons by protons would yield valuable information. Neither experiment has yet been performed.

2.2 *Time-reversal*

The violation of CP invariance observed in the decay of neutral kaons, together with the CPT theorem, implies violation of time-reversal invariance in at least one of the fundamental interactions. Since the nature of this violation is still not very clear, many experiments in nuclear and particle physics have been performed with the aim of obtaining further information on the violation of T.

The discussion of T violation is more complicated than that of P violation. Invariance of the Hamiltonian with respect to P leads to certain selection rules: certain reactions are P-forbidden, as discussed in sect. 2.1. In contrast to P, the operator T is antilinear. For this reason, time-reversal invariance does not in general lead to selection rules, but only specifies the relative phases of nuclear matrix elements. (The expectation values of odd L electric moments form an exception; they must vanish if T invariance holds.) Therefore, tests of T invariance usually consist in a comparison between two reactions or two matrix elements rather than in a search for forbidden processes. Detailed-balance tests for T provide a typical example of this statement.

A second difficulty with T-invariance tests is that there does not exist as yet a satisfactory and accepted theory of T violation in the $K \rightarrow 2\pi$ decay. In the parity experiments, the order of magnitude of the expected effects could be estimated on the basis of the current-current hypothesis (1). Electromagnetic, weak, and superweak interactions have been proposed as the source of the observed T violation in kaon decays. Of these theories, only the suggestion by Bernstein, Feinberg, and Lee [9] leads to detectable effects in nuclear physics. The lack of a generally accepted theoretical framework for T violation makes it difficult to evaluate the significance of the various experimental tests of T violation. For this reason, I shall confine myself in the following to a summary of the experiments and of the upper limit on T violation given by the data, without attempting any interpretation. In the classification of the experiments, I shall follow the scheme suggested by Henley [2].

As mentioned above, T invariance makes it possible to specify the relative phases between nuclear matrix elements. In the experiments to be discussed, the relevant nuclear matrix elements can always be chosen real. Violations of time-reversal invariance can be measured by the deviation of an angle η from 0° or 180°.

Table 3. Tests of T invariance in nuclear and particle physics

		Refs.
1) Beta-decay tests		
Neutron decay: $\eta = (178.7 \pm 1.3)°$		[51]
^{19}Ne decay $\eta = (180.2 \pm 1.6)°$		[48,49]
2) Electromagnetic tests		
a) high energies		
$e + p \to N^* + e$	$|\eta| < 5 - 10°$	[42,58]
$\gamma + d \rightleftarrows n + p$	see text	[44,45]
b) nuclear physics		
^{56}Mn (beta-decay)	$\eta = (1.5 \pm 0.8)°$	
	$\eta = (-2.6 \pm 1.5)°$	[52]
	$\triangle\|\delta\cdot\eta\|/\delta_{average} = (0.23 \pm 1.48)°$	
^{106}Pd (beta-decay)	$\eta = (0.2 \pm 1.0)°$	[57]
^{49}Ti (pol. neutrons)	$\eta = (1.0 \pm 1.4)°$	[54,55]
^{36}Cl (pol. neutrons)	$\eta = (0.05 \pm 0.13)°$	[50]
^{99}Ru (Mössbauer)	$\eta_{uncorr} = (0.06 \pm 0.10)°$	[46,56]
^{183}Ir (Mössbauer)	$\eta_{uncorr} = (0.06 \pm 0.21)°$	[43]
3) Comparison of polarization and asymmetry in p-p scattering		
	$\|\eta\| < 3.4°$	[53]
	$\|\eta\| < 1-2°$	[63]
4) Detailed balance in nuclear reactions		
^{24}Mg + d $\rightleftarrows$ ^{25}Mg + p	$\|\eta\| < 0.17°$	[47,62]
^{24}Mg + α $\rightleftarrows$ ^{27}Al + p	$\|\eta\| < 0.17°$	[59,60, 61]

Table 3 lists the experiments carried through up to the present time. Beta-decay tests consist in measuring the triple correlation $(\vec{J}\cdot[\vec{k}_e \times \vec{k}_\nu])$ between the polarization $\vec{J}$ of the decaying nucleus (n or ^{19}Ne), the direction $\vec{k}_e$ of the emitted electron, and the direction $\vec{k}_\nu$ of the emitted neutrino, measured through the recoil of the residual nucleus. These tests give an upper limit of about 3% on the deviation of the phase angle from zero or π. In the electromagnetic tests, the seat of T-violating effects is sought in the electromagnetic interaction, in accord with the proposal by Bernstein et al. [9]. One may look for an asymmetry in the inelastic scattering of electrons on protons, polarized perpendicularly to the reaction plane, or for a violation of detailed balance in the reaction $\gamma + d \rightleftarrows n + p$. The first experiment was carried out by a Harvard group [42], who found no deviation from T invariance and who established an upper limit ten times smaller than the maximum T violation expected on the basis of ref. [9]. The experiment was carried to significantly higher energies (18 GeV maximum electron bombarding energy) by a Stanford-Berkeley group [58]. The significance of these high-energy experiments, the accuracy of which remains considerably below some of the nuclear physics experiments, is that they involve matrix elements of the electromagnetic current far off the energy shell. Detailed balance in the reaction $\gamma + d \rightleftharpoons n + p$ has been tested by two groups in collaboration [44,45]. The Princeton-Pennsylvania group finds agreement with the inverse reaction within the experimental accuracy at all neutron energies except the highest one, $E_n = 580 \pm 60$ MeV, where the χ^2 deviation of their results (6 data points) from the inverse reaction is $\chi^2 = 10$. It is, however, probably premature to ascribe this finding to T violation. The nuclear physics experiments either use a γ-γ angular correlation, measured on a polarized initial nucleus (the polarization may be due to a preceding beta decay or to the capture of polarized neutrons), or measure the angular correlation of elliptically polarized light with oriented nuclei through Mössbauer techniques. The latter method is the most accurate one. It yields an upper limit of 3 parts in 10^{-3} for T violation. When the accuracy attains such a high value, several corrections due to final-state interactions must be considered. They put the mean value of η listed in table 3 much closer to zero without affecting the error.

T invariance can also be tested by comparing the polarization of elastically scattered protons with the asymmetry produced by the scattering of polarized protons. With T invariance, polarization and asymmetry should be equal. One uses a polarized target of H and scatters either a polarized or an unpolarized beam of protons elastically. These experiments yield an upper limit of several percent on η.

Very high accuracy can also be attained in nuclear reaction tests of detailed balance, which establish an upper bound of 3 parts of 10^{-3} on η.

In conclusion, we see that nuclear physics tests have established an upper bound on η which is about 3×10^{-3}, while high-energy experiments yield a bound of a few percent. It should be added that searches for an electric dipole moment of the neutron [10] or the proton [11] have failed to give any evidence for a simultaneous violation of T and P.

In view of this situation and the fact that parity is violated in nuclei only to one part in 10^{-6} or so, we shall in the remainder of this paper assume that P and T are both conserved. This is justified because the accuracy of typical nuclear polarization experiments using polarized beams and/or targets is considerably smaller than the numbers discussed above. It should be emphasized, however, that nuclear polarization experiments to measure P violation and to detect T violation remain of central interest for the understanding of the fundamental interactions. Measurements of parity violation in nuclear physics provide the only means of finding out certain properties of the weak interactions. If it were possible to decrease the upper bound on T violation in nuclei significantly, this would also help in our understanding of the $K \to 2\pi$ decay.

3. CONSEQUENCES OF P AND T INVARIANCE

In order to discuss the influence of parity and time-reversal invariance on polarization phenomena of reactions involving particles with non-zero spins in the entrance and exit channels, one needs a suitable parameterization of all possible scattering experiments. In a very general frame, such a parameterization has been given by Csonka, Moravcsik, and Scadron in a series of papers [12]. These authors also discuss the restrictions imposed by P and T invariance in such reactions. In less general terms, problems of this type have also been studied in ref. [13].

Due to lack of space, the general case, i.e., a reaction of the type

$$A + B \to A' + B' \tag{3}$$

with A, B, A', B' all having non-zero spin, cannot be fully discussed here. We confine ourselves to a brief sketch of the theoretical method employed in some treatments, and to illustrative examples which show typical limitations imposed by P and T invariance.

Let s_1, s_2, s_1', s_2' denote the spins of the particles in reaction (3), and m_1, m_2, m_1', m_2' their magnetic quantum numbers. For a fixed scattering angle θ, the elements M of the reaction matrix corresponding to the reaction (3) can be written in the form $M_{s_1'm_1's_2'm_2', s_1m_1s_2m_2}$. Thus, $\underset{\approx}{M}$ is a rectangular matrix with $(2s_1+1)(2s_2+1)$ rows and $(2s_1'+1)(2s_2'+1)$ columns. This matrix can be written as a linear combination of spin operators, acting in the product of the spin spaces of the four particles. A simple example is given by the elastic scattering of a particle of spin 1/2 by a particle of spin 0, where $\underset{\approx}{M}$ takes the form ($\underset{\approx}{\vec{\sigma}}$ is the vector consisting of the three Pauli matrices)

$$\underset{\approx}{M} = \underset{\approx}{1}A + (\vec{B} \cdot \vec{\sigma}). \tag{4}$$

The fact that $\vec{B}(\theta)$ must be a vector follows from rotational invariance. An expression for $\underset{\approx}{M}$ similar to (4), although more complicated, can be found in the general case. Restrictions on the coefficients in this general expansion of $\underset{\approx}{M}$ follow from invariance under P and T. We demonstrate this, using the form (4) as an example. Let $\vec{k}$ and $\vec{k}'$ be the relative momenta of the two fragments before and after the collision. We define three linearly independent unit vectors $\vec{e}_i$ by

$$\vec{e}_1 = \vec{k} \times \vec{k}' / |\vec{k} \times \vec{k}'|, \tag{5a}$$

$$\vec{e}_2 = (\vec{k}' - \vec{k})/|(\vec{k} - \vec{k}')| , \tag{5b}$$

$$\vec{e}_3 = \vec{e}_1 \times \vec{e}_2 . \tag{5c}$$

$\underset{\approx}{M}$ can be written in the form

$$\underset{\approx}{M} = \underset{\approx}{1} A(\theta) + \sum_{j=1}^{3} B_j(\theta) (\vec{e}_j \cdot \underset{\approx}{\vec{\sigma}}_1) . \tag{6}$$

Under the parity operation, $\vec{k} \rightarrow -\vec{k}$, $\vec{k}' \rightarrow -\vec{k}'$, and $\underset{\approx}{\vec{\sigma}} \rightarrow \underset{\approx}{\vec{\sigma}}$. Parity invariance therefore requires that the invariant amplitudes $B_2(\theta)$ and $B_3(\theta)$ in eq. (6) vanish. Under time-reversal, $\vec{k} \rightarrow -\vec{k}'$, $\vec{k}' \rightarrow -\vec{k}$, and $\underset{\approx}{\vec{\sigma}} \rightarrow -\underset{\approx}{\vec{\sigma}}$. Time-reversal invariance therefore requires $B_2(\theta) = 0$. In this case, the requirement of parity invariance alone implies the results of T invariance. This is not generally the case, however.

After the restrictions imposed upon $\underset{\approx}{M}$ by invariance under P and T have been worked out, one can discuss the relations between observables implied by these invariance requirements. Let $\underset{\approx}{\rho}$ denote the density matrix for the system before the reaction. Tensor polarizations of target and projectile of arbitrary rank can be specified in terms of $\underset{\approx}{\rho}$. The density matrix $\underset{\approx}{\rho}'$ in spin space after the reaction is given by the matrix product

$$\underset{\approx}{\rho}' = \underset{\approx}{M} \underset{\approx}{\rho} \underset{\approx}{M}^{\dagger} . \tag{7}$$

Expectation values of observables can be calculated by taking the trace of the product of the corresponding operator with $\underset{\approx}{\rho}'$. For instance, the cross section corresponding to arbitrary polarization of target and projectile (specified by $\underset{\approx}{\rho}$) is given by $\mathrm{Tr} \{\underset{\approx}{\rho}'\}$. Similarly, the vector polarization $\vec{P}_{A'}$, of particle A' is given by

$$\vec{P}_{A'} = \mathrm{Tr} \{\underset{\approx}{\vec{s}}_1' \underset{\approx}{\rho}'\}/\mathrm{Tr} \{\underset{\approx}{\rho}'\} , \tag{8}$$

where $\underset{\approx}{s}'$ is the spin operator of particle A'. Relations between observables follow from the restrictions imposed upon $\underset{\approx}{M}$ discussed earlier. In the elastic scattering of an unpolarized spin-1/2 particle by a spin-0 particle, for instance, parity invariance ensures that the polarization after the scattering must be perpendicular to the reaction plane. In the elastic scattering of particles of arbitrary spin, time-reversal invariance guarantees the equality of polarization and asymmetry used in sect. 2.2 above. More generally speaking, time-reversal invariance also implies the principle of detailed balance, etc.

Before we leave this subject, a few special features deserve attention. Taking into account the restrictions imposed by invariance under P and T, one may ask for a set of scattering experiments which completely determines the elements of the scattering matrix in a particular reaction, except for a common phase. It is pointed out in ref. [14] that in reactions involving the spins $1/2 + 1/2 \to 0 + 0$ and $1 + 1 \to 0 + 0$ such a complete set can only be obtained if one uses polarized beams *and* targets. This demonstrates the need for polarized targets. In a more general context, this question has been discussed in ref. [15], where the need for polarization correlation experiments is also emphasized. In practice, such correlation experiments imply the necessity of using polarized targets and of performing polarization transfer experiments.

In reactions involving compound nuclear resonances, it is advantageous to write the scattering matrix in angular momentum representation. In an experiment involving an unpolarized beam and target, a non-vanishing vector polarization for the emitted particles can only occur if the scattering matrix has at least two interfering non-vanishing amplitudes in the exit channels. This statement [16] holds independently of the coupling scheme used, and applies more generally to all odd-rank polarization tensors of the emitted particles. The statement is important [17] for reactions through isobaric analog resonances. In such cases, one often has only one amplitude in the final channels. Then the polarization of the emitted nucleons vanishes, unless a direct term interferes with the compound amplitude.

In the case of inelastic scattering or transfer reactions, it is also well to remember that the equality of polarization and asymmetry valid in the elastic scattering of spin-1/2 particles does not apply, and that polarization and asymmetry yield different pieces of information. The use of polarimeters and the use of polarized beams (or targets) are not equivalent, not even from a theoretical point of view.

4. CONSEQUENCES OF ISOSPIN INVARIANCE

Isospin is conserved by the strong, and violated by the electromagnetic and weak, interactions. As pointed out by Barshay and Temmer [18], this fact leads to approximate symmetries in nuclear reactions. We consider the reaction

$$A + B \to C + C'. \tag{9}$$

If the isospin has a definite value in the entrance channel, if it is conserved in the reaction, and if C and C' are members of an isospin multiplet, the cross section for the reaction (9) is symmetric about 90 degrees in the c.m. system. This statement holds indepedently of the reaction mechanism and rests on the symmetry (or antisymmetry) of the wave function with respect to an interchange of C and C'. Experiments [19] testing this hypothesis contribute to our understanding of the isospin purity of light nuclei. In ref. [20], further consequences of isospin conservation are investigated. In particular, it is shown that for the polarizations P_c and $P_{c'}$ of the two reaction products, isospin conservation implies the relation

$$P_c(\theta) = P_{c'}(\pi - \theta) \tag{10}$$

where θ is the c.m. scattering angle. This shows that polarization experiments can also be used to test isospin symmetry in nuclear reactions.

In conclusion, we have seen that nuclear polarization experiments are of great importance in testing symmetry properties of the fundamental interactions, and that in turn symmetry principles are vital in correlating and analyzing nuclear polarization experiments.

REFERENCES

[1]   D. H. Wilkinson, Phys. Lett. 31B (1970) 447; D. E. Alburger and D. H. Wilkinson, Phys. Lett. 32B (1970) 190.
[2]   E. M. Henley, Ann. Rev. Nucl. Sci. 19 (1969) 367.
[3]   Sov. Phys.—Usp. 11, no. 4 (1969).
[4]   H. Wäffler, cited as private communication by M. Gari, Phys. Lett. 31B (1970) 627.
[5]   K. D. Baker and W. D. Hamilton, Phys. Lett. 31B (1970) 557.
[6]   H. A. Weidenmüller, Second Polarization Symp., p. 219.
[7]   V. M. Lobashov et al., Sov. Phys.—JETP Lett. 3 (1966) 76.
[8]   E. W. Koopmann et al., Nucl. Phys. B19 (1970) 107.
[9]   J. Bernstein, G. Feinberg, and T. D. Lee, Phys. Rev. 139 (1965) B1650.
[10] P. D. Miller et al., Phys. Rev. Lett. 19 (1967) 381; Phys. Rev. 170 (1968) 1200; C. G. Shull and R. Nathans, ibid., p. 384; V. W. Cohen et al., Phys. Rev. 177 (1969) 1942.
[11] G. E. Harrison, P. G. H. Sandars, and S. J. Wright, Phys. Rev. Lett. 22 (1969) 1263.
[12] P. L. Csonka, M. J. Moravcsik, and M. D. Scadron, Ann. Phys. (N.Y.) 40 (1966) 100, ibid. 41 (1967) 1, Nuovo Cim. 42 (1966) 743, Phys. Rev. 143 (1966) 1324, ibid. 152 (1966) 1310 and earlier references therein.
[13] H. Arenhövel, Phys. Rev. 171 (1968) 1212; G. Bergdolt, Ann. Phys. (Paris) 10 (1965) 857; P. Dumontet, P. Gaillard, and M. Lambert, Nucl. Phys. 83 (1966) 169; I. Duck, Nucl. Phys. 80 (1966) 617; K. Hehl, Fortschr. d. Phys. 13 (1965) 557; P. W. Keaton, Jr., Los Alamos preprint LA 4373 MS (1970); N. P. Klepikov, Z. Sh. Kogan, and S. V. Shamanin, Sov. J. Nucl. Phys. 5 (1967) 928; R. G. Seyler, Nucl. Phys. A124 (1969) 253.
[14] D. Fick, Phys. Lett. 24B (1967) 13.

[15] M. Simonius, Phys. Rev. Lett. 19 (1967) 279.

[16] H. L. Harney, Phys. Lett. 28B (1969) 249.

[17] H. L. Harney et al., Nucl. Phys. A145 (1970) 282.

[18] S. Barshay and G. M. Temmer, Phys. Rev. Lett. 12 (1964) 728.

[19] W. von Oertzen et al., Phys. Lett. 28B (1969) 482; H. T. Fortune, A. Richter, and B. Zeidman, Phys. Lett. 30B (1969) 175.

[20] D. Robson and A. Richter, to be published in Ann. Phys. (N.Y.)

[21] V. M. Lobashov et al., Phys. Lett. 30B (1969) 39.

[22] E. Kankeleit, Congr. Int. Phys. Nucl. 2 (CNRS, Paris, 1964) p. 1206.

[23] V. M. Lobashov et al., Sov. Phys.—JETP Lett. 3 (1966) 173.

[24] F. Boehm and E. Kankeleit, Nucl. Phys. A109 (1968) 457.

[25] H. Blumberg et al., Phys. Lett. 22 (1966) 328.

[26] P. Bock, B. Jenschke, and H. Schopper, Phys. Lett. 22 (1966) 316.

[27] B. Jenschke and P. Bock, Phys. Lett. 31B (1970) 65.

[28] P. Bock and H. Schopper, Phys. Lett. 6 (1965) 284.

[29] B. Jenschke and P. Bock, to be published.

[30] V. M. Lobashov et al., Sov. Phys.—JETP Lett. 5 (1967) 59, and Phys. Lett. 25B (1967) 104.

[31] V. M. Lobashov et al., Proc. Conf. Electron Capture and High-order Processes in Nuclear Decays, vol. III (Eötvös Lorand Phys. Soc., Budapest, 1968).

[32] J. C. Vanderleeden and F. Boehm, Bull. Am. Phys. Soc. 14 (1969) 587.

[33] H. Diehl et al., to be published.

[34] D. W. Cruse and W. D. Hamilton, Nucl. Phys. A125 (1969) 241.

[35] E. Bodenstedt et al., Phys. Lett. 29B (1969) 165, and Nucl. Phys. A137 (1969) 33.

[36] R. Haas, L. B. Leipuner, and R. K. Adair, Phys. Rev. 116 (1959) 221.

[37] Yu. G. Abov, P. A. Krupchitsky, and Yu. A. Oratovskii, Phys. Lett. 12 (1964) 25.

[38] Yu. G. Abov, P. A. Krupchitsky, and Yu. A. Oratovskii, Sov. J. Nucl. Phys. 1 (1965) 341.

[39] Yu. G. Abov et al., Phys. Lett. 27B (1968) 16.

[40] E. Warming et al., Phys. Lett. 25B (1967) 200.

[41] E. Warming, Phys. Lett. 29B (1969) 564.

[42] J. A. Appel et al., Phys. Rev. D1 (1970) 1285.

[43] M. Atac et al., Phys. Rev. Lett. 20 (1968) 691.

[44] R. L. Anderson, R. Prepost, and B. H. Wiik, Phys. Rev. Lett. 22 (1969) 651.

[45] D. F. Bartlett et al., Phys. Rev. Lett. 23 (1969) 893.

[46] M. Blume and O. C. Kistner, Phys. Rev. 171 (1968) 417.

[47] D. Bodansky et al., Phys. Rev. Lett. 17 (1966) 589.

[48] F. P. Calaprice et al., Phys. Rev. Lett. 18 (1967) 918.

[49] F. P. Calaprice et al., Phys. Rev. 184 (1969) 1117.

[50] J. Eichler, Nucl. Phys. A 120 (1968) 535.

[51] B. G. Erozolimsky et al., Phys. Lett. 27B (1968) 557.

[52] M. H. Garrell et al., Phys. Rev. 187 (1969) 1410.

[53] R. Handler et al., Phys. Rev. Lett. 19 (1967) 933.

[54] J. Kaifosz, J. Kopecky, and J. Honzatko, Phys. Lett. 20 (1965) 284.

[55] J. Kaifosz, J. Kopecky, and J. Honzatko, Nucl. Phys. A120 (1968) 225.

[56] O. C. Kistner, Phys. Rev. Lett. 19 (1967) 872.

[57] R. B. Perkins and E. T. Ritter, Phys. Rev. 174 (1968) 1426.

[58] S. Rock et al., Phys. Rev. Lett. 24 (1970) 748.

[59] W. von Witsch, A. Richter, and P. von Brentano, Phys. Lett. 22 (1966) 631.

[60] W. von Witsch, A. Richter, and P. von Brentano, Phys. Rev. Lett. 19 (1967) 524.

[61] W. von Witsch, A. Richter, and P. von Brentano, Phys. Rev. 169 (1968) 923.

[62] W. G. Weitkamp et al., Phys. Rev. 165 (1968) 1233.

[63] R. Ya. Zul'karneev, V. S. Nadeshdin, and V. I. Satarov, Sov. J. Nucl. Phys. 10 (1970) 559.

DISCUSSION

Segel:

There are two cases in which the parity experiments are getting close to the fundamental issues. One is a preliminary result reported at the 1969 Columbia meeting by Lobashev on the γ-ray circular polarization following thermal n-p capture. Here there is virtually no nuclear physics to worry about. The other case is ^{41}K where gamma-ray circular polarization has been reported, also by Lobashev et al. Since ^{41}K is well described by the shell model, it should be possible to test the correctness of the parity violating potential in this case.

Keaton:

You mentioned that a complete set of measurements to test symmetry relations requires polarized beam *and* polarized targets. Could one also use polarization transfer experiments?

Weidenmüller:

No.

Polarization Phenomena in
Nuclear Reactions

HERMAN FESHBACH, Massachusetts Institute of Technology,
Cambridge, USA*

My paper will be in the nature of a review [1]. Our topic is polarization in nuclear reactions. And we shall briefly discuss two aspects.
The first is the important technical problem of the polarization observables and of their representation in terms of partial waves. The other
is the use of polarization experiments for the probing of nuclear structure and nuclear reaction mechanisms.

With respect to the first problem, recall that the polarization properties of a particle state can be described in terms of the spherical
tensors $T_{k\kappa}(\vec{S})$, where $\vec{S}$ is the spin operator for the particle in question.
The reduced width of $T_{k\kappa}(S)$ will be taken to be $\sqrt{2S+1}$. For a spin of
S, the maximum value of k is 2S. Any operator in the Hilbert space
associated with the spin S can be expressed in terms of $T_{k\kappa}$, and in
particular the density matrix ρ may be so decomposed. This should be
considered as a "multipole" expansion of ρ in spin space:

$$\rho(S) = \frac{1}{2S+1} \sum T_{k\kappa}(\hbar\, T_{k\kappa}{}^\dagger \rho) = \sum T_{k\kappa}(S)\langle T_{k\kappa}^\dagger(S)\rangle. \tag{1}$$

Thus the values of $\langle T_{kq}\rangle$ specify the "polarization" character of the
beam, and we may therefore restrict our attention to T_{kq} and its expectation values whether the initial or final beam is being considered.

Not only do we need to describe the polarization properties of
initial and final beams, but we are concerned with the connection between these initial and final states. This is furnished by the scattering matrix or scattering operator, which for most of the problems of
interest today (see preceding paper by Weidenmüller [2]) must be a rotational, reflection and time-reversal invariant, corresponding to the
conservation of angular momentum, to parity, and to the symmetry between initial and final states. These symmetries are realized by taking
proper combinations of the observables. This is a familiar operation
and its simplest example is in the scattering between a particle of
spin 0 and a particle of spin S. The process can be elastic or inelastic
as long as these spins are maintained. Then the scattering amplitude

operator can be expanded in the tensor operator $T_{kq}(S)$. These must be combined with the three kinematical quantities, the three vectors

$$\hat{n} = \frac{\vec{k} \times \vec{k}'}{|\vec{k} \times \vec{k}'|} \,, \quad \hat{q} = \frac{\vec{k}' - \vec{k}}{|\vec{k}' - \vec{k}|} \,, \quad \hat{p} = \frac{\vec{k}' + \vec{k}}{|\vec{k}' + \vec{k}|} \tag{2}$$

to form the appropriate variants. Here $\vec{k}$ is the incident momentum, $\vec{k}'$ the final momentum in the c.m. frame. If $S = 1/2$, the familiar form is obtained

$$M = f(\vec{k} \cdot \vec{k}', k, k') + g(\vec{k} \cdot \vec{k}', k, k')\vec{\sigma} \cdot \hat{n}, \tag{3}$$

where M is the scattering operator. For spin 1 we obtain

$$M = A + B\,\vec{S} \cdot \hat{n} + C\,\vec{S} \cdot \hat{n}\,\vec{S} \cdot \hat{n} + D[(\vec{S} \cdot \hat{p})(\vec{S} \cdot \hat{p}) - (\vec{S} \cdot \hat{q})(\vec{S} \cdot \hat{q})]. \tag{4}$$

Here A, B, C, D are invariant functions like f and g in eq. (3). The form of the last term indicates one of the problems with this method, namely, that all the possible invariants are not independent. In this case $(\vec{S} \cdot \hat{n}\ \vec{S} \cdot \hat{n})$, $(\vec{S} \cdot \hat{p}\ \vec{S} \cdot \hat{p})$, and $(\vec{S} \cdot \hat{q}\ \vec{S} \cdot \hat{q})$ add up to a "c number." This method has been much used in the analysis of nucleon–nucleon scattering and more recently for the scattering of spin-1/2 systems, for example, p on ^{3}He. For the general case of a spin-0 target,

$$M = \sum (2k + 1) A_{kq}\,(\hat{k}',\hat{k}) T_{kq}(\vec{S}). \tag{5}$$

This expansion as well as the general properties of the coefficients A_{kq}, and the more general case in which both particles have spin, has been discussed by Kerman [3]. He has also obtained expansions of the coefficients A_{kq} in a partial wave series. However, it turns out to be just as easy to obtain these and more general results without using the tensor formalism. To illustrate, I shall give the result for arbitrary spins of the particles for both initial and final systems, using the channel spin representation. (Similar expressions for the spin-orbit coupling scheme have also been obtained.)

$$\langle S'\mu', s'\nu' | M | S\mu, s\nu \rangle = 4\pi \sum \langle S'\mu', s'\nu' | S'm' \rangle \langle S'm', LM | Sm \rangle \cdot$$

$$\cdot (-)^{J+S'+\ell'-\ell} (2J+1) \sqrt{\frac{2J'+1}{2S+1}} \begin{Bmatrix} S' & S & L \\ \ell & \ell' & J \end{Bmatrix} \cdot$$

$$\cdot i^{\ell'-\ell} \mathcal{B}_{LM} |\ell',\ell| f_J[(S's')S'\ell'; (Ss)S\ell] \tag{6}$$

where

$$\mathcal{B}_{LM} = \sum_{mm'} Y_{\ell'm'}(\Omega')Y_{\ell m}(\vec{\Omega})\langle \ell'm'\ell m|LM\rangle \tag{7}$$

$$= \sqrt{\frac{2\ell+1}{4\pi}}\, Y_{\ell'M}(\vec{\Omega}')\langle \ell'M\,\ell\,0|LM\rangle. \tag{8}$$

In this formula, unprimed quantities refer to the initial beam, prime quantities to the emergent beam. Thus S, s, S', ℓ, $\vec{\Omega}$ give the initial spin of the target, the spin of the projectile, the channel spin, the orbital angular momentum of the projectile, and the incident direction of motion, respectively. Since angular momentum is conserved, M can be decomposed into partial wave amplitudes of a given total angular momentum J. The quantity f_J is the transition amplitude for this partial wave. From this formula it is a simple matter to compute the expectation value of any of the tensors $T_{K\kappa}$ of the various spins S, s, S', s'. The angular dependence is carried by $\mathcal{B}_{LM}$. Eq. (8) applies when $\vec{\Omega} = 0$, that is, when the direction of motion of the incident beam is taken to be z-axis. These functions were discussed by Rose [4], Brink and Satchler [5], and Kerman [3]. They transform as spherical tensors of rank and order κ. A few examples will be helpful:

$$\mathcal{B}_{00} = \frac{(-)^{\ell}\delta_{\ell\ell'}}{4\pi}\,\sqrt{2\ell+1}\;P_{\ell}(\vec{\Omega}\cdot\vec{\Omega}') \tag{9a}$$

$$\mathcal{B}_{1\mu} = i\,(-)^{\ell}\sqrt{\frac{2\ell+1}{4\pi\ell(\ell+1)}}\;P_{\ell}^{(1)}(\vec{\Omega}\cdot\vec{\Omega}')Y_{1\mu}(\hat{n}) \quad \text{for } \ell = \ell' \tag{9b}$$

$$\mathcal{B}_{2\mu} = c(\ell,\ell')Y_{2\mu}(\hat{n}) + b(\ell,\ell')[Y_{2\mu}(\hat{p}) - Y_{2\mu}(\hat{q})], \tag{9c}$$

where b and c are known functions of ℓ and ℓ'.

The maximum value of L is the minimum of $S + S'$ and $\ell + \ell'$, and thus gives the number of independent functions of $\vec{\Omega}'$ which the matrix element (eq. (6)) contains. Thus if the channel spins are 0, only $\mathcal{B}_{00}$ is involved, if $S + S'$ equals 1/2, $\mathcal{B}_{00}$ and $\mathcal{B}_{1\mu}$ are involved, while $\mathcal{B}_{2\mu}$ is involved as well if its value is 1. It is interesting to compare eqs. (4) and (9c). It is to be emphasized that eq. (6) holds for any reaction. When spin-orbit forces are important, it is more convenient to expand in terms of $f_J((s'\ell')j'S', (s\ell)jS)$. The expansion for this case has also been obtained.

Remarkably simple expressions for the polarization cross sections for many cases have been given by Kerman [3] and can be easily extended to situations he did not happen to consider. As a first example, consider the cross section for the polarization tensor $T_{K\kappa}(S)$:

$$\langle T_{\kappa\kappa}\rangle \frac{d\sigma}{d\Omega} = \frac{4\pi^2}{(2S+1)(2S'+1)} \sum (f_{a_1'a_1}^{J_1})^* f_{a_2'a_2}^{J_2} \langle a_2 J_2 || Y_L || a_1 J_1 \rangle \cdot$$

$$\cdot \langle a_1'J_1 || T_L(L',k) || a_2'J_2 \rangle \frac{(-)^{J_2-J_1}}{\sqrt{(2k+1)(2L+1)}} \mathcal{B}_{\kappa\kappa}(L,L'). \qquad (10)$$

Here $a_{1,2}$ represent the quantum numbers in the entrance channel, coupling scheme unspecified; the primed quantities apply to the exit channel. $T_L(L',k)$ is the spherical tensor formed from $Y_{L'm'}(\Omega')$ and $T_{\kappa\kappa}(S')$:

$$Y_{L'M'} T_{\kappa\kappa} = \sum \langle L'M', k\kappa | LM \rangle T_{LM}(L',k). \qquad (11)$$

The angular distribution is carried by $\mathcal{B}_{\kappa\kappa}$, which is defined by eq. (6). $\mathcal{B}_{\kappa\kappa}$ transforms, as it should, like a tensor of rank k order μ. To fill out this formula one needs the reduced matrix elements which do depend upon the coupling scheme adopted for the description of the reaction. However, regardless of the coupling scheme chosen, it is possible to derive, for example, complexity theorems. Let ℓ_i and ℓ_i' be two possible incident or emergent orbital angular momenta, respectively. Then from the reduced matrix elements it follows that

$$\vec{\ell}_1 + \vec{\ell}_2 = \vec{L}$$

$$\vec{\ell}_1' + \vec{\ell}_2' = \vec{L}'. \qquad (12)$$

For a given J, $\ell_i' - \ell_i'$ is fixed by parity; hence $L - L'$ is even. It also follows that

$$\vec{L} + \vec{L}' = \vec{k}$$

$$\vec{J}_1 + \vec{J}_2 = \vec{L}. \qquad (13)$$

It is now a simple matter to develop a "complexity" theorem. As an illustration take $k = 1$, the vector polarization. We look for the maximum value of L' given a maximum value of either ℓ_i or ℓ_i' or both. From eq. (12)

$$(L')_{max} = 2 \min \ell'.$$

From eq. (13) $L' = L-1, L, L+1$. Because of parity conservation, only the term $L = L'$ survives:

$$(L')_{max} = 2 \min \ell.$$

Similarly, from eq. (13)

$$(L')_{max} = 2 \min J.$$

From all these results it follows that

$$(L')_{max} = 2 \min (J, \ell, \ell'). \tag{14}$$

It is easy to generalize eq. (11) to the case for which the incident beam is also polarized. Consider the case for which $T_{k\mu}$ has non-zero expectation value for the incident beam. Then the two reduced matrix elements in (10) are replaced by

$$\langle a_2 J_2 || T_p(L,k) || a_1 J_1 \rangle \langle a_1' J_1' || T_p(L',k') || a_2' J_2' \rangle$$

while $\mathcal{B}$ is replaced by

$$\sum_{mm'} \langle L'm', k'\kappa' | P p \rangle \langle LM, k\kappa | P p \rangle Y_{L'm'} Y_{Lm} = \mathcal{B}_{Pp}(Lk, L'k'). \tag{15}$$

It is now an easy matter to derive "complexity" rules for this case. Again eq. (12) and the parity rule on $L - L'$ hold. Eq. (13) is replaced by

$$\vec{L} + \vec{P} = \vec{k}$$

$$\vec{L}' + \vec{P} = \vec{k}'$$

$$\vec{J}_1 + \vec{J}_2 = \vec{P}. \tag{16}$$

The general rules immediately follow from this equation.

We turn now to the use of polarization experiments for the probing of nuclear structure and nuclear reaction mechanisms. We divide the discussion into two parts, one dealing with the compound nucleus and resonance reactions; the other, with direct reactions and potential scattering. Regardless of which of these areas is under consideration, polarization experiments are useful because they help determine (1) the spin-dependent parts of the nuclear Hamiltonian and (2) the quantum numbers, particularly the spin quantum numbers of excited states. In addition, because polarization is an interference phenomenon, the effect of related but often small components of the nuclear wavefunction become more visible. These points should be borne in mind in connection with the discussions that follow.

Let us begin with the compound nucleus and resonance reactions. The use of polarization experiments to help determine the spin of a compound nuclear state is familiar. For vector polarization the effect comes from the interference of the resonant with the non-resonant background. For target spin-0 and spin-1/2 particles the sign of the polarization is sufficient to determine whether the spin of the state being excited is $\ell + 1/2$ or $\ell - 1/2$. This procedure has recently become of greater interest because of its use in connection with analog resonances [6].

Another phenomenon of interest has to do with the statistical theory of nuclear reactions. In this theory the energy average of interference terms is zero, so that the average vector polarization is zero. There will be fluctuations away from the average. It would be interesting to compare the fluctuations of expectation values of various tensor polarizations. These results will bear upon the interpretation of the width of their self-correlations as a lifetime. In addition, if there are cross-correlations in these cross sections, particularly when their width is much larger than that of the fluctuations, doorway states may be indicated. Evidence relating to the existence of doorway states is given in the contribution of Ruh and Marmier [7] to this conference. It would be very interesting to investigate the fluctuations and regularities of the entire spectrum for reactions, ranging from the energy region in which statistical nuclear reactions prevail to the region dominated by direct processes.

Let us turn now to the direct reactions and potential scattering. Here it is useful to use fig. 1. The diagram on the right describes elastic scattering in the presence of an external field provided by the target nucleus. Only external quantum numbers of the target nucleus, such as mass and spin, can be involved. The model corresponding to this model is the optical model. The diagram to the left describes the direct reaction

$$A + a \rightarrow A' + a'. \tag{17}$$

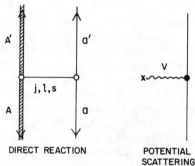

DIRECT REACTION   POTENTIAL SCATTERING

Fig. 1

The target nucleus A undergoes a transition to A' transferring angular momentum spin, parity, isospin, and baryon number to the incident projectile a at the vertex converting the latter with a'. The input at this vertex is referred to as the form factor.

The spin-dependence of the optical model potential has been thoroughly documented, particularly for protons [8]. There for spin-0 nuclei the spin-dependent term is of the spin-orbit type. Measurement of the polarization of unpolarized incident beams (or the related asymmetry for polarized incident beams) as well as the differential scattering cross section is not generally sufficient to determine the phase shifts. In addition, measurement of R or A is required, that is, measurement involving a polarized incident beam is needed. However, even in the absence of these more difficult experiments, much information is obtained just from the polarization experiments. For small spin-orbit effects

$$\sigma \sim |f|^2 \qquad P\sigma \sim |g| \operatorname{Re}\frac{1}{f} \tag{18}$$

where g is the spin-flip amplitude. For protons one has in addition the interference of the known Coulomb amplitude with the nuclear amplitude. It then becomes possible to determine all the nuclear amplitudes.

When the target nucleus has spin, additional optical model interactions are possible. These include $\vec{S} \cdot \vec{\sigma}$, $\vec{S} \cdot \vec{r}\ \vec{\sigma} \cdot \vec{r}$, and so on [9]. The determination of the strengths of these terms appears to be quite difficult because of their weakness. One method involves looking at the depolarization parameter D, which equals unity for spin-0 targets. An analysis of this possibility has been performed by Stamp [10]. Other procedures involve the scattering of polarized beams by oriented nuclei. Recent measurements at Stanford have detected such spin-spin effects [11].

Another interesting modification of the nucleon optical potential has involved the inclusion of deformation in the spin-orbit term [12]. This was required for the understanding of the polarization induced in the scattering of protons by some vibrational nuclei [13].

The development of the deuteron optical potential has not, I believe, reached this level of sophistication. In addition to the spin-orbit term, there are other possibilities [14] which occur because the deuteron has spin of 1 rather than 1/2. These give rise to the tensor type coupling exemplified by $(\vec{s} \cdot \vec{r})\,(\vec{s} \cdot \vec{r})$, $(\vec{s} \cdot \vec{p})\,(\vec{s} \cdot \vec{p})$, $(\vec{s} \cdot \vec{\ell})\,(\vec{s} \cdot \vec{\ell})$. Of course there are correspondingly more complex polarization phenomena which can be investigated, for example, the average value of the second rank tensors in spin space in addition to the first rank which is just the vector polarization. These quantities can be investigated both with polarized and with unpolarized incident beams. Present indications are that the tensor terms are small [15]. And there are additional terms in the optical model potential, if the target nucleus has spin.

A good understanding of these optical wavefunctions is essential for the unraveling of the direct reactions. The optical model can be extended to take the *a-a'* vertex into account by the method of coupled channels, which includes the possibility of elastic scattering as well as the various reaction channels. The vertex is then represented by a potential matrix. DWBA consists in treating the non-diagonal parts of the matrix by perturbation theory, and it is here that the optical model wavefunctions become important. The potential matrix depends upon the nucleon transition *A-A'*.

How do polarizations arise? For nucleon polarizations it is essential that some asymmetry with respect to reaction planes exist. In inelastic nucleon scattering the absence of such an asymmetry would make the inelastic scattering of nucleons with spin oriented up indistinguishable from the inelastic scattering with spin oriented down. Another way to say the same thing is that the potential matrix must depend upon $\vec{\sigma}$ for polarization to exist. This has the corollary that spin-flip transitions and therefore $\ell = 1$ transfers must exist. There are two possible sources for asymmetry [16]. One is the spin-dependent distortions of the incident and emergent *a* and *a'*. The other may be provided by the multipole of the nucleon-nucleus interaction which induces the transition from *A* to *A'*. For this purpose it is essential that the form factor for this transition, and thus the potential matrix describing the transition, should not be trivially dependent on the "m" value, i.e., upon the orientation of the multipole. Investigation of the polarization in inelastic transitions when combined with elastic scattering thus can provide definite information on these multipole moments and on the nature of the nucleon-nucleus interaction and the nuclear transition it induces. The need for a vibrating spin-orbit term was discovered in this fashion.

Aspects of a more microscopic model can be investigated. One can employ nuclear wavefunctions which depend on a more detailed way on the nuclear Hamiltonian [17]. One can also relate the nucleon-nucleus interaction to the two-body residual interaction and attempt to use polarization experiments to investigate its spin dependence.

Turning now to the stripping reaction, we hardly need to be reminded of the experiments performed by Yule and Haeberli [18], who by using polarized deuteron beams were able to determine the j-values transferred in the reaction. The asymmetry with respect to the reaction plane in this case, as pointed out by Huby, Refai, and Satchler [19], can be described by giving the average value of the component of the transferred orbital angular momentum normal to the reaction plane. Because of this asymmetry the proton will have opposite polarizations for $j = \ell + 1/2$ and $\ell - 1/2$. Such an average non-zero value in the absence of spin-dependence can be present only because of distortions [20]. Plane wave descriptions for the incident deuteron and final proton will give no polarization since, as it is very easy to show, they will give no net asymmetry of the transferred angular momentum. For $\ell = 0$ transfers no such asymmetry can appear, and these transitions are there-fore sensitive to spin-dependent distortions. By and large, one has a qualitative understanding of polarization in (d,p) reactions, but some effects such as the effect of the D-state of the deuteron seem to be poorly understood [21].

Even the cursory review offered here shows how much has been learned and, more importantly, how much more can be learned from ex-periments using polarized beams. The growing use of polarized deu-terons, tritons, and ^{3}He beams, particularly in particle transfer experi-ments, will undoubtedly give us many new insights into the nature of these reactions and the nuclear transitions they induce.

REFERENCES

*Work supported in part by the U.S. Atomic Energy Commission.

[1]   For recent reviews see: N. Austern, Direct Nuclear Reactions (Wiley-Interscience, New York, 1970) chap. 9; C. Glashausser and J. Thirion, Advances in Nuclear Physics, eds., M. Baranger and E. Vogt (Plenum Press, New York 1969) p. 79.

[2]   H. Weidenmüller, Third Polarization Symp.

[3]   A. K. Kerman, U.S.A.E.C. Report LAMS 2740, TID-4500 (1961).

[4]   M. E. Rose, Oak Ridge Report ORNL-2516 (1958).

[5]   D. Brink and G. R. Satchler, Angular Momentum (Oxford Univer-sity Press, Oxford, 1962) p. 55.

[6]   J. L. Adams, W. J. Thompson, and D. Robson, Nucl. Phys. 89 (1966) 377; G. Graw, Third Polarization Symp.

[7]   A. Ruh and P. Marmier, Third Polarization Symp.

[8]   L. Rosen, J. G. Beery, A. S. Goldhaber, and E. H. Auerbach, Ann. Phys. (N.Y.) 34 (1965); 96; F. Perey, Nucleon Elastic-Scattering, Third Polarization Symp.

[9]   H. Feshbach, Ann. Rev. Nucl. Sci. 8 (1958) 49.

[10] A. Stamp, Phys. Rev. 153 (1967) 1052.

[11] D. C. Healey, J. S. McCarthy, D. Parks, and T. R. Fisher, Phys. Rev. Lett. 25 (1970) 117; P. Catillon, Third Polarization Symp.

[12] H. Sherif and J. Blair, Phys. Lett. 26B (1968) 489; J. S. Blair and H. Sherif, Third Polarization Symp.

[13] M. Fricke, E. E. Gross, and A. Zucker, Phys. Rev. 163 (1967) 1113; C. Glashausser, R. de Swiniarski, J. Thirion, and A. D. Hill, Phys. Rev. 164 (1967) 164.

[14] G. R. Satchler, Nucl. Phys. 21 (1960) 116; R. C. Johnson, Third Polarization Symp.

[15] P. Schwandt and W. Haeberli, Nucl. Phys. A110 (1968) 585.

[16] G. R. Satchler, Third Polarization Symp.

[17] W. G. Love and G. R. Satchler, Nucl. Phys. A101 (1967) 424; J. Raynal, Third Polarization Symp.

[18]   T. J. Yule and W. Haeberli, Phys. Rev. Lett. 19 (1967) 756;
       Nucl. Phys. A117 (1968) 1.
[19]   R. Huby, M. Y. Refai, and G. R. Satchler, Nucl. Phys. 9 (1968)
       94; L. J. B. Goldfarb, Third Polarization Symp.
[20]   H. C. Newns, Proc. Phys. Soc. (London) A66 (1953) 477.
[21]   R. C. Johnson, Nucl. Phys. A90 (1967) 289; F. D. Santon, Third
       Polarization Symp.

Description of Polarization and
Suggestions for Additional Conventions

S. E. DARDEN, University of Notre Dame, Indiana, USA*

Abstract: The expressions relating polarization, cross sections, and
analyzing power are summarized, using spherical and Cartesian
tensors most frequently employed in the literature. Emphasis
is placed on coordinate systems and the relation between analyzing power and polarization produced in the inverse reaction.
For the spherical tensors the expressions are generalized to
include the effect of third-rank polarization. Simple surfaces
containing multipole deformations are used to aid in visualizing
second- and third-rank polarization.

1. INTRODUCTION

The parameters used to describe the polarization of beams of particles and the expressions for cross sections of processes induced by
polarized beams have been the subject of many papers [1–32]. These
topics have been rather thoroughly discussed for spin-one-half and
spin-one particles. In the case of spin-one particles, where several
parameters are required for a complete description of polarization,
various authors have used different parameters, coordinate systems,
notation, and, in a few instances, different normalizations. Within the
past decade, experimental work on deuteron polarization has increased
to the point where the number of publications and reports relating to
this topic is well over one hundred. In the majority of papers, either the
spherical tensor moments $\langle T_{kq} \rangle$, with normalization as given, for example, by Lakin [8], or the Cartesian tensor moments, P_i, P_{ij}, as
normalized by Goldfarb [20], have been used. Occasionally confusion
arises as to the correct sign of certain tensor moments, as well as
over coordinate systems to which the polarization or analyzing power
is referred. At the very least, a certain amount of discussion of these
matters is required in a carefully prepared publication on deuteron
polarization experiments. In the hope of helping to minimize both the
confusion and the amount of such discussion required in future publications, this paper is presented. Essentially all of the formulas and

expressions required to extract and present experimental results on spin-one polarization or analyzing power are given in the following sections. Particular emphasis is placed on coordinate systems.

2. TENSOR MOMENTS

Listed below are the tensor moments most commonly used in discussing spin-one polarization. The spherical tensors have been chosen to satisfy (for spin-one) the normalization condition [8].

$$\mathrm{Tr}\, T_{kq} T_{k'q'}^{\dagger} = 3\,\delta_{kq,k'q'} \tag{1}$$

$$k \leq 2, \; -k \leq q \leq k.$$

For the Cartesian tensors, the normalization is that given by Goldfarb [20].

In terms of the angular momentum operators for spin-one particles these tensor moments are:

$$t_{00} = \langle T_{00} \rangle = 1$$

$$t_{10} = \langle T_{10} \rangle = \sqrt{3/2}\,\langle S_z \rangle \qquad t_{1\pm1} = \langle T_{1\pm1} \rangle = \mp(\sqrt{3}/2)\,\langle S_x \pm i S_y \rangle$$

$$t_{20} = \langle T_{20} \rangle = (1/\sqrt{2})\,\langle 3 S_z^2 - 2 \rangle$$

$$t_{2\pm1} = \langle T_{2\pm1} \rangle = \mp(\sqrt{3}/2)\,\langle (S_x \pm i S_y) S_z + S_z (S_x \pm i S_y) \rangle$$

$$t_{2\pm2} = \langle T_{2\pm2} \rangle = (\sqrt{3}/2)\,\langle (S_x \pm i S_y)^2 \rangle$$

$$P_i = \langle S_i \rangle \qquad i = x, y, z$$

$$P_{ij} = \langle S_{ij} \rangle = \frac{3}{2}\,\langle S_i S_j + S_j S_i \rangle - 2\,\delta_{ij} \tag{2}$$

with the spin-one operators given by

$$S_x = (1/\sqrt{2}) \begin{pmatrix} 0 & 1 & 0 \\ 1 & 0 & 1 \\ 0 & 1 & 0 \end{pmatrix} \quad S_y = (1/\sqrt{2}) \begin{pmatrix} 0 & -i & 0 \\ i & 0 & -i \\ 0 & i & 0 \end{pmatrix} \quad S_z = \begin{pmatrix} 1 & 0 & 0 \\ 0 & 0 & 0 \\ 0 & 0 & -1 \end{pmatrix}.$$

The brackets represent the expectation value averaged over all the particles in the beam.

The two systems of tensor moments are related as follows:

$$t_{10} = (\sqrt{3/2}) P_z \qquad\qquad t_{1\pm1} = \mp(\sqrt{3/2})(P_x \pm i P_y)$$

$$t_{20} = P_{zz}/\sqrt{2} \qquad\qquad t_{2\pm1} = \mp(1/\sqrt{3})(P_{xz} \pm i P_{yz})$$

$$t_{2\pm2} = (1/2\sqrt{3})(P_{xx} - P_{yy} \pm 2 i P_{xy})$$

$$P_x = (-1/\sqrt{3})(t_{11} - t_{1-1})$$

$$P_y = (i/\sqrt{3})(t_{11} + t_{1-1}) \tag{3}$$

$$P_z = (\sqrt{2/3}) t_{10}$$

$$P_{xx} = (\sqrt{3}/2)(t_{22} + t_{2-2}) - (t_{20}/\sqrt{2})$$

$$P_{yy} = -(\sqrt{3}/2)(t_{22} + t_{2-2}) - (t_{20}/\sqrt{2})$$

$$P_{xy} = P_{yx} = -i(\sqrt{3}/2)(t_{22} - t_{2-2})$$

$$P_{xz} = P_{zx} = -(\sqrt{3}/2)(t_{21} - t_{2-1})$$

$$P_{yz} = P_{zy} = i(\sqrt{3}/2)(t_{21} + t_{2-1})$$

$$P_{zz} = \sqrt{2}\, t_{20}$$

The t_{kq} are in general complex, while the Cartesian moments are all real.

The spherical tensors satisfy the relation

$$T_{k-q} = (-1)^q T_{kq}^\dagger \tag{4}$$

and the corresponding relation for the spin moments t_{kq} is

$$t_{k-q} = (-1)^q t_{kq}^* . \tag{5}$$

Since the t_{kq} are defined so as to be analogous to the spherical harmonics $Y_k^q(\theta,\phi)$, they can be easily generalized for higher spins. For example, beams of particles of spin-3/2 will require, in addition to t_{1q} and t_{2q} similar to those in eq. (2), third-rank spin moments, t_{3q}, for their description. From the similarity between the t_{kq} and the Y_k^q and using the commutation relations [33] for T_{kq} these can be seen to be

$$t_{30} = (1/6\sqrt{5}) \langle 20 S_z^3 - 41 S_z \rangle$$

$$t_{3\pm1} = \mp(1/24\sqrt{15}) \langle (60 S_z^2 - 51)(S_x \pm i S_y) + (S_x \pm i S_y)(60 S_z^2 - 51) \rangle$$

$$t_{3\pm2} = (1/\sqrt{6}) \langle S_z(S_x \pm iS_y)^2 + (S_x \pm iS_y)^2 S_z \rangle$$

$$t_{3\pm3} = \mp(1/3)\langle((S_x \pm iS_y)^3\rangle. \tag{6}$$

Polarization of a beam of particles is also conveniently expressed by the density matrix [15], ρ. For spin-one particles, ρ is given in terms of the T_{kq} by [8]

$$\rho = 1/3 \sum_{k,q} t_{kq} T_{kq}^\dagger. \tag{7}$$

For any spin operator the averaged expectation value is given in terms of ρ by

$$\langle O \rangle = \frac{Tr(\rho O)}{Tr\rho}. \tag{8}$$

It is sometimes useful for purposes of visualizing the higher rank spin moments to make use of analogies which exist between spin moments and other quantities [11, 18, 30, 34]. One such analogy is that between spin moments and multipole moments of a charge distribution [18, 34]. A similar classical pictorialization involves the use of closed surfaces containing quadrupole, octupole, etc., deformations [35] to symbolize second, third, and higher rank polarizations, respectively. Such a surface can be represented by an equation giving the length of the radius vector, $\vec{R}$, from the origin to the surface as a function of angular position of $\vec{R}$.

$$R(\theta,\phi) = R_0 \left\{ 1 + \sum_q a_{kq} Y_k^{q*}(\theta,\phi) \right\}. \qquad k \geq 2 \tag{9}$$

The requirement that R be real yields the condition

$$a_{k-q} = (-1)^q a_{kq}^*. \tag{10}$$

In this scheme, an unpolarized beam would be represented by a sphere, with all $a_{kq} = 0$. If the length of the radius $R(\theta,\phi)$ in a given direction is taken to be proportional to the number of particles in the beam having spins pointing in that direction, then a_{kq} have a simple interpretation, viz., they are essentially the t_{kq}. This can be seen by noting that in this classical analogy, the averaged expectation value of T_{kq} is represented by $Y_k^q(\theta,\phi)$ averaged over the beam.

$$t_{kq} = \frac{1}{R_0} \int Y_k^q (\theta,\phi) \, R(\theta,\phi) \, d\Omega$$

$$= \int Y_k^q \left\{ 1 + \sum_q a_{kq} Y_k^{q*} \right\} d\Omega = a_{kq} \qquad (11)$$

Eq. (9) can then be rewritten

$$R(\theta,\phi) = R_0 \left\{ 1 + \sum_q t_{kq} Y_k^{q*} (\theta,\phi) \right\} . \qquad (9a)$$

Some examples of surfaces symbolizing various second- and third-rank moments are shown in fig. 1. On the left side of the figure are shown surfaces corresponding to beams having only the non-vanishing moments t_{20}, R.P. t_{21}, and R.P. t_{22}, respectively. On the right are given surfaces for t_{30}, I.P. t_{31}, I.P. t_{32}, and I.P. t_{33}, respectively. Figures corresponding to complex values of t_{kq} are obtained by rotating the surfaces in fig. 1 about the z-axis.

3. POLARIZATION PRODUCED IN REACTIONS INITIATED BY UNPOLARIZED PARTICLES

For a two-body reaction or scattering process characterized by the scattering matrix M, the density matrix ρ_f, for the outgoing particles is given in terms of the density matrix for the incident beam by [8]

$$\rho_f = M \rho_i M^\dagger . \qquad (12)$$

If the incident beam is unpolarized, $\rho_i = 1/3$, and

$$\rho_f = 1/3 \, MM^\dagger . \qquad (12a)$$

Spin moments for the outgoing beam are given, according to eq. (8), by

$$t_{kq} = \frac{Tr(\rho_f T_{kq})}{Tr\rho_f} = \frac{Tr(MM^\dagger T_{kq})}{Tr(MM^\dagger)} \qquad (13)$$

or, using Cartesian tensors,

$$\begin{Bmatrix} P_i \\ P_{ij} \end{Bmatrix} = \frac{Tr \left(MM^\dagger \begin{Bmatrix} S_i \\ S_{ij} \end{Bmatrix} \right)}{Tr(MM^\dagger)} . \qquad (13a)$$

SECOND RANK MOMENTS   THIRD RANK MOMENTS

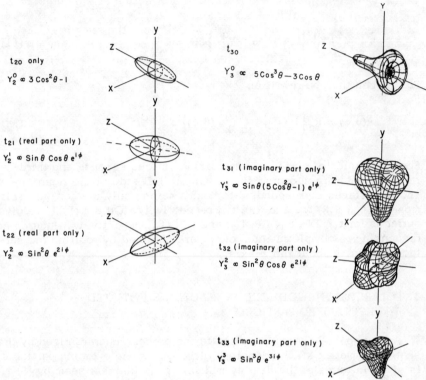

t_{20} only
$Y_2^0 \propto 3\cos^2\theta - 1$

t_{21} (real part only)
$Y_2^1 \propto \sin\theta\cos\theta\, e^{i\phi}$

t_{22} (real part only)
$Y_2^2 \propto \sin^2\theta\, e^{2i\phi}$

t_{30}
$Y_3^0 \propto 5\cos^3\theta - 3\cos\theta$

t_{31} (imaginary part only)
$Y_3^1 \propto \sin\theta(5\cos^2\theta - 1)\, e^{i\phi}$

t_{32} (imaginary part only)
$Y_3^2 \propto \sin^2\theta\cos\theta\, e^{2i\phi}$

t_{33} (imaginary part only)
$Y_3^3 \propto \sin^3\theta\, e^{3i\phi}$

Fig. 1. Examples of surfaces symbolizing second- and third-rank spin moments. On the left are shown surfaces having quadrupole deformations corresponding to t_{20}, R.P. t_{21}, and R.P. t_{22}, respectively. On the right are shown surfaces representing t_{30}, I.P. t_{31}, I.P. t_{32}, and I.P. t_{33}, respectively.

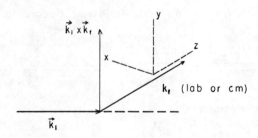

Fig. 2. Coordinate system for spin moments describing polarization produced in a reaction or scattering.

It is convenient to refer the t_{kq} or P_i, P_{ij} to a right-handed co-ordinate system having z-axis along the direction of the outgoing particle momentum, $\vec{k}_f$ (lab or c.m.), and y-axis along $\vec{k}_i \times \vec{k}_f$. This coordinate system is shown in fig. 2. Throughout the remainder of this paper, whenever the symbols t_{kq} or P_i, P_{ij} are used to denote the polarization of particles produced in a reaction or scattering, it will be understood that they are referred to the coordinate system of fig. 2 unless explicitly stated otherwise. Using the coordinate system of fig. 2, the restrictions imposed by parity conservation [8] on the complexity of polarization which can be produced in a reaction with initially unpolarized particles can be expressed as

$$t_{k-q} = (-1)^{k+q} t_{kq}$$

$$P_x = P_z = P_{xy} = P_{yz} = 0 \tag{14}$$

$$P_{xxx} = P_{zzz} = P_{zzx} = P_{xxz} = P_{yyz} = P_{xyy} = 0.$$

These relations can be visualized with the help of the polarization surfaces (9) as illustrated in fig. 3. A given reaction or scattering is symbolized by the momentum vectors $\vec{k}_i$ and $\vec{k}_f$, and a surface representing the polarization of the outgoing beam. Vector polarization (rank one) is represented by a double arrow. As viewed in a mirror, the figures symbolizing odd-rank polarization appear inverted compared to ordinary mirror images as a consequence of the axial vector nature of spin. Parity conservation requires that the reaction and its image as seen in a mirror be identical. As a result, only polarization corresponding to the real parts of t_{2q} and the imaginary parts of t_{1q} and t_{3q} can be produced. These are just the polarizations which were symbolized by the surfaces shown in fig. 1 (except for t_{30}). For spin-3/2 particles, an additional three parameters beyond the four needed for spin-one particles are required.

4. ANALYZING POWER (EFFICIENCY TENSORS) FOR REACTIONS INITIATED BY POLARIZED BEAMS

If a two-body reaction is initiated by a beam of polarized particles, the intensity of the outgoing particles is given by

$$I = Tr\rho_f = Tr M \rho_i M^\dagger. \tag{15}$$

Substitution of ρ_i from eq. (7) yields

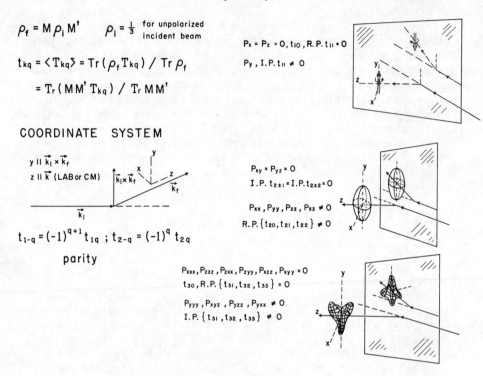

$$\rho_f = M \rho_i M^\dagger \qquad \rho_i = \tfrac{1}{3} \text{ for unpolarized incident beam}$$

$$t_{kq} = \langle T_{kq} \rangle = Tr(\rho_f T_{kq}) / Tr \rho_f$$

$$= T_r(MM^\dagger T_{kq}) / T_r MM^\dagger$$

$P_x = P_z = 0, \, t_{10}, \, R.P. \, t_{11} = 0$

$P_y, \, I.P. \, t_{11} \neq 0$

COORDINATE SYSTEM

$y \parallel \vec{k}_i \times \vec{k}_f$

$z \parallel \vec{k}$ (LAB or CM)

$P_{xy} = P_{yz} = 0$

$I.P. \, t_{2\pm1} = I.P. \, t_{2\pm2} = 0$

$P_{xx}, P_{yy}, P_{zz}, P_{xz} \neq 0$

$R.P. \{t_{20}, t_{21}, t_{22}\} \neq 0$

$$t_{1-q} = (-1)^{q+1} t_{1q} \; ; \; t_{2-q} = (-1)^q t_{2q}$$

parity

$P_{xxx}, P_{zzz}, P_{zxx}, P_{zyy}, P_{xzz}, P_{xyy} = 0$

$t_{30}, R.P. \{t_{31}, t_{32}, t_{33}\} = 0$

$P_{yyy}, P_{xyz}, P_{yzz}, P_{yxx} \neq 0$

$I.P. \{t_{31}, t_{32}, t_{33}\} \neq 0$

Fig. 3. Restrictions on the polarization which can be produced in a reaction or scattering with initially unpolarized particles. As seen in the mirror, the reaction must appear unchanged.

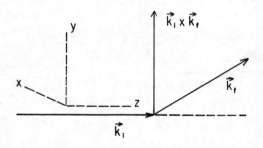

Fig. 4. Coordinate system to which the analyzing powers ε_{kq} and A_i, A_{ij} are referred.

$$I = \frac{1}{3} \sum_{k,q} t_{kq} \, Tr(M T_{kq}^{\dagger} M^{\dagger}), \tag{15a}$$

where the t_{kq} describe the polarization of the incident beam. The intensity for the case in which the incident beam is unpolarized is given from (12a) by $I_0 = Tr(1/3 \, MM^{\dagger})$. When this is inserted in (15a), one obtains

$$I = I_0 \left\{ 1 + \sum_{\substack{k,q \\ k \neq 0}} t_{kq} \frac{Tr(M T_{kq}^{\dagger} M^{\dagger})}{Tr(MM^{\dagger})} \right\}$$

$$= I_0 \left\{ 1 + \sum_{\substack{k,q \\ k \neq 0}} t_{k,q} \, \varepsilon_{kq}^{*} \right\}. \tag{16}$$

The analyzing power or efficiency tensors, ε_{kq}, are defined by

$$\varepsilon_{kq} \equiv \frac{Tr(M T_{kq} M^{\dagger})}{Tr(MM^{\dagger})} \tag{17}$$

and, in terms of the Cartesian tensors,

$$\begin{Bmatrix} A_i \\ A_{ij} \end{Bmatrix} \equiv \frac{Tr\left(M \begin{Bmatrix} S_i \\ S_{ij} \end{Bmatrix} M^{\dagger}\right)}{Tr(MM^{\dagger})}. \tag{17a}$$

In eq. (16), both the t_{kq} and the ε_{kq} must be referred to the same coordinate system. A convenient and frequently used coordinate system (fig. 4) which will be used throughout this paper for the analyzing power is one in which the z-axis is along the momentum vector $\vec{k}_i$, of the incident beam, and the y-axis is along $\vec{k}_i \times \vec{k}_f$, for the reaction in question. With this choice of coordinate system, the ε_{kq} and A_i, A_{ij}, etc., satisfy the relations (14) for parity conserving interactions.

$$\varepsilon_{k-q} = (-1)^{k+q} \varepsilon_{kq}$$

$$A_x = A_z = 0 \qquad\qquad A_{xy} = A_{yz} = 0 \tag{18}$$

$$A_{xxx} = A_{zzz} = A_{zzx} = A_{xxz} = A_{yyz} = A_{xyy} = 0.$$

Using eq. (18), the expression (16) for the intensity can be written out explicitly

$$I = I_0 \left\{ 1 + 2\,(i\,\varepsilon_{11})\,R.P.\,(i\,t_{11}) + \varepsilon_{20}\,t_{20} + 2\,\varepsilon_{21}\,R.P.\,t_{21} \right.$$

$$+ 2\,\varepsilon_{22}\,R.P.\,t_{22} + 2(i\,\varepsilon_{31})\,R.P.\,(i\,t_{31})$$

$$\left. + 2\,(i\,\varepsilon_{32})\,R.P.\,(i\,t_{32}) + 2(i\,\varepsilon_{33})\,R.P.\,(i\,t_{33}) \right\}. \qquad (19)$$

In eq. (19) the third-rank terms have been included, making this expression valid[†] for particles of spin $\leq 3/2$. For spin-one particles, the third-rank terms are absent, and for spin-one-half particles, only the $(i\,\varepsilon_{11})$ term remains. With the Cartesian tensors, the intensity expression (spin-one only) is

$$I = I_0 \left\{ 1 + \frac{3}{2} P_y A_y + \frac{1}{2} P_{zz} A_{zz} + \frac{2}{3} P_{xz} A_{xz} + \frac{1}{6}(P_{xx} - P_{yy})(A_{xx} - A_{yy}) \right\}.$$

$$(19a)$$

At this point it should again be stressed that in (19) *both* the $t_{kq}(P_i, P_{ij})$ and the $\varepsilon_{kq}(A_i, A_{ij})$ are referred to the coordinate system of fig. 4.

5. RELATION BETWEEN ANALYZING POWER AND POLARIZATION PRODUCED IN THE INVERSE REACTION

For time-reversal-invariant interactions the efficiency tensors are simply related to the polarizations produced in the inverse reaction [23]. This relationship is symbolized in fig. 5. In the left portion of the figure are shown surfaces representing separately the various first- and second-rank analyzing tensors, i.e., the quantity which connects polarization of the incident beam with intensity (and *only* intensity) of the outgoing beam. All of the analyzing tensors in the figure have been chosen to have positive values relative to the coordinate system of fig. 4. The interpretation of these diagrams is as follows: if an incident beam is polarized in such a way that the surface representing a particular component of its polarization (referred to the coordinate system of fig. 4) matches the surface symbolizing the corresponding analyzing power, then the related term in eq. (19) is positive. For example, if the incident beam has R.P. $t_{22} > 0$, then $\langle S_x^2 - S_y^2 \rangle > 0$ and

---

[†]When eq. (19) is used for spin-3/2 particles, the t_{1q} and t_{2q} will differ slightly from those given in eq. (2), since for spin-3/2 the normalization condition (eq. (1)) becomes $\mathrm{Tr}\,T_{kq}\,T_{k'q'}^\dagger = 4\,\delta_{kq,k'q'}$.

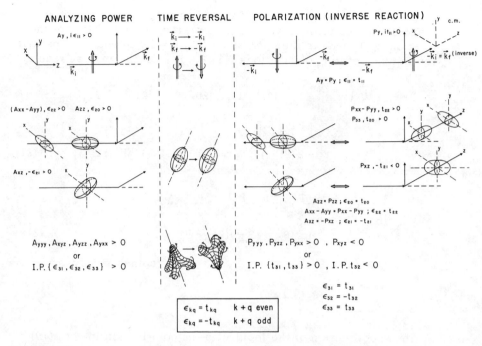

Fig. 5. Illustration of the connection between analyzing power and polarization parameters for the inverse reaction.

the polarization surface is elongated along the x-axis. Since this is the case for the surface symbolizing ε_{22} in fig. 5, the $k = 2$, $q = 2$ term in eq. (19) will make a positive contribution to the intensity. The effect of time reversal on these diagrams is illustrated in the center portion of fig. 5. In addition to reversing the directions of $\vec{k}_i$ and $\vec{k}_f$, time reversal requires the inversion (through their origin) of figures symbolizing odd-rank efficiency tensors as shown. The results of this operation is given by the diagrams on the right. These represent, for time-reversal-invariant interactions, the connection between *intensity* of the incident beam and polarization of the outgoing beam, for the inverse reaction. At the far right in fig. 5 these diagrams are inverted so as to exhibit the polarizations referred to the coordinate system of fig. 2. All of the first- and second-rank polarizations for the inverse reactions are seen to be equal to the corresponding efficiency tensors for the forward reaction except for t_{21} (P_{xz}), for which a sign change occurs. When higher rank polarization is considered, the general relation between analyzing power and polarization in the inverse reaction emerges

$$\varepsilon_{kq} = (-1)^{k+q} t_{kq} \qquad (20)$$

where the coordinate systems of the ε_{kq} and t_{kq} are as shown in figs. 4 and 2 (applied to the inverse reaction), respectively. Use of the relations (20) permits the expression (19) for the intensity to be written in terms of the polarizations t'_{kq}, P'_i, P'_{ij} for the inverse reaction

$$I = I_0 \Big\{ 1 + 2\,(R.P.\,i\,t_{11})(i\,t'_{11}) + t_{20}\,t'_{20} - 2\,t'_{21}\,R.P.\,t_{21}$$

$$+ 2\,t'_{22}\,R.P.\,t_{22} + 2\,(i\,t'_{31})\,R.P.\,(i\,t_{31})$$

$$- 2\,(i\,t'_{32})\,R.P.\,(i\,t_{32}) + 2\,(i\,t'_{33})\,R.P.\,(i\,t_{33}) \Big\} \qquad (21)$$

$$I = I_0 \Big\{ 1 + \frac{3}{2} P_y\,P'_y + \frac{1}{2} P_{zz}\,P'_{zz} - \frac{2}{3} P_{xz}\,P'_{xz}$$

$$+ \frac{1}{6}(P_{xx} - P_{yy})(P'_{xx} - P'_{yy}) \Big\}. \qquad (21a)$$

(spin-one only)

6. SPECIAL CASES

The specification of the incident beam polarization in eqs. (19) and (21) commonly involves two cases of interest:

1. *Incident beam polarized as a result of a previous scattering or reaction involving initially unpolarized particles.* In this case it is frequently desirable to write the intensity (21) with the incident beam polarization referred to the coordinate system of fig. 2 applied to the first reaction. Designating the polarizations so referred as $t_{kq}(\theta_1)$, $P_i(\theta_1)$, $P_{ij}(\theta_1)$, they can be shown (eq. 24)) to be related to the t_{kq}, P_i, P_{ij} of eq. (21) by

$$t_{kq} = t_{kq}(\theta_1)\,e^{-iq\phi} \qquad (22)$$

where ϕ is the angle between the two reaction planes; $\cos\phi = \hat{n}_1 \cdot \hat{n}_2$:

$$\hat{n}_{1,2} = \left\{ \frac{\left(\vec{k}_i \times \vec{k}_f\right)}{\left|\vec{k}_i \times \vec{k}_f\right|} \right\}_{1,2} .$$

With the use of the relations (14) and (22) in eq. (21), the intensity for the second of two scattering or reaction processes reduces to

$$I = I_0 \left\{ 1 + 2\,i t_{11}(\theta_1)\,i t_{11}(\theta_2) \cos\phi + t_{20}(\theta_1)\,t_{20}(\theta_2) \right.$$

$$- 2 t_{21}(\theta_1)\,t_{21}(\theta_2) \cos\phi + 2 t_{22}(\theta_1)\,t_{22}(\theta_2) \cos 2\phi$$

$$+ 2\,i t_{31}(\theta_1)\,i t_{31}(\theta_2) - 2\,i t_{32}(\theta_1)\,i t_{32}(\theta_2)$$

$$\left. + 2\,i t_{33}(\theta_1)\,i t_{33}(\theta_2) \right\}$$

(23a)

$$I = I_0 \left\{ 1 + \frac{3}{2} P_y(\theta_1)\,P_y(\theta_2) \cos\phi + \frac{1}{2} P_{zz}(\theta_1)\,P_{zz}(\theta_2) \right.$$

$$- \frac{2}{3} P_{xz}(\theta_1)\,P_{xz}(\theta_2) \cos\phi + \frac{1}{6}\left[P_{xx}(\theta_1) - P_{yy}(\theta_1) \right]$$

$$\left[P_{xx}(\theta_2) - P_{yy}(\theta_2) \right] \right\}$$

(spin-one only)

where the polarizations for the inverse second reaction have been denoted by $t_{kq}(\theta_2)$, $P_i(\theta_2)$, $P_{ij}(\theta_2)$.

2. *Incident beam polarized in a polarized ion source.* Here it is normally convenient to specify the incident beam polarization parameters in eqs. (19) and (21) in terms of the parameters giving the beam polarization relative to the symmetry axis of the spin system and the angles which specify the orientation of this axis with respect to the coordinate system of fig. 4. In the following discussion, the coordinate system of fig. 4 will be referred to as S. Denoting the tensor moments which express the beam polarization referred to the spin-symmetry axis as z-axis by τ_{10}, τ_{20}, τ_{30} (p_z, p_{zz}), the t_{kq} are given by the transformation equations [36]

$$t_{kq} = \sum_{q'} \tau_{kq'} D^k_{q'q}(\Phi, \theta, \Psi) = \tau_{k0} D^k_{0q}(\Phi, \theta, \Psi)$$

(24)

where Φ, θ, and Ψ are the Euler angles rotating the coordinate system of the τ_{k0} into S. These are, in succession, right-handed rotations around the old z-axis, new y-axis, and final z-axis, respectively. Angles specifying the orientation of the spin symmetry axis (Z) with respect to S are shown in fig. 6a. The angle between Z and $\vec{k}_i$ is β, and ϕ denotes the angle between the projection of Z on the x-y plane of S and $\vec{k}_i \times \vec{k}_f$. In terms of β and ϕ, the Euler angles which carry X Y Z into S (fig. 6b) are $\Phi = 0$, $\theta = -\beta$, $\Psi = -(90° + \phi)$. Using these angles in eq. (24) results in the expressions for the t_{kq} of eqs. (19) and (21):

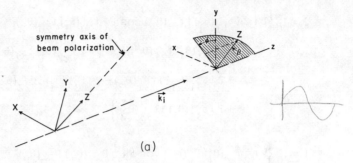

(a)

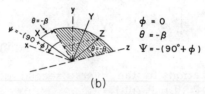

$$\phi = 0$$
$$\theta = -\beta$$
$$\Psi = -(90° + \phi)$$

(b)

Fig. 6. Rotation angles relating the spin symmetry axis of a beam from a polarized-ion
source to the coordinate system of fig. 4.
(a) Polar and azimuthal angles specifying the
orientation of the spin symmetry axis relative
to S. (b) Euler angles which carry X, Y, Z
into S.

$$t_{11} = -i\,\tau_{10}\,\frac{\sin\beta\,e^{i\phi}}{\sqrt{2}}$$

$$t_{20} = (\tau_{20}/2)\,(3\cos^2\beta - 1)$$

$$t_{21} = -i\,\tau_{20}\,\sqrt{3/2}\,\sin\beta\cos\beta\,e^{i\phi}$$

$$t_{22} = -\tau_{20}\,\sqrt{3/8}\,\sin^2\beta\,e^{2i\phi}$$

$$t_{31} = -i\,\tau_{30}\,\sqrt{3/16}\,(5\cos^2\beta - 1)\sin\beta\,e^{i\phi}$$

$$t_{32} = -\tau_{30}\,\sqrt{15/8}\,\cos\beta\sin^2\beta\,e^{2i\phi}$$

$$t_{33} = i\,\tau_{30}\,\sqrt{5/16}\,\sin^3\beta\,e^{3i\phi} \qquad (25)$$

Finally, substitution of the relations (25) into eqs. (19) and (21) yield expressions useful for extracting either polarizations (t'_{kq}, P'_i, P'_{ij}, inverse reaction) or efficiency tensors (ε_{kq}, A_i, A_{ij}) from measured particle intensities.

$$I = I_0 \Big\{ 1 + \sqrt{2}\, \tau_{10}(i t'_{11}) \sin\beta \cos\phi + \frac{1}{2} \tau_{20} t'_{20} (3\cos^2\beta - 1)$$

$$- \sqrt{6}\, \tau_{20} t'_{21} \sin\beta \cos\beta \sin\phi - \sqrt{3/2}\, \tau_{20} t'_{22} \sin^2\beta \cos 2\phi$$

$$+ (\sqrt{3}/2) \tau_{30} (i t'_{31}) (5\cos^2\beta - 1) \sin\beta \cos\phi$$

$$- \sqrt{15/2}\, \tau_{30} (i t'_{32}) \cos\beta \sin^2\beta \sin 2\phi$$

$$- (\sqrt{5}/2) \tau_{30} (i t'_{33}) \sin^3\beta \cos 3\phi \Big\} . \tag{26}$$

$$I = I_0 \Big\{ 1 + \frac{3}{2} p_z P'_y \sin\beta \cos\phi + \frac{1}{4} p_{zz} P'_{zz} (3\cos^2\beta - 1)$$

$$+ p_{zz} P'_{xz} \sin\beta \cos\beta \sin\phi - \frac{1}{4} p_{zz} (P'_{xx} - P'_{yy}) \sin^2\beta \cos 2\phi \Big\} .$$

(spin-one only)

In terms of efficiency tensors:

$$I = I_0 \Big\{ 1 + \sqrt{2}\, \tau_{10} (i \varepsilon_{11}) \sin\beta \cos\phi + \frac{1}{2} \tau_{20} \varepsilon_{20} (3\cos^2\beta - 1)$$

$$+ \sqrt{6}\, \tau_{20} \varepsilon_{21} \sin\beta \cos\beta \sin\phi - \sqrt{3/2}\, \tau_{20} \varepsilon_{22} \sin^2\beta \cos 2\phi$$

$$+ (\sqrt{3}/2)\tau_{30} (i \varepsilon_{31}) (5\cos^2\beta - 1) \sin\beta \cos\phi$$

$$+ \sqrt{15/2}\, \tau_{30} (i \varepsilon_{32}) \cos\beta \sin^2\beta \sin 2\phi$$

$$- (\sqrt{5}/2) \tau_{30} (i \varepsilon_{33}) \sin^3\beta \cos 3\phi \Big\} . \tag{27}$$

$$I = I_0 \Big\{ 1 + \frac{3}{2} p_z A_y \sin\beta \cos\phi + \frac{1}{4} p_{zz} A_{zz} (3\cos^2\beta - 1)$$

$$- p_{zz} A_{xz} \sin\beta \cos\beta \sin\phi - \frac{1}{4} p_{zz} (A_{xx} - A_{yy}) \sin^2\beta \cos 2\phi \Big\} .$$

(spin-one only)

7. RECOMMENDATIONS

A. In describing polarization of spin-one particles, either spherical tensor moments, denoted by t_{kq} (or some other symbol), or Cartesian moments, denoted by P_i, P_{ij}, should be used. The normalization should be as in eqs. (1) and (2).

Tensor moments describing the polarization of spin-one particles emitted in a scattering or reaction should, where possible, be referred to the coordinate system of fig. 2.

B. In reporting results of intensity or cross-section measurements for reactions initiated by polarized spin-one particles, it is recommended that *either* analyzing powers (efficiency tensors) denoted by some definite symbol, such as ε_{kq} (spherical) or A_i, A_{ij} (Cartesian), and referred to the coordinate system of fig. 4 be used, *or* polarization parameters (as defined in eq. (13) for the inverse reaction, with notation and coordinate system as in A) be used.[†]

The author would like to thank Professor W. Haeberli and Drs. G. Ohlsen and P. W. Keaton for extremely helpful conversations and suggestions.

[†]*Editors' note:* Following the recommendations made in this paper the Symposium adopted unanimously the "Madison Convention" which is described elsewhere in these Proceedings. According to the convention, ε_{kq} of this paper should be replaced by T_{kq}.

REFERENCES

*Research supported in part by the National Science Foundation.

[1]   L. Wolfenstein, Phys. Rev. 75 (1949) 1664.
[2]   R. J. Blin-Stoyle, Proc. Phys. Soc. A64 (1951) 700.
[3]   R. J. Blin-Stoyle, Proc. Phys. Soc. A65 (1952) 452.
[4]   R. H. Dalitz, Proc. Phys. Soc. A65 (1952) 175.
[5]   L. Wolfenstein, Phys. Rev. 92 (1953) 123.
[6]   A. Simon and T. A. Welton, Phys. Rev. 90 (1953) 1036.
[7]   A. Simon, Phys. Rev. 92 (1953) 1050.
[8]   W. Lakin, Phys. Rev. 98 (1955) 139.
[9]   H. Stapp, University of California Radiation Laboratory, Report UCRL-3098 (1955) (unpublished).
[10] G. R. Satchler, Proc. Phys. Soc. A68 (1955) 1041.
[11] S. W. MacDowell, Anais da Academia Brasileira de Ciencias 28 (1956) 71.
[12] S. W. MacDowell and J. Tiomno, Anais da Academia Brasileira de Ciencias 28 (1956) 157.

[13] L. Wolfenstein, Ann. Rev. Nucl. Sci. 6 (1956) 43.
[14] O. D. Cheishvili, Sov. Phys.—JETP 3 (1957) 974.
[15] U. Fano, Rev. Mod. Phys. 29 (1957) 74.
[16] S. Devons and L. J. B. Goldfarb, Handbuch der Physik 42 (1957) 362.
[17] H. Stapp, Phys. Rev. 107 (1957) 607.
[18] R. J. Blin-Stoyle and M. A. Grace, Handbuch der Physik 42 (1957) 558.
[19] O. D. Cheishvili, Sov. Phys.—JETP 5 (1957) 1009.
[20] L. J. B. Goldfarb, Nucl. Phys. 7 (1958) 622.
[21] C. B. Van Wyk, Nuovo Cim. 9 (1958) 270.
[22] L. C. Biedenharn, Ann. Phys. (N.Y.) 4 (1958) 104, and 6 (1959) 399.
[23] G. R. Satchler, Nucl. Phys. 8 (1958) 65.
[24] H. Faissner, Ergeb. Exakt. Naturw. 32 (1959) 180.
[25] L. J. B. Goldfarb and J. R. Rook, Nucl. Phys. 12 (1959) 494.
[26] L. J. B. Goldfarb, Nucl. Phys. 12 (1959) 657.
[27] J. Raynal, J. Phys. Radium 21 (1960) 373; Thesis, Paris (1964).
[28] G. R. Satchler, Oak Ridge National Laboratory, Report ORNL-2861 (1960) (unpublished).
[29] T. A. Welton, Fast Neutron Physics, vol. II (Interscience, 1963) p. 1317.
[30] J. Button and R. Mermod, Phys. Rev. 118 (1960) 1333.
[31] E. V. Inopin, Nucl. Phys. 48 (1963) 517.
[32] C. J. Mullin, J. M. Keller, C. L. Hammer, and R. H. Good, Jr., Ann. Phys. (N.Y.) 37 (1966) 55.
[33] G. Racah, Phys. Rev. 62 (1942) 438.
[34] S. E. Darden, Am. J. Phys. 35 (1967) 727.
[35] M. A. Preston, Physics of the Nucleus (Addison-Wesley Co., 1962) p. 230.
[36] M. Rose, Elementary Theory of Angular Momentum (John Wiley and Sons, 1957).

DISCUSSION

Keaton:

I believe you have done the Physics community a service by bringing to everyone's attention the important difference between *efficiency tensors* and the polarization produced in the inverse reaction, which, for lack of a better term, I will call *inverse polarization*. There are, however, several points about your proposal which I would like to stress. First, the simple relationship of a sign change under time reversal between efficiency tensors and inverse polarizations. An expression of the cross section in the laboratory in terms of inverse polarization would have a complicated appearance. Second, I find it most unsatisfying that the tensor moments of the beam are expressed in one coordinate system, and the tensor moments describing the reaction (inverse polarization) are expressed in another coordinate system. Third, I would like to see cross sections written simply as an invariant contraction of tensors. This form would hold in the c.m. or the lab system. This becomes more important as the formalism is generalized to include the polarization transfer experiments which are now being done.

Breit:

Whatever convention is adopted it appears desirable that in its presentation the earlier work on nucleon-nucleon scattering be taken into account by stating the relationship or equivalence of the new to the old symbols. I am referring to the Wolfenstein triple-scattering parameters D, R, R', A, A' and the correlation coefficients C_{nn}, etc.

Darden:

Dr. Ohlsen will probably comment on this. I believe the parameters you have mentioned are just the polarization transfer coefficients. I think it may be premature to establish conventions for these.

Breit:

Cartesian tensors are not really a bad representation, especially in the case of scattering of spin 1/2 particles by each other. The total spin is either zero or one. The former is very easy to deal with. In the latter case, spin rotations may be described in the Cartesian system as those of a vector, and the equations are readily understood by anyone knowing the elements of vector analysis.

Darden:

I am sorry my prejudices showed through so strongly. I did not mean to imply that one should not use Cartesian moments.

The Nucleon-Nucleon Interaction

MALCOLM H. Mac GREGOR, Lawrence Radiation Laboratory, Livermore,
California, USA[*]

Nucleon-nucleon scattering is the best example we can name of an
experiment that requires polarized sources or targets. On the one
hand, we have the fact that there are five scattering amplitudes pres-
ent for all values of $\vec{J} = \vec{\ell} + \vec{S}$, so we need five kinds of experiments
to make any kind of a decent analysis. (If inelastic channels are open,
the situation gets even worse; we then need nine kinds of experiments.)
On the other hand, we have the fact that nucleon-nucleon phases are
very democratic. There are no resonances to be found, at least in the
elastic energy region, and no phase tries to dominate the other phases.
Hence a rigorous analysis is required to obtain a definitive result.
Fig. 1 illustrates the standard kinds of polarization experiments that
are performed.

As a means of summarizing the experiments that have been carried
out, I have shown in figs. 2a, 2b, and 2c the p,p and n,p experiments
that were carried out in the years up to the times of the First, Second,

N–N EXPERIMENTS IN THE LAB FRAME

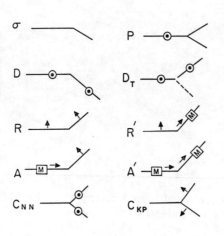

Fig. 1. Standard nucleon-
nucleon scattering experiments. Dots and arrows
indicate polarization vectors, and M indicates precession by 90° in a magnetic
field. Other observables
include $A_{yy} \equiv C_{nn}$, and A_{xx},
which is similar to C_{kp}.

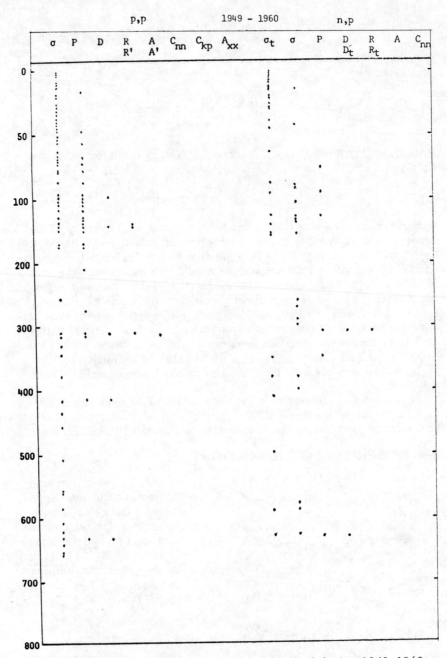

Fig. 2a. Nucleon-nucleon experiments published during 1949-1960, the interval corresponding to the First Polarization Symposium.

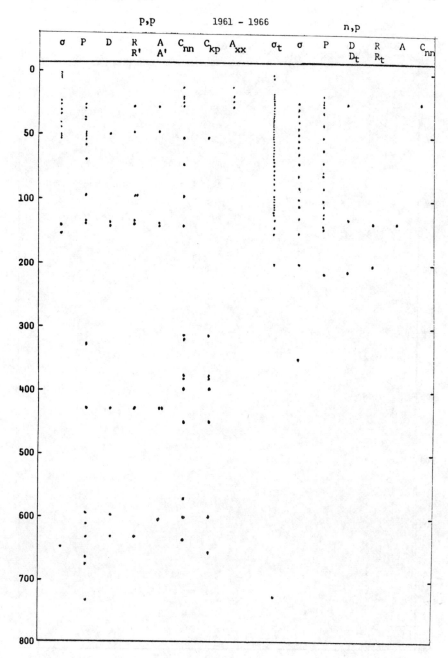

Fig. 2b. Nucleon-nucleon experiments published during 1961-1966, the interval corresponding to the Second Polarization Symposium.

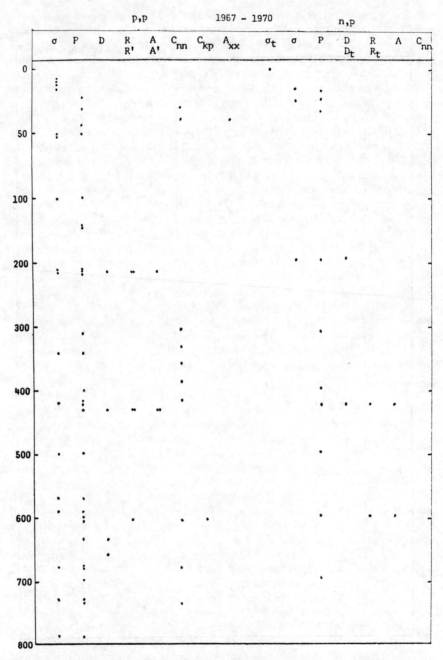

Fig. 2c. Nucleon–nucleon experiments published during 1967–1970,
the interval corresponding to the Third Polarization Symposium.

and Third Polarization Symposia. We can see that whereas the early work was limited mainly to differential cross-section and polarization data, later measurements include the more involved so-called double-scattering and triple-scattering experiments, names that modern polarization technology has fortunately rendered somewhat obsolete. What these figures do not show is that many of the early measurements included only one or a few scattering angles, whereas the later measurements generally include many angles and very good statistics, although absolute normalization continues in some cases to be a problem.

If we want to represent the information contained in the data of figs. 2a—2c in some kind of useful condensed form, the one way that suggests itself is phase shift analysis. We can reproduce the store of information contained in these data by specifying about two dozen phase shifts as a function of energy. At the time of the First Polarization Symposium at Basel in 1960, modern phase shift analyses were just getting started. Stapp at Berkeley had completed his classic analysis of the 310 MeV Berkeley triple-scattering data, and the Livermore group was starting to extend Stapp's results. Breit at Yale had carried out the first energy-dependent analyses. By the time of the Second Polarization Symposium at Karlsruhe in 1965, both Yale and Livermore had comprehensive phase shift programs under way, and a number of more specialized phase shift analyses were being carried out at various laboratories and universities. However, the data were not yet complete enough, and in some cases not accurate enough, and the computing techniques were not yet refined enough to permit really definitive analyses.

As we meet here for this Third Polarization Symposium at Madison, one major phase of this work has come to fruition. The experimental groups have resolved their major differences and have done an admirable job of providing a complete set of p,p and n,p measurements over the whole elastic energy range from about 1 to 450 MeV. The Yale [1] and Livermore [2—5] groups have completed analyses of these data that give precise fits to essentially all of the p,p and n,p data below 450 MeV, a total of some 2000 data points. I think it is a fair statement to say that any differences between the Yale phases and the Livermore phases are smaller than the uncertainties we know still exist because of inconsistencies or incompleteness in the data selection and because of certain small but important theoretical corrections that must be included in the parameterization of the phase shifts.

I would now like to mention the areas in which difficulties still reside in the p,p analyses. One area is the region above 450 MeV, where inelastic effects become important. By the time we reach an energy of 650 MeV (the Dubna cyclotron energy), almost half of the total cross section is inelastic. In our high-energy Livermore analyses [3], we found a strong coupling between the elastic and the inelastic matrix elements. Fig. 3 shows how two of the uncoupled phase shifts, $^1D_2$ and $^3F_3$, vary as we add more and more inelastic freedom. Our

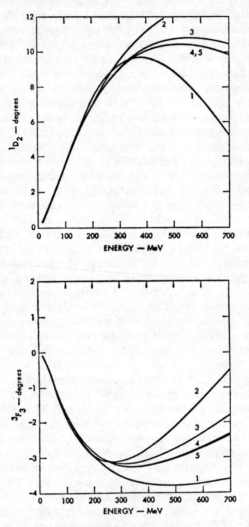

Fig. 3. Variations in the phases $^1D_2$ and $^3F_3$ from a Livermore solution at high energies [3] as the inelastic freedom is increased: curve (1), no inelasticity; (2) $^1D_2$ inelastic; (3) $^3P_1$ and $^3F_3$ inelasticity added; (4) $^3P_2$ and $^3F_2$ inelasticity added; (5) $^3P_0$ and $^1G_4$ inelasticity added. Curves (4) and (5) indicate some stabilizing of the solution.

conclusion [3] is that since we have no definitive way of handling inelastic effects, we cannot carry out a definitive analysis of the elastic phases. Of course, if we had a complete set of nine kinds of p,p experiments for all scattering angles at a single energy, we could in principle determine the elastic scattering matrix without knowing anything about the inelasticity. But such data do not yet exist. However, our conclusion about the difficulties at high energies may be somewhat pessimistic. The inelasticity can be inserted in a fairly reasonable manner by adding imaginary components to a few selected phases and constraining them by fitting to measured total and reaction cross sections. Experimental work at 650 MeV is being extended [6], and high-energy phase shift analyses are also being extended [7]. Fig. 4 shows a comparison of recent CERN p,p differential cross-section and polarization data [8] with earlier predictions from the Livermore high-energy phase shift analyses [3]. As can be seen, the agreement is quite close.

Some recent measurements of p,p scattering in the 35−50 MeV range have been made by groups at Saclay and Grenoble. Fig. 5 shows new polarization data [9] at 30 MeV, together with the phase shift predictions from Yale [1] and Livermore [5]. Fig. 6 shows new A_{xx} and A_{yy} measurements [10] at 37.2 MeV. These are compared to predictions from the Livermore analyses [5] and from the Tamagaki potential [11]. A_{yy} accurately fits the Livermore prediction, whereas A_{xx} does not, which suggests [10] that the Livermore P-phases may need some adjustment in this energy region.

At low energies, p,p experiments have recently centered around attempts to clear up difficulties with the differential cross section at 10 MeV [12−16]. These difficulties have to do with absolute normalizations and the complications concomitant with making measurements at small angles in the region of the steeply-rising Coulomb peak. The most recent Los Alamos measurements [16] appear to have cleared up the discrepancies. Recent work at Michigan State and Los Alamos [17] has included a careful re-examination of all the p,p data below 30 MeV and a detailed analysis of the various small but important electromagnetic corrections that must be applied to phase shift analyses in this energy region. In carrying out their own phase shift analysis of these data [17], these workers conclude that inconsistencies still exist in some of the low-energy p,p data.

The situation with respect to n,p phase shift analyses, while in reasonably satisfactory shape, is understandably not as favorable as for p,p analyses. In n,p scattering, both isotopic spin $I = 1$ and $I = 0$ components are present. Since the data are not sufficient to specify both the $I = 1$ and the $I = 0$ scattering matrices (this would require five experiments at each energy measured over the complete angular range), we must take the $I = 1$ scattering matrix from p,p analyses and use this plus the n,p data to determine the $I = 0$ matrix. Thus any uncertainties in the p,p results get coupled directly into the n,p analyses. Also,

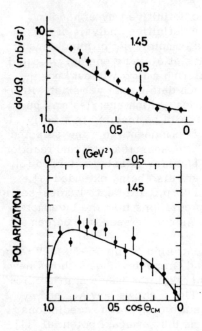

Fig. 4. p,p differential cross-section and polarization fits of recent CERN measurements [8] at 1.45 GeV/c ($T_{lab} = 0.79$ GeV) to the Livermore [3] solution 4 of fig. 3 in the present paper.

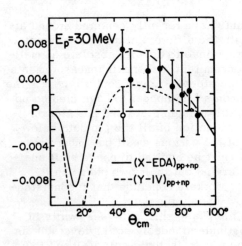

Fig. 5. Recent p,p differential cross-section and polarization data from Grenoble [9] fitted to the Yale [1] and Livermore [5] p,p predictions. The predictions bracket the experimental values.

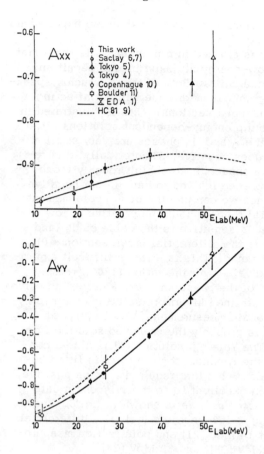

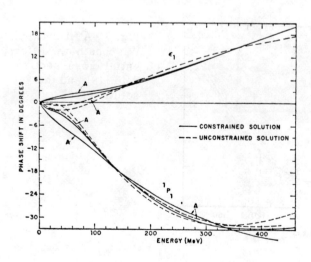

Fig. 6. Recent p,p A_{xx} and A_{yy} data at 37 MeV [10] fitted to the Livermore [5] p,p phase shift solution and the predictions of the Tamagaki potential [11]. The Livermore predictions fit A_{yy} accurately, but not A_{xx}.

Fig. 7. Livermore energy-dependent n,p phases ϵ_1 and $^1P_1$ [5]. The unconstrained solution goes to unphysical values for these phases at low energies. The curves labeled A are a forced fit to Wisconsin data [19] and give a more realistic solution.

n,p experiments are in general harder to carry out than p,p experiments to the same level of accuracy.

The main uncertainty that has existed in n,p analyses is somewhat of a fluke. It turns out that low-energy n,p experiments permit an ambiguity in the ϵ_1 and $^1P_1$ phase shifts. Both of these, roughly speaking, can get large or small together in magnitude without affecting fits to quantities such as total cross sections. This is illustrated in fig. 7, which shows Livermore n,p energy-dependent solutions [5]. With no constraints applied, the ϵ_1 and $^1P_1$ phases are very small at low energies, and ϵ_1 actually goes negative. Theoretically, this result seems incorrect, since we expect $^1P_1$ to exhibit roughly a one-pion-exchange behavior at low energies and to have larger negative values near threshold, and since we can use the quadrupole moment of the deuteron to conclude [18] that ϵ_1 should be positive at low energies. One measurement that is sensitive to the value of $^1P_1$ (and hence also to the value of ϵ_1) is the differential cross section $\sigma(\theta)_{np}$, measured over a wide angular range. This is a rather difficult experiment, and the lack of accurate values for the ratio $\sigma(180°)/\sigma(90°)$ has been responsible for most of the phase shift ambiguity. The curves labeled A in fig. 7 show a constrained Livermore solution obtained by forcing a fit to a recent Wisconsin measurement [19] of $\sigma(\theta)_{np}$ at 24 MeV. This constrained fit is probably the preferred solution from the Livermore n,p analyses. The Yale n,p solution [1] is also a constrained fit in the sense that the ϵ_1 phase, for example, is fitted below 10 MeV to a potential model value that required ϵ_1 to be positive. The Yale solution [20] gives an excellent fit to the Wisconsin data at 24 MeV, which was added *a posteriori*, as is shown in fig. 8. Recent measurements of n,p scattering at low energies include a differential cross-section measurement at 14 MeV [21] and polarization measurements at 16.1 MeV [22], 21.6 MeV [23], and 32 MeV [24].

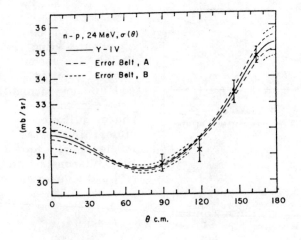

Fig. 8. A fit between Wisconsin n,p differential cross-section data [19] and the Yale solution [20].

With respect to nucleon-nucleon experiments that are still in the preliminary stages, I should mention a Harwell measurement of small angle p,p differential elastic scattering at about 159 MeV [25]. An interesting feature of this work is the use of a free-standing solid hydrogen target 1 cm in diameter and 5 cm high. In general, emphasis has shifted from nucleon-nucleon experiments to experiments involving three or more nucleons [26, 27]. In some cases, with their nucleon-nucleon missions accomplished, cyclotrons have been dismantled [28].

Turning to theoretical work, we first mention some of the efforts that have been made to improve phase shift analyses. One of the most interesting results comes from Carnegie-Mellon [29]. In the Yale and Livermore p,p analyses, each of 14 phases was parameterized separately. However, in a dispersion relation sense, these phases should be regarded as arising from partial wave projections over a common set of left-hand discontinuities. Hence their parameterizations should be related. By using relationships of this general nature, the Carnegie-Mellon workers [29] were able to reduce the required number of free phases from 14 to 9 or 10 while maintaining the same statistical accuracy. In a different context, workers from Dubna, Saclay, and CERN have been developing statistical techniques for removing ambiguities from phase shift solutions [30], including predictions of the most useful experiments to perform.

In direct attempts to solve the nuclear force problem, we have the by now familiar situation that the one-pion-exchange diagram gives an important and easy to calculate contribution to the nuclear force, whereas two- and higher-pion-exchange diagrams present extremely formidable calculational problems. The higher-pion contributions can be tackled directly by evaluating the left-hand discontinuities in dispersion integrals, or they can be handled more tractably, but also more approximately, by using one-boson-exchange models. Probably the most determined efforts in this area have come from the Japanese physicists. Without attempting to sort out the papers as to the particular type of calculation being performed, we give in ref. [31] a list of papers that were supplied to the speaker in connection with this symposium. With respect to calculations of a nucleon-nucleon potential, we mention research efforts that have been carried out recently at Zürich [32] and at Dubna and Prague [33]. Some work has also been done recently at Michigan State [34], and somewhat earlier at Cornell [35].

One well-known theoretical result that gives a striking (if somewhat limited) success is the simple one-boson-exchange model. If, in addition to the one-pion-exchange diagram, we add in exchanges of σ, ρ, and ω "particles," we obtain good qualitative values for the phase shifts, even for P-waves, as shown in fig. 9 [28]. Other "particles" can be added to this model, but these seem to be the important ones. (Since the masses of these "particles" are not really determined by fitting to nucleon-nucleon phases, we can, if we like, use the ϵ resonance instead of the σ resonance.) As examples of recent work in this

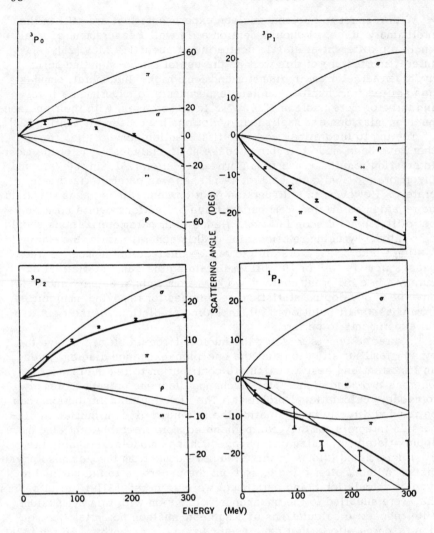

Fig. 9. Contributions of π, σ, ρ, and ω one-boson-exchange terms to nucleon-nucleon P-phases. This figure was taken from ref. [28].

area, in addition to the references [31] of the Japanese workers, we
cite papers from Texas A & M [36] and a review article from Hamburg
[37]. The comment about the somewhat limited success of one-boson-
exchange models at the beginning of this paragraph means that although
this model gives good qualitative results, it is difficult to know just
how one should proceed in attempting to provide a more rigorous treatment than is supplied by the use of simple one-particle-exchange
Feynman diagrams. These, among other things, are manifestly non-
unitary, since they contribute only real amplitudes. Recent reviews
of the general problems associated with potential models have come
from Buffalo [38]. For a recent theoretical study involving a single
nucleon, see ref. [39].

REFERENCES

*Work performed under the auspices of the U.S. Atomic Energy
Commission.

[1] R. E. Seamon et al., Phys. Rev. 165 (1968) 1579.
[2] M. H. Mac Gregor, R. A. Arndt, and R. M. Wright, Phys. Rev.
     169 (1968) 1128.
[3] M. H. Mac Gregor, R. A. Arndt, and R. M. Wright, Phys. Rev.
     169 (1968) 1149.
[4] M. H. Mac Gregor, R. A. Arndt, and R. M. Wright, Phys. Rev.
     173 (1968) 1272.
[5] M. H. Mac Gregor, R. A. Arndt, and R. M. Wright, Phys. Rev.
     182 (1969) 1714.
[6] J. Bystrický et al., Phys. Lett. 28B (1969) 572; L. P. Glonti, Yu.
     M. Kazarinov, and M. R. Khayatov, Dubna preprint; S. I. Bilen-
     kaya, L. N. Glonti, Yu. M. Kazarinov, and V. S. Kiselev, JINR
     preprint Pl-4960 (1970).
[7] Preprint Pl-4960 cited in ref. [7] gives the latest results of the
     continuing Dubna phase shift analyses at 650 MeV; see also
     N. Hoshizaki and T. Kadota, Prog. Theor. Phys. 42 (1969) 815,
     826 for phase shift analyses at even higher energies.
[8] M. G. Albrow et al., to be published.
[9] J. Arvieux et al., Third Polarization Symp.
[10] D. Garreta, K. Nisimura, and M. Fruneau, Phys. Lett. 31B (1970)
      363; a similar result has also been obtained at 46.9 MeV.
[11] R. Tamagaki and W. Watari, Progr. Theoret. Phys. (Kyoto), Suppl.
      39 (1967) 23.
[12] R. J. Slobodrian, H. E. Conzett, E. Shield, and W. F. Trivol,
      Phys. Rev. 174 (1968) 1122.
[13] M. H. Mac Gregor, R. A. Arndt, and R. M. Wright, Phys. Rev.
      179 (1969) 1624.

[14]   N. Jarmie, R. E. Brown, R. L. Hutson, and J. L. Detch, Jr., Phys.
        Rev. Lett. 24 (1970) 240.

[15]   J. Holdeman, P. Signell, and M. Sher, Phys. Rev. Lett. 24 (1970)
        243.

[16]   N. Jarmie, J. H. Lett, J. L. Detch, Jr., and R. L. Hutson, Phys.
        Rev. Lett. 25 (1970) 34.

[17]   M. S. Sher, P. Signell, and L. Heller, Ann. Phys. (N.Y.) 58
        (1970) 1.

[18]   G. Breit and R. D. Haracz, in High Energy Physics, ed. E. H. S.
        Burhop (Academic Press, New York, 1967), vol. I, p. 127, n. 15.

[19]   L. N. Rothenberg, Phys. Rev. C1 (1970) 1226.

[20]   G. Breit, J. Lucas, and M. Tischler, Phys. Rev. 184 (1969) 1668.

[21]   M. Tanaka, N. Koori, and S. Shirato, J. Phys. Soc. Japan 28
        (1970) 11. Fig. 4 of this ref. shows the data as compared to
        the unconstrained Livermore solution. The constrained Liver-
        more solution of course gives a better fit, although the experi-
        mental results appear to indicate more anisotropy than indicated
        by the phase shift analyses.

[22]   R. Garrett et al., Third Polarization Symp.

[23]   F. D. Brooks and D. T. L. Jones, Third Polarization Symp.

[24]   N. Ryu et al., Third Polarization Symp.

[25]   O. N. Jarvis, private communication.

[26]   C. J. Batty, private communication.

[27]   N. Booth and C. Dolnick, to be published; N. E. Booth et al.,
        to be published.

[28]   See M. H. Mac Gregor, Phys. Today 22 (1969) 21.

[29]   Yung-An Chao, Phys. Rev. Lett. 25 (1970) 309.

[30]   A. Pazman, JINR preprint E5-3775, Dubna, 1968; A. Pazman et
        al., Czech. J. Phys. B19 (1969) 882; J. Bystrický and F. Lehar,
        to be published; Dubna preprint P1-4960 (ref. [6]).

[31]   S. Furuichi and M. Yonezawa, Progr. Theoret. Phys. (Kyoto) 38
        (1967) 1200; S. Furuichi, ibid. 42 (1969) 837; S. Furuichi and T.
        Sokawa, ibid. 41 (1969) 1504 and 42 (1969) 146 E; S. Furuichi,
        H. Suemitsu, W. Watari, and M. Yonezawa, ibid. 40 (1968) 523;
        S. Furuichi, T. Ueda, W. Watari, and M. Yonezawa, ibid. 41
        (1969) 131; S. Furuichi, H. Suemitsu, W. Watari, and M. Yone-
        zawa, ibid. 41 (1969) 461; S. Furuichi, W. Watari, and M. Yone-
        zawa, ibid. 41 (1969) 850; S. Furuichi, H. Kanada and K. Wata-
        nabe, RUP-70-2, to be published in ibid.; K. Takada, S. Takagi,
        and W. Watari, ibid. 38 (1967) 144; Y. Kishi, S. Sawada, and W.
        Watari, ibid. 38 (1967) 892; M. Kikugawa, W. Watari, and M.
        Yonezawa, ibid., Suppl., Extra Number (1968) 170; M. Kikugawa,
        W. Watari, and M. Yonezawa, ibid. 43 (1970) 407; S. Otsuki,
        ibid., Suppl. 42 (1968) 39; T. Ueda, ibid. 43 (1970) 696; T. Ueda,
        to be published in ibid.; S. Sato and T. Ueda, to be published
        in ibid.; T. Ueda and A. E. S. Green, Nucl. Phys. B10 (1969) 289;

T. Ueda, Progr. Theoret. Phys. (Kyoto) 41 (1969) 131; T. Ueda
and A. E. S. Green, Phys. Rev. 174 (1968) 1304.

[32]   J. Benn and G. Scharf, Nucl. Phys. A134 (1969) 481.
[33]   J. Bystrický, F. Lehar, and I. Úlehla, Phys. Lett. 20 (1966) 186;
       I. Úlehla, Czech. Phys. B19 (1969) 553; I. Úlehla et al., ibid.
       B19 (1969) 1570; I. Úlehla, ibid. B20 (1970) 401.
[34]   P. Signell, to be published.
[35]   R. V. Reid, Jr., Ann. Phys. (N.Y.) 50 (1968) 411.
[36]   R. A. Bryan, Nucl. Phys. A146 (1970) 359; R. A. Bryan, in press.
[37]   G. Kramer, DESY T-70/3, July (1970).
[38]   G. Breit, Nucl. Phys. B14 (1969) 507; G. Breit, Int. Conf. on
       Properties of Nuclear States," Montreal (1969).
[39]   M. H. Mac Gregor, UCRL-72736 and UCRL-72747, to be pub-
       lished in Lett. Nuovo Cim.

DISCUSSION

Simmons:

I have a comment on the low energy p-p system, namely about
the Δ^p_{LS} phase discussed by Signell earlier. Near 11 MeV the
p-p experiments do not affect this quantity much. However, the
n-p polarization at 90° c.m. does relate to it directly. Improved
measurements to a precision of ±0.002 or so, would represent a
definite contribution to knowledge of this quantity.

Slobodrian:

Your phase shift analyses go up to 700 MeV, and you use the
Pauli spin formalism. Would it not be more adequate to use
Dirac wave functions for the phase shift parameterization of
nucleon-nucleon experimental data when one is reaching energies of the order of half the nucleon rest mass?

Mac Gregor:

We rely on the experimentalists to reduce their experimental
data to the center of mass. Then the Pauli formalism is all right.
You are correct that there are important corrections at high energies.

Feshbach:

The calculation of a nucleon-nucleon potential so that unitarity
is automatically satisfied has been made by Partovi and Lomon.
The various vector and scalar mesons are included and recoil
effects are consistently taken into account. The method used
is based on the Blankenbecler and Sugar approximation to the
Bethe-Salpeter equation. A second remark has to do with electron-

deuteron scattering which can be exploited to yield the charge form factor of the neutron. Until recently the low-q slope of the form factor was known and some values at high q, where q is the momentum transfer. Recent experiments have been at the low momentum transfer end, so the coefficient of the q^2 term and q^4 terms have been measured. These are consistent with the boundary-condition model of nucleon forces and are in distinct disagreement with the Hamada-Johnston potential.

The Three- and Four-Nucleon System

J. S. C. McKEE, University of Birmingham, England

1. NUCLEON-DEUTERON SCATTERING

Polarization effects in n-d and p-d scattering have become clearly established over the past few years. The contour map prepared by Haeberli [1], which is shown in fig. 1, is a comprehensive survey of the most recent p-d polarization data in the range from 0 to 50-MeV incident proton energy. It shows distinctly that the polarization becomes negative at backward angles in the vicinity of 19-MeV incident proton energy, a fact first established at Berkeley [2] and later confirmed in a more precise experiment carried out by Faivre [3]. The

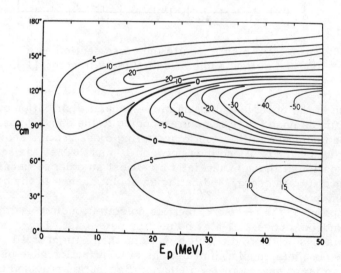

Fig. 1. Contour map of the polarization of protons scattered from deuterium. Curves of constant polarization are shown as a function of proton laboratory energy [1].

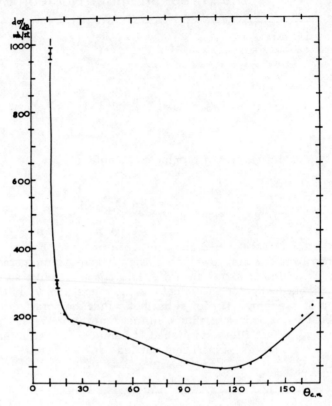

Fig. 2. P-d elastic scattering cross section as a
function of c.m. angle at 8 MeV (data from ref. [1]).

position of the observed minimum in the angular distribution of the
asymmetry is near 120° c.m., corresponding to the angular position of
the broad minimum in the elastic scattering cross section shown in
fig. 2. In the range of energies between 10 and 20 MeV the scattering
cross section at the minimum falls by almost an order of magnitude for
a factor of two in energy, indicating an abrupt change in the complexity of the problem.

Compared to the p-d work, there is no corresponding wealth of n-d
data. Indeed, above 22.7 MeV no neutron data exist. Experiments
with neutrons are more difficult to perform than with charged particles,
and few complete angular distributions of polarization have been obtained so far at any energy. An angular distribution provided by Walter
[4] is shown in fig. 3. Clearly it shows the same general shape as
p-d scattering and is unremarkable.

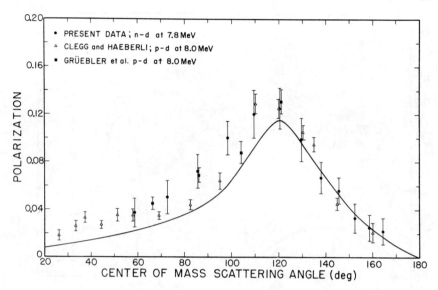

Fig. 3. Angular distribution of the polarization of neutrons elastically scattered from deuterium at 7.8 MeV-incident energy [4].

The only point worth noting in the n-d scattering is the suggestion that the dip at backward angles may be deeper at some energies for neutrons than for protons. This is illustrated in fig. 4, which shows data obtained at 23 MeV for both elastic processes. Accurate measurements of n-d elastic scattering and polarization above 23 MeV would be useful. Bonner [5] is currently engaged on such a program at the Rutherford laboratory.

The three-body problem concerning two bound nucleons and one free nucleon is extremely complex, but capable of eventual theoretical solution in terms of the Faddeev equations. The spin structure of the problem is complicated, and the only useful calculations made to date have involved approximate methods for dealing with it. The most popular of these has been the impulse approximation.

A calculation of the polarization in p-d scattering at 150 MeV using the impulse approximation made by Kottler and Kowalski [6] was remarkably successful at reproducing the p-d data of Postma [7]. The method was, however, unsuccessful at reproducing the 40-MeV data of Conzett [8] for the same reaction. Until recently only the simple model of Hüfner and de Shalit [9] has been successful in fitting the lower energy data in a convincing way. They interpreted the rapid variation of $P(\theta)$ with angle θ around the minimum of $d\sigma/d\Omega$ in terms of a diffraction mechanism. Their fit (fig. 5) is remarkable, as it involves only one free parameter.

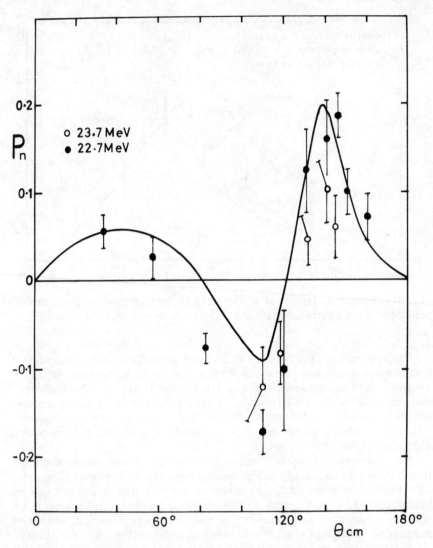

Fig. 4. Angular distribution of the polarization of neutrons elastically scattered from deuterium at 22.7 MeV (full circles [51]) and 23.7 MeV (open circles [52]). The solid line shows the trend of p-d data for comparison purposes.

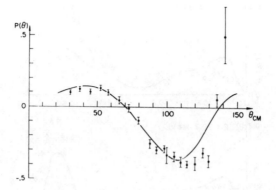

Fig. 5. Fit (Hüfner and de Shalit [9]) to the
angular distribution of polarization in p-d
scattering at 40-MeV incident laboratory
energy [8].

Krauss and Kowalski [10] suggest that whereas the study of the
spin-dependent properties of the three-nucleon system may be a sensitive probe of the underlying dynamics, clearly a study of the cross
section is not. They make the point that most calculations of low
energy nucleon-deuteron scattering have been performed under the
assumption of central two-body interactions which render them in-
capable of predicting the polarization of the scattered nucleon. The
"exact" solution to the spin-dependent problem is of great complexity, but Krauss and Kowalski have reduced this complexity by means
of the approximate procedure of Sloan [11, 12]. In this method the
integral equations for three-particle scattering are written in the form
of a Heitler equation which is solved by partial-wave analysis. The
approximation lies in replacing the source term in this equation by the
sum of a nucleon-exchange term and a impulse graph. The two-
nucleon forces are introduced through simple separable Yamaguchi
terms, and only the singlet S-wave and the triplet S- and D-waves
are included in the calculation. The polarizations predicted by this
model (which fits cross-section data reasonably well) are shown in
fig. 6 for proton energies of 14.4 and 22.7 MeV. The data are those
of Faivre [3]. The method has succeeded in predicting the negative
dip at backward angles in the 22.7-MeV data and the absence of such
a dip at 14.4 MeV. Considering the crudeness of the two-nucleon
force used, the agreement with experiment is remarkable and represents a considerable step forward in our understanding of the important features of the problem. While reviewing elastic scattering,
it is worth drawing attention to the contributed paper from the Zürich
group [13] in which new data on the tensor moments $\langle iT_{11} \rangle$, $\langle T_{20} \rangle$,

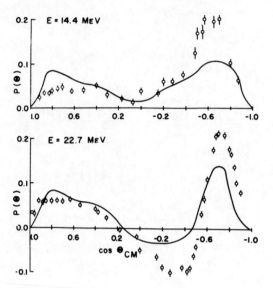

Fig. 6. Fits (Krauss and Kowalski [10]) to
the p-d polarization data of ref. [3] at
energies of 14.4 and 22.7 MeV.

$\langle T_{21} \rangle$, and $\langle T_{22} \rangle$ of polarized deuterons elastically scattered from protons are presented. These data cover a considerable angular range
and show significant tensor polarization at 6-MeV incident deuteron
energy. The conclusion that spin-dependent forces are required in
the consideration of three-nucleon systems at c.m. energies as low
as 2 MeV illustrates the complexity of the problem which requires
solution. Some work on the same reaction at 12.2-MeV deuteron energy, communicated to me privately by Griffith (Birmingham), confirms
these findings and is shown in fig. 7, together with predictions from
a phase-shift analysis of Berovic and Clews [14]. There has so far
been no measurement of the nucleon-deuteron spin correlation parameters, although Berovic hopes to perform such a measurement in the
near future by scattering polarized deuterons from a polarized proton
target. No further data on the triple scattering parameters have been
obtained since Haeberli [1] reviewed the subject in 1969.

I will now discuss the polarization effects observed in the breakup
of the deuteron by nucleons. Asymmetry and polarization are not necessarily the same for inelastic scattering processes. Darden [15] has
reviewed this topic admirably. Clegg [16] has shown the asymmetry
of the total number of breakup neutrons to be small over a considerable
forward angular range at energies of 10 and 12 MeV, and Arvieux [17]
has suggested that the asymmetry in quasi-scattering of protons from
the virtual singlet deuteron is similar to that obtained in scattering

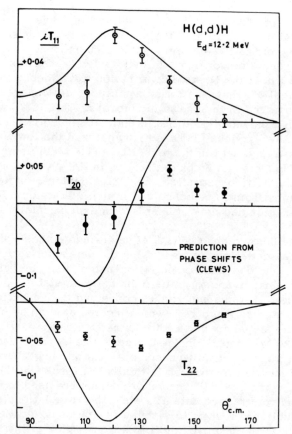

Fig. 7. Angular distibutions for the tensor
moments iT_{11}, T_{20}, and T_{22} in the elastic
scattering of polarized deuterons by hydrogen
at 12.2 MeV [53].

from the real triplet deuteron. This suggestion is supported by his
experimental data. Other experiments by Kuroda [18–21] and Spalding
[22] examine the nucleon-nucleon interaction potential through polarization effects in inelastic scattering processes.

   It is interesting to examine various analyses of p–d elastic scattering data which have been performed in recent years and to assess
their contribution to the overall understanding of the problem. Seyler
[23] has derived fromulae relating the phase shifts in proton-deuteron
scattering to the differential cross section and polarization. The number of parameters involved in this analysis is formidable, and it is
clearly easier to calculate phase shifts from an approximate theory,
if such is available, rather than to use what Seagrave has described

as the complete "bone-crushing formalism" of Seyler. Purrington and Gammel [24] have calculated the nucleon-deuteron phase shifts using the Born approximation and introducing a nucleon-nucleon interaction which includes a tensor term. In their analysis, they couple S- and D-states and adjust the D-state contribution to the deuteron wave function empirically. Their scattering amplitudes are complex, and channel spin is conserved in the scattering process. This is a low-energy analysis, but produces good fits to both cross-section and polarization data at 9 MeV. Arvieux [25] has also analyzed the low-energy problem (in this case, p-d data in the region 1.5–11.5 MeV) using the phase shifts of Van Oers and Brockman [26] as initial parameters. Arvieux has used the helicity formalism, neglected mixing of states, and assumed channel spin to be conserved in the interaction. This frequently made assumption has not yet been shown experimentally to be valid [27].

Recently Berovic and Clews [14] have made a phase-shift analysis of the same scattering process, using the parameterization of Seyler [23] for the nucleon-nucleon collision matrix, and have allowed for splitting of phases and for transitions between states. Again they have assumed that channel spin is conserved throughout the scattering. The phase shift program (which incorporates an automatic search routine) was tested against known Rutherford cross sections up to an energy of 20 MeV for the incident proton, and angular distributions calculated from the unsplit phases of Christian and Gammel and Van Oers and Brockman reproduced faithfully the results of the original analyses. When the split phases of Seyler were used to predict cross-section and polarization data at 3 MeV, the calculations again agreed with the published results to within a few percent, a difference readily attributable to the improved facilities for complex arithmetic and double precision available on the IBM 360/44 computer used in this work. The phase parameters of Arvieux failed, however, to reproduce published polarization when used in the program, although the cross-section data were readily and faithfully computed. Two typical cases are shown in fig. 8. The phase shifts of Arvieux [25] are compared with those of Clews [14] in table 1. Differences in the sets of phases are quite small and involve mainly the splitting of ^2P phases and the inclusion of $^4D_{3/2}$, $^4D_{5/2}$, and $^4D_{7/2}$ contributions in all but the Arvieux set. Berovic concludes that an error in phase has occurred in the Arvieux parameterization, as the differential cross sections (phase insensitive) are well reproduced whereas the polarizations (phase sensitive) are not. The Berovic-Clews analysis has been used to fit cross sections and polarizations up to 6 MeV using thirteen free parameters associated with S-, P-, and D-waves which all vary smoothly with energy (fig. 9). At present this analysis is being extended to search in the mixing parameters, with the hope of removing the restriction of channel spin conservation. The additional parameters intro-

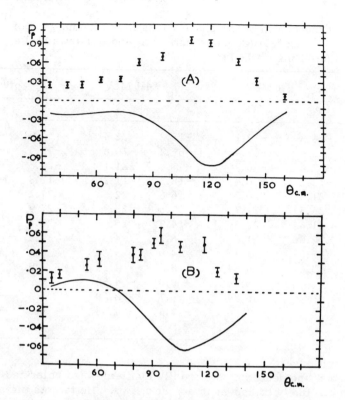

Fig. 8. Fit by Berovic and Clews [14] using the phase
parameters of Arvieux to p-d polarization data at
6 MeV (curve A) and 3 MeV (curve B).

duced by this extension demand further experimental data and, in
particular, the expectation values of additional tensor moments, such
as the spin correlation operators $\langle T_{11,11} \rangle$ and $\langle T_{11,1-1} \rangle$. Undoubtedly
this additional information will shortly be obtained, as both polarized
beams and targets are becoming readily available.

Additional facts on the three-body problem may be found in the review articles by Haeberli [1] and myself [28].

2. POLARIZATION IN n-T and p-³He SCATTERING

Experiments completed recently on the four-nucleon system include the elastic scattering of nucleons by ³He and tritium, the elastic
scattering of deuterons by deuterons, and the various breakup reactions possible with these combinations of particles. Polarized beams

Table 1. Comparison of the phase shifts from the analyses of
Arvieux [25] (columns marked A) and of Clews [14]
(columns marked C)

	E_p = 3.0 MeV		E_p = 4.0 MeV		E_p = 6.0 MeV	
	A	C	A	C	A	C
$^2S_{1/2}$	−14.3	−22.1	−15.2	−22.4	−46.9	−28.8
$^4S_{3/2}$	−65.6	−67.4	−69.2	−70.8	−78.5	−80.2
$^2P_{1/2}$	− 7.4	− 6.9	− 7.4	−21.1	−21.8	−25.0
$^2P_{3/2}$	−13.2	− 6.4	−18.3	−11.2	−12.8	−18.0
$^4P_{1/2}$	21.8	20.8	16.6	22.5	27.2	23.9
$^4P_{3/2}$	25.3	21.7	17.5	23.9	27.9	25.7
$^4P_{5/2}$	21.6	23.3	14.9	25.7	22.6	28.0
$^2D_{3/2}$	2.06	1.6	6.2	4.3	1.2	8.32
$^2D_{5/2}$	2.06	1.4	4.6	5.64	3.3	9.76
$^4D_{1/2}$	− 3.67	− 4.64	− 8.4	− 4.29	− 6.9	4.6
$^4D_{3/2}$		− 5.26	− 9.0	− 4.56	− 7.2	− 4.78
$^4D_{5/2}$		− 6.84	− 6.7	− 8.60	− 5.4	−15.11
$^4D_{7/2}$		− 4.66	− 7.3	− 4.53	− 6.5	− 1.34
$^4F_{3/2}$					2.0	

of nucleons and deuterons are readily available, and polarized ^{3}He
sources are likely to appear in a year or two. ^{3}He targets already
exist. Measurements of the polarization and of the reaction asymmetry
of outgoing particles from reactions involving four nucleons are being
made, and an impressive amount of undigested and largely uncompre-
hended data has been accumulated.

The recent measurement of the polarization of neutrons elastically
scattered from tritium [29] yielded unexpected results. The Los Alamos
group, using 40% polarized neutrons, was able to measure neutron polarization for lab angles from 40° to 118.5° (136° c.m.). The scatterer was
one mole of liquid tritium. The energy of the neutrons was 22.1 MeV,
only slightly greater than that at which Tivol [30] had earlier studied
the charge-conjugate reaction ^{3}He($\vec{p}$,p)^{3}He (21.3 MeV). The unexpected
and significant feature of the Los Alamos data is that the neutron
polarizations are greater than the corresponding proton polarizations
at all angles except for the crossover point at 112° c.m. where both
are zero. In all other studies of charge-conjugate reactions the neutron polarizations are found to be either equal to or significantly less
than the corresponding proton values. (Note, however, the earlier comment on n-d scattering at 22 MeV.) Accurate angular distributions of
scattering cross sections for these two processes at this energy could
be instructive.

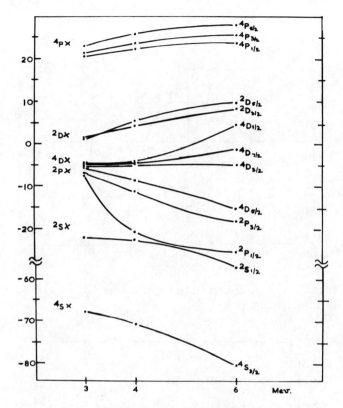

Fig. 9. Energy variation of the phases of Berovic
and Clews [14]. The unsplit phases of Christian
and Gammel are indicated by the symbol ×.

Earlier data on polarization effects in p-^{3}He scattering have been
reviewed by Tombrello [31], who made a phase-shift analysis of data
obtained at energies between 1.0 and 11.5 MeV. Tivol [30] has extended the range of the data in five steps to 21.3 MeV, and Griffiths
[32] has recently published results obtained at energies of 30 and
50 MeV at the Rutherford laboratory. Work is in progress on a phase-
shift analysis of these data. A phase-shift analysis carried out by
Morrow and Haeberli [33] of all data between 1 and 11 MeV has enabled a broad region of solutions to be defined, including the earlier
solutions of Tombrello [31]. The spin-correlation A_{xz} measurement
of McSherry [34] at 8.8 MeV has significantly reduced the uncertainty
with which the p-^{3}He phase shifts can be determined (see fig. 10),
in particular the $^3P_0$ phase and the coupling parameter ϵ_0 for states of
$J^\pi = 1^-$ in the p-^{3}He system. No extension of this analysis to include
the higher energy data has yet been reported.

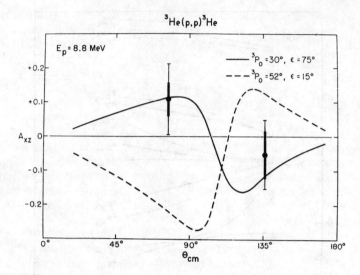

Fig. 10. The spin correlation parameter A_{xz} measured at 8.8 MeV by McSherry [34].

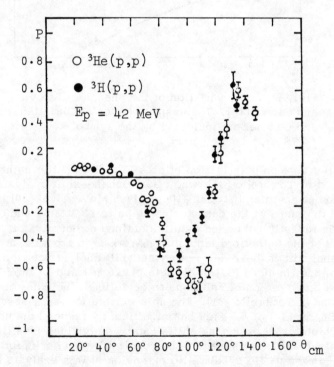

Fig. 11. Comparison of the polarization in $^3\text{He}(p,p)^3\text{He}$ and $T(p,p)T$.

Some new data from Grenoble (Arvieux et al.) are available in which the p-^{3}He polarizations at 42 MeV are shown to be higher than those found by Griffith [32] at 30 and 50 MeV, and also greater than p-T values at the same energy (see fig. 11).

3. POLARIZATION EXPERIMENTS INVOLVING p + T AND n + ^{3}He FINAL STATES

Keaton [35] has described work carried out at Los Alamos using the polarized triton facility in a study of the elastic scattering of polarized tritons by hydrogen. This experiment was performed at incident triton energies between 6.4 and 9.7 MeV. The only reaction channels open are to (T + p) and (n + ^{3}He) final states. Triton polarizations were calculated from a knowledge of the states in the mass-4 system obtained by Werntz and Meyerhof [36] from measurements of the T(p,n)^{3}He and T(p,n)^{3}He processes. The values obtained are shown as dashed lines in fig. 12, and the agreement with experimental results is good. The details of the calculation of the polarization are soon to be published by Dodder. Cross sections and polarizations are calculated using the formulae of Lane and Thomas [37]. So far it has not been

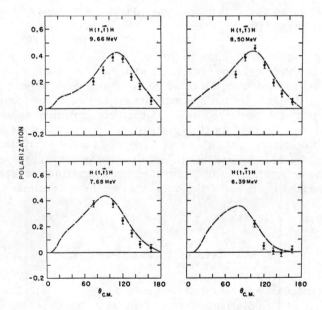

Fig. 12. Calculated triton polarizations compared with the experimental results of Keaton [35] in elastic scattering of tritons by hydrogen at incident laboratory energies between 6.4 and 9.7 MeV.

possible to find a notably better set of energy levels than those already deduced by Werntz and Meyerhof.

The reactions T(p,p)T, ^{3}He(n,p)T, T(p,n)^{3}He, and ^{3}He(n,n)^{3}He are all useful for the investigation of possible excited states of the ^{4}He system. Evidence for a 0^+ state at around 20-MeV incident neutron energy in the ^{3}He(n,p)T reaction was first published some time ago [38], and later confirmed by Passell and Schermer [39] in a study of the transmission of polarized thermal neutrons through polarized ^{3}He. Also, low-energy T(p,n)^{3}He polarization and cross-section data indicate 0^- and 2^-, T = 0 states in ^{4}He at energies around 21.4 and 22.4 MeV, and this indication is reflected in the spectrum of protons from the ^{3}He-d reaction studied by Jarmie [40]. Further work on these reactions will undoubtedly involve precise polarization measurements.

Cranberg (Virginia) has some new data on the polarization of neutrons from the T(p,n)^{3}He reaction, as measured by scattering from ^{4}He using the time-of-flight method for protons of up to 2.2-MeV incident energy. Comparison with the predictions of the theory of Werntz and Meyerhof [36] indicates the need for adjustment of the energies of the virtual 0^- and 2^- states in ^{4}He at around 22-MeV excitation energy in this case.

4. DEUTERONS AS PROJECTILES—D(d,d)D, D(d,p)T, AND D(d,n)^{3}He REACTIONS

It is impossible to discuss in this paper all experiments relevant to a study of the four-body system. I will therefore limit myself largely to experiments which use polarized deuterons as the incident particles. In elastic scattering, for each scattering angle θ, the dependence of the intensity of the scattered beam on the polarization of the incident beam $\langle T_{qk} \rangle$ is described by five parameters. These are the vector analyzing power $\langle iT_{11} \rangle$, the three second-rank tensor analyzing powers $\langle T_{20} \rangle$, $\langle T_{21} \rangle$, and $\langle T_{22} \rangle$, and the differential cross section for an unpolarized beam $\sigma_0(\theta)$. If $\sigma_p(\theta,\phi)$ denotes the scattering cross section for the polarized beam and ϕ the azimuthal angle of scattering, then

$$\sigma(\theta,\phi) = \sigma_0(\theta) \left\{ 1 + \langle t_{20} \rangle \langle T_{20} \rangle + 2 \langle t_{21} \rangle \langle T_{21} \rangle \cos\phi \right.$$
$$\left. + 2 \langle it_{11} \rangle \langle iT_{11} \rangle \cos\phi + 2 \langle t_{22} \rangle \langle T_{22} \rangle \cos 2\phi \right\}$$

gives the cross section required (see refs. [41] and [42]). The sign convention adopted for polarization is analogous to the Basel convention for spin-1/2 particles, and $\vec{k}_{in} \times \vec{k}_{out}$ is taken as positive. If the reaction is time-reversed, the vector analyzing power $\langle iT_{11} \rangle$ is identical with the vector polarization of deuterons in the inverse reaction using unpolarized particles at the same c.m. energy.

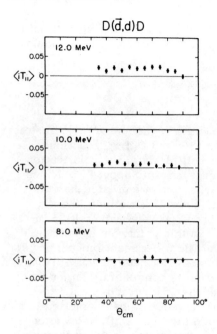

Fig. 13. Vector analyzing power for the reaction $D(\vec{d},d)d$ at 8.0, 10.0, and 12.0 MeV [44].

Bernstein [43] has measured the vector analyzing power for the elastic scattering of deuterons by deuterium at 10-MeV incident energy and has found values that were very small or zero at all angles measured. The polarized deuterons were generated by bombarding a deuterium target with alpha particles, and the errors on the measurements were substantial. Plattner and Keller [44] have examined the same problem at 8, 10, and 12 MeV using the polarized beam from the Wisconsin tandem accelerator. These data are of excellent quality and also indicate that the vector polarization is very small in the energy interval studied. At 10 MeV and 12 MeV, however, the data are not compatible with zero, although the polarizations are all less than 3% (fig. 13). Plattner and Keller point out that the low polarizations are surprising, since it is known that orbital momenta greater than zero are important in the scattering cross section. They also comment that polarization in d-d elastic scattering is almost an order of magnitude smaller than that found in other elastic scattering processes from targets consisting of a few nucleons, whether they be fermions or bosons. Vector and tensor polarizations have also been measured at 21.4 MeV by Arvieux [45] using the polarized beam at Saclay. Here again the vector polarization is small, not substantially greater than that observed at 12 MeV. In summary then, the d-d system is not well understood. The latest elastic cross-section data [46—47] show the angular distributions to be featureless over the energy range between

3 and 12.25 MeV and to vary smoothly with energy. The spin structure of the problem is known to be complex, but as yet we have found no handle to the solution.

5. D(d,$\vec{n}$)^{3}He and D(d,$\vec{p}$)T REACTIONS

Families of polarization curves compiled by Seagrave [29] illustrate clearly the data available on the D(d,n)^{3}He and D(d,p)T reactions. Because of the identical particles in the initial state, the polarization curves for the outgoing nucleon are antisymmetric about 90°. Above 6 MeV it is noted that the proton polarizations are greater than the charge-conjugate neutron values by a factor of something like $\sqrt{2}$. This fact is still a puzzle. Recent work from Duke University communicated to me by Walter is relevant to this point. An experimental polarization excitation function for the D(d,$\vec{n}$)^{3}He reaction near 45° c.m. is compared with proton data as well as earlier neutron work. From fig. 14 it is clear that proton polarization is consistently larger than neutron polarization at all angles over the energy range from 4 to 14 MeV. The advent of polarized deuteron sources has permitted the reactions mentioned above to be studied in more detail, and vector analyzing power and polarization have already been studied to some extent. Bernstein [43] detected protons from the D($\vec{d}$,p)^{3}H reaction, and ^{3}He from the D($\vec{d}$,n)^{3}He reaction at 10 MeV, thus deriving information on the analyzing power for protons at forward angles and neutrons at backward angles (fig. 15). Rather optimistically, on the basis of two overlapping points, it was then suggested that the charge independence of nuclear forces was satisfied in these reactions. Recent neutron data from Birmingham [48] at the same energy tend to reinforce this view. There is certainly no evidence for a difference in the vector analyzing powers of these two reactions.

Huber [49] has presented work on the same pair of reactions carried out at low energy, between 150 and 460 keV. In this energy region the vector analyzing power changes very slightly, and the same angular dependence is observed in both reactions. Two tensor moments vary smoothly with energy, but the depolarization parameter shows a change in both magnitude and sign over this interval. The fact that there is no difference in the vector analyzing powers of the two reactions at this energy is not too significant, as polarizations in the D(d,$\vec{n}$)^{3}He and D(d,$\vec{p}$)T reactions are also indistinguishable at these energies.

Berovic and Clews [14] have measured the vector analyzing power of the reaction D($\vec{d}$,p)T at 12 MeV. The results are consistent with the neutron data of Blyth [48] at slightly lower energy, but may indicate that the points of Bernstein [43] at backward angles are larger than expected. The "best fit" curve of Blyth would accommodate the points of Berovic quite readily on the basis of charge independence in these reactions.

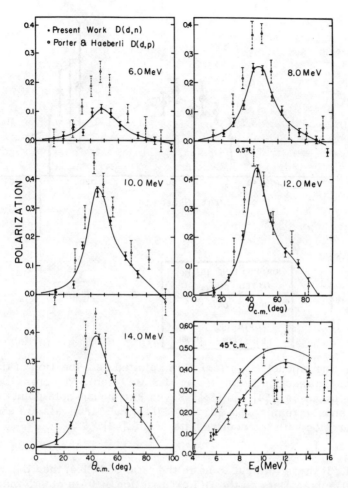

Fig. 14. The reactions $D(\vec{d},\vec{n})^3He$ and $D(\vec{d},\vec{p})T$ are compared as a function of energy and angle up to 14.5 MeV incident energy [4].

Finally, a remark about polarization transfer in few-body reactions. Measurements by Simmons [50] and Blyth [48] have been made on the 0° transfer of polarization from deuteron to neutron in the $^3H(\vec{d},\vec{n})\alpha$ and $D(\vec{d},\vec{n})^3He$ reactions, respectively. In each case substantial neutron polarizations have been observed, amounting to more than 2/3 of the vector polarization of the deuteron beam. Blyth has described a simple model which fits the $H(\vec{d},\vec{n})2p$ and $D(\vec{d},\vec{n})^3He$ reactions at 0°, but predicts slightly larger neutron polarization in the $T(\vec{d},\vec{n})\alpha$ case than is observed. No simple model of this kind will predict different polarization for the neutron and the proton in the outgoing

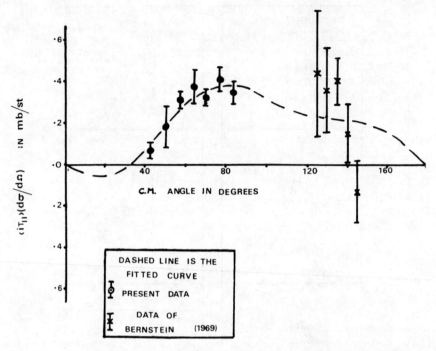

Fig. 15. The product $\langle it_{11} \rangle$ $(d\sigma/d\Omega)$ is plotted as a function of the c.m. angle for the reaction $D(d,n)^3He$ at 11.5 MeV (lab). The cross-section data are from refs. [54] and [55]. Seven Birmingham data points covering the angular region from 30° to 6° (lab) in 5° intervals are shown, in conjunction with the neutron data of ref. [43].

channel. If these turn out to be in the ratio of 1 : 1.4, then the model is of little use. Data on the $D(\vec{d},\vec{p})T$ reaction at 0° is clearly needed.

I am well aware that I have omitted from this paper data considered by many to be important, and that I have included some experimental work already quite familiar. My aim has been to spotlight those areas of the subject in which substantial advances have been made in the recent past, and to underline discrepancies, ambiguities, and enigmas wherever they have occurred. Much has yet to be done before the four-body problem, quite apart from its solution, is clearly defined.

REFERENCES

[1] W. Haeberli, Proc. Int. Conf. on Three-Body Problem (North-Holland Publishing Co., Amsterdam, 1970) p. 188.

[2] J. S. C. McKee, D. J. Clark, R. J. Slobodrian, and W. F. Tivol, Phys. Lett. 24B (1967) 240.

[3] J. C. Faivre et al., Nucl. Phys. A127 (1969) 169.

[4] R. L. Walter, private communication; R. K. Walter et al., Phys. Lett. 30B (1969) 626.

[5] B. E. Bonner, private communication.

[6] H. Kottler and K. L. Kowalski, Phys. Rev. 138 (1965) 619.

[7] H. Postma and R. Wilson, Phys. Rev. 121 (1961) 1229.

[8] H. E. Conzett et al., Phys. Lett. 11 (1964) 68.

[9] A. de Shalit, Preludes in Theoretical Physics (North-Holland Publishing Co., Amsterdam, 1966) p. 35.

[10] J. Krauss and K. L. Kowalski, Phys. Lett. 31B (1970) 263.

[11] I. H. Sloan, Phys. Rev. 165 (1968) 1587.

[12] I. H. Sloan, Phys. Lett. 25B (1967) 84.

[13] R. E. White et al., Third Polarization Symp.

[14] N. Berovic, private communication; C. Clews, M.Sc. Thesis, University of Birmingham (1970).

[15] S. E. Darden, Am. J. Phys. 35 (1967) 727.

[16] T. B. Clegg, G. R. Plattner, and W. Haeberli, Nucl. Phys. A119 (1968) 238.

[17] J. Arvieux, J. L. Durand, A. Papineau, and A. Tarrats, Proc. Int. Conf. on Three-Body Problem (North-Holland Publishing Co., Amsterdam, 1970) p. 212.

[18] K. Kuroda, A. Michalowich, M. Poulet, and F. Gomez-Gimeno, Phys. Lett. 16 (1965) 133.

[19] K. Kuroda, A. Michalowich, and M. Poulet, J. Phys. (Paris) 26 (1965) 303.

[20] K. Kuroda, Can. J. Phys. 20 (1966) 277.

[21] K. Kuroda, A. Michalowich, and M. Poulet, Nucl. Phys. 88 (1966) 33.

[22] D. Spalding, A. R. Thomas, N. W. Reay, and E. H. Thorndike, Phys. Rev. 150 (1966) 806.

[23] R. G. Seyler, Nucl. Phys. A124 (1969) 253.

[24] R. D. Purrington and J. L. Gammel, Phys. Rev. 168 (1968) 1174.

[25] J. Arvieux et al., Nucl. Phys. A102 (1967) 503.

[26] W. T. H. Van Oers and K. W. Brockman, Jr., Nucl. Phys. A92 (1967) 561.

[27] J. Raynal, Nucl. Phys. 100 (1967) 473.

[28] J. S. C. McKee, Reports on Progress in Physics (1970), in press.

[29] J. D. Seagrave, Proc. Int. Symp. on Few-Body Problems (Gordon & Breach, New York, 1968) p. 40, and Third Polarization Symp.

[30] W. F. Tivol et al., University of California Report 17299 (1966) B1.

[31] T. A. Tombrello, Phys. Rev. 138 (1965) 41, and University of California Report UCRL-17299 (1967) p. 83.

[32] R. J. Griffith et al., PLA Progress Report (1968) p. 22.

[33] L. W. Morrow and W. Haeberli, Nucl. Phys. A126 (1969) 225.
[34] D. S. McSherry, S. D. Baker, G. R. Plattner, and T. B. Clegg,
      Nucl. Phys. A126 (1969) 233.
[35] P. W. Keaton, Jr., Proc. Symp. on Nuclear Reaction Mechanisms
      (Quebec, 1969) p. 331.
[36] C. Werntz and W. E. Meyerhof, Nucl. Phys. A121 (1968) 38.
[37] A. M. Lane and R. G. Thomas, Rev. Mod. Phys. 30 (1958) 257.
[38] A. A. Bergman and F. L. Shapiro, Sov. Phys.—JETP 13 (1961) 895.
[39] L. Passell and R. I. Schermer, Phys. Rev. 150 (1966) 146.
[40] J. N. Jarmie, R. N. Stokes, G. G. Ohlsen, and R. W. Newsome, Jr.,
      Phys. Rev. 161 (1967) 1050.
[41] W. Lakin, Phys. Rev. 98 (1955) 139.
[42] G. R. Satchler, Nucl. Phys. 8 (1958) 65.
[43] E. M. Bernstein, G. G. Ohlsen, V. S. Starkovich, and W. G. Simon,
      Nucl. Phys. A126 (1969) 641.
[44] G. R. Plattner and L. G. Keller, Phys. Lett. 30B (1969) 327.
[45] J. Arvieux, J. Goudergue, B. Mayer, and A. Papineau, Phys. Lett.
      22 (1966) 610.
[46] A. S. Wilson, M. C. Taylor, J. C. Legg, and G. C. Phillips,
      Nucl. Phys. A126 (1969) 193.
[47] A. D. Bacher and T. A. Tombrello, Nucl. Phys. A113 (1968) 557.
[48] C. O. Blyth, P. B. Dunscombe, J. S. C. McKee, and C. Pope,
      Third Polarization Symp.
[49] P. Huber, Proc. Symp. on Nuclear Reaction Mechanisms (Quebec,
      1969) p. 287.
[50] J. E. Simmons et al., Bull. Am. Phys. Soc. 14 (1969) 1230.
[51] J. J. Malanify et al., Phys. Rev. 146 (1966) 632.
[52] R. L. Walter and C. A. Kelsey, Nucl. Phys. 46 (1963) 66.
[53] J. A. R. Griffith (Birmingham), private communication.
[54] M. D. Goldberg and J. M. LeBlanc, Phys. Rev. 119 (1960) 1992.
[55] S. T. Thornton, Nucl. Phys. A136 (1969) 25.

DISCUSSION

Baker:

Data now available from Saclay on p + ^{3}He include proton polarization, ^{3}He polarization, and the spin-correlation parameters
A_{yy} and A_{xx}, measured at $E_p = 19.4$ MeV. It appears that at this
energy, despite inelastic processes whose magnitudes have not
been directly measured, some of the phase shifts are better determined than at energies below 12 MeV.

Cahill:

Differential cross sections of ^{3}He(p,p)^{3}He have been measured
between 18 and 57 MeV. The phase-shift analysis referred to
by Baker has been extended to these data and makes use of the

Rutherford Laboratory polarization data at 30 MeV, and the
Grenoble polarization data at 42.7 MeV presented at this meet-
ing. The analysis presents a smooth behavior in energy to 30
MeV and extends easily to 42.7 MeV.

Ohlsen:

We are repeating the experiment of Bernstein et al. using the
Los Alamos polarized source. The work is about half completed
and includes measurements of all four analyzing tensors for the
$D(\vec{d},t)p$, $D(\vec{d},{}^3He)n$, $D(\vec{d},p){}^3He$, and $D(\vec{d},d)D$ reactions at 10 MeV.
We confirm that there is no significant difference between the
p-t and the n-^{3}He outgoing particle channels.

Arvieux:

There is a discrepancy in the sign of the proton polarization and
the vector polarization of the deuteron, when one analyzes p-d
elastic scattering with Seyler's and with my phase-shift program.
This was brought to my attention by Weitrutz and Brüning from
Hamburg when they compared both formalisms some months ago.
My error, which occurred in computing the rotation matrix elements, has been corrected. A new analysis, which supersedes
the old one, has been made by the Hamburg group for both p-d
and n-d scattering up to E_{lab} = 8 MeV. The results of this
analysis are included in the Proceedings of this symposium.

The Five-Nucleon System and the Calibration of Polarization of Beams of Neutrons, Protons, and Deuterons

H. R. STRIEBEL, University of Basel, Switzerland

1. INTRODUCTION

The main purpose of this paper is to survey the recent experimental work on polarization effects in five-nucleon systems and to discuss the possibilities of calibrating the polarization of nucleons and deuterons. At the Karlsruhe conference, five years ago, Barschall [1] summarized the results in this field very carefully. Since then efficient sources of polarized protons and deuterons have been installed in many laboratories or existing sources have been improved by increasing the beam current and polarization. This has revived the interest in interactions suitable to calibrate nuclear polarizations and at the same time has allowed polarization measurements of high precision.

At moderate energies the most widely used monitor for nucleon polarization is the nucleon-helium scattering, while the tensor polarization of deuterons is preferably calibrated by the reactions ^{3}He(d,p)^{4}He or T(d,n)^{4}He.

The subject of the first section of this paper is the elastic scattering of nucleons from ^{4}He. In the second section, the ^{3}He(d,p)- and T(d,n)-reaction and the scattering of deuterons from ^{3}He and ^{3}H are discussed.

2. ELASTIC SCATTERING OF NUCLEONS FROM ^{4}He-NUCLEI

2.1. *Proton-α scattering*

Since the Karlsruhe conference, differential cross-section and polarization measurements for p-α scattering have been performed over an energy range from 1 to about 70 MeV. Only data published since 1965 will be mentioned here (for earlier references see [1]).

Polarization angular distributions at ten proton energies between 0.9 and 3.2 MeV have been obtained [2] at the DTM in Washington and analyzed [3] satisfactorily, using S- and P-wave phase shifts only. Weitkamp and Haeberli [4] have analyzed their polarization measurements in the energy region from 17.5 to 27 MeV together with the cross section data of Allison and Smythe [5]. They included S-, P-, D-, and F-waves and arrived at a fair representation of the experimental polarization, but the calculated differential cross sections near the $3/2^+$ resonance exceeded the experimental values by about 10%.

At Saclay [6] the differential cross section and the polarization have been remeasured with 12-, 14.3- and 17.5-MeV protons and at twelve scattering angles for thirteen proton energies between 22 and 25 MeV, i.e., near the $3/2^+$ resonance. The phase-shift analysis included the data of refs. [4] and [5]. The analysis is very similar to that of ref. [4], but resulted in a better fit of the differential cross section.

Jahns and Bernstein [7] measured the proton polarization at eleven points, including six different angles and four energies between 6 and 11 MeV. Their results are in good agreement with previous experiments. In order to fill a gap, Schwandt et al. [8] studied p-α scattering between 4.6 and 12 MeV. They measured polarization angular distributions at five energies and represented the phases by a generalized effective-range expansion. Plummer et al. [9] deduced phase shifts for the 10-MeV region from their polarization measurements at 10-MeV proton energy and from some earlier data.

Polarization and cross-section measurements carried out at the Lawrence Radiation Laboratory [10] at proton energies of 20 to 40 MeV are reported elsewhere in these Proceedings. An analysis of these latter experiments is under way.

The average experimental error of most of the recent polarization data on p-α scattering is less than ± 0.02 and the deviations between comparable experimental values obtained by different authors are in general not much larger. Fig. 1 is a contour plot of the analyzing power of p-α elastic scattering. The data are mainly from refs. [3, 4, 6, 7, 9].

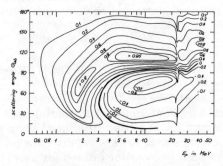

Fig. 1. Contour plot of the analyzing power of p-α scattering. Energies and scattering angles are given in the lab system.

An inspection of fig. 1 shows clearly that p-α scattering is an excellent analyzer for lab energies between 2 and 22 MeV as well as above the $3/2^+$ resonance near 23 MeV. For proton energies larger than 4 MeV and scattering angles around 120°, analyzing powers of nearly 1 are available and the values are very insensitive to proton energy and scattering angle. In the low energy region (1.5—4 MeV) scattering angles between 60° and 100° are preferable.

2.2. *Neutron-α scattering*

The papers concerning experimental cross sections and polarizations in n-α scattering are collected in ref. [11]. Publications of the last six years on this topic cover mainly the energy interval between 0.2 and 34 MeV.

Since the analyzing power of helium for neutron polarization is usually deduced from the n-α scattering phase shifts obtained from differential cross-section data, such measurements have been performed with high precision. Hoop and Barschall [12] fitted their experimental differential cross sections at 25 neutron energies between 2 and 30.7 MeV by a set of phase shifts including S-, P-, D-, and F-waves. Above 22 MeV, where the t + d channel opens, inelastic parameters were included also. In the energy region from 0.2 to 7 MeV Morgan and Walter [13] have measured 22 angular distributions of neutrons scattered from He. They corrected the data for the finite energy resolution of the He recoil detector. Their phase-shift analysis, which allows for S- and P-waves, is in fair agreement with ref. [12]. The energy dependence of the phases corresponds to single-level dispersion theory. Time-of-flight data of the angular distributions [14] at energies between 0.5 and 1.4 MeV include small angles and were fitted by phases similar to those of ref. [13]. Niiler et al. [15] report on differential cross-section measurements and a phase-shift analysis near 20-MeV neutron energy. Their S- and $P_{1/2}$-phase shifts are appreciably smaller than those of ref. [12], while the D- and F-phases are larger. Furthermore, Niiler et al. also considered G-waves.

Polarization measurements for n-α scattering are inevitably less reliable than for p-α scattering. In the neutron case the polarization of the incident particles, arising from a nuclear reaction, varies rapidly with emission angle and energy. Protons from a source of polarized ions, on the other hand, are available at all energies in a well-collimated monoergic beam and have constant polarization. Thus, in analyzing left-right asymmetries of scattered neutrons, the polarization of the incident neutrons is handled as an adjustable parameter.

Angular distributions of the asymmetry in the scattering of polarized neutrons from ^{4}He have been measured with high accuracy at 1 and 2.4 MeV by Sawers et al. [16]. An energy dependent phase-shift analysis of these results and of the angular distribution data is given in ref. [13]. At Duke University, Stammbach et al. [17] have measured and analyzed polarization angular distributions at 7.80 and 3.38 MeV.

Polarization data have been obtained at five neutron energies between 2 and 24 MeV by May et al. [18] and corrected for geometrical and multiple scattering effects in refs. [13, 16]. These data are well fitted by the phase shifts of Hoop and Barschall [12]. The polarization measurements of Büsser et al. [19] at 12 and 16.2 MeV confirm the observations of ref. [18].

Broste et al. [20, 21] have measured the analyzing power of n-α scattering at six energies between 11 and 30 MeV. Their results are well fitted by slightly adjusted Hoop-Barschall phase shifts. Polarization measurements of Arifkhanov et al. [22] with 27.8- and 34.1-MeV neutrons are well confirmed by the results of ref. [21].

The statistical uncertainties of the analyzing power of n-α scattering are usually of the order of 0.05 with the exception of the Duke measurements [16, 17], where an error < 0.02 is given. In addition, unknown systematic errors arising from P_1, the polarization of the incident neutrons, might be important. In view of these uncertainties the recent data agree well. A contour plot of the analyzing power of ^{4}He for the polarization of elastically scattered neutrons is shown in fig. 2, which has been drawn on the basis of refs. [12, 13, 16—22].

The plot exhibits a wide region with analyzing power P > 0.9 between 3 and 20 MeV and for angles 105° < θ_{lab} < 120°. At energies above 23 MeV the preferable scattering angle is 130°. A high analyzing power is expected near E_n = 600 keV for a scattering angle of 60°, but in this region no polarization measurements have been performed. An isolated measurement at E_n = 262 keV and θ_{lab} = 75° by Jewell et al. [23] gave P_2 = +0.21 ± 0.06.

2.3. *Comparison of the nucleon-α scattering analyses*

Several recent papers are concerned with a comparison of different sets of phase shifts, with a comparison of n-α and p-α scattering, or with nuclear model calculations.

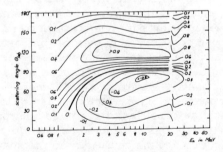

Fig. 2. Contour plot of the analyzing power of n-α scattering. Energies and scattering angles are given in the lab system.

Satchler et al. [24] used a real optical model potential of Woods-Saxon form with a spin-orbit coupling term peaking inside the central potential. The radius of the central well and the strength of the spin-orbit term were allowed to depend linearly on energy. For the best fit of the data the potential strengths were slightly different for protons and neutrons. The phase shifts, polarizations, and cross sections computed from the potential model for nucleon energies between 0 and 20 MeV agree well with the experimental values.

In their n-α interaction study, Morgan and Walter [13] calculated phase shifts based on the single-level dispersion theory. Stammbach and Walter [25] have reinvestigated this problem, fitting the actual data instead of empirically derived phase shifts. They list resonance parameters for S-, P-, D-, and F-waves. For the best fit to the experimental data (0–20 MeV) slightly different reduced widths were necessary for p-α and n-α scattering. The agreement with experimental data in general is quite good.

Arndt and Roper [11] analyzed selected n-α scattering measurements using an effective range parameterization of the phase shifts. They were not able to determine reliably D- or F-waves in the energy range between 0 and 20 MeV. Reasonable fits to the 0–16.4 MeV data were obtained, but beyond this range the fits are poorer.

A two-channel (^{3}He-d and ^{4}He-p) calculation based on a refined cluster model was performed by Heiss and Hackenbroich for the ^{5}Li [26] and the ^{5}He [27] scattering states. Their results agree qualitatively with the experimental values. In particular the $3/2^+$ resonant state is explained as a consequence of the strong coupling between the $D_{3/2}$ nucleon-α channel and the $S_{3/2}$ T-d channel.

Kraus et al. [28] have satisfactorily described polarizations and differential cross sections of the p-α scattering from 1 to 12 MeV in the formalism of Humblet and Rosenfeld using complex resonant terms for the $P_{3/2}$ and $P_{1/2}$ resonances.

Since the n-α analyzing power is usually calculated from phase shifts, it is interesting to compare the polarizations computed from the various sets of phase shifts. Rhea et al. [29] show that in the energy region from 0.5 MeV to 14 MeV the analyzing power for the scattering angle where $P(\theta)$ reaches its maximum depends little on the chosen set of phase shifts (see figs. 1 and 2 of ref. [29]).

3. FIVE-NUCLEON INTERACTIONS INDUCED BY POLARIZED DEUTERONS

3.1. $^3He(d,p)^4He$ reaction

As proposed by Galonski, Willard, and Welton [30] the ^{3}He(d,p)^{4}He reaction as well as its mirror reaction have been intensively used at energies below 1 MeV to calibrate the tensor polarization of beams from polarized deuteron sources. Only recently have these reactions been studied with polarized deuterons above 1 MeV as well as with polarized ³He targets. About twenty papers on this subject were submitted to this conference.

Starkovich et al. [31] obtained vector polarized deuterons by scattering deuterons from helium and used them to study the vector analyzing power of the ³He(d,p) reaction at 10, 12, and 14 MeV. Plattner and Keller [32] performed measurements of the angular distribution of the vector analyzing power $\langle iT_{11}\rangle$ for the ³He(d,p) reaction between 4 and 10 MeV, and Klingler and Dusch [33] report measurements between 2 and 13 MeV. These three experiments agree generally to better than ±10%. At 430 keV Forssmann et al. [34] have obtained data on the vector analyzing power.

Grüebler et al. [35] measured the angular distribution of all four tensor moments $\langle iT_{11}\rangle$, $\langle T_{20}\rangle$, $\langle T_{21}\rangle$, and $\langle T_{22}\rangle$ at energies between 2.8 and 10 MeV. Their measurements of $\langle iT_{11}\rangle$ confirm the other measurements [31−33]. The vector and tensor analyzing power measured by Keaton et al. [36] at 10 MeV is described elsewhere in these Proceedings. The general features of the results of refs. [35] and [36] are the same, but the values of $\langle T_{20}\rangle$ differ roughly by 15%.

Fig. 3 shows the vector analyzing power of refs. [31−35] and the tensor polarization data of [35, 36]. The drastic changes of the analyzing power with energy near 6-MeV lab energy must be attributed to the known state of ⁵Li near 20 MeV. The analyzing power between 0 and 2 MeV is mainly determined by the S-wave resonance corresponding to the 3/2⁺ state in ⁵Li at 16.65 MeV. The dashed portion of the

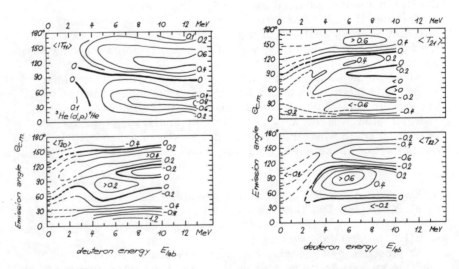

Fig. 3. Tensor moments $\langle iT_{11}\rangle$, $\langle T_{20}\rangle$, $\langle T_{21}\rangle$, and $\langle T_{22}\rangle$ of the ³He(d,p)⁴He reaction.

curves connects the measured data with the values [30] of the analyzing power that would result from a purely resonant S-wave interaction leading to the $3/2^+$-state.

From fig. 3 it is evident that all components of the analyzing power
of the ^{3}He(d,p)^4 reaction in the energy region at least up to 12 MeV
are large enough to mark this reaction as an excellent analyzer for
deuteron polarization.

Since polarized ^{3}He targets are available at several laboratories,
the ^{3}He(d,p)^{4}He reaction has repeatedly been studied with polarized
^{3}He nuclei. The first measurements of this sort are those of Baker et
al. [37] at several deuteron energies up to 10 MeV. Brückmann and
Schmidt [38] observed this reaction with a polarized target at 52-MeV
deuteron energy. Watt and Leland [33] are reporting measurements at
6, 8, and 10 MeV; Huber et al. [40], experimental data at energies
below 2.5 MeV. Measurements between 2.8 and 10 MeV have been
submitted by Leemann, Grüebler, et al. [41] and data at 10 MeV by
Baker [42].

Fig. 4 depicts the analyzing power $A(\theta)$ of the ^{3}He(d,p) reaction for
^{3}He polarization. The data, taken from refs. [37, 39—41], agree in general within the limits of error.

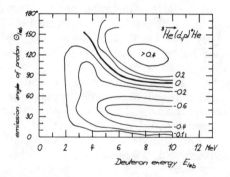

Fig. 4. Analyzing power of the
^{3}He(d,p)^{4}He reaction for the ^{3}He
polarization.

Between the analyzing power $A(\theta)$ for the ^{3}He polarization and the
proton polarization $P(\theta)$ of the ^{3}He(d,p) reaction induced with unpolarized nuclei there seems to exist the approximate relationship $P = -A$
(e.g., see refs. [32, 37—39]). This has been discussed in several theoretical papers (see [43] for references). Experimentally these connec-
tions are valid only qualitatively.

In order to find at low deuteron energies contributions to the
^{3}He(d,p)^{4}He reaction other than those of S-waves leading to the $3/2^+$
compound state, Leemann et al. [44] studied the reaction with polar-

ized beam and target. Their results show that higher angular momenta are needed for good fits. A limited set of matrix elements has been proposed by Seiler and Baumgartner [45].

3.2. $T(d,n)^4He$ reaction

Since the T(d,n) and ^{3}He(d,p) reaction are charge-symmetric, their characteristics are expected to be the same except for Coulomb effects. Comparing experimental polarizations of the emitted nucleons, the neutron polarization is often found systematically lower than the proton polarization, occasionally by as much as 25%. In the most recent experiment [46] at 7- and 11.4-MeV deuteron energy, however, the differences between the nucleon polarizations are appreciably smaller.

As a further contribution to the question of polarization in mirror reactions, Davis et al. [47] have compared the vector analyzing power $\langle iT_{11} \rangle$ of the T(d,n) and its mirror reaction at 6, 10, and 11.3 MeV. The differences are small. For deuteron energies below 1 MeV the ^{3}He(d,p) data of refs. [34] and [44] (the quantities in fig. 1 of [44] are proportional to the D and D_{ik} used in [48]) can be compared with the T(d,n) results of Grunder et al. [48]: The tensor analyzing powers of the two reactions agree well, while the vector effects, arising only from the non-dominant matrix elements, are very small but completely different. An analysis of the T(d,n) experiment [48] is given in [45].

Ohlsen et al. [49] have studied the analyzing power of the T(d,n) reaction at energies below 100 keV. They found, under the assumption of pure S-wave interaction, a sensitivity to P_{zz} of 0.95 P^0_{zz}, where $P^0_{zz} = 3/2 \cos^2 \theta_{c.m.} - 1/2$ is the analyzing power resulting from a pure S-wave $3/2^+$ interaction. Because many deuteron polarization experiments at low energies are based on the tensor calibration with the T(d,n) or ^{3}He(d,p) reaction, these measurements are of great importance.

Experimental [50] and theoretical [51] investigations on polarization transfer processes $T(\vec{d},\vec{n})$ are presented elsewhere in these Proceedings.

3.3. Elastic scattering of deuterons from ^3He and Tritium

The vector and tensor analyzing power of the elastic deuteron scattering from ^{3}He has been measured by König et al. [52] between 4 and 10 MeV and by Dodder et al. [53] at 10 and 12 MeV deuteron energy. These data together with those of Plattner and Keller [32] of the vector analyzing power are shown in the contour plots of fig. 5. Since at forward angles and over a wide range of energy the vector analyzing power is appreciably larger than the tensor analyzing power, this process may become a useful analyzer for the vector polarization of deuterons. Watt and Leland [54] have observed asymmetries in scattering 9.9- and 11.9-MeV unpolarized deuterons from polarized ^{3}He nuclei.

Results on the mirror process $T(\vec{d},d)T$ are reported by Treiber et al. [55] for energies between 3.3 and 12.3 MeV.

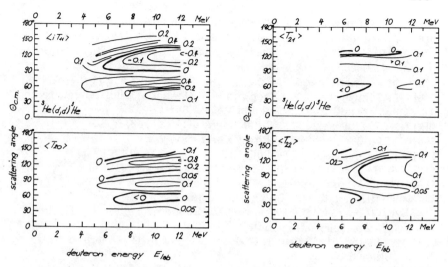

Fig. 5. Tensor moments $\langle iT_{11}\rangle$, $\langle T_{20}\rangle$, $\langle T_{21}\rangle$, and $\langle T_{22}\rangle$ of the elastic scattering process $^3\mathrm{He}(d,d)^3\mathrm{He}$.

4. CONCLUSIONS

For the calibration of nucleon polarization in the energy region below 100 MeV the elastic scattering from $^4\mathrm{He}$ is still the preferred process. The uncertainty of the proton data is in general about 3 to 5 times smaller than that of the neutron data.

For deuterons the $^3\mathrm{He}(d,p)^4\mathrm{He}$ and $\mathrm{T}(d,n)^4\mathrm{He}$ reactions allow an easy determination of the tensor polarization by observing the angular distribution of the emitted nucleons or α-particles. However, only the relative analyzing power is well known. Therefore an absolute calibration is desirable. Prior et al. [56] have done this, following a proposal by Jacobsohn and Ryndin [57]. Two-body nuclear reactions involving one spin-1 and three spin-0 particles have an absolute analyzing power independent of the specific reaction and independent of angle and energy [58]. Prior et al. used the $^{16}\mathrm{O}(d,\alpha_1)^{14}\mathrm{N}*(2.31)$ reaction for this calibration.

REFERENCES

[1] H. H. Barschall, Second Polarization Symp.
[2] L. Brown and W. Trächslin, Nucl. Phys. A90 (1967) 334.
[3] L. Brown, W. Haeberli, and W. Trächslin, Nucl. Phys. A90 (1967) 339.
[4] W. G. Weitkamp and W. Haeberli, Nucl. Phys. 83 (1966) 46.

[5]   P. W. Allison and R. Smythe, Nucl. Phys. A121 (1968) 97.

[6]   P. Darriulat, D. Garreta, A. Tarrats, and J. Testoni, Nucl. Phys.
      A108 (1968) 316; D. Garreta, J. Sura, and A. Tarrats, ibid. A132
      (1969) 204.

[7]   M. F. Jahns and E. M. Bernstein, Phys. Rev. 162 (1967) 871.

[8]   P. Schwandt, T. B. Clegg, and W. Haeberli, Nucl. Phys. (to be
      published), and Third Polarization Symp.

[9]   D. J. Plummer et al., Nucl. Phys. A115 (1968) 253.

[10] A. D. Bacher et al., Third Polarization Symp.

[11] R. A. Arndt and L. D. Roper, Phys. Rev. C1 (1970) 903.

[12] B. Hoop and H. H. Barschall, Nucl. Phys. 83 (1966) 65.

[13] G. L. Morgan and R. L. Walter, Phys. Rev. 168 (1968) 1114. The
      modified data will be published later.

[14] D. Cramer, D. Simmons, and L. Cranberg, University of Virginia
      (private communication).

[15] A. Niiler, M. Drosg, J. C. Hopkins, and J. D. Seagrave, Third
      Polarization Symp.

[16] J. R. Sawers, G. L. Morgan, L. A. Schaller, and R. L. Walter,
      Phys. Rev. 168 (1968) 1102.

[18] T. H. May, R. L. Walter, and H. H. Barschall, Nucl. Phys. 45
      (1963) 17.

[19] F. W. Büsser, F. Niebergall, and G. Söhngen, Nucl. Phys. 88
      (1966) 593.

[20] W. B. Broste, J. E. Simmons, and G. S. Mutchler, Bull. Am. Phys.
      Soc. 14 (1969) 123.

[21] W. B. Broste and J. E. Simmons, Third Polarization Symp.

[22] U. R. Arifkhanov, N. A. Vlasov, V. V. Davydov, and L. N.
      Samoilov, Sov. J. Nucl. Phys. 2 (1966) 170.

[23] R. W. Jewell, W. John, J. E. Sherwood, and D. H. White, Phys.
      Rev. 142 (1966) 687.

[24] G. R. Satchler et al., Nucl. Phys. A112 (1968) 1.

[25] Th. Stammbach and R. L. Walter, to be published.

[26] P. Heiss and H. H. Hackenbroich, Phys. Lett. 30B (1969) 373.

[27] P. Heiss and H. H. Hackenbroich, Z. Physik 231 (1970) 230.

[28] L. Kraus, I. Linck, and D. Magnac-Valette, to be published.

[29] T. C. Rhea, Th. Stammbach, and R. L. Walter, Third Polarization
      Symp.

[30] A. Galonski, H. B. Willard, and T. A. Welton, Phys. Rev. Lett. 2
      (1959) 349.

[31] V. S. Starkovich et al., Bull. Am. Phys. Soc. 13 (1968) 1448.

[32] G. R. Plattner and L. G. Keller, Phys. Lett. 29B (1969) 301.

[33] W. Klingler and F. Dusch, Third Polarization Symp.

[34] B. Forssmann, G. Graf, H. P. Jochim, and H. Schober, Third
      Polarization Symp.

[35] W. Grüebler et al., Third Polarization Symp.

[36] P. W. Keaton et al., Third Polarization Symp.

[37] S. D. Baker, G. Roy, G. C. Philips, and G. K. Walters, Phys. Rev. Lett. 15 (1965) 115.

[38] H. Brückmann and F. K. Schmidt, Nucl. Phys. A136 (1969) 81.

[39] B. E. Watt and W. T. Leland, Third Polarization Symp.

[40] P. Huber et al., Third Polarization Symp.

[41] Ch. Leemann et al., Third Polarization Symp.

[42] H. Brückmann and F. K. Schmidt, Nucl. Phys. A136 (1969) 81.

[43] B. DeFacio and R. K. Umerjee, Third Polarization Symp.

[44] Ch. Leemann et al., Third Polarization Symp.

[45] F. Seiler and E. Baumgartner, Third Polarization Symp.

[46] W. B. Broste and J. E. Simmons, Third Polarization Symp.

[47] J. C. Davis, D. Hilscher, and P. A. Quin, Third Polarization Symp.

[48] H. A. Grunder et al., Third Polarization Symp.

[49] G. G. Ohlsen, J. L. McKibben, and G. P. Lawrence, Third Polarization Symp., and Bull. Am. Phys. Soc. 13 (1968) 1443.

[50] W. B. Brost et al., Third Polarization Symp.

[51] G. G. Ohlsen, P. W. Keaton, and J. L. Gammel, Third Polarization Symp.

[52] V. König et al., Third Polarization Symp.

[53] D. C. Dodder et al., Third Polarization Symp.

[54] B. E. Watt and W. T. Leland, Third Polarization Symp.

[55] H. Treiber, K. Kilian, R. Strauss, and D. Fick, Third Polarization Symp.

[56] R. M. Prior, K. W. Corrigan, and S. E. Darden, Bull. Am. Phys. Soc. 15 (1970) 35.

[57] B. A. Jacobsohn and R. M. Ryndin, Nucl. Phys. 24 (1961) 505.

[58] P. W. Keaton et al., Third Polarization Symp.

DISCUSSION

L. Brown:

The reaction $^6Li(p,\alpha)^3He$ has large analyzing power and cross section for proton energies of a few hundred keV. This is convenient, perhaps unique, for protons with energies below 1 MeV. Results have been published recently by Petitjean and Brown in Nuclear Physics.

Slobodrian:

On the slide for n-^{4}He scattering polarization patterns, what were the dashed lines? Were they calculated from existing phase shifts? Were any phase-shift fits shown on the slide? Do such fits exist?

Striebel:

The dashed lines were just hand extrapolations.

Cranberg:

Almost all data reported on n-^{4}He scattering have been based on observations of alpha recoils. The interpretation of recoil data present resolution problems at back angles; small-angle data are hard to obtain; and, in general, at low energies there are questions of linearity of pulse-height versus energy of the recoils. It is therefore of interest that data on the scattered neutrons have been obtained by time-of-flight methods (Cramer, Simmons, Cranberg) for incident neutron energies from 0.5 to 1.4 MeV, where comparison may be made with predictions from phase-shift analyses based on the recoil data of Morgan and Walter. It is gratifying to note that the agreement between the results obtained by these very different methods is excellent.

Polarization and Compound
Nuclear States in Light Nuclei

G. R. PLATTNER, University of Basel, Switzerland

1. INTRODUCTION

An extensive paper covering the same topic was given by Darden [1] at the Second Symposium on Polarization Phenomena, held at Karlsruhe in 1965. In it the use of polarization work in compound level spectroscopy was reviewed and documented very thoroughly, so that in this paper all the work which was done in the period before 1965 need not be dealt with again. Even so, the number of results obtained since that time exceeds by far the level at which a comprehensive review would be possible in the space allotted to the subject. I will therefore try to present a detailed survey of the present state of knowledge only for those five or six scatterings or reactions which have attracted the strongest interest in the years since 1965. For a few other processes it will be possible to refer to results similar to those described in more detail; but I will have to limit myself to a mere bookkeeping job with respect to a large amount of interesting work in the case of those processes where a reasonably complete and consistent picture cannot yet be obtained from the polarization experiments performed up to now.

From a survey of the related literature, one fact emerges quite clearly and should, I feel, be stated in the beginning. It is the fact that in spite of the tremendous progress made since 1965 with sources of polarized ions, the number of cases in which polarization experiments have been instrumental in determining compound level parameters is still rather limited. This is clearly a consequence of certain basic problems associated with this type of nuclear spectroscopy; and in my concluding remarks, when some of the difficulties will have become more evident, I shall try to sum them up and to offer an opinion about how best to overcome them.

2. ELASTIC SCATTERING FROM SPIN-ZERO NUCLEI

The simplest process which one can study in order to obtain spectroscopic information about compound nuclear states from polarization

experiments is the elastic scattering of nucleons from spin-zero nuclei. It is well known that in this case there exist only three independent observables if parity conservation is assumed. In view of
this inherent simplicity and the additional constraint that due to the
short range of the nuclear forces only a few orbital angular momenta
participate in the process, one can hope to analyze uniquely even an
incomplete set of experimental results like cross-section and polarization data in terms of partial wave amplitudes or phase shifts. In
practice this procedure works well even at energies where processes
other than elastic scattering can occur and the phase shifts become
complex rather than real, thus effectively doubling the number of
parameters to be determined.

Apart from nucleon-alpha scattering, which is discussed in another
invited paper [2] at this conference, the scattering of protons and, to
a lesser degree, of neutrons from ^{12}C have been the prime objects of
interest for this kind of study.

Since 1965 several groups have measured [3—13] the polarization
of protons elastically scattered from ^{12}C. New and more precise measurements of the differential elastic cross section have also been reported [14—19]. The present state of knowledge is summarized in fig. 1.

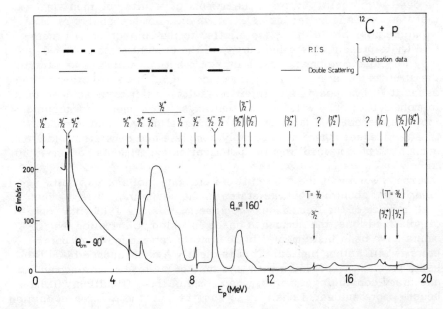

Fig. 1. Summary of ^{12}C(p,p)^{12}C elastic scattering. Above an excitation
function of the cross section, level assignments to states in ^{13}N are
shown. The availability of polarization data obtained by double scattering methods or with polarized-ion sources (P.I.S.) is marked by
horizontal lines at the top of the figure. Heavy bars indicate that
complete angular distributions were measured.

The excitation function of the cross section shows the fluctuations due to the many compound states excited in ^{13}N up to 20 MeV. The positions of the various levels and, if known, their total angular momenta and parities are marked. The horizontal lines at the top of the figure indicate the energy regions where polarization experiments with distinct bearing on the level structure of ^{13}N have been performed, either by double scattering or by using a beam from a polarized-ion source (P.I.S.). Heavy bars serve to indicate that rather complete angular distributions have been obtained.

With one exception, no changes in the J^π-assignments to levels in ^{13}N have been found necessary since 1965. The one change concerns the resonance structure at 9.15 MeV proton energy, which already at the Karlsruhe conference was suspected to be caused by two levels in ^{13}N. The work of three groups at Rice [14, 20], University of Texas (7, 8], and Erlangen [13] has since demonstrated the validity of this suspicion; and definite assignments have been made to the two levels in question. In fig. 2, some of the relevant results are shown. The cross-section data in the upper left-hand corner clearly show the double structure of the anomaly. The University of Texas double scattering polarization data then led to the unambiguous assignment of 5/2- and 7/2- to the two states; and the still unpublished, precise results obtained with the Erlangen polarized-ion source finally allowed an accurate phase shift analysis and determination of the level parameters, as shown on the right.

It would seem that below 10 MeV proton energy the scattering of protons on ^{12}C is very well understood. Phase shifts of little ambiguity are available [6, 8, 15–17, 20–23] over the entire energy range, and though they are not always based on cross-section *and* polarization data there seem to be no more doubts about the gross structure of ^{13}N as manifest in p-^{12}C elastic scattering. This is not to say, however, that with one or two more polarization experiments in the appropriate energy regions a significant increase in the accuracy of the level parameters could not be achieved.

In the field of analyses, a simple optical model calculation has been reported [22], which satisfactorily generates the three levels with dominant single particle character, namely, the 1/2$^+$ at 0.46 MeV, the 5/2$^+$ at 1.75 MeV, and the 3/2$^+$ at approximately 6.7 MeV. If a simple form of coupling to the first excited state of ^{12}C is included [24], two other levels—the 5/2$^+$ at 4.81 MeV and the 3/2$^+$ at 5.3 MeV—appear in agreement with experiment. More sophisticated coupled channel approaches have since been developed, however, and ought to be applied to p + ^{12}C scattering.

If below 10 MeV the work seems almost over for the experimentalist, this is certainly not the case above that energy. Between 10 and 20 MeV there is still considerable structure in the cross section, but very little polarization data are available. Apart from an excitation function at four angles up to 11.5 MeV from Erlangen [13] and a few double

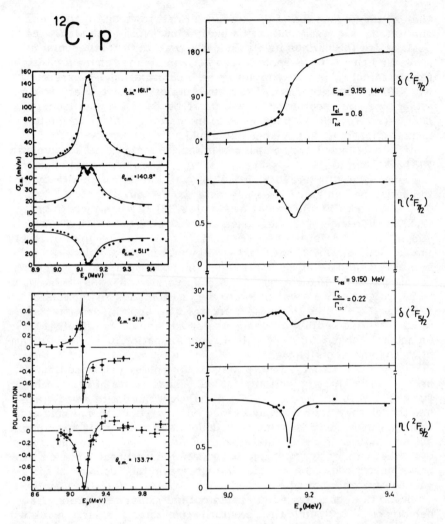

Fig. 2. The 9.15 MeV doublet in ^{12}C(p,p)^{12}C. The cross-section and polarization data from refs. [7, 8] and results of the phase shift analysis from ref. [13] are shown. The solid curves through the data are phase shift fits, the curves connecting the phase shift results serve only to guide the eye.

scattering results obtained around 13 MeV at Urbana [5], there exist
only two angular distributions [10] at 16.5 and 19 MeV and an excitation function at one angle in 0.5 MeV steps [11] between 10 and 21
MeV, both measured at Saclay. Clearly this is not enough to study
the structure of ^{13}N in a region where the level density is still not too
large for reasonably well isolated resonances to be seen. Measurements of good quality are needed to clarify the structure just above
10 MeV, and also to check the assignments (see fig. 1) obtained by
the groups at Stanford [18] and Yale [19] from their cross-section
measurements. Of particular interest are the T = 3/2 states belonging
to the A = 13 isospin multiplet.

Above 20 MeV we reach the region where the optical model should
become more and more appropriate to describe the scattering. But
even here it has been shown [25] that more or less pronounced resonances cannot be entirely excluded in an analysis. Cross-section
data [26] and polarization data from the Rutherford High Energy Laboratory (RHEL) [4] have been quite successfully interpreted in terms of
the optical model with three compound resonances added [4, 27] (see
also [28]). The uniqueness of the level assignments thus obtained,
however, is questionable.

For the benefit of future analyses it should be mentioned that new
measurements of the total reaction cross section have also been reported [29, 30].

Proton elastic scattering from ^{16}O has not received the same attention as p-^{12}C scattering in the last years. Up to the time of this conference almost all the new polarization data were taken below 3 MeV
by the group at the Carnegie Institution's Department of Terrestrial
Magnetism (DTM) in Washington [6] and between 16 and 40 MeV at
RHEL [31] and UCLA [32] (see also [33]) with the exception of a few
results with lesser bearing on the level structure of ^{17}F [34, 35]. Since
Darden's survey [1] of this scattering in 1965, virtually no new spectroscopic information has been gained. The level at 2.66 MeV proton energy has been re-analyzed by the group at São Paulo [36], using their
cross-section data and the old off-resonance polarization measurements from Wisconsin [37]. The assignment of 1/2$^-$ for this level has
been confirmed (not 1/2$^+$ as erroneously marked in [1]) and the São
Paulo phase shifts predict the later DTM polarizations [6] very well.
At the high energies, the RHEL group analyzed their cross-section and
polarization data again in the same manner as with p-^{12}C elastic scattering and found it necessary to add four compound levels between 17
and 29 MeV to the optical model calculation [31]. The quality of their
fits and with it the reliability of their assignments leaves something
to be desired, however.

The large gap in polarization measurements between 5 and 15 MeV
has only recently been attacked, and first results are presented in a
contribution to this conference [38].

The study of neutron induced processes has of course not profited from a technical progress similar to the one achieved for charged particles with polarized-ion sources. This fact is clearly reflected in the relative paucity of new results on scattering of neutrons from spin-zero nuclei. Whereas to my knowledge for the scattering of neutrons from ^{16}O only cross-section measurements [39, 40] have been performed since 1965 (and there were not many results from polarization experiments at that time, either), four groups have investigated the polarization of neutrons scattered from ^{12}C. The Tübingen experiment [41, 42] at 16 MeV neutron energy has not led to an analysis in terms of ^{13}C levels; and the results by Miller and Biggerstaff [43], consisting of a polarization excitation function at one angle between 3 and 4 MeV, agree quite well with existing phase shifts by Wills et al. [44] and need not be discussed further here, particularly since $n + {}^{12}C$ elastic scattering between 3 and 7 MeV is treated in detail in another contribution to this conference [45]. The other two publications by Aspelund [46] and by the Argonne group [47] deal with the energy region below 2 MeV neutron energy. In particular, the Argonne data by Lane et al. [47] have been analyzed extensively and have led to a considerably refined knowledge of the ^{13}C level structure. In fig. 3 are shown the results of both their measurements and their analysis parameterized in terms of the Legendre polynomial expansion coefficients for the cross section and for the neutron polarization. The solid curve shown in the figure has been obtained by using a two-level R-matrix analysis to reproduce three compound states at approximately 3.3 and 2.9 MeV (both $3/2^+$), 2.1 MeV ($5/2^+$), and a bound $1/2^-$ level. Distant levels were taken into account by adding background terms to the R-function expansion. A second analysis, shown in fig. 3 as a dashed line, was then undertaken to try to reproduce the levels with dominant single particle character ($3/2^+$ at 3.3 MeV, $5/2^+$ and $1/2^-$ bound) and the background contributions with a simple spherical square well potential. As can be seen in fig. 3, the attempt was quite successful and a great amount of well-based spectroscopic information has been obtained, though of course none of the general features of ^{13}C were ever in doubt.

This might be the place to note that two additional levels of ^{13}C which were at one time claimed to be present in this region of excitation energy [48] have since been shown not to exist [49] and indeed do not show up in the work of the Argonne group.

Several theoretical analyses [50—52] of n-^{12}C-elastic scattering below 5 MeV using a deformed, diffuse potential to generate all the observed resonances by coupling to one or several excited states of collective nature in ^{12}C have also been reported. The calculations generally show a considerable amount of agreement with the experimental results, though quantitatively there is still room for improvements, particularly at the higher energies. Yet another analysis by Lovas [53], using the Feshbach unified reaction model, agrees in part

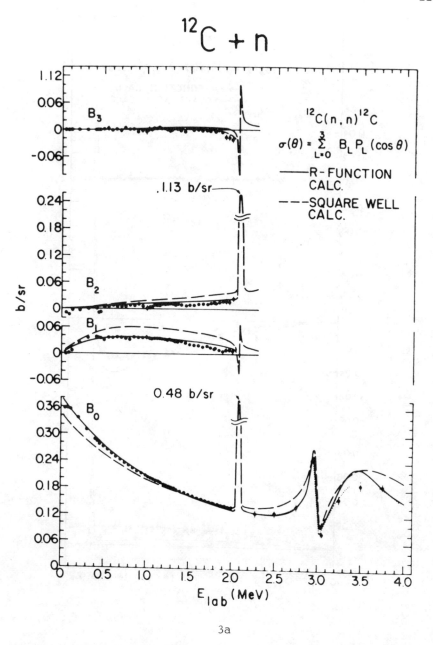

$$^{12}C + n$$

3a

Fig. 3. Results of measurements and analysis of $^{12}C(n,n)^{12}C$ below a few MeV. Both parts of the figure (see following page for 3b) are from ref. [47] and show the Legendre expansion coefficients of cross-section and polarization data. The solid and dashed curves are results of two analyses described in the text.

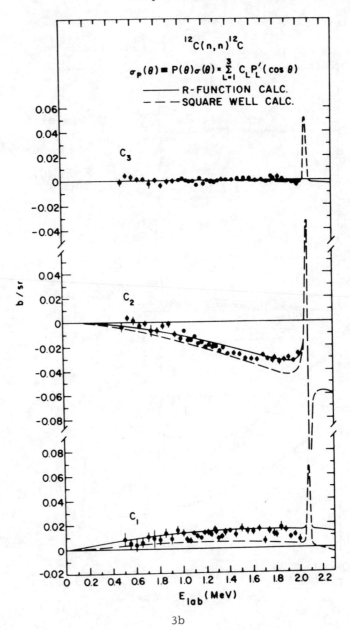

3b

with experimental results below 4.5 MeV, although the parity of one of the levels generated disagrees with experiment.

Again it should be mentioned that new total cross-section measurements for neutrons on ^{12}C have been reported [54, 55].

Still a relatively simple process in terms of the complexity of the spin structure is deuteron elastic scattering from spin-zero nuclei. Though the number of observables which constitute a complete set assuming time reversal invariance and parity conservation is seven in this case, five independent quantities are relatively easy to measure once a good polarized-ion source is available. These five include the cross section, the vector polarization iT_{11}, and the three second-rank tensor parameters T_{20}, T_{21}, and T_{22} which are needed to completely describe the polarization of a polarized deuteron beam. Thus a fairly comprehensive set of observables is open to experimental investigation, and spectroscopic information about compound levels can be obtained without an undue amount of ambiguity.

The only such process for which all five parameters just mentioned have been measured with sufficient accuracy to encourage a thorough analysis is deuteron-alpha elastic scattering. The present state of knowledge is summarized in fig. 4, where the same notation as in fig. 1

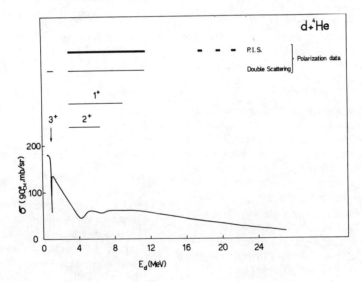

Fig. 4. Summary of ^{4}He(d,d)^{4}He. Above an excitation function of the cross section, level assignments to states in ^{6}Li are shown. The availability of polarization data obtained by double scattering methods or with polarized-ion sources (P.I.S.) is marked by horizontal lines at the top of the figure. Heavy bars indicate that complete angular distributions of more than two deuteron polarization parameters were measured.

has been used. No new anomalies
have been found below 24 MeV
nor have any of the known levels
been given different spin or parity
assignments. Deuteron–alpha
scattering has nevertheless attracted enormous interest, mostly
because of its potential use as a
convenient deuteron polarization
analyzer and because of an old
disagreement on whether negative
parity states in ^{6}Li exist below
12 MeV deuteron energy [1].

Double scattering measurements
of the deuteron polarization were
made across the 1.07 MeV 3^+ level
[56] and from 3 to 12 MeV [57–
60]. With polarized deuterons the
process was extensively studied
between 3 and 12 MeV by the groups
at Wisconsin [61, 62] and at Zürich
[63–65]; and at three energies
around 20 MeV quite complete data
were obtained at Saclay [66]. One
new cross-section experiment between 3 and 14 MeV has also been
reported [67].

As is beautifully demonstrated
in two contributions to this conference [68, 69], deuteron–alpha scattering has now been shown to provide an efficient analyzer for deuteron polarization over most of the
energy range below 12 MeV [62–
65]. Particularly noteworthy in
this context are the measurements
of the second-rank tensor parameters in d–α scattering by the group
from the ETH in Zürich [64, 65],
since to my knowledge this is the
first time results of this quality
and scope have been obtained, a
fact which dramatically demonstrates the progress in polarized-
ion source technology.

That these measurements yielded
more than just a refinement of already known facts is shown in fig. 5,

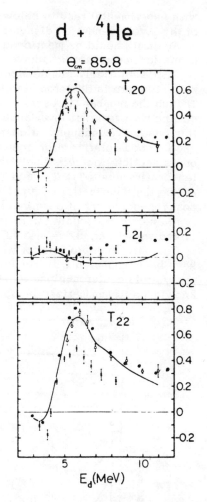

Fig. 5. Comparison of old and
recent results of deuteron polarization measurements in d–α
elastic scattering. Crosses
are double scattering results
from ref. [58]; open circles, results obtained with a polarized-
ion source [61]; and dots, the
most recent results from refs.
[64, 65]. The solid curve is the
result of a phase shift analysis
by Keller and Haeberli [62]. The
definitions of the T_{qk} are as in
refs. [64, 65].

where some old and new results are compared. The double scattering results from Wisconsin [58] are shown as crosses; those obtained there later [61] with their first polarized-ion source, as open circles. The black dots mark the new Zürich measurements [64, 65]. It can be seen that large discrepancies exist, particularly between the double scattering results obtained using the ^{3}He(d,p)^{4}He reaction as a deuteron polarization analyzer and the results obtained with a polarized-ion source. To what extent these discrepancies are due to an overestimate of the analyzing power of the ^{3}He(d,p)^{4}He reaction, to calibration problems with the beam from polarized-ion sources, or to other experimental errors is unclear. Several contributed papers at this conference deal with this complex of questions and it must be hoped that it can be cleared up soon.

The solid line in fig. 5 shows the phase shift fit obtained by the Wisconsin group [62] using their recent vector polarization data [62] and their old measurements of the second-rank tensor parameters [58, 61]. The main disagreement of this fit with the new Zürich data is in the parameter T_{21}, a fact which for a short time even conjured the ghosts of Senhouse and Tombrello's [70] odd parity levels. Thanks to the cooperation of the Zürich group who obligingly furnished a copy of the results of their preliminary phase shift analysis [69], it can be shown in fig. 6 that there is no need for such an unexpected reversal, and that the main difference between the Wisconsin phases [62], shown as dotted lines, and the Zürich phases, shown as solid lines, lies in the $^3P_0$ phase shift. This parameter now follows a hard sphere energy dependence more closely than before, certainly no indication of a negative parity level. The error bars shown at 9 MeV mark the uncertainty in the phase shifts as determined at Wisconsin.

Another noteworthy development since 1965 concerns the f-waves. It has been found [62] that the discrepancy between the old McIntyre and Haeberli phase shift predictions [58] for the vector polarization at forward angles and corresponding measurements of that quantity [60–62] can be made to disappear by including small f-wave contributions even below 10 MeV.

The low energy phase shifts have been extended up to 21 MeV by the Saclay group [71]. These phases however are based only on cross-section measurements up to 17 MeV and therefore cannot be regarded as very accurate.

A different parameterization of the experimental data in terms of Humblet-Rosenfeld S-matrix theory has been carried out at Strasbourg [72] with the intent to bypass the well-known ambiguities of R-matrix calculations and to provide a consistent set of level parameters for the three known resonances in d-α scattering. The authors of that paper, unfortunately, disregarded the wealth of polarization data and thus the quality of their parameterization cannot be properly judged.

Many model calculations [73–78] have been performed since 1965, extending the already considerable list of publications on this subject.

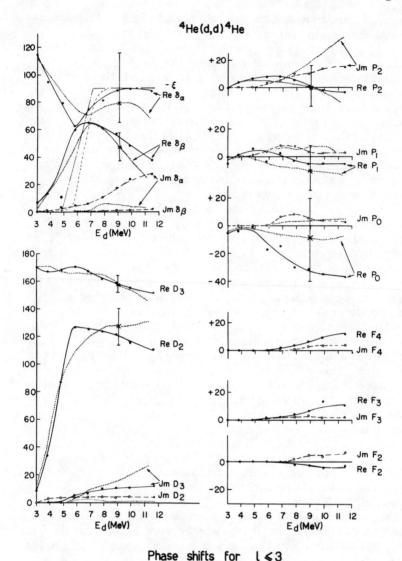

Phase shifts for $\ell \leqslant 3$

Fig. 6. Comparison of the d-α phase-shifts of Keller and Haeberli [62] (dotted lines) with the recent phase shifts of the Zürich group (solid lines) which included accurate measurements of second-rank tensor parameters [63, 65]. The f-waves look quite similar in both cases and are shown only once. The error bars at 9 MeV characterize the uncertainty in the phase-shifts [62] as determined without the new data. δ_α, δ_β, and ξ are the three parameters which replace the S_1 and D_1 phase shift in the case of ℓ-mixing (see [58]).

Resonating group calculations [73, 74, 78] have given good fits to the phase shifts determined from experiment with no other adjustable parameter than the ^{6}Li binding energy.

Another approach has been to reproduce d-α scattering by separating it into two two-body nucleon-alpha interactions plus a simple nucleon-nucleon interaction [75—77]. Particularly in the most recent of these calculations by Shanley [76, 77] remarkable fits to all the experimentally determined quantities are obtained when nucleon-alpha interaction, as taken from n-alpha phase shifts, and s-wave nucleon-nucleon interaction are properly taken into account.

Though a large body of data, including polarization measurements, has been accumulated since 1965 for deuteron-^{12}C [79—88] and deuteron-^{16}O [89] elastic scattering, very little has been learned from this about the level structure of the nuclei ^{14}N and ^{18}F.

Experiments with polarized particles other than protons or deuterons have become more and more attractive in the last few years. In our context, the scattering of ^{3}He (and also tritium) from ^{4}He [90—97] and ^{12}C [98—100] are particularly noteworthy since several experiments have been reported. Again the spectroscopic information about compound levels obtained from these investigations is too sparse to give a detailed treatment here.

3. ELASTIC SCATTERING FROM NUCLEI WITH SPIN

In contrast to scattering from spin-zero nuclei, the complexity of the spin structure and with it the number of observables in a complete set increases greatly as soon as the target has spin. It is this fact which makes an unambiguous analysis of experimental results almost impossible in this case, unless either very difficult experiments are performed or rather special circumstances mitigate somewhat the complexity arising from the spin of the target. In this context I will concentrate only on the elastic scattering of protons from ^{6}Li, though it is by no means the only process of this kind for which polarization experiments have been done, nor is it the one for which the most experimental information has become available. The reason for my preference lies in the fact that at energies below a few MeV this process seems to be dominated by two well-separated resonances and therefore an analysis in terms of phase shifts has been tried [101], even though only a very limited amount of experimental information was available.

In fig. 7, a summary of the problem is given. The experimental data consist of cross-section measurements [102—104] and a determination of the proton polarization [101] by the group at the DTM in the energy range indicated. The phase shift analysis performed by the DTM group in collaboration with Seyler from Ohio State University gave the results shown in the upper half of the figure. The two reso-

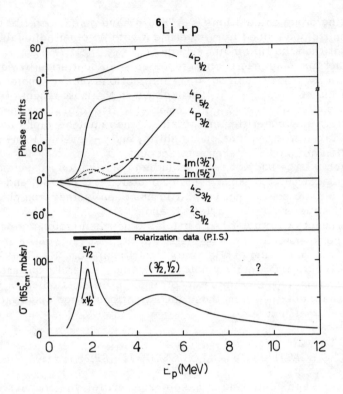

Fig. 7. Summary of ^{6}Li(p,p)^{6}Li. Above an excitation
function of the cross section, level assignments to
states in ^{7}Be are shown. The region where complete
angular distributions of the proton polarization,
taken with a polarized-ion source (P.I.S.), are available is marked with a heavy horizontal bar. The upper
part of the figure shows the results of a phase shift
analysis discussed in the text.

nances are identified as a 5/2$^-$ level in ^{7}Be at approximately 2 MeV
proton energy and a 3/2$^-$ level at around 5 MeV. In order to illustrate
the difficulties inherent in such an analysis, some of the assumptions
that were made more or less a priori will have to be enumerated. All
the mixing parameters (describing to what extent two different values
of the orbital angular momentum or of the channel spin participate in
the formation and decay of a given state) as well as the d-phases and
the doublet p-phases were set to zero after some attempts to include
them had shown no positive indication of their necessity. The absorption parameters or imaginary parts of the phase shifts were fixed at
physically reasonable values, as shown in the figure for the two p-

waves. Smoothly energy dependent s-wave absorption was also included but is not shown in fig. 6. The only constraint on the absorption parameters was that they should reproduce the experimental total cross section for the $^6Li(p,^3He)^4He$ reaction, the only inelastic channel of importance at these energies.

Even with all these simplifying assumptions the J-value of the upper resonance could not be uniquely determined. It was found that if the $3/2^-$ and the $1/2^-$ p-wave phase shifts are interchanged, virtually no detectable difference in the quality of the fits is produced at any energy. Only a previous study of the same level [105] in ^7Be via inelastic scattering saved the situation, since these measurements exclude the $1/2^-$ assignment.

From this brief description of the problems involved in such an analysis, it can be seen that even in a very favorable case the inclusion of measurements of just one more parameter (here the proton polarization) helps little to provide more insight than if cross-section data alone were used. In fact both level assignments had been determined earlier with not much more ambiguity [104, 105] and without the benefit of the polarization measurements. The authors of the $p + ^6Li$ analysis do indeed point out that according to their calculations only a measurement of the ^6Li second-rank tensor polarization produced in the scattering could really bring about a significant reduction in ambiguity.

In the study of other processes of similar character, the Argonne group, whose results on $n-^{12}C$ elastic scattering have already been discussed, seems to have had a great deal of success with R-matrix analysis of cross-section and polarization data for neutron scattering from ^{10}B and ^{11}B at energies below 2 MeV [106, 107]. At the time of this writing, no detailed report is yet available.

In $p-^{14}N$ scattering, a few experiments including some with polarized particles have been reported [5, 108—110], but far more information will be needed for a meaningful analysis in terms of levels in ^{15}O.

An experimentally rather complete study of $p + ^7Li$ elastic and inelastic scattering with polarized protons has been performed at Erlangen [111], but no analysis was tried. Quite a bit of work has been done [112—117] on the elastic scattering of protons from ^9Be, and some spectroscopic information on ^{10}B has been obtained. Polarization experiments, though quite complete and accurate, played a minor role in these analyses.

4. REACTIONS

Polarization experiments involving the study of reactions rather than elastic scattering have yielded much less information about nuclear levels. This is not due to a paucity or poor quality of such experiments, but is caused by the fact that an analysis in terms of a model independent formalism, like a phase shift analysis or, more properly, a parameterization in terms of reaction amplitudes, is in general much more complicated for a reaction than for elastic scattering because of the lack of symmetry between the initial and final states. In addition, direct reaction mechanisms often play an important role and obscure the contributions from compound levels.

An excellent example of a frustrating experience in reaction analysis—and the only one I am going to talk about in detail—is the study of the even spin and parity levels in ^{8}Be via the two reactions ^{6}Li(d,a)^{4}He and ^{7}Li(p,a)^{4}He. Both processes have long been favorites of experimentalists and theoreticians alike, mostly because they seem to be so clearly dominated by compound nuclear mechanisms and because their spin structure is relatively simple due to the two identical spinless particles in the exit channel. Fig. 8 shows the total cross sections for both reactions on a common scale giving the excitation energy in ^{8}Be. One can be fairly certain, it would seem, that at most three or four levels in ^{8}Be dominate this energy region. Since there were initial difficulties in determining the level parameters, a true information explosion has taken place since 1965. In rapid succession there have become available new cross-section measurements for both reactions [118—122], many rather complete polarization experiments performed with double scattering techniques and polarized-ion sources, with protons [111, 123—127] and with deuterons [128—132] as marked in fig. 8; and about as many different interpretations in terms of ^{8}Be levels [111, 120, 124, 128, 129, 131—136] as there are publications. In order to illustrate this profusion of proposed levels in ^{8}Be, the upper half of fig. 8 represents the results obtained by the different groups. It should be said of course that these interpretations have been obtained on very different levels of sophistication, ranging from probably erroneous computer calculations over rather qualitative arguments all the way to mutlilevel R- and S-matrix fits. Unfortunately these efforts did not lead to a quantiative representation of the experimental results. While the agreement between the different groups has been quite satisfactory as far as the direct experimental results are concerned, the overlap between the different interpretations is much less convincing.

It is indeed quite ironical that at a polarization conference I should have to take recourse to one of the few processes for which polarization experiments will never be done, namely, alpha-alpha elastic scattering, in order to try to shed some light on the confusion left by the polarization work. The very same even spin and parity levels of ^{8}Be are also seen in this simple scattering, and the recent accurate data [137] obtained at the LRL in Berkeley will lead to a well established knowledge of the existence and positions of these levels of ^{8}Be in the energy region of interest. The preliminary results of a phase shift analysis of the data are shown at the bottom of the upper part in fig. 8. There is some degree of coincidence between the levels proposed from the study of the two reactions discussed above and those

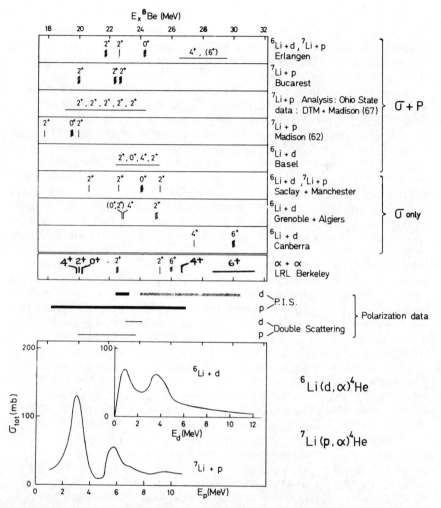

Fig. 8. Summary of the state of knowledge about the even spin and parity states in ^{8}Be in connection with the study of ^{6}Li(d,a)^{4}He and ^{7}Li(p,a)^{4}He. In the lower part of the figure, the total cross sections for the two processes are shown on a common scale giving the excitation energy in ^{8}Be. The availability of polarization data for ^{6}Li(d,a)^{4}He and ^{7}Li(p,a)^{4}He is shown with horizontal lines marked d and p, respectively. Heavy bars indicate that complete angular distributions have been measured. The cross-hatched marking signifies a measurement of complete angular distributions of the deuteron vector polarization only. In the upper part of the figure, a table of level assignments to states in ^{8}Be is shown. The assignments were deduced by different groups from a study of the reaction indicated. At the bottom of this table, the level assignments obtained from a-a elastic scattering [137] are shown for comparison.

found in alpha-alpha scattering. The most unexpected new levels are the 4^+ and 0^+ near 20 MeV which had never been suspected to exist. These levels overlap with the previously known 2^+ level in that energy region, and their combined effects produce the strong single peak in the $^7Li(p,\alpha)^4He$ total cross section.

Though we are still far from having exhausted all sources of experimental information on the two reactions $^6Li(d,\alpha)^4He$ and $^7Li(p,\alpha)^4He$, in my opinion the task before us is not to obtain more data but to try to understand the reaction mechanism. Probably more than just formation and decay of compound nuclear states are involved in this case, and an attempt should be made to include direct reaction contributions in the calculations while using the alpha-alpha results to properly predict the compound formation amplitudes. A successful first step in that direction, introducing the deuteron-pickup concept into the discussion of $^6Li(d,\alpha)^4He$, has very recently been reported [122], and one can hope that more progress will be made if work continues along such lines.

A great number of other reactions have been studied with polarized particles, but in the great majority of cases a clear connection to the topic of this paper is not present. Only in a few papers are attempts made to assign level parameters to observed anomalies, whereas in most cases an analysis in terms of a direct reaction mechanism is proposed and observed anomalies are regarded as bad luck. Those experiments where an analysis in terms of compound level contributions has been tried, have mostly been performed at very low energies, where low orbital angular momenta dominate strongly. Very often reactions involving polarized deuterons were studied, so that a substantial number of observables were open for investigation. Into this class fall the studies performed by the Basel group on $^6Li(d,p)^7Li$ and $^6Li(d,p')^7Li^*$ [138], $^7Li(d,n)^8Be$ [139], $^{11}B(d,n)^{12}C$ [140], and $^{12}C(d,p)^{13}C$ [141]; and, to a lesser extent, studies at other laboratories on $^9Be(p,d)^8Be$ [142, 143], $^{10}B(d,p)^{11}B$ [144], $^{11}B(d,p)^{12}B$ [145], $^{11}B(d,n)^{12}C$ [146, 147], $^9Be(\alpha,n)^{12}C$ [148, 149], $^{13}C(\alpha,n)^{16}O$ [150—152], $^6Li(p,\alpha)^3He$ [153], and $^{15}N(p,\alpha)^{12}C$ [154]. In most of the cases just mentioned level parameters were determined or checked, though not always with sufficient lack of ambiguity and enough precision to regard the assignments as final. I regret not being able to deal with some of these reports a little bit more specifically, but this would clearly exceed the possibilities of this short review.

This brings me to the end of my paper. As promised in the introduction, I would like to use the concluding remarks to try to understand why so little new spectroscopic information about compound levels has been obtained from so many experiments with such sophisticated equipment. The reason seems to be that it is indeed a small class of processes for which a simple polarization experiment is at the same time indispensable and, together with cross-section data, sufficient to allow an intelligent analysis to be performed. Another class—and not a large one, either—is made up of very simply structured reactions or scatterings for which cross-section data alone have already provided most of the essential information. But by far the largest class is taken up by all those processes for which cross-section and polarization data in the normal sense do not provide enough information to disentangle nature's complexities.

Basically, the only way out of these problems is the measurement of additional parameters. In the case of nucleon-nucleon scattering this has been recognized long ago, and because of the basic importance of that interaction, experimentalists have invested a great deal of hard work in measuring triple-scattering (Wolfenstein) and spin-correlation parameters. With the advent of intense polarized beams from polarized-ion sources, such measurements have become much less demanding and should be done even in cases where less fundamental insights are to be gained. Double scattering experiments with polarized beams can provide information on some of these additional parameters for many of the processes I have listed.

Could it be that after the long, lean years of hard double scattering experiments with unpolarized beams, experimentalists are enjoying too much the ease and elegance with which such experiments can be done with today's highly efficient experimental technology? It seems to me that a step back into those hard times, doing double scattering experiments but now with polarized beams, could provide the stimulus needed to make the use of polarization experiments for compound level spectroscopy a flourishing branch of low energy nuclear physics again.

REFERENCES

[1]   S. E. Darden, Second Polarization Symp., p. 433.
[2]   H. R. Striebel, Third Polarization Symp.
[3]   I. Bondouk, F. Asfour, V. J. Gontchar, and F. Machali, Nucl. Phys. 65 (1965) 490.
[4]   R. M. Craig et al., Nucl. Phys. 79 (1966) 177.
[5]   J. D. Steben and M. K. Brussel, Phys. Rev. 146 (1966) 780.
[6]   W. Trächslin and L. Brown, Nucl. Phys. A101 (1967) 273.
[7]   G. E. Terrell, M. F. Jahns, M. R. Kostoff, and E. M. Bernstein, Phys. Rev. 173 (1968) 931.
[8]   E. M. Bernstein and G. E. Terrell, Phys. Rev. 173 (1968) 937.
[9]   R. P. Slabospitskii et al., Sov. J. Nucl. Phys. 8 (1965) 502.
[10] P. Darriulat, J. M. Fowler, R. de Swiniarski, and J. Thirion, Second Polarization Symp., p. 342.
[11] J. C. Faivre et al., Nucl. Phys. A127 (1969) 169.
[12] H. Guratzsch, G. Hofmann, H. Müller, and G. Stiller, Nucl. Phys. A129 (1969) 405.
[13] G. Hartmann, University of Erlangen, Ph.D. Thesis, 1968 (unpublished), and private communication.

[14] J. B. Swint, A. C. L. Barnard, T. B. Clegg, and J. L. Weil, Nucl. Phys. 86 (1966) 119.

[15] A. C. L. Barnard, J. B. Swint, and T. B. Clegg, Nucl. Phys. 86 (1966) 130.

[16] J. C. Armstrong, M. J. Baggett, W. R. Harris, and V. A. Latorre, Phys. Rev. 144 (1966) 823.

[17] G. S. Mani, A. Sadeghi, and A. Tarrats, J. Phys. (Paris) Colloq. 1 (1966) 130.

[18] H. M. Kuan and S. S. Hanna, Phys. Lett. B24 (1967) 566.

[19] M. J. LeVine and P. D. Parker, Phys. Rev. 186 (1969) 1021.

[20] J. B. Swint, J. S. Duval, Jr., A. C. L. Barnard, and T. B. Clegg, Nucl. Phys. A93 (1967) 177.

[21] J. S. Duval, Jr., A. C. L. Barnard, and J. B. Swint, Nucl. Phys. A93 (1967) 164.

[22] S. J. Moss and W. Haeberli, Nucl. Phys. 72 (1965) 417.

[23] A. C. L. Barnard, J. S. Duval, Jr., and J. B. Swint, Phys. Lett. 20 (1966) 412.

[24] A. C. L. Barnard, Phys. Rev. 155 (1967) 1135.

[25] T. Tamura and T. Teresawa, Phys. Lett. 8 (1964) 41.

[26] J. K. Dickens, D. A. Haner, and Ch. N. Waddell, Phys. Rev. 132 (1963) 2159.

[27] J. Lowe and D. L. Watson, Phys. Lett. 23 (1966) 261, and B24 (1967) 174.

[28] D. K. Scott, P. S. Fisher, and N. S. Chant, Nucl. Phys. A99 (1967) 177.

[29] J. F. Dicello, Jr. and G. Igo, Phys. Rev. C2 (1970) 488.

[30] M. Q. Makino, C. N. Waddell, and R. M. Eisberg, Nucl. Phys. 68 (1969) 378.

[31] O. Karban et al., Nucl. Phys. A132 (1969) 548.

[32] J. M. Cameron, J. R. Richardson, W. T. H. van Oers, and J. W. Verba, Phys. Rev. 167 (1968) 908.

[33] H. B. Eldridge et al., Phys. Rev. 167 (1968) 915.

[34] L. Drigo et al., Nuovo Cim. 37 (1965) 1766.

[35] I. Boca, M. Cenja, E. Iliescu, and N. Martalogu, Rev. Roum. Phys. 10 (1969) 415.

[36] V. Gomes, R. A. Douglas, T. Polga, and O. Sala, Nucl. Phys. 68 (1969) 417.

[37] R. A. Blue and W. Haeberli, Phys. Rev. 137 (1965) B284.

[38] R. M. Prior, K. W. Corrigan, and S. E. Darden, Third Polarization Symp.

[39] C. H. Johnson and J. L. Fowler, Phys. Rev. 162 (1967) 890.

[40] J. L. Fowler and C. H. Johnson, Phys. Rev. C2 (1970) 124.

[41] G. Mack and G. Mertens, Z. Naturf. 22a (1967) 1640.

[42] G. Mack, Z. Physik 212 (1968) 365.

[43] T. G. Miller and J. A. Biggerstaff, Nucl. Phys. A124 (1969) 637.

[44] J. E. Wills, Jr., J. K. Bair, H. O. Cohen, and H. B. Willard, Phys. Rev. 109 (1958) 891.

[45] J. L. Weil and W. Galati, Third Polarization Symp.

[46] O. Aspelund, Phys. Norvegica 3 (1968) 43.

[47] R. O. Lane, R. D. Kashel, and J. E. Monahan, Phys. Rev. 188 (1969) 1618.

[48] C. D. Moak, A. Galonsky, R. L. Traughber, and C. M. Jones, Phys. Rev. 110 (1958) 1369.

[49] P. R. Christensen and C. L. Cocke, Phys. Lett. 22 (1966) 503.

[50] G. Pisent and A. M. Saruis, Nucl. Phys. A91 (1967) 561.

[51] J. T. Reynolds, C. J. Slavik, C. R. Lubitz, and N. C. Francis, Phys. Rev. 176 (1968) 1213.

[52] J. L. Roeder, Ann. Phys. (N.Y.) 43 (1967) 382.

[53] I. Lovas, Nucl. Phys. 81 (1966) 353.

[54] J. C. Davis and F. T. Noda, Nucl. Phys. A134 (1969) 361.

[55] R. B. Schwartz, H. T. Heaton II, and R. A. Schrack, Bull. Am. Phys. Soc. 15 (1970) 567.

[56] H. Meiner et al., Helv. Phys. Acta 40 (1967) 483.

[57] P. G. Young, G. G. Ohlsen, and M. Ivanovich, Nucl. Phys. A90 (1967) 41.

[58] L. C. McIntyre and W. Haeberli, Nucl. Phys. A91 (1967) 369, 382.

[59] G. G. Ohlsen, V. S. Starkovich, W. G. Simon, and E. M. Bernstein, Phys. Lett. B28 (1969) 404.

[60] E. M. Bernstein, G. G. Ohlsen, V. S. Starkovich, and W. G. Simon, Phys. Rev. Lett. 18 (1967) 966.

[61] A. Trier and W. Haeberli, Phys. Rev. Lett. 18 (1967) 915, and Ph.D. Thesis, University of Wisconsin, 1966 (available from University Microfilms, Ann Arbor, Michigan, USA).

[62] L. G. Keller and W. Haeberli, to be published in Nucl. Phys.

[63] W. Grüebler, V. Koenig, P. A. Schmelzbach, and P. Marmier, Nucl. Phys. A134 (1969) 686.

[64] V. Koenig, W. Grüebler, P. A. Schmelzbach, and P. Marmier, Nucl. Phys. A148 (1970) 380.

[65] W. Grüebler, V. Koenig, P. A. Schmelzbach, and P. Marmier, Nucl. Phys. A148 (1970) 391.

[66] P. Darriulat, D. Garreta, A. Tarrats, and J. Arvieux, Nucl. Phys. A94 (1967) 653.

[67] G. S. Mani and A. Tarrats, Nucl. Phys. A107 (1968) 624.

[68] V. Koenig, W. Grüebler, P. A. Schmelzbach, and P. Marmier, Third Polarization Symp.

[69] P. A. Schmelzbach, W. Grüebler, V. Koenig, and P. Marmier, Third Polarization Symp.

[70] L. S. Senhouse and T. A. Tombrello, Nucl. Phys. 57 (1964) 624.

[71] J. Arvieux et al., Nucl. Phys. A94 (1967) 663.

[72] L. Kraus, I. Linck, and D. Magnac-Valette, Nucl. Phys. A136 (1969) 301.

[73] D. R. Thompson and Y. C. Tang, Phys. Lett. B26 (1968) 194.

[74] D. R. Thompson and Y. C. Tang, Phys. Rev. 179 (1969) 971.

[75] P. M. Fishbane and J. V. Noble, Phys. Rev. 171 (1968) 1190.

[76] P. E. Shanley, Phys. Rev. Lett. 21 (1968) 627.

[77] P. E. Shanley, Phys. Rev. 187 (1969) 1328.

[78] W. Laskar and R. LeMaitre, J. Phys. (Paris) 29 (1968) 409.

[79] J. F. Arvieux et al., Phys. Lett. 16 (1965) 149.

[80] Yu. V. Gofman et al., Sov. J. Nucl. Phys. 5 (1967) 510.

[81] N. I. Zaika et al., Sov. J. Nucl. Phys. 7 (1968) 460.

[82] G. Clausnitzer, R. Fleischmann, and W. Wilsch, Phys. Lett. B29 (1967) 466.

[83] R. L. A. Cottrell, J. C. Lisle, and J. O. Newton Nucl. Phys. A109 (1968) 288.

[84] H. Cords, G. U. Din, M. Ivanovich, and B. A. Robson, Nucl. Phys. A113 (1968) 608.

[85] D. G. Gerke, D. R. Tilley, N. R. Fletcher, and R. M. Williamson, Nucl. Phys. 75 (1966) 609.

[86] H. O. Meyer, University of Basel, Ph.D. Thesis (1970), and private communication.

[87] H. Paetz gen. Schieck et al., Bull. Am. Phys. Soc. 14 (1969) 507.

[88] T. S. Katman, N. R. Fletcher, D. R. Tilley, and R. M. Williamson, Nucl. Phys. 80 (1966) 449.

[89] H. Cords, G. U. Din, and B. A. Robson, Nucl. Phys. A134 (1969) 561.

[90] R. J. Spiger and T. A. Tombrello, Phys. Rev. 163 (1967) 964.

[91] D. D. Armstrong, L. L. Catlin, P. W. Keaton, Jr., and L. R. Veeser, Phys. Rev. Lett. 23 (1969) 135.

[92] P. Schwandt et al., Phys. Lett. B30 (1969) 30.

[93] D. M. Hardy et al., Phys. Lett. B31 (1970) 355.

[94] W. S. McEver et al., Phys. Lett. B31 (1970) 560.

[95] M. Ivanovich, P. G. Young, and G. G. Ohlsen, Nucl. Phys. A110 (1968) 441.

[96] F. Dunnill, T. I. Gray, H. T. Fortune, and N. R. Fletcher, Nucl. Phys. A93 (1967) 201.

[97] P. W. Keaton, Jr., D. D. Armstrong, and L. R. Veeser, Phys. Rev. Lett. 20 (1968) 1392.

[98] R. L. Hutson et al., Phys. Lett. B27 (1968) 153.

[99] J. B. A. England et al., Phys. Lett. B30 (1969) 476.

[100] W. S. McEver et al., Phys. Rev. Lett. 24 (1970) 1123.

[101] C. Petitjean, L. Brown, and R. G. Seyler, Nucl. Phys. A129 (1969) 209.

[102] U. Fasoli, E. A. Silverstein, D. Toniolo, and G. Zago, Nuovo Cim. 34 (1964) 1832.

[103] J. A. McCray, Phys. Rev. 130 (1963) 2034.

[104] W. D. Harrison and A. B. Whitehead, Phys. Rev. 132 (1963) 2607.

[105] W. D. Harrison, Nucl. Phys. A92 (1967) 253, 260.

[106] S. L. Hausladen et al., Bull. Am. Phys. Soc. 19 (1970) 567.

[107] C. E. Nelson et al., Bull. Am. Phys. Soc. 15 (1970) 567.

[108] R. I. Brown, Nucl. Phys. 78 (1966) 492.

[109] L. Drigo et al., Nuovo Cim. B45 (1966) 206.
[110] F. Boreli, P. N. Shrivastava, B. B. Kinsey, and V. D. Mistry, Phys. Rev. 174 (1968) 1221.
[111] K. Kilian et al., Nucl. Phys. A126 (1969) 529.
[112] I. Bondouk, F. Asfour, V. J. Gontchar, and F. Machali, Nucl. Phys. 65 (1965) 490.
[113] K. Fukunaga, J. Phys. Soc. Japan 20 (1965) 1.
[114] V. S. Siskin et al., Sov. J. Nucl. Phys. 7 (1968) 571.
[115] G. B. Andreev et al., Sov. J. Nucl. Phys. 10 (1970) 521.
[116] D. H. Loyd and W. Haeberli, Nucl. Phys. A148 (1970) 236.
[117] T. Mo and W. F. Hornyak, Phys. Rev. 187 (1969) 1220.
[118] G. Bruno et al., J. Phys. (Paris) 27 (1966) 517.
[119] N. Longequeue et al., J. Phys. (Paris) 27 (1966) 649.
[120] G. J. Clark, D. J. Sullivan, and P. B. Treacy, Nucl. Phys. A98 (1967) 473.
[121] G. S. Mani and R. M. Freeman, Proc. Phys. Soc. (London) 85 (1965) 281.
[122] Y. F. Antouviev et al., Nucl. Sci. Abstr. 15792 (1970).
[123] I. Boca, M. Borsaru, M. Cenja, and E. Iliescu, Nucl. Phys. 62 (1965) 75.
[124] I. Boca et al., Phys. Lett. 22 (1966) 76.
[125] J. Arvieux and S. Roman, Phys. Lett. B26 (1968) 153.
[126] C. Petitjean and L. Brown, Nucl. Phys. A111 (1968) 177.
[127] G. R. Plattner, T. B. Clegg, and L. G. Keller, Nucl. Phys. A111 (1968) 481.
[128] H. Bürgisser et al., Helv. Phys. Acta 40 (1967) 185.
[129] G. R. Plattner et al., Helv. Phys. Acta 40 (1967) 465.
[130] W. E. Burcham et al., Nucl. Phys. A120 (1968) 145.
[131] W. Dürr et al., Nucl. Phys. A120 (1968) 678.
[132] R. Neff, P. Huber, H. P. Naegele, and H. Rudin, Third Polarization Symp., and Helv. Phys. Acta 42 (1969) 915.
[133] R. M. Freeman et al., Proc. Phys. Soc. (London) 85 (1965) 267.
[134] T. Chan, J. P. Longequeue, and H. Beaumevieille, Nucl. Phys. A124 (1969) 449.
[135] M. Borsaru, M. Cenja, C. Hategan, and E. Iliescu, Rev. Roum. Phys. 12 (1967) 661.
[136] M. Dayhuff and R. G. Seyler, Bull. Am. Phys. Soc. 14 (1969) 1213.
[137] A. D. Bacher et al., Bull. Am. Phys. Soc. 14 (1969) 1218.
[138] H. P. Naegele et al., Helv. Phys. Acta 42 (1969) 566.
[139] U. von Moellendorff et al., Helv. Phys. Acta 43 (1970), in press.
[140] S. M. Rizvi et al., Helv. Phys. Acta 43 (1970), in press.
[141] R. Gleyvod et al., Helv. Phys. Acta 41 (1968) 442.
[142] D. C. Robinson, Nucl. Phys. A95 (1967) 663.
[143] A. J. Froelich and S. E. Darden, Nucl. Phys. A119 (1968) 97.
[144] G. Dietze and K. D. Stahl, Z. Physik 228 (1969) 172.
[145] L. Pfeiffer and L. Madansky, Phys. Rev. 163 (1967) 999.
[146] G. R. Mason and J. T. Sample, Nucl. Phys. 82 (1966) 635.

[147] T. G. Miller and J. A. Biggerstaff, Phys. Rev. 187 (1969) 1266.
[148] G. P. Lietz, S. F. Trevino, A. F. Behof, and S. E. Darden, Nucl.
      Phys. 67 (1965) 193.
[149] H. O. Klages and H. Schoelermann, Z. Physik 227 (1969) 344.
[150] H. Schoelermann, Z. Physik 220 (1969) 211.
[151] W. L. Baker et al., Bull. Am. Phys. Soc. 14 (1969) 1230.
[152] T. R. Donoghue et al., Third Polarization Symp.
[153] L. Brown and C. Petitjean, Nucl. Phys. A117 (1968) 343.
[154] B. P. Ad'yasevich et al., Sov. J. Nucl. Phys. 3 (1966) 208.

DISCUSSION

Tanifuji:

In my opinion, the deformation parameter of ^{12}C obtained by
Reynolds et al. by n + ^{12}C scattering is too small. We calculated proton-^{12}C scattering and found that the parameter is about
-0.5, the magnitude of which is larger than that given by Reynolds
by an order of magnitude.

Pisent:

I will only observe that in coupled-channel calculations of
nucleon-^{12}C scattering, we have also obtained (like Tanifuji) a
good fit with a higher deformation value (-0.9), which seems to
be more correct.

Slobodrian:

You showed a pair of states in the p + ^{12}C system (5/2$^-$ and 7/2$^-$)
with a remarkably symmetric double hump in the cross section.
In the book by Goldberger and Watson on collision theory it is
shown that a simple pole in the S-matrix may produce a "double
hump" resonance. Was this possibility looked into?

Plattner:

I do not know whether that was considered, but it would seem to
me that the parameterization of the Erlangen data in terms of
phase shifts clearly shows that there are two distinct states of
very different widths present.

Weil:

The doublet at 9.14 MeV in ^{12}C(p,p)^{12}C was first noticed at Rice
in the excitation function of the inelastic scattering to the first
excited state. Hence, there is evidence for the doublet in both
the elastic and the inelastic channels.

Clausnitzer:

Complete angular distributions of the cross section and analyzing
power were measured at Erlangen in 2 keV steps through the resonance. The representation of the resonant partial amplitudes in
the complex plane as a function of energy exhibits typical resonance behavior (circles), from which width ratios were determined.

Nucleon-Nucleus Optical Model Potential

F. G. PEREY, Oak Ridge National Laboratory, Tennessee, USA*

At the Second International Symposium on Polarization Phenomena, the interpretation of data on polarization in elastic scattering of nucleons from nuclei in terms of the optical model required that the spin-orbit potential have, in general, a smaller radius than the real part of the potential. There is no similar important contribution to the knowledge of the optical model potential resulting from polarization measurements to be reported at this conference. However, most optical model analyses in the past few years have made extensive use of polarization data. I will not review the present situation regarding the optical model potential for nucleons, a task far too big for this short paper, but will restrict myself to comments on a few areas of optical model studies.

Let me begin by making a somewhat artificial grouping of optical model analyses. In general one can classify the various analyses in different categories according to the aims of these analyses. First there are the "global fits." By these I mean attempts at describing the optical model potential parameters as a function of energy and some nuclear properties such as charge, neutron excess, etc., to obtain a good fit to large quantities of elastic scattering data. A typical example of such studies is the recent work of Becchetti and Greenlees [1]. One of the remarkable features of such analyses is that it is in general possible to fit a very large class of data over a large energy range using some very simple dependence of the parameters on mass, energy, neutron excess, etc. It is never clear, however, in such analyses to what extent the shape of the potentials selected, the functional dependence used in the variable parameters, the selection of data, etc., affect the end results. It is generally clear from the many different analyses which have been made that it is possible to perform them in different ways, almost always with equivalent success, but also with different parameter variations. As one attempts to fit more data, the description of the parameters usually gets more complicated, and it is often not clear what physical significance can be attached to some of them. In any event, these studies are very useful in giving recipes for finding optical model parameters for use in nuclear reaction analyses such as the DWBA.

The second class of analyses is more restricted in the sense that a much more limited data set is investigated. For instance, one studies one element as a function of energy or a series of nuclei at a given energy. An underlying assumption of these analyses is that by restricting the data sets in such a manner one may get at the energy dependence of the parameters or at the neutron excess dependence in a more unambiguous way. From this second class of studies, where much better fits are obtained, it seems clear that the effective one-body potential, the optical model potential, changes in shape, as well as in strength, as a function of energy and mass number. This is not at all surprising, since the effective one-body potential has a very complicated relationship to the simple concepts of nucleon-nucleon potentials and matter distribution.

The third class of analyses deals with what one would call a more fundamental approach to the optical model potential. The so-called "reformulated optical model" potential of Greenlees and collaborators [2] is typical of this class. Because the many-body problem involved is formidable, one does not attempt to solve it directly but tries to go back to more fundamental concepts, such as neutron and proton density distributions and effective nucleon-nucleon forces. The "reformulated optical model" potential, for instance, is a folding model of the matter distribution in the nucleus and an effective nucleon-nucleon force. Such a method has some advantages. It allows one to investigate some hopefully general property of the potential, such as its various moments, and relate them to more fundamental quantities, such as the root-mean-square radius of the matter distribution and the strength of the effective nucleon-nucleon interaction. What is not clear at this stage is the meaning of these more fundamental quantities, outside of the context of the model chosen, in view of the gross oversimplifications still made in these models. In particular, we should note the neglect of exchange, the effects of core polarization, and the use of a phenomenological imaginary potential. Some of these effects have been looked at and found to be not negligible. As an example, exchange contributions have been found [3] to be appreciable and strongly energy dependent. From a practical viewpoint such analyses are very different from the other ones, since the desired information is obtained by fitting one nucleus at a time and they do not constrain some of the parameters of the optical model potential to have a particular functional dependence, on mass number, for instance. To be more explicit, one assumes that the proton distribution in the target nucleus is known from electron scattering or mu-mesic X-ray measurements, and by varying the shape of the potential one determines the neutron distribution. One of the remarkable features of this model is that if one fixes the range of the effective nucleon-nucleon potential at a reasonable value, then the neutron distributions and the strength of the effective nucleon-nucleon potential assume very plausible values. Looking at the various moments of the optical model potential is just another mathematical way of describing the potential and is not related to some a priori model for this potential.

For almost twenty years we have made optical model analyses, and we can fit a large body of data, including differential cross sections, polarizations, and reaction cross sections. Our potentials have become standardized to a large degree, and we know with some degree of confidence the behavior of the standardized parameters as a function of mass number and energy. It is not surprising that we have not yet been able to calculate the optical model potential from a very fundamental approach because it is a complex many-body problem. However, I am amazed to see that in spite of the improvements in our ability to take better and more complete data and to perform more complex global fits on this great mass of data, we have not been compelled to modify the shapes of our phenomenological potentials for more than fifteen years. A number of different attempts have been made to change them, but basically the Woods-Saxon shape for the real potential has been very successful, and we have not come up with a modification of its shape which would systematically improve the fits. The Woods-Saxon shape may not be the best way to parameterize the potential, and, as emphasized by Greenlees and collaborators, quantities such as the volume integral and the root-mean-square radius of the potential may have a deeper meaning than the usual parameterization. It is, however, obvious that much more than these two moments are important, otherwise a square-well potential would do just as well. Maybe a greater effort made on analyzing data for nuclei where the usual optical model analyses have not been too successful would yield some interesting answers. The two nuclei which immediately come to mind are carbon and calcium. In the case of carbon, the strong coupling of the first excited state will undoubtedly complicate matters. But, from our understanding of the effects of strong coupling, the case of elastic scattering from carbon at 40 to 50 MeV should not be very different from the one for many other nuclei at lower energies. The case of calcium is more puzzling, since strong coupling cannot be involved. It is worth noting that at a given incident bombarding energy the quality of the fits improves as the nuclei get heavier, and the quality of the fits worsens for a given nucleus as the energy increases. The quality of the fits is given by the usual chi-square value. These last two remarks apply also to the elastic scattering of more complex particles. One could conjecture that we are trying to get too much out of such a simple local potential. The actual potential may be very complicated and angular momentum dependent. However, our simple optical model potentials may have been successful because we have been reproducing only an important aspect of the phase shift behavior as a function of angular momentum and energy. Elastic scattering is dominated by a strong diffraction pattern, and the answer to the questions we might want to ask of the potential may be buried in small deviations from this diffraction pattern. My feeling at this stage of the development of the optical model potential is that not only will we have to be asking the right questions in the right way, but very likely we will have to obtain better data than were thought adequate in the past. I do not know what the right questions are, but I have a feeling that even in today's type of analyses we have not investigated our methodology thoroughly enough. We have only a very limited understanding of how our simple chi-square criteria, used in automatic computer searches, are affected by the particular sampling of angles where data are available. We indiscriminately treat the data points without respect for their relative position on the diffraction pattern. We do not fully understand the importance of relative weighting of data points as a function of angle. All these problems prevent us from comparing meaningfully the quality of fits for two different angular distributions, whether for different nuclei or different energies. I know from many personal experiences that these factors play an important role in specific cases, but I am at a loss to provide any useful guidelines in general terms. There are many other similar problems, such as the relative weighting of differential cross-section, polarization, and reaction cross-section data, or the influence of absolute normalization of the data. All these problems partially explain the multitude of seemingly different results obtained from analyses of the same set of data, and I feel that the "jungle" of the optical model parameter space is in great part due to our lack of understanding of our methodology.

Another aspect of our problem is the accuracy of the data with which we work. I do not think that the experimentalists are to blame for this. They will usually take data of the required accuracy when the effort is worthwhile. However, optical modelists have not conveyed in a convincing manner their requirements on accuracy for a meaningful analysis. As a result, the experimentalists have been faced with the task of determining on their own how accurately to take data which might be useful for such analyses. I am not thinking of the accuracy of elastic scattering data which come as a by-product of an experiment designed for other purposes, but of the accuracy of data whose main objective is the determination of optical model parameters. There are a number of experiments which have been performed on the same nuclei at very nearly the same energies, but very few of them involve the measurement of both the differential cross section and polarization. The two examples shown are probably illustrative of the present data situation. Fig. 1 shows the differential cross-section data for scattering of 40-MeV protons from ^{60}Ni. The energies at which the ORNL [4] and the Minnesota [5] experiments were performed are slightly different. Some of the effects of this difference in energy have been compensated for by plotting the ratio of the cross section to the Rutherford cross section. What is rewarding is the very good agreement between the two sets of data up to about 70°. Beyond 70°, however, except for a few points, the two sets of data differ by about 15%. The ORNL data were renormalized upward by 7.7% as a result of

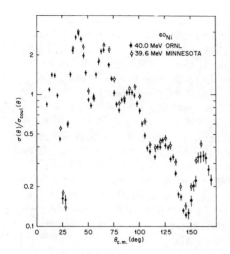

Fig. 1. Differential cross-section data for 40-MeV protons on ^{60}Ni from refs. [4] and [5].

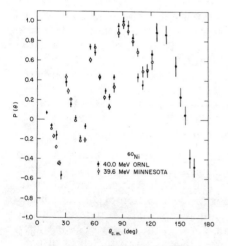

Fig. 2. Polarization data for 40-MeV protons on ^{60}Ni from refs. [4] and [5].

a subsequent remeasurement of the target thickness [6]. We now have these two data sets, not quite at the same energy, which should be good individually to about 3% in absolute magnitude, but which differ over nearly two-thirds of the measured angular distributions by about 15%. There are, of course, several possible reasons for such a difference, including the one that this difference is real and due to the slightly different energies. The polarization data from these measurements are shown in fig. 2 and except for a few points are in good agreement. This situation is very disturbing and causes one to wonder how far one should attempt to push the optical model analyses. I would not like to give the impression that this is an isolated instance, as I am aware of several other similar cases. Although such differences in the data will give pretty much the same value for the lower moments of the potential, this is not true for the higher ones from which we obtain information on the shape of the potentials.

I would like now to turn to a related problem which deals more directly with the subject of this conference. Recently Sinha and Edwards [7] have reported that much better fits could be obtained to the elastic proton scattering from Pb, in the range of 30 to 60 MeV, if one adds to the Woods-Saxon real potential a hump inside the surface. The improvements in the quality of fits are of the order of two or three in chi-square. We were extremely interested in their results, but a little puzzled by a footnote indicating that the quality of the fits was not sensitive to the width of the hump. Therefore, we decided to parameterize the shape of the real potential in a slightly different manner and repeat the analysis. There were several other reasons why we wanted to repeat the study, the two main ones being as follows: (1) it was not obvious that polarization data, which were available at 30 and 40 MeV, had been used in the fits and (2) we knew [6] that the elastic scattering data at 40 MeV had to be renormalized by 10%. The shape of the real potential we used is a Woods-Saxon potential, but we had the possibility of using a different diffuseness parameter inside and outside the halfway point. We investigated, in addition, the inclusion of a Gaussian depression at the center of the potential of the form

$$b V_s \exp\left[-(r/a_g)^2\right]$$

where a_g is the width parameter of the Gaussian and b expresses the depression as a fraction of the potential depth. When b is negative, the real potential has a wine-bottle shape. This parameterization of the real potential permits independent investigation of an asymmetric surface region with only one more parameter, and of a depression in the center with two additional parameters. The results of Sinha and Edwards translated in terms of these parameters are that b is of the order of -0.1 and the internal diffuseness is about half the external diffuseness. For Pb, a_g is 4–5 fm. We include a standard volume and surface imaginary potential and a Thomas-type spin-orbit potential. The differential cross-section data used at 30.3 MeV were taken from Ridley and Turner [8], and the polarization data at the same energy were obtained by Greenlees et al. [9]. The errors were the same as given in the experimental papers.

Six analyses were made at 30 MeV, and the results are shown in table 1. We fitted the differential cross sections alone, or simultaneously with the polarization. Type A potential is the standard optical

Table 1. Results of the search on fitting angular distributions of 30-MeV protons scattered by ^{208}Pb

	Best fits to $\sigma(\theta)$ only			Best fits to $\sigma(\theta)$ and $P(\theta)$		
	A	B	C	A	B	C
V_S (MeV)	54.54	51.88	54.88	52.25	51.77	54.15
r_S (fm)	1.171	1.175	1.167	1.190	1.190	1.180
a_S int (fm)	0.713	0.387	0.429	0.703	0.622	0.629
a_S ext (fm)	0.713	0.733	0.721	0.703	0.710	0.702
b (fm)			-0.066			-0.077
a_g (fm)			6.01			4.88
W_S (MeV)	1.49	1.18	0.734	2.32	2.52	2.17
W_D (MeV)	8.78	10.26	10.16	8.08	8.16	8.25
r_I (fm)	1.280	1.270	1.261	1.282	1.274	1.268
a_I (fm)	0.767	0.660	0.704	0.743	0.712	0.737
V_{SO} (MeV)	6.02	5.91	5.77	6.02	6.01	6.03
r_{SO} (fm)	1.111	0.919	0.890	1.18	1.181	1.176
a_{SO} (fm)	1.241	1.098	0.911	0.602	0.591	0.595
χ_σ^2/N_σ	1.4	0.89	0.85	5.7	5.7	5.5
χ_P^2/N_P	(482)	(316)	(275)	13.8	13.0	12.3
χ^2	103	64	61	958	926	884
σ_R (mb)	1982	1838	1861	1943	1893	1901
J/A (MeV fm^3/ nucl)	404.7	416.1	413.8	404.8	407.9	405.5
$\langle r^2 \rangle^{1/2}$ (fm)	5.99	6.04	6.00	6.05	6.06	6.04
$\langle r^4 \rangle^{1/4}$ (fm)	6.44	6.48	6.44	6.50	6.50	6.47

model potential. Type B potential has only the asymmetric surface in
the Woods-Saxon potential. Type C potential is a type B potential
with the addition of a depression in the center. Comparison between
columns 1 and 3 shows the effect observed by Sinha and Edwards.
The chi-square for the differential cross section is improved by a
factor of about two. The agreement with the polarization is improved
also by a factor of two, but it is very bad, chi-square being of several
hundred per point. Recall that for these three columns we did not
search on the polarization data. The internal diffuseness is about
half the external diffuseness, and the depression in the center is of
the right magnitude. The second column indicates that all the improvement was brought about by the asymmetric Woods-Saxon form and not
the central depression. This fact may explain the lack of sensitivity
which they found to the width of their hump. If we look at the last
three columns, where polarization data are also used in the fitting,
we see that the effect disappears. It is very doubtful, from a purely
phenomenological point of view, if an improvement of a few percent
in chi-square can justify the addition of one or three parameters. We
can only conclude that the data neither support nor exclude such a
change in potential shape. The reaction cross section given at the
bottom of the columns is not affected; neither are the first three even
moments of the real part of the potential.

For the 40-MeV analyses we used the differential cross-section and
polarization data of Blumberg et al. [4]. The differential cross-section
data were renormalized upward by 10%. We used the errors given in
the paper except that for angles less than 30° we assigned 5% error,
and between 30 and 48° errors of 3%; these data had less than 1% statistical error (the cross section of the 135° angle was also given 20%
error). For the polarization data we took into account the finite angular aperture of the detectors of ±1.2°. Table 2 shows the results of
the four fits which were made. Whereas we were able to reproduce
the results of Sinha and Edwards at 30 MeV using the differential cross
section data only, we failed to do so at 40 MeV. The change in normalization and/or the different weighting scheme which we used may
explain this difference. In any event, the use of an asymmetric Woods-
Saxon potential does not improve the fits significantly whether we
use polarization data or not. The reaction cross section has been
measured [6] at 40 MeV and is 2023 ± 100 mb, and this is in good
agreement with these results.

For even-even nuclei there can be no additional terms in the optical model potential. In the case of nuclei with spin different from
zero we can expect a term in the potential to represent the spin-spin
interaction. Measurements of total cross section of neutrons on
oriented nuclei have indicated that the strength of this term is very
small. Of course, such a term will have to be included to explain
depolarization in elastic scattering. There are two contributions to
this conference on these measurements. It is very likely that for the

Table 2. Results of the search on fitting angular distributions of 40-MeV protons scattered by ^{208}Pb

	Best fits to $\sigma(\theta)$ only		Best fits to $\sigma(\theta)$ and $P(\theta)$	
	A	B	A	B
V_S (MeV)	53.82	51.16	50.14	49.45
r_S (fm)	1.148	1.172	1.193	1.192
a_S int (fm)	0.755	0.621	0.717	0.688
a_S ext (fm)	0.755	0.762	0.717	0.762
W_S (MeV)	5.20	2.22	4.38	3.03
W_D (MeV)	5.39	8.76	6.56	10.09
r_I (fm)	1.254	1.244	1.303	1.290
a_I (fm)	0.754	0.700	0.680	0.618
V_{so} (MeV)	6.77	6.67	5.89	5.43
r_{so} (fm)	1.261	1.165	1.119	1.136
a_{so} (fm)	1.093	1.079	0.741	0.658
χ_σ^2/N_σ	5.5	4.7	7.6	7.8
χ_P^2/N_P	(152)	(56)	10.0	7.3
χ^2	379	310	927	814
σ_R (mb)	2051	2033	2095	2081
J/A (MeV fm^3/ nucl)	382.3	394.6	392.6	395.3
$\langle r^2 \rangle^{1/2}$ (fm)	5.97	6.07	6.09	6.15
$\langle r^4 \rangle^{1/4}$ (fm)	6.44	6.54	6.54	6.62

$\sigma(\theta)$: Data renormalized by $+10\%$; $\theta < 30$: 5% errors;
$\theta \geq 50$ as in ref. [4] except for $\theta = 135°$; 20% error.
$P(\theta)$: $\Delta\theta = \pm 1.2°$.

proton measurements a straight-forward analysis, including a spin-spin term in the potential, is possible. In the case of the neutron measurements, more analyses will have to be made in view of the presence of compound elastic contributions. It seems evident, however, that the contribution of this term to both the differential cross section and polarization is small.

It is a pleasure to acknowledge the help of Claire Perey in performing the Pb optical model analysis.

REFERENCES

*Research sponsored by the U.S. Atomic Energy Commission under contract with Union Carbide Corporation.

[1] F. D. Becchetti, Jr., and G. W. Greenlees, Phys. Rev. 182 (1969) 1916.
[2] G. W. Greenlees, G. J. Pyle, and Y. C. Tang, Phys. Rev. C1 (1970) 1145.
[3] W. G. Love and G. R. Satchler, to be published.
[4] L. N. Blumberg et al., Phys. Rev. 147 (1966) 812.
[5] H. S. Liers et al., Phys. Rev. C2 (1970) 1399.
[6] J. Menet, E. E. Gross, A. Van der Woude, and A. Zucker, private communication.
[7] G. C. Sinha and V. R. W. Edwards, Phys. Lett. 31B (1970) 273.
[8] B. W. Ridley and J. F. Turner, Nucl. Phys. 58 (1964) 497.
[9] G. W. Greenlees et al., Report No. COO-1265-88, University of Minnesota (unpublished).

DISCUSSION

Sinha:

Ambiguities in optical model analyses arise from the parameters used in the model which are not known a priori. If ρ_n, the matter distribution, is known a priori by nuclear matter calculations, it is much easier to get the effective interaction correctly. We find that in this way one can determine the effective interaction more precisely in the case of ^{40}Ca. Measurements of the cross section and polarization at backward angles should be done with more precision because that determines the optical parameters more accurately.

Greenlees:

In our work we also find that the volume integral per nucleon and the r.m.s. radius of the real central potential increase as the energy decreases. This is not unexpected, since second-order effects become more important at lower energies. For example, the effects of antisymmetrization will increase as the incident energy gets closer to the Fermi energy; this will reduce the potential in the interaction and will appear as an apparent increase in the halfway radius. The problem is to find a way of introducing such effects into the model which has a physical basis and is not simply an increased mathematical freedom due to the use of additional parameters. It is not clear to me that Perey's approach has such a physical basis.

Perey:

I agree with the objections. The present approach attempts to delineate what changes in the potential are needed.

Greenlees:

It is perhaps unfortunate that we conventionally use a particular functional form (the Fermi shape) for such work. I think we tend to overemphasize halfway radii and diffuseness values in this context.

Benenson:

Earlier work reported a proton potential anomaly. Does one still see this in optical model analysis?

Perey:

Yes, the depth is greater for protons than for neutrons.

Greenlees:

It is perhaps dangerous to attach too much importance to this difference since there is a scarcity of neutron data, and the available data tend to be for different nuclei and at different energies from the proton data.

Perey:

In the Becchetti global analysis, which examined essentially all the proton data and independently all the neutron data for $A > 40$ and $E < 50$ MeV, it was evident that the proton analysis was dominated by data with $A < 90$ and $E > 10$ MeV, whereas the neutron analysis was largely determined by data for $A > 90$ and $E < 14$ MeV.

Wigner:

If the number of neutrons in the nucleus is larger than the number of protons, it is reasonable to have a stronger attraction for protons because the effective proton-neutron interaction is greater than the effective neutron-neutron interaction. The question which one would like to have answered, therefore, is whether the difference is present also in nuclei with equal numbers of protons and neutrons such as ^{16}O or ^{40}Ca. That would be surprising and important.

Greenlees:

After one allows for the n-p difference via an isospin term, there still seems to be an effect.

Breit:

Just what was done in the calculations for protons on ^{16}O as compared with neutrons on ^{16}O and in other similar cases? How was the Coulomb potential in the transition region and in the nuclear interaction corrected for in the comparisons?

Perey:

Those particular nuclei are difficult to examine in such detail since it is in general difficult to get a good fit with an optical potential.

Breit:

In that case one cannot speak of a comparison.

Perey:

True. However, a systematic analysis of data for a range of nuclei seems to show a difference, but it is not positively established.

Wigner:

If this could be established, it would be revolutionary and would really upset theoreticians.

Elastic Scattering of
Deuterons and Heavier Projectiles

R. C. JOHNSON, University of Surrey, Guildford, England

1. INTRODUCTION

I have interpreted my brief here as a discussion of the way in which knowledge of the optical potentials for deuterons, ^{3}He, and tritons has been extended by the great increase of experiments involving polarized projectiles since the last polarization symposium.

In the first place, of course, polarization experiments provide unique information concerning the spin-dependence of the optical potentials. However, polarization experiments can also lead to improved knowledge of the parameters of the central part of the potentials. An example of this is the analysis by Yule and Haeberli [1] of their results on deuteron scattering by ^{40}Ca at 11 MeV, including both differential cross-section and vector polarization data. The parameters they obtained for the deuteron optical potential led to an improved fit to (d,p) cross sections [2]. They found that the improved fit they obtained at large angles is due to small differences in the parameters of their central potential compared with parameters obtained without fitting to elastic polarization data.

2. POLARIZATION EFFECTS IN ELASTIC DEUTERON SCATTERING

The situation up to about 1967 has been reviewed by Hodgson [3]. At that time the only relevant polarization data were the 21 MeV results obtained with ^{40}Ca targets at Saclay [4]. These data consisted of the angular distributions of the differential cross section, the vector polarization (iT_{11}), and two of the three real numbers (T_{22}, T_{20}, T_{21}) that specify the second-rank tensor components of the deuteron polarization produced in the scattering. These data were successfully fitted by Raynal [5], who showed that they were consistent with an optical potential of the form

$$V_d = V \text{ (central)} + V(\vec{L} \cdot \vec{S}), \tag{1}$$

143

with V (central) and $V(\vec{L}\cdot\vec{S})$ having roughly the magnitudes one would expect to be given by the Watanabe model [6], i.e.,

$$V_d \approx \langle \Phi_d | V_n + V_p | \Phi_d \rangle, \tag{2}$$

where V_n and V_p are nucleon optical potentials. A feature of Raynal's work was that he found no evidence from the data for any term in the deuteron optical potential having a more complicated dependence on the deuteron spin. I refer here to the second-rank tensor potentials first written down by Satchler [7] and now conventionally denoted by

$$T_R(r) = (\vec{S}_d \cdot \vec{r})^2 / r^2 - \frac{2}{3}, \tag{3a}$$

$$T_L(r) = (\vec{L}\cdot\vec{S}_d)^2 + \frac{1}{2}(\vec{L}\cdot\vec{S}_d) - 2\vec{L}^2, \tag{3b}$$

$$T_p(r) = (\vec{S}_d \cdot \vec{p})^2 - \frac{2}{3}\vec{p}^2 \tag{3c}$$

where in each case a multiplicative radial factor has been omitted. I shall refer to these potentials collectively as V (tensor). Although such potentials appear very exotic, even the simplest models of the deuteron optical potential bring them in immediately. Satchler [7] and, in more detail, Raynal [8] and Testoni and Gomes [9] have shown that when the Watanabe model (2) is taken seriously, the D-state of the deuteron gives rise to a potential of the T_R type, while potentials of the T_P and T_L type can arise in the same model from the non-locality of the nucleon optical potentials [10]. It was therefore most satis- factory when later experiments between 5 and 11 MeV on ^{40}Ca [11], and other targets [12], measuring the same quantities as in the Saclay experiment, showed the necessity for a small term in the deuteron optical potential of the T_R type. Raynal [13] pointed out that in fact the T_R potential obtained by Schwandt and Haeberli [11] by straight- forward fitting procedures was close (within a factor of 2) to the T_R potential predicted by Raynal [8] on the basis of eq. (2) and reasonable deuteron internal wave functions.

Tensor terms in qualitative agreement with those found by Schwandt and Haeberli [11, 12] have also been reported as the result of the analysis of polarization effects in deuteron elastic scattering from ^{12}C at 3.5 to 7.1 MeV [14] and by Mg at 7 MeV [15]. Both these ex- periments differed from the experiments of Schwandt and Haeberli [11, 12] in that a source of polarized ions was not used and that the T_{21} component of the deuteron polarization was also measured. These authors report [14, 15] that T_{21} is particularly sensitive to the presence of V (tensor) terms in the deuteron potential. I shall return again to this point.

In order to understand some of the features of these data, it is useful to expand the transition matrix $M(\theta)$, for elastic deuteron scattering from a spin-zero target, as a linear combination of the same tensor operators T_{KQ} whose average values are the quantities measured in polarization experiments.

$$\langle \sigma' | M | \sigma \rangle = M^0(\theta) + \sum_Q M^1_Q(\theta) \langle \sigma' | T^\dagger_{1Q} | \sigma \rangle + \sum_Q M^2_Q(\theta) \langle \sigma' | T^\dagger_{2Q} | \sigma \rangle.$$

(4)

The usefulness of this expansion for our purposes lies in the approximate correspondence, shown in eq. (5), between the different terms in the expansion and various components of the deuteron optical potential

$$M^0(\theta) \rightarrow V(\text{central}) + \dots,$$

(5a)

$$M^1_Q(\theta) \rightarrow V(\vec{L}\cdot\vec{S}) + \dots\dots\dots,$$

(5b)

$$M^2_Q(\theta) \rightarrow (V(\vec{L}\cdot\vec{S}))^2 + V(\text{tensor}) + \dots\dots$$

(5c)

I am appealing here to the relative weakness of the spin-dependent terms, so that a perturbation treatment should give a good idea of the way spin-dependent terms influence observed quantities [16, 17]. Thus eq. (5) is meant to indicate that, for example, the terms M^2_Q in eq. (4) only enter through quantities which are at least second order in $V(\vec{L}\cdot\vec{S})$ or first order in $V(\text{tensor})$.

Expressions for observable quantities in terms of the functions M^K_Q are shown in eq. (6).

$$\sigma_0 \, i T_{11} = -2\,\text{Im}\left\{M^1_1 M^{0*} - \frac{1}{4}(2)^{1/2} M^1_1 (M^{2*}_0 + (6)^{1/2} M^{2*}_2)\right\},$$

(6a)

$$\sigma_0 \bar{T}_{20} = -\frac{1}{2}(2)^{1/2} |M^1_1|^2 + 2\,\text{Re}\{M^0 M^{2*}_0\}$$

$$+ \frac{1}{2}(2)^{1/2} |M^2_0|^2 - (2)^{1/2} |M^2_2|^2,$$

(6b)

$$\sigma_0 \bar{T}_{22} = -\frac{1}{2}(3)^{1/2} |M^1_1|^2 + 2\,\text{Re}\{M^0 M^{2*}_2\} + \text{Re}\{M^2_0 M^{2*}_2\},$$

(6c)

$$\sigma_0 \bar{T}_{21} = -(3)^{1/2}\,\text{Re}\left\{M^1_1 (M^{2*}_2 - (3/2)^{1/2} M^{2*}_0)\right\},$$

(6d)

$$M^1_0 = M^2_1 = 0, \qquad M^1_1 = M^1_{-1}, \qquad M^2_2 = M^2_{-2}.$$

(6e)

The quantities $\overline{T}_{KQ}$ describe the polarization of deuterons outgoing along $\vec{k}_2$ in the c.m. system in an elastic scattering experiment with unpolarized deuterons incident along $\vec{k}_1$. The bar notation denotes that these tensors refer to a y-axis along $\vec{k}_1 \times \vec{k}_2$ and a z-axis along $(\vec{k}_1 + \vec{k}_2)$. Of course the components of the tensors describing the polarization state in any other coordinate system are simply related to these quantities; in particular the components T_{KQ} referred to a co-ordinate system with z-axis along $\vec{k}_2$ are linear combinations of the $\overline{T}_{KQ}$'s, with simple functions of the scattering angle θ as coefficients. If terms involving M^2 are neglected, eq. (6) predicts

$$\overline{T}_{22} = \left(\frac{3}{2}\right)^{1/2} \overline{T}_{20}, \tag{7a}$$

$$\overline{T}_{21} = 0, \tag{7b}$$

and

$$\overline{T}_{20}, \overline{T}_{22} \leq 0. \tag{7c}$$

The same equations apply with the bars dropped in this approximation. These predictions appear to have been satisfied qualitatively in the Saclay [4] experiment (fig. 1), although T_{21} was not measured. Hence any contributions from the M_Q^2 terms must lead to a net negative con-tribution, and it would not be surprising in this case if effects caused by tensor potentials could be masked by a suitably chosen $\vec{L} \cdot \vec{S}$ force. In this sort of situation it is clearly important that T_{21} be measured. At the lower energies used by Schwandt and Haeberli [11, 12], although again T_{21} was not measured, the measured T_{20} and T_{22} yielded both positive and negative values (see fig. 2), indicating that M^2 terms must be present; and, in fact, they are able to show that some V (tensor) was necessary, as already mentioned.

It is crucial to note that in all these experiments V (central) and $V(\vec{L} \cdot \vec{S})$ were tied down pretty closely by measurements of the differen-tial cross section and iT_{11}, respectively. The accurate measurements of this quantity which are now becoming available for a wide range of nuclei from ^{12}C to ^{208}Pb are therefore extremely valuable [18–22].

Eq. (6a) shows that the dominant term in $i\overline{T}_{11}$ (= iT_{11}) involves the term M^0, which is dominated by the large V (central), interfering with the term M_1^1, which is linear in $V(\vec{L} \cdot \vec{S})$ in lowest order. In order to look closely at the possible tensor potential contributions, we would like to look closely at the terms $M^0 M^{2*}$ in eq. (6b, c). These can be dis-entangled from the other terms by simple subtraction. In terms of the T_{2Q}, we have

$$\cos \theta (T_{22} - (3/2)^{1/2} T_{20}) - 2 \sin \theta\, T_{21} = \overline{T}_{22} - (3/2)^{1/2} \overline{T}_{20}, \tag{8a}$$

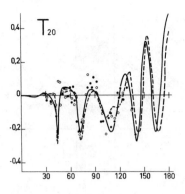

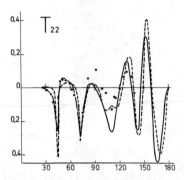

Fig. 1. Second-rank deuteron
tensor polarization components
for Ca(d,d) Ca^{40} at 21.7 MeV.
The data are from Beurtey et al.
[4] and the curves are optical
model fits [5] obtained with a
potential of the form of eq. (1).

$$\sin \theta (T_{22} - (3/2)^{1/2} T_{20}) + 2 \cos \theta \, T_{21} = 2 \overline{T}_{21}, \tag{8b}$$

$$T_{22} + (6)^{-1/2} T_{20} = \overline{T}_{22} + (6)^{-1/2} \overline{T}_{20}, \tag{8c}$$

$$\overline{T}_{22} - (3/2)^{1/2} \overline{T}_{20} \sim M^0 M^{2*}, \tag{8d}$$

$$\overline{T}_{21} \sim M^1 M^{2*} \tag{8e}$$

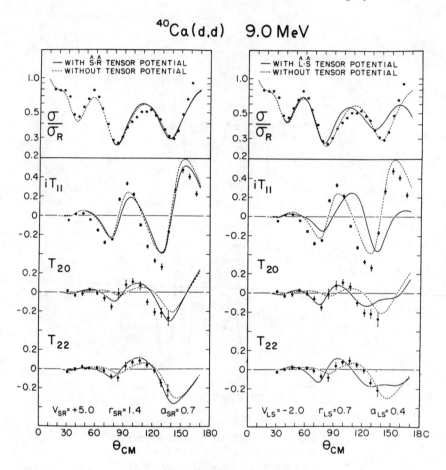

Fig. 2. Angular distributions for ^{40}Ca(d,d) at 9 MeV. Both the experimental points and the optical model calculations are from Schwandt and Haeberli [11].

The left hand side of eq. (8a) involves a knowledge of all three components T_{2Q}. Also shown in eq. (8b) is the relationship between the T_{2Q} and $\overline{T}_{21}$. It can be seen from eq. (6d) that $\overline{T}_{21}$ arises from terms of at least third order in $V(\vec{L}\cdot\vec{S})$ and therefore would normally be expected to be small. It is also clear from eq. (8) that in agreement with the observations of refs. [14, 15] the component T_{21} is likely to be very sensitive to the presence of V (tensor) because it is a linear combination of quantities involving $M^0(\theta)$ and M_Q^2 with no dependence on the $|M^1|^2$ terms.

An illustration of the usefulness of this type of analysis is shown

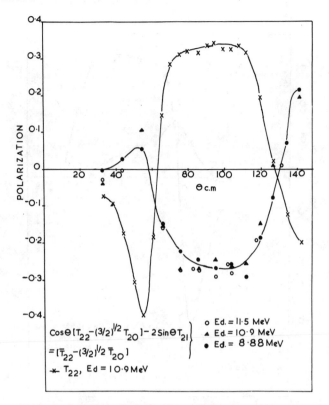

Fig. 3. Angular distribution of the quantity
$(\bar{T}_{22} - (3/2)^{1/2}\ \bar{T}_{20})$ (see eq. (8a)) for the d-α data
of König et al. [24]. (Figure prepared by V. O'Hara-
Boyce, University of Surrey.)

in figs. 3 and 4. (Similar results at 12 MeV are presented at this conference by G. P. Laurence et al. [23].) Here accurate data of König et
al. [24] have been used to evaluate the left-hand sides of eq. (8a, b).
Fig. 3 therefore shows directly the second-order $V(\vec{L}\cdot\vec{S})$ terms and
first-order tensor potential terms interfering with the central force
contributions, while fig. 4 shows effects which would be at least third
order in $V(\vec{L}\cdot\vec{S})$ in the absence of any V (tensor). I feel that this way
[16] of looking at these rather complicated polarizations data should
prove very useful; and as more accurate measurements of this type
become available, it should be possible to tie down the spin dependence of the deuteron optical potential.

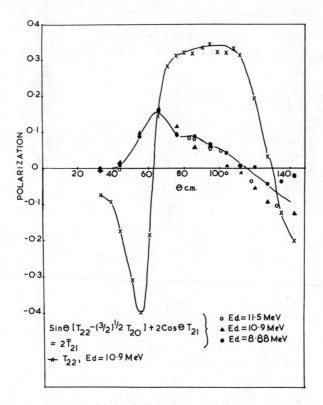

Fig. 4. Angular distribution of the quantity $2\bar{T}_{21}$ (see eq. (8b)) for the d-α data of König et al. [24]. (Figure prepared by V. O'Hara-Boyce, University of Surrey.)

3. POLARIZATION EFFECTS IN ELASTIC SCATTERING OF ^{3}He AND TRITON

The small amount of polarization data, of relevance to the ^{3}He and triton optical potentials, collected up to 1968 has been reviewed by Hodgson [25], together with the considerably greater information on the central parts of the ^{3}He and triton optical potentials obtained by that date. The most extensive polarization results so far are reported at this conference by McEver et al. [26], using ^{4}He scattering as an analyzer. These authors report that values of the ^{3}He spin-orbit force in the region of 4 to 5 MeV are needed in order to fit their data on ^{12}C, ^{9}Be, and ^{16}O targets with an 18 MeV bombarding energy. These values appear to be somewhat higher than the value of 2.7 ± 0.7 MeV obtained by Patterson and Cramer [27] on the basis of spin-flip measurements in inelastic ^{3}He scattering, and theoretical estimates [28] of about 2 MeV based on the generalization [29] of the Watanabe model for deuterons (eq. (2)). However, calculation shows that the spin-orbit potentials obtained by McEver et al. [26] in the region of space just outside the range of the absorptive potentials take on values which are very similar to those of a spin-orbit potential of depth 2.5 MeV, with a radius and diffuseness compatible with averaging nucleon optical potentials over the ^{3}He internal wave function. These results suggest that Born approximation estimates of ^{3}He and triton polarization, which give

$$P\left(^3He\right) \sim \frac{1}{9} P \text{ (nucleon)},$$

may be as misleading for medium mass nuclei as they obviously are for light nuclei. Full elucidation of this point will probably have to await the development of polarized ^{3}He and triton sources [30].

4. CONCLUSIONS

Apart from their intrinsic interest, it appears to me that experiments involving polarized deuterons, ^{3}He, and tritons derive their importance from the implications they may have for other direct reaction studies involving the same projectiles. As already mentioned in the introduction, the effects of additional elastic data can show up in a direct-reaction calculation simply through the improved optical model parameters that are obtained. However, these data can also be relevant at a more basic level. Thus, for example, it is well known that the imaginary part of the deuteron optical potential predicted by eq. (2) is very inaccurate [31] (the situation is similar in the ^{3}He case [28, 32]). Some suggestions have been made [33, 34, 35] recently concerning the way these discrepancies can be accounted for in terms of coupling to definite channels. The ability of these theories to give a good account of elastic polarization measurements would give additional credence to the implications they have for the analysis of stripping and pickup reactions [36].

REFERENCES

[1] T. J. Yule and W. Haeberli, Nucl. Phys. A117 (1968) 1.
[2] B. A. Robson, Phys. Lett. 26B (1968) 501.
[3] P. E. Hodgson, Adv. Phys. 15 (1966) 329.
[4] R. Beurtey et al., Compt. Rend. 256 (1963) 922.
[5] J. Raynal, Phys. Lett. 7 (1964) 281.

[6]   S. Watanabe, Nucl. Phys. 8 (1958) 484.

[7]   G. R. Satchler, Nucl. Phys. 21 (1960) 116.

[8]   J. Raynal, Thesis, Paris, 1964 (unpublished).

[9]   J. Testoni and L. C. Gomes, Nucl. Phys. 89 (1966) 288.

[10]   E. B. Lyovshin, Ukr. Fiz. Zh. 13 (1968) 1939.

[11]   P. Schwandt and W. Haeberli, Nucl. Phys. A123 (1969) 401.

[12]   P. Schwandt and W. Haeberli, Nucl. Phys. A110 (1968) 585.

[13]   J. Raynal, Phys. Lett. 29B (1969) 93.

[14]   H. Cords et al., Nucl. Phys. A113 (1968) 608.

[15]   A. Djaloeis and J. Nurzynski, Third Polarization Symp.

[16]   R. C. Johnson and D. J. Hooton, to be published.

[17]   E. B. Lyovshin, Ukr. Fiz. Zh. 13 (1968) 975.

[18]   A. M. Baxter et al., Nucl. Phys. A112 (1968) 209.

[19]   J. A. R. Griffith et al., Nucl. Phys. A146 (1970) 193.

[20]   J. A. R. Griffith et al., to be published.

[21]   R. C. Brown et al., Third Polarization Symp.

[22]   J. Lohr, Third Polarization Symp.

[23]   G. P. Lawrence et al., Third Polarization Symp.

[24]   V. König, W. Grüebler, D. A. Schmelzbach, and P. Marmier,
       Nucl. Phys. A134 (1969) 686, A148 (1970) 38, and A148 (1970)
       391; Third Polarization Symp.

[25]   P. E. Hodgson, Adv. Phys. 17 (1968) 563.

[26]   W. S. McEver et al., Third Polarization Symp.

[27]   D. M. Patterson and J. G. Cramer, Phys. Lett. 27B (1968) 373.

[28]   A. Y. Abul-Magd and M. El-Nadi, Prog. Theoret. Phys. (Kyoto)
       35 (1966) 798.

[29]   J. R. Rook, Nucl. Phys. 61 (1965) 219.

[30]   D. D. Armstrong, P. W. Keaton, Jr., and L. R. Veeser, Third
       Polarization Symp.

[31]   F. G. Perey and G. R. Satchler, Nucl. Phys. A97 (1967) 515.

[32]   F. D. Becchetti, Jr. and G. W. Greenlees, Third Polarization
       Symp.

[33]   R. C. Johnson and P. J. R. Soper, Phys. Rev. C1 (1970) 976.

[34]   G. H. Rawitscher, Phys. Rev. 181 (1969) 1518, and 163 (1967)
       1223.

[35]   M. Ichimura, M. Kawai, T. Ohmura, and B. Imanishi, Phys. Lett.
       30B (1969) 143, and to be published.

[36]   J. D. Harvey and R. C. Johnson, to be published.

DISCUSSION

Keaton:

The form factor of the tensor potential used by Schwandt and
Haeberli is quite different from that predicted by the calculations of Raynal where he included the D-state of the deuteron.
Do you know how sensitive the fits are to the shape of the tensor potential?

Raynal:

The form factor of Schwandt and Haeberli has a smaller range but is two times deeper than the form factor I calculated. If they increase the parameter of their Gaussian form factor, they would have to decrease its depth. So I think that it is in quantitative agreement.

Johnson:

We have found a phenomenological form factor which fits the shape of the potentials predicted by Raynal when he folds the deuteron wave function with the nucleon optical potential. It is simple and can be included in optical model computer calculations as a choice of form factors along with the usual ones.

Keaton:

We have measured all the efficiency tensors including T_{21} for $^{56}Fe(\vec{d},d)^{56}Fe$ at 15 MeV. We are planning a similar study of many nuclei. If I recall correctly, we do find T_{21} to be very small, as you suggested. It is larger than about 0.03 only beyond 100°. Perhaps, since you included ^3He, it should be mentioned that we have also used polarized tritons, and this too will provide a sensitive test of the spin-orbit potential for composite particles.

Raynal:

In the old analysis of d-^{40}Ca scattering, the conclusion was that the effect of tensor interaction is less than the experimental errors, and that one must decrease the radius of the $\vec{L} \cdot \vec{S}$ interaction. In the later analysis of Schwandt and Haeberli at 9 MeV, which showed a tensor potential effect, I think that a smaller spin-orbit radius would improve the agreement.

Proton Inelastic Scattering

G. R. SATCHLER, Oak Ridge National Laboratory, Tennessee, USA[*]

Abstract: The results of recent polarization measurements, including the so-called spin-flip measurements, on the inelastic scattering of protons are reviewed. These indicate that a "collective" or deformed potential model gives a good account of most of the data, provided both the absorptive and the spin-orbit coupling components of the optical potential are deformed. For the latter, partly on experimental and partly on theoretical grounds, the so-called Thomas form of deformed spin-orbit coupling is to be preferred over the simplified form used earlier. Further, there are indications that the spin-orbit part should have a greater deformation than the central part. A microscopic description of the polarization in terms of shell-model wave functions and effective nucleon-nucleon interactions has not so far been successful. It is most likely that nucleon-nucleon spin-orbit and tensor interactions will have to be introduced.

As the rest of this conference testifies, there has been a considerable increase in the number of polarization measurements since the last conference in 1965. The area of proton inelastic scattering presents no exception. This subject was reviewed by F. G. Perey at the 1965 conference, and at that time he was only able to give a foretaste of the inelastic data which are now available.

Table 1 lists the currently available data, mostly obtained with polarized ion sources. I have not made a completely thorough search of the literature, so I apologize to any authors whose work I have inadvertently overlooked. (It was almost by accident that I discovered a large number of data from the last flurry of activity at the Rutherford Laboratory PLA which were hiding in progress reports!) Contributions to this meeting are not included.

Inelastic scattering in general leaves the spin of the excited nucleus with a non-random orientation whose polarization can then be measured by observing the angular distribution of some decay product (such as a γ ray) or its angular correlation with the outgoing proton.

Table 1. Polarization measurements on proton inelastic scattering

Targets	Measured[†]	E_p (MeV)	Refs.
^{24}Mg, ^{52}Cr, ^{58}Ni	P	5.38–5.95	[1]
^{24}Mg	P	5.8	[2]
^{48}Ti, ^{58}Ni	P	7–8	[3]
^{54}Fe, ^{60}Ni	A	10	[4]
^{65}Cu	A	15.7	[5]
^{12}C, ^{60,62}Ni	A	16.5	[6]
^{48,50}Ti, ^{52}Cr, ^{54,56}Fe, ^{58,62,64}Ni, ^{63}Cu	A	18.6	[7]
^{12}C	A	19	[6]
^{12}C	A	20.2–28.3	[8]
^{12}C, ^{16}O, ^{24,25,26}Mg, ^{27}Al, ^{28}Si, ^{40}Ca	A	20.3	[9]
^{90,92}Zr, ^{92}Mo	A	20.3	[10]
C, Si	A	29,49	[11]
^{9}Be	A	30.3	[12]
^{10}B	A	30	[13]
^{11}B, ^{13}C, ^{16}O	A	30.3	[14]
^{40}Ar	A	30,50	[15]
^{54,56}Fe, ^{58}Ni, ^{120}Sn, ^{208}Pb	A	30.3	[16]
^{54}Fe	A	30.4	[17]
^{64,66,68}Zn	A	30	[18]
^{92,96,100}Mo	A	30	[19]
^{28}Si, ^{54}Fe, ^{58,60}Ni	A	40	[20,21]
^{6,7}Li, ^{9}Be	A	50	[22]
^{12}C	A	49.5	[23]
^{24}Mg, ^{64,68}Zn	A	50	[24]
^{54,56,58}Fe	A	50	[25]
^{89}Y, ^{90,91,92,94,96}Zr	A	50	[26]
^{114}Cd	A	50	[27]
^{148}Sm	A	50	[28]
^{12}C	P	135	[29]
^{12}C	A	145	[30]
^{12}C	P	152	[31]
^{24}Mg, ^{28}Si, ^{32}S, ^{40}Ca	P	152	[32]
^{6}Li	P	155	[33]
^{12}C	P	155	[34]
^{208}Pb	P	155	[35]
^{12}C, ^{16}O, ^{40}Ca	P	173	[36]
^{24}Mg	P	185	[37]

[†] P denotes measurement of polarization of outgoing protons, A denotes asymmetry from polarized beam.

Table 2. Spin-flip measurements on proton inelastic scattering

Targets	E_p (MeV)	Refs.
^{12}C, $^{58,60,64}Ni$	9.25–20	[38,39]
^{12}C, ^{24}Mg, ^{58}Ni	~ 10	[40]
^{54}Fe, ^{60}Ni	10	[4]
$^{60,62,64}Ni$, $^{64,66}Zn$	10,11,12,14.5	[41]
^{52}Cr, ^{54}Fe	11	[42]
$^{148,152}Sm$	12	[43]
$^{54,56}Fe$	19.6	[44]
^{12}C	26.2,40	[45]
^{12}C†	30	[46]
^{28}Si	30,40	[47]

†Measured using polarized protons.

I exclude discussion of these observations except for the so-called spin-flip measurements which are intimately related to proton polarization measurements. Table 2 lists the available spin-flip data.

Figs. 1 and 2 show samples of the asymmetry data taken with polarized protons of 30 and 40 MeV, respectively. The curves are theoretical predictions obtained using the simple collective or deformed potential model to which I will return later. A wide range of targets has been studied. However, an interesting feature of the experiments listed in Table 1 is that they include not only these strong, "collective" transitions to the lowest 2^+ and 3^- states in even nuclei, but also transitions in odd nuclei and excitation of unnatural parity 2^- and 3^+ states and states with spins up to 5^-. The targets studied include those like ^{90}Zr which exhibit spectra well described by the simple shell model.

Most of the measurements were made with protons with energies of about 20, 30, 40, and 50 MeV, with small samples taken either at lower energies or between 100 and 200 MeV. The higher energy measurements are concentrated at forward scattering angles and in general have been plagued by difficulties with energy resolution. However, as we shall see, they are capable of yielding very interesting information.

The use of energies much below about 20 MeV encounters compound nucleus contributions from light nuclei and the Coulomb barrier in heavy nuclei. The former is well illustrated by the work of Ahmed et al. [4], whose calculations based on Hauser-Feshbach theory indicate appreciable contributions from the compound nucleus (fig. 3). The amount by which the bombarding energy exceeds the lowest (p,n) threshold is a better measure of the importance of the compound nucleus than the energy itself. Thus the compound nucleus is more important [4]

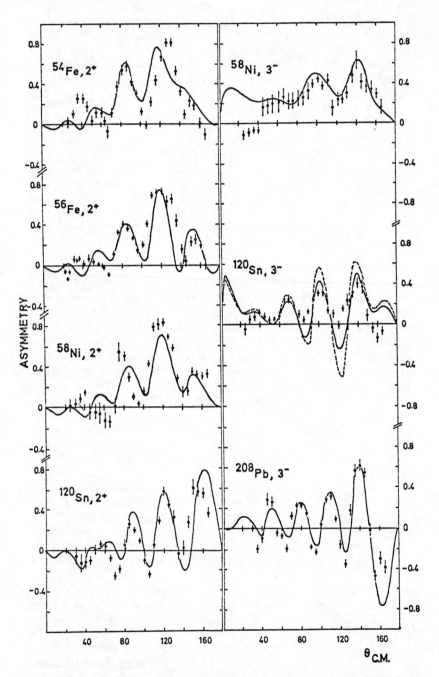

Fig. 1. Asymmetries from the inelastic scattering of 30-MeV protons
[16]. The curves are for DWBA calculations.

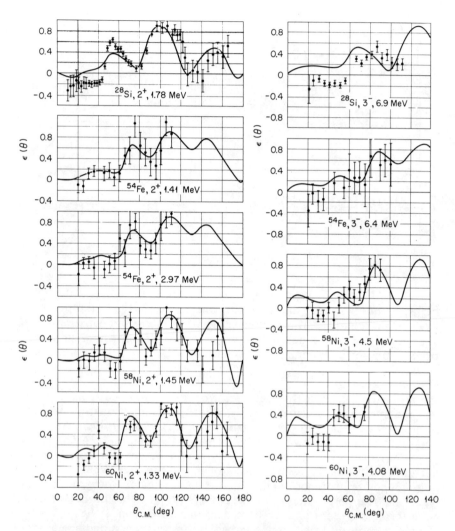

Fig. 2. Asymmetries from the inelastic scattering of 40-MeV protons
[21]. The curves are for DWBA calculations.

for 10-MeV protons on ^{54}Fe ($Q_0(p,n)$ = -9.6 MeV) than on ^{60}Ni ($Q_0(p,n)$ = -6.9 MeV).

If we believe the statistical model, we expect the compound component to be unpolarized [48] and simply to reduce in proportion the polarization from the direct component. The measurement in some cases [1,3] of appreciable polarizations of the outgoing protons, together with cross sections which vary rapidly with energy, indicates

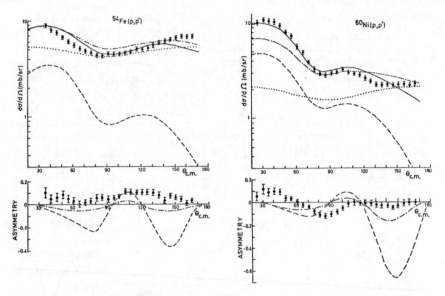

Fig 3. Cross sections and asymmetries from exciting 2^+ states with 10-MeV protons [4]. The dashed curves are the direct reaction contributions, the dotted curves are statistical compound nuclear predictions. The solid and dash-dot curves result from combining these.

the presence either of fluctuations or of non-statistical compound phenomena such as doorway states. Statistical fluctuations are amplified in polarization measurements because of their greater dependence upon interferences between different partial waves [48]. A polarized beam facility would be useful for studying doorway states except for certain limitations [1]. An asymmetry in the inelastic scattering would be produced only if more than one entrance channel wave participates and there is interference between them [49]. Such would not be the case for a spin-zero target which exhibits a resonance with definite J^π unless there is interference between the resonance and some background amplitude (whether compound or direct). If the background were of a statistical compound nature, one would again expect this interference to vanish. However, doorway states induced by polarized protons incident on *odd* targets with non-zero spin could be studied in this way in general.

Analogous conditions for the exit channel apply for the detection of the polarization of the outgoing protons. However, this is much less likely to be a restriction in practice because the residual nuclear spin is seldom zero.

We may ask the general question, What differences may we expect between measurements of polarization and asymmetry for a given inelastic transition? I am not aware of any experimental results. We have just seen that under certain special conditions the two measurements may give quite different results. For direct reactions, there is a theorem that applies in the adiabatic limit of negligible energy loss [50]. The effective interaction operator acting on the scattering proton can be classified according to whether it is even or odd under time reversal; for example, $\vec{\sigma}$ is odd while the product $\vec{L} \cdot \vec{\sigma}$ is even. Correspondingly, the inelastic transition operator is a sum of two parts with differing asymmetry. In a first-order theory, such as the distorted-wave Born or impulse approximations [51], each part of the interaction gives rise to the corresponding part of the transition amplitude. If only even, or only odd, parts of the transition amplitude contribute to a given transition, then the polarization and asymmetry are equal in this adiabatic limit. However, the contributions from interference between these two terms to the polarization and asymmetry are of equal magnitude but opposite sign. Hence for differences to appear we need both even and odd interactions to contribute and also to interfere.

The collective or deformed potential model, even when the spin-orbit coupling term is deformed, gives an even interaction in this sense [50], so would yield polarizations and asymmetries which are similar. A microscopic description of the interaction is more likely to produce differences, both from exchange effects and from spin-spin and tensor terms in the nucleon-nucleon interaction. Calculations of the DWBA type which I have seen indicate that the adiabatic condition is reasonably well satisfied for the excitation of low-lying levels and with energies of about 20 MeV and above. These calculations also seem to support the notion that larger polarization-asymmetry differences are to be found in microscopic calculations, although I have not noticed *very* large differences (even with mixtures of spin-spin and central interactions) except in some cases where the so-called "unmatural-parity" exchange terms contribute (an example being the excitation of the 2^- level in ^{16}O to which unnatural parity terms with $L = 2$ may contribute). I have not seen comparisons for such calculations which have included tensor or spin-orbit forces. It would be interesting to look into this more carefully; it is possible that a comparison of the two measurements for selected transitions could yield useful information on the nature of the effective interaction.

Let us now survey what we have learned from the mass of inelastic polarization data that has accumulated. The simplest picture emerges when we use the collective model. Strictly, this should only be applicable to the strong, collective excitations of vibrational or rotational states, but in practice it gives a good account of many weaker, non-collective transitions. The model is applied by assuming that the optical potential follows the nuclear shape when it is deformed away from sphericity. The usual prescription [51,52] is to deform the surface $r = R$ according to

$$R \rightarrow R + \sum_L \delta_L Y_L(\theta), \tag{1}$$

where for simplicity I have only written the expression for an axially symmetric deformation. δ_L is the deformation length for the 2^L-th multipole, often written as $\delta_L = \beta_L R$, and $Y_L(\theta)$ is the corresponding spherical harmonic. This radius R appears as a parameter in the optical potential U, which we now expand in Taylor series,

$$U(r,R) \rightarrow U(r,R) + \sum_L \delta_L Y_L(\theta) \frac{\partial U(r,R)}{\partial R} + \ldots \tag{2}$$

The non-spherical terms may then induce inelastic scattering. Often only the term which is first order in δ_L is kept and then its effects calculated with the DWBA, although higher order terms will be included in a coupled-equations calculation [51,52].

The optical potential contains a number of distinct terms, and it is reasonable that each should follow the shape of the nuclear density distribution when it is deformed. These deformations have been introduced, historically more-or-less in the following order, into the

1. real part of the central nuclear potential,
2. Coulomb potential,
3. imaginary part of the central nuclear potential,
4. spin-orbit potential.

(This evolution occurred simply as a matter of computational convenience.) The biggest contribution to the transition amplitude for protons comes from deforming the real potential. As is to be expected, the deformation of the charge distribution (which gives rise to Coulomb excitation) is more important at the lower proton energies. The effect of the non-spherical part of the imaginary or absorptive potential on the differential cross sections tends to be more important as the energy is increased. It was also found to have very important effects on the polarization or asymmetry. Fig. 4 illustrates these features for a quadrupole excitation in ^{60}Ni by 40-MeV protons. First we note that deforming the imaginary part as well as the real (denoted "complex") makes significant changes in the differential cross section at this energy. The effect on the predicted asymmetry is even more striking. The curve from the real part alone is fairly structureless. The absorptive term introduces the diffraction-like oscillations which are seen experimentally, although the agreement with experiment is still only qualitative.

Deforming the spin-orbit coupling term in the optical potential also affects the inelastic cross sections at large scattering angles. From its very nature one would expect this term to have a strong influence on the asymmetries, and we see that it does indeed produce large changes. With the real part of the deformed central potential it still

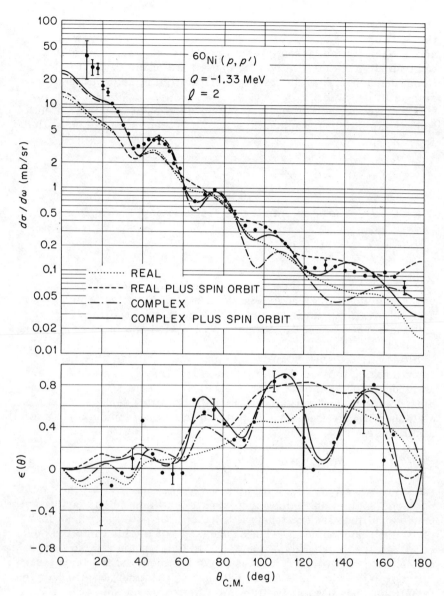

Fig. 4. Cross sections and asymmetries for inelastic scattering of
40-MeV protons [21]. The curves are for DWBA calculations as described in the text.

does not reproduce the structure seen experimentally, but in conjunction with the full complex potential we see that it gives very nice agreement with the measurements except at the most forward scattering angles. This last was a common feature of the earliest calculations, and I must now discuss more carefully the deformation of the spin-orbit potential.

According to the Dirac equation, an electron moving in a potential field U experiences a spin-orbit coupling potential of the so-called Thomas type,

$$-i\left(\frac{\hbar}{2mc}\right)^2 \vec{\sigma}\cdot(\vec{\nabla}U \times \vec{\nabla}). \tag{3}$$

When U is a central potential, this reduces to

$$\left(\frac{\hbar}{2mc}\right)^2 \frac{1}{r} \frac{\partial U(r)}{\partial r} \vec{\sigma}\cdot\vec{L}. \tag{4}$$

The same form is usually employed in the single-particle potentials of the nuclear optical model and shell model, but it certainly does not arise in this case as a simple relativistic term. Empirically we find that the coupling needed has the opposite sign and is 20 to 30 times stronger.

However, the same form is obtained when one uses the impulse approximation to describe the interaction. This seems to have been derived first by Fernbach et al. [53] some 15 years ago. In its simplest form, one regards the proton scattering from each target nucleon as though it were free. We write the free nucleon-nucleon scattering amplitude as

$$t(\vec{k},\vec{k}') = a(\vec{k},\vec{k}') + b(\vec{k},\vec{k}')(\vec{\sigma}_t + \vec{\sigma})\cdot\vec{n}, \tag{5}$$

(where $\vec{n}$ is the unit vector along $\vec{k} \times \vec{k}'$), and sum over all the target nucleons. With some minor approximations, the result is equivalent to the amplitude for scattering from an optical potential of the form (3), with U now being the Fourier transform of b folded into the nuclear density distribution ρ [53,54]. This derivation is also applicable to inelastic scattering if it takes place between states with the same intrinsic wave function. Somewhat more sophisticated "effective interactions" may also be written in the form (5) and the same results follow. Quite early it was noted [55] that this approach could explain the striking similarity at small angles between elastic and inelastic polarizations observed for high energy protons.

The phenomenological potential used assumes the form (3) has a more general validity and allows the potential function U_S to be parameterized freely, although retaining a shape similar to that of the density (almost invariably a Woods-Saxon form is chosen). In par-

ticular, analyses of elastic scattering indicate that it should have a
somewhat smaller radius and possibly a smaller surface diffuseness
than the real part of the central potential. That such differences
should occur is not surprising within the context of the impulse approximation. The central potential is then given by the Fourier transform of a in eq. (5) folded into the density distribution ρ and the observed differences could be ascribed to the spin-orbit part of the
effective interaction (5) having a shorter range than the central part.

When the surface deformation (1) is introduced, we obtain to first
order in δ

$$-i\,\vec{\sigma}\cdot(\vec{\nabla}U_S \times \vec{\nabla}) = \left[\frac{1}{r}\frac{\partial U_S}{\partial r}\,\vec{\sigma}\cdot\vec{L}\right] + \delta_L\left[\frac{1}{r}\frac{\partial}{\partial r}\frac{\partial U_S}{\partial R}Y_L(\theta)\vec{\sigma}\cdot\vec{L}\right.$$

$$\left. - \frac{i}{r}\frac{\partial U_S}{\partial R}\frac{\partial Y_L}{\partial \theta}\,\vec{\sigma}\cdot(\vec{a}_\theta \times \vec{\nabla})\right] + \ldots \qquad (6)$$

The first part just gives the usual spin-orbit term (4) for spherical
potentials, and the parameters of U_S may be determined from analyses
of elastic scattering. The second part gives the deformed part which
can contribute to inelastic scattering. (For simplicity, we have again
written (6) for an axially symmetric deformation. Also, $\vec{a}_\theta$ is a unit
vector normal to $\vec{r}$ in the direction of increasing θ.) The second term
of this part contains the gradient operator which, for example, in a
DWBA will act on the incident distorted wave. This requires some
fairly extensive modification of current DWBA computer codes, where-
as the first term, proportional to $\vec{\sigma}\cdot\vec{L}$, requires only trivial changes.
Hence, in accordance with the philosophy that one should do the
simplest thing first, the early calculations [20,21] were made with the
phenomenological interaction [51]

$$\frac{1}{2}h_L(r)\,[Y_L(\theta)\,\vec{\sigma}\cdot\vec{L} + \vec{\sigma}\cdot\vec{L}\,Y_L(\theta)], \qquad (7)$$

where, guided by eq. (6), one chose

$$h_L(r) = \delta_L\frac{1}{r}\frac{\partial}{\partial r}\frac{\partial U_S}{\partial R}, \qquad (8)$$

with U_S given by the elastic scattering. This form, sometimes called
the "Oak Ridge type," was used for the calculated curves shown in
figs. 2 and 4. (Refs. [7,9] use a radial form slightly different from
eq. (8).)

The symmetrization in eq. (7) is required to make the interaction
hermitian. Except for this, with eq. (8) it is the same as the first part

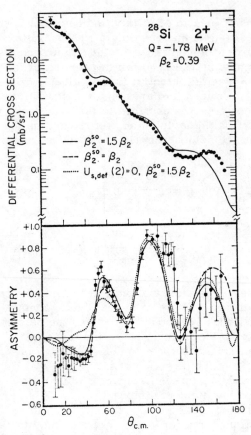

Fig. 5. Cross sections and asymmetries
for inelastic scattering of 40-MeV protons
[21]. The curves are for DWBA calculations
as described in the text.

of the deformed interaction in eq. (6). However, as I mentioned above,
characteristically this form fails to reproduce the measured asymmetries
at the forward angles. It has now been shown by Sherif [56], using the
DWBA, that the additional term in the so-called "full Thomas" form (6)
gives better agreement with the data in this region. (Rook [57] has
shown that in a plane-wave treatment this extra term contributes significantly only for small momentum transfers. However, it is not clear
at present whether this is relevant to the DWBA calculations.) Examples are shown in figs. 5—7 for 40-MeV protons. Except at forward
angles, the two interactions give similar results. ($U_{S,def}(2) = 0$ means
the second part of (6) is *not* included.) There is also some indication

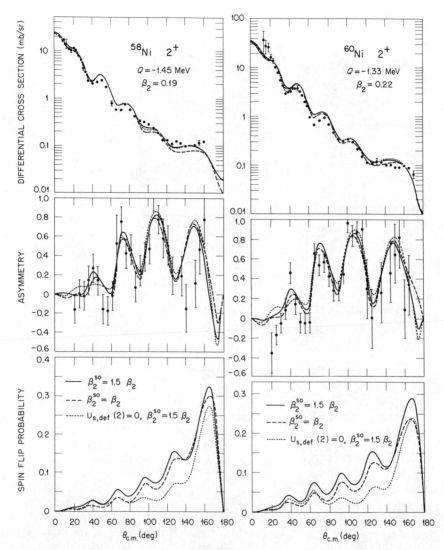

Fig. 6. Cross sections, asymmetries, and spin-flip probabilities for inelastic scattering of 40-MeV protons [21]. The curves are for DWBA calculations as described in the text.

of improved fits when the deformation β^{SO} of the spin-orbit term is made larger than that, β, for the central potential. This again is a general characteristic; it may be due in part to the physically significant quantity being the deformation length $\delta = \beta R$ rather than β itself. Calculations with the full Thomas term (6) have also been made recently by Raynal [58], using the coupled-equations technique.

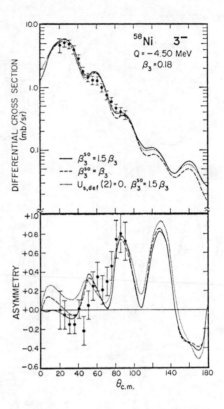

Fig. 7. Cross sections and asymmetries
for inelastic scattering of 40-MeV protons [21]. The curves are for DWBA calculations as described in the text.

Perhaps the most striking evidence for the validity of the form (6)
comes from analysis [59] of the polarization from high energy (155 MeV)
proton scattering. Fig. 8 shows this for ^{12}C. These data are confined
to the forward angles and the full Thomas interaction gives an excellent fit. Fig. 9 for ^{24}Mg shows the effect of not including the deformed
spin-orbit at all ("no DSO") and also the results of using the distorted-
wave impulse approximation (DWIA) (interaction as in eq. (5) and deformed Hartree-Fock wave functions). Except at the few largest angles,
the macroscopic collective model is markedly superior to the microscopic description. Fig. 10 shows that the deformed spin-orbit potential also has an appreciable effect on the differential cross sections
both in magnitude (30%) and in shape. The results of further analyses
of this type have been submitted to this conference (J. S. Blair and
H. Sherif).

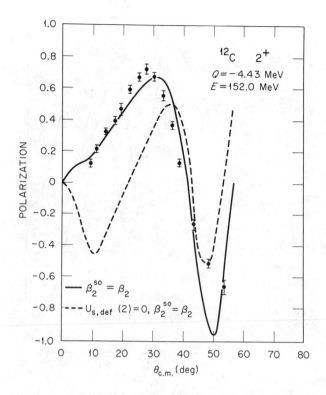

Fig. 8. Polarization of 152-MeV protons inelastically scattered [31] from ^{12}C. The curves are for
DWBA calculations as described in the text.

Before leaving the collective model, we should discuss the spin-
flip measurements. A symmetry property of the transition amplitude
[60] determines that a 2^+ state excited in a spin-zero target can emit
γ-rays in a direction perpendicular to the scattering plane only if the
scattered proton has its spin flipped with respect to this direction [40,
61]. Hence observation of γ-rays in this direction in coincidence with
the outgoing protons provides a measure of this spin-flip probability.
There are two sources of this spin flip: it may occur (1) during the
elastic scattering due to the spherical spin-orbit potential, and (2) during the inelastic events if there is a spin-dependent coupling term
such as the deformed spin-orbit. The lowest part of fig. 6 shows some
predictions of the collective model as a function of proton scattering
angle. Here we see that the extra term of the full Thomas prescription
has its largest effects at angles of 90° and more. Fig. 11 shows some
comparisons with data for 30- and 40-MeV protons on ^{28}Si. The extra

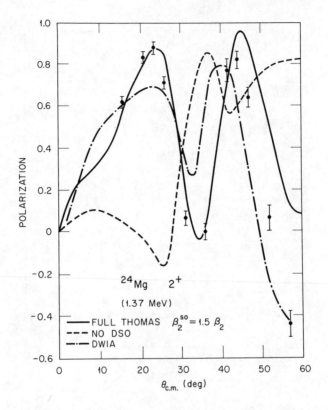

Fig. 9. Polarization of 155-MeV protons inelastically scattered [32] from ^{24}Mg. The curves are for
DWBA calculations as described in the text.

probability predicted by the full Thomas coupling is in better agreement with the 30 MeV measurements between 60° and 120°. The same
kind of result is obtained at lower energies [38,39]. Again there is
some preference for $\beta^{SO} > \beta$. Altogether, though, one would find it
difficult to make a strong case for the full Thomas coupling based upon
the present spin-flip data alone. The effects of the finite size of the
solid angle subtended by the γ detector, for example, are comparable
[61] to the differences between the full Thomas and the simpler form
of coupling. These are difficult experiments, but it would be interesting to have some precise data at a moderately high energy (like 40
MeV or above).

An even more difficult experiment was made [46] at the Rutherford
Laboratory PLA: spin-flip measurements on ^{12}C using polarized protons
of 30 MeV. The analysis is not yet complete, but such measurements

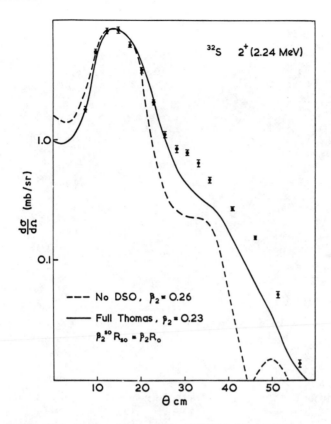

Fig. 10. Cross sections for 155-MeV protons inelastically scattered [32] from ^{32}S. The curves
are for DWBA calculations as described in the text.

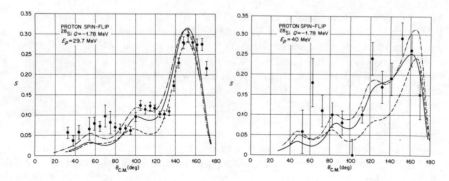

Fig. 11. Spin-flip probabilities for 30- and 40-MeV protons [47] on
^{28}Si. The curves are for DWBA calculations; the full curve is for the
full Thomas term, the dashed curve is for $U_{S,def}(2) = 0$, and the dot-
dash curve is for $\beta_2^{SO} = 1.5\beta_2$.

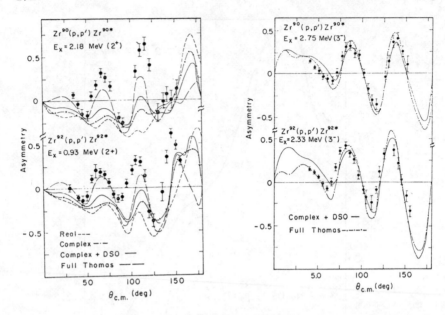

Fig. 12. Asymmetries for inelastic scattering of 20.3-MeV protons [10] from ^{90,92}Zr. The curves are for DWBA calculations as described in the text.

should throw further light on the nature of the spin-dependent interaction [61,62].

So far we have been discussing strong transitions which excite the lowest 2^+ and 3^- states in even nuclei and for which it is not unreasonable that a collective model might work. There are data for many other transitions (table 1) for which we might expect to need a more detailed or microscopic description in terms of nucleon-nucleon effective interactions and wave functions in a shell model representation. The interactions are to be obtained, in one way or another, from nucleon-nucleon scattering data, and are not to be treated as adjustable parameters. Fig. 12 shows on the right the very similar data for two strong octupole transitions, together with the collective model predictions. Again the full Thomas coupling gives a better fit. On the left are data for two relatively weak quadrupole excitations in the same nuclei. First we notice significant differences between the measured angular distributions in the two cases. Second, the collective model does not give a good account of these data, although again the calculations with the full Thomas term come closest. Now the level in ^{90}Zr is thought to be well described by the recoupling of two valence protons in the $1g_{9/2}$ orbit, while that in ^{92}Zr is due to the recoupling of two neutrons in the $2d_{5/2}$ orbit. (In addition, ^{92}Mo has a structure sim-

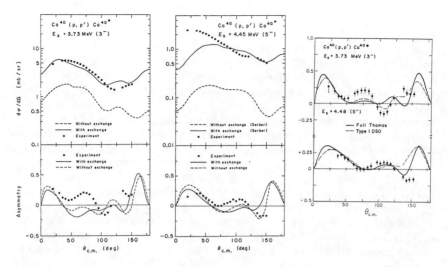

Fig. 13. Cross sections and asymmetries for 20.3-MeV protons [9] scattered from ^{40}Ca. The curves are for DWBA calculations as described in the text.

ilar to ^{90}Zr, and the measured asymmetries are also similar.) This difference in structure alone would give rise to some differences between their excitation in a microscopic model. For example, the nucleon-nucleon spin-orbit interaction is expected to be stronger between two protons than between a proton and a neutron. However, a microscopic analysis of such reactions is still in its infancy. We know that exchange between the incident proton and the target nucleons must be included [63] and that core polarization (virtual excitation of the closed shells) may often dominate the transition [64]. We can expect the tensor force to be important in some cases. Especially, in view of the arguments given above for the origin of the deformed spin-orbit potential, we must expect the nucleon-nucleon spin-orbit force to be important.

Fig. 13 shows results [10] for ^{40}Ca using particle-hole wave functions and a central force with the spin-spin term implied by a Serber exchange mixture (see also Escudie et al., this conference). In neither case are the asymmetries fitted as well as with the collective model. Indeed even the cross sections are not reproduced very well. Both transitions here are quite collective, so it is clear we must include spin-orbit forces. In addition, we noted earlier that in the collective model it is important to deform the imaginary potential also. The interactions used so far in microscopic calculations are real (except for the impulse approximation used at high energies), which leads one to speculate that it may be necessary to include an imaginary component here also for a complete description.

   Calculations [32,65] with the distorted-wave impulse approximation include both spin-orbit and tensor terms in the effective nucleon-
nucleon interaction (only the former is indicated in eq. (5)). Fig. 9
includes an example of its use; the results are generally similar to
those for the collective model. At lower energies, only a few microscopic calculations, including exchange with a tensor [66] or spin-
orbit [58] force, have been reported so far. However, now that the
computational facilities are available we may expect to learn better
to what extent a microscopic description of these reactions is feasible.
   Finally, I should comment on two points of interest. First, there
is a suggestion that while the collective model almost always seems
to give a good account of the data at energies of 30 MeV and above,
it often fails for energies near 20 MeV, sometimes with the same target. The classic example is ^{54}Fe which at 40 MeV shows asymmetries
very similar to those of ^{56}Fe, yet the two nuclei give different results
at 18.6 MeV. (The data for 30 MeV shown in fig. 1 also seem to indicate some differences. Further, the errors on the 40-MeV measurements are somewhat larger.) The situation is further confused by the
result that at 19.6 MeV the spin-flip distributions for the two nuclei
are very similar [44].
   The second point is that some of the spin-flip measurements have
indicated a peak in the spin-flip probability for scattering angles
near 70° which is *not* predicted by the collective model in its present
form. Some indication of this is seen for ^{28}Si in fig. 11 and it has
been seen [44,67] most clearly for 19.6-MeV protons on ^{54}Fe and ^{56}Fe.
On the other hand, it does not appear for ^{58}Ni at the same energy [67].
   Asymmetry (or polarization) and spin-flip measurements are more
sensitive to small components in the transition amplitudes (such as
those due to spin-spin forces, etc.) than is the differential cross section, so if these apparent discrepancies persevere they may yield
valuable information about these components.

REFERENCES

   *Research sponsored by the U.S. Atomic Energy Commission under
contract with the Union Carbide Corporation.

[1]   A. Ruh and P. Marmier, Nucl. Phys. A151 (1970) 479.
[2]   I. Boca et al., Nucl. Phys. 79 (1966) 188.
[3]   E. E. Griffin and W. Parker Alford, Nucl. Phys. 36 (1962) 305.
[4]   M. Ahmed, J. Lowe, P. M. Rolph, and O. Karban, Nucl. Phys.
      A147 (1970) 273.
[5]   C. Glashausser et al., Bull. Am. Phys. Soc. 15 (1970) 497.
[6]   P. Darriulat, J. M. Fowler, R. de Swiniarski, and J. Thirion,
      Second Polarization Symp., p. 342.

[7]   C. Glashausser, R. de Swiniarski, J. Thirion, and A. D. Hill,
        Phys. Rev. 164 (1967) 1437.
[8]   R. M. Craig et al., Nucl. Phys. 79 (1966) 177.
[9]   A. G. Blair et al., Phys. Rev. C1 (1970) 444.
[10]   C. Glashausser et al., Phys. Rev. 184 (1969) 1217.
[11]   R. M. Craig et al., Nucl. Phys. 83 (1966) 493.
[12]   R. J. Batten et al., Rutherford High Energy Lab. Report RHEL/R-
        156 (1967).
[13]   P. D. Greaves et al., Rutherford High Energy Lab. Report RHEL/
        R-170 (1968).
[14]   O. Karban, J. Lowe, P. D. Greaves, and V. Hnizdo, Nucl. Phys.
        A133 (1969) 255 and A141 (1970) 675; Rutherford High Energy
        Lab. Report RHEL/R-187 (1969).
[15]   A. A. Rush, E. J. Burge, and D. A. Smith, Rutherford High Energy
        Lab. Report RHEL/R-170 (1968).
[16]   O. Karban et al., Nucl. Phys. A147 (1970) 461.
[17]   D. J. Baugh et al., Nucl. Phys. A99 (1967) 203.
[18]   W. H. Tait, E. J. Burge, M. Calderbank, and D. A. Smith, Ruther-
        ford High Energy Lab. Report RHEL/R-170 (1968).
[19]   W. H. Tait and V. R. W. Edwards, Rutherford High Energy Lab.
        Report RHEL/R-187 (1969).
[20]   M. P. Fricke, E. E. Gross, and A. Zucker, Phys. Rev. 163 (1967)
        1153.
[21]   M. P. Fricke et al., Phys. Rev. Lett. 16 (1966) 746.
[22]   G. S. Mani, A. D. B. Dix, D. T. Jones, and D. Jacques, Ruther-
        ford High Energy Lab. Reports RHEL/R-156 (1967) and RHEL/R-
        170 (1968).
[23]   J. A. Fannon, E. J. Burge, D. A. Smith, and N. K. Ganguly, Nucl.
        Phys. A97 (1967) 263.
[24]   V. E. Lewis, E. J. Burge, A. A. Rush, and D. A. Smith, Nucl. Phys.
        A101 (1967) 589.
[25]   G. S. Mani, D. T. Jones, D. Jacques, and A. Dix, Rutherford High
        Energy Lab. Report RHEL/R-187 (1969).'
[26]   G. S. Mani et al., Rutherford High Energy Lab. Report RHEL/R-
        187 (1969).
[27]   V. E. Lewis et al., Rutherford High Energy Lab. Report RHEL/R-
        156 (1967).
[28]   Joan F. Grace and V. E. Lewis, Rutherford High Energy Lab.
        Report RHEL/R-187 (1969).
[29]   J. M. Dickson and D. C. Salter, Nuovo Cim. 6 (1957) 235.
[30]   J. McL. Emmerson et al., Nucl. Phys. 77 (1966) 305.
[31]   B. Tatischeff et al., Phys. Lett. 16 (1965) 282.
[32]   A. Willis et al., Nucl. Phys. A112 (1968) 417.
[33]   N. Marty et al., Proc. Int. Congress on Nuclear Physics, ed.
        P. Gugenberger (Paris, 1964).
[34]   R. Alphonce, A. Johansson, and G. Tibell, Nucl. Phys. 3 (1957)
        185 and 4 (1957) 672.

[35]   A. Willis et al., J. de Phys. 30 (1969) 13.
[36]   P. Hillman, A. Johansson, and H. Tyren, Nucl. Phys. 4 (1957)
       648.
[37]   B. Hoistad, A. Ingemarsson, A. Johansson, and G. Tibell, Nucl.
       Phys. A119 (1968) 290.
[38]   W. A. Kolasinski et al., Phys. Rev. 180 (1969) 1006.
[39]   J. Eenmaa, F. H. Schmidt, and J. R. Tesmer, Phys. Lett. 28B
       (1968) 321.
[40]   F. H. Schmidt, R. E. Brown, J. B. Gerhart, and W. A. Kolasinski,
       Nucl. Phys. 52 (1964) 353.
[41]   J. Delaunay, P. Passerieux, and F. G. Perey, reported by F. G.
       Perey, Second Polarization Symp., p. 191.
[42]   R. Ballini et al., Nucl. Phys. A97 (1967) 561.
[43]   A. B. Kurepin, P. R. Christensen, and N. Trautner, Nucl. Phys.
       A115 (1968) 471.
[44]   D. L. Hendrie, C. Glashausser, J. M. Moss, and J. Thirion, Phys.
       Rev. 186 (1969) 1188.
[45]   J. J. Kolata and A. Galonsky, Phys. Rev. 182 (1969) 1073.
[46]   M. Ahmed, J. Lowe, P. M. Rolph, and V. Hnizdo, Rutherford High
       Energy Lab. Report RHEL/R-187 (1969).
[47]   R. O. Ginaven, E. E. Gross, J. J. Malanify, and A. Zucker, Phys.
       Rev. Lett. 21 (1968) 552.
[48]   M. Lambert and G. Dumazet, Nucl. Phys. 83 (1966) 181.
[49]   H. L. Harney, Phys. Lett. 28B (1968) 249.
[50]   H. Sherif, to be published.
[51]   G. R. Satchler, Lectures in Theoretical Physics, 1965, eds.
       P. D. Kunz, D. A. Lind, and W. E. Brittin (University of Colorado
       Press, Boulder, 1966).
[52]   T. Tamura, Revs. Mod. Phys. 37 (1965) 679.
[53]   S. Fernbach, W. Heckrotte, and J. V. Lepore, Phys. Rev. 97 (1955)
       1059.
[54]   G. E. Brown, Proc. Phys. Soc. (London) 70A (1957) 361.
[55]   Th. A. J. Maris, Nucl. Phys. 3 (1957) 213; K. Nishimura and M.
       Ruderman, Phys. Rev. 106 (1957) 558; H. S. Kohler, Nucl. Phys.
       9 (1958) 49.
[56]   H. Sherif and J. S. Blair, Phys. Lett. 26B (1968) 489; H. Sherif
       and R. de Swiniarski, Phys. Lett. 28B (1968) 96; H. Sherif, Nucl.
       Phys. A131 (1969) 532.
[57]   J. R. Rook, Phys. Lett. 29B (1969) 86.
[58]   J. Raynal, to be published.
[59]   H. Sherif and J. S. Blair, Nucl. Phys. A140 (1970) 33.
[60]   A. Bohr, Nucl. Phys. 10 (1959) 486.
[61]   F. Rybicki, T. Tamura, and G. R. Satchler, Nucl. Phys. A146
       (1970) 659.
[62]   A. B. Clegg and G. R. Satchler, Nucl. Phys. 27 (1961) 431.
[63]   J. Atkinson and V. Madsen, Phys. Rev. C1 (1970) 1377; R. Schaeffer,

Nucl. Phys. A132 (1969) 186; W. G. Love and G. R. Satchler, Nucl. Phys., to be published.

[64] W. G. Love and G. R. Satchler, Nucl. Phys. A101 (1967) 424; R. Schaeffer, Nucl. Phys. A135 (1969) 231; F. Petrovich, H. McManus, and J. Borysowicz, Proc. Int. Conf. on Properties of Nuclear States (University of Montreal Press, 1969) p. 716.

[65] R. M. Haybron and H. McManus, Phys. Rev. 140 (1965) B638.

[66] W. G. Love, L. J. Parish, and A. Richter, Phys. Lett. 31B (1970) 167.

[67] F. H. Schmidt et al., Progress Report, University of Washington (1970).

DISCUSSION

Raynal:

The forward asymmetry of ^{90}Zr can be understood in the microscopic model, using a two-body $\vec{L} \cdot \vec{S}$ interaction. The excitation must include more than a simple proton configuration. The best test case is ^{54}Fe in which there are two 2^+ states, one with large and the other with small asymmetry. Wave functions of Gillet et al. reproduce these features in the forward direction. One obtains large asymmetries where there is an open shell of protons and a closed shell of neutrons; this effect seems to be related to the quantum numbers of the open shell. As I pointed out in a contribution to the Montreal conference and in the symposium at Quebec which followed it, the *full Thomas* and the *phenomenological* spin-orbit deformation have opposite effects on the S matrix. These effects add coherently in the forward direction.

Breit:

The full Thomas treatment multiplies the term found long ago by Thomas by a factor of 20 to 30 so as to use the phenomenological N–N potential V_{LS}. It appears that the terminology is misleading because it gives the wrong impression. The physical origin of the N–N effect is caused largely by vector meson exchange. So much for semantics. If the presence of the extra terms (the ones caused by deformations) can be established, it appears to be a very important finding. It would then demonstrate the *same* exchange of vector mesons between nucleons when they are in a nucleus as when they are free.

Isobaric Analog Resonances

G. GRAW, University of Erlangen-Nürnberg, Germany

1. SUMMARY

Since the discovery of isobaric analog resonances (IAR) as compound nuclear resonances in proton scattering by Fox et al. in 1964 [1], IAR's have been investigated extensively. Progress in this field has been summarized in the Proceedings of the first and second conferences on nuclear isospin in 1966 and 1969 [2, 3]; a discussion of the general properties of IAR's may be found in these reports. The present paper is confined to polarization experiments.

Measurements of polarization or analyzing power in proton scattering in the region of IAR's have been performed on a number of nuclei. In elastic scattering, most of these experiments were done on spin-zero target nuclei near closed neutron shells [4−14], but some experiments on Sn [15] and Te [16] isotopes have also been reported. The only odd-mass target nucleus investigated until now has been ^{89}Y [17] which has spin $I = 1/2$.

The main purpose of these experiments was to determine the total angular momentum j of IAR's with known orbital angular momentum ℓ. Furthermore, resonance parameters can be determined more accurately, particularly in those cases where the compound elastic cross section cannot be neglected.

In inelastic scattering some experimental data on analyzing power [18, 19] are now available. These experiments, though not yet completely analyzed, will provide information on the complete wave function of the IAR's as well as on the off-resonant direct scattering amplitudes.

2. INTRODUCTION

2.1. *The wave function of an IAR*
Isobaric analog resonances in heavy nuclei are a consequence of the charge independence of nuclear forces. In a simple picture, they are members of isobaric spin multiplets and are created from the parent

179

bound states (with isospin $T = T_z = (N - Z)/2 = T_>$) by application of the isospin lowering operator T^-, which transforms neutrons into protons without essential change of the configuration:

$$\Psi_{IAR} = \frac{T^-}{\sqrt{2T_0 + 1}} \, \Phi_{Parent}.$$

(1)

Because of the increase in Coulomb energy for these members of the multiplet, they are shifted in energy and occur as resonances observable in proton scattering.

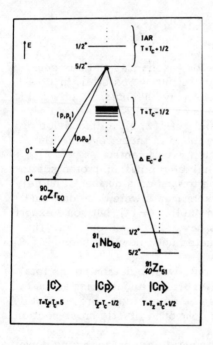

Fig. 1. Energy diagram of levels participating in the IAR. Target (= core state): ^{90}Zr with isospin $T = T_z = (N - Z)/2 = T_c$. The IAR in ^{91}Nb and the parent-analog-state have isospin $T = T_c + 1/2 = T_>$. ΔE_c = Coulomb-shift. $\delta = n - p$ mass-difference.

The properties of these isospin multiplet states may be investigated either as bound states in the parent nucleus, e.g., via (d,p) reactions (see the invited paper to this conference by Haeberli [45]), or as IAR's in elastic or inelastic proton scattering. To extract spectroscopic information from the data, a reaction theory [21—24] is needed in both cases. For the determination of energy, parity, and—from an experiment with polarized particles—the total angular momentum, the details of these theories are in many cases unimportant. However, the accuracy of spectroscopic factors depends strongly on the quality of the reaction theory used.

Inelastic scattering near IAR's may offer a possibility to determine the complete wave function of the parent state. The wave function of a nucleus with an odd neutron and even proton number may be expanded in a linear superposition of single neutron states n^j coupled to unexcited (spin 0^+) and to excited (spin I) collective core states:

$$\Phi^{J^\pi} = \sqrt{S}\, |n^{J^\pi} \times C(0^+)\rangle + \sum_{jI} b_{jI} |n^j \times C^*(I)\rangle^{J^\pi}. \tag{2}$$

The first term is proportional to the square root of the spectroscopic factor S and may be determined from a (d,p) experiment; the b_{jI} are the coefficients of parentage and cannot be determined directly from (d,p) experiments. The wave function of the IAR is obtained from eqs. (1) and (2) by application of the T^- operator (T_0 is the total isospin quantum number of the core):

$$\Psi_{IAR}^{J^\pi} = \frac{1}{\sqrt{2T_0+1}} \left\{ \sqrt{S}\left[|p^{J^\pi} \times C(0^+)\rangle + \sqrt{2T_0}\,|n^{J^\pi} \times A(0^+)\rangle \right] \right.$$

$$\left. + \sum_{jI} b_{jI}\left[|p^j \times C^*(I)\rangle^{J^\pi} + \sqrt{2T_0}\,|n^j \times A^*(I)\rangle^{J^\pi} \right] \right\}. \tag{3}$$

The quantity p^j is the wave function of a single proton (scattering) state, the isobaric analog to n^j, and the A(I) are the IAR's of the core states C(I).

The first term describes the elastic proton channel, the others describe inelastic proton channels leaving the target nucleus in a collective excited state (3rd term) or in neutron particle-hole states (2nd and 4th terms). The strength of these various exit channels of the IAR depends on the expansion coefficients bearing the spectroscopic information. The quantity S is determined from elastic scattering; the parentage coefficients are determined by inelastic scattering to collectively excited target states.

2.2. Influence of compound-nucleus states

IAR's in heavy nuclei are complicated resonance phenomena: they occur in an energy region where the density of compound nuclear (CN) states with natural isospin $T_< = T_> - 1$ is very high. Within the width of the resonance, there is a large number of strongly overlapping CN states, for which a statistical description is assumed to be valid. The IAR ($T_>$) and the CN states are coupled mainly by Coulomb forces [21]. Because of this coupling, the strength of $T_<$ states with the same spin and parity as the IAR is enhanced and the total width Γ of the IAR is spread (Γ_c = partial width for channel c, W = spreading width describing strength of the coupling between the IAR and the $T_<$ states):

$$\Gamma = \sum_{c} \Gamma_c + W_c. \tag{4}$$

For the description of proton scattering experiments in terms of an S-Matrix, the S-Matrix is separated into an energy averaged part $\langle S \rangle$ and a fluctuating part.

$$S_{cc'} = \langle S_{cc'} \rangle + (S - \langle S_{cc'} \rangle). \tag{5}$$

The averaging interval may be considered to be the experimental energy resolution; it must be small compared to the total resonance width but large enough so that the statistical assumption for the $T_<$ states is valid.

The energy averaged part is usually called "direct" scattering, the fluctuating part "compound" (CN) scattering

$$S_{cc'}^{DI} = \langle S_{cc'} \rangle, \tag{6}$$

$$S_{cc'}^{CN} = S_{cc'} - \langle S_{cc'} \rangle. \tag{7}$$

The energy averaged differential cross section $\langle \sigma(\theta) \rangle$ is the sum of two terms, one arising from direct terms only and one from fluctuating terms only

$$\langle \sigma(\theta) \rangle = \sigma(\theta)^{DI} + \sigma(\theta)^{CN}. \tag{8}$$

The energy averaged product of analyzing power and differential cross section depends only on direct terms [25]

$$\langle \sigma(\theta)A \rangle = \sigma(\theta)^{DI} A^{DI}. \tag{9}$$

To overcome all difficulties with compound nuclear scattering in the analysis of resonance effects, it is obviously advantageous to determine $\sigma(\theta)A$.

Often the IAR occurs well above (p,n) thresholds; then many channels are open for the decay of the CN states. In the case of elastic scattering, it may then be assumed that

$$\sigma(\theta)^{DI} \gg \sigma(\theta)^{CN}. \tag{10}$$

However, if the (p,n) channels are closed, or if only a few channels are open, CN scattering may not be neglected in the cross section.

If it is possible to fit the direct scattering using $\sigma(\theta)A$, then σ_{DI}

can be calculated and the compound elastic cross section is obtained directly, since it is the difference

$$\sigma(\theta)^{CN} = \sigma(\theta)^{exp} - \sigma(\theta)^{DI}. \tag{11}$$

3. ANALYSIS OF ELASTIC PROTON SCATTERING

The direct elastic scattering may be described by phase shifts. We separate the off-resonance scattering and the resonance scattering [22–24]

$$S_{\ell ij} = \exp[2 i \operatorname{Re}(\delta_{\ell j})] \left[\exp[-2 \operatorname{Im}(\delta_{\ell j})] + i \sum_R \frac{\exp[2 i \phi_R] \Gamma_{pR}}{E_R - E - i/2 \Gamma_R} \right]. \tag{12}$$

In the case of a spin–zero target, the cross section and analyzing power are obtained by the usual formulas

$$\sigma(\theta) = |g|^2 + |h|^2 \tag{13}$$

$$\sigma(\theta)A = 2 \operatorname{Re}(gh^*), \tag{14}$$

where g and h are given by

$$g = \text{Coulomb} + \frac{i}{2k} \sum_{\ell j} (j + 1/2)(1 - S_{\ell j}) e^{2i\omega_\ell} P_\ell^0 (\cos \theta), \tag{15}$$

$$h = \frac{1}{2k} \sum_{\ell j} (-1)^{j+\ell+1/2} (1 - S_{\ell j}) e^{2i\omega_\ell} P_\ell^1 (\cos \theta). \tag{16}$$

The complex phase shifts $\delta_{\ell j}$ describe the off-resonant scattering; they should be consistent with an optical model description. The IAR with quantum numbers ℓ and $j = \ell \pm 1/2$ is described by a Breit–Wigner term containing the resonance energy E_R, total width Γ_R, partial width Γ_{pR}, and a resonance mixing phase ϕ_R. The sum of Breit–Wigner terms for several IAR's with the same spin and parity is correct only if the energy separation of these IAR's is much larger than their total width. The case of overlapping resonances with the same spin and parity is not yet completely solved [22, 23].

The energies of most of the pronounced IAR's are near the Coulomb barrier for proton scattering. Then, the off-resonance scattering is dominated by Coulomb scattering and therefore the optical model phase

shifts are not large. The resonance mixing phase ϕ_R, which depends
on the influence of the imaginary part of the optical potential [22–24]
is also small. For the purpose of spin determination only, if the ℓ-
value and the resonance energy are known, the details of the optical
model parameters are unimportant. This is supported by the experimental fact that the shapes of the excitation functions for cross section and analyzing power over IAR's with the same ℓ- and j-values
are quite similar in a given energy and mass region.

The determination of the total angular momentum is quite easy from
a polarization measurement since the resonance effect of σA differs
in sign depending on whether $j = \ell \pm 1/2$. This is because σA is linear
in h (eq. (14)) and the resonance contributes to h with a different sign
(eq. (16)) depending on whether $j = \ell \pm 1/2$. In principle, a j-determina-
tion is possible from $\sigma(\theta)$ only, because the resonance term has different weight in g depending on whether $j = \ell \pm 1/2$, whereas in h there
is no change in weight. In elastic scattering, however, there is a
strong off-resonant scattering amplitude in g (mainly Coulomb scattering). Therefore, the main resonance contribution to $\sigma(\theta)$ is due to interference between the resonance term in g and the Coulomb amplitude.
Because of the relatively small pure resonance effect, the angular
distribution is very similar for both j-values so that the j-determina-
tion from elastic cross sections only is rather uncertain.

Lieb et al. [27], however, have shown that in inelastic scattering
which leaves the target nucleus in an excited 0^+ state, the direct off-
resonance amplitude is small. They determined reliably the spins of
a number of states in the N = 50 region from the angular distribution of
$\sigma(\theta)$ only.

4. SPIN DETERMINATION FROM
ELASTIC SCATTERING EXPERIMENTS

Fig. 2 shows the first polarization experiment on an IAR, performed
by Moore and Terrell [4] in 1966. In a double scattering experiment
they measured the polarization of protons elastically scattered from a
^{90}Zr target. The polarization data in the energy region of the IAR of
the second excited state of ^{91}Zr (E_x = 2.06 MeV) clearly confirms the
$3/2^+$ assignment obtained from the shell model for this d-wave resonance. The solid and the dashed curves are calculated for the two
alternatives $j = \ell \pm 1/2$ and with resonance parameters obtained from
fitting the differential cross section.

Fig. 3 shows a careful investigation of low lying states of ^{89}Sr,
performed by Ellis and Haeberli [6] with a polarized beam. The IAR
of the ground state (g.s.) and of a number of excited states were fitted
using resonance parameters obtained by Cosman et al. [28] from the
analysis of the differential cross section. To reproduce the off-
resonance analyzing power observed at higher energies, the phase

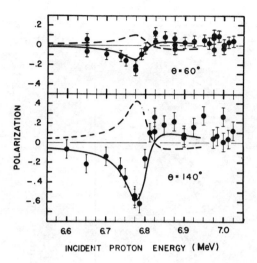

Fig. 2. Polarization mea-
surement in $^{90}Zr(p,p_0)^{90}Zr$
[4] near the $d_{3/2}$ IAR at
E_p = 6.8 MeV. Dashed
curve: spin assignment
$d_{5/2}$.

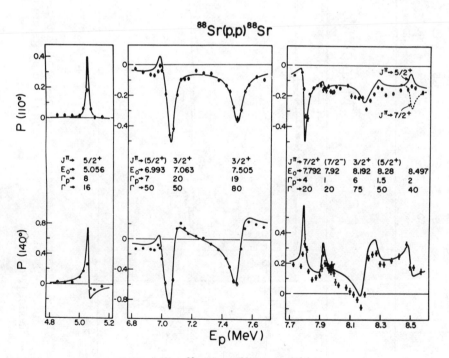

Fig. 3. Analyzing power for $^{88}Sr(p,p_0)^{88}Sr$ [6]. The lines are calculated
using the resonance parameters shown. Assignments in parentheses
and the assignment for the E_p = 8.497 MeV resonances are uncertain.

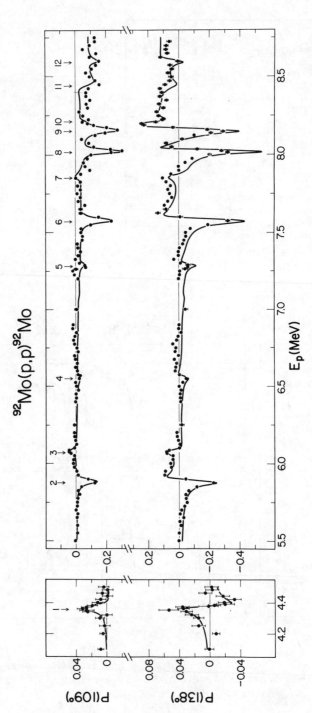

	1	2	3	4	5	6	7	8	9	10	11	12
J^π	$5/2^+$	$3/2^+$	$5/2^+$	$3/2^+$	$(3/2^+)$	$3/2^+$	$(5/2^+)$	$3/2^+$	$3/2^+$	$(5/2^-)$	$(3/2^-)$	$(3/2^+)$
E_0	4.355	5.875	6.07	6.55	7.28	7.57	7.85	8.015	8.15	8.21	8.44	8.59
Γ_p	1.5	5	1.5	1	1	6.5	1.5	5.5	7.5	1	5	1.5
Γ	30	27	30	30	20	30	40	20	35	25	70	30
ϕ	10°	5°	0°	10°	0°	0°	-10°	0°	0°	0°	10°	-10°

^{92}Mo(p,p)^{92}Mo

E_p(MeV)

Fig. 4. Analyzing power for ^{92}Mo(p,p$_0$)^{92}Mo in the region of the g.s. and of excited-state IAR's [6]. The off-resonance phase shifts are calculated from an optical model [39].

shifts were calculated from an optical potential. The resonance mixing phases were found to be near zero. The shape of the resonances
is reproduced well, and the spins of the strong resonances are determined reliably.

At higher energies, the analysis becomes difficult because the contributions of weak resonances are not known well enough to be included in the calculation. This fact explains some of the discrepancies between the data and the fit. A similar measurement by the same
group is shown in fig. 4. The partial width of the $d_{5/2}$ g.s. IAR in
^{92}Mo(p,p$_0$)^{92}Mo at E_p = 4.355 MeV is strongly reduced by the Coulomb
barrier. This resonance has not been observed in the cross section,
but its existence is clearly established from the analyzing power
measurements. A second $d_{5/2}$ resonance has been identified near the
lowest $d_{3/2}$ resonance, similar to the case of ^{88}Sr. Again, several
$d_{3/2}$ resonances are observed. A measurement on the low lying states
of ^{91}Zr was performed by Wienhard et al. [8] with the Erlangen polarized beam [20] (fig. 5). Five resonances, all $d_{3/2}$, were analyzed. The
off-resonance analyzing power was calculated from phase shift analyses at several off-resonance energies. All $3/2^+$ resonances have
very similar shapes, even at energies where the off-resonance analyzing

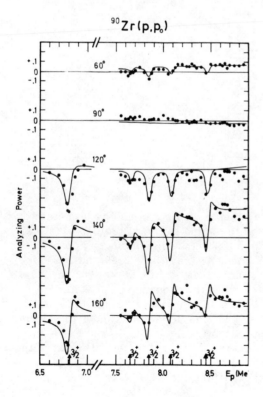

Fig. 5. Analyzing power
for ^{90}Zr(p,p$_0$)^{90}Zr [8]. The
line is calculated for $d_{3/2}$
resonances only; some
small resonances are not
included. The off-resonance
amplitudes are obtained
from a phase shift analysis.

power is 0.2. The deviations are again due mainly to weak resonances not included in the fit.

The results of the experiments on states in the Zr isotopes ^{91}Zr, ^{93}Zr, ^{95}Zr, and ^{97}Zr are shown in fig. 6 [8, 29]. In the ^{91}Zr g.s. there is one $2d_{5/2}$ neutron above the closed shell; in ^{96}Zr, the $2d_{5/2}$ neutron shell is closed. The number of observed $d_{3/2}$ states decreases very rapidly with increasing neutron number as one approaches the closure of the $d_{5/2}$ neutron shell. This supports the idea that the origin of most of these $d_{3/2}$ states is the coupling of the $d_{5/2}$ states with excited core states, mainly the 2^+ state. These states, of course, do not occur in ^{97}Zr, where the $d_{5/2}$ shell is closed. The lowest observed $d_{3/2}$ resonance in ^{96}Zr$(p,p_0)^{96}$Zr is stronger than in the other Zr isotopes, because the $2d_{3/2}$ single-particle strength is no longer spread over so many states.

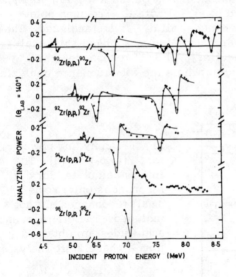

Fig. 6. Analyzing power for elastic scattering from ^{90}Zr, ^{92}Zr, ^{94}Zr, and ^{96}Zr at $\theta_{lab} = 140°$ [8, 29].

An experiment on a target with spin was performed by Bernstein et al. [17] (fig. 7). They measured the polarization of protons elastically scattered from the spin-$1/2^-$ nucleus ^{89}Y in the energy region of two neighboring IAR's at $E_p = 7.29$ and 7.46 MeV ($E_x = 2.48$ and 2.63 MeV). The fit to the cross section and polarization proved that the resonance effect is due only to $d_{3/2}$ waves for both resonances. Possible $d_{5/2}$ and $s_{1/2}$ admixtures to these 2^- and 1^- states were found to be quite small. The spin sequence 2^- and 1^- was determined under the assumption that the two states are created by the coupling of the same $2d_{3/2}$ neutron to the $1/2^-$ core. In this case, the proton partial width of the IAR should be the same for both cases if correct statistical factors are used. For the 2^-, 1^- sequence, partial widths of 32 keV were obtained

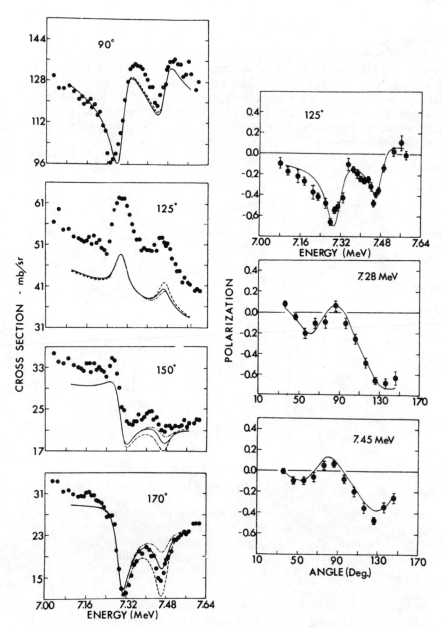

Fig. 7. Polarization and cross section for $^{89}Y(p,p_0)^{89}Y$ [17]. The 2^- and 1^- IAR at E_p = 7.29 and 7.46 MeV arise from the coupling of the $d_{3/2}$ neutron state to the $J^\pi = 1/2^-$ target state only. Contributions of $s_{1/2}$ (dashed line) and $d_{5/2}$ are very small.

for both states, whereas for the 1^-, 2^- sequence the partial widths
were 40 and 20 keV. Hence 2^-, 1^- is the strongly preferred sequence.

Nuclei far from the closed neutron shell have been investigated by
Veeser and Ellis [15]. In a study of the elastic scattering of polarized
protons by the isotopes ^{116}Sn, ^{118}Sn, and ^{120}Sn (fig. 8), they found excellent agreement with the spin assignments $d_{3/2}$, $s_{1/2}$, $d_{5/2}$, and $d_{5/2}$
of Schneid et al. [32] for the four lowest states. There is also a contribution to this symposium by Roth et al. [16] who report spin determinations for states in ^{129}Te.

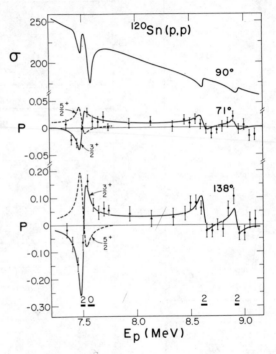

Fig. 8. Cross section and analyzing power
for ^{120}Sn$(p,p_0)^{120}$Sn [15]. The lines are calculated curves for spin assignments $d_{3/2}$, $s_{1/2}$,
$d_{5/2}$, and $d_{5/2}$. Near the g.s. IAR the dashed
curve is for $d_{5/2}$, the solid curve for $d_{3/2}$.

In the region of the $N = 82$ shell some spin assignments had to be
revised. Fig. 9 shows a recent Erlangen measurement of elastic scattering of polarized protons by ^{140}Ce. The data have not been completely analyzed yet, however [28, 30].

In the $N = 83$ nuclei ^{139}Ba, ^{141}Ce, ^{143}Nd, and ^{145}Sm, Veeser et al. [9,
10], Clausnitzer et al. [11], and Fiarman et al. [13] found that the
second excited states with large spectroscopic factors were $p_{1/2}$ and
not $p_{3/2}$ as determined earlier from the analysis of cross sections only.

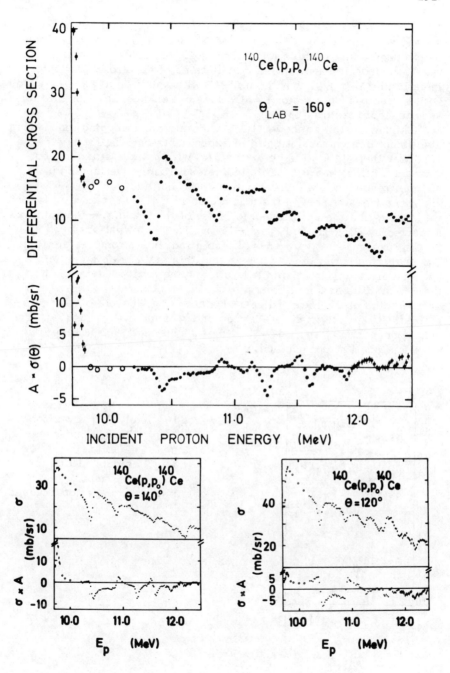

Fig. 9. $\sigma(\theta)A$ and $\sigma(\theta)$ for $^{140}Ce(p,p_0)^{140}Ce$ [29, 30].

In a preliminary analysis we find an $f_{7/2}$ resonance just below the
second $f_{5/2}$ resonance at $E_p = 11.5$ MeV. Such an $f_{7/2}$ state has never
before been found in this region of nuclei. The existence of a second
$f_{7/2}$ resonance is important for calculations of spectroscopic factors.
In particular it proves that the g.s. IAR with spin $f_{7/2}$ at $E_p = 9.74$ MeV
with a spectroscopic factor $S \approx 0.9$ does not exhaust the sum rule for
spectroscopic factors [33−35].

At higher energies (above 11 MeV) there are overlapping resonances,
also found in high resolution (d,p) experiments on ^{138}Ba [36]. At $E_p =$
12.2 MeV there is a strong effect in $\sigma(\theta)$ with p-wave structure, but
very little effect in $\sigma(\theta)A$. This indicates strongly overlapping $p_{3/2}$
and $p_{1/2}$ resonances which cancel in $\sigma(\theta)A$, but add in $\sigma(\theta)$.

From a measurement in the g.s. region of ^{147}Nd [12] shown in fig. 10,
we learned that the spectroscopic factor of the 5/2 g.s.—the spin was
measured by NMR-methods [37]—is small. The lowest observed state
in (d,p) experiments (with 15-keV resolution) is a strong f-state [33].
The analyzing power clearly shows that this state is an $f_{7/2}$-state.
Because of the different spin it is a low lying excited state and not
the 5/2 g.s., as assumed before.

The spins of the lowest d- and g-states of ^{209}Pb were confirmed by
a polarization experiment of Fiarman and Marquardt (fig. 11) [14]. The
spin sequence $g_{9/2}$, $d_{5/2}$, $g_{7/2}$, and $d_{3/2}$ shows that ^{209}Pb is really an
excellent shell model nucleus.

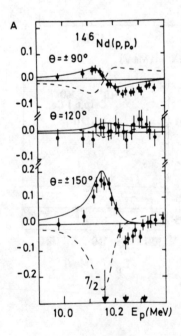

Fig. 10. Analyzing power for
^{146}Nd(p,p$_0$)^{146}Nd [12] near the
g.s. IAR. The solid line is for
spin $f_{7/2}$, the dashed line for
$f_{5/2}$. Small resonances indicated by arrows are not included.

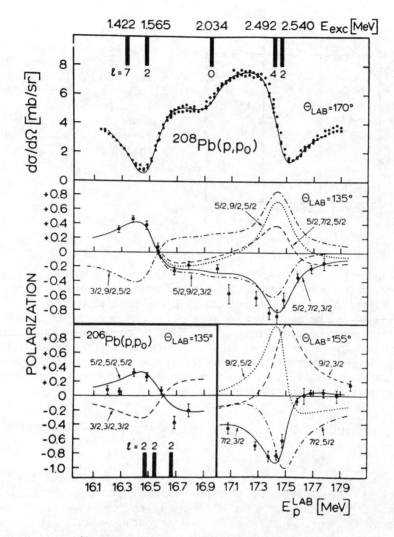

Fig. 11. $^{208}Pb(p,p_0)^{208}Pb$ and $^{206}Pb(p,p_0)^{206}Pb$, $\sigma(\theta)$ and P [14].
The solid line is calculated for the shell model spin se-
quence $d_{5/2}$, $g_{7/2}$, $d_{5/2}$.

5. COMPOUND ELASTIC SCATTERING

As pointed out above, the product $\sigma(\theta)A$ is independent of compound
elastic scattering amplitudes [25], i.e., statistically fluctuating amplitudes. The cross section, however, is a sum of a direct and a compound term. To investigate $\sigma(\theta)^{CE}$, we have performed measurements
on N = 50 nuclei near the $2d_{5/2}$ resonance at $E_p \approx 5$ MeV [31].

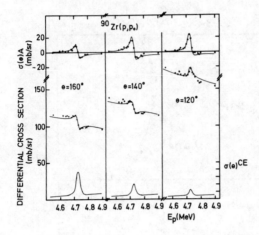

Fig. 12. $\sigma(\theta)$ and $\sigma(\theta)A$ for $^{90}Zr(p,p_0)^{90}Zr$ [31] near the $d_{5/2}$ g.s. IAR. The solid curve is calculated with the resonance parameters indicated: $\sigma(\theta) = \sigma(\theta)^{DI} + \sigma(\theta)^{CE}$. At the bottom of the figure $\sigma(\theta)^{CE}$ is shown.

For the target nucleus ^{90}Zr (fig. 12) this energy is below the (p,n) threshold ($E_p = 6.97$ MeV). The elastic scattering channel may be considered to be the only open channel; inelastic scattering ($E_x = 1.75$ MeV) can be neglected because of the Coulomb barrier. In such a one-channel case, the compound elastic cross section can be calculated from the parameters of direct scattering. From the unitarity of the "energy-unaveraged" S-matrix it follows that the reaction cross section, calculated from the energy-averaged amplitudes, is equal to the compound elastic cross section [40]

$$\sigma^{CE}(\theta) = \frac{1}{4k^2} \sum_{\ell j} (1 - |\langle S_{\ell j}\rangle|^2) \left\{ [(j + \tfrac{1}{2})P_\ell^0 (\cos\theta)]^2 + [P_\ell^1(\cos\theta)]^2 \right\}. \tag{17}$$

The measurement of σA and $\sigma(\theta)$ is fitted with these formulas; the full line in fig. 12 is calculated with the resonance parameters indicated. At the bottom of the figure the compound elastic cross section is shown. Its contribution is 30 mb/sr at $\theta = 160°$, about one-third of $\sigma(\theta)$. The resonance parameters, extracted from the fit to both analyzing power and differential cross section, are

$$E_R = 4.73 \text{ MeV}, \quad J^\pi = 5/2^+, \quad \Gamma = 17 \pm 2 \text{ keV}, \quad \Gamma_p = 3.1 \pm 0.3 \text{ keV},$$

$$\text{and } \phi_R = 0.05.$$

The off-resonance scattering is derived from an optical model using Perey's program [38] and the parameters of Becchetti et al. [39], with a value of 4 MeV for W.

The resonance mixing phase $\phi_R = 0.05$, extracted from this analysis,

has positive sign and is very small. This supports the reaction theory
of Thompson-Adams-Robson [22] and Weidenmüller [24] which proposes
a positive sign for the mixing phase.

In this single channel case the spreading width W [21] is defined
by

$$W = \Gamma - \Gamma_p. \tag{18}$$

The quantity W is determined to be 14 keV, which is more than 4 times
the proton partial width. This indicates a strong coupling between
the IAR and the CN states.

An example of compound elastic scattering in the case of open
neutron channels is the measurement near the $d_{5/2}$ IAR in ^{88}Sr(p,p_0)^{88}Sr
at E_p = 5.075 MeV shown in fig. 13. The $T_<$ states with the same
value of j^π as the IAR may decay by elastic scattering or by $p_{3/2}$ neutron waves to the 4^- ground state of ^{88}Y (E_n = 0.6 MeV). Dutt and
Schrills [41] showed that the n-decay to the 0.395 keV 1^- state can
be neglected.

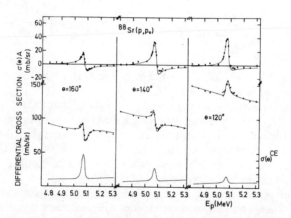

Fig. 13. Cross section and $\sigma(\theta)A$ for ^{88}Sr(p,p_0)^{88}Sr
[31] near the $d_{5/2}$ g.s. IAR. The solid line is
calculated for $\sigma(\theta)A$ and $\sigma^{DI}(\theta)$, adding $\sigma^{CE}(\theta)$ as
indicated in the figure and calculated from for-
mulas (19) and (20).

Our measurement shows strong compound elastic scattering. The
solid curves are calculated for the direct contribution only. The resonance parameters are E_p = 5.076 MeV; Γ = 19 keV, Γ_p = 6.0 keV, and
ϕ_R = 0.08. The partial width Γ_p = 6.0 keV is smaller than the value
Γ_p = 8 keV obtained by Cosman [28] from the analysis of the differential cross section only. The difference could result from neglecting
fine structure effects in the cross section.

The (p,n) measurement of Dutt and this measurement show the same order of magnitude for the (p,n) and compound elastic cross sections.

The CE-cross section is fitted using transmission coefficients T_c [23]. The proton channel with the same quantum numbers ℓ, j as the IAR is enhanced.

$$T_p^{\ell,j} = [1 - \exp(-4 \operatorname{Im} \delta_{\ell j})] \frac{(E - E_R + \Delta)^2 + B}{(E - E_R)^2 + \Gamma^2/4}. \tag{19}$$

For simplicity B has been set equal to zero, as derived earlier by Robson [21]. Then the magnitude of the resonance effect of $\sigma(\theta)^{CE}$ is determined only by the imaginary part of the optical model phase shift $\delta_{\ell j}$ and a resonance energy shift Δ.

If one neutron channel is open and all inelastic proton channels are closed, the compound elastic cross section for the resonant partial wave is

$$\sigma^{CE}(\theta) = \frac{1}{4k^2} \left\{ [(j + \frac{1}{2}) P_\ell^0(\cos\theta)]^2 + [P_\ell^1(\cos\theta)]^2 \right\} \frac{(T_p^{\ell j})^2}{T_p^{\ell j} + T_n}. \tag{20}$$

Using a shift $\Delta = -52$ keV, we obtain a good fit with phase shifts from an optical model with W = 4 MeV.

The measurement (fig. 14a) on the $J^\pi = 1/2^-$ nucleus ^{89}Y in the region of the $d_{5/2}$ analog resonances is a further example of compound elastic scattering. These states are assumed to originate from the coupling of a $d_{5/2}$ neutron state to the $1/2^-$ target state to 2^- (g.s.) and 3^-. Both resonances show strong CE-scattering. Fig. 14b shows the total (p,n) cross section obtained by Johnson et al. [42]. The ratio $\sigma_{CE}/\sigma(p,n)$ is larger for the lower 2^- resonance than for the higher 3^- resonance. This shows the decrease of CE scattering when more neutron channels are open. However, the CE cross section at the 3^- resonance is large even though the $p_{3/2}$ neutron decay to the $9/2^+$ g.s. of ^{89}Zr is strongly preferred.

This example demonstrates the importance of compound elastic contributions for low lying resonances, and suggests that partial widths extracted from the good polarization measurements that are now possible with the available intense polarized beams will be more reliable.

6. INELASTIC SCATTERING

In elastic scattering the polarization of protons in the exit channel, obtained from scattering of an unpolarized beam by an unpolarized

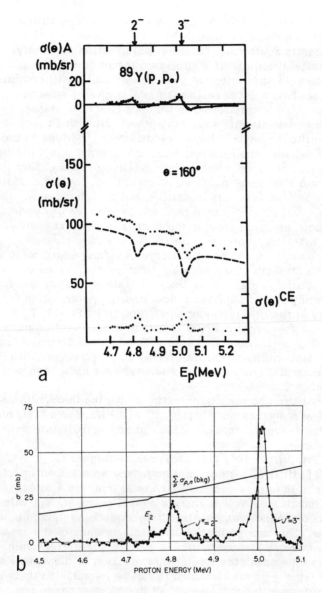

Fig. 14. a: ^{89}Y(p,p$_0$)^{89}Y; $\sigma(\theta)$A and $\sigma(\theta)$ [31] near the 2$^-$ g.s. and 3$^-$ IAR. The curves are calculated for direct scattering only. At the bottom is shown $\sigma(\theta)^{CE} = \sigma(\theta)^{exp} - \sigma(\theta)^{DI}$. b: Total neutron cross section for ^{89}Y(p,n)^{89}Zr [42].

target, is identical to the analyzing power obtained with a polarized
beam in the entrance channel.

In inelastic scattering, however, polarization and analyzing power
are generally different. If a spin-zero target nucleus is left in an excited state with spin different from zero, P and A will contain different
information, because the resonance may decay by several partial waves
with different spin interfering in the exit channel. Harney [43] pointed
out that a non-vanishing analyzing power is obtained only if the transition amplitudes have at least two different J^π values in the entrance
channel. Because the resonant entrance channel has a distinct J^π
value, at least one more scattering amplitude is necessary in order
to produce a non-zero analyzing power. Thus, inelastic scattering of
polarized protons is a very sensitive tool to determine whether or not
there are small off-resonance reaction amplitudes or overlapping resonances, both of which provide the necessary second channel.

If the reaction proceeds via a single entrance channel, e.g., at an
isolated resonance, the polarization provides spectroscopic information on the strength of the various decay partial waves of the IAR, which
leave the target nucleus in an excited state with spin different from
zero. Harney et al. [19] have calculated the polarization for inelastic
scattering at the $g_{9/2}$ resonance in $^{208}Pb(p,p')^{208}Pb$ (5^-, $E_x = 3.702$ MeV).
In this case the target nucleus is left in a pure neutron particle-hole
state. The polarization can be as high as 0.6 and depends strongly
on the partial widths of the interfering outgoing waves with different
angular momenta. However, no measurements have been performed
until now.

The analyzing power for inelastic scattering through IAR's has been
observed in reactions on ^{90}Zr [18], ^{140}Ce, ^{142}Nd, ^{144}Nd [28], and ^{208}Pb
[19], which leave the target nucleus in an excited state with collec-
tive character.

The $^{90}Zr(p,p_1)^{90}Zr$ (0^+, 1.75 MeV) reaction (near the $d_{3/2}$ IAR at $E_p = 8.45$ MeV [18]) (fig. 15) does not show any pronounced interference
character in the cross section that would indicate a direct off-resonance
reaction amplitude. The analyzing power, however, has values up to
0.3 in the resonance region; the off-resonance analyzing power is almost zero. This proves that there is interference of the resonance with
at least one more direct reaction amplitude of different spin.

Because the spin structure of this reaction is the same as in the
elastic scattering case, we used the same formulae to fit the data. For
an order-of-magnitude fit, we chose the off-resonance scattering amplitude g to depend only on the scattering angle θ and neglected the
weak energy dependence. The "spin-flip" amplitude h is set equal to
zero because of the vanishing off-resonance analyzing power. To obtain a satisfactory fit to both $\sigma(\theta)$ and A, the main part of the off-
resonant cross section (90%) is assumed to be compound, e.g., incoherent; only 5 to 10% of the off-resonant cross section is direct. When
only the direct part is analyzed, p-waves seem to dominate, as is

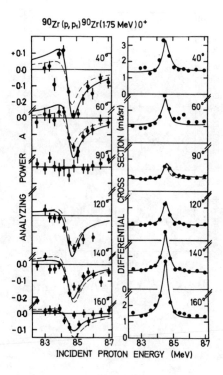

Fig. 15. Analyzing power and differential cross section for $^{90}Zr(p,p_1)^{90}Zr$ (0^+, E_p = 1.75 MeV) [18]. The solid curve has been calculated with an off-resonance amplitude in g only: $\rho e^{i\phi}$ and a resonance amplitude. A CN cross section σ_{inc} is added. The dashed curve is calculated assuming direct p-waves only.

indicated by the nearly symmetric pattern of A about θ = 90°. The dashed curve is calculated assuming that the direct background originates from p-waves only. The larger direct reaction amplitudes observed at forward angles indicate the influence of further reaction amplitudes. The main result of this measurement is that the off-resonant cross section is not purely compound, as had been assumed earlier [44].

The inelastic scattering of polarized protons from N = 82 nuclei leaving the target nucleus in the first excited 2^+ state again shows resonance effects in the analyzing power (fig. 16). For the well separated resonances in $^{140}Ce(p,p')^{140}Ce$ (2^+) at lower energies, these effects are smaller than in the case of some overlapping resonances at higher energies. The pattern at a scattering angle θ = 160° is very similar to that at θ = 140°. This experiment shows again that there is an off-resonant direct reaction amplitude populating the 2^+ state. Furthermore, the measurement is very sensitive to strongly overlapping resonances; in those cases the analyzing power may have appreciable values. This latter fact is also borne out by measurements on ^{142}Nd and ^{144}Nd in the energy region of overlapping resonances [30].

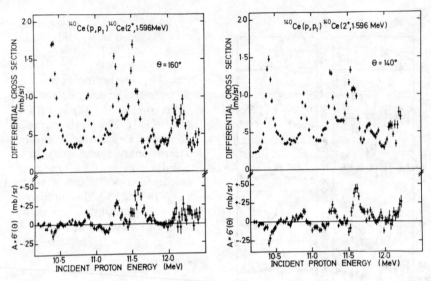

Fig. 16. $\sigma(\theta)$ and $\sigma(\theta)A$ for ^{140}Ce(p,p$_1$)^{140}Ce (2$^+$, 1.6 MeV) [30] in the region above the g.s. IAR.

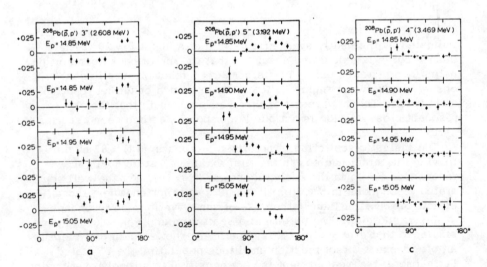

Fig. 17. Analyzing power for the reaction ^{208}Pb(p,p') near the $g_{9/2}$ IAR [19]. The excited target states are collective states (a, b) or states of pure neutron particle-hole character (c).

A comparison of inelastic scattering near IAR's leaving the target nucleus in excited states of collective or pure neutron particle-hole configuration has been performed by Harney et al. [19]. They observed the ^{208}Pb(p,p') reaction in the region of the $g_{9/2}$ IAR at E_p = 14.91 MeV to the collective 3^- (2.608 MeV) and 5^- (3.192 MeV) states and also to the 4^- (3.469 MeV) state, known to have very pure neutron particle-hole character (fig. 17). In the last case, zero analyzing power is observed, indicating that this excited state is populated only through the IAR, whereas the collective states are populated by various reaction channels and therefore show considerable analyzing power.

7. CONCLUSION

The purpose of this paper is to point out that polarization measurements on IAR's offer information important for the understanding of nuclear structure and of the reaction mechanism. In elastic scattering spins can be determined uniquely; and in cases where incoherent scattering is important, or in cases of overlapping resonances, resonance parameters can be determined more reliably.

In inelastic scattering, the configuration of particle-hole states, populated only through the IAR, should be investigated by double scattering experiments.

The analyzing power, less difficult to measure, offers information in all those cases where at least two channels interfere in the entrance channel, as is the case for non-resonant direct scattering, overlapping resonances, or targets with spin different from zero.

REFERENCES

[1] J. D. Fox, C. F. Moore, and D. Robson, Phys. Rev. Lett. 12 (1964) 198.
[2] Isobaric Spin in Nuclear Physics, eds. J. D. Fox and D. Robson (Academic Press, New York, 1966).
[3] Nuclear Isospin, eds. J. D. Anderson et al. (Academic Press, New York, 1969).
[4] C. F. Moore and C. E. Terrell, Phys. Rev. Lett. 16 (1966) 804.
[5] L. Brown and W. Trächslin, Nucl. Phys. A96 (1967) 238.
[6] J. L. Ellis and W. Haeberli, in Nuclear Isospin, eds. J. D. Anderson et al. (Academic Press, New York, 1969) pp. 585, 651; J. L. Ellis, Ph.D. Thesis, University of Wisconsin (1970).
[7] G. Clausnitzer, R. Fleischmann, G. Graw, K. Wienhard, ibid., p. 629.
[8] K. Wienhard, G. Clausnitzer, G. Graw, and R. Fleischmann, Z. Physik 228 (1969) 129.

[9]   L. Veeser, J. Ellis, and W. Haeberli, Phys. Rev. Lett. 18 (1967) 1063.

[10] L. Veeser and W. Haeberli, Nucl. Phys. A115 (1968) 172.

[11] G. Clausnitzer et al., Nucl. Phys. A106 (1967) 99.

[12] G. Graw, G. Clausnitzer, R. Fleischmann, and K. Wienhard, Phys. Lett. B28 (1969) 583.

[13] S. Fiarman, E. J. Ludwig, L. Michelman, and A. B. Robbins, Phys. Lett. 22 (1966) 175, and Nucl. Phys. A131 (1969) 267.

[14] S. Fiarman and N. Marquardt, Z. Physik 221 (1969) 494.

[15] L. Veeser and J. Ellis, Nucl. Phys. A115 (1968) 185.

[16] M. Roth et al., Third Polarization Symp.

[17] E. M. Bernstein et al., Phys. Rev. Lett. 22 (1969) 476, and Phys. Rev. 185 (1969) 1485.

[18] G. Graw et al., Phys. Lett. B30 (1969) 465.

[19] H. L. Harney et al., Nucl. Phys. A145 (1970) 282.

[20] G. Clausnitzer et al., Nucl. Instr. 80 (1970) 245.

[21] D. Robson, Phys. Rev. 137 (1965) 535.

[22] W. J. Thompson, J. L. Adams, and D. Robson, Phys. Rev. 173 (1968) 975.

[23] H. A. Weidenmüller, in Nuclear Isospin, eds. J. D. Anderson et al. (Academic Press, New York, 1969) p. 361.

[24] H. L. Harney and H. A. Weidenmüller, Nucl. Phys. 139 (1969) 241.

[25] W. J. Thompson, Phys. Lett. B25 (1967) 454.

[26] J. L. Adams, W. J. Thompson, and D. Robson, Nucl. Phys. 89 (1966) 377.

[27] K. P. Lieb, J. J. Kent, T. Hausmann, and C. E. Watson, Phys. Lett. B32 (1970) 273.

[28] E. R. Cosman, H. A. Enge, and A. Sperduto, Phys. Rev. 165 (1968) 1175.

[29] P. Schulze-Döbold, University of Erlangen, Diplomarbeit (1970).

[30] H. Clement, University of Erlangen, Diplomarbeit (1970).

[31] W. Kretschmer, University of Erlangen, Ph.D. Thesis, to be published.

[32] E. J. Schneid, A. Prakash, and B. L. Cohen, Phys. Rev. 156 (1967) 1316.

[33] C. A. Wiedner, A. Hensler, J. Solf, and J. P. Wurm, Nucl. Phys. A103 (1967) 433.

[34] H. Seitz et al., Nucl. Phys. A140 (1970) 673.

[35] E. Grosse et al., Nucl. Phys. A142 (1970) 345.

[36] D. Von Ehrenstein, G. C. Morrison, J. A. Nolen, and N. Williams, Phys. Rev. C1 (1970) 2066.

[37] R. W. Kedzie, M. Abraham, and C. D. Jeffries, Phys. Rev. 108 (1957) 54.

[38] F. G. Perey, Phys. Rev. 131 (1963) 745.

[39] F. D. Becchetti and G. W. Greenlees, Phys. Rev. 182 (1969) 1190.

[40]   D. Robson and A. M. Lane, Phys. Rev. 161 (1967) 982.
[41]   G. C. Dutt and R. Schrils, Phys. Rev. 175 (1968) 1413.
[42]   C. H. Johnson, R. L. Kernell, and S. Ramavataram, Nucl. Phys. 107 (1968) 21.
[43]   H. L. Harney, Phys. Lett. B28 (1968) 249.
[44]   K. P. Lieb, J. J. Kent, and C. F. Moore, Phys. Rev. 175 (1968) 1482.
[45]   W. Haeberli, Third Polarization Symp.

DISCUSSION

Weidenmüller

The mixing between the analog resonance and the surrounding $T_<$ states plays a role at three places in the analysis. (1) A resonance mixing phase appears which, as Dr. Graw pointed out, is usually found to be quite small. (2) The total width is not assumed to be equal to the sum of the partial widths. (3) There appears a compound part of the cross section which tends to reduce the polarization in elastic scattering through the analog resonance. This, however, comes into play only for nuclei in the Zr region.

Von Ehrenstein:

Dr. Graw said that there are probably two unresolved f-resonances in the ^{140}Ce(p,p$_0$) excitation function around 1.8 MeV. These are above the lowest $f_{7/2}$-resonance which is the analog of the ground state of ^{141}Ce. In our investigation of one of the isotones with the ^{138}Ba(d,p)^{139}Ba reaction, we have indeed resolved two close resonances ($\sim$ 18 keV separation) at the corresponding excitation energy of 1.7 MeV. Both are strongly populated and have $\ell = 3$. The previous spin assignment of 5/2 was obtained by scattering polarized protons from the analog resonance, and it assumed only one level. More polarization data are desirable to identify a possible 7/2$^-$ level.

Theory of Transfer Reactions

L. J. B. GOLDFARB, The University of Manchester, England

1. INTRODUCTION

Two aspects come to mind when we question the point of polarization studies in transfer reactions. First, we are provided with an effective probe to some of the spectroscopic features in a stripping reaction. We are familiar with the successes in determining the ℓ-values of states resulting from the transfer process simply through an analysis of the shapes of angular distributions. Measurement of the proton polarization or of the asymmetry associated with vector polarized deuterons in the neighborhood of the first peak in a (d,p) reaction points decisively to the j-value. This has been demonstrated by the experiments of Yule and Haeberli [1].

The second aspect, which is a common feature of most polarization studies, is where we seek to supplement information already gained from the study of differential cross sections in order to feel more confident of our description of the process. Polarization measurements tend not to be as well-fitted as differential cross sections. The motivation is ultimately to learn about nuclear spectroscopic factors, quantities which are determinable only with limited confidence owing to the sensitivity of calculations to a considerable number of features.

Practical considerations require that I confine my remarks to the (d,p) process. These can be extended with little modification so as to relate to the inverse pickup process and to proton transfers. Two-nucleon transfer or processes involving ^{3}He, tritons, alpha particles or, indeed, heavy ions are more difficult to describe. There are few known polarization measurements, although several contributions to this symposium report on (p,t) and (p, ^{3}He) transfers, and I shall say no more about these reactions.

The idea of acquiring more information about the stripping process is most relevant today, when much more critical thought is being given to the question of the nature of the reaction mechanism. Contrast the present situation with that at our earlier symposium at Karlsruhe [2] when great optimism was expressed about the applicability of the distorted-wave Born approximation (DWBA) [3]. Stress was placed particularly on the need to collect reliable information about the nature of the distorted waves. The DWBA implies consistency between the wave functions deduced from elastic scattering studies and those needed for the DWBA analysis—both in the entrance and in the exit channel. Few reactions have so far met this test over an extensive energy range.

Current theory of the (d,p) process attempts to go beyond the simplest characterization of the deuteron and of the target nucleus. The conventional DWBA treatment leads to integrals involving products of wave functions for the deuteron and proton, in addition to which there is a neutron form factor. Viewing the target as completely inert, this factor is simply the wave function of the transferred neutron. Otherwise, we must treat consistently both the neutron stripped from the deuteron and the other neutrons that are outside the target core [4]. This leads to the requirement that we solve a set of coupled equations for these neutrons, the outcome of which is a more realistic form factor. Several calculations have been performed along these lines [4], but little if any attention has been given to date to the consequences of polarization.

The deuteron likewise is a composite system, but this particular aspect has been dealt with only recently. The transition amplitude according to the DWBA is principally the matrix element of V_{np} and consequently only those components of the deuteron with neutron and proton in close proximity enter into the calculation. This has the effect of minimizing the importance of finite-range effects. Almost all calculations based on the DWBA rest on the "zero-range" approximation or on some simple amendments which bear relation to what is sometimes called the local-energy approximation [5]. The aim is the reduction of the problem to the solution of standard three-dimensional integrals. Johnson and Santos [6, 7] have utilized this feature to handle contributions from the d-state of the deuteron, while Delic and Robson [8] resort to the use of a full finite-range treatment to deal with d-state effects. Johnson and Soper [9] have gone further, with the deuteron still viewed as in a compact configuration, and have indicated how to deal simultaneously with the possibilities that the deuteron will break up, strip, or scatter elastically. Somewhat related is a coupled-channel procedure developed by Rawitscher [10].

Doubts have been expressed about the reliability of depicting the deuteron motion even in terms of an optical model, particularly when close to the nuclear surface, and the approach of Johnson and Soper [9], in fact, avoids this need. This feature is shared by two other parallel developments involving one school associated with Butler [11] and another connected with Pearson [12]. I shall dwell on these innovations in more detail later, but I should not neglect pointing to other developments which try to deal more rigorously with the far from trivial three-body problem [13]. Most of this work is in the developmental

state, so there has been little inclination to refine the theory to
handle polarization phenomena.

   The bulk of the analysis of transfer processes has been made in
the spirit of the DWBA, and I should like to review some of the consequences of the theory.

2. THE DWBA THEORY

   It is realized by now [14] that for a proper description of polarization it is crucial to give account to the spin-dependence of the optical-
model potentials used to generate the distorted waves. Nevertheless,
situations remain where the consequences of spin-independent distortion (SID) are sufficient. These might be at angles near the first
stripping peak or in Coulomb stripping reactions, that is, at energies
well below the Coulomb barrier, where nuclear distortion effects, even
those arising from central interactions, are of minor importance. Differential cross sections, on the other hand, are fairly insensitive to
spin-dependent distortion (SDD) apart from the large angles where the
yields are relatively small.

   Polarization measurements are of two varieties. First, one is provided with a polarized-ion source, say, which for deuterons involves
the vector-polarization component P(d) that is normal to the reaction
plane and tensor polarizations $P_{ij}(d)$, (i,j = 1, 2, or 3). The angular
distribution associated with such deuterons is of the form

$$\frac{d\sigma}{d\Omega} = \left(\frac{d\sigma}{d\Omega}\right)_0 \left[1 + \frac{3}{2}P(d)A(d,\theta) + \sum_{i,j} n_{ij} P_{ij}(d)A_{ij}(d,\theta)\right], \qquad (1)$$

where $(d\sigma/d\Omega)_0$ arises from no polarization and n_{ij} are numbers dependent on the choice of axes used to describe the deuteron polarization. I shall not be concerned with their values as I shall say little
about tensor polarization. The quantity $A(d,\theta)$ can be referred to as
the vector-polarization analyzing power.

   Secondly, the proton polarization $P(p,\theta)$, normal to the reaction
plane, is in question, and we infer this through measurements of the
angular distribution in a second reaction which involves the proton
analyzing power $A(p,\bar{\theta})$ and is of the form

$$\frac{d\sigma}{d\Omega} = \left(\frac{d\sigma}{d\Omega}\right)_0 [1 + P(p,\theta)A(p,\bar{\theta})]. \qquad (2)$$

Usually $A(p,\bar{\theta})$ is known. The next order of complication involves
polarization transfer where measurement is made of $P(p,\theta)$ in associa-
tion with incident deuterons of polarization $P(d)$ and $P_{ij}(d)$. This
particular field of enquiry might be thought of as being fairly remote;

nevertheless there are contributions to the symposium which do relate to this type of experiment. I shall say no more about the matter here.

For the present, SID will be taken to include neglect of the deuteron d-state as well as the non-central part of V_{np}. In such situations, we find [15]

$$A(d,\theta) = 2P(p,\theta),\tag{3}$$

i.e.,

$$\frac{d\sigma}{d\Omega} = \left(\frac{d\sigma}{d\Omega}\right)_0 [1 + 3P(d)P(p,\theta)].\tag{4}$$

There is no sensitivity [15] to $P_{ij}(d)$, no polarization effect when $\ell = 0$, and a correlation [16] of polarizations when $\ell \neq 0$ to the j-value. Thus,

$$x(j = \ell + \frac{1}{2}; \theta) = -\frac{\ell}{\ell + 1} x(j = \ell - \frac{1}{2}; \theta)\tag{5}$$

where $x(\theta)$ is either $A(d,\theta)$ or $P(p,\theta)$. Note that identical distributions are predicted for the deuteron analyzing power and the proton polarization, no matter what the j-value is. Further [16], there are bounds to the polarization magnitude. Thus, $P(p,\theta)$ is no greater than 1/3 or $\ell[3(\ell+1)]^{-1}$ according as j is $\ell + 1/2$ or $\ell - 1/2$, respectively. The successes of Yule and Haeberli [1] in identifying j-values by the signs of $A(d,\theta)$ at small angles near the first stripping peak are a consequence of the applicability of eq. (5). One can justify the operation of SID by numerical calculation.

With the introduction of SDD, few qualitative statements can be made. Both $P(p,\theta)$ and $A(d,\theta)$ are found to be large and near the maximum values of unity in some instances. The quantities $A_{ij}(d,\theta)$, on the other hand, tend to be small, but there are few examples of such measurements. The experiments of the type performed by Yule and Haeberli [1] bear some interest in the sense that they can shed light on the nature of SDD. Seyler and I [17] have recently extended eq. (3) in order to unravel the separate effects of proton spin-dependent distortions (PSDD) and deuteron spin-dependent distortions (DSDD) by procedures valid to first and higher orders in the strength of SDD. Neglecting the effect of SDD on the neutron, which numerical calculation shows to be fairly small, the following formulae hold:

$$[X(\theta)P(p,\theta)]_j = [X(\theta)P(p,\theta)]_j^{(0)} + a_d(\theta) + \sqrt{3/2}\, a_p(\theta) - [\beta_p(\theta)]_j$$

$$[X(\theta)A(d,\theta)]_j = 2[X(\theta)P(p,\theta\phi)]_j^{(0)} + a_d(\theta) + \sqrt{2/3}\, a_p(\theta) + \sqrt{3/2}\,[\beta_d(\theta)]_j$$

$$\left(\frac{d\sigma}{d\Omega}\right)_j \propto [X(\theta)]_j = [X(\theta)]^{(0)} + [\gamma_p(\theta)]_j + \sqrt{6}\,[\gamma_d(\theta)]_j.\tag{6}$$

We assume that the incoming deuterons have no polarization of the tensor variety. The exact form of the quantities on the right-hand side can be stated, but this will not be done here; however, we can note that terms with the suffix j have a j-dependence precisely as in eq. (5). The functions $a_d(\theta)$ and $a_p(\theta)$, on the other hand, are j-independent and are the only non-vanishing quantities for the case of $\ell = 0$ aside from the quantity $[X(\theta)]^{(0)}$. Further, they show the same parametric dependence on the optical model except that they refer to different channels; the same is true for the $\beta(\theta)$ and $\gamma(\theta)$ functions.

The formulae, which are valid only to first order in SDD, have shown their applicability in calculations for ^{40}Ca and ^{90}Zr targets, deuteron energies near 10 MeV, and angles $\lesssim 30°$. The inclusion of SDD adds to the stripping amplitude tensors in spin-space of rank $t = 1$ and $t = 2$ (only in the case of DSDD) and these enter into the polarization expressions in bilinear combination. Fig. 1 shows the effect of PSDD on $P(p,\theta)$ for the ^{40}Ca(d,p)^{41}Ca reaction (E_d = 7 MeV) and an excitation energy of 2.46 MeV. The $t = 1$ contributions should be relatively small if a first-order treatment is to be valid, and this is certainly not the case over a good part of the angular range. The $t = 2$ terms are not included in eq. (6), but they are small at angles

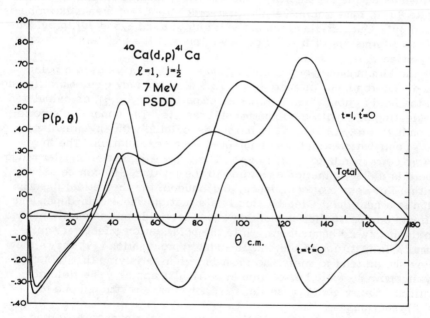

Fig. 1. Proton polarization, $P(p,\theta)$, for a hypothetical transition ^{40}Ca(d,p)^{41}Ca (E_d = 7 MeV) and an excitation energy of 2.46 MeV. Only SDD effects in the proton channel are taken into account and these are treated as tensors of rank 0 and 1 which are taken in bilinear combination (from ref. [17]).

$\lesssim 30°$ and at energies above 7 MeV. The analysis is no longer simple when the first-order treatment is invalid, as the $t = 0$ terms, indicated by zero superscripts, become spin-dependent.

With such formulae, we have the possibility of unraveling features concerning SDD effects in each channel. This is particularly useful if it is only the deuteron channel that is in doubt. The formulae apply reasonably well within limited energy ranges and at small angles. Even with the breakdown of a first-order treatment, the equations have validity owing to a phenomenon we term separability. Thus, if $x(d,p)$ refers to $P(p,\theta)$ or $A(d,\theta)$ as calculated with SDD in each channel, then separability requires that

$$x(d,p) = x(d,0) + x(0,p) - x(0,0) \qquad (7)$$

where 0 refers to SID in a particular channel. In the case of $\ell = 0$, $x(0,0)$ vanishes and one just has to add contributions from SDD in each channel. This seems to be valid over a considerable angular integral and over a wide range of angles, particularly if $\ell \neq 0$. Fig. 2 shows for the $^{90}Zr(d,p)^{91}Zr$ ($E_d = 11$ MeV) reaction a comparison of results of a full numerical treatment of $P(p,\theta)$, labeled 2, with that obtained using the separability approximation, labeled 1. The agreement up to 90° is remarkable; yet the first-order treatment breaks down beyond 30°. Our formulae thus are valid up to and beyond 90°, provided the $t = 2$ terms are of little importance, even if the $t = 0$ terms are spin-dependent.

The Birmingham group of Debenham et al. [18] have used a polarized deuteron beam of energy $E_d = 12.3$ MeV to study Coulomb stripping in the lead region. Fig. 3 shows a comparison of $A(d,\theta)$, as calculated neglecting SDD with the experimental results. The upper curve corresponds to $\ell = 2$, $j = 3/2$ and is further associated with an unresolved $g_{7/2}$ contribution which was neglected in the calculation. The lower curve corresponds to $\ell = 2$, $j = 5/2$. Throughout the entire angular range there is a discrimination according to the j-value, SID can be assumed with a fair degree of confidence, and, indeed, the theoretical description is a neat one. Clearly, this is a situation where spin-dependent effects can be studied in more detail. The yields are generally peaked in the backward hemisphere, while the polarization effects are quite small. In fact, use of the saddle-point approximation [19] shows that they should tend to vanish here. It is rather satisfying that the analyzing power should be so large at particular angles. The field of fruitful enquiry along these lines is, however, a restricted one for (d,p) reactions.

A good deal of attention has been given in the past to $\ell = 0$ reactions. Both $P(p,\theta)$ and $A(d,\theta)$ are consequences of SDD in either channel. If we ignore SDD but include effects of the deuteron d-state and a non-central V_{np}, the only non-vanishing polarization are those of

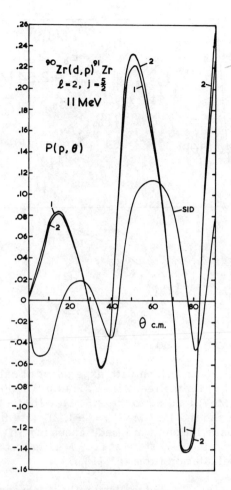

Fig. 2. Comparison of calculations of
P(p,θ) for a hypothetical ground-state
^{90}Zr(d,p)^{91}Zr transition corresponding to
E_d = 11 MeV as performed using the
separability approximation (labeled 1)
with the results of a full numerical
treatment (labeled 2). The SID results
are also shown (from ref. [17]).

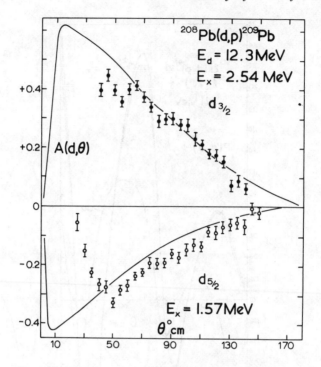

Fig. 3. Comparison with experiment of calculations
of $A(d,\theta)$ for the reactions ^{208}Pb(d,p)^{209}Pb (E_d =
12.3 MeV) leading to the 2.54 MeV ($j = \ell - 1/2$)
state and the 1.57 MeV ($j = \ell + 1/2$) state in ^{209}Pb.
The calculations were made assuming SID. The
neighboring $g_{7/2}$ state at 2.49 MeV is ignored in
the calculation (from ref. [18]).

the tensor variety [6]. Assuming separability, Johnson's results [6]
become

$$X(\theta)P(p,\theta) = a_d(\theta) + \sqrt{3/2}\; a_p(\theta) + \delta_d(SD;\theta)$$

$$X(\theta)A(d,\theta) = a_d(\theta) + \sqrt{2/3}\; a_p(\theta) + \frac{3}{2}\delta_d(SD;\theta) + \delta_p(SD;\theta), \quad (8)$$

where the extra contributions δ_p and δ_d arise from a first-order treatment of the effects of the d-state and a non-central V_{np}. Such contributions are, however, probably less important here than they are for
larger ℓ-values [7]. Neglecting them, the contributions for SDD for
$\ell = 0$ can be assessed directly from separate measurements of $P(p,\theta)$
and $A(d,\theta)$, provided that separability holds. Seyler and I [20] analyzed
the $\ell = 0$ ^{28}Si(d,p) ground-state reaction, and fig. 4 shows a plot of

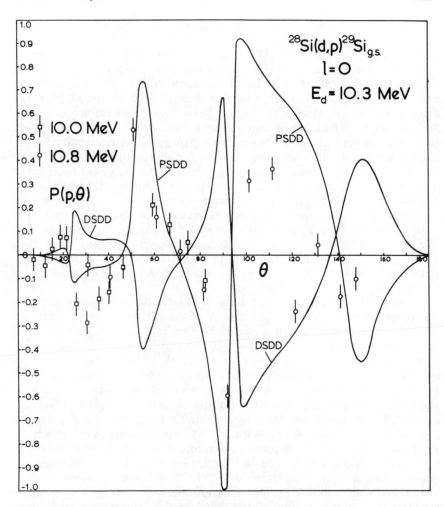

Fig. 4. Contributions of $\alpha_d(\theta)$ and $\alpha_p(\theta)$ (labeled DSDD and PSDD, respectively) to $P(p,\theta)$ for the ground-state $^{28}Si(d,p)$ transition. Comparison is to be made with the experimental results which were found for $E_d = 10.0$ and 10.8 MeV (from ref. [20]).

$\alpha_d(\theta)$ and $\alpha_p(\theta)$. These were computed using published optical-model parameters at deuteron energies near 10 MeV, but the strengths of the spin-orbit interactions were taken to be equal. We see here an odd anticorrelation of the two functions, something which might be fortuitous if it were not for the suggestion of the same feature operating with the $^{118}Sn(d,p)$ ground-state reaction [21] and the $^{12}C(d,p)$ transition to the first excited state of ^{13}C as reported by the Cracow workers [23].

Fig. 4 also suggests that proton SDD is dominant; although if this is the case, $P(p,\theta)$ would be nearly equal to $3/2$ $A(d,\theta)$ and this is not borne out by experiments [24]. Thus, the ^{28}Si-transition is somewhat obscure. The dominance of PSDD was expected by Hooper [25] in his analysis of $\ell = 0$ transfers. The deuteron channel is normally associated with large absorption effects and this tends to diminish $a_d(\theta)$. Hooper's analysis also explains the "saw-tooth" behavior observed for various $\ell = 0$ reactions. With strong absorption, a small spectrum of partial waves dominates the stripping integral. Since the latter is expressed simply as a weighted sum of Legendre polynomials, there is expected to be a change in sign of the polarization near the zero of a dominant Legendre polynomial. Relying on a first-order treatment, the stripping amplitude is at a minimum near this angle and the polarization is therefore near a maximum magnitude; yet it is here that the polarization switches sign. This applies both to $P(p,\theta)$ and to $A(d,\theta)$ and the rapid swing in these magnitudes at specific angles gives rise to the "saw-tooth" behavior which is well corroborated by recent polarization measurements. The feature is, however, restricted to s-wave transfers.

Perhaps the reason for the failure of the DWBA in the case of the ^{28}Si reaction is owing to the lightness of the target. Compare, for example, the successful fit reported by Kamitsubo and Mayer [21] of their experiment involving the $\ell = 0$ ^{118}Sn(p,d)^{117}Sn ground-state reaction. Here, the analyzing power $A(p,\theta)$ for 24.5 MeV protons was measured and compared with DWBA predictions. The fit, as shown in fig. 5, is quite impressive, although it is not clear whether the optical-model parameters account for the elastic scattering. Again, it appears that PSDD dominates and that its contribution is out of phase with that associated with DSDD. With the ^{28}Si lesson still in our minds, we should like to know about the deuteron polarization for this reaction, and, indeed, we have some information about this from recent experiments of Kocher and Haeberli [22]. This involves measurement of the deuteron analyzing power for the inverse ^{117}Sn(d,p)^{118}Sn reaction (which equals the deuteron polarization for the original reaction), but at a somewhat lower energy. Both sets of results are shown in fig. 5, and a comparison shows that, aside from angles $\lesssim 50°$, they seem to point consistently to the working of PSDD. Note also the rapid oscillations to be found in the figure. This feature is characteristic of high-energy $\ell = 0$ reactions. Taking into account the saw-tooth behavior, we realize the need for careful measurements. Transitions to the excited states of ^{117}Sn were also measured, but the fits were not too impressive. Experimental effort in this region of the periodic table nevertheless is to be applauded.

A poor quality in fits to experiment might point to the operation of factors excluded in the usual DWBA analyses. The spin-dependence of the deuteron optical-model interaction can include terms of the type

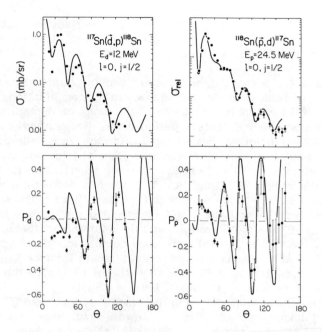

Fig. 5. Comparison with experiment of calculations
of $(d\sigma/d\Omega)_0$ and $A(p,\theta)$ for the $\ell = 0$ ground-state
reaction ^{118}Sn(p,d)^{117}Sn (E_p = 24.5 MeV) (from Escudié
et al., Saclay Annual Report CEA-N-1232, p. 103), and
of $(d\sigma/d\Omega)_0$ and $A(d,\theta)$ for the inverse reaction
^{117}Sn(d,p)^{118}Sn but where E_d = 12 MeV (from ref. [22]).

$$T_R = r^{-2}(\vec{S} \cdot \vec{r})^2 - \frac{2}{3}$$

$$T_L = (\vec{L} \cdot \vec{S})^2 + \frac{1}{2}(\vec{L} \cdot \vec{S}) - \frac{2}{3}L^2, \qquad (9)$$

where S is the spin-one operator. Their effect on the elastic scattering of polarized deuterons was investigated by Schwandt and Haeberli
[26], and the general conclusion is that terms of the T_R-type, characterized by a long-range attractive interaction, could effect an improvement but that the T_L-terms failed to improve matters. This is particularly satisfying for stripping reactions, if we refer to the calculations
of Delic and Robson [27]. They dealt with the ^{40}Ca(d,p)^{41}Ca (E_d =
9 MeV) reaction leading to states with ℓ = 1, j = 1/2 and 3/2 corresponding to excitations of 2.47 and 3.95 MeV, respectively. The T_L-term
seriously altered the differential cross section and all polarizations
were greatly affected. The T_R-term, on the other hand, hardly changed

T_{20} , T_{21}

the differential cross section and the vector polarization entities; but the tensor analyzing-power terms $A_{33}(d,\theta)$ and $A_{13}(d,\theta)$, which are normally small, showed large changes, albeit only for the $j = 1/2$ transition.

If, indeed, the tensor interactions are generally unimportant for vector polarization, we must look elsewhere for complications. Delic and Robson [8] have included effects of the deuteron d-state in what appears to me to be a Herculean performance. Not only have they included the T_R-interaction, but no approximations were made concerning finite-range features. Two reactions were considered. In one case, calculations were made of the deuteron polarization for the ground-state transition $^{16}O(p,d)^{15}O$ (E_p = 30 MeV) and for a transition to an excited state. These were both $\ell = 1$ transitions with $j = 3/2$ and $1/2$, respectively. Another set of $\ell = 1$ transitions relates to the $^{52}Cr(d,p)^{53}Cr$ (E_d = 8 MeV) reaction and the deuteron analyzing powers were calculated. Just as in the previous calculation relating to the effect of T_R, only the tensor polarization terms of the type $P_{13}(d,\theta)$, $P_{33}(d,\theta)$, $A_{13}(d,\theta)$, and $A_{33}(d,\theta)$ were appreciably affected; even so, the effect is most evident for the lower j-value. Santos [28] also has presented a contribution following the simplified treatment [7] developed with Johnson. The effect of the d-state tends to be small, particularly if $\ell = 0$, it seems to increase with the Q-value, and it applies more to the proton polarization than to the deuteron analyzing power.

3. CURRENT THEORETICAL DEVELOPMENTS

Differential cross sections are, as a rule, much more easily described by theory than are polarization phenomena. This perhaps is owing to some ignorance of the details of the spin-dependent parts of the optical-model interactions, particularly for the deuterons, although we must also be certain about the nature of the deuteron central interaction. Differential cross sections are sensitive to spin-dependent terms mostly at those angles where the yield is relatively weak, at large angles, for example, where the Lee–Schiffer effect [29] was observed for $\ell = 1$ transitions. Successes in DWBA analysis are generally less easily attained if the deuteron energy exceeds 20 MeV [30]. Johnson and Soper [9] and Rawitscher [10] have suggested the need to account for deuteron breakup as a likely possibility at these energies. In support of this we can refer to the recent measurements of Udo et al. [31] and May et al. [32], which report on measurements of nucleon pairs as a consequence of deuteron breakup. Johnson and Soper [9] suggest a refinement to the DWBA which is based on an adiabatic treatment of the deuteron, that is, one which views the deuteron as moving slowly (in relative s-state configurations) compared to the motion of its center-of-mass. The transition amplitude is as it is in the DWBA except for the need to replace $\chi_0(R)$, the eigenfunction of the deuteron center-of-mass which characterizes elastic scattering, by $\bar{\chi}(R)$ which satisfies

$$[E + B_d - T_R - V_c(R) - \bar{V}(R)] \, \bar{\chi}(R) = 0 \tag{10}$$

where B_d is the deuteron binding-energy, V_c is the Coulomb potential, and $\vec{R}$ is the vector joining the deuteron center-of-mass to the target nucleus. The interaction term $\bar{V}(R)$, in a zero-range treatment, is merely the sum $V_n(R) + V_p(R)$, and this is to be used instead of the deuteron model potential. Further, if the local-energy approximation is used, then a better approximation for $\bar{V}(R)$ is

$$\bar{V}(R) = \frac{\int d\vec{r} \, V_{np}(r) \, [V_n(|\vec{R} + \frac{1}{2}\vec{r}|) + V_p(|\vec{R} - \frac{1}{2}\vec{r}|)] \phi_d(r)}{\int d\vec{r} \, V_{np}(r) \phi_d(r)}, \tag{11}$$

which is the sum of the neutron and proton potentials averaged over the range of the neutron-proton interaction. The potentials V_n and V_p are then replaced by Woods-Saxon terms corresponding to individual motion at half the deuteron energy. Also, $\bar{\chi}(R)$ differs from $\chi_0(R)$ as follows:

$$\bar{\chi}(R) = \chi_0(R) + \int d\vec{k} \, \frac{\phi^{(+)}(\varepsilon_{\vec{k}}, r = 0)}{\phi_d(r = 0)} \, \chi(\varepsilon_{\vec{k}}, R) \tag{12}$$

where $\phi^{(+)}(\varepsilon_{\vec{k}}, r)$ denotes an $\ell = 0$ scattering state of the neutron-proton pair moving with energy $\varepsilon_{\vec{k}}$. Thus, in contrast to earlier assertions, the wave function needed for the stripping calculation is not simply that associated with elastic scattering, for eq. (12) presents further contributions associated with deuteron breakup with characteristic energies which are asserted to be no more than 10 MeV.

Calculations by Harvey and Johnson [33] indicate that the radial extension of the real and imaginary parts of $V(R)$ are similar, in contrast to the conventional parameterization which favors a larger extent for the imaginary part. The case of $^{54}Fe(d,p)^{55}Fe$ studied experimentally by Yntema and Ohnuma [30] at $E_d = 23$ MeV illustrates a characteristic failing of the DWBA. Fig. 6 shows the angular distribution leading to the first excited state and the poor account of the DWBA beyond the second stripping peak. The needed falloff was only obtainable if non-locality features of unusually large range were ascribed to the deuteron. The angular distribution falls off naturally here owing to a pronounced ℓ-space localization, which Harvey and Johnson show arises as a consequence of the convergence of the radii of the real and imaginary potentials.

No polarization measurements cover this reaction, however. More relevant are the asymmetry measurements of Chant and Nelson [30] in

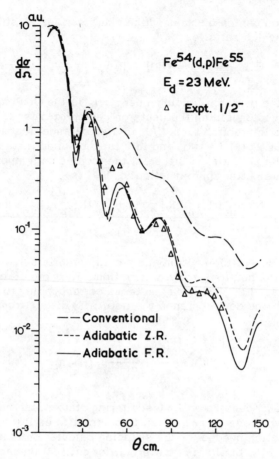

Fig. 6. Comparison with experiment of calculations of the differential cross section for
$^{54}Fe(d,p)^{55}Fe$ reaction leading to the first excited state of ^{55}Fe, (E_d = 23 MeV). The conventional DWBA calculation takes into account
finite-range effects (from ref. [33]).

connection with the $^{40}Ca(p,d)$ ground-state pickup reaction where
30.5 MeV polarized protons were used. Differential cross sections
again were poorly fitted, as is seen in fig. 7; however, some improvement was gained by damping deuteron contributions from the interior
through the use of an imaginary potential of a very large magnitude.
This is in line with Rawitscher's [10] work, and the effect is similar
to that achieved by introducing highly non-local interactions. The
adiabatic treatment goes a long way in accounting for the differential
cross section; however, the proton analyzing power is poorly repro-

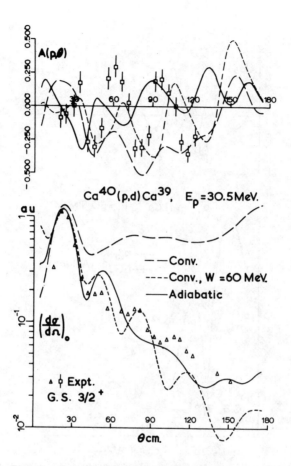

Fig. 7. Comparison with experiment of calcu-
lations of $(d\sigma/d\Omega)_0$ and $A(p,\theta)$ for the ground-
state $^{40}Ca(p,d)^{41}Ca$ reaction (E_p = 30.5 MeV).
The parameterization for the adiabatic treatment
is as in refs. [34, 35] (from ref. [33]).

duced. A better fit to the latter measurements is achieved by altering
the parameterization from that suggested by Perey [34] and by Perey
and Buck (see ref. [35]) to a more recent set proposed by Becchetti
and Greenlees [36]. This is seen in fig. 8; although there is a deterioration in the fit to the differential cross section. Perhaps some
reason for the overall discrepancy is owing to anomalies already seen
in attempting to fit the elastic scattering of protons by ^{40}Ca at high
energies. Attempts by Harvey and Johnson [33] to improve the fit to

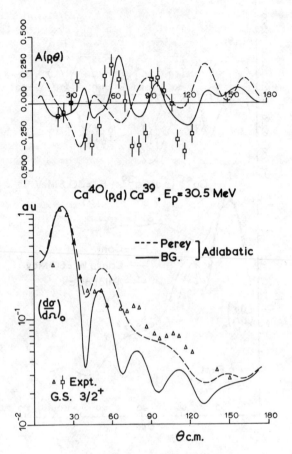

Fig. 8. As in fig. 7 but with a different set
of distorting parameters corresponding to
that given in ref. [36] (from ref. [33]).

the analyzing power by changing the parameterization of the optical
models invariably resulted in a worsening of the description of the
differential cross sections. Care must be given when applying the
adiabatic theory to deal with transfers where the deuteron energy is
less than 20 MeV. Most polarization studies lie outside the range
covered by this new view of stripping, and one welcomes more experi-
mentation at higher energies to learn more about the reaction mechanism.

     There are two other new versions of the stripping process which
bear strong similarity to each other and which also relate to the adiabatic approach. Perhaps another characteristic is the fact that they

have not met with too ready an acceptance. One [11] will be labeled
the BHMM treatment, which is based on the sudden approximation for
the deuteron. Here, the description of the deuteron state function is
intended to cover periods that are short compared to the natural period
of the deuteron. The motion is then in terms of separate wave packets
appropriate to the neutron and proton with their momenta correlated
according to the Fourier transform of the deuteron internal wave function. In the weakly-bound projectile (WBP) model [12], there is a
further correlation of the neutron with its proton pair in the description of the neutron interaction with the target.

There is an essential difference between the two descriptions, however. In the BHMM theory, the transition amplitude is cast in terms
of a linear combination of neutron and proton states in the continuum,
and instead of calculating for transitions leading to a final neutron
bound state, the theory deals with matrix elements of V_{np} *leading to*
neutron-proton breakup which are further to be multiplied by factors
expressing the overlap of the neutron-proton system with the final
state of outgoing proton and bound neutron. This leads to a peculiar
dependence of the cross section in terms of the quantity $S(1-S)^{-2}$
where S is the spectroscopic factor. The WBP theory avoids the introduction of a complete set of neutron states and the cross section
merely depends linearly on S.

On the other hand, the formal expressions for the transition amplitudes, as given by the two theories, show great resemblance. They
are:

$$
T_{fi}^{BHMM} x(1-S) = \int_{k_n' \neq i\kappa} d\vec{k}_p' d\vec{k}_n' \langle \psi^{(-)}(\kappa,\vec{k}_p)|\psi^{(+)}(\vec{k}_n',\vec{k}_p')\rangle \langle \psi^{(+)}(\vec{k}_n',\vec{k}_p')|V_{np}|\psi_d^{(+)}\rangle
$$

$$
= \int_{k_p'=k_p} d\vec{k}_p' \, S(\vec{k}_p,\vec{k}_p')\langle F(\vec{r}_n)|\psi^{(+)}(\vec{k}_d-\vec{k}_p')\rangle \, g(\vec{k}_d,\vec{k}_p')
$$

$$(13)$$

and

$$
T_{fi}^{WBP} = \int dk_p' \langle \psi^{(-)}(\kappa,\vec{k}_p)|\psi^{(+)}(\vec{k}_p')\rangle \langle \psi^{(+)}(\vec{k}_p')|V_{np}|\psi_d^{(+)}\rangle
$$

$$
= \int_{k_p'=k_p} d\vec{k}_p' \, S(\vec{k}_p,\vec{k}_p')\langle F(\vec{r}_n)|\mathscr{Y}(\vec{r}_n,\vec{k}_p')|\psi^{(+)}(\vec{k}_d-\vec{k}_p')\rangle \, G(\vec{k}_d,\vec{k}_p')
$$

$$(14)$$

where

$$G(\vec{k}_d, \vec{k}'_p) = \left[\left(\frac{1}{2}\vec{k}_d - \vec{k}'_p\right)^2 + \alpha^2\right]^{-1} g(\vec{k}_d, \vec{k}'_p)$$

$$= \left[\left(\frac{1}{2}\vec{k}_d - \vec{k}'_p\right)^2 + \alpha^2\right]^{-1} - \left[\left(\frac{1}{2}\vec{k}_d - \vec{k}'_p\right)^2 + \beta^2\right]^{-1} \tag{15}$$

$$\psi_d(r) \propto r^{-1}[\exp(-\alpha r) - \exp(-\beta r)] \tag{16}$$

$$\mathcal{V}(\vec{r}_n, \vec{k}'_p) = \int |\chi_p^{(+)}(\vec{k}_p, \vec{r}_p)|^2 \, V_{np}(\vec{r}_n - \vec{r}_p) \, d\vec{r}_p$$

$$\cong \begin{cases} \left[\left(\frac{1}{2}\vec{k}_d - \vec{k}'_p\right)^2 + \alpha^2\right] \ldots, \; r_n \geq R - 1 \text{ fm} \\ \\ 0 \ldots \ldots \ldots \ldots, \; r_n < R - 1 \text{ fm} . \end{cases} \tag{17}$$

In the above expressions, $F(r_n)$ denotes the neutron bound-state function and $S(\vec{k}_p, \vec{k}'_p)$ is the scattering matrix for the elastic scattering of protons at an energy equal to the energy of the outgoing protons. The capturing interaction $\mathcal{V}(\vec{r}_n, \vec{k}'_p)$ depends on the proton density function and in the zero-range approximation for V_{np} leads to a cutoff in the r_n-integration. More refined calculations by Pearson et al. [12] have taken into account the non-spherical nature of the capturing interaction, and this has the effect of greatly complicating the numerical analysis. The differential cross sections may not be much altered, but polarization effects are considerably modified in some instances.

Baker and I [37] attempted to assess the energy dependence of calculations based on the BHMM theory. The approximation obviously is better the higher the energy. In fact, one can show that the theory diverges as the deuteron energy tends to zero. The study was confined to targets already considered earlier by Butler et al. [11], but the energy was allowed to vary. In all the cases considered, there was a marked deterioration on altering the energy. Any sensible fit to experiment required modification of the spectroscopic factor. The "energy dependence" of S is shown in fig. 9 for ground-state transitions with ^{40}Ca, ^{52}Cr, and ^{90}Zr as targets. Since S should be energy independent, the theory faces a severe shortcoming. Further, the angular distributions present poor fits to experiments, although the picture improves at energies beyond 100 MeV. The calculations proved to be very sensitive to the parameterization of the optical model used to generate the neutron wave functions. Fig. 10 shows as an example the effect of altering the real part of the neutron potential strength by just ±1 MeV for the ^{90}Zr(d,p)^{91}Zr ground-state reaction ($E_d = 15$ MeV). Such a sensitivity would seem to be removed from the WBP model owing to the presence of the capturing interaction which cuts out a

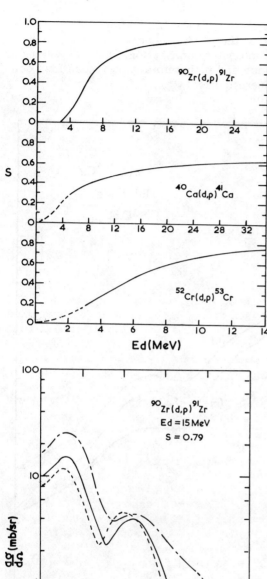

Fig. 9. The apparent
energy dependence of
spectroscopic factors
for the ground-state re-
actions ^{90}Zr(d,p)^{91}Zr,
^{52}Cr(d,p)^{53}Cr, and
^{40}Ca(d,p)^{41}Ca as a
function of the deuteron
energy (from ref. [37]).

Fig. 10. Effect on the
differential cross sec-
tion for the ground-state
^{90}Zr(d,p)^{91}Zr reaction
(E_d = 15 MeV) of a change
in the magnitude of the
real part of the neutron
optical potential. The
full curve is calculated
using the parameters of
ref. [35]; the dashed
curve corresponds to an
increase of the magni-
tude of the real potential
by 1 MeV and the other
corresponds to a de-
crease by 1 MeV (from
ref. [37]).

good deal of the inner contributions in the integration over r_n. Bang
and Pearson [12] point out that the BHMM theory shows important
cancellation between inner and outer contributions, but the WBP model
is relieved of this interference and it therefore shows much less sensitivity to the neutron parameterization.

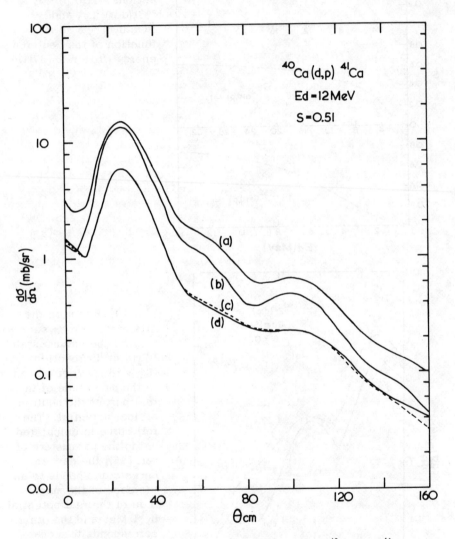

Fig. 11. Angular distributions for the ground-state ^{40}Ca(d,p)^{41}Ca reaction (E_d = 12 MeV). Curve (a) corresponds to SID, curve (b) allows
for the spin-orbit term for the bound state of the neutron, curve (c)
includes SDD for the neutron both for the bound state and for the continuum, and curve (d) allows fully for SDD (from ref. [37]).

Our calculations also showed the BHMM theory to be fairly independent of the proton spin-dependent distortion (PSDD). This is indicated for the ^{40}Ca(d,p) ground-state reaction in fig. 11 where several curves are shown. Curves (a) and (b) correspond to SID, but in the latter case the neutron bound-state function involves a spin-orbit interaction. With SDD acting on the neutron function in the continuum, we arrive at curve (c) and this is practically indistinguishable from curve (d), which is the result of a full SDD treatment. This insensitivity to PSDD does not seem to be suggested by the WBM model. Although tempted to study the systematics of polarization effects within the framework of the BHMM theory, we abandoned the project owing to the very unsatisfactory fits to differential cross sections.

The WBP model, on the other hand, has managed to fit quite well both angular distributions and polarization measurements. As an example [38] of the high quality of the fit that has been obtained, we see in fig. 12 a comparison of A(d,θ) with experiment for the ^{12}C(d,p)^{13}C ground-state reaction at energies varying from 7 MeV to 22 MeV. The dashed curves differ from the continuous curves in that the proton radius-parameter is 1.05 fm instead of 1.15 fm. The dot-dashed curve at 12 MeV indicates the effect of increasing the strength of the imaginary neutron potential from 8 MeV to 11 MeV. The experimental fits are indeed impressive, particularly for such a light nucleus as ^{12}C.

The transition amplitude as written in eq. (14) can be cast in a different form if we add and subtract unity from the proton scattering amplitude $S(\vec{k}_p, \vec{k}'_p)$, in which case we have

$$T_{fi}(\vec{k}_d, \vec{k}_p) = T_{fi}^{(u)}(\vec{k}_d, \vec{k}_d - \vec{k}_p) + \int d\cos\theta'_p \, f_{sc}(\theta_{pp'}) \tau(\vec{k}_d, \vec{k}_d - \vec{k}'_p). \quad (18)$$

Here, $T_{fi}^{(u)}$ is the unscattered contribution which corresponds to free motion for the stripped proton and it is symmetric about the recoil direction defined by $\vec{k}_n = \vec{k}_d - \vec{k}_p$. The quantity $f_{sc}(\theta_{pp'})$ is meant to represent the proton elastic scattering amplitude at an energy associated with k_p, while $\tau(\vec{k}_d, \vec{k}_d - \vec{k}'_p)$ is an unspecified function which shows a dependence on $\vec{k}'_n = \vec{k}_d - \vec{k}'_p$. Note however that $k_p = k'_p$. Both terms in the expression for T_{fi} relate to neutron motion in the continuum with kinetic energy $E_n = (\hbar k_n)^2 (2m)^{-1}$ and $(\hbar k'_n)^2 (2m)^{-1}$, respectively. The range of values of E_n is fairly great. Fig. 13 shows how it varies with θ'_p for the ^{40}Ca(d,p) ground-state reaction where $E_d = 12$ MeV. Most (d,p) ground-state reactions are characterized by values of k_p and k_d which are not too different. Thus, at 0°, we find E_n is just 390 keV while it is 76 MeV at 180°, and yet the deuteron energy is just 12 MeV. Beyond 50° the neutron energy exceeds that of the deuteron. The one-to-one correspondence of E_n with angle θ, however, is lost on account of the scattered contribution. Thus, for a particular angle θ there is an averaging over energy, the weighting

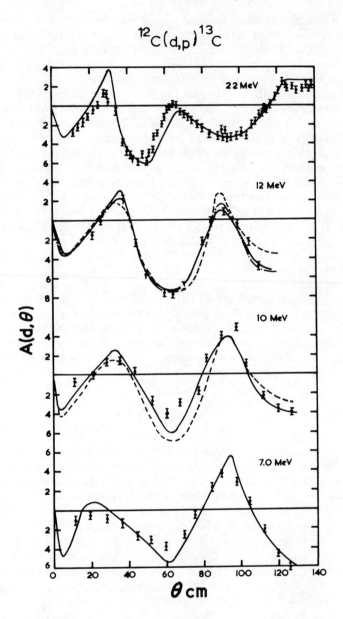

Fig. 12. Comparison of calculations of $A(d,\theta)$ for the ground-state $^{12}C(d,p)^{13}C$ reaction for various values of E_d. The dashed curves show the effect of lowering the radius parameter r_0 for the proton from the value 1.15 fm to 1.05 fm. The dot-dashed curve shows the effect of increasing the imaginary part of the neutron optical potential from 8 MeV to 11 MeV (from ref. [38]).

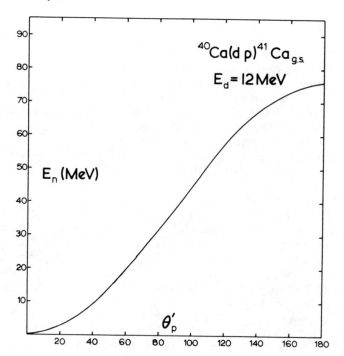

Fig. 13. The variation of the neutron energy in the continuum E_n as a function of the angle θ'_p for the ground-state ^{40}Ca(d,p) reaction where E_d = 12 MeV.

being determined by $f_{sc}(\theta_{pp'})$, the elastic scattering amplitude. Difficulties in the depiction of an optical model at low energies are avoided by this energy averaging, but this is not the case with the contributions from $T_{fi}^{(u)}$. This can be a problem as the unscattered contribution is found to account for the overall shape of the differential cross section. The detailed structure depends on the scattered contribution. Bang and Pearson find that $\tau(\vec{k}_d - \vec{k}'_p)$ is predominant near the first stripping peak, and this feature prevails even at large reaction angles θ. Thus stripping, according to this view, is characterized first by protons emerging at an angle near that of the stripping peak from which the protons scatter elastically. This would be through large angles if θ is large. If the stripping peak is at a very small angle $\theta^{(0)}$, we can identify $\theta_{pp'}$ with θ and confine the θ'_p integration to involve just a small-angle interval about $\theta^{(0)}$. The unscattered contribution has a form identical to the one associated with SID. Consequently, if $\ell = 0$, we are drawn to an interesting correlation of polarization effects in stripping reactions with those seen in elastic scattering processes. Here, $T_{fi}^{(u)}$ can be shown to contribute nothing, the second term involves $f_{sc}(\theta)$, and the correlation easily follows. Pearson, Bang, and Pocs [12] bear this out in their calculations for the $\ell = 0$ reactions $^{28}Si(d,p)^{29}Si$ and $^{88}Sr(d,p)^{89}Sr$, as is illustrated in figs. 14 and 15, respectively. In the first case the dashed curve represents the proton polarization associated with elastic scattering. In fig. 15 the dashed curve shows the effect of doubling the strength of the imaginary part of the neutron potential, while the dot-dashed curve shows the elastic scattering polarization. Even if $\ell \neq 0$, there should be a trace of this correlation.

Successful accounts were also given of the proton polarization for the $^{12}C(d,p)^{13}C$ ground-state reaction ($E_d = 10$ MeV and 15 MeV) in addition to the $^{40}Ca(d,p)^{41}Ca$ ground-state reaction at $E_d \approx 10$ MeV, the $^{24}Mg(d,p)^{25}Mg$ ground-state reaction ($E_d = 15$ MeV), and the $^{28}Si(d,p)^{29}Si$ $\ell = 2$ reaction at 15 MeV leading to the state of 1.86 MeV excitation. Further successes were recorded by Pearson, Willcott, and McIntyre [38] in calculations of $A(d,\theta)$ for reactions on ^{12}C (both the ground state and first excited state) and for ^{24}Mg, ^{40}Ca, and ^{52}Cr targets leading to various final states. The achievement of this fairly impressive feat required certain small-scale changes in the values of the distorting parameters. For example, a smaller radius parameter was chosen for the reaction with the ^{12}C target. There was also reasonable agreement with experimental values of $(d\sigma/d\Omega)_0$.

Further semi-qualitative results come out of this approach. In particular,

$$\left(\frac{d\sigma}{d\Omega}\right)_0^{j = \ell + \frac{1}{2}} - \left(\frac{d\sigma}{d\Omega}\right)_0^{j = \ell - \frac{1}{2}} = BP_{el}(\theta) \tag{19}$$

where $B > 0$ and $P_{el}(\theta)$ represents the proton elastic scattering polarization. Some of the qualitative notions first introduced by Newns [39] are applicable if we take account of the non-spherical nature of the capturing interaction, $\mathcal{V}(\vec{r}_n, \vec{k}'_p)$. This has the effect of highlighting one hemisphere presented to the neutron as against the other, with the final outcome that the proton polarization $P(p,\theta)$ is expected to be positive or negative near the first stripping rank according as j is $\ell + 1/2$ or $\ell - 1/2$, respectively. This is entirely in agreement with Newns' prediction, if proton absorption is taken into account. Although this feature is verified in many cases, it seems to be somewhat dubious and one finds exceptions—at least the DWBA results are not generally in accord with this result, as can be checked in fig. 2.

Just as with the DWBA, selection rules for polarization effects can be given with the WBP model [38]. These may be summarized as follows: If $\ell = 0$, then $A(d,\theta)$ arises mainly from PSDD, and if the contribution of neutron spin-dependent distortions (NSDD) is negligible, we have the correlation

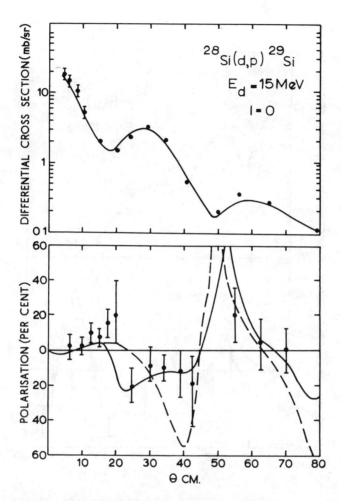

Fig. 14. Comparison with experiment of calculations
of angular distributions and P(p,θ) for the ground-state
^{28}Si(d,p)^{29}Si reaction (E$_d$ = 15 MeV). The neutron parameters are as in ref. [35] except for a doubling of the
strength of the imaginary potential in the case of the
continuous curve. The dashed curve represents the
proton elastic scattering polarization (from ref. [12],
Ann. Phys. 52 (1969) 33).

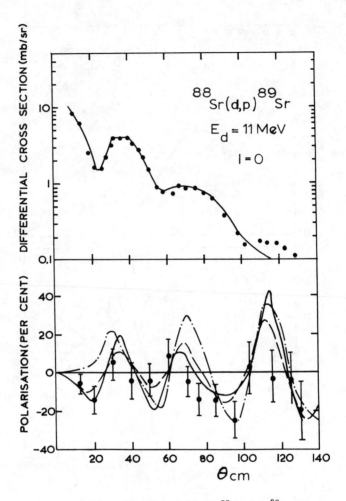

Fig. 15. As in fig. 14 but for the ^{88}Sr(d,p)^{89}Sr (Q = 3.11 MeV) reaction (E$_d$ = 11 MeV). The dashed curve now shows the effect of multiplying the strength of the imaginary potential while the dot-dashed curve represents the proton elastic scattering polarization (from ref. [12], Ann. Phys. 52 (1969) 33).

$$A(d,\theta) = \frac{2}{3} P(p,\theta),\tag{20}$$

which is precisely what we find from the DWBA, if there is no DSDD (cf. eq. (6)). On the other hand, if $\ell \neq 0$, then $A(d,\theta)$ is mainly associated with NSDD, and if we neglect PSDD, we find

$$A(d,\theta) = 2 P(p,\theta),\tag{21}$$

which corresponds to the SID prediction in the DWBA treatment even when the neutron bound-state function is subject to spin-orbit interactions. Further, if the neutron scattering states are not subject to SDD, we find that the correlation according to the j-value represented by eq. (5) holds. Indeed, the WBP model presents no new results.

Consideration has also been given to the consequences of the d-state of the deuteron [40]. Calculations show that proton polarization is more affected than the deuteron analyzing power, that the changes become more marked with increasing angle, as in the DWBA treatment [28], and that they are greater than has recently been found using the DWBA. The suggestion is also made that d-state effects are best considered at a later stage when the effects of the nuclear distortion are better understood.

A great deal of effort has been put into this new formulation by Pearson and collaborators [12], but perhaps a proper evaluation is somewhat premature at the present time. Much more remains to be done in assessing the importance of the various parameters that are needed for the calculation and in extending these calculations over a wider spectrum of energies and nuclei. Very little has been done with regard to the calculation of spectroscopic factors. Further, one hopes that some of the mathematical complication will prove to be unnecessary. This would allow for a freer acceptance and use of the theory. This is not to belittle the many successes reported using the conventional DWBA, which covers a wider class of transfer processes than just (d,p) reactions. Polarization phenomena present a difficult challenge, connecting theory and experiment, and the successes of the new formulation in accounting for polarization effects merit that more attention be given to it. This symposium can play an important role in the encouragement of further polarization studies and thus contribute in an important manner towards the ultimate attainment of a trustworthy theoretical understanding of the transfer mechanism.

REFERENCES

[1]   T. J. Yule and W. Haeberli, Nucl. Phys. A117 (1968) 1.
[2]   Second Polarization Symp.

[3] N. Austern, in Selected Topics in Nuclear Theory, ed. F. Janouch (International Atomic Energy Agency, Vienna, 1963); L. L. Lee, Jr., et al., Phys. Rev. 136 (1964) B971.

[4] See, for example, R. J. Philpott, W. T. Pinkston, and G. R. Satchler, Nucl. Phys. A119 (1968) 241.

[5] P. J. A. Buttle and L. J. B. Goldfarb, Proc. Phys. Soc. 83 (1964) 701; G. Bencze and J. Zimányi, Phys. Lett. 9 (1964) 246.

[6] R. C. Johnson, Nucl. Phys. A90 (1966) 289.

[7] R. C. Johnson and F. Santos, Phys. Rev. Lett. 19 (1967) 364.

[8] G. Delic and B. A. Robson, Third Polarization Symp.

[9] R. C. Johnson and P. J. R. Soper, Phys. Rev. C1 (1970) 976.

[10] G. H. Rawitscher, Phys. Rev. 163 (1967) 1223.

[11] S. T. Butler, R. G. L. Hewitt, B. H. J. McKellar, and R. M. May, Ann. Phys. (N.Y.) 43 (1967) 282; R. May and J. S. Truelove, ibid., 43 (1967) 322.

[12] C. A. Pearson and M. Coz, Nucl. Phys. 82 (1966) 533, 545; J. M. Bang, C. A. Pearson, and L. Pocs, ibid., A100 (1967) 1, 24; C. A. Pearson, J. M. Bang, and L. Pocs, Ann. Phys. (N.Y.) 52 (1969) 33.

[13] See, for example, K. R. Greider and L. R. Dodd, Phys. Rev. 146 (1966) 671; A. M. Mitra, Nucl. Phys. 32 (1962) 529; R. Aaron and P. E. Shanley, Phys. Rev. 142 (1966) 508; A. I. Baz', V. F. Demin, and I. I. Kuz'min, Sov. J. Nucl. Phys. 4 (1967) 815; A. J. Kromminga, K. L. Lim, and I. E. McCarthy, Phys. Rev. 157 (1967) 770.

[14] L. J. B. Goldfarb and R. C. Johnson, Nucl. Phys. 18 (1960) 353; see also, L. J. B. Goldfarb, Second Polarization Symp.

[15] G. R. Satchler, Nucl. Phys. 3 (1958) 67.

[16] R. Huby, M. Y. Refai, and G. R. Satchler, Nucl. Phys. 9 (1958) 4.

[17] L. J. B. Goldfarb and R. G. Seyler, Nucl. Phys. A149 (1970) 545.

[18] A. A. Debenham, J. A. R. Griffith, M. Irshad, and S. Roman, Nucl. Phys. A151 (1970) 81; J. A. R. Griffith and J. Roman, Phys. Rev. Lett. 24 (1970) 1496.

[19] L. J. B. Goldfarb and K. K. Wong, Nucl. Phys. A90 (1967) 361.

[20] R. G. Seyler and L. J. B. Goldfarb, Third Polarization Symp.

[21] H. Kamitsubo and B. Mayer, Third Polarization Symp.

[22] D. Kocher and W. Haeberli, private communication.

[23] A. Budzanowski et al., to be published.

[24] W. E. Maddox, C. T. Killey, Jr., and D. W. Miller, Phys. Rev. C1 (1970) 476; H. H. Cuno, G. Clausnitzer, and R. Fleischmann, Nucl. Phys. A139 (1969) 657; J. A. R. Griffith, M. Irshad, O. Karban, and S. Roman, Proc. Symp. on Nuclear Reaction Mechanisms, Quebec, 1969, to be published.

[25] M. B. Hooper, Nucl. Phys. 76 (1966) 449.

[26] P. Schwandt and W. Haeberli, Nucl. Phys. A110 (1968) 585 and A123 (1969) 401; H. Cords, G. U. Din, M. Ivanovitch, and B. A. Robson, ibid., A113 (1968) 606.

[27]   G. Delic and B. A. Robson, Nucl. Phys. A127 (1969) 234.
[28]   F. D. Santos, Third Polarization Symp.
[29]   L. L. Lee, Jr., and J. P. Schiffer, Phys. Rev. Lett. 12 (1964) 108.
[30]   J. L. Yntema and H. Ohnuma, Phys. Rev. Lett. 19 (1967) 1341;
       N. S. Chant and J. M. Nelson, Nucl. Phys. A117 (1968) 385.
[31]   F. Udo, Rev. Mod. Phys. 37 (1965) 365.
[32]   E. C. May, B. L. Cohen, and T. M. O'Keefe, Phys. Rev. 164
       (1967) 1253.
[33]   J. D. Harvey, Ph.D. Thesis, Surrey University, 1970 (unpublished).
[34]   F. G. Perey, Phys. Rev. 131 (1963) 745.
[35]   L. Rosen, J. G. Beery, A. S. Goldhaber, and E. H. Auerbach, Ann.
       Phys. (N.Y.) 34 (1965) 96.
[36]   F. D. Becchetti, Jr., and G. W. Greenlees, Phys. Rev. 182 (1969)
       1190.
[37]   T. F. Baker and L. J. B. Goldfarb, Nucl. Phys. A146 (1970) 577.
[38]   C. A. Pearson, J. C. Wilcott, and L. C. McIntyre, Nucl. Phys.
       A125 (1969) 111.
[39]   H. C. Newns, Proc. Phys. Soc. A66 (1953) 477.
[40]   C. A. Pearson, D. Rickel, and D. Zissermann, Nucl. Phys. A148
       (1970) 273.

DISCUSSION

Pearson:

Spectroscopic factors have recently been obtained using the
WBP model for several of the low-lying states in ^{41}Ca. These
spectroscopic factors overlap those obtained by Seth and
Satchler for the same states.

Rawitscher:

We are now including spins into our coupled equations, and we
are concentrating our efforts on the polarization tensors in the
deuteron channel and the stripping channel. We find that the
polarization functions do indeed provide very stringent tests
of the reaction theory.

Glashausser:

There has been a considerable controversy in the literature concerning the mathematical and physical validity of both the Butler
model and the Pearson model. Are there still problems with
either of the models?

Goldfarb:

Perhaps of an emotional kind. I think it is fair to say that the
proponents of these new models have answered those objections which are of a mathematical nature. The final test is
reasonable agreement with experimental data.

Pearson:

I know of no justification for any of the present stripping calculations in the sense that their formulae have been shown to be *valid* approximations to some exact result. They should all be regarded as models whose acceptance depends on their usefulness for reproducing data over a wide range of target nuclei, energy, and angular momentum transfer.

Tolsma:

When you look at the experimental results for the analyzing power A_d in (d,p) reactions, for example the results of Yule and Haeberli, then you see that for the forward angles there is no influence on A_d of spin-orbit distortions of the channels. Is this why you say that for the forward angles the reaction takes place on the surface of the target nucleus?

Goldfarb:

I cannot see why the effects on the stripping amplitude owing to spin-orbit distortion seem not to be different over the whole angular range. It is the part of the stripping amplitude that is calculated without consideration of spin-dependent distortion that is so large in the region of the stripping peak.

Schwandt:

In connection with the conventional form of the DW treatment of (d,p) stripping, you indicated that the use of elastic-scattering wave functions in the deuteron channel may not be terribly realistic. Would it make any sense, then, to depart from the common usage of elastic-scattering distorting potentials and adjust potential parameters to fit the stripping measurements without worrying too much about the validity of such potentials for the description of the elastic scattering?

Goldfarb:

No, certainly not, if you are referring to the proton channel. Also, this procedure is not what is suggested by Johnson and Soper for the deuteron channel.

D. W. Miller:

You indicated that it seems important to extend the measurements to higher energies. Do you have a particular energy range in mind which might be optimum for testing these models? Is there an upper energy limit useful for this purpose?

Goldfarb:

I was more interested in seeing that measurements are extended towards a wider spectrum of targets and, in particular, away from the light targets. I do not mean to say that we have to extend studies beyond approximately 20 or 30 MeV.

Experiments on Transfer Reactions

W. HAEBERLI, University of Wisconsin, Madison, USA

1. INTRODUCTION

In the discussion of stripping reactions at the Second Polarization
Symposium, Miller [1] pointed out the lack of a consistent set of experimental observations. The main body of information at that time
consisted of a number of measurements of the polarization of protons
from (d,p) reactions. For light target nuclei (A ≤ 16), the results showed
strong fluctuations with energy and thus were difficult to interpret.
For heavier target nuclei, where the separation between the final states
is smaller, the experiments were terribly difficult so that the measured effects were not much larger than the error bars. Some comparisons were made with calculations based on the distorted wave Born
approximation (DWBA). The fact that measurable polarization was
found for transitions with orbital angular momentum transfer $\ell = 0$ indicated the need to take spin-dependent forces into account, but there
were not enough experimental data to test reaction theories. Besides
the interest in reaction mechanisms, the study of polarization in stripping reactions has for a long time been motivated by the proposal of
Newns [2] that the sign of the polarization should depend on the j-value
of the captured particle, $j = \ell + 1/2$ or $j = \ell - 1/2$. The Pittsburgh
group [3] five years ago presented some evidence that such a sign rule
holds near the stripping peak, but the lack of accurate information prevented clear conclusions.

In the last three years a large number of accurate experiments on
polarization effects in transfer reactions have become available. Most
of these measurements have made use of polarized particle beams from
accelerators. The studies are primarily of two kinds. In the first type
of experiment, polarized deuterons are used to initiate a (d,p) reaction,
and the analyzing power P_d of the reaction is observed. The first
measurements of this kind were carried out as early as 1963 at Saclay
(see ref. [1]) where the $\ell = 1$ reaction $^{12}C(d,p)^{13}C$ (g.s.) and the $\ell = 0$
reaction $^{28}Si(d,p)^{29}Si$ (g.s.) were investigated at 22-MeV deuteron energy. The first systematic study of transitions with different values
of ℓ and j was reported in 1967 by the group at Wisconsin [4]. These

results showed a clear j-dependence for transitions with ℓ = 1, 2, and
3. By comparison with measurements of the proton polarization, which
requires a double-scattering technique, the experiments are very easy.
With polarized-beam intensities of a few nA and solid state detectors
of good resolution, angular distributions of the analyzing power can
be obtained simultaneously for a large number of transitions in a few
hours. More than a hundred transitions have been studied by the
groups at Birmingham, Erlangen, and Wisconsin. The second type of
polarized-beam experiment concerns the study of (p,d) reactions with
polarized protons. The information obtained is equivalent to a measurement of the proton polarization P_p in the time-reversed reaction
(d,p). Again, because polarized beams can be used, accurate results
over a wide range of angles can be obtained. Results of this type have
been reported recently, mostly by the group at Saclay. Since time-
reversal invariance provides an intimate connection between (d,p) and
(p,d) reactions the present summary presents both types of results together in sect. 2, separated according to the ℓ-value of the transferred
nucleon.

Another area where considerable progress has been made since the
last symposium is the measurement of the neutron polarization from
(d,n) reactions. These measurements were made possible by development of helium polarimeters (gas or liquid) of relatively high efficiency
and by the application of time-of-flight methods. Most results reported so far have the drawback that they were done at relatively low
bombarding energies and for light target nuclei where resonance effects may play an important role. The neutron polarization measurements are summarized in sect. 3. For other transfer reactions the results still are very sparse (sects. 4 and 5).

Table 1 gives a summary of results (refs. [4—95]) reported since the
last symposium; work reported at that symposium is not included. The
tabulation is limited to one- and two-nucleon transfer reactions for
$A \geq 6$ and bombarding energies greater than 1 MeV. The discussion
does not cover all reactions, but concentrates on studies of medium
and heavy nuclei and bombarding energies of several MeV where one
can expect that the stripping process is the dominant reaction mechanism.

A review of the theory of transfer reactions is presented by Goldfarb [96] in the preceding paper. A recent short summary of experimental results is contained in a review by Glashausser and Thirion
[97], where references to earlier reviews are also given.

2. NEUTRON TRANSFER REACTIONS (d,p) AND (p,d)

In the following, the quantity P_p refers to the proton polarization
in a (d,$\vec{p}$) reaction or to the analyzing power of the inverse reaction
($\vec{p}$,d) induced by polarized protons. These two measurements are

Table 1. Polarization measurements in transfer reactions

A	$E_x(B)$ MeV	E_{beam} MeV	Ref.	A	$E_x(B)$ MeV	E_{beam} MeV	Ref.
I. Reactions* $A(\vec{d},p)B$				^{52}Cr	5 states	8	[4]
				^{52}Cr	19 states	10	[19]
^{6}Li	g.s.; 0.48	2.1−10.9	[5]	^{52}Cr	g.s.; 0.57	10	[24]
^{6}Li	g.s.; 0.48	10, 12	[6]	^{52}Cr	5 states	11	[25]
^{9}Be	g.s.	7	[4]	^{52}Cr	g.s.	11	[26]
^{9}Be	g.s.; 3.37	9	[7]	^{52}Cr	5 states	12.3	[13]
^{9}Be	g.s.	10	[8]	^{53}Cr	18 states	10	[27]
^{9}Be	g.s.; 3.37	10, 12	[6]	^{53}Cr	0.83	10	[24]
^{9}Be	g.s.	12.3	[9]	^{54}Fe	4 states	8	[25]
^{9}Be			[10,11]	^{54}Fe	g.s.; 0.41	10	[24]
^{10}B	g.s.	10	[8]	^{54}Fe	17 states	10	[19]
^{10}B	g.s.	10, 12	[6]	^{57}Fe	15 states	10	[27]
^{10}B			[11]	^{88}Sr	g.s.; 1.04	12	[13]
^{11}B	g.s.; 0.95	10, 12	[6]	^{90}Zr	g.s.	6−9	[28]
^{12}C	g.s.; 3.09	7−10	[4]	^{90}Zr	g.s.; 1.21	9.9−10.1	[8]
^{12}C	g.s.; 3.09	7.7−10	[8]	^{90}Zr	4 states	11	[29]
^{12}C †	g.s.	12.3	[12]	^{117}Sn	g.s.; 1.22	12	[27]
^{12}C	g.s.; 3.09	12.3	[9,13]	^{119}Sn	g.s.; 1.17	12	[27]
^{12}C	g.s.	12.4	[14,15]	^{207}Pb	g.s.	12.3	[13,30]
^{14}N	g.s.	10, 12	[6]	^{208}Pb	4 states	12.3	[13,30]
^{16}O	g.s.; 0.87	9−10.3	[8]				
^{16}O	6 states	9.3−13.3	[17]				
^{16}O	g.s.; 0.87	8; 12.3	[16,18]	II. Reactions $A(d,\vec{p})B$			
^{28}Si	7 states	9	[16]	^{9}Be	g.s.; 3.37	1−6	[31]
^{28}Si	3 states	9−10.3	[8]	^{10}B	g.s.	1.2−1.9	[32]
^{28}Si	g.s.; 1.27	12.3	[13]	^{10}B	g.s.	13	[15]
^{24}Mg	5 states	8	[4]	^{12}C	3.09	2.8−3.2	[33]
^{24}Mg	6 states	12.3	[18]	^{12}C	g.s.	3.1; 3.3	[34]
^{40}Ca	g.s.	5−10.1	[8]	^{12}C	g.s.; 3.09	12.4	[35]
^{40}Ca	4 states	7	[4]	^{12}C	g.s.		[36]
^{40}Ca	14 states	11	[19]	^{14}N	g.s.	13	[15]
^{40}Ca	1.95−3.95	11	[20]	^{28}Si	g.s.	5	[37]
^{40}Ca	4 states	5; 9; 11	[21]	^{28}Si	4 states	10.8	[38]
^{40}Ca	g.s.	11.2	[22]	^{40}Ca	1.95	4.8	[39]
^{48}Ca	g.s.; 2.03	5.5−11	[23]	^{40}Ca	g.s.	5	[40]
^{46}Ti	16 states	10	[19]	^{40}Ca	g.s.	5.6; 6.5	[41]
^{48}Ti	20 states	10	[19]	^{40}Ca	3.95	11	[42]
^{50}Ti	15 states	10	[19]	^{40}Ca	3 states	10.8	[43]

*Vector analyzing power except where noted

†Tensor analyzing power

Table 1 *(continued)*

A	$E_x(B)$ MeV	E_{beam} MeV	Ref.
^{40}Ca		≈ 10	[22]
^{40}Ca	g.s.		[36]
^{88}Sr	1.05	11	[44]
^{90}Zr	g.s.; 1.2	11	[45]

III. Reactions A($\vec{p}$,d)B

A	$E_x(B)$ MeV	E_{beam} MeV	Ref.
^{9}Be	g.s.	5–12	[7]
^{12}C	4 states	30.3	[46]
^{16}O	g.s.; 0.16	30.3	[46]
^{28}Si	g.s.	30.5	[47]
^{40}Ca	g.s.	30.5	[47]
^{48}Ca	g.s.	22.9	[48]
^{49}Ti	g.s.; 0.98	20.9	[48]
^{53}Cr	g.s.	16.6	[26]
^{57}Fe	g.s.; 0.85	17.3	[49,50,51]
^{61}Ni	3 states	16.6	[49,50,51]
^{90}Zr	3 states	22.9	[48]
^{91}Zr	g.s.	24.5	[49,50]
^{92}Mo	3 states	24.5	[48]
^{118}Sn	6 states	24.5	[52]
^{119}Sn	g.s.; 1.23	24.5	[52]

IV. Reactions A(p,$\vec{d}$)B

A	$E_x(B)$ MeV	E_{beam} MeV	Ref.
^{9}Be‡	g.s.	1.6–3.8	[53]
^{9}Be‡	g.s.	2.5; 3.7	[54]
^{9}Be‡	g.s.	3; 4; 5	[55]
^{9}Be‡	g.s.	4.9–9.8	[56]
^{9}Be, ^{12}C, ^{28}Si**		185	[57]

V. Reactions A($\vec{d}$,n)B

A	$E_x(B)$ MeV	E_{beam} MeV	Ref.
^{11}B	g.s.; 4.43	10; 11.8	[58]
^{14}N	g.s.	10; 11.8	[58]
^{89}Y	g.s.	11	[58]

VI. Reactions A(d,$\vec{n}$)B

A	$E_x(B)$ MeV	E_{beam} MeV	Ref.
^{6}Li	(g.s. + 0.43)	2.5–3.8	[59]
^{7}Li	g.s.; 2.9	2.5–3.8	[59]
^{9}Be	5 states	0.9–2.5	[60]
^{9}Be	5 states	3; 3.5	[61]
^{10}B	g.s.	1.2–2.9	[61]
^{10}B	3 states	2.5–4	[63]
^{11}B	g.s.; 4.43	1.5	[64]
^{11}B	5 states	2	[65]
^{11}B	g.s.; 4.43	2.8; 4	[66]
^{11}B	g.s.; 4.43	5.5	[67]
^{11}B	g.s.; 4.43	7.6–11.7	[68]
^{12}C	g.s.	1.7–2.8	[69]
^{12}C	g.s.	2.8–4.2	[70]
^{12}C	g.s.; 2.37	3.9–5	[71]
^{12}C	g.s.; 2.37	4–6.2	[72]
^{12}C	g.s.	4–7.5	[73]
^{12}C	g.s.	51.5	[74]
^{13}C	3 states	2.5–4	[63]
^{14}N	g.s.	1.65–2.9	[75]
^{14}N	g.s.	3.1–3.7	[76]
^{14}N	g.s.	3.7	[77]
^{14}N	g.s.	4.2–6.0	[67]
^{14}N	g.s.	5.35	[78]
^{15}N	g.s.	1.6–3.0	[79]
^{15}N		3.1–3.8	[80]
^{15}N	g.s.	4.41; 5.5	[81]
^{15}N	g.s.	2.5–3.8	[59]
^{16}O	g.s.	5.35	[78]
^{28}Si	3 states	5; 3.8	[82]
^{40}Ca	3 states	6	[83]
^{56}Fe, ^{59}Co, ^{60,62,64}Ni		7.7–11.5	[84]

‡Tensor polarization
**Vector polarization

Table 1 *(continued)*

A	$E_x(B)$ MeV	E_{beam} MeV	Ref.	A	$E_x(B)$ MeV	E_{beam} MeV	Ref.
VII. Reactions A($\vec{d}$,t)B				^{16}O	7.03	43.8	[88]
				^{16}O	g.s.; 2.31	49.5	[87]
^9Be	g.s.	9	[7]				
^{13}C	g.s.; 4.43	12	[85]	X. Reactions A(^{3}He,$\vec{n}$)B			
^{208}Pb		12.3	[86]				
				^{12}C	g.s.	2.2−3.7	[89]
VIII. Reactions A($\vec{p}$,t)B				^{13}C	g.s.	2.9−3.9	[90]
				^{13}C	g.s.	4.2−5.7	[91]
^{12}C	g.s.; 3.36	49.5	[87]	^{24}Mg	g.s.	5; 5.8	[92]
^{15}N	3.51; 7.38	43.8	[88]				
^{16}O	4 states	43.8	[88]	XI. Reactions A(^{3}He,$\vec{p}$)B			
^{16}O	g.s.	49.5	[87]				
^{28}Si	g.s.	49.5	[87]	^{11}B	4.43	1.8−2.8	[93]
				^{12}C	g.s.	2.95	[94]
IX. Reactions A($\vec{p}$,^{3}He)B				^{12}C	g.s.; 2.31	2.5−5.5	[95]
^{15}N	3.68; 7.55	43.8	[88]				

equivalent except that in practice the inverse reaction cannot be observed if the final nucleus is in an excited state, since this would require a target of nuclei in the excited state. The quantity P_d refers to the analyzing power of a $(\vec{d},p)$ reaction or to the deuteron polarization in the inverse $(p,\vec{d})$ reaction. The measurements of P_p and P_d are not equivalent, except in special cases (see ref. [96]), although they are related in a complicated way through the spin-dependence of the reaction matrix elements. A difficulty arises in the description of deuteron polarization and analyzing power, because spherical and Cartesian tensors are used [98]. To avoid changes in the original drawings from the literature, iT_{11} will be used besides P_d to denote the deuteron vector analyzing power. The relationship is $P_d = iT_{11}\sqrt{3}/2$.

2.1. *Transitions with $\ell = 0$*

Transitions with $\ell = 0$ are thought to be particularly simple, since the polarization effects arise from spin-dependent distortions only. Measurements have been made of the proton polarization [38] and the deuteron vector analyzing power [13, 16] for the strong $\ell = 0$ transition in $^{28}Si(d,p)^{29}Si$ (g.s.). These results, however, are not particularly suited as a test of reaction theories, because recent measurements at Erlangen [8] for six different energies between 9 and 10.3 MeV show considerable fluctuation of the analyzing power with energy.

Some measurements on heavier nuclei are now available. Fig. 1

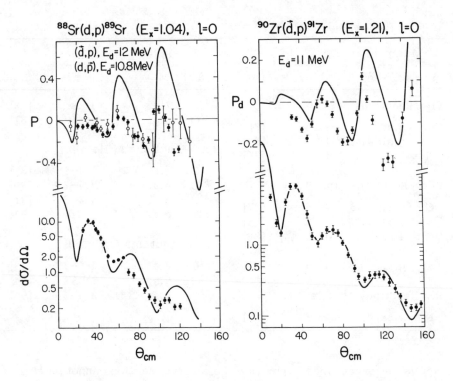

Fig. 1. Cross section and deuteron vector analyzing power (solid dots) for $\ell = 0$ transitions. For ^{88}Sr(d,p)^{89}Sr the proton polarization (open circles) is also shown. The curves are the results of DWBA calculations. The results for ^{88}Sr(d,p)^{89}Sr are from ref. [13], those for ^{90}Zr(d,p)^{91}Zr from ref. [29].

shows the analyzing power for vector-polarized deuterons in the reactions ^{88}Sr(d,p)^{89}Sr (E_x = 1.04 MeV) at 12 MeV measured at Birmingham [13] and similar results from Wisconsin [29] for ^{90}Zr(d,p)^{91}Zr (E_x = 1.21 MeV) at 11 MeV. In both cases the neutron capture is to a good single-particle state. The parameters used for the DWBA calculations on ^{88}Sr(d,p) have not been published. For ^{90}Zr(d,p), the parameters were obtained from analyses of cross section and polarization for deuteron elastic scattering at 11 MeV and for proton elastic scattering at the energy appropriate to the exit channel (14.8 MeV). In view of the fact that no parameters were adjusted to fit the reaction data, the agreement with the measurements is reasonable. For both reactions, the polarization of the protons has also been measured [44, 45]. The polarization P_p (open circles, fig. 1) shows a great deal of similarity to the deuteron analyzing power P_d.

As mentioned earlier, one expects much better measurements of P_p if one measures the asymmetry in the inverse reaction with a polarized beam. In a contribution to this symposium [52] the group at Saclay reports two observations of $1/2^+$ transitions in $(\vec{p},d)$ reactions, namely, the g.s. transitions on $^{118}Sn(\vec{p},d)^{117}Sn$ and $^{119}Sn(\vec{p},d)^{118}Sn$ at $E_p = 24.5$ MeV. The results show a pronounced oscillatory pattern which agrees very well with DWBA calculations. Since the final nucleus is stable, it is feasible to study also the analyzing power in the inverse reaction $^{117}Sn(\vec{d},p)^{118}Sn$. This has been done recently by Kocher [27] at Wisconsin (see fig. 5 of the preceding paper). The bombarding energy was 12 MeV, while 17.5 MeV would have been necessary to match the c.m. energy of the Saclay experiment. This is unfortunate because the measurements would have been of sufficient accuracy to extract separately the polarization effects caused by the spin-dependence in the proton and the deuteron channel.

2.2 *Transitions with $\ell = 1$*

For transitions with $\ell > 0$, one of the possible applications of polarization experiments is the determination of j-values. It is assumed that we would normally determine ℓ from the cross section and only wish to distinguish between the two possible j-values, $j = \ell + 1/2$ and $j = \ell - 1/2$. The first unambiguous j-dependent polarization effect was discovered [4] in measurements of the vector analyzing power for the strong $\ell = 1$ transitions in $^{40}Ca(d,p)^{41}Ca$ to the states at $E_x = 1.95$ MeV ($j = 3/2^-$) and 3.95 MeV ($j = 1/2^-$). As fig. 2 shows, the vector analyzing power for $j = 3/2^-$ is of opposite sign and half as large in magnitude as that for $j = 1/2^-$. The relationship $P_d (j = 1/2^-) = -2 P_d (j = 3/2^-)$ (solid and dashed curves in fig. 2) is expected if spin-dependent distortions contribute little to the analyzing power.

Some examples of more recent results which are of higher accuracy and extend over a much wider range of angles are shown in fig. 3, which illustrates the j-dependence for several strong $\ell = 1$ transitions in the 2p-shell (^{50}Ti, ^{52}Cr, ^{54}Fe) for 10-MeV deuterons [19]. Transitions with $j = 1/2^-$ are characterized by a maximum in iT_{11} at 35°, $j = 3/2^-$ by a minimum. The results on more than 40 transitions in the 2p-shell [19] show that the analyzing power between 15° and 60° is sufficiently insensitive to target mass, bombarding energy, Q-value, and spectroscopic factor that no theory is required to make j-assignments. Transitions have been studied with Q-values between 1 and 7 MeV and spectroscopic factors between 0.02 (1.59-MeV state in ^{49}Ti) and 0.96 (g.s. ^{51}Ti). Measurements for $^{40}Ca(d,p)$ at 5, 7, 9, and 11 MeV [4, 21] show the same kind of j-dependence at forward angles for all bombarding energies, although for reaction angles larger than 90° the behavior differs from one energy to the next.

The j-assignments based on the measured vector analyzing power agree in general with those obtained from other experiments. Some exceptions are discussed in refs. [4, 20]. For the state in ^{41}Ca at

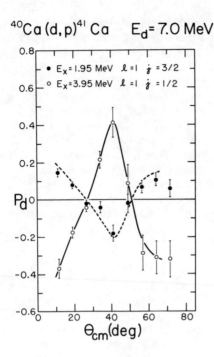

^{40}Ca (d,p)^{41} Ca E_d= 7.0 MeV

- E_x= 1.95 MeV ℓ =1 j = 3/2
- E_x=3.95 MeV ℓ =1 j = 1/2

Fig. 2. First observation of the strong j-dependence in a ($\vec{d}$,p) reaction [4]. The solid and dashed curves are related by P_d (j = 1/2$^-$) = $-2P_d$(j = 3/2$^-$).

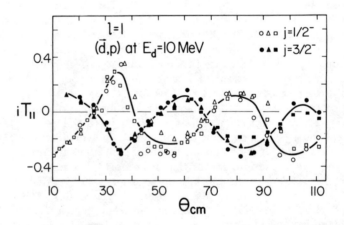

l=1
($\vec{d}$,p) at E_d=10 MeV

o △ □ j=1/2$^-$
● ▲ ■ j=3/2$^-$

Fig. 3. Illustration of the j-dependence of the vector analyzing power in ($\vec{d}$,p) reactions with ℓ = 1. The measurements are for the target nuclei ^{50}Ti, ^{52}Cr, ^{54}Fe [19].

3.62 MeV where (d,pγ) measurements reported j = 3/2⁻, the j = 1/2⁻
assignment based on the vector analyzing power is now supported by
a measurement [99] of the circular polarization of γ-rays following
the capture of polarized thermal neutrons. Measurements of the vector analyzing power thus appear to provide a reliable method of determining j-values. With presently available beam currents (10–100 nA)
the polarized-beam experiments are actually less time consuming than
other methods (e.g., the Lee-Schiffer j-dependence in the cross section at back angles), because the j-dependence in the analyzing power
is very large and can be observed near the stripping peak where the
cross section is about a hundred times larger than at back angles.

   In fig. 4, the cross section and vector analyzing power for three
ℓ = 1 transitions in ^{52}Cr(d,p)^{53}Cr is compared to DWBA calculations [19].

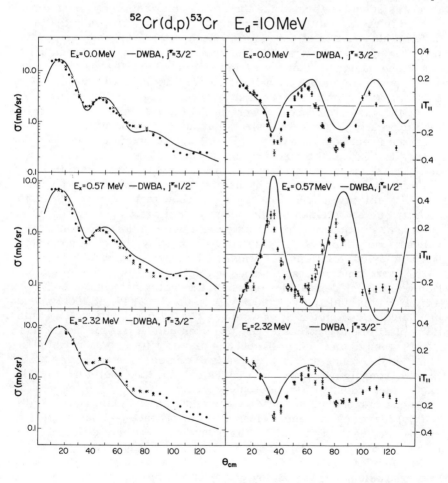

Fig. 4. Comparison of DWBA calculations with the observed cross section and analyzing power for three ℓ = 1 transitions in ^{52}Cr(d,p)^{53}Cr.

In view of the fact that no parameters were adjusted (they were determined from analyses of elastic scattering measurements) the agreement with the measurements is reasonable. However, the deviations between observed and calculated analyzing power are significant. Quite universally, for instance, the first maximum in the calculated $j = 1/2^-$ analyzing power is too large for nuclei in this mass region. Yule [4] has shown that the calculations are surprisingly insensitive to variations in the optical model parameters. In particular, the results are not altered appreciably if the spin-orbit coupling in the deuteron and the proton potential are entirely neglected. For lighter target nuclei the sensitivity to the potential parameters is greater.

The next topic of discussion concerns the polarization of protons from $(d,\vec{p})$ reactions. For many years the question whether DWBA theory is adequate to describe the proton polarization has been discussed in the literature. Pearson et al. [100] have proposed that a different description of the reaction mechanism (WBP) is more promising. The apparent failure of DWBA seemed particularly surprising after it had been found [4] that the same theory explains the deuteron analyzing power adequately. The problem was further investigated recently in a cooperative experiment between Saclay and Wisconsin [26] where the vector analyzing powers in $^{52}Cr(\vec{d},p)^{53}Cr$ and in the inverse reaction $^{53}Cr(\vec{p},d)^{52}Cr$ were measured at the same c.m. energy, the second experiment being simply an elegant way to obtain the proton polarization in $^{52}Cr(d,\vec{p})^{53}Cr$. Scattering of polarized deuterons from ^{52}Cr and of polarized protons from ^{53}Cr was observed at the same time, and the results were used to obtain the optical model parameters for the DWBA calculations. The results (fig. 5) show that a reasonable fit to all data can be obtained simultaneously. A deuteron potential with surface absorption gave better agreement with the measurements than volume absorption, but this may not be conclusive because no systematic study of variations of optical model parameters was made.

The work mentioned so far does not yet answer the other old question whether a j-dependence exists in the proton polarization. The answer was provided a year ago by Escudie et al. [49] at Saclay, who used a polarized proton beam of about 17 MeV to study two $\ell = 1$ transitions of different j-values. Only one such pair of measurements was made ($^{57}Fe(\vec{p},d)$ and $^{61}Ni(\vec{p},d)$), but the results (fig. 6) demonstrated a clear j-dependence over a wide range of angles. The analyzing power is in reasonable agreement with DWBA calculations (solid lines). The angular dependence of the proton analyzing power $P_p(\theta)$ in these (p,d) reactions is remarkably similar to the deuteron analyzing power $P_d(\theta)$ in $(\vec{d},p)$ reactions on nearby elements (fig. 7), but P_p is considerably smaller in magnitude than P_d. The behavior is not very different from that expected if spin-dependent forces are negligible (i.e., $P_d = 2P_p$). The same relationship can also be seen in fig. 5. In a contribution to this symposium [48] the Saclay group reports further measurements for $\ell = 1$ transitions for somewhat heavier targets (^{90}Zr, ^{92}Mo) at about

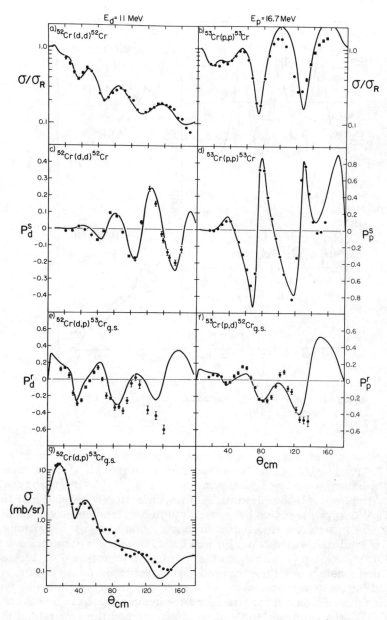

Fig. 5. Cross section and analyzing power for $^{52}Cr(\vec{d},p)^{53}Cr$ and for the inverse reaction $^{53}Cr(\vec{p},d)^{52}Cr$. The curves are DWBA calculations. The optical model parameters were obtained from an analysis of the elastic scattering measurements shown in the upper part of the figure [26].

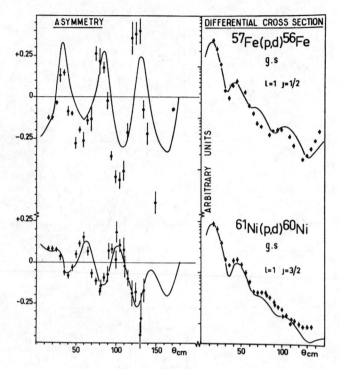

Fig. 6. Illustration of the j-dependence of the proton analyzing power in ($\vec{p}$,d) reactions with ℓ = 1. The figure is from ref. [49].

23-MeV proton energy. In these cases a pronounced minimum in P_d near 35° for j = 3/2⁻ is no longer present (P_d remains positive forward of 60°), but a pronounced maximum for j = 1/2⁻ at 30° is still found.

For light nuclei, the $^9Be(d,p)^{10}Be$ (g.s.) reaction (j = 3/2⁻) and the $^{12}C(d,p)^{13}C$ (g.s.) reaction have been studied extensively (see table 1). Many of the features are similar to those found in ℓ = 1 transitions on heavier nuclei. Baxter et al. [9] and Budzanowski [35] pointed out the similarity between P_d and P_p in the $^{12}C(d,p)^{13}C$ reaction near 12 MeV. The Birmingham group [9] compared the deuteron vector analyzing powers for the two reactions at 12.4 MeV and observed the same type of j-dependence near 35° as is found for heavier nuclei. However, for $^{12}C(d,p)^{13}C$ the maximum in P_d at 35° is lower and the excursions to negative values are much more pronounced (fig. 8) than for other j = 1/2⁻ transitions. The angular distributions measured at Erlangen [8] change quite slowly with energy between 7.7 and 10 MeV, but DWBA calculations at 10 and 12.4 MeV [4, 13, 35] reproduce the measurements not even qualitatively. A comparison with WBP calculations

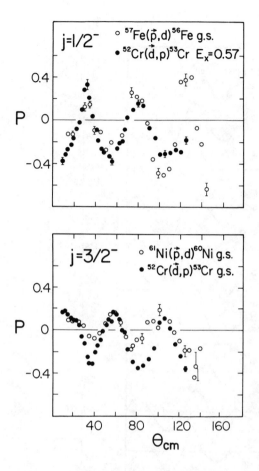

Fig. 7. Comparison of P_p and P_d for ℓ = 1 transitions of j = 1/2⁻ and j = 3/2⁻ in the 2p-shell. The measurements of P_p [50] are for a proton energy of about 17 MeV. The measurements of P_d [19] are for a correspondingly lower deuteron energy of 10 MeV.

has been discussed by Goldfarb in the preceding paper. For ⁹Be(d,p), Loyd [7] has reported qualitative agreement of DWBA calculations with his measurements at 9 MeV for angles forward of 90°.

It is usually pointed out that DWBA may be unsuccessful for light nuclei because of resonance effects in the experiments. However, the problem is in part that the predicted A-dependence is much stronger

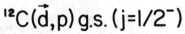

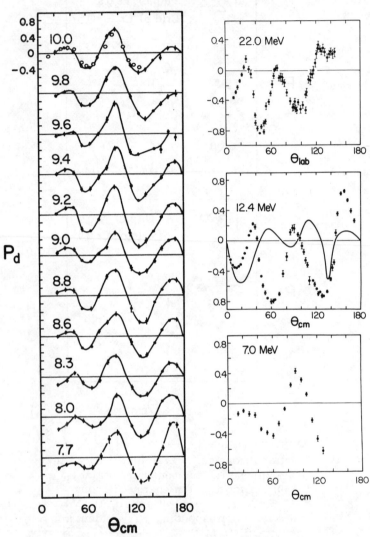

Fig. 8. Deuteron vector analyzing power P_d for $^{12}C(d,p)^{13}C$
(g.s.). The measurements shown on the left are from ref.
[8], except for the open circles, which are from ref. [4]. The
results shown on the right are from refs. [7, 13, 101]. The
curve for 12.4 MeV is from a DWBA calculation [13].

than the observed one. In fact, the observed positions of maxima and minima in the analyzing power are remarkably insensitive to A (compare 12.4-MeV data in fig. 8 with the j = 1/2 transition in fig. 4).

2.3 *Transitions with* ℓ_n = 2

The j-dependence for ℓ = 2 transitions was first observed [4] by comparing the deuteron vector analyzing power for a $3/2^+$ and $5/2^+$ final state in $^{24}Mg(d,p)^{25}Mg$. It was concluded that P_d was of opposite sign ($P_d \simeq \pm 0.2$ for j = $\ell \pm 1/2$) near the stripping peak ($\sim 30°$). Recent studies [13, 16] of ℓ = 2 transitions in $^{16}O(d,p)^{17}O$ and $^{28}Si(d,p)^{29}Si$ show the same characteristic behavior at forward angles, but at larger angles the analyzing power for transitions of the same j shows no close similarity. Measurements at Erlangen [8] on the same reactions between 9 and 10.3 MeV indicate considerable fluctuation with energy. The only heavier target nucleus for which a j = $3/2^+$ and a j = $5/2^+$ transition can be compared is ^{90}Zr where very recent measurements are available from the Wisconsin group [102] at a deuteron energy of 12 MeV (fig. 9). The simple sign rule mentioned above no longer applies. The oscillations for the two j-values are opposite in phase at the forward angles and move progressively closer at large angles. No analysis of these results has yet been performed, but for j = $5/2^+$ transitions DWBA calculations are available from Birmingham for $^{88}Sr(d,p)^{89}Sr$ (g.s.) and from Wisconsin for $^{90}Zr(d,p)^{91}Zr$ (g.s.) (fig. 10).

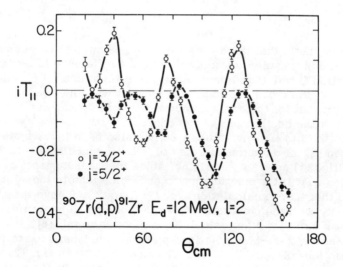

Fig. 9. Illustration of the j-dependence of the vector analyzing power for ℓ = 2 transitions in $^{90}Zr(d,p)^{91}Zr$ (g.s. and E_x = 2.06 MeV). The measurements are from ref. [102].

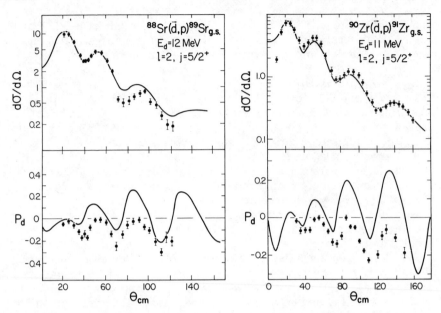

Fig. 10. Comparison of measured cross sections and vector analyzing powers for $\ell = 2$, $j = 5/2^+$ transitions with DWBA calculations. The results for ^{88}Sr(d,p)^{89}Sr are from ref. [13], those for ^{90}Zr(d,p)^{91}Zr from ref. [29].

It is very likely that the difference in the calculated curves for these two reactions results not so much from the difference in target mass or reaction Q-value as from the use of different optical model parameters. For the cross section the agreement with the measurements is excellent, but for the analyzing power only the phase but not the amplitude of the oscillations is reproduced.

The ^{90}Zr(d,p)^{91}Zr (g.s.) transition has also been investigated at Erlangen [8] for deuteron energies of 9.9, 10, and 10.1 MeV. The results (fig. 11) are very surprising, since the measurements at these energies bear little resemblance to each other and to those of fig. 10. Nor do their DWBA calculations for $\ell = 2$ and $\ell = 0$ have the slightest similarity with the calculations by other groups (figs. 1, 10). Similar problems exist for other calculations reported in ref. [8].

At Saclay, 24.5-MeV polarized protons have been used to induce the reaction ^{91}Zr($\vec{p}$,d)^{90}Zr (g.s.). The measured $P_p(\theta)$ (fig. 12) shows similarity with $P_d(\theta)$ of the inverse reaction (figs. 9, 10) except that the maxima and minima are shifted toward smaller angles (e.g., third minimum at 100° and 110° for P_p and P_d, respectively). This shift presumably is a consequence of the higher bombarding energy in the measurements of the proton analyzing power ($E_p = 24.5$ MeV corresponds to $E_d = 19.5$ MeV in the inverse reaction).

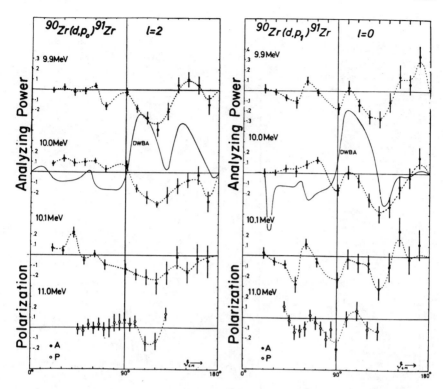

Fig. 11. Vector analyzing power (solid dots) and proton polarization (open circles) for the $\ell = 2$, $j = 5/2^+$ g.s. transition and the $\ell = 0$ ($E_x = 1.21$ MeV) transition in $^{90}Zr(d,p)^{91}Zr$. The figure is from ref. [8].

In a contribution to this symposium [52] the Saclay group compared for the first time proton analyzing powers in $(\vec{p},d)$ reactions for $j = 3/2^+$ and $j = 5/2^+$. The results show a surprisingly weak j-dependence, except in a small angular range near the main stripping peak.

2.4. *Transitions with $\ell \geq 3$*

The strong $\ell = 3$, $j = 5/2^-$ transitions in $^{54}Fe(\vec{d},p)^{55}Fe$ [19, 25] and $^{52}Cr(\vec{d},p)^{53}Cr$ [19] can be compared to the strong $\ell = 3$, $j = 7/2^-$ g.s. transition in $^{40}Ca(\vec{d},p)^{41}Ca$ [21]. The results are shown in figs. 13 and 14. The j-dependence of the deuteron vector analyzing power is evident, particularly near the stripping peak where iT_{11} has a broad, pronounced minimum ($iT_{11} \approx -0.2$) for $j = 5/2^-$ and a maximum for $j = 7/2^-$. The results for $^{40}Ca(d,p)^{41}Ca$ suggest that at least at forward angles ($\theta < 80°$) the analyzing power is fairly insensitive to bombarding energy (however, see ref. [8]). Weaker $\ell = 3$ transitions to many higher excited states in ^{53}Cr and ^{55}Fe and to several states in ^{47}Ti, ^{48}Ti, and

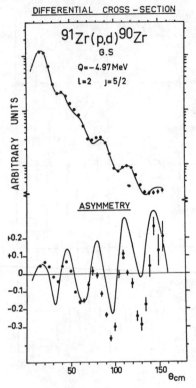

DIFFERENTIAL CROSS – SECTION

^{91}Zr(p,d)^{90}Zr
G.S
Q = – 4.97 MeV
l = 2   j = 5/2

ARBITRARY UNITS

ASYMMETRY

Fig. 12 *(left)*. Vector analyzing power for a $(\vec{p},d)$ reaction with $\ell = 2$. The figure is from ref. [49].

Fig. 13 *(below)*. Cross section and vector analyzing power for $\ell = 3$, $j = 5/2^-$ transitions.

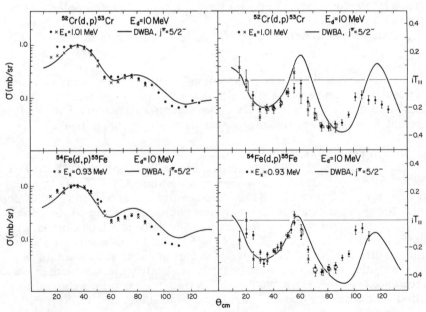

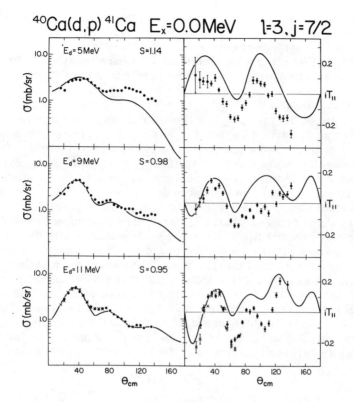

^{40}Ca(d,p)^{41}Ca   E$_x$=0.0MeV   l=3, j=7/2

Fig. 14. Cross section and vector analyzing power
for the ℓ = 3, j = 7/2$^-$ transition.

^{51}Ti [19] show the same characteristics as the data in figs. 13 and 14.
Measurements of the analyzing power for $(\vec{p},d)$ reactions on ^{48}Ca and
^{49}Ti with orbital angular momentum transfer ℓ = 3, j = 7/2$^-$ are reported
in ref. [48]. The angular dependence of P$_d$ at forward angles is similar
to that of P$_d$ for a 7/2$^-$ transition.

A contributed paper by Kocher et al. [103] discusses the analyzing
power in $(\vec{d},p)$ reactions with orbital angular momentum transfer ℓ = 4.
For the five transitions studied, the analyzing power is small ($|iT_{11}| <$
0.15) and the angular distributions show no pronounced features. For
^{90}Zr$(\vec{d},p)^{91}$Zr (E$_x$ = 2.21, j = 7/2$^+$) the agreement with DWBA calculations
is very poor (see fig. 1 of ref. [103]). Transitions with ℓ = 4 have also
been studied at Birmingham for ^{208}Pb$(\vec{d},p)^{209}$Pb [30]. The proton analyzing
power for transitions with ℓ = 4, j = 7/2$^+$ and ℓ = 5, j = 11/2$^-$ have been
measured in ^{118}Sn$(\vec{p},d)$. The agreement with DWBA calculations is quite
unsatisfactory, particularly for the ℓ = 4 transition.

Inspection of all DWBA calculations of deuteron and proton analyzing
powers in $(\vec{d},p)$ and $(\vec{p},d)$ reactions gives one the impression that the

agreement with measurement is worse for even than for odd ℓ-values, and for a given ℓ-value is worse for $j = \ell - 1/2$ than for $j = \ell + 1/2$. As a whole, the agreement gets worse with increasing ℓ. Of course, results from different groups should not be compared directly, since often the philosophy used in adjusting parameters is different.

2.5 *Mixed j-values*

If the spin of the target I_i and the spin of the final state I_f are different from zero, several values of ℓ and j of the captured neutron are consistent with conservation of angular momentum and parity. The range of possible values of j is $|I_i - I_f| \le j \le (I_i + I_f)$. The cross sections from the different ℓ, j add incoherently [5, 24]. In the study of nuclear structure one is interested to determine separately the transition amplitudes for different values of ℓ, j. This is possible because the shape of the cross-section angular distribution depends primarily on the relative contribution from different ℓ-values, while the analyzing power (or polarization) is sensitive to the j-values present. The method was first applied by the Erlangen group [5] to study the reaction ^{6}Li(d,p)^{7}Li. The cross section indicates that the reaction to the g.s. and first excited state of ^{7}Li proceeds by capture of an $\ell = 1$ neutron. The problem then is to determine the relative transition amplitude for $j = 3/2^-$ and $j = 1/2^-$. Measurements of the vector analyzing power were made at 6.7 and 7.0 MeV (fig. 15). For the transition to the first excited state $P_d(\theta)$ is similar to that for pure $3/2^-$ transitions observed

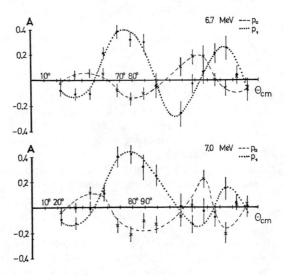

Fig. 15. Analyzing power for the reaction ^{6}Li($\vec{\mathrm{d}}$,p)^{7}Li (g.s. and first excited state) for deuteron energies 6.7 and 7.0 MeV. The figure is from ref. [5].

on spin-zero nuclei (i.e., minimum at $\theta \sim 35°$). For the g.s. transition the analyzing power is intermediate between that for $1/2^-$ and $3/2^-$ transitions, indicating that both j-values contribute noticeably to the configuration of the ground state. Quantitative conclusions could not be reached, since detailed studies of the analyzing power as a function of energy between 2.9 and 10.9 MeV showed considerable fluctuations in the analyzing power from one energy to another (compound nucleus formation). The energies for the measurements of fig. 15 were carefully chosen to be away from prominent resonances.

The same group [6] has recently studied many other $\ell = 1$ transitions on nuclei in the 1p-shell. However, observations were made only at two bombarding energies (10 and 12 MeV). In some cases (e.g., $^{13}C(d,p)^{14}C$) the analyzing power was drastically different at these two energies. The values of the mixing ratio derived from these measurements must be considered with caution until more is known about fluctuations with energy and about the dependence of the analyzing power on Q-value and on nuclear mass.

For heavier nuclei (2p-shell) recent work [24] has shown that it is possible in some cases to determine the mixing ratio quantitatively. Fig. 16 shows the cross section and vector analyzing power for $^{53}Cr(d,p)^{54}Cr$ leading to the 2^+ first excited state of ^{54}Cr. Since the spin of the target nucleus is $3/2^-$, the permitted values of ℓ and j are $\ell = 1 (j = 1/2, 3/2)$ and $\ell = 3 (j = 5/2, 7/2)$. For comparison, the measurements on pure $j = 1/2^-$ and $j = 3/2^-$ transitions of about the same Q-values on ^{54}Fe are shown as dashed lines. The results for the transition on ^{53}Cr can be explained quantitatively (solid curves) as an incoherent superposition of contributions with $\ell = 1$, $j = 1/2^-$, and $j = 3/2^-$. The ratio of the spectroscopic factors was found to be $S_{3/2}/S_{1/2} = 1.48$. The cross section indicates that the $\ell = 3$ contribution is negligible. The use of comparison measurements on a nearby target with zero spin (i.e., single value of j) is necessary because DWBA calculations do not predict the vector analyzing powers with sufficient accuracy.

A similar analysis has been carried out by the Wisconsin group [27] for many transitions on $^{53}Cr(d,p)^{54}Cr$ and $^{57}Fe(d,p)^{58}Fe$. Transitions in $^{117}Sr(\vec{d},p)^{118}Sr$ and $^{119}Sr(\vec{d},p)^{120}Sr$ to the first excited state (2^+) have been investigated at 10-MeV deuteron energy. Since the target spin is $1/2^+$, the permitted j-values are $3/2^+$ and $5/2^+$. In this case no measurements on nearby spin-zero nuclei were available for comparison. An attempt to explain the results by comparison with DWBA calculations for $j = 3/2^+$ and $5/2^+$ was not successful.

A paper submitted to this symposium [51] discusses the determination of mixing ratios from asymmetry measurements in $(\vec{p},d)$ reactions.

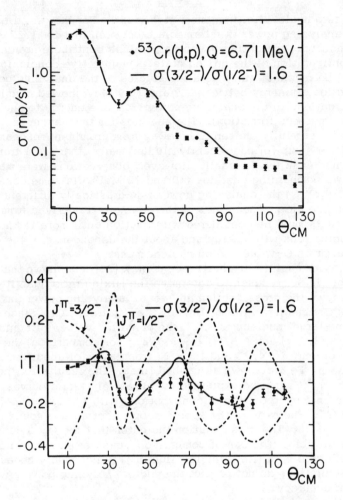

Fig. 16. Analyzing power and cross sections for the
reaction $^{53}Cr(\vec{d},p)^{54}Cr$ (E_x = 0.83) at E_d = 10 MeV.
For comparison, the dashed curves show the analyzing
powers for j = 1/2$^-$ and j = 3/2$^-$ transitions of about
the same Q-value in $^{54}Fe(d,p)^{55}Fe$ (see fig. 3). The
solid curves are calculated from an incoherent super-
position of $\sigma(\theta)$ and $iT_{11}(\theta) \cdot \sigma(\theta)$ (see ref. [24]).

3. POLARIZATION EFFECTS IN (d,n) REACTIONS

3.1. *Polarization of neutrons from (d,n) reactions*

Since the last symposium, accurate measurements of the polarization of neutrons from (d,n) reactions have become available. Most of the results are for light target nuclei and for bombarding energies below 6 MeV (see table 1). Results for many of these reactions have been summarized by Walter [104] at this symposium. As an example of recent measurements, the neutron polarization from $^{15}N(d,n)^{16}O$ is shown in fig. 17. The angular dependence of P_n changes considerably with bombarding energy, so that DWBA analysis is of doubtful value. In general, little success with DWBA calculations has been reported for (d,n) reactions on light nuclei. For comparison with stripping theory, heavier target nuclei are of more interest. The Alberta group [82] has studied $^{28}Si(d,n)^{29}P$ at $E_d = 5$ MeV. The results for the $\ell = 0$ g.s. transition are reasonably well reproduced by DWBA calculations (fig. 18). For the $\ell = 2$ transitions ($E_x = 1.38$ MeV and $E_x = 1.96$ MeV, $j = 5/2^+$) the measured polarization shows no clear j-dependence and the calculated polarizations bear little resemblance to the measurements. The Alberta group [83] also measured the polarization of the neutrons from $^{40}Ca(d,n)^{41}Sc$ for the $\ell = 3$ g.s. transition and for the $\ell = 1$ transitions to the states at $E_x = 1.72$ and 2.42 MeV. The measurements are of relatively high accuracy ($\Delta P \approx 0.03$) and extend over a wide range of angles. The polarization is found to be small ($|P| \approx 0.1$). For the g.s. and 1.72-MeV state, DWBA calculations show similarity with the measurements only at forward angles. The measurements for heavier nuclei in the 2p-shell [84] are very limited in angular range and energy resolution.

3.2. *Analyzing power in (d,n) reactions induced with vector-polarized deuterons*

The intensity of beams from polarized-ion sources has reached the point where $(\vec{d},n)$ reactions can be investigated. Results of this type have been presented to this symposium [58]. In addition to two light target nuclei, the reaction $^{89}Y(\vec{d},n)^{90}Zr$ at $E_d = 11$ MeV was also studied (fig. 19). These results are of particular interest because bombarding energy and target mass are in the region where one expects stripping theory to apply. The observed analyzing power for $^{89}Y(d,n)^{90}Zr$ reaches values as large as $iT_{11} = 0.45$ (i.e., $P_d = 0.5$). Comparison with fig. 3 shows the surprising result that the analyzing power for $^{89}Y(\vec{d},n)^{90}Zr$, for which $j = 1/2^-$, has the characteristics of $(\vec{d},p)$ transition with $j = 3/2^-$ rather than $1/2^-$. The right-hand side of fig. 19 shows DWBA calculations for $(\vec{d},n)$ and $(\vec{d},p)$ of the same j- and Q-value. The calculations reproduce qualitatively the large difference in the observed analyzing power between (d,p) and (d,n) transitions. Coulomb effects in stripping reactions apparently are more important than is generally assumed.

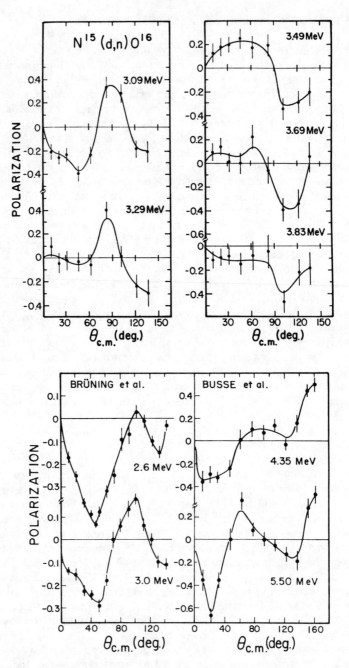

Fig. 17. Polarization of neutrons from the ^{15}N(d,n)^{16}O reaction between 2.6 and 5.5 MeV. The figure is from ref. [80].

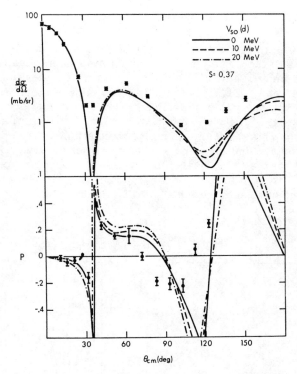

Fig. 18. Comparison of the cross section and neutron polarization in ^{28}Si(d,n)^{29}P (g.s.) with DWBA calculations. The deuteron energy was 5 MeV [82].

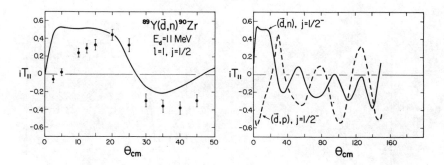

Fig. 19. Vector analyzing power of the ^{89}Y(d,n)^{90}Zr reaction [58]. The solid curve is the result of a DWBA calculation. The curves on the right demonstrate the large difference between predicted ($\vec{d}$,n) and ($\vec{d}$,p) analyzing powers. The same optical model parameters [58] were used for both calculations.

4. ANALYZING POWER IN $(\vec{d},t)$ REACTIONS

Only three $(\vec{d},t)$ reactions have been studied so far (see table 1). Transitions of the same ℓ but different j have been observed in $^{13}C(\vec{d},t)^{12}C$ and $^{208}Pb(\vec{d},t)^{207}Pb$. The results for angular momentum transfer j = 3/2⁻ and j = 1/2⁻ in $^{208}Pb(d,t)^{207}Pb$ show a clear j-dependence (fig. 20). The analyzing power for j = 1/2⁻ has a variation with angle similar to that of the j = 1/2 g.s. transition in $^{207}Pb(\vec{d},p)$ at the same bombarding energy, but the sign of the analyzing power is opposite in the two cases. Similarly, the analyzing power observed by Debenham et al. [85] for the $^{13}C(\vec{d},t)^{12}C$ g.s. reaction (j = 1/2⁻) at forward angles is opposite in sign to that for the $^{12}C(\vec{d},p)^{13}C$ g.s. reaction of the same j. Both measurements were at E_d = 12.3 MeV.

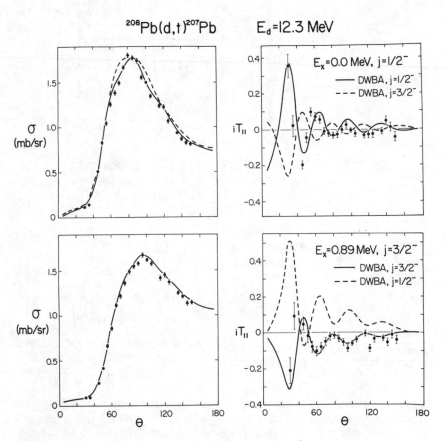

Fig. 20. Analyzing power in the reaction $^{208}Pb(\vec{d},t)^{207}Pb$ for ℓ = 1 transitions with j = 1/2⁻ and j = 3/2⁻ [86].

5. TWO-NUCLEON TRANSFER REACTIONS

Several (^{3}He,$\vec{p}$) and (^{3}He,$\vec{n}$) experiments have been reported for light target nuclei and incident energies of a few MeV (see table 1). Neutron polarizations of large magnitude are observed (e.g., P = 0.9 ± 0.1 for ^{12}C(^{3}He,n)^{14}O at 2.4 MeV, θ_{lab} = 50°, ref. [89]), but the low cross section ($\approx$ 1 mb/sr) reduces the usefulness of these reactions as sources of polarized neutrons. Attempts to describe the results by DWBA calculations have been only moderately successful (e.g., refs. [89, 90]).

The groups at Oxford and Berkeley have used polarized proton beams of 49.5 MeV and 43.8 MeV, respectively, to induce (p,t) and (p,^{3}He) reactions on light nuclei. Nelson et al. [87] conclude that the analyzing powers of ^{16}O(p,t)^{14}O (g.s.) and ^{16}O(p,^{3}He)^{14}N (E_x = 2.31 MeV) are the same within the experimental errors, as expected for reactions leading to mirror final states. They find satisfactory agreement with DWBA only if finite-range corrections are included in the calculations. The Berkeley group studied five ($\vec{p}$,t) transitions for which the transferred quantum numbers are the same (L = 2, S = 0). The measured analyzing powers seem to fall into two distinct classes (see the contribution to this symposium by Hardy et al. [88]).

6. CONCLUSIONS

The development of polarization studies in transfer reactions since the last symposium is an example of the fruitful interplay of experiment and theory. The analyzing power of ($\vec{d}$,p) and ($\vec{p}$,d) reactions has been shown to exhibit a clear j-dependence, and the main features are understood in terms of the DWBA. For the first time enough is known about polarization measurements to apply them as a reliable spectroscopic tool. The feasibility of extending the same method to targets with spin, where mixing of j-values occurs, has been demonstrated. However, in this case the accuracy of DWBA predictions is not yet sufficient to obtain reliable results. Particularly for high ℓ-values, the present theory is far from quantitative, and even for transitions with ℓ = 1 substantial systematic differences between calculations and experiment are found. However, these problems offer interesting possibilities for further work. Theories of transfer reactions other than DWBA (e.g., WBP; see Goldfarb's paper [96]) have to be explored further. For DWBA calculations, the effects of the deuteron d-state and of the inclusion of tensor forces in the deuteron-nucleus interaction will have to be studied more closely. This will require measurements of the deuteron tensor analyzing power in elastic scattering and in deuteron induced reactions. The state of technology of polarized-ion sources has reached the point where such measurements can be carried out in a routine manner. A first result has been reported to this symposium [12]. One question still of much concern is the choice of deuteron optical model parameters in DWBA calculations, since the observed cross sections and vector polarizations in the elastic channel permit substantial correlated changes in the parameters. Here again measurements of the three tensor polarizations for deuteron scattering are expected to lead to more realistic potentials. These results will be of real value only if at the same time on the theoretical side the effect of channel coupling is studied further. Another type of experiment which is becoming feasible is the measurement of the outgoing nucleon polarization in (d,p) reactions induced by polarized deuterons (polarization transfer). A first experiment has been reported by the Birmingham group [105]. The proton polarimeter developed at Montreal [106] offers interesting possibilities for such experiments.

Much work is still required to understand the difficulties that have been experienced in the interpretation of (d,p) and (d,n) reactions on light nuclei. Since measurements at isolated energies may be influenced by compound nuclear resonances it is important that the experiments cover a wide range of energies. This point is illustrated, for instance, by the Erlangen measurements on $^6\text{Li}(\vec{d},p)^7\text{Li}$ where 27 angular distributions between 2.9 and 10.9 MeV were observed. An energy interval of one or two MeV is not sufficient because the resonances may be broad. There is general agreement that DWBA has not been successful for stripping on light nuclei, but it is not clear that the problem is a fundamental one. Numerical calculations clearly show that for light targets the calculated polarization effects are more sensitive to the choice of optical model parameters than for the heavier targets. The practice of determining the optical parameters for the DWBA calculations from observations on the elastic cross section and polarization is a reasonable first step, but often this approach is followed too literally. The optical model may well be made to fit the elastic scattering cross section and polarization even in the presence of a compound nuclear resonance, but the internal wave function may not be at all realistic. The use of some kind of average parameters obtained from studies at many energies may in fact be more realistic.

The vigorous activity which has developed over the last few years in neutron polarization measurements is now being extended to higher bombarding energies [69, 74] and heavier target nuclei. The first measurements of the analyzing power in $(\vec{d},n)$ reactions have been made only recently [58]. The systematic exploration of $(\vec{d},n)$ reactions will require the development of pulsed polarized beams for application of the neutron time-of-flight method.

The discussion in the present paper has concentrated on the deuteron stripping reaction and the inverse nucleon pickup reaction. Work on other types of transfer reactions, including two-nucleon transfer reactions, has only just begun. No doubt by the time the next symposium is held many of the problems which still exist in the understanding of transfer reactions will have been solved.

REFERENCES

[1] D. W. Miller, Second Polarization Symp., p. 410.

[2] H. C. Newns, Proc. Phys. Soc. 66 (1953) 477; S. T. Butler, Proc. Rutherford Int. Conf. (Heywood Co., London, 1961) p. 492.

[3] J. X. Saladin et al., Second Polarization Symp., p. 507.

[4] T. J. Yule and W. Haeberli, Phys. Rev. Lett. 19 (1967) 756; T. J. Yule and W. Haeberli, Nucl. Phys. A117 (1968) 1.

[5] W. Dürr et al., Nucl. Phys. A122 (1968) 153; D. Fick et al., Phys. Lett. 27B (1968) 366.

[6] D. Fick, R. Kankowsky, K. Kilian, and E. Salzborn, Phys. Rev. Lett. 24 (1970) 1503, and Third Polarization Symp.

[7] D. H. Loyd and W. Haeberli, to be published; D. H. Loyd and W. Haeberli, Bull. Am. Phys. Soc. 14 (1969) 37.

[8] H. H. Cuno, G. Clausnitzer, and R. Fleischmann, Nucl. Phys. A139 (1969) 657.

[9] A. M. Baxter, J. A. R. Griffith, and S. Roman, Phys. Rev. Lett. 20 (1968) 1114.

[10] L. S. Saltykov, Izv. Akad. Nauk SSSR, Ser. Fiz. 31 (1967) 323.

[11] L. S. Saltykov, Izv. Akad. Nauk SSSR, Ser. Fiz. 31 (1967) 260.

[12] A. A. Debenham et al., Third Polarization Symp.

[13] S. Roman, Symp. on Nuclear Reaction Mechanisms (Quebec, 1969) p. 351.

[14] A. I. Malko, N. N. Pucherov, and L. S. Saltykov, Sov. J. Nucl. Phys. 3 (1966) 221.

[15] Yu. Gofman et al., Sov. J. Nucl. Phys. 5 (1967) 510.

[16] D. C. Kocher, P. J. Bjorkholm, and W. Haeberli, to be published in Nucl. Phys.

[17] K. W. Corrigan, R. M. Prior, and S. E. Darden, Bull. Am. Phys. Soc. 15 (1970) 483.

[18] G. Hudson et al., Third Polarization Symp.

[19] D. C. Kocher and W. Haeberli, Third Polarization Symp.

[20] D. C. Kocher and W. Haeberli, Phys. Rev. Lett. 25 (1970) 36.

[21] D. C. Kocher and W. Haeberli, to be published in Nucl. Phys.

[22] A. S. Klimenko, A. I. Mal'ko, L. S. Saltykov, and V. I. Chirko, Ukr. Fiz. Zh. 13 (1968) 334.

[23] P. A. Quin and J. J. Bogaards, Bull. Am. Phys. Soc. 15 (1970) 48.

[24] D. C. Kocher and W. Haeberli, Phys. Rev. Lett. 23 (1969) 315.

[25] P. J. Bjorkholm and W. Haeberli, Bull. Am. Phys. Soc. 13 (1968) 723, and to be published in Nucl. Phys.

[26] P. J. Bjorkholm, W. Haeberli, and B. Mayer, Phys. Rev. Lett. 22 (1969) 955.

[27] D. C. Kocher and W. Haeberli, Bull. Am. Phys. Soc. 14 (1969) 602; D. C. Kocher and W. Haeberli, to be published.

[28] G. Clausnitzer, G. Graw, C. F. Moore, and K. Wienhard, Phys. Rev. Lett. 22 (1969) 793.

[29] P. J. Bjorkholm, Thesis, University of Wisconsin (1969).

[30] A. A. Debenham, J. A. R. Griffith, M. Irshad, and S. Roman, Nucl. Phys. A151 (1970) 81; R. D. Rathmell and W. Haeberli, Third Polarization Symp.

[31] R. A. Blue, K. J. Stout, and G. Marr, Nucl. Phys. A90 (1967) 601.

[32] G. Dietze and K. D. Stahl, Z. Phys. 228 (1969) 172.

[33] N. C. Herbert, Thesis, Stanford University (1967).

[34] M. Sosnowski, Nucl. Phys. A133 (1969) 266.

[35] A. Budzanowski et al., Nucl. Phys. (to be published), and Third Polarization Symp.

[36] Ch. Chiang et al., Hsueh Pao 22 (1966) 554.

[37] D. P. Gurd, G. Roy, and H. G. Leighton, Nucl. Phys. A120 (1968) 94.

[38] W. E. Maddox, C. T. Kelley, Jr., and D. W. Miller, Phys. Rev. C1 (1970) 476.

[39] G. Roy, H. Sherif, G. Moss, and W. Saunders, Bull. Am. Phys. Soc. 15 (1970) 484.

[40] H. G. Leighton, G. Roy, and D. P. Gurd, Nucl. Phys. A115 (1968) 108.

[41] K. A. Kuenhold and T. R. Donoghue, to be published in Nucl. Phys.

[42] C. C. Foster, W. E. Maddox, and D. W. Miller, Phys. Rev. 181 (1969) 1529.

[43] C. T. Kelley, Jr., W. E. Maddox, and D. W. Miller, Phys. Rev. C1 (1970) 488.

[44] E. J. Ludwig and D. W. Miller, Phys. Rev. 138 (1965) B364.

[45] L. S. Michelman, S. Friarman, E. J. Ludwig, and A. B. Robbins, Phys. Rev. 180 (1969) 1114.

[46] N. S. Chant, P. S. Fisher, and D. K. Scott, Nucl. Phys. A99 (1967) 669.

[47] N. S. Chant and J. M. Nelson, Nucl. Phys. A117 (1968) 385.

[48] J. L. Escudie, J. Gosset, H. Kamitsubo, and B. Mayer, Third Polarization Symp.

[49] J. L. Escudie et al., Phys. Rev. Lett. 23 (1969) 1251.

[50] L. Escudie et al., Int. Conf. on Properties of Nuclear States (University of Montreal Press, 1969) addendum p. 11.

[51] J. Gosset, H. Kamitsubo, and B. Mayer, Third Polarization Symp.

[52] J. L. Escudie et al., Third Polarization Symp.; H. Kamitsubo and B. Mayer, Third Polarization Symp.

[53] A. J. Froelich and S. E. Darden, Nucl. Phys. A119 (1968) 97.

[54] S. E. Darden and A. J. Froelich, Phys. Rev. 140 (1965) 69.

[55] V. V. Verbinski and M. S. Bokhari, Phys. Rev. 143 (1966) 688.

[56] M. Ivanovich, H. Cords, and G. U. Din, Nucl. Phys. A97 (1967) 177.

[57] A. Ingemarsson, A. Johansson, and G. Tibell, Phys. Lett. 27B (1968) 439.

[58] D. Hilscher, P. A. Quin, and J. C. Davis, Third Polarization Symp.

[59] R. S. Thomason, M. M. Meier, J. Taylor, and R. L. Walter, Bull. Am. Phys. Soc. 13 (1968) 603.

[60] T. G. Miller and J. A. Biggerstaff, Phys. Rev. C1 (1970) 763.
[61] G. Spalek et al., Third Polarization Symp.
[62] R. Brüning, F. W. Busser, H. Dubenkropp, and F. Niebergall, Nucl. Phys. 121 (1968) 224.
[63] M. M. Meier, R. S. Thomason, and R. L. Walter, Bull. Am. Phys. Soc. 12 (1967) 1197.
[64] G. R. Mason and J. T. Sample, Nucl. Phys. 82 (1966) 635.
[65] T. G. Miller, Bull. Am. Phys. Soc. 12 (1967) 1143.
[66] M. M. Meier, F. O. Purser, Jr., G. L. Morgan, and R. L. Walter, Symp. on Nuclear Reaction Mechanisms (Quebec, 1969) p. 409.
[67] W. Busse et al., to be published in Nucl. Phys.
[68] J. Taylor et al., Third Polarization Symp.
[69] M. M. Meier, L. A. Schaller, and R. L. Walter, Phys. Rev. 150 (1966) 825.
[70] J. R. Sawers, Jr., F. O. Purser, Jr., and R. L. Walter, Phys. Rev. 141 (1966) 825.
[71] P. D. Parker, Phys. Rev. 150 (1966) 830.
[72] T. R. Donoghue et al., Phys. Rev. 173 (1968) 952.
[73] C. A. Kelsey and A. S. Mahajan, Nucl. Phys. 71 (1965) 157.
[74] H. Brückmann, W. Kluge, and L. Schaenzler, Z. Physik 221 (1969) 379.
[75] F. W. Buesser, J. Christiansen, F. Niebergall, and G. Soehngen, Nucl. Phys. 69 (1965) 103.
[76] M. M. Meier, F. O. Purser, Jr., and R. L. Walter, Phys. Rev. 163 (1967) 1056.
[77] N. P. Babenko et al., Sov. J. Nucl. Phys. 1 (1965) 323.
[78] S. T. Thornton et al., Third Polarization Symp.
[79] R. Brüning, F. W. Buesser, F. Niebergall, and J. Christiansen, Phys. Lett. 21 (1966) 435.
[80] M. M. Meier, R. S. Thomason, and R. L. Walter, Nucl. Phys. 115 (1968) 540.
[81] W. Busse et al., Z. Physik 206 (1967) 404.
[82] S. T. Lam et al., Nucl. Phys. A119 (1968) 146; J. Taylor, Jr., G. Spalek, Th. Stammbach, and R. L. Walter, Symp. on Nuclear Reaction Mechanisms (Quebec, 1969) p. 403.
[83] D. A. Gedcke et al., Nucl. Phys. A134 (1969) 141.
[84] I. I. Levintov et al., Sov. J. Nucl. Phys. 8 (1969) 8.
[85] A. A. Debenham, J. A. R. Griffith, M. Irshad, and S. Roman, Int. Conf. on Properties of Nuclear States (University of Montreal Press, 1969) p. 274.
[86] H. S. Liers, R. D. Rathmell, S. E. Vigdor, and W. Haeberli, to be published in Phys. Rev. Lett.
[87] J. M. Nelson, N. S. Chant, and P. S. Fisher, Third Polarization Symp., and Phys. Lett. B31 (1970) 445; J. M. Nelson, N. S. Chant, and P. S. Fisher, to be published.
[88] J. C. Hardy et al., Third Polarization Symp., and Phys. Rev. Lett. 25 (1970) 298.

[89] L. A. Schaller et al., Phys. Rev. 163 (1967) 1034.

[90] Th. Stammbach, R. S. Thomason, J. Taylor, Jr., and R. L. Walter,
      Phys. Rev. 174 (1968) 1119; R. S. Thomason, Th. Stammbach,
      J. Taylor, Jr., and R. L. Walter, Symp. on Nuclear Reaction
      Mechanisms (Quebec, 1969) p. 413.

[91] D. C. DeMartini and T. R. Donoghue, Ohio State University
      Progress Report 1968/1970, p. 5.

[92] D. C. DeMartini, W. L. Baker, C. E. Busch, and T. R. Donoghue,
      Ohio State University Progress Report 1968/1970, p. 9.

[93] D. G. Simons, Phys. Rev. 155 (1967) 1132; D. G. Simons and
      R. W. Detenbeck, Phys. Rev. 137 (1965) 1471.

[94] M. Krivopustov, G. Schirmer, I. V. Sizov, and H. Oehler, Joint
      Institute for Nuclear Research, Dubna (1969), quoted in Nuclear
      Science Abstracts No. 21261 (23) 1969.

[95] G. Marr, Thesis, Ohio State University (1968).

[96] L. J. B. Goldfarb, Third Polarization Symp.

[97] G. Glashausser and J. Thirion, Adv. Nucl. Phys. 2 (1969) 79.

[98] S. E. Darden, Third Polarization Symp.

[99] K. Abrahams, private communication.

[100] C. A. Pearson, J. C. Wilcott, and L. C. McIntyre, Nucl. Phys.
      A125 (1969) 111.

[101] R. Beurtey et al., J. Phys. (Paris) 24 (1963) 1038.

[102] R. D. Rathmell and W. Haeberli (unpublished).

[103] D. C. Kocher, R. D. Rathmell, and W. Haeberli, Third Polariza-
      tion Symp.

[104] R. L. Walter, Third Polarization Symp.

[105] R. C. Brown, J. A. R. Griffith, O. Karban, and S. Roman, Third
      Polarization Symp.

[106] R. Bangert et al., Third Polarization Symp.

DISCUSSION

Roman:

DW calculations for (d,p) reactions often give a good description
of polarization and cross section, but not always, as Haeberli
has shown. As a rule one uses the distorting optical model
potentials which describe best the elastic scattering and no
adjustments are made to these parameters. For the incident
channel, however, ambiguous sets of optical model parameters
often exist which describe equally well the deuteron elastic
scattering. It is gratifying to find, when these sets are inserted
into DWBA, that the best results give sets of deuteron potentials
with central strengths equal to the strength obtained by folding
the proton and neutron potential.

Atomic Beam Sources

H. F. GLAVISH, Stanford University, California, USA

1. INTRODUCTION

In the past ten years there has been repeated demonstration that
the atomic beam principle initiated by Clausnitzer, Fleischmann, and
Schopper [1] can be used successfully for the production of polarized
positive and negative ions. The refinement of rf transitions and strong
field ionization to enhance polarization and improve flexibility is now
used almost without exception. The subject of atomic beam sources
was discussed at some length in the Proceedings of the previous two
symposia on polarization phenomena. In addition there have been
several other reports and reviews in the literature: Fleischmann [2],
Baumgartner [3], Daniels [4], Dickson [5], Beurtey [6], Craddock [7],
Haeberli [8], Drake [9], and Allen [10]. In particular, the accounts
of Dickson [5] and Haeberli [8] are extensive and quite detailed. A
recent conference was also held at ORTEC Inc. [11].

Recently there has been a demand for an increase in beam current
to offset the substantial loss which occurs upon injection and acceleration. This may be appreciated by noting the output and target currents of some representative sources as set out in table 1.

Table 1. Comparison of atomic beam sources (polarizations 60–
100% of ideal values)

Place	Accelerator		Output beam	Target beam
ETH Zürich [29]	Tandem	p,d	15 nA neg	1.5 nA
Erlangen [30]	Tandem	p,d	10 nA neg	1.5 nA
Berkeley [12]	88-inch Cyclotron	p,d	2.5 μA pos	100 nA
Birmingham [31]	40-inch Cyclotron	p,d	100 nA pos	1 nA
CEN Saclay [32]	28-MeV Cyclotron	p,d	4 μA pos	100 nA
Rutherford Lab [33]	Linac*	p	1 μA pos	15 nA
Basel [34]	Cockcroft-Walton	p,d	100 nA pos	15 nA

*No longer in operation.

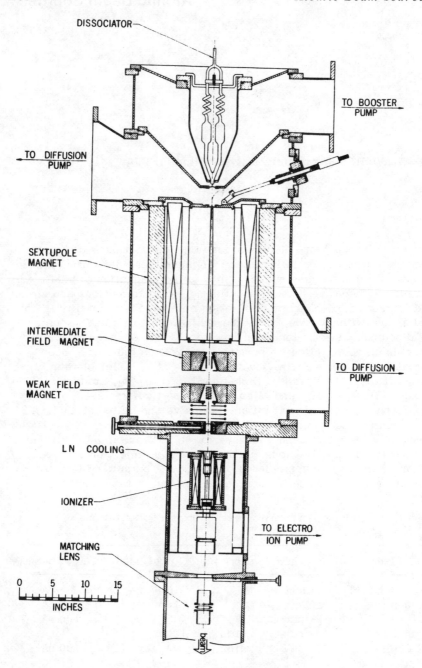

Fig. 1. Atomic beam polarized positive ion source on the Berkeley 88-inch cyclotron.

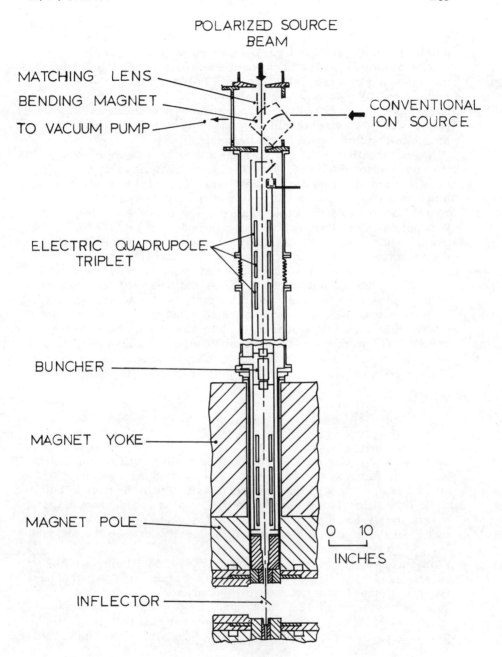

Fig. 2. Axial injection system for the Berkeley cyclotron.

An example of an atomic beam source of positive ions constructed for the Berkeley 88-inch cyclotron [12] is shown in fig. 1, and the axial injection system in fig. 2. This source is based on the design used at Saclay [13, 14] and employs a dissociator scheme similar to that proposed by Keller [15]. The $m_J = 1/2$ hyperfine states are separated from the $m_J = -1/2$ states in a sextupole magnet. Ions are formed by electron bombardment in a strong magnetic field. Various polarization states are obtained by inducing rf transitions in the atomic beam after separation in the sextupole.

Quite recently [16, 17, 18] several groups have become engaged in the construction of compact polarized sources to fit in the terminal of single-ended electrostatic accelerators. Paetz [19] has reported the progress of the Ohio terminal source at this symposium. This type of source has to operate from the limited power of 4 kVA available in the terminal. The sublimation pumps are an important part of the development, as they must have sufficient capacity for the gas load and at the same time operate for long periods without recharging.

Ebinghaus [20] has discussed the application of the atomic beam principle to the production of a 1 μA beam of polarized ^{6}Li ions. An interesting feature is the almost 100% ionization efficiency obtained by surface ionization on heated tungsten using the Langmuir-Taylor effect. A more complete account of this work may be found in the literature [21]. Work on polarized Li and Cs ions has also been carried out at Wisconsin [22]. The development of an atomic beam polarized ^{3}He source using nuclear magnetic separation is continuing [23, 24].

2. NOZZLE DESIGN

Nozzle design is always a problem in atomic beam sources. Almost every type of nozzle has been used, from a bundle of fine capillaries to a single large hole. The properties of nozzles are now better understood, but there is still uncertainty, particularly in the case of supersonic flow at low pressures. Since the pumping speed between the nozzle and separating magnet is limited by geometry, it is desirable that a nozzle produce a high axial directionality to keep the residual gas pressure low.

At dissociator pressures of 0.5 Torr or less, capillaries are frequently used. The flow through a single canal of length ℓ and diameter d is illustrated in fig. 3. Flow through canals is discussed in detail in the literature [25, 26, 27] and Lew [28] has provided a useful summary. At low pressures, when the mean free path is greater than ℓ, simple molecular effusion takes place. In terms of the source density n_0 and mean velocity $\bar{v}$:

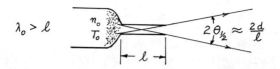

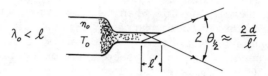

Fig. 3. Molecular effusion from a canal
for two different pressure regimes.

Center-line intensity $\quad I(0) = \dfrac{n_0 \bar{v}}{4\pi} \cdot \dfrac{\pi d^2}{4} \quad$ atoms sec^{-1} sr^{-1};

Total flow $\qquad\qquad F = \dfrac{4d}{3\ell} \cdot \dfrac{n_0 \bar{v}}{4} \cdot \dfrac{\pi d^2}{4} \quad$ atoms sec^{-1};

Half-width $\qquad\qquad \theta_{1/2} \approx \dfrac{d}{\ell} \ $ rad. $\qquad\qquad\qquad$ (1)

At higher pressures, when the mean free path becomes less than ℓ, the
total flux is still given by eq. (1), but free molecular effusion does
not occur until a point along the canal is reached where the mean free
path is equal to the remaining length of the canal, called the "effec-
tive length" ℓ'. In this pressure region $\theta_{1/2} \approx d/\ell' \propto d(n_0/\ell)^{1/2}$ and
$I(0) \propto d^2(n_0/\ell)^{1/2}$. Thus we observe that with increasing pressure the
directionality deteriorates and the center-line intensity increases.
This description applies provided there is a sufficiently low pressure
in the nozzle exhaust chamber. In the case of a bundle of capillaries,
$I(0)$ should be enhanced by a factor equal to the number of capillaries
present. However another effect is present which can produce attenuation regardless of the background pressure. As the source pressure
increases, the jets from each capillary broaden and overlap. A region
of constant gas density exists [27] downstream from the nozzle for a
length $L \sim D/\theta_{1/2}$, where D is the diameter of the bundle. Because of
this effect it is found in practice that a single or few canals often
produce a slightly higher intensity (flux per sr). Of course the pumping capacity must be sufficient to handle the larger gas loads which
arise from the inferior directionality. A single canal probably permits
a higher dissociation to be obtained, since there is an absence of
large surface areas in the vicinity of the nozzle. Another advantage

is that the atoms are concentrated near the axis, where the acceptance
of the magnetic separator is largest. Also, the beam is more compact
at the ionizer. When there is limited pumping capacity (e.g., terminal
source for electrostatic accelerator), the multi-capillary nozzle has
an advantage. The canal diameter d should be as small as possible
within practical limits, but d/ℓ needs to be greater than the acceptance
half-angle of the magnetic separator. Very fine capillaries are observed [35] to have a short life when the discharge is on. At the
Rutherford laboratory the performances of various multi-capillary
nozzles [36] have been compared. Better results were obtained in the
source if the bundle diameter was decreased from 8mm to 6 mm.
Clausnitzer [30] has compared the directionalities of various nozzles
and obtained the results shown in fig. 4.

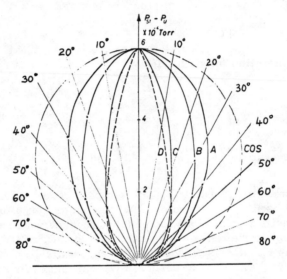

Fig. 4. Directionality for various nozzles
(ref. [30]). A = single 3mm opening; B =
8 × 1.5 mm diam openings; C = 230 × 0.34 mm
diam capillaries; D = 1000 × 0.04 mm diam
capillaries. Canal lengths for A and B were
5 mm; and for B and C, 2 mm. The total gas
flow was 2 Torr liters sec^{-1}.

Kantrowitz and Grey [37] proposed that intense, well-directed,
neutral beams might be obtained by supersonic flow from a nozzle.
The application to atomic beam sources was suggested by Keller [15].
An arrangement of this type used at Saclay [14] produced an atomic
beam flux after the sextupole of 5×10^{16} atoms sec^{-1}. The best fluxes
obtained with other nozzles are $5-10 \times 10^{15}$ atoms sec^{-1}. In the last

decade supersonic flow at low pressures has been studied at some
length. Anderson, Andres, and Fenn [38] have provided an excellent
summary. An extensive empirical investigation of the effect of the
various physical parameters of the nozzle has been made by Campargue
[39].

When gas flows adiabatically through a nozzle, random Maxwellian
motion is traded off for forward directed mass-flow [40], and the internal temperature decreases according to the equation

$$T = T_0 \left(1 + \frac{\gamma - 1}{2} M^2\right)^{-1},$$

(2)

where T_0 is the stagnation temperature, M the Mach number, and γ
the specific heat ratio. An increase in M is accompanied by a reduction in the width of the velocity distribution. If the downstream background pressure is low and the nozzle suitably designed, a value M =
1 can be obtained at the nozzle throat. If there is expansion after the
throat, the Mach number continues to increase downstream until either
the density becomes so low that there is transition to free flow or a
normal shock (Mach disk) occurs which thermalizes the jet. An empirical relation [41] expresses the location L of the Mach disk downstream from a nozzle of diameter D, as a function of stagnation pressure p_0 and background pressure p_{bg}

$$L/D = 0.67(p_0/p_{bg})^{1/2}.$$

(3)

At the low pressures typical of a polarized proton or deuteron source
transition to free flow can be assumed to occur before the Mach disk
is formed. This is experimentally verified [38, 42] in supersonic
molecular beam devices. Transition to free flow and freezing of the
Mach number is expected to occur when the mean free path

$$\lambda = \lambda_0 \left(1 + \frac{\gamma - 1}{2} M^2\right)^{1/(\gamma - 1)}$$

(4)

becomes comparable to the jet diameter. Thus for $p_0 = 3$ Torr, the terminal value for M should be about 4, occurring at a distance of about
5D from the nozzle. The dependence of M on distance is given by
Ashkenas and Sherman [43]. To test this prediction, M would have to
be found from an experimental velocity distribution measurement [38].
At pressures $p_0 = 10$ Torr and greater there is agreement with the above
description, but there are few data at lower pressures. It would be
very enlightening to have the results of similar measurements and
empirical studies at lower pressures. This would help to resolve the
long-standing question of whether or not supersonic flow is applicable
to an atomic beam source for hydrogen or deuterium. While the inten-

sities measured at Saclay are comparatively high, there is no clear
evidence that a supersonic beam is obtained. Most groups find little
benefit in dissociator pressures greater than about 3 Torr because of
the low dissociations obtained, presumably as a consequence of volume recombination. This restriction does not apply to a polarized ^{3}He
source [24] where a value $M = 10$, determined from a velocity spectrum,
has been obtained for $p_0 = 36$ Torr and $T_0 = 5.9$ K.

In the original proposal for a supersonic beam [37] a converging-
diverging Laval nozzle was suggested. It is now generally agreed
that at low densities the diverging part of the Laval nozzle has little
effect [38]. In fact, if a true boundary layer does form in the diverging
part, it would be very thick because of the low density and would en-
croach on the isentropic core and reduce the area of the jet.

Once transition to free flow has occurred, the jet becomes porous
and feels the background gas for the first time. This can cause significant attenuation [44, 45, 46]. It is therefore advisable to take
advantage of differential pumping by inserting a skimmer near the
transition point, as indicated in fig. 5. The skimmer should be of
sufficient diameter to accept the whole of the isentropic core. The
area of this core as a function of M and nozzle throat area A^* has a
theoretical value of

$$A_s = \frac{A^*}{M} \left\{ \frac{1 + \frac{\gamma - 1}{2} M^2}{\frac{\gamma + 1}{2}} \right\}^{\frac{\gamma + 1}{2(\gamma - 1)}} . \tag{5}$$

Empirical investigation of the effect of the skimmer diameter might
help to determine how applicable this result is at low pressures. A
surprising result observed at $p_0 \sim 2$ Torr in the Australian National
University source [47], the Berkeley source [12, 48], and the apparatus
of Campargue [39] is the insensitivity to skimmer position and therefore to solid angle subtended at the nozzle. At high pressures the
intensity shows a characteristic sensitive dependence on skimmer
position, determined by shock formation at the skimmer inlet when the
skimmer is close to the nozzle, and by gas scattering attenuation when
it is further downstream. The posssibility that a normal detached
shock may form at the skimmer entrance is the reason for the characteristic conical shape and the sharp leading edge of the skimmer.

The center-line intensity and total flux of a supersonic jet are
given in theory as [38, 40]

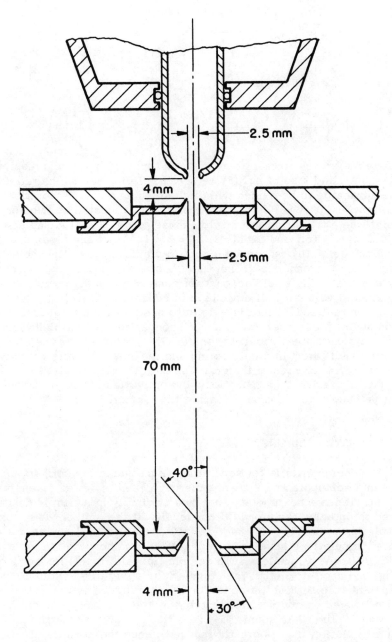

Fig. 5. Nozzle, skimmer, and collimator on the Berkeley source.

$$I(0) = A_s a_0 n_0 \frac{\gamma(\gamma/2)^{1/2}}{2\pi} \cdot \frac{M^3}{\left(1 + \frac{\gamma-1}{2} M^2\right)^{\frac{\gamma+1}{2(\gamma-1)}}} \quad (M > 3)$$

$$F = \left(\frac{2}{\gamma+1}\right)^{\frac{1}{\gamma-1}} \left(\frac{\gamma}{\gamma+1}\right)^{1/2} a_0 n_0 A^* ; \quad a_0 = \left(\frac{2kT_0}{m}\right)^{1/2} \quad (6)$$

where A_s is the skimmer area. If A_s is replaced by the expression on the right-hand side of eq. (5) we see that $I(0) \propto M^2$, which reflects the advantage to be gained with high Mach numbers. Theoretically the last result is valid only if the skimmer is located at or before the point where transition to free flow takes place. For H atoms at $T_0 = 400$ K, $p_0 = 3$ Torr and for $M = 3$, $A^* = 3 \times 10^{-2}$ cm^2 and a sextupole acceptance of 10^{-3} sr, eqs. (5) and (6) predict a flux after the sextupole of $\approx 5 \times 10^{17}$ atoms sec^{-1}. Low degree of dissociation and attentuation by gas scattering would account for some of the discrepancy with experimental values. Full advantage of supersonic flow has probably not been realized. There is a definite need for more empirical information at low pressures. A popular misconception is that high Mach number beams are difficult to ionize. This is not so, as the high M value results from internal cooling and the most probable velocity is never very much greater than a_0. The narrow velocity distribution might be an advantage if the focusing properties of the sextupole magnet produce a very compact beam at the ionizer.

3. SEPARATING MAGNET

The design criteria for separating magnets and the comparison between a sextupole and a quadrupole have been reviewed extensively [5, 8]. Tapered sextupoles, analyzed originally by Keller [49], are used to optimize the atomic beam density at the ionizer, as for example in the Saclay [15], Berkeley [12], and ANU [47] sources. Computer calculations have been made [50] to find how the ion yield varies with different parameters such as pole tip field, taper, ionizer location, and geometry, etc. The calculations apply to a sextupole with constant pole tip field, a condition easily fulfilled in practice even for a tapered sextupole, as supported by the measurements on the magnet for the ORIC source [51] (see fig. 6). For this particular case the calculation time is greatly reduced, since the atom trajectories have an exact solution [50], with the reasonable approximation of a constant atomic dipole moment. The exact solution applies even to atoms which carry angular momentum with respect to the sextupole axis. By considering some 10^5 different trajectories, integration over

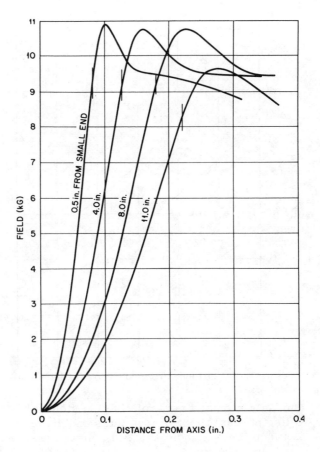

Fig. 6. Variation of sextupole field with distance
from the axis and position along the length (ref. [51]).

the entrance parameters and velocity distribution is accurate. Ray
tracing calculations have been carried out by other groups [30, 52],
but only for constant aperture sextupoles. The calculations show that
for a Glavish [53] type ionizer at 35 cm from the sextupole exit, the
ion yield is improved by a factor of 2 for a taper (exit/entrance aperture) of 2:1. This result has wide applicability and is not very sensitive to the sextupole length, pole tip field, or entrance aperture
diameter if these are in the respective ranges 25–50 cm, 6–10 kG,
and 0.5–0.8 cm. In the Auckland source, the poles of an untapered
sextupole were re-machined with a 2:1 taper and the ion current was
found to increase by a factor of 2.5, all other parameters remaining
fixed.

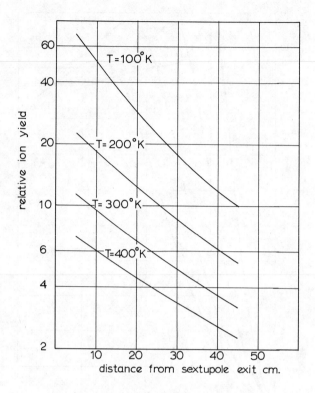

Fig. 7. Calculations of relative ion yield for
different dissociator temperatures and distances
between the sextupole exit and ionizer entrance.

   The calculations have been extended to consider the effect of a
cooled dissociator. This can lead to an increase in ion yield in two
ways: the ionizer efficiency varies as $T^{-1/2}$ and the sextupole acceptance as T^{-1}, where T is the dissociator temperature. The theoretical
ion yield as a function of ionizer-sextupole separation is shown in
fig. 7 for various temperatures. At large separation the gain is not as
good as $T^{-3/2}$ because of the divergence of the atomic beam after it
leaves the sextupole. The curves in fig. 7 are normalized to constant
gas flow through the nozzle. At a low temperature the dissociation
might be reduced, particularly as the pressure must be higher to maintain a constant gas flow. A factor of 2 has been reported for cooling
to liquid nitrogen temperatures [54, 55]. A small microwave dissociator
has been used successfully at the Rutherford laboratory [56]. A discharge was sustained with only 50 W input power compared with 300 W

for the conventional 20 MHz system. The small size and low power might be useful for studies of liquid-nitrogen cooled discharges.

The focusing action of sextupole magnets has been investigated [30, 49, 57], but it is not clear that this property can be used to great advantage because of the large velocity spread in the atoms, the extended length of most ionizers, and the large ionizer-sextupole separation. The situation might be very different with high Mach number beams and different style ionizers.

4. RADIO FREQUENCY TRANSITIONS

Rf transitions, when used in conjunction with strong field ionization, permit the selection of various polarization states. This is particularly valuable in the case of deuterium. The adiabatic passage method of Abragam and Winter [58] is used, probably without exception. Beurtey [59] and Haeberli [8] discuss the application to atomic beam sources at some length. Sequences of transitions are used and different polarizations are obtained by appropriate switching. In the Saclay source the schemes (b), (c), (e), (f), (g), and (h) in table 2 are employed, producing six different polarization states for deuterons. Three transition units are used in the sequence weak field, strong field, weak field, the strong field unit being tuned to 2 ⟷ 5 for vector and 2 ⟷ 6 for tensor polarization measurements. A more compact

Table 2. Possible combinations of radio frequency transitions

		A Weak field	B Two-level tuned to	C Weak field	P_z $=\sqrt{2/3}\,\tau_{10}$	P_{zz} $=\sqrt{2}\,\tau_{20}$
Deuterium	a	on			− 2/3	0
	b	on	2 ⟷ 5		− 2/3	0
	c	on	2 ⟷ 5	on	+ 2/3	0
Deuterium	d		3 ⟷ 5		+ 1/3	− 1
	e		2 ⟷ 6		+ 1/3	+ 1
	f	on	2 ⟷ 6		− 1/3	+ 1
	g	on	2 ⟷ 6	on	+ 1/3	− 1
	h		2 ⟷ 6	on	− 1/3	− 1
	i		3 ⟷ 5	on	− 1/3	+ 1
Hydrogen	j	on			− 1	
	k		2 ⟷ 4		+ 1	

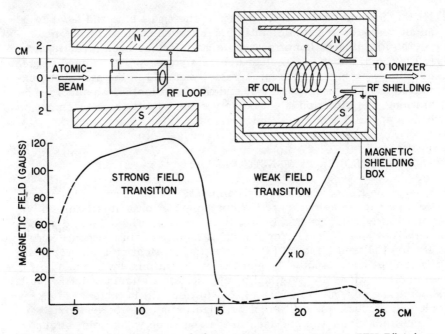

Fig. 8. Schematic diagram of rf transition apparatus on ETH Zürich source.

system omits the first weak field unit and corresponds to (a), (c), (d), (e), (h), and (i). This scheme is used at Basel [60] and ETH Zürich [29]. A schematic diagram of the Zürich apparatus is shown in fig. 8. This method requires the strong field transition to be tuned to each of the two-level transitions 2 ⟷ 5, 2 ⟷ 6, 3 ⟷ 5.

An alternative method with one weak field unit and two strong field units, as shown in table 3, does not require a strong field unit to be

Table 3. Deuterium transitions with one weak field unit and two strong field units

Weak field	Strong field 3 → 5	Strong field 2 → 6	P_z	P_{zz}
off	on	off	+ 1/3	− 1
off	off	on	+ 1/3	+ 1
off	on	on	+ 2/3	0
on	off	off	− 2/3	0
on	on	off	− 1/3	− 1
on	off	on	− 1/3	+ 1

tuned to different transitions. The role of the weak field unit is to reverse the sign of P_z, which of course can also be achieved with a magnetic field during or after ionization. This scheme is used in the ANU [47] and Minnesota [61] sources.

The case of protons is more straightforward: the weak field transition produces $P_z = -1$ and the strong field $(2 \leftrightarrow 4)$ $P_z = +1$.

To minimize the ionizer-sextupole separation it is important to make the rf units compact. For strong field transitions in hydrogen as well as in deuterium, the system shown in fig. 9 is employed on the ANU and Minnesota sources [62]. Two $\lambda/2$ resonant elements are enclosed in a cavity and are excited with anti-parallel currents by choosing the plane of the drive loop as indicated in the figure. A very

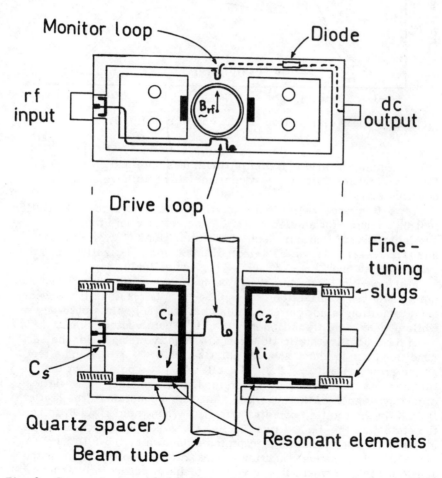

Fig. 9. Compact strong field rf transition units with twin $\lambda/2$ resonant elements enclosed in a cavity.

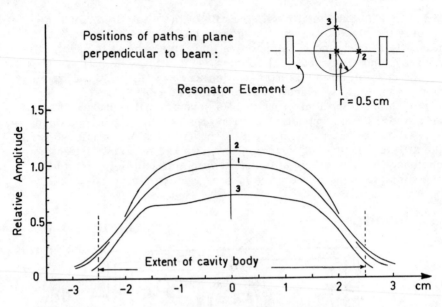

Fig. 10. Rf field distribution for the cavities of fig. 9.

uniform and symmetric rf field distribution is obtained as shown in
fig. 10. Cavities with twin λ resonating elements have also been
constructed [51].

A two-frequency weak field transition scheme is used by McCullen
[63] in a source for an electrostatic accelerator where the space
available is very limited. Each of the transitions $(1,2,3) \rightarrow (2,3,4)$
and $(1,2,3) \rightarrow (1,3,4)$ can be selected, producing polarizations of $P_z =
-2/3$, $P_{zz} = 0$ and $P_z = -1/3$, $P_{zz} = +1$, respectively.

In the case of deuterium it has been observed [64, 29, 65] that substantial tensor polarization ($P_{zz} \sim 0.34-0.4$) is obtained if the weak
field transition is not correctly adjusted. Oh [66] points out that
while a weak field transition may lead to a pure vector-polarized
beam for a positive static field ramp, a negative ramp under the same
conditions can produce a vector polarization of $P_z = -0.4$ and a tensor polarization of $P_{zz} = 0.3$. McCullen [67] has pointed out that
this behavior is most easily understood if we consider the variation
of energy of each of the hyperfine states as a function of the static
field B_0 in the rotating coordinate frame. When the rf field B_1 is zero,
the states are the dashed lines in fig. 11. When B_1 is turned on, the
states split away from the asymptotes as shown by the full curves.
The adiabatic passage condition is most difficult to satisfy near the
$1 \leftrightarrow 4$ resonance where the curvature is sharp, particularly at low
field strengths for B_1. Now consider a positive ramp and a value for
B_1 which is too small to satisfy the adiabatic condition in the $1 \leftrightarrow 4$

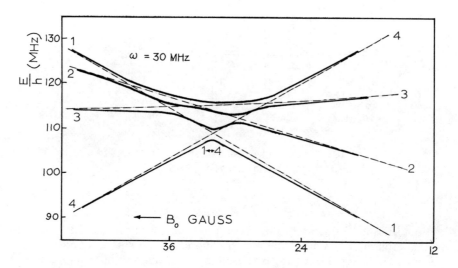

Fig. 11. Energy levels of the first four hyperfine states of deuterium in a rotating frame (30 MHz).

region. As we pass from right to left, state (3) will pass smoothly to state (2). Near the 1 ⟷ 4 resonance, part of (1) will spill over and go to (3), but this is largely compensated by part of state (2) going to (4). The net result is (1,2,3) → (2,3,4) (more or less) even though the adiabatic passage condition is not satisfied. For a negative ramp we pass from left to right, and the transitions 1 → 4, 2 → 3 occur smoothly, but state (3) goes to both (1) and (2). A compensation mechanism does not occur. The above description is an over-simplification, as each state will to some extent spread over all of the other states. The most important condition is that B_1 must be strong enough to satisfy the adiabatic passage condition as originally suggested by Abragam and Winter [58].

5. IONIZATION

We can sum up the situation on ionization in just one statement: ionizers in current use, which employ electron bombardment, have an efficiency of the order of 10^{-3} or less. Beurtey no longer considers this to be sufficient, and he is experimenting with new methods of ionization [68]. In the case of negative ion production, Haeberli [69] has suggested that large yields might be obtained if the polarized thermal beam is crossed with a fast beam of unpolarized negative hydrogen ions or neutral atoms such as Cs (fig. 12). Similarly, one might use fast H^+ ions in a process such as

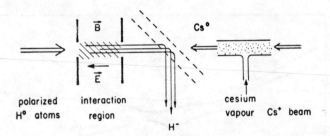

Fig. 12. Colliding beam scheme suggested for
producing high negative ion yields.

$$\vec{H}^0 + H^+ \rightarrow \vec{H}^+ + H^+ + \vec{e}$$

to obtain useful beams of polarized positive ions, although a colliding
beam technique does not appear to have much advantage over electron
bombardment for this case, if the cross sections and obtainable primary beam currents are considered. Some systematic experiments need
to be conducted to see if these new techniques can be utilized in a
practical source.

In the conventional electron bombardment ionizer, strong field
ionization is used in almost every case, but weak field ionizers of
comparable efficiency have been constructed [70, 71, 72]. However,
larger electron emitting surfaces are required.

Most strong field ionizers use the same method; electrons are injected along the atomic beam and are confined by the strong magnetic
field which is coaxial with the atomic beam. In this way a large ionization volume can be obtained and the electrons can make several
transits, if negative potentials are present at each end of the ionization volume. A typical example [53] is shown in fig. 13. In this
ionizer the active ionization volume is 15 cm long. The ions are extracted from one end by a sequence of electrodes with negative potentials. Before the ions leave the magnetic field, which is produced by
a solenoid, they reach about 2 keV energy. A gridded lens is used to
focus and accelerate the beam to about 5 keV. The atomic beam must
pass through the filament, which is immersed in the magnetic field,
before the ionization volume is reached. By careful choice of configurations of the electron gun it is also possible to inject the electrons
with a filament off to one side and out of the path of the atomic beam
[17]. The intensity of unpolarized ions from atoms or molecules which
collide with high temperature surfaces might be reduced with this
arrangement.

In addition to the discussions on strong field ionization in the Proceedings of the second polarization symposium, there are numerous
other accounts in the literature [11, 29, 30, 73, 74, 75]. Space charge
effects appear to be present in most strong field ionizers in that two

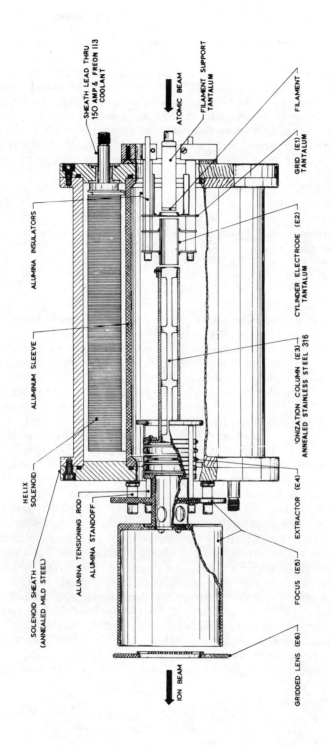

SHEATH LEAD THRU
150 AMP & FREON 113
COOLANT

ATOMIC BEAM

FILAMENT SUPPORT
TANTALUM

FILAMENT

GRID (E1)
TANTALUM

CYLINDER ELECTRODE (E2)
TANTALUM

ALUMINA INSULATORS

ALUMINUM SLEEVE

IONIZATION COLUMN (E3)
ANNEALED STAINLESS STEEL 316

EXTRACTOR (E4)

HELIX
SOLENOID

ALUMINA TENSIONING ROD

ALUMINA STANDOFF

SOLENOID SHEATH
(ANNEALED MILD STEEL)

FOCUS (E5)

GRIDDED LENS (E6)

ION BEAM

ONE INCH

Fig. 13. Auckland strong field ionizer.

operating states are observed, one state being more critical and also more efficient than the other. This is noticed in the Auckland ionizer [53], particularly at pressures above 2×10^{-6} Torr. Trapping of positive ions was discussed by Powell [76], but the conclusions were not definite and efficiencies higher than found in other ionizers were not obtained. Space charge neutralization does not seem to be consistent with positive ion extraction when the residual pressures are low (see also ref. [53]). The observed efficiencies are substantially lower [53] than one would expect even for the case when space charge neutralization is absent.

There is a possibility that higher efficiencies might be obtained by ionization in an rf discharge using a buffer gas such as He [77, 78]. It is not yet clear, however, that a strong guide field is sufficient to preserve the polarization. A design for an ionizer operating in the pulsed mode has been discussed [79].

6. NEGATIVE IONS

Polarized negative ions were first produced with an atomic beam source by passing polarized positive ions through a thin foil [80]. Now it is more common to use charge exchange in gas or vapor, since the thickness can be adjusted and the quality of the ion beam does not deteriorate as with foils. Grüebler [81] has shown that high negative yields are obtained with alkali vapors (see fig. 14). In particular, Na is attractive, since the high yields are realized at higher ion energies which are preferred from the point of view of beam optics.

The charge exchange canal on the source for the ETH Zürich tandem [29] generates Na vapor by heating an internal reservoir containing metallic Na, as shown in fig. 15. A strong solenoid field coaxial with the beam and spin axis is applied to decouple the nuclear spin and the unpolarized donor electron spins of the neutral states. Otherwise there is substantial depolarization. The guide field has been repeatedly demonstrated [29, 47, 30] to be effective, and its influence can be accounted for quantitatively [30] by considering the spin state structure of the hyperfine states as a function of field (see fig. 16).

Since the particles change charge, the magnetic field is able to impart angular momentum to the emerging ions [82]. For small radius beams this does not significantly affect the beam quality, provided the magnetic field is not unnecessarily high. A similar effect occurs in the ionizer, and partial compensation results if the ionizer and canal fields point in the same direction [83]. In connection with the choice of canal dimensions, it should be noted that the solenoid focuses the beam and increases the acceptance of the canal over its geometrical value [83].

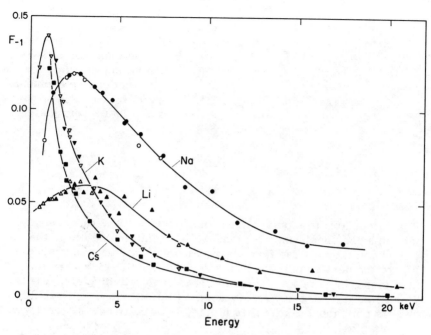

Fig. 14. Conversion fractions for the conversion of H$^+$ ions to H$^-$ ions as a function of H$^+$ ion energy.

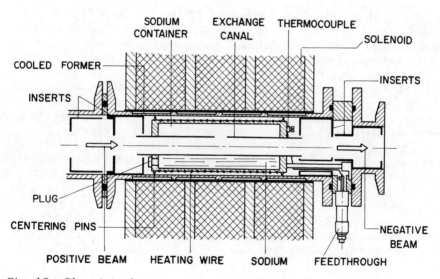

Fig. 15. Charge exchange canal on the ETH Zürich tandem source.

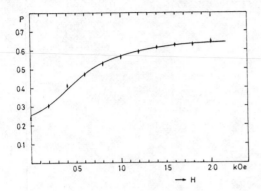

Fig. 16. Measurements of the proton polarization on the Erlangen source for different charge exchange guide fields. The solid curve is a theoretical fit.

The output from the ionizer contains a significant H_2^+ component which produces unpolarized H^+ ions by dissociative capture in the canal. When they emerge from the canal these ions possess half the energy of the desired polarized ions derived from H^+ ions. The unwanted ions can be removed from the beam by a retarding einzel lens, a Wien filter [29], or a deflection magnet [30]. Since the energy difference is only about 2 keV, it is preferable to carry out this operation at low energies.

7. SPIN PRECESSION

For many polarization experiments it is an advantage to be able to change the direction of the spin quantization axis independently of the beam direction. Wien filters (i.e., crossed field analyzers) or mirror-solenoid systems are used. In a Wien filter, a magnetic field produces the spin precession and a compensating electric field prevents deflection of the beam. In the other system, a 90° electrostatic deflection after the strong field ionizer or charge exchange canal leaves the spin axis at 90° to the beam direction. A solenoid coaxial with the deflected beam then precesses the spin axis (see fig. 17). Any direction for the spin axis is possible if the Wien filter can rotate around the beam axis, or if two successive mirror-solenoid systems are used. Both schemes must be designed with care to avoid loss of beam quality (and perhaps beam). In a cyclotron, the spin axis must be aligned along the cyclotron field direction to avoid depolarization during acceleration. In this case the atomic beam source axis is made parallel to the field (see figs. 1 and 2), and spin precession devices are not required.

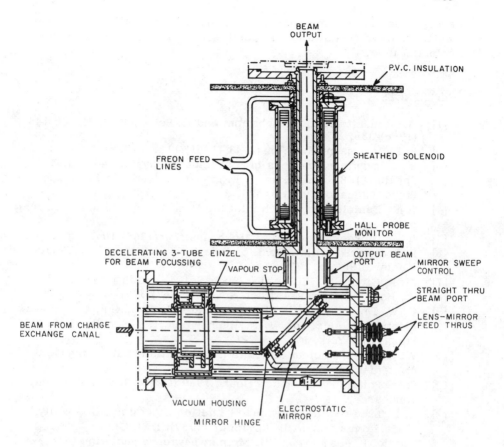

Fig. 17. Auckland mirror-solenoid system and retarding einzel lens for stopping half-energy unpolarized H⁻ ions.

8. INJECTION AND ACCELERATION

Axial injection into cyclotrons and acceleration without depolarization is discussed by Powell [76, 84] and more recently by Clark [85]. The Saclay source employs radial injection along the median plane [86], since the magnet pole does not possess a suitable axial hole. Typical target currents and transmission factors are shown in table 1.

The situation for tandems is at present not so satisfactory, as table 1 indicates. A better performance is expected from the ANU source [47] for a number of reasons: the output current is substantially improved, the beam emittance is comparable with the tandem acceptance, and the injection scheme will enable small or large beam waists to be formed at different positions before the low-energy accelerator tube.

REFERENCES

[1]   G. Clausnitzer, R. Fleischmann, and H. Schopper, Z. Physik 144 (1956) 336.
[2]   R. Fleischmann, Nucl. Instr. 11 (1961) 112.
[3]   E. Baumgartner, Progress in Fast Neutron Physics, eds. G. C. Phillips, J. B. Marion, and J. R. Risser (University of Chicago Press, Chicago, 1963) p. 323.
[4]   J. M. Daniels, Oriented Nuclei (Academic Press, New York, 1965) p. 165.
[5]   J. M. Dickson, Progr. Nucl. Tech. Instr. 1 (1965) 105.
[6]   R. Beurtey, Proc. Int. Conf. on Polarized Targets and Ion Sources (Saclay, 1966) p. 177.
[7]   M. K. Craddock, IEEE Trans. Nucl. Sci. 13, no. 4 (1966) 127.
[8]   W. Haeberli, Ann. Rev. Nucl. Sci. 17 (1967) 373.
[9]   C. W. Drake, Methods of Experimental Physics, 4-B, Atomic and Electron Physics, eds. V. W. Hughes and H. L. Schultz (Academic Press, New York, 1967) p. 226.
[10] W. D. Allen, Nuclear Structure (North-Holland Publishing Co., Amsterdam, 1967) p. 259.
[11] Proc. Polarized Ion Source Conf., 1969 (ORTEC Inc., Oak Ridge, Tennessee).
[12] A. U. Luccio et al., Lawrence Radiation Laboratory, University of California, Report UCRL-18607 (1969); D. J. Clark, A. U. Luccio, F. Resmini, and H. Meiner, Lawrence Radiation Laboratory, University of California, Report UCRL-18934 (1969).
[13] R. Beurtey, Ph.D. Thesis, University of Paris (1963), and Report CEA-R2366, CEA France (1964).
[14] R. Beurtey et al., Saclay Progr. Report CEA-N 621 (1966) 81.
[15] R. Keller, L. Dick, and M. Fidecaro, Report CERN 60-2 (Geneva, 1960), and First Polarization Symp., p. 48.
[16] T. Donoghue, Proc. Polarized Ion Source Conf., 1969 (ORTEC, Inc., Oak Ridge, Tennessee) sect. D.
[17] J. McCullen, Proc. Polarized Ion Source Conf., 1969 (ORTEC, Inc., Oak Ridge, Tennessee) sect. C.
[18] I. J. Barit, G. A. Vasil'ev, E. A. Glasov, and V. V. Jolkin, Nucl. Instr. 57 (1967) 160.
[19] H. Paetz gen. Schieck, C. E. Busch, J. A. Keane, and T. R. Donoghue, Third Polarization Symp.
[20] H. Ebinghaus, Third Polarization Symp.
[21] H. Ebinghaus, U. Holm, H. V. Klapdor, and H. Neuert, Z. Physik 199 (1967) 68; U. Holm et al., ibid. 233 (1970) 415.

[22] R. E. Miers and L. W. Anderson, Rev. Sci. Instr. 39 (1967) 336.
[23] R. Vyse, J. C. Heggie, and M. K. Craddock, Proc. Conf. Rarefied
      Gas Dynamics 6th, 2 (1969) 939.
[24] R. Vyse, D. Axen, and M. K. Craddock, Rev. Sci. Instr. 41 (1970)
      87.
[25] J. A. Giordmaine and T. C. Wang, J. Appl. Phys. 31 (1960) 463.
[26] G. R. Hanes, J. Appl. Phys. 31 (1960) 2171.
[27] J. C. Helmer, F. B. Jacobus, and P. A. Sturrock, J. Appl. Phys. 31
      (1960) 458.
[28] H. Lew, Methods of Experimental Physics, 4-A, Atomic and Elec-
      tron Physics, eds. V. W. Hughes and H. L. Schultz (Academic
      Press, New York, 1967) p. 155.
[29] W. Grüebler, V. König, and P. Marmier, Nucl. Instr. 62 (1968)
      115; W. Grüebler, V. König, and P. A. Schmelzbach, Report ETH
      Zurich (1970); W. Grüebler, V. König, and P. A. Schmelzbach,
      Third Polarization Symp.
[30] G. Clausnitzer et al., Nucl. Instr. 80 (1970) 245.
[31] W. B. Powell, Second Polarization Symp., p. 47.
[32] R. Beurtey, private communication. See also refs. [13, 14].
[33] F. J. Swales and A. G. D. Payne, Rutherford High Energy Labora-
      tory, PLA Ann. Progr. Report RHEL/R187 (1969) 8.
[34] H. Rudin, private communication.
[35] C. D. Moak, private communication.
[36] A. P. Banford, Rutherford High Energy Laboratory, PLA Ann. Progr.
      Report RHEL/R136 (1966) 17.
[37] A. Kantrowitz and J. Grey, Rev. Sci. Instr. 22 (1951) 328.
[38] J. B. Anderson, R. P. Andres, and J. B. Fenn, Adv. Chem. Phys.
      10 (1966) 275.
[39] R. Campargue, Proc. Conf. Rarefied Gas Dynamics 4th, 2 (1966)
      279.
[40] A. G. Hansen, Fluid Mechanics (Wiley, New York, 1967) ch. 7.
[41] J. B. Fenn and J. B. Anderson, Proc. Conf. Rarefied Gas Dynamics
      4th, 2 (1966) 311.
[42] J. B. Anderson and J. B. Fenn, Phys. Fluids 8 (1965) 780.
[43] H. Ashkenas and F. S. Sherman, Proc. Conf. Rarefied Gas Dy-
      namics 2 (1965) 85.
[44] J. B. Fenn and J. Deckers, Proc. Conf. Rarefied Gas Dynamics
      1 (1963) 497.
[45] J. E. Scott and J. E. Drewry, Proc. Conf. Rarefied Gas Dynamics
      1 (1963) 516.
[46] T. R. Govers, R. L. LeRoy, and J. M. Deckers, Proc. Conf. Rarefied
      Gas Dynamics 6th, 2 (1969) 985.
[47] H. F. Glavish et al., Third Polarization Symp.
[48] D. J. Clark, private communication.
[49] R. Keller, Report CERN 57-30 (1957).
[50] H. F. Glavish, Ph.D. Thesis, University of Auckland (1968) (un-
      published).

[51] R. S. Lord et al., Oak Ridge National Laboratory, Electronuclear Div. Ann. Progr. Report (1968) 96.

[52] D. F. Jones, Ohio State University, Physics Dept. Report (1969) (unpublished).

[53] H. F. Glavish, Nucl. Instr. 65 (1968) 1.

[54] B. P. Ad'yasevich, V. G. Antonenko, Yu. P. Polunin, and D. E. Fomenko, Atomnaya Energiya 17 (1964) 17, and Plasma Phys. (J. Nucl. Energy) C7 (1965) 187.

[55] W. Grüebler, private communication.

[56] A. P. Banford, N. J. Diserens, R. C. Carter, and A. D. G. Payne, Rutherford High Energy Laboratory, PLA Ann. Progr. Report RHEL/ R156 (1967) 9.

[57] A. I. Livshits, Sov. Phys.—Tech. Phys. 12 (1967) 650.

[58] A. Abragam and J. M. Winter, Phys. Rev. Lett. 1 (1958) 374, and Compt. Rend. 255 (1962) 1099.

[59] R. Beurtey, Second Polarization Symp., p. 33.

[60] H. Rudin, E. Steiner, and H. R. Striebel, Second Polarization Symp., p. 101.

[61] C. H. Poppe, V. Shkolnik, and D. L. Watson, Third Polarization Symp.

[62] M. A. Stentiford, to be published.

[63] J. D. McCullen, W. S. Smith, and W. J. Tipton, Phys. Rev. Lett. 23 (1969) 457.

[64] K. Jeltsch, P. Huber, A. Janett, and H. R. Striebel, Helv. Phys. Acta 439 (1970) 279.

[65] H. Paetz gen. Schieck, private communication.

[66] S. Oh, Nucl. Instr. 82 (1970) 189.

[67] J. D. McCullen, private communication.

[68] R. Beurtey, Saclay Progr. Report CEA-N 1032 (1968) 70.

[69] W. Haeberli, Nucl. Instr. 62 (1968) 355.

[70] D. A. G. Broad, A. P. Banford, and J. M. Dickson, Second Polarization Symp., p. 76.

[71] G. A. Vasil'ev and E. A. Glasov, Nucl. Instr. 58 (1968) 303.

[72] F. Haring and D. Fick, Nucl. Instr. 64 (1968) 285.

[73] J. F. Bruandet, F. Ripouteau, and M. Fruneau, Rev. Phys. Appliqué 4 (1969) 169.

[74] R. N. Boyd et al., Nucl. Instr. 63 (1968) 210.

[75] P. Birien, H. Poussard, and P. Roche, Saclay Progr. Report CEA-N 844 (1967) 77.

[76] W. B. Powell, Second Polarization Symp., p. 47.

[77] M. Heyman, P. Delpierre, and R. Sene, Second Polarization Symp., p. 97.

[78] B. P. Ad'yasevich, V. D. Aleshin, and G. V. Smirnov, Pribor̄y Éksp. 4 (1965) 32.

[79] P. Coiffet, Compt. Rend. 270 (1970) 343.

[80] W. Haeberli, Second Polarization Symp., p. 64.

[81]   W. Grüebler, P. A. Schemlzbach, V. König, and P. Marmier,
       Helv. Phys. Acta 43 (1970) 254.
[82]   G. G. Ohlsen, J. L. McKibben, R. R. Stevens, and G. P. Lawrence,
       Nucl. Instr. 73 (1969) 45.
[83]   H. F. Glavish, B. A. MacKinnon, and J. P. Ruffell, to be published.
[84]   W. B. Powell, IEEE Trans. Nucl. Sci. 13, no. 4 (1966) 147.
[85]   D. J. Clark, Lawrence Radiation Laboratory, University of Cali-
       fornia, Report UCRL-18980 (1969).
[86]   R. Beurtey and J. M. Durand, Nucl. Instr. 57 (1967) 313.

Lamb-Shift Sources

BAILEY L. DONNALLY, Lake Forest College, Illinois, USA

1. INTRODUCTION

Polarized ion sources utilizing the special properties of hydrogen and deuterium atoms in the $2\,^2S_{1/2}$ state are becoming increasingly common at accelerator installations. Although only six years have elapsed since the discovery of charge exchange processes which made the method seem attractive, polarized proton and deuteron sources of this type have been used in conjunction with several tandem accelerators to produce a considerable amount of useful information. Some of these "Lamb-shift" sources, as they have come to be called, are notable for their simplicity and low cost. Others are distinguished by high intensity and high polarization available on target. They are particularly attractive when high quality polarized negative ions are required, but recent studies indicate that positive ion beams may be obtained as well. Several review articles contain information about Lamb-shift sources [1, 2, 3, 4].

Fig. 1 shows, in general terms, how the source works. Although protons are shown here, deuterons or tritons may be used. Protons from some more or less conventional ion source enter a cell containing cesium vapor where many of them capture electrons, preferentially forming n = 2 states of the hydrogen atom. All excited states of the hydrogen atom except the $2\,^2S_{1/2}$ state quickly decay to the ground state, so that what emerges from the cesium cell is some mixture of H(2S), H(1S), H⁻, and H⁺. Hydrogen atoms in the $2\,^2S_{1/2}$ state cannot decay to the ground state by electric dipole radiation and are, therefore, metastable. Spontaneous decay to the $2\,^2P_{1/2}$ state is negligible

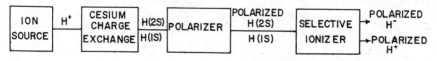

Fig. 1. Schematic diagram of Lamb-shift polarized ion sources.

because of the small energy separation. The dominant decay mode of
the upperturbed H(2S) is a two-quantum transition to the ground state,
giving it a lifetime of about 1/7 s. However, electric fields will
cause Stark mixing with the short-lived 2P states nearby and produce
decay to the ground state (quenching). The H^+ and H^- must be removed from the beam by a transverse electric field small enough not
to cause significant quenching of H(2S). The H(2S) beam is then
polarized by selective quenching of undesired hyperfine states, using
techniques described in more detail in sect. 4. Finally, the electrically
neutral H(2S) atoms either capture electrons or lose electrons in collisions with other atoms. In this paper we will (incorrectly, of course)
call both these processes ionization. The collision process which
forms ions must strongly discriminate against the abundant ground
state atoms which accompany the H(2S). These ionization processes
are discussed in sect. 5.

2. ION SOURCES

The most crucial component in determining the intensity of the
polarized ion beam is the ion source and the associated focusing system. Both rf ion sources and duoplasmatrons have been used. There
is little doubt that the duoplasmatron is superior for this application
in virtually every aspect except cost. There are many disadvantages
of the rf ion source:

(1) A fairly intense beam of neutral atoms emerges from the source
along with the proton or deuteron beam and with the same energy as
the ion beam. These neutral atoms are not removed from the beam by
the deflection plates which remove the charged components of the
beam. Thus they appear in the ionizing gas cell and contribute quite
significantly to the unpolarized background current which contam-
inates the desired polarized beam. Of course if the protons are accelerated or decelerated after they are extracted from the source and
before they enter the cesium cell, the metastable hydrogen atoms
entering the ionizing gas cell will have a different energy from the
nuetrals which emerge directly from the source and, after capturing
or losing an electron in the gas cell, can be separated from them.
These neutrals can also be eliminated by bending the ion beam before
it enters the cesium cell.

(2) The ion beam emerging from the rf ion source may not have an
energy exactly equal to the voltage used to extract the beam. With
a fixed extraction voltage, the energy of the ions can vary considerably with changes in the rf discharge [5]. This problem makes it difficult to make sure that the ion energy is the optimal one for the
charge exchange processes.

(3) The ion beam has a considerable spread in energy which in-

creases with extraction voltage, making it difficult to focus the beam
well, and making it particularly difficult to extract with a large voltage
and decelerate to the energy desired to end up with a well collimated
beam.

Most of the Lamb-shift ion sources now in operation use a duoplasmatron as the proton source. The protons are usually extracted
from the duoplasmatron with a high voltage and are subsequently decelerated to the required energy. The configuration of source and
lenses which has been most successful thus far [6] was developed at
Los Alamos Scientific Laboratory (LASL) and has been adopted by several other groups. It is shown in fig. 2. The source employs a plasma

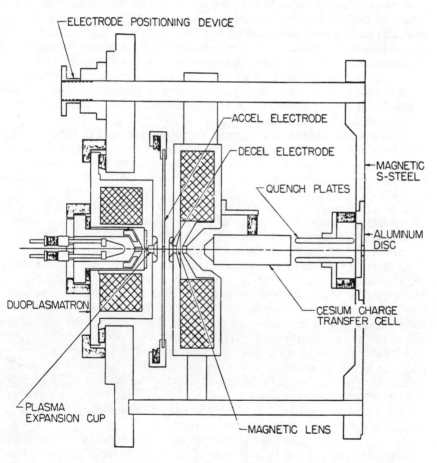

Fig. 2. Cross-sectional view of the Los Alamos Scientific Laboratory
ion source, lens system, and cesium charge transfer cell. The position
of the accel electrode is adjustable from outside the vacuum chamber,
an important feature for getting good ion beam alignment.

expansion cup from which ions are extracted by an electrode at about
-15 kV consisting of a plate with a hole in it. After passing through
the hole the protons are decelerated and immediately pass into a
strong magnetic lens which collimates them. The cesium cell in which
the protons are converted to metastable hydrogen atoms is placed very
near to the magnetic lens so that the protons are neutralized before
the space charge can expand and diverge the beam, and so that the
cesium emerging from the channel toward the ion source can help
neutralize space charge in the acceleration-deceleration region. Space
charge neutralization occurs because the cesium is ionized by the
proton beam, producing low velocity electrons, and because electrodes
coated with cesium readily eject electrons when hit by other particles.
It is extremely important to provide space charge neutralization in the
region of the ion source. Various schemes for producing space charge
neutralization have been tried by the LASL group, including thermionic
electron emitters in the source region, but no method has proved to
be better than merely turning on the cesium oven. Proton currents of
up to 0.5 mA have been measured 60 cm from the deceleration electrode with a Faraday cup having an aperture of diameter 2.5 cm. Without the cesium to provide space charge neutralization, the current is
only about 1/10 of this value.

It seems reasonable to expect substantial increase in beam currents
from improved ion sources which are being developed and studied.
For example, Morgan [7] has designed a low energy ion source for use
in fusion research that gives a beam current of 20-25 mA of 1 keV ions
on a cup 5 cm in diameter at a distance of about 50 cm from the lens.
This source uses a duoplasmatron with a plasma expansion cup of 2-
cm diameter covered with a grid at the exit end to define the plasma
boundary. The ion beam is extracted and accelerated by a gridded
electrode at high negative voltage, and is quickly decelerated to the
desired energy by another gridded electrode. Morgan's source also
uses a magnetic lens, which he finds less useful at low energies.
Space charge neutralization proved to be extremely important, and it
was accomplished by heating the thoriated tungsten grids in the ion
beam. This way of injecting electrons for space charge neutralization
was much more effective than using ring emitters around the beam. It
was found that if the magnetic lens was used, a space charge neu-
tralizing grid was necessary at the center of the magnetic lens because low energy electrons produced outside the magnetic lens were
reflected by the magnetic mirror and could not neutralize the beam in
the magnetic lens region. The critical defect in the Morgan source is
the unacceptable short life of the grids.

To preserve the excellent beam quality of the duoplasmatron, it is
important to extract the ions from a magnetic field-free region and to
do the cesium charge exchange in a magnetic field-free region [8].
This point will be discussed again in sect. 5.

3. PRODUCTION OF H(2S) ATOMS

The electron bombardment sources of H(2S) such as the ones used by Lamb and Retherford [9] yield rather low intensity, slow beams. Vályi [10] has built a 0.6 nA polarized deuteron source using electron bombardment excitation of ground state deuterium atoms and selective ionization of the D(2S) by low energy electrons. However, his method does not seem to be capable of enough improvement to make it attractive. Madansky and Owen [11] proposed that H(2S) atoms be produced by passing protons through a gas. The idea was that when the protons captured electrons, some of the atoms formed would be in the 2S state. They were able to show experimentally that H(2S) atoms were formed when 10 keV protons were sent through molecular hydrogen gas, but the fraction of atoms in the 2S state was small. Subsequently, other workers [12] have verified the finding of Madansky and Owen and have studied these processes in detail, both for molecular hydrogen and for other gases as target. Donnally et al. [13] argued that a more suitable atom from which to capture electrons was cesium, and found experimentally that the cross section for the reaction $H^+ + Cs \rightarrow H(2S) + Ca^+$ is large for protons with energies between 160 eV and 3000 eV. Fig. 3 shows the measured cross sections. Of course, some electron capture in cesium vapor leads to hydrogen atoms in the ground state or to

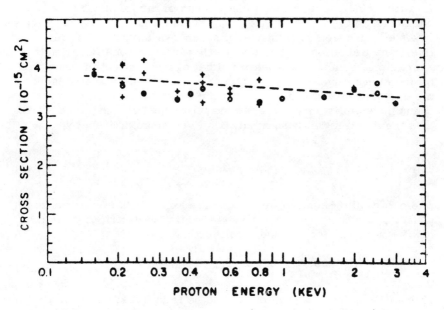

Fig. 3. Cross section for the reaction $H^+ + Cs \rightarrow H(2S) + Cs^+$ as a function of energy. From ref. [13].

excited states which quickly decay to the ground state. For 500 eV
protons, the fraction of atoms which emerge from the cesium vapor
cell in the 2S state is 0.32 [14]. If the cesium vapor thickness is
high, some of the 2S atoms will be quenched by collisions with cesium
atoms and the fraction of atoms in the 2S state in the beam emerging
from the vapor cell will be less. The cesium vapor thickness which
gives the optimum compromise between intensity and high H(2S) fraction has been studied by several groups [15, 16, 17] and has been found
to be about 2.5 micron-cm. It has also been found that H_2^+ ions and
H_3^+ ions produce negligible amounts of H(2S) in the cesium cell [15].
This is a crucial point since it means that these ions need not be
separated from the H^+ beam before the beam enters the cesium cell.

4. POLARIZATION SCHEMES

One of the principle technical problems in every polarization scheme
currently being used is changing in magnitude, and perhaps in direction,
the magnetic field seen by metastable hydrogen atoms without causing
transitions to different states. This change in magnetic field is always
made by having the atom move from one part of a static magnetic field
to another part. The problem, then, is to shape the magnetic field properly to obtain adiabatic passage. There are two separate parts to this
problem: (1) to avoid quenching the metastable atoms by the motional
electric field associated with components of the magnetic field perpendicular to the velocity, and (2) to avoid transitions between hyperfine states induced by rotating components of the magnetic field
(Majorana transitions). For a magnetic field rotationally symmetric
about the direction of the beam, neither of these effects is important
for the atoms traveling precisely along the axis of symmetry, since
there is no transverse field. But if the magnetic field on the axis is
changing with distance, z, along the beam, then at a distance, r, off
axis there is a radial component, B_R, given approximately by

$$B_R = -\frac{r}{2}\frac{\partial B_z}{\partial z}, \tag{1}$$

where B_z is the magnetic field along the axis. This transverse component of the magnetic field causes a motional electric field in the
rest frame of an atom moving parallel to the axis of

$$E = vB_R \times 10^{-4} \tag{2}$$

where E is in V/cm, v is in cm/s, and B_R is in Gauss. The lifetime for
a metastable hydrogen atom in an electric field is approximately [1]

$$\tau = \frac{1.13}{E^2} \ [\ (574 \mp B)^2 + 716] \times 10^{-9} \ s. \tag{3}$$

Clearly these equations show that there is a limit to how fast the magnetic field can be changed along the axis if excessive quenching is to be avoided in those parts of the beam at the greatest distance from the axis. On the other hand, for a fixed magnetic field gradient, eqs. (1), (2), and (3) set an upper limit to the acceptable beam diameter. For example, if we take a rather large gradient, $\partial B/\partial z = 100$ G/cm, with B near zero, and we consider atoms of velocity 3×10^7 cm/s at a distance of 1 cm from the axis, about 2% of the atoms are quenched per centimeter of travel. This example shows that although this quenching effect should be considered in designing a source, it is unlikely to be a major problem. It should be remembered that a diverging atom beam will be partially quenched, even in a uniform axial magnetic field, by the motional electric field. In most polarized ion sources Majorana depolarization is a more serious problem. Ohlsen [18] has considered this problem in considerable detail, calculating the expected depletion of particular hyperfine states for some specified field configurations. At points off the axis of a cylindrically symmetric magnetic field there is a radial component of the magnetic field given by eq. (1) and an axial component of approximately B_z. The total magnetic field, as seen by the atom as the atom moves parallel to the axis, will change its direction unless the axial field varies with z as $B_z = B_0 e^{Kz}$, where B_0 and K are constants. To avoid depolarization due to Majorana transitions, the rate at which the magnetic field rotates, as viewed by the atom, should be small compared to the frequency required to induce transitions between the initial hyperfine state and the adjacent hyperfine states. Thus it is more difficult to avoid Majorana transitions at low magnetic fields, where the hyperfine states are close together, than at higher fields. Again, it is not the atoms traveling exactly along the axis that are affected, but the atoms off axis. Some of Ohlsen's results are shown in fig. 4.

4.1. *Adiabatic field reduction method*

This method for polarizing the nuclei of metastable hydrogen atoms was derived from the observation of Lamb and Retherford [9] that passage through a 575 G magnetic field in which there was a small electric field perpendicular to the magnetic field produces electronic polarization. Fig. 5 shows why this is so. Near 575 G, the $2\,{}^2S_{1/2}$ ($m_J = -1/2$) states (β states) of the atoms cross the $2\,{}^2P_{1/2}$ ($m_J = 1/2$) states of the atom and the electric dipole coupling between the states produced by the small electric field causes these states to decay because of the admixture of the short-lived 2P state. Because of their short lifetime (1.6×10^{-9} s), the natural width of the 2P states is large (about 100 MHz wide), so exact crossing of the states is not required.

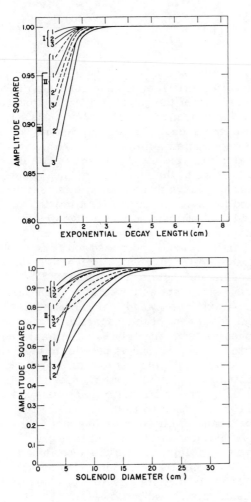

Fig. 4. Retained fraction of hydrogen atoms in $\alpha\,(m_I = 1/2)$ state (I), deuterium atoms in $\alpha\,(m_I = 1)$ state (II), and deuterium atoms in $\alpha\,(m_I = 0)$ state (III). The top graph is for a field which decays exponentially from 580 G to 5 G $(B_z = 5 + 575 \exp.\ (-z/Z))$. The abscissa is the "1/e" length Z. The curves marked 1, 2, and 3 correspond to particles which travel 1.25, 2.50, and 3.75 cm from the axis. A velocity of 3×10^7 cm/s is assumed. The lower graph is for a magnetic field approximating that which may be obtained by a solenoid in an iron cylinder with a small oppositely directed correction coil about one diameter from the solenoid end. This field falls from 575 G to about 5 G in about 1.25 solenoid diameters. From ref. [18].

Both β_+ and β_- hyperfine states are quenched. The remaining H(2S) atoms are in the $m_J = 1/2$ states (α states) and therefore have 100% electronic polarization. Zavoiskii [19] suggested that if these electronically polarized atoms were taken adiabatically to a weak field, the state α_+ would retain its nuclear polarization with $m_I = 1/2$, and state α_- would become an equal admixture of $m_I = 1/2$ and $m_I = -1/2$, yielding a total nuclear polarization for the atoms of 50%.[†] If deuterons

---

[†] In this section the polarizations quoted refer to the theoretical limit of either extremely low field or extremely high field. In a prac-

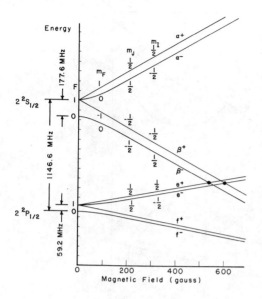

Fig. 5. Energy level dia-
gram for the hydrogen atom
with n = 2. The other fine
structure state, $2\,^2P_{3/2}$, lies
about 10,000 MHz above
the $2\,^2S_{1/2}$ state and is not
shown here. Both the low
field quantum numbers (F,
m_F) and the high field quantum numbers (m_J, m_I) are
shown. The designations
α, β, e, and f for the states
as shown are due to Lamb
and Retherford [9].

are used instead of protons, the $m_J = -1/2$ states would be quenched,
and when the remaining atoms are taken adiabatically to a low field,
the nuclei would have $P_z = -1/3$ and $P_{zz} = -1/3$. The first Lamb-shift
polarized ion sources used this method of polarization and were built
at Yale [20, 21] and Milan [3, 22]. Subsequently this type of source
was constructed at Wisconsin [23] and at Notre Dame [24, 25].

4.2 *Spin filter*

McKibben, Lawrence, and Ohlsen [26] have developed a very versatile scheme for polarizing H(2S) or D(2S) based on an effect observed
and explained by Lamb and Retherford [9, 27]. It utilizes a three-level
interaction involving α, β, and e states, all with the same nuclear
polarization. Consider metastable deuterium atoms in a magnetic field
of 565 G where the β($m_I = 1$) state crosses the e($m_I = 1$) state as shown
in fig. 6. A small transverse d.c. electric field would then quench β

---

tical device these limits may not be reached but the correction to be
applied can easily be calculated from formulas listed in ref. [1]. The
stated polarizations obviously also do not take into account a contamination of the polarized ion beam by ions resulting from ground
state atom collisions. Perhaps it should also be mentioned here that
the ground state atoms which arise from quenching of some hyperfine
states of the metastable atoms will usually have some nuclear polarization.

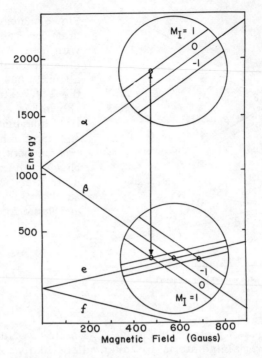

Fig. 6. Energy level diagram for the
deuterium atom in the n = 2 state. Only
the fine structure states are shown except inside the circles, where a magni-
fied view of the hyperfine states is shown.

states. On the other hand, if instead of applying the transverse static
electric field we applied an axial rf electric field of frequency about
1600 MHz, just corresponding to the energy separation of the $a(m_I = 1)$ and the $e(m_I = 1)$ states, the a states would be quenched. The situation is very different if both electric fields are applied simultane-
ously. In particular, if we start with a beam of unpolarized D(2S)
atoms in a 565 G axial magnetic field and take them into a static transverse electric field, all the β states will be quenched, but all the a
states will be preserved. Now with the transverse static field still
applied, if the atoms are taken into a slowly increasing axial rf field,
the atoms in the $a(m_I = 1)$ state will become an admixture of a and β
which does not decay. As these atoms emerge slowly from the rf field,
they revert to the $a(m_I = 1)$ state. The effect of having the two interactions applied together is to preserve this single hyperfine state of
D(2S). All other states are quenched by one of the electric fields. If
the magnetic field is increased to 575 G, the $a(m_I = 0)$ state would be

preserved, and at 585 G, the $\alpha(m_I = -1)$ state would be preserved. If the $m_I = 1$ state is preserved, it can be ionized subsequently in either a strong field or a weak field, but if either $m_I = 0$ or $m_I = -1$ states are selected, the atoms must be ionized in a strong field (i.e., large compared to the "critical field" of 14.6 G for deuterium) in order to retain the nuclear polarization. The deuteron polarizations obtainable with this source are $(P_z = 1, P_{zz} = 1)$, $(P_z = 0, P_{zz} = -2)$, or $(P_z = -1, P_{zz} = 1)$, corresponding to selecting $m_I = 1$, 0, or -1, respectively. If protons are used with the spin filter, α states with either $m_I = 1/2$ or $m_I = -1/2$ may be selected. But if $m_I = -1/2$ is selected the atoms must be ionized in a strong field (i.e., large compared to 63.4 G, the critical field for H(2S)). The charm of the spin filter lies in its ability to produce beams of high polarization, the ease with which it can be switched from one polarization to another, and its ability to polarize beams of large diameter, resulting in good intensity. Its main disadvantages are its cost and the technical difficulty in setting it up and making it work initially. Ohlsen and McKibben [28] have studied the theory of the spin filter, giving careful attention to how the fields must be varied to avoid transient 2P state amplitudes which would cause the desired spin states to decay. McKibben, Lawrence, and Ohlsen [26] reported the successful operation of a spin filter, and have used it in a very high performance polarized proton and deuteron source now operating on the LASL tandem accelerator [29]. The design of a cavity, split into quadrants which allows application of a transverse d.c. electric field and a slowly increasing and decreasing longitudinal rf electric field is shown in fig. 7. Clegg and Bissinger [30] have also constructed a source of this type and have installed it on the TUNL accelerator.

4.3 *Diabatic field reversal*

In the simple adiabatic field reduction method of polarizing H(2S), the beam polarization is low. It was Sona [31] who pointed out that higher polarization could be obtained by passing the beam through a simple configuration of static magnetic fields. Sona's scheme involves taking the α-state metastable atoms, which remain after passage through a 575 G field, adiabatically to a weak field and then through a zero field to a field reversed in direction. The appropriate axial field strength variation is shown in fig. 8. The crucial diabatic process occurs during the field reversal. At this time the states 1 and 2 of fig. 9 become states 1' and 2', respectively. When these two new states are taken adiabatically to a strong (negative) field, the nuclear polarization is 100%. Note that the m_I indicated in fig. 9 is with respect to the direction of the positive magnetic field (i.e., the 575 G field that quenched states 3 and 4). Fig. 10 shows what happens when deuterium atoms are used. The states 1, 2, and 3 which remain after the 575 G passage are transformed to the 1', 2', and 3' states, respectively, by the field reversal. In a strong field these states give $P_z = 2/3$, $P_{zz} = 0$.

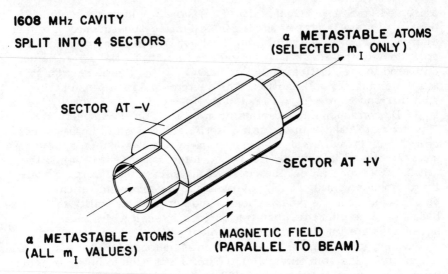

I608 MHz CAVITY SPLIT INTO 4 SECTORS

a METASTABLE ATOMS (SELECTED m_I ONLY)

SECTOR AT −V

SECTOR AT +V

a METASTABLE ATOMS (ALL m_I VALUES)

MAGNETIC FIELD (PARALLEL TO BEAM)

Fig. 7. Schematic diagram of the LASL nuclear spin filter. The overall length of the device is 34 cm. From ref. [26].

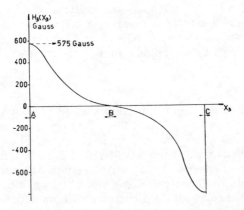

Fig. 8. A qualitative representation of the magnetic field distribution required for the diabatic field reversal scheme. From ref. [31].

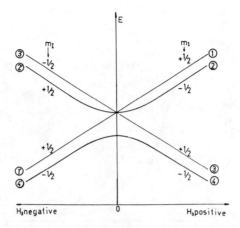

Fig. 9. Zeeman hyperfine splitting for hydrogen atoms in the $2\,^2S_{1/2}$ state. From ref. [31].

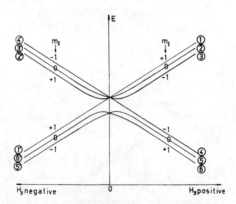

Fig. 10. Zeeman hyperfine splitting for deuterium atoms in the $2\,^2S_{1/2}$ state. From ref. [31].

If the 1' state is quenched by a small electric field at negative 575 G, the remaining states 2' and 3' would yield $P_z = 1/2$, $P_{zz} = -1/2$. In order to get the desired transitions during the field reversal, the principal physical condition that must be satisfied is that the rate of rotation of the magnetic field seen by the atom must be much larger than the Larmor presession frequency. It is easy to insure this condition for atoms traveling along the axis. But where the axial field is changing,

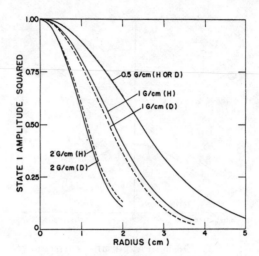

Fig. 11. Fraction of the initial state-1
atoms making the desired transition when
the field is linearly reversed at the indicated rate. The abscissa is the distance
of the atom from the (cylindrically symmetric) field axis. From ref. [18].

the off-axis field has a radial component and the net magnetic field
seen by an atom traveling off axis is rotating less rapidly the farther
from the axis the atom travels. To insure that atoms traveling at a
distance r from the axis undergo the desired transitions, Sona showed
that the condition $H'r^2 \ll 14$ G cm must be satisfied for hydrogen atoms,
where H' is the gradient of the field on the axis. Ohlsen [18] has
made more detailed calculations of what happens to state 1 for various gradients and radii. Fig. 11 shows his results for an axial field
varying linearly near the zero crossing. Clearly, for a given gradient
there is a limit to the radius of the beam if the desired transitions
are to occur in the atoms at the edge of the beam. Both Sona [31] and
Bechtold et al. [32] find that even small transverse fields in the region
of the zero crossing seriously diminish the polarization. Bechtold et
al. experimentally determined that with a deuteron beam of diameter
1.6 cm no serious depolarization occurred for a gradient as large as
1.9 G/cm, but that an externally applied transverse field of 0.5 G reduced the polarization to 80% of its maximum value. Most groups
have used smaller gradients. Clegg et al. [33] pointed out that if the
deuterium beam is taken adiabatically to a weak field after passing
through the second β-state quenching region the polarization is $P_z =$
0, $P_{zz} = -1$. Except for the slightly tedious job of initially adjusting
the magnetic field so that it is properly shaped in the vicinity of the

zero crossing, sources based on this scheme are technically relatively simple. It has the feature that it can provide purely tensor
polarized deuteron beams (i.e., with vector polarization zero) or purely
vector polarized beams. That the diabatic field reversal method works
was first shown experimentally by Meiner, Michel, and Corrigan [34]
at Notre Dame. Since then, several groups have successfully utilized
this polarization scheme [33, 32].

4.4. *Weak field rf state selection*
Electric dipole transitions between 2S and 2P states may be induced by small rf electric fields of the appropriate frequency. At magnetic fields above about 50 G it is not possible to polarize H(2S)
highly by selective quenching of particular hyperfine states using rf
electric dipole transitions, because the frequency required to quench
an unwanted state is only about 1.2 line widths different from a frequency that quenches the desired state. Donnally [35] pointed out
that at very low magnetic field the $F = 0$ hyperfine state (see fig. 5)
is sufficiently separated from the three $F = 1$ hyperfine states to allow
rf state selection. An 1147 MHz rf field will quench the $F = 1$ states
by inducing transitions between the $2\,{}^2S_{1/2}$ ($F = 1$, $m_F = \pm\, 1$, 0) states,
and the $2\,{}^2P_{1/2}$ ($F = 1$, $m_F = \pm\, 1$, 0) states. This frequency is separated
from the frequency that would quench the $2\,{}^2S_{1/2}$ ($F = 0$, $m_F = 0$) state
(by transitions to the $2\,{}^2P_{1/2}$ ($F = 1$, $m_F = \pm\, 1$, 0) states) by 2.4 line
widths. The $F = 0$ to $F = 0$ transition is forbidden. The weak magnetic
field should be at an angle of about 45° to the rf electric field so that
all the $F = 1$ states will be quenched. If the surviving $2\,{}^2S_{1/2}$ ($F = 0$,
$m_F = 0$) atoms are ionized in a strong magnetic field, the proton polarization will be 100%. This method is not applicable to deuterium atoms
because of their small hyperfine splitting. It is the simplest way to
produce highly polarized protons. Because the $F = 0$ state is well separated from other states at all magnetic fields of interest, it is uniquely
immune to Majorana transitions. This suggests the possibility of constructing a very short polarized proton source with a large beam current. Boyd et al. [36] have built a polarized proton source based on
this method.

4.5. *rf state selection with diabatic field reversal*
In the diabatic field reversal scheme β states are quenched before,
and sometimes also after, the field reversal by passage through a
575 G magnetic field with a transverse static electric field. Donnally
[37] has suggested that the diabatic field reversal scheme is more
versatile if the rf state selection is employed. For example at a magnetic field of about 120 G, the α–f transition frequency and the β–e
transition frequency differ by about 4 line widths and therefore a transverse rf electric field of the appropriate frequency can be used to
quench either the α states or the β states. If an rf state selector is
available both before and after the field reversal region and the atoms

are ionized in a weak magnetic field, the polarization combinations possible for a deuterium beam are ($P_z = 0$, $P_{zz} = -1$), ($P_z = -1$, $P_{zz} = 1$), ($P_z = 0$, $P_{zz} = 0$), and ($P_z = 1$, $P_{zz} = 1$). If the beam is ionized in a strong field other combinations are possible, including ($P_z = -2/3$, $P_{zz} = 0$) and ($P_z = 2/3$, $P_{zz} = 0$). With hydrogen beams, polarizations of $+1$ and -1 are possible using either strong field or weak field ionization. Because the magnetic field required for the state selection is only a little above 100 G, the field reversal region can be made shorter, resulting in larger beam currents.

4.6. *2364 G crossing method*

Leventhal [38] demonstrated that at a magnetic field of 2364 G, where the a_- state crosses the $2\,^2P_{3/2}$ ($m_J = -3/2$, $m_I = +1/2$) state, a transverse electric field preferentially quenches the a_- states. The electric field also partially quenches a_+ states by coupling to other 2P states. The β states have to be removed by a 575 G crossing. This method is not attractive for use in a practical polarized ion source because of the large magnetic field required and because it would not yield a particularly high polarization.

If the states selected by one of these polarization schemes must be ionized in a "strong" magnetic field, the beam quality will be degraded as it emerges from the magnetic field. Ohlsen et al. [8] have shown that for a parallel, cylindrical beam of radius R inside a magnetic field, the effective transverse emittance for the beam after it emerges from the field will be $\eta = 3.46\, B_z R^2 / M^{1/2}$ cm rad $(eV)^{1/2}$, where B_z is in kG, R is in cm, and M is the mass of the particles in amu. As an example, consider a hydrogen negative ion beam of diameter 3 cm emerging from a "strong" field 3 times the critical field (the critical field for H(2S) is 63.4 G). The emittance would be increased by 3 cm rad $(eV)^{1/2}$, still well within the acceptance of typical commercial tandem accelerators. For deuterium negative ions the situation is even better since it is more massive and the critical field for D(2S) is only 14.6 G. However, this problem would become serious if beams of larger radius were to be used.

5. SELECTIVE IONIZATION

After the nuclei of the metastable atoms are polarized, the atoms must either capture an electron or lose an electron so that it can be accelerated. Furthermore, since there are considerably more ground state atoms than metastable atoms in the beam, the process by which the metastable atoms gain or lose electrons must discriminate sharply against ground state atoms. Madansky and Owen [11] suggested photoionization with photons of energy insufficient to ionize ground state atoms. The huge photon fluxes required to get good yields make this method unsatisfactory. Alexeff [39] attempted to detect preferential

stripping of H(2S) in atomic collisions but failed to detect any effect
when the metastable atoms were quenched. Donnally et al. [13] suggested the nearly resonant charge transfer reaction H(2S) + Cs$^+$ →
H$^+$ + Cs, which should be quite selective. Although plasmas containing sufficiently high densities of Cs$^+$ to carry out this reaction can
probably be provided, the suggestion is not feasible because the
metastable atoms would be quenched in plasmas of adequate density
[40]. The first successful solution to this problem was demonstrated
by Donnally and Sawyer [41]. They argued that because the potential
energy of H$^-$ + A$^+$ is a few electron volts above H(2S) + A for large
internuclear separations, the Coulomb interaction would pull the two
energies together as the internuclear separation decreases. A pseudo-
crossing of potential energy curves would cause the reaction H(2S) + Ar
→ H$^-$ + Ar$^+$ to have a substantial cross section for some H(2S) velocity.
Because the potential energy curve for H(1S) + Ar would not come near
that for H$^-$ + Ar$^+$ except, perhaps, at very small internuclear separation, the cross section for H(1S) + Ar → H$^-$ + Ar$^+$ should be smaller.
Fig. 12 shows Donnally and Sawyer's experimental verification of
these ideas. For energies around 500 eV for protons (1000 eV for deuterons), the negative ion current is much larger when the beam contains H(2S) than when the H(2S) is quenched. Various groups have
studied the energy and argon gas thickness that are optimal for the
reaction H(2S) + Ar → H$^-$ + Ar$^+$ when used in polarized particle sources
[15, 16, 17]. Bechtold et al. [42] do not observe the peak at about
500 eV in the selectivity. One possible reason for the different selectivities observed at different laboratories is that the H$^-$ produced
from H(1S) is scattered more in the collision with argon than those

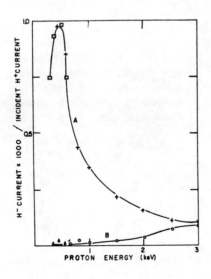

Fig. 12. The yield of H$^-$
with argon as the target gas.
Curve A is the fraction of
protons converted to H$^-$ by
all processes. Curve B is
the fraction of protons converted to H$^-$ when the metastable atoms are quenched
before they enter the gas
cell. From ref. [41].

produced from H(2S), and therefore the angular acceptance of the H⁻
detector plays a significant role. Furthermore, in most laboratories
the fast neutral particles emerging from the source region are not
separated from the protons. Lawrence, Ohlsen, and McKibben [29]
have found that if the H⁻ is focused by a lens with small aberration
and passed through a small hole, the quenchable fraction is much
larger than if no angle-defining apertures are used. Several target
gases will selectively attach electrons to H(2S) [15] but none have
been found that are as satisfactory as argon.

Since many accelerators require positive ions, it is desirable to
find a reaction which will selectively strip electrons from H(2S) to
give protons. Brückmann, Finken, and Friedrich [43] and Donnally
and Sawyer [44] showed that H_2 and He gas strip H(2S) more readily
than H(1S), but the maximum tensor polarization of 1 keV D^+ produced
in these reactions [43] was only -0.45. A more satisfactory reaction
is H(2S) + I_2 → H^+ + I_2——(or perhaps H(2S) + I_2 → H^+ + I^- + I)——which
has been studied by Knutson [45] and by Brückmann, Finken, and
Friedrich [46]. Fig. 13 shows the very encouraging results obtained
by Knutson. The selectivity is much better than for the reaction involving H_2 or He. Brückmann, Finken, and Friedrich [46] measured
the tensor polarization of 1 keV D^+ formed from collisions with I_2 and
found P_{zz} = 0.64. The intensity of the polarized beam of D^+ is about
two times the intensity of the beam of D^- obtained by collisions with
argon.

Since none of the charge transfer reactions discussed here involve
significant momentum transfer, the emittance of polarized ion sources
using them is consistently less than 1 cm rad $(eV)^{1/2}$.

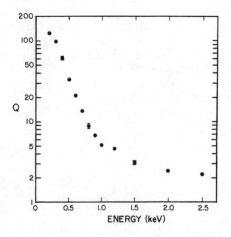

Fig. 13. The ratio, Q, of H^+
ion current for unquenched
atom beam to that for a
quenched atom beam. I_2 is
the target gas. From ref. [45].

Table 1

	Polarization scheme	Hydrogen			Deuterium			
		Beam out of source, nA	Beam on target, nA	P	Beam out of source, nA	Beam on target, nA	P_z	P_{zz}
Los Alamos	SF	120	60	0.87	160	80	0.78	0.78
					160	80	0.01	-1.5
Notre Dame	DFR		0.5	0.6	8	1.5	0.49	-0.62
Rutgers	LFRF	12*	1.5	0.50				
TUNL	SF	5.7	0.7	75	2.8		0.67	0.67
					2.8			-1.0
Wisconsin	DFR	10	0.75	0.75	200*	60*	0.5	
					70	20		-0.75

Table 2

| | Polarization scheme | Deuterium | |
		Beam out of source, nA	P_{zz}
Erlangen	DFR		
Karlsruhe	DFR	150	−0.70
		300 (D^+)	−0.64
Kyoto	DFR		
Milan	AFR	0.8	−0.24
Tokyo	SF		
Washington	DFR	300	−0.76
Yale	AFR	0.7	−0.19
Zürich	AFR		

6. PERFORMANCE OF LAMB-SHIFT SOURCES

Several papers have been submitted to this symposium describing the constructional features of some of the best existing Lamb-shift sources. These papers include information about beam handling, spin precession devices, and injection into accelerators. It would be redundant to include that material here. Instead, tables 1 and 2 list the performance figures for Lamb-shift sources. Table 1 represents those sources that have already been installed on accelerators. Table 2 shows the rest of the sources. Where no data are presented, the sources are still under construction. In table 1 an effort has been made to present average values of currents and polarizations, rather than the maximum values obtained. The numbers which are peak values rather than typical values are marked with an asterisk (*). No attempt was made to present typical values in table 2. Under the heading, polarization scheme, AFR means the adiabatic field reduction method as described in sect. 4, DFR means diabatic field reversal, LFRF means low field rf, and SF means spin filter.

Up to this time, considerable effort has been expended to perfect the polarization schemes. Almost surely the focus of work in the future will be on increasing the intensity. Several groups predict that very high quality microampere beams will become available soon.

REFERENCES

[1]   W. Haeberli, Ann. Rev. Nucl. Sci. 17 (1967) 373.
[2]   A. Cesati, F. Cristofori, L. Milazzo Colli, and P. G. Sona, Progr. Nucl. Phys. 10 (1969) 119.
[3]   A. Cesati, F. Cristofori, L. Milazzo Colli, and P. G. Sona, Energia Nucleare 13 (1966) 649.
[4]   C. W. Drake, in Methods of Experimental Physics, IVB, Atomic and Electron Physics, eds. V. W. Hughes and M. Schultz (Academic Press).
[5]   C. V. Cook, O. Heinz, D. C. Lorents, and J. R. Peterson, Rev. Sci. Instr. 33 (1962) 649.
[6]   G. P. Lawrence, G. G. Ohlsen, and J. L. McKibben, Phys. Lett. 28B (1969) 594.
[7]   O. B. Morgan, Thesis, University of Wisconsin (1970).
[8]   G. G. Ohlsen, J. L. McKibben, R. R. Stevens, Jr., and G. P. Lawrence, Nucl. Instr. 73 (1969) 45.
[9]   W. E. Lamb, Jr., and R. C. Retherford, Phys. Rev. 79 (1950) 549.
[10] L. Vályi, Nucl. Instr. 58 (1968) 21.
[11] L. Madansky and G. E. Owen, Phys. Rev. Lett. 2 (1959) 209.
[12] J. E. Bayfield, Phys. Rev. 182 (1969) 115. This paper contains references to earlier work.
[13] B. L. Donnally, T. Clapp, W. Sawyer, and M. Schultz, Phys. Rev. Lett. 12 (1964) 502.
[14] B. L. Donnally, J. Odell, and R. Becker, to be published.
[15] B. Donnally and W. Sawyer, Second Polarization Symp., p. 71.
[16] H. Brückmann, D. Finken, and L. Friedrich, Z. Physik 224 (1969) 586.
[17] H. Treiber, Thesis, University of Erlangen–Nürnberg (1969).
[18] G. Ohlsen, Los Alamos Scientific Laboratory Report LA-3949.
[19] E. K. Zavoiskii, Soviet Phys.— JETP 5 (1957) 603.
[20] C. W. Drake and R. Krotkov, Phys. Rev. Lett. 16 (1966) 848.
[21] C. W. Drake and R. V. Krotkov, IEEE Trans. Nucl. Sci. 13, no. 4 (1966) 142.
[22] A. Cesati, F. Cristofori, and L. Milazzo Colli, Energia Nucleare 13 (1966) 328.
[23] T. B. Clegg, G. R. Plattner, L. G. Keller, and W. Haeberli, Nucl. Instr. 57 (1967) 167.
[24] G. Michel, K. Corrigan, and S. E. Darden, Bull. Am. Phys. Soc. 12 (1967) 1204.
[25] G. Michel, K. Corrigan, H. Meiner, R. M. Prior, and S. E. Darden, Nucl. Instr. 78 (1970) 261.
[26] J. L. McKibben, G. P. Lawrence, and G. G. Olhsen, Phys. Rev. Lett. 20 (1968) 1180.
[27] W. E. Lamb, Jr., Phys. Rev. 85 (1952) 259.
[28] G. G. Ohlsen and J. L. McKibben, Los Alamos Scientific Laboratory Report LA-3725.

[29]   G. P. Lawrence, G. G. Ohlsen, and J. L. McKibben, Phys. Lett.
       28B (1969) 594.
[30]   T. B. Clegg, G. A. Bissinger, W. Haeberli, and P. A. Quin, Third
       Polarization Symp.
[31]   P. G. Sona, Energia Nucleare 14 (1967) 295.
[32]   V. Bechtold, H. Brückmann, D. Finken, and L. Friedrich, Z. Physik
       231 (1970) 98.
[33]   T. B. Clegg, G. R. Plattner, and W. Haeberli, Nucl. Instr. 62
       (1968) 343.
[34]   H. Meiner, G. Michel, and K. Carrigan, Nucl. Instr. 62 (1968)
       203.
[35]   B. Donnally, Bull. Am. Phys. Soc. 12 (1967) 509.
[36]   R. N. Boyd, J. C. Lombardi, A. B. Robbins, and D. E. Schechter,
       Nucl. Instr. 81 (1970) 149.
[37]   B. Donnally, to be published.
[38]   M. Leventhal, Phys. Lett. 20 (1966) 625.
[39]   I. Alexeff, First Polarization Symp.
[40]   E. M. Purcell, Astrophys. J. 116 (1952) 457.
[41]   B. L. Donnally and W. Sawyer, Phys. Rev. Lett. 15 (1965) 439.
[42]   V. Bechtold et al., Third Polarization Symp.
[43]   H. Brückmann, D. Finken, and L. Friedrich, Phys. Lett. 29B (1969)
       223.
[44]   B. Donnally and W. Sawyer, Sixth Int. Conf. Physics of Elect.
       and Atomic Collisions: Abstracts of Papers (M.I.T. Press, Cam-
       bridge, 1969), p. 488.
[45]   L. D. Knutson, Phys. Rev., to be published.
[46]   H. Brückmann, D. Finken, and L. Friedrich, Third Polarization
       Symp., and private communication.

Sources of Polarized Fast Neutrons

RICHARD L. WALTER, Duke University and Triangle Universities
Nuclear Laboratory, Durham, North Carolina, USA*

1. INTRODUCTION

Since the Karlsruhe conference, a large amount of data on neutron
polarization produced in reactions has been recorded, but usually with
a view toward learning more about the reaction mechanism or about
resonance parameters than about providing a useful, calibrated source
of polarized neutrons. On the other hand, a surprisingly small number of laboratories have been investigating the interaction of polarized
neutrons with targets even though there is a great need for good, elastic scattering data for polarized neutrons. Clearly the latter situation
exists because clean, polarized neutron beams have not been readily
available. The purpose of this paper is to discuss some of the methods
for obtaining polarized neutrons, the current best sources, and some
promising sources for the future. I have tried to be thorough in my
search for data, and since the reaction compilation will continue, in
collaboration with T. Donoghue, we would appreciate receiving information about omitted useful polarization data. Time will not permit me
to detail all the sources of polarized neutrons. I will neglect the
planned program of J. Dabbs at the Oak Ridge Electron Linear Accelerator using the transmission of neutron beams through polarized targets
to produce a high flux of polarized neutrons, but some comments
about a similar method at Los Alamos will be inserted for interest.
I have not been able to make a fair evaluation of the method [1] of
using 150-keV polarized deuteron beams to provide 14-MeV polarized
neutron beams through the T(d,n) reaction, but a quick glance suggests
that this method may be useful, given sufficient deuteron intensity.

2. PRODUCTION OF POLARIZED NEUTRONS THROUGH
REACTIONS USING UNPOLARIZED INCIDENT BEAMS

2.1. *Introduction*

Most scattering experiments with polarized neutrons to date have
used reactions as sources, usually $(p,\vec{n})$, or $(d,\vec{n})$ reactions. The ideal

neutron-producing reaction would be monoenergetic and would have
a large differential cross section at the reaction angle employed relative to the total neutron production cross section. It would have self-
supporting, isotopically enriched targets which withstand high beam
currents and would yield a large polarization which has been accurately calibrated. In reality, most reactions give several neutron
groups, so that their usefulness may be limited to laboratories with
pulsed beam, time-of-flight apparatus. Several laboratories [2] have
been studying the use of the associated particle method for discrim-
inating against unwanted groups or other backgrounds. If the counters
selected can survive the high neutron flux, this technique will bypass
the pulsed beam requirement, as fast time-of-flight measurements are
then possible by employing the associated particle for a timing pulse.
Although the associated-particle method looks encouraging, I won't
be able to discuss these advantages further. The final introductory
remark is that deuterons are typically harder to work with because
^{12}C, which appears on collimators and on the target itself, has a large
(d,n) cross section and the D(d,n) neutrons from the beam stop produce
a large background flux. Proton reactions, on the other hand, often
have large negative Q-values, so that the upper limit of the neutron
range is quite low, even for tandem Van de Graaff accelerators.

   With the ideal reaction in mind, let me present the current knowledge, first describing the calibration, then the results for a few important reactions, and concluding with several comparisons of all the
reactions which have been studied.

2.2. *Calibration*

The greatest difficulty encountered in preparing meaningful polarization plots was that the analyzing power employed several years
ago differs from the one favored today. Those groups who have used
^{12}C as an analyzer have based their values on either the phase shifts
of Meier et al. [3] or Wills et al. [4] or both. Reynolds et al. [5] have
contributed another n-^{12}C analysis. For data reported using ^{12}C as
an analyzer, I have adjusted the data to correspond to an analyzing
power given by the phase shifts of Wills et al., except in one instance where I have noted a Reynolds et al. analysis. ^{4}He has most
often been employed because the background problems can be reduced
by using the ^{4}He recoils in a coincidence arrangement. One of the
most successful methods for measuring neutron polarization accurately
has been to insert a spin-precession solenoid between the ^{4}He scatterer and the target to interchange the role of the detectors. Such an
arrangement is shown in fig. 1. The neutrons are emitted from the
reaction at an angle θ_1 and pass along the axis of the solenoid. Those
neutrons which scatter through the angle θ_2 are detected in proton-
recoil counters, usually plastic scintillators. Typically, the ^{4}He is
contained at about 150 atm in a thin-walled vessel which also acts
as a scintillation chamber. The He recoil pulses, which occur in

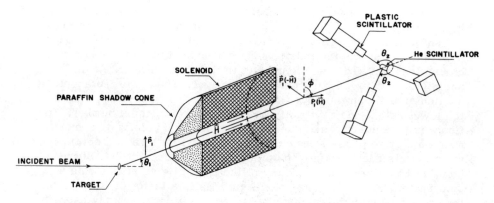

Fig. 1. Typical polarimeter.

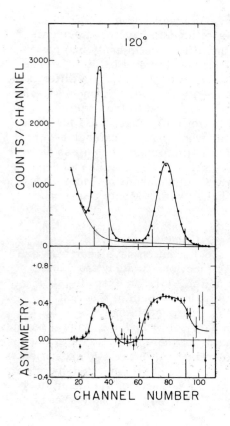

Fig. 2. Recoil ^{4}He spectrum for pulses in coincidence with pulses in the side detector at 120°. Lower half shows the asymmetry as a function of channel number. The curves are calculated fits to the data.

coincidence with neutrons detected at θ_2, are then studied in a pulse
height analyzer. In fig. 2 a helium-recoil spectrum generated in such
a fashion is shown. This is from the ^{9}Be$(\alpha,n)^{12}$C study of Stammbach
et al. [6] and shows the recoil peaks corresponding to the neutron
groups leaving ^{12}C in the ground and first excited states. At this
angle (120°) the analyzing power is nearly 1. The curve represents
a calculated fit, in the upper half to the measured spectrum, and in
the lower half to the asymmetry as a function of channel number. The
figure is presented to indicate that clean neutron polarization experiments are quite possible now and that reliable data can be obtained
when the residual background can be evaluated accurately. Simmons'
group at Los Alamos uses a thin-walled liquid He scintillator as a
polarimeter. Since the density of liquid He is a factor of four greater
than that for the highest pressure gas scintillators commonly used,
the group has been able to perform some unusual experiments within
a reasonable amount of beam time. With the improvement [7] of the
liquid He scintillator resolution, this method is becoming very attractive. Other methods have used gas cells with vanes and detected the
α-recoils [8]. This is a fairly efficient method, but it sometimes suf-
fers from resolution problems.

The n-^{4}He analyzing power is now very well known below 15 MeV,
but above this energy the n-^{4}He scattering data have been accurate
only to ±5%. Rhea et al. [9] show the difference in analyzing power
for the various sets of n-^{4}He phase shifts. There is little difference
below 15 MeV near 120° lab, but at forward angles, sizable differences
exist. For the data shown in this paper, when the difference was significant, the values were adjusted to correspond to the analyzing
power given by the Stammbach and Walter [10] phase shifts for all
reactions except for the T + d neutrons whose energy extends beyond
the range of these phase shifts. In the T(d,n) case, the values are
based on Hoop and Barschall's phase shifts [11]. In all cases where
adjustments were made, the new value includes multiple-scattering
and finite-geometry corrections computed with a Monte Carlo code
available at our laboratory.

2.3 *Values obtained from studying* $P_{He}(\theta)$

The best determinations of source polarization have been from experiments which used neutron source reactions to study the angular
dependence of the n-^{4}He analyzing power $P_2(\theta)$. An example [6] is
shown in fig. 3 for 7.8 MeV neutrons from the ^{9}Be(α,n) reaction. By
renormalizing such asymmetry data in a resonance analysis of all the
n-^{4}He asymmetry and cross-section data below 21 MeV, we [10] have
been able to determine the scale factor for the asymmetry data, the
factor being the incident neutron polarization. In table 1, experiments
of this type are listed, along with the reactions, the neutron energies,
the extracted polarization values, and an estimate of the uncertainty
in the polarization. The uncertainty at 17.7 MeV is in parentheses,

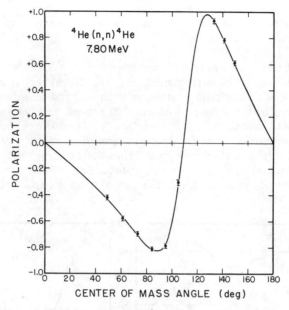

Fig. 3. Polarization for ^{4}He(n,n) for 7.8 MeV
neutrons. The measured asymmetry is scaled
by $1/P_1$ where $P_1 = 0.53$.

Table 1. Sources of polarized neutrons calibrated through
$P_1 P_2(\theta)$ fitting

E_n (MeV)	Reaction	θ_{lab} (deg)	E_{inc} (MeV)	ΔE_{inc} (MeV)	P_1	ΔP_1	Ref.
1.01	Li(p,n)	50°	2.91	0.033	0.310	±0.004	[13]
2.00	T(p,n)	33°	3.10	0.20	0.228	±0.010	[14]
2.43	^{12}C(d,n)	25°	2.82	0.175	0.469	±0.006	[13]
3.38	^{9}Be(α,n$_1$)	45°	2.55	0.30	0.535	±0.003	[6]
5.92	T(p,n)	40°	7.80	0.20	0.198	±0.007	[14]
7.80	^{9}Be(α,n$_0$)	45°	2.55	0.30	0.530	±0.005	[6]
10.0	D(d,n)	32°	8.4	0.50	0.226	±0.010	[14]
11.0	D(t,n)	133.4°	11.53	0.30	0.605	±0.01	[15]
16.4	T(d,n)	90°	6.0	0.50	0.464	±0.02	[14]
17.7	T(d,n)	82°	7.00	0.30	0.413	(±0.02)	[15]

as there probably will be some adjustments to the phase shifts in
this energy region based on new n-^{4}He cross-section data [12] reported at this conference.

2.4. *The more important reactions*

For high-energy polarized neutrons, the T(d,n) reaction is clearly
the best because the polarization is high and the lower energy neutrons arising from the n + p + t channel are weak and are well separated
from the main group. Many of the data [8, 16—20] are represented in
fig. 4. Here we see that for deuteron energies above 4 MeV, the T(d,n)
reaction yields polarizations exceeding 0.4 at back angles, and above
7 MeV, polarizations near 0.6 at 30° lab. Because of the uncertainty
in analyzing power at 23 MeV, the Los Alamos group bombarded a

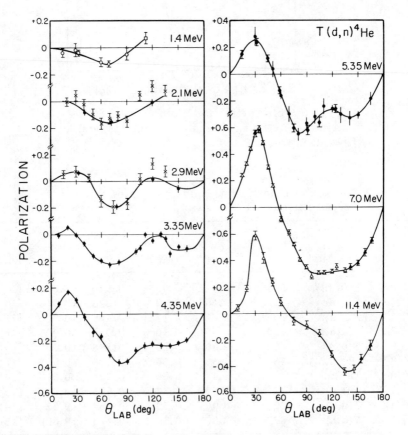

Fig. 4. Neutron polarization for the T(d,n) reaction. The refs.
are the following: △ [15], ▲ [20], ◆[19], ●[18], x [8], ◇[16],
○[16], □[17].

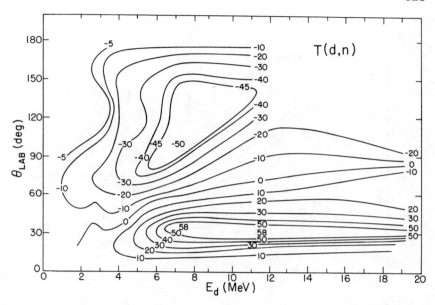

Fig. 5. Contour plot of the polarization from the T(d,n) reaction.

deuterium target with about 10.5-MeV tritons and measured the polarization at 143°, which corresponds to 30° in the T(d,n) reaction at
7 MeV. In the D(t,n) reaction the neutron energy was 13 MeV, an energy of fairly well known ⁴He analyzing power. The solid triangles
at 7.0 MeV represent these data transformed into the T(d,n) system.
The agreement is excellent. In Fig. 5 a contour plot produced from
the curves in fig. 4 and other data (refs. [21, 22] and table 1) show
the wide range for which the T(d,n) will provide highly polarized
neutrons.

For neutron energies below 18 MeV, the D(d,n)³He reaction is a
useful source because of its large cross sections and because of the
considerable energy difference between the breakup neutrons and the
main neutron group. This reaction has been reinvestigated above
6 MeV to test whether the differences between it and the D(d,p) reaction are real. The results of Spalek et al. [23] are compared to the
work of Porter and Haeberli [24] in fig. 6. The solid curves are derived from Legendre polynomial fits to the product P(θ)σ(θ). Indeed
a difference does still exist. This also is apparently true for the mirror reactions, ³He(d,p) and T(d,n) [20]. In fig. 7, the available polarization data [14, 25–28] at 45° c.m. are plotted. My best estimate of
the polarization at this angle is given by the solid curve. According
to the Pσ fit of Spalek et al., the polarization peaks close to 45° c.m.
for energies above 6 MeV. This reaction is quite useful, the polariza-

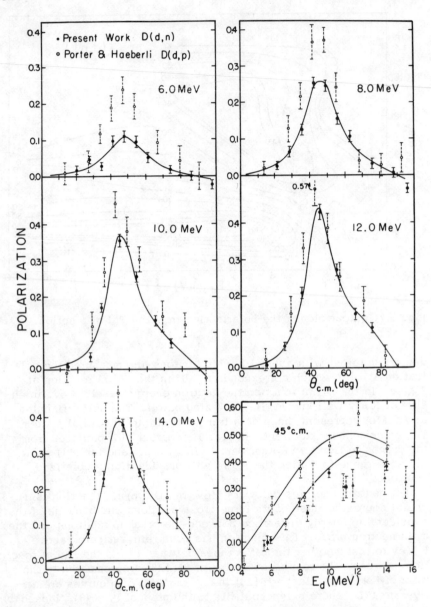

Fig. 6. Polarization of D(d,n) reaction at 6−14 MeV from ref. [23] and D(d,p) from ref. [24].

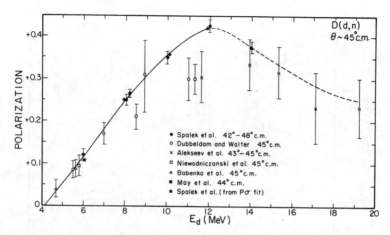

Fig. 7. Polarization near 45° c.m. for the D(d,n) reaction. The data are from refs. [25—28], [23], and [14].

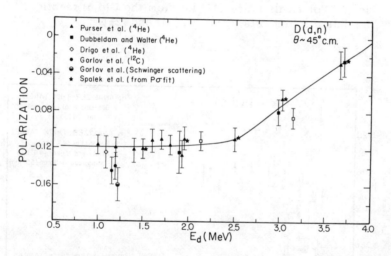

Fig. 8. Polarization near 45° c.m. for the D(d,n) reaction. The data are from refs. [29—31], [23], and [21].

tion attaining a value of 0.42 at 12 MeV. In fig. 8 we see that below 4 MeV, the polarization [29—31] at 45° c.m. is low but probably fairly well determined. The maximum polarization in this region occurs near 49° lab, but as can be seen in fig. 9 the data [3, 32, 33] scatter considerably. Here again the solid curve represents my best estimate. The datum of Bodarenko and Ot-Stavnov [32] was taken with a thick target and probably is meaningless. The data of Levintov et al.

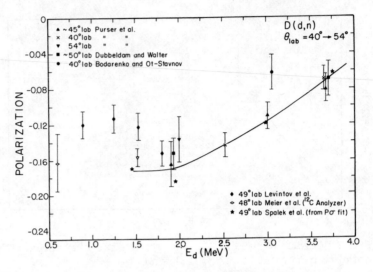

Fig. 9. Polarization near 49° lab from the D(d,n) reaction.
The data are from refs. [31–33], [3], and [25].

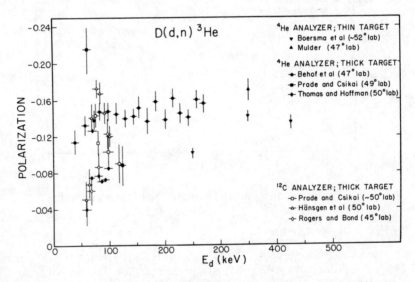

Fig. 10. Polarization near 49° lab. The data are from refs. [35–
41]. (See text regarding the data of Behof et al.)

[33] taken in 1956 do not appear consistent although the new values of the n-⁴He analyzing power were used to obtain the values plotted.

For accelerators whose maximum voltage is less than 500 keV, the only possible source reaction known to me is D(d,n). At the Karlsruhe conference, Barschall [34] indicated that more data at low energies were needed to clarify the situation. Fig. 10 shows the deluge of new data [35—41]. Here we have the current status for 49° lab. A few of the early results [42, 43] have been removed for clarity, as the backgrounds may have been improperly taken into account. (The data of Behof et al. should be increased by 0.06 as I erred in making corrections to these data.) I have shifted the energies of the thick target data, corrected for the n-⁴He and n-¹²C phase shifts, and applied the corrections to the data of Boersma et al. for instrumental asymmetries as suggested by Boersma [44]. After all this effort, I believe that we have eliminated the debated resonance [45] at 90 keV. My feeling is that no solid evidence of the resonance exists in these data, and I expect that the polarization ranges between 0.14 and 0.15 from 400 keV down to 60 keV and then drops gradually as the data of Behof et al. indicate.

The T(p,n) reaction is probably the most convenient, clean source to use in that the neutron contribution from breakup is very low and the protons do not produce much background in the beam stop. Unfortunately, as is shown in fig. 11, the magnitude of the polarization [14, 26, 46—49] seems to be below 0.2 for proton energies above 3 MeV. At 2.3 MeV, this reaction supplies a 40% polarized neutron beam of 1.2-MeV energy. Recent data by Cramer and Cranberg [49] indicate

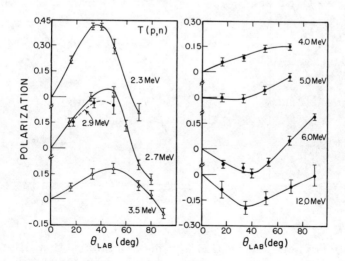

Fig. 11. Polarization from the T(p,n) reaction. The data are from refs. [46] (●) and [47] (△).

that the polarization rises almost linearly from zero at threshold to
0.4 at 2 MeV and 32° lab. When other factors are taken into account,
this is the most efficient reaction source known.

As one would expect, the target most studied [13, 50, 51] for reaction polarization is obviously "hard-to-fit" ^{12}C. An example of the
size of the ^{12}C(d,n) polarization and of the amount of our knowledge
is given in fig. 12 which was prepared by Donoghue from the data of
refs. [51–53]. The statistical uncertainties are typically less than
±0.05. At first glance it appears that the polarization does not vary

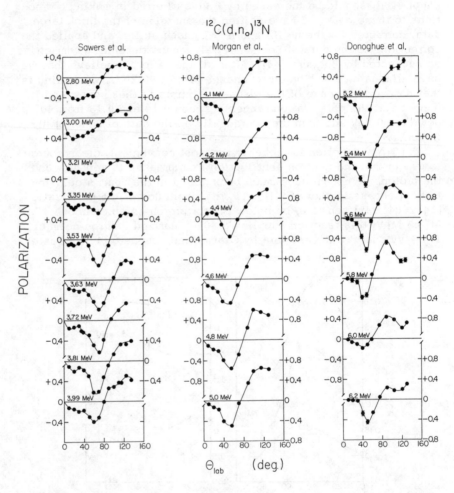

Fig. 12. Example of polarization from the ^{12}C(d,n) reaction. The
uncertainties are smaller than the point size except where indicated.
The data are from refs. [51–53].

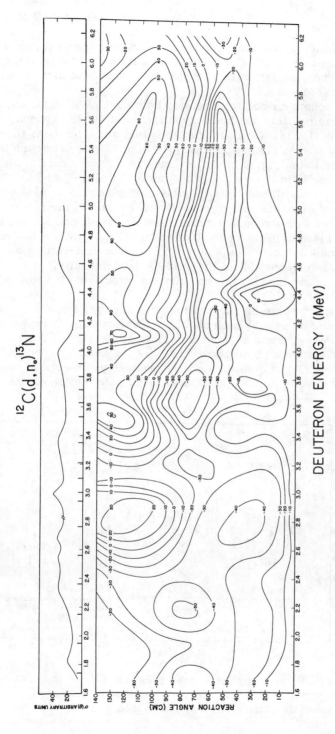

$^{12}C(d,n_o)^{13}N$

Fig. 13. Polarization contour plot of the $^{12}C(d,n)$ reaction.

much with energy, but from a contour plot of this reaction (fig. 13) one realizes that to use this source, he must select target thicknesses and beam energies carefully. Several long islands of high polarization are available, however.

A good source for neutron energies below 2 MeV is the ^{7}Li(p,n) reaction. This reaction has been investigated at 50° lab by many groups, as shown in fig. 14. (See Kuchnir et al. [54] and references therein.) Below E_p = 3 MeV, Elwyn and Lane [55] were able to determine the polarization using Mg, O, and C as analyzers. Hardekopf et al. [56] attempted to fill in the gaps left by Andress et al. [57] and to eliminate some of the difficulty in selecting the real polarization value between 3 and 4 MeV from this collection of data. In fig. 15, the new results are compared with the earlier data of Andress et al. and Sawers et al. [13]. The lower half shows the calibration of the first-excited-state neutron group which is only 430 keV from the ground-state group, but which has only about 10% of the ground-state intensity.

Fig. 16 shows the ^{14}N(d,n) contour plot of polarization from a paper by Meier et al. [58].

Before we compare all the available reactions, I would like to give an example of an effort which should be duplicated in order to enhance the usefulness of some of the more practical reactions. The results of a determination by Stammbach et al. for the ^{9}Be(α,n_0) reactions are shown in fig. 17. In order to make resonance-type reactions useful for targets several hundred keV thick, or for polarization excitation

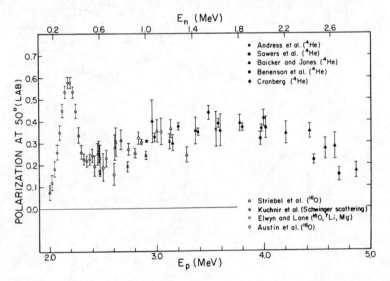

Fig. 14. Polarization at 51° lab from the ^{7}Li(p,n) reaction. The sources of the data are given in refs. [54] and [57]. The datum of Sawers et al. is from table 1.

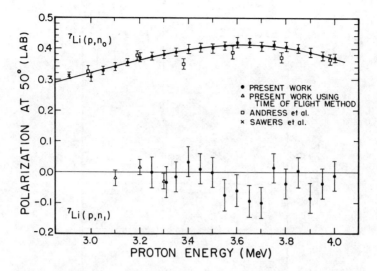

Fig. 15. Recent polarization data for the ^{7}Li(p,n_0) and
Li(p,n_1) reactions from refs. [56, 57] and table 1.

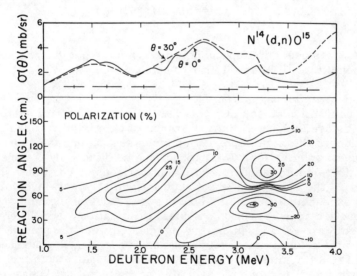

Fig. 16. Polarization contour plot of the ^{14}N(d,n) reaction
from ref. [58].

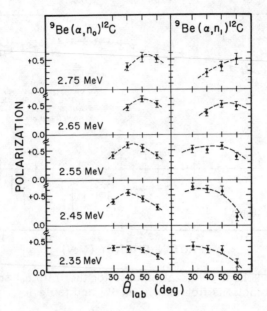

Fig. 17. Polarization from the ^{9}Be(α,n$_0$) and
^{9}Be(α,n$_1$) reactions from ref. [59].

functions, one needs to map the polarization quite carefully. These
experiments take time, but are most useful.

2.5. *Mountains and molehills*

The maximum accuracy in a polarization measurement by scattering that is available in a minimum amount of time is proportional to
the product of the neutron production cross section and the square of
the reaction polarization. This product can be divided by the energy
loss in the target to standardize flux comparisons. Listed in tables 2
and 3 is all the information about neutron sources which have polarizations exceeding about 0.2. Given are the ranges of incident energies, reaction angles, neutron energies, polarizations, and differential
cross sections. Also given are the energies between the ground-state
and the first-excited-state groups, the energy loss S, the maximum
$P^2\sigma$ and $P^2\sigma/S$ (in units of 10^{-12}/eV sr). The ΔE_n indicates either the
range over which neutrons are available or the target thickness used
in the measurement. The last column gives a reference either for the
data themselves or for a work which cites the polarization data. An
attempt was made to summarize this information graphically so that
comparisons could be made at a glance. The only satisfactory scheme
I found was the "mountains and molehills" one shown in fig. 18. Here
is displayed $P^2\sigma/S$ versus the outgoing neutron energy for incident

Table 2. Projectile beam energies from 1 to 4 MeV

Target + Proj.	E_{inc} (MeV)	θ_{lab} (deg)	E_{n_0} (MeV)	$E_{n_0}-E_{n_1}$ (MeV)	$|P|$	σ_{lab} (mb/sr)	S (10^{-15} eV-cm^2)	$P^2\sigma$ (mb/sr)	$P^2\sigma/S$ (10^{-12}/eV-sr)	ΔE_n (keV)	Target Form	Ref. for P
T+p	1.5-2.3	32	0.6- 1.3	–	0.2 -0.4	45	0.8-0.6	8.8	15	100	GAS, T-Zr	46,47,49
	2.3	33-40	1.3- 1.2	–	0.40	55	0.6	5.1	10	300		
	2.7	33-50	1.7- 1.4	–	0.32	50	0.5	3.1	6	200		
		33-50	1.8- 1.6	–	0.25	50	0.5	1.3	3	>300		
	2.9-3.5	50	1.6- 2.0	–	0.18	40	0.4					
^{7}Li+p	2.1-2.3	50	0.3- 0.5	–	0.26-0.6	25-60	1.6	18.0	11.0	<200	SOLID	13, 54-56
	2.3-3.0	50	0.5- 1.1	0.45	0.22-0.30	60-30	1.6-1.4	2.7	1.9	300		
	3.0-4.0	50	1.1- 2.0	0.44	0.30	25	1.4-1.0	2.3	2.3	300		
^{9}Be+p	2.6	30	0.6	CONT.	0.25	7	1.6	0.43	0.25	100	SOLID	61
	2.7-3.4	90	0.5- 1.1	"	0.20	5-11	1.6-1.2	0.44	0.37	>300		
D+d	1.0-2.5	35-50	3.6- 5.1	–	0.12-0.16	11	2-0.9	0.25	0.28	>300	GAS, D-Zr	TEXT
T+d	2.1-4.0	70	16.0-17.2	–	0.18-0.30	10-6	1.1-0.6	0.5	0.8	>300	GAS, T-Zr	TEXT
^{11}B+d	1.8-2.0	20-40	14.9-15.3	4.2	0.20	3	4	0.12	0.03	200	SOLID	62
	2.0	90-110	14.2-13.8	4.0	0.18	1	4	0.04	0.01	300		
	2.0-2.4	30	15.2-15.6	4.2	0.30	1.5	4	0.14	0.04	>300		
	3.0-3.4	45	16.0-16.4	4.2	0.35-0.40	1-2	3-2.5	0.32	0.13	>300		
	3.2-4.0	20	16.6-17.5	4.2	0.20-0.15	2	3-2.5	0.08	0.03	>300		
^{12}C+d	2.6-3.2	25	2.1-2.8	2.6	0.30-0.46	30-40	4	8.5	2.1	>300	SOLID FOIL	50
	3.7	25	3.3	2.5	0.20	20	4	0.8	0.7	50		
	1.7-2.6	40	1.3-2.1	–	0.30	12-40	5.4	3.6	0.5	>300		
	1.7-2.4	70	1.2-1.4	–	0.3-0.4	6-17	5	2.6	0.6	200		
	2.1-2.4	70	1.5-1.8	2.3	0.4-0.5	16-8	5	2.5	0.1	300		
	2.4-3.9	70	1.8-2.6	2.3	0.4-0.2	6-4	4	0.5	0.2	>300		
			2.6-3.1		0.3-0.5	4-2.5	3	0.6		>300		
^{14}N+d	2.1-3.3	60	6.8	–	0.25	1.6	5	0.10	0.02	300	GAS, Si$_3$N$_4$	58
	3.1-3.3	45	7.9-8.1	–	0.4-0.35	1	4	0.16	0.04	300		
	3.5	45	8.3	–	0.25	1		0.06	0.02	300		
	3.1-3.7	85-105	7.0-7.8	–	0.20-0.33	0.4	4-3	0.04	0.01	>300		
^{15}N+d	1.6	40-70	11.0-11.4	–	0.32	3-1.5	6	0.3	0.05	300	GAS, Si$_3$N$_4$	65
	2.2-2.6	30-70	11.9-12.3	–	0.3-0.5	1.5-0.3	5	0.14	0.03	>300		
	2.6	45	12.1	–	0.48	1.5	5	0.35	0.07	300		
	3.1	45	12.6	–	0.40	1.5	4	0.24	0.06	300		
	3.1-3.3	70	12.2-12.4	–	0.30	0.4	3	0.04	0.01	300		
	3.8	100	12.2	–	0.47	0.3		0.07	0.02	300		
^{12}C+^{3}He	2.2-2.4	50-90	0.5-0.9	–	0.50-0.9	0.9	21	0.7	0.03	300	SOLID	66
	2.2-2.4	60	0.7-0.9	–	0.77	0.9	21	0.5	0.02	300		
	2.9	100-140	1.0-0.8	–	0.37	1.5-2.2	18	0.3	0.02	200		
	3.5-3.7	30-67	1.7-2.2	–	0.50-0.72	2.2-1.5	16	0.7	0.05	300		

Table 2 *(continued)*

Target + Proj.	E_{inc} (MeV)	θ_{lab} (deg)	E_{n_0} (MeV)	$E_{n_0}-E_{n_1}$ (MeV)	$|P|$	σ_{lab} (mb/sr)	S (10^{-15} eV-cm^2)	$P^2\sigma$ (mb/sr)	$P^2\sigma/S$ (10^{-12}/eV-sr)	ΔE_n (keV)	Target Form	Ref. for P
^{9}Be+^{4}He	2.0	105	6.2	4.0	0.3	3	20	2.7	0.14	450	SOLID	67,68
	2.4-2.7	40-60	7.4-8.0	4.4	0.30-0.63	18-8	18	3.0	0.16	>300		59
	2.5	45	7.8	4.4	0.63	16	18	6.3	0.35	100		
^{11}B+α	3.6	30	3.3	2.6	0.7	10†	18	4.9	0.27	150t	SOLID	69
	3.8	30-60	3.5-3.0	2.6	0.9-0.5	6†	18	4.8	0.27	150t	FOIL	
^{13}C+^{4}He	1.65	30-150	3.8-2.8	--	0.3-0.7	5c	32	2.5	0.08	900t		
	2.07	10-130	4.2-3.2	--	0.3-0.8	3-7	29	3.8	0.13	60t		
	2.24	55-90	4.1-3.7	--	0.3-0.5	1.5-5	28	0.75	0.03	60t		
	2.43	60-90	4.2-3.8	--	0.3-0.56	2-3	27	0.62	0.02	60t		
	3.36	90	4.4	--	0.40	5	23	0.8	0.04	<50	SOLID	69,70
	3.44	25	5.4	--	0.30	25	23	2.3	0.10	<40	FOIL	

† guess, t thickness used in measurement

Table 3. Projectile beam energies above 4.0 MeV

Target + Proj.	E_{inc} (MeV)	θ_{lab} (deg)	E_{n_0} (MeV)	$E_{n_0}-E_{n_1}$ (MeV)	$|P|$	σ_{lab} (mb/sr)	S (10^{-15} eV-cm^2)	$P^2\sigma$ (mb/sr)	$P^2\sigma/S$ (10^{-12}/eV-sr)	ΔE_n (keV)	Target Form	Ref. for P
T+p	4-5	90	1.4- 1.9	–	0.14	23	0.4 -0.3	0.45	1.5	>500	GAS,	TEXT,
	6-12	40-60	3.6- 9.5	–	0.12-0.20	28-20	0.3 -0.14	0.8	5.7	>500	T-Zr	47,48,14
	12-17	45	9.1-13.2	–	0.20	19-12*	0.14-0.11	0.8	5.7	>500		
Li+p	4-5	30-40	2.1- 3.2	0.43	0.3 -0.2	20*	1	1.8	1.8	>300	SOLID	60
	7	20-110	5.2- 3.4	0.43	0.18	15+	0.6	0.5	0.9	>300		
^{9}Be+p	4.7-6	30	2.7- 4.0	CONT.	0.4 -0.3	15-10	0.8	2.4	3.0	>500	SOLID,	61,71
	5-7	50	2.8- 4.7	"	0.3 -0.45	12-10	0.8 -0.7	2.1	2.7	>500	FOIL	
	7	20-90	5.1- 3.9	"	0.4 -0.7	10-5	0.7	2.5	3.6	100		
	7-8	65	4.4- 5.4	"	0.7 -0.55	6	0.6	3.0	4.3	>500		
	8-11	65	5.4- 7.9	"	0.55-0.25	6-5	0.7 -0.5	1.8	2.6	>500		
^{11}B+p	7.6	20-50	4.7- 4.4	2.1	0.25-0.30	56-12	0.9	3.5	3.9	100t	SOLID	71
	8.5	10-40	5.7- 5.4	2.1	0.30-0.55	31-17	0.8	7.5	9.5	100	FOIL	
	9.4	20-70	6.5- 5.8	2.0	0.30-0.5	50-10	0.7	4.5	6.4	100		
	11.4	20-30	8.5- 8.4	2.0	0.25	20+	0.6	1.3	2.2	100		
^{13}C+p	6.9	20-60	3.8- 3.5	1.3	0.35-0.7	23	1.2	12	10	480t	SOLID,	72
	8.0	20-80	4.9- 4.2	1.3	0.3 -0.5	20-11	1.1	3.2	3.0	420	FOIL	
	8.8	20-80	5.7- 5.0	1.3	0.2	20-10	1.0	0.8	0.8	400		
	10.9	40	7.5	1.4	0.4	5	0.8	0.8	1.0	340		
^{15}N+p	7.9	10-60	4.3- 3.9	–	0.20-0.35	14	1.2	1.7	1.4	280t	GAS	72
	9.2	40-70	5.4- 4.8	5.2	0.35-0.55	12-4	1.1	1.8	1.6	240		
	10.3	30-80	6.6- 5.8	5.2	0.35-0.75	11-2	1.0	2.2	2.2	200		
	11.3	30-60	7.5- 7.1	5.2	0.65-0.90	6-3	0.9	3.5	4.0	200		
	12.3	40-70	8.4- 7.9	6	0.3 -0.6	6	0.9	2.2	2.4	200		
D+d	7-12	32	8.9-12.9	CONT.	0.18-0.42	5.5-4.1	0.4 -0.3	0.65	2.2	>500	GAS,	TEXT
	12-19	32	12.9-18.5	"	0.42-0.25	4.1-3	0.3 -0.2	0.65	2.2	>500	D-Zr	
T+d	4.4	60-150	18.3-12.2	–	0.2 -0.37	5+	0.6	0.5	0.9	>500	GAS,	TEXT
	5-7	70-150	12.2-19.0	–	0.3 -0.5	5-3	0.5	1.0	2.6	>500	T-Zr	
	5-11	30	21.1-26.9	–	0.25-0.59	6-3	0.5 -0.3	1.1	3.6	>500		
	11-19	30	26.9-34.1	–	0.5	3-2+	0.3 -0.2	0.7	2.5	>500		
	11.4	120-160	14.8-12.1	–	0.30-0.44	2-4	0.3	0.6	1.8	>500		
^{11}B+d	5.5	60	18.0	4.0	0.22	0.3	1.9	0.02	0.01	>300	SOLID	63,64
	5.5	100	16.5	4.0	0.3	0.2	1.9	0.02	0.01	>300	FOIL	
	11.7	70	22.9	4.2	0.35	0.3+	1.1	0.04	0.04	>300		
^{12}C+d	4.0-4.4	45-65	3.1- 3.8	2.4	0.3 -0.5	3	2.7	0.75	0.28	>300	SOLID	51,52
	4.4-5.7	50	3.7- 4.9	2.4	0.4 -0.5	3-5+	2.7 -2.2	1.25	0.57	>300	FOIL	
	4.0-4.6	120	2.4- 3.0	2.0	0.4 -0.7	3	2.7	1.5	0.55	>300		
	4.6-6.0	105	3.2- 4.2	2.1	0.3 -0.6	2+	2.5 -2.0	0.72	0.36	>300		

Table 3 (*continued*)

Target + Proj.	E_inc (MeV)	θ_lab (deg)	E_n0 (MeV)	E_n0 - E_n1 (MeV)	|P|	σ_lab (mb/sr)	S (10^-15 eV-cm)	p²σ (mb/sr)	p²σ/S (10^-12/eV-sr)	ΔE_n (keV)	Target Form	Ref. for P
^{14}N+d	5.5	10-30	10.5-10.4	5.2	0.25-0.3	4-5	2.5	0.36	0.14	>200	GAS, Si$_3$N$_4$	63
	5.5	95	9.1	4.8	0.2	1.4	2.5	0.06	0.02	>200		
^{15}N+d	4.4	10-30	14.2-14.1	6.1	0.3	3	3.0	0.27	0.09	>200	GAS, Si$_3$N$_4$	63
	4.4	150-160	11.8	5.5	0.5 -0.6	1.4	3.0	0.35	0.12	>200		
	5.5	10-30	15.4-15.1	6.0	0.4 -0.4	4	2.5	1.4	0.42	>200		
	5.5	150-160	12.6-12.5	5.4	0.3 -0.4	1.3	2.5	0.21	0.08	>200		
^{13}C+^{3}He	4.2	50-60	10.8-10.5	5.1	0.65	0.7	14	0.30	0.02	250	SOLID FOIL	73
	4.2	110	9.4	4.7	0.4	1.0	14	0.16	0.01	250		
	4.7-5.7	70	10.7-11.5	5.0	0.4 -0.7	0.6-0.7c	13-11	0.35	0.03	>500		
^{9}Be+^{4}He	4.2	60	8.7	4.4	0.45	6	13	1.2	0.09	200t	SOLID FOIL	74,68
	4.5-4.8	60	9.0-9.3	4.4	0.5 -0.3	5-3	12	1.3	0.11	>300		
	4.8-5.5	30	10.1-10.8	4.6	0.25-0.65	13-20	11	8.4	0.75	>500		
	5.25	120	7.5	3.8	0.75	3+	11	1.6	0.15	200t		
	5.5	105-130	8.1- 7.4	3.8	0.6 -0.5	3	10	1.1	0.11	200t		
	5.85	60-70	10.0- 9.5	4.4	0.6	2	9	0.7	0.06	200t		
	5.85	120-130	7.8- 7.5	3.8	0.4 -0.6	4	9	1.4	0.15	200t		
^{13}C+^{4}He	4.36	35-55	6.2- 5.8	- -	0.6 -0.7	4+	18	2.0	0.11	40t	SOLID FOIL	70
	4.70	60-80	6.0- 5.6	- -	0.5 -0.6	2-4	17	1.4	0.08	50t		
	4.80	30-60	6.6- 6.1	- -	0.4 -0.5	2	16	0.5	0.3	60t		

* extrapolation or interpolation, † guess, t thickness used in measurement

336

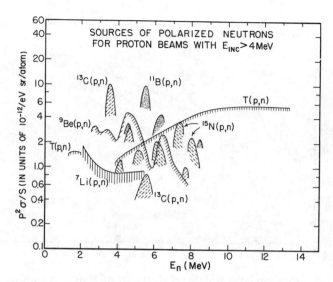

Fig. 18. Comparison of the usefulness of reactions as sources of polarized neutrons in terms of $P^2\sigma/S$ vs. neutron energy for proton beams with energies above 4 MeV. (See table 3.)

proton energies above 4 MeV. Except for the T(p,n) reaction, these reactions have considerable resonance structure, enough in some cases to make them less favorable if the scattering experiment permits a wide neutron energy spread, as in n-p or n-d scattering, for example. The $P^2\sigma/S$ for the T(p,n) reaction looks very good but, in addition to the ordinary gas target problems, one may not be able to work easily with a neutron beam emitted at 45° if a long gas target is necessary to take advantage of the low energy loss. Additionally, if the differential cross section is at a minimum of the angle of use for any of the reactions, considerable extra time might be spent measuring backgrounds. Lastly, the lower the source polarization, the more sensitive the experiment becomes to instrumental asymmetries. Thus, one should use these graphs with care, keeping in mind tables 2 and 3 and the experimental conditions. In fig. 19 are shown the $P^2\sigma/S$ curves for other reactions for beams above 4 MeV. Shown in fig. 20 are similar curves for reactions with beams below 4 MeV. Here, one sees that a wide neutron energy range is available with quite low energy accelerators. A final comment about this tabulation. Excluded were (d,n) reactions on ^{6,7}Li, ^{9}Be, ^{10}B, and ^{13}C, even though data exist, because the separation between the neutron groups was too small to make them valuable. Also, no data on targets heavier than ^{15}N were given because the existing polarization is typically low and because the energy loss is large.

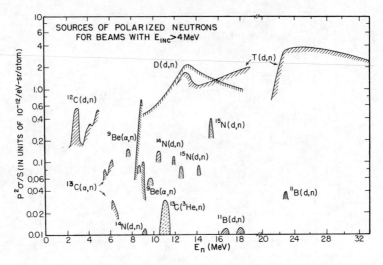

Fig. 19. Comparison of the usefulness of reactions as sources
of polarized neutrons in terms of $P^2\sigma/S$ vs. neutron energy for
deuteron and helium beams with energies above 4 MeV. (See
table 3.)

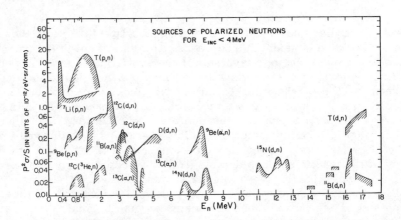

Fig. 20. Comparison of the usefulness of reactions as sources
of polarized neutrons in terms of $P^2\sigma/S$ vs. neutron energy for
incident beams below 4 MeV. (See table 2.)

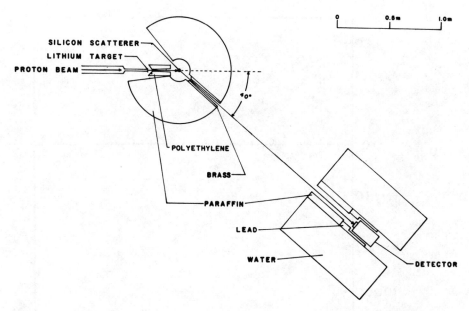

Fig. 21. Arrangement of Lunnon and Lefevre for scattering unpolarized (or polarized) neutrons from Si.

2.6. *Thick targets*

Before leaving this topic, I would like to mention a promising method of combining very thick targets with high-resolution time-of-flight techniques to perform neutron scattering studies which has been initiated by Lunnon and Lefevre [75]. Shown in fig. 21 is a top view of their setup for measuring the scattering cross section at 40° for Si. For the polarization experiment the apparatus is rotated so that neutrons emitted at 50° lab from the Li target strike the Si. When the pulsed, 3.5-MeV proton beam bombards the thick ^{7}Li target, polarized neutrons from 1.5 MeV down to 0 are emitted at 50°. By measuring the time-of-flight spectra for scattering from Si on each side of the neutron beam axis, the analyzing power of Si over a wide energy range is determined. An example of some preliminary data is shown in fig. 22, where the lower plot represents the differential cross section for 30° lab scattering from Si, and the upper curve, the right-left asymmetry for the same scattering angle. The data have not been corrected for the ^{7}Li(p,n$_1$) group nor have they been divided by P_1 (E_n) which ranges from 0.3 to 0.4. The flight path was about 2.5 m and it took approximately 30 hours to obtain the data shown. One limitation of this method is that the useful target thickness can be no greater than the separation to the first excited state of the scatterer, unless the inelastic scattering can be neglected.

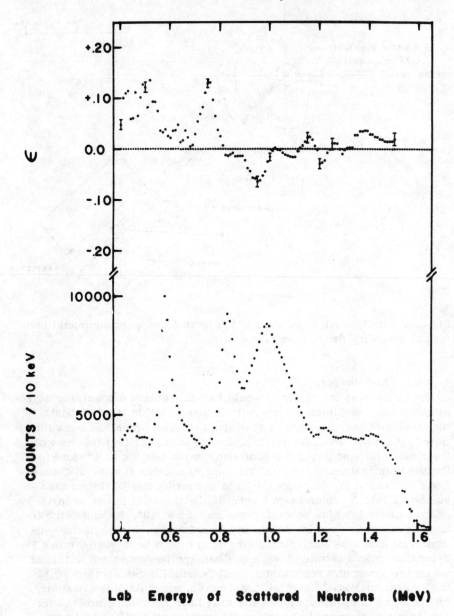

Fig. 22. Results from Lunnon and Lefevre of thick target Li + p neutrons (emitted at 51° lab) scattered from Si. Data in the lower half are proportional to the differential cross section. Data in the upper half are proportional to the polarization. (See text.)

3. POLARIZATION TRANSFER REACTIONS

3.1. $(\vec{p},\vec{n})$ *reactions*

During the last five years, beams of polarized neutrons have been produced at high energies using reactions of the type X(p,n)Y. For neutrons emitted close to 0, the polarization of the neutron beam is related to the incident proton polarization through the polarization transfer coefficient a_t by $P_n = a_t P_p$. In a 1965 Rutherford Laboratory progress report, R. C. Hanna and co-workers give a value of -0.3 for a_t for neutrons emitted at 0° from the interaction of 50-MeV polarized protons with a liquid deuterium target. Follow-up experiments [76] with D and Li targets were reported in 1969 when a_t at 0° for the D target was given as -0.13 and -0.23 at 30 and 50 MeV, respectively. Thomas et al. [77] have studied the neutron spectrum from the bombardment of D_2 with 200-MeV protons. They report that the high-energy component due to charge exchange has a half-width of 12 MeV at 10° where they measure a_t to be -0.84. In both of these experiments, the polarized proton source was very weak. The usefulness of such polarized neutron beams has not yet been exploited. (Perhaps the cleanest source of 100–200-MeV neutrons will be stripping reactions on light nuclei with polarized deuteron beams. I have not located any reports concerning this type of experiment at high energies.)

3.2. $(\vec{d},\vec{n})$ *reactions*

At the other end of the energy scale, Ohlsen [78] proposes use of intense polarized deuteron beams below 500 keV to produce highly polarized 14-MeV neutrons through the T(d,n) reaction. For a deuteron beam which is 100% in the $m_I = +1$ state, he suggests that a neutron beam with a longitudinal polarization of 1.0 or a transverse polarization of 0.8 might be achieved. An experiment [79] was performed in Basel in 1962 to measure the neutron polarization, but the vector polarization was relatively low. As the intensity of the polarized deuteron sources increases, the practicality of this source of 14-MeV neutrons grows.

In a beautiful set of experiments, the Los Alamos group has been measuring the neutron polarization produced from the T($\vec{d},\vec{n}$) reaction from 4 to 15 MeV. Their results, which are reported elsewhere in these Proceedings, are shown in fig. 23. Polarized deuterons with $\langle P_y \rangle = \langle P_{yy} \rangle = 0.76$ were incident on a tritium target and the neutron polarization was measured by scattering from liquid helium. Below 6 MeV, the polarization drops because the influence of the 107-keV $3/2^+$ resonance is strong in this region. The prediction of the maximum neutron polarization (for an incident beam in the $m_I = +1$ state) is represented by the dashed curve. For their deuteron beam, a polarization of 0.565 ± 0.036 at 15 MeV was observed.

Since the D(d,n) reaction at 0° provides a clean beam of unpolarized neutrons from 5 to 18 MeV (with tandem accelerators) one expects

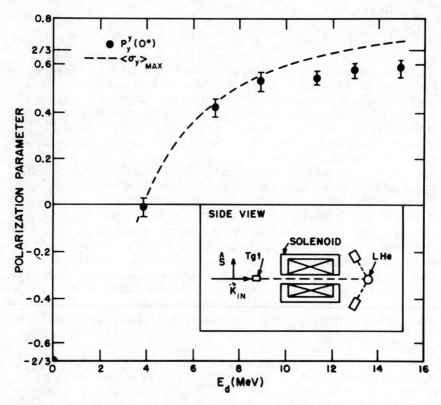

Fig. 23. Neutron polarization produced in the T(d,n) reaction with polarized deuterons. (Taken from ref. [80].)

that this would be the most promising source of polarized neutrons in this range if polarized deuteron beams were employed. As this reaction appears to occur predominantly by ℓ = 0 stripping, one naively expects that the neutron of the polarized deuteron would be little perturbed in a 0° reaction, retaining most of its original spin orientation. According to the data reported by Simmons et al. [81], this indeed appears to be true for deuterons from 4 to 13 MeV. For a 100% polarized deuteron beam (m_I = 1), their experiments indicate that the neutron polarization would be about 0.90. The significance of these data on the T($\vec{d}$,n) and D($\vec{d}$,n) reactions will become very apparent as more polarized-ion sources in the 100-nA range become available. For example, from 7 to 12 MeV, the D(d,$\vec{n}$) reaction with unpolarized deuterons has an average $P^2\sigma$ of about 0.3 mb/sr at 32° lab, the angle of maximum P^2. The D($\vec{d}$,n) reaction at 0° gives a $P^2\sigma$ of about 80 mb/sr for the 100% polarized deuteron beam. Obviously then, compared to a normal 5 μA beam the cross-over point from $P^2\sigma$ considerations alone for switching to the transfer reaction would be when a beam of 20 nA became available. (For the $T(\vec{d},\vec{n})$ reaction, P^2 from the unpolarized reaction is quite large, so the cross over would be at an appreciably higher current.) Considering (1) background problems when one works with reaction neutrons produced in a cross-section minimum, as in the $D(d,\vec{n})$ or $T(d,\vec{n})$ case, (2) the ease of changing the spin direction with a polarized-ion source, and (3) the increase in the $1/r^2$ factor if there is no spin-precession solenoid between the neutron source and a scatterer, the $D(\vec{d},\vec{n})$ transfer reaction probably competes favorably when more than 3—5 nA of 100% m_I = +1 beam is available.

4. SMALL-ANGLE SCATTERING

The possibility of using small-angle scattering or Schwinger scattering [83] as a source of polarized neutrons is worth discussing. Here the polarization is dominantly produced by the interaction of the neutron's magnetic moment with the Coulomb field of the nucleus. Since the magnitude of the polarization is insensitive to reasonable variations in optical model parameters, the polarization can be accurately computed. Also appealing is the fact that the polarization varies relatively little with energy permitting the use of thick targets for neutron production. As seen in fig. 24, from the recent work of Hogan and Seyler [84] the calculated polarization for 7-MeV neutrons scattered from Li reaches a magnitude of 1.0 near 0.5° and stays

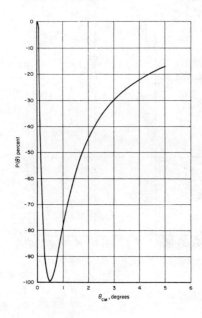

Fig. 24. Polarization produced in elastic neutron scattering through the interaction of the magnetic moment of the neutron with the Coulomb field of the nucleus.

above 0.5 out to 2°. The experimental difficulty here is that one must first collimate a neutron beam (produced in some reaction) to a cone of about 0.5° half angle and then may utilize only a small cone of the scattered flux. One of the nicest experiments to study polarization effects in small angle scattering was performed by Kuchnir et al. [54] whose results are shown in fig. 25. These data were obtained by measuring the right-left scattering asymmetry for polarized neutrons produced in the ^{7}Li(p,n) reaction at 50° lab. The beam was relatively large (20 μA) and the target was about 70 keV thick. The shape of the data follows that of the calculated curves, but if the data are normalized to the curves, the extracted polarization values for the ^{7}Li(p,n) reaction are about 15–20% below other values, some of which have been obtained recently. Apparently, even with the extreme care taken

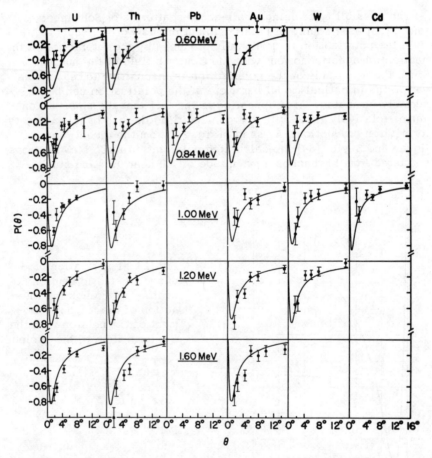

Fig. 25. Polarization results of Kuchnir et al. [54] for small angle scattering of polarized neutrons.

by the Argonne group, it seems difficult to avoid inherent problems connected with air scattering and with the fine collimation required in these experiments. Perhaps with very high flux devices, Schwinger scattering will be a useful method of producing polarized neutron beams, but I believe that other techniques will probably always outweigh this method.

I owe a great deal of gratitude to T. R. Donoghue and G. L. Morgan for their helpful discussions and suggestions. To my students R. Hardekopf, T. Rhea, G. Spalek, and J. Taylor, who gave unselfishly of their time, "thanks." J. E. Simmons and others at LASL have been most helpful in communicating their results and tutoring me about their work. Sessions with G. G. Ohlsen, P. W. Keaton, and R. Finlay are gratefully appreciated.

REFERENCES

*Work supported by the U.S. Atomic Energy Commission.

[1]   P. Delpierre, R. Sene, J. Kahane, and M. de Crespin de Billy, Third Polarization Symp.; J. F. Bruandet, F. Ripouteau, and M. Fruneau, Rev. Phys. Appl. 4 (1969) 169.

[2]   H. Prade and Gy. Máthé, Acta Phys. Acad. Sci. Hung. 25 (1968) 83; G. E. Tripard, L. F. C. Monier, B. L. White, and P. W. Martin, Nucl. Instr. 66 (1968) 261; D. G. Schuster and R. L. Hagengruber, Third Polarization Symp.

[3]   R. W. Meier, P. Scherrer, and G. Trumpy, Helv. Phys. Acta 27 (1954) 577.

[4]   J. E. Wills, J. K. Blair, H. O. Cohn, and H. B. Willard, Phys. Rev. 109 (1958) 891.

[5]   J. T. Reynolds, C. J. Slavik, C. R. Lubitz, and N. C. Francis, Phys. Rev. 176 (1968) 1213.

[6]   Th. Stammbach, J. Taylor, G. Spalek, and R. L. Walter, Phys. Rev. C2 (1970) 434.

[7]   J. C. Martin, W. B. Broste, and J. E. Simmons, Third Polarization Symp.

[8]   See, for example, J. Christiansen, F. W. Büsser, F. Niebergall, and G. Söhngen, Nucl. Phys. 67 (1965) 133.

[9]   T. C. Rhea, Th. Stammbach, and R. L. Walter, Third Polarization Symp.

[10]   Th. Stammbach and R. L. Walter, to be published.

[11]   B. Hoop, Jr., and H. H. Barschall, Nucl. Phys. 83 (1966) 65.

[12]   A. Niiler, M. Drosg, J. C. Hopkins, and J. D. Seagrave, Third Polarization Symp.

[13]   J. R. Sawers, G. L. Morgan, L. A. Schaller, and R. L. Walter, Phys. Rev. 168 (1968) 1102.

[14] T. H. May, R. L. Walter, and H. H. Barschall, Nucl. Phys. 45 (1963) 17.

[15] W. B. Broste, J. E. Simmons, and G. S. Mutchler, Bull. Am. Phys. Soc. 14 (1969) 1230.

[16] Z. Wilhelmi et al., Int. Conf. on Properties of Nuclear States, 1969 (Univ. of Montreal Press) p. 284; Günther Hentschel, Z. Physik 219 (1969) 32.

[17] F. Boreli, V. Lazarević, and N. Radišić, Nucl. Phys. 66 (1965) 301.

[18] W. Busse et al., Nucl. Phys. A100 (1967) 490.

[19] R. B. Perkins and J. E. Simmons, Phys. Rev. 124 (1961) 1153.

[20] W. B. Broste and J. E. Simmons, Third Polarization Symp.

[21] N. V. Alekseev et al., Sov. Phys.—JETP 20 (1965) 287.

[22] R. L. Walter, W. Benenson, T. H. May, and C. A. Kelsey, Nucl. Phys. 59 (1964) 235.

[23] G. Spalek et al., Third Polarization Symp.

[24] L. E. Potter and W. Haeberli, Phys. Rev. 164 (1967) 164.

[25] P. S. Dubbeldam and R. L. Walter, Nucl. Phys. 28 (1961) 414.

[26] N. V. Alekseev et al., Sov. Phys.—JETP 18 (1964) 979.

[27] N. P. Babenko, I. O. Konstantinov, A. P. Moskalev, and Y. A. Nemilov, Sov. Phys.—JETP 20 (1965) 512.

[28] H. Niewodniczanski, J. Szmider, and J. Szymankowski, J. Phys. 24 (1963) 871.

[29] L. Drigo et al., Nuovo Cim. 1 (1969) 237.

[30] G. V. Gorlov, N. S. Lebedeva, and V. M. Morozov, Sov. Phys.—Doklady 9 (1965) 806; Sov. J. Nucl. Phys. 8 (1969) 630. See also ref. [5].

[31] F. O. Purser, Jr., J. R. Sawers, Jr., and R. L. Walter, Phys. Rev. 140 (1965) 870; F. O. Purser, Ph.D. Dissertation, Duke University (1966).

[32] I. I. Bondarenko and P. S. Ot-Stavnov, Sov. Phys.—JETP 20 (1965) 67.

[33] I. I. Levintov, A. V. Miller, E. Z. Tarumov, and V. N. Shamshev, Nucl. Phys. 3 (1957) 237.

[34] H. H. Barschall, Second Polarization Symp., p. 399.

[35] H. Hänsgen, H. Pose, G. Schirmer, and D. Seeliger, Nucl. Phys. 73 (1965) 417.

[36] H. J. Boersma, C. C. Jonker, J. G. Nijenhuis, and P. J. van Hall, Nucl. Phys. 46 (1963) 660.

[37] J. T. Rogers and C. D. Bond, Nucl. Phys. 53 (1964) 297.

[38] A. F. Behof, T. H. May, and W. I. McGarry, Nucl. Phys. A108 (1968) 250.

[39] K. Thomas and A. Hoffman, Z. Physik 217 (1968) 128.

[40] J. P. Mulder, Doctoral Dissertation, University of Groningen (1968).

[41] H. Prade and J. Csikai, Nucl. Phys. A123 (1969) 365.

[42] P. J. Pasma, Nucl. Phys. 6 (1958) 141, and private communication.

[43] P. P. Kane, Nucl. Phys. 10 (1959) 429.

[44] H. J. Boersma, private communication.

[45] See T. R. Donoghue, H. Paetz gen. Shiek, C. E. Busch, and J. A. Keane, Third Polarization Symp.

[46] R. L. Walter, W. Benenson, P. S. Dubbeldam, and T. H. May, Nucl. Phys. 30 (1962) 292.

[47] C. A. Kelsey, B. Hoop, and P. Van der Maat, Nucl. Phys. 51 (1964) 395.

[48] K. P. Artemov, N. A. Vlasov, and L. N. Samoilov, Sov. Phys.— JETP 10 (1960) 841.

[49] D. S. Cramer and L. Cranberg, Bull. Am. Phys. Soc. 14 (1969) 553.

[50] M. M. Meier, L. A. Schaller, and R. L. Walter, Phys. Rev. 150 (1966) 821, and references cited therein.

[51] T. R. Donoghue et al., Phys. Rev. 173 (1968) 952, and references therein.

[52] G. L. Morgan, R. L. Walter, C. S. Soltez, and T. R. Donoghue, Phys. Rev. 150 (1966) 830.

[53] J. R. Sawers, F. O. Purser, and R. L. Walter, Phys. Rev. 141 (1966) 825.

[54] F. T. Kuchnir et al., Phys. Rev. 176 (1968) 1405.

[55] A. J. Elwyn and R. O. Lane, Nucl. Phys. 31 (1962) 78.

[56] R. A. Hardekopf, R. L. Walter, J. M. Joyce, and G. L. Morgan, Third Polarization Symp.

[57] W. D. Andress, F. O. Purser, J. R. Sawers, and R. L. Walter, Nucl. Phys. 70 (1965) 313.

[58] M. M. Meier, F. O. Purser, and R. L. Walter, Phys. Rev. 163 (1967) 1056.

[59] Th. Stammbach, G. Spalek, J. Taylor, and R. L. Walter, Nucl. Instr. 80 (1970) 304.

[60] W. Benenson, T. H. May, and R. L. Walter, Nucl. Phys. 32 (1962) 510.

[61] C. A. Kelsey, Nucl. Phys. 45 (1963) 235.

[62] T. G. Miller and J. A. Biggerstaff, Phys. Rev. 187 (1969) 1266; F. W. Büsser, private communication; R. Brüning et al., Second Polarization Symp., p. 141; M. M. Meier, F. O. Purser, G. L. Morgan, and R. L. Walter, Symp. on Nuclear Reaction Mechanisms (Quebec, 1969) p. 409.

[63] D. Hilscher, private communication.

[64] J. Taylor et al., Third Polarization Symp.

[65] M. M. Meier, R. S. Thomason, and R. L. Walter, Nucl. Phys. A115 (1968) 540; R. Brüning, F. W. Büsser, F. Niebergall, and J. Cristiansen, Phys. Lett. 21 (1966) 435.

[66] L. A. Schaller et al., Phys. Rev. 163 (1967) 1034.

[67] H. O. Klages and H. Schölermann, Z. Phys. 227 (1969) 344.

[68] T. R. Donoghue, private communication.

[69] A. Ciocanel, M. Molea, C. Pencea, and O. Salagean, Report of Inst. de Fizica Atomica, Romania (1969).

[70]   W. L. Baker, Ph.D. dissertation, Ohio State University (1969);
       T. R. Donoghue, C. E. Busch, J. A. Keane, and H. Paetz gen.
       Schieck, Third Polarization Symp.; H. Schölermann, Z. Phys. 220
       (1969) 211.

[71]   B. D. Walker, C. Wong, J. D. Anderson, and J. W. McClure,
       Phys. Rev. 137 (1965) 1504.

[72]   B. D. Walker et al., Phys. Rev. 137 (1965) B347.

[73]   D. De Martini, Ph.D. dissertation, Ohio State Univ. (1969).

[74]   G. P. Lietz, S. F. Trevino, A. F. Behof, and S. E. Darden, Nucl.
       Phys. 67 (1965) 193.

[75]   H. Lefevre, private communication.

[76]   L. P. Robertson et al., Nucl. Phys. A134 (1969) 545.

[77]   A. R. Thomas, D. Spalding, and E. H. Thorndike, Phys. Rev. 167
       (1968) 1240.

[78]   G. G. Ohlsen, Phys. Rev. 164 (1967) 1268.

[79]   F. Seiler et al., Helv. Phys. Acta 35 (1962) 385.

[80]   W. B. Broste et al., Third Polarization Symp.

[81]   J. E. Simmons, W. B. Broste, George P. Lawrence, and G. G.
       Ohlsen, Third Polarization Symp.

[82]   C. O. Blyth, P. B. Dunscombe, J. S. C. McKee, and C. Pope,
       Third Polarization Symp.

[83]   J. Schwinger, Phys. Rev. 73 (1948) 407.

[84]   W. S. Hogan and R. G. Seyler, Phys. Rev. 177 (1969) 1706.

DISCUSSION

McKee:

It is important in using superconducting spin-precession solenoids to be able to check experimentally on depolarization effects introduced by the magnet. At Birmingham we have the
facility to precess neutron spin through 360°, which enables us
to make this check for neutron energies up to 12 MeV. Few if
any angular distributions of polarization of neutrons from the
reactions mentioned have been published recently. Is there not
a need for such data? At Birmingham we have also worked at
polarization transfer in the D($\vec{\text{d}}$,n)^{3}He reaction at 2°. Polarization transfer of ~ 80% is found at 12 MeV deuteron energy in
agreement with Simmons' observations at the same energy. The
fact that polarization transfer in this reaction is high and apparently energy independent below 15 MeV seems to me a fasci-
nating discovery and of great use for the future.

Simmons:

In the use of polarized neutrons from polarized beams at Los
Alamos, we have found it helpful to reverse the neutron spin
direction by spin selection at the ion source. For certain pur-

poses the usual precession solenoid could be eliminated, with gain of counting rate by an order of magnitude.

Cranberg:

It seems that pulsed-beam time-of-flight methods will have to be used when sources of polarized neutrons are applied to investigations of elastic and inelastic scattering. Thus it seems reasonable to consider this when assigning a figure of merit to sources of polarized neutrons. One should quote $P^2 \bar{I}_{pulsed}$ where $\bar{I}_{pulsed}$ is proportional to the time-averaged intensity of the neutron source operated on a pulsed basis at some appropriate duty cycle (e.g., 1%).

Polarized Solid Targets

C. D. JEFFRIES, University of California, Berkeley, USA*

1. INTRODUCTION

Nuclei in solids may be oriented by establishing thermal equilibrium at low temperatures T in large external magnetic fields H or with strong hyperfine interactions in magnetic ions or atoms. These methods [1] have been widely used in orienting microscopic quantities of radionuclei, but only in a very few cases, e.g., in ^{165}Ho [2] has it been practical to obtain sizable orientations in bulk targets for scattering experiments, which is the main subject of interest at this conference. For protons, the thermal equilibrium polarization is $p_0 \approx (H/T) \times 10^{-7}$ gauss/K, and in the absence of strong hyperfine interaction, one can obtain only 1% polarization in an external field of 100 kG at 1 K. One can resort to a kind of trickery, known as dynamic nuclear polarization, in which the polarization is enhanced many orders over the thermal equilibrium value. In this paper we discuss these dynamic methods under three broad categories: (1) microwave dynamic polarization; (2) optical pumping in solids; (3) spin refrigerators. Only the first category has so far been used in actual targets, but we will discuss the physical principles of all types. We conclude with a brief discussion of recent advances in thermal equilibrium methods.

2. MICROWAVE DYNAMIC POLARIZATION [3, 4, 5]

This method (the "solid effect") is applicable to any nucleus of spin I of a diamagnetic atom in a solid containing about 1% of intentionally added electron "spins" S, e.g., paramagnetic ions, atoms, or molecules. To fix ideas, consider a crystal of the double nitrate $(Nd,La)_2 Mg_3(NO_3)_{12} \cdot 24 H_2O$, known as Nd:LMN, where the protons in the waters of hydration are the nuclear spins $I = 1/2$, and the Nd^{3+} paramagnetic ions are the electron spins $S = 1/2$. This substance was actually the first in which large polarizations (70%) were obtained [6], in Berkeley in 1962. In the usual experimental arrangement (fig. 1), the crystal is placed in a microwave cavity containing a microwave field

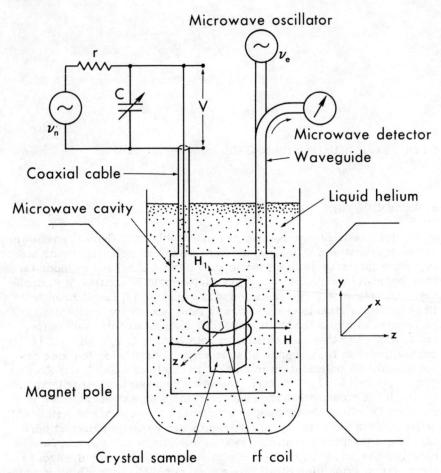

Fig. 1. Basic experimental arrangement for microwave dynamic polarization.

H_1 at ν_e, immersed in liquid ^{4}He ($\approx$ 1 K) or ^{3}He ($\approx$ 0.4 K), and in a uniform external d.c. field H $\approx$ 15 to 50 kG. An rf coil is wound around or near the crystal, to be used for measuring the proton nuclear magnetic resonance (NMR) at $\nu_n = g_n \beta H/h$. The NMR signal is proportional to the proton polarization p. The protons are dynamically polarized by the combined actions of the dipole-dipole interaction $\mathcal{H}_{I,S}$, the thermal lattice vibrations, and the applied d.c. anu microwave magnetic fields. For a typical I,S pair of neighbors separated by distance r the static Hamiltonian is

$$\mathcal{H} = g\beta\vec{H}\cdot\vec{S} - g_n\beta\vec{H}\cdot\vec{I} - \frac{gg_n\beta^2}{r^3}\left[\vec{I}\cdot\vec{S} - \frac{3(\vec{I}\cdot\vec{r})(\vec{I}\cdot\vec{S})}{r^2}\right], \tag{1}$$

where the terms are the electron Zeeman, nuclear Zeeman and dipole-dipole, respectively; β is the Bohr magneton; the effective g-factor for Nd^{3+} is $g = 2.7$; and for protons $g_n = 0.003$. The energy levels (fig. 2)

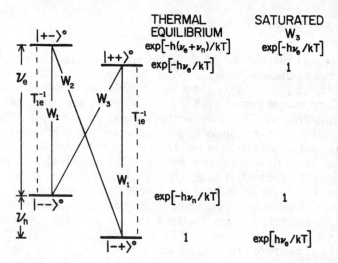

Fig. 2. Energy levels, transition rates, and populations for dynamic polarization.

are characterized in zero order by $|M,m\rangle^0$, where $M = \langle S_z \rangle = \pm 1/2$, and $m = \langle I_z \rangle = \pm 1/2$. However the term I_+S_z in $\mathcal{H}_{I,S}$ admixes these slightly, so that in addition to the usual allowed electron paramagnetic resonance (EPR) transition W_1 at $\nu_e = g\beta H/h$ one observes two weak forbidden transitions W_2 and W_3 at $(\nu_e + \nu_n)$ and $(\nu_e - \nu_n)$, weaker than the central line by the small factor

$$\sigma \approx \left[\frac{\text{dipolar energy}}{\text{nuclear Zeeman energy}}\right]^2 \approx \frac{3}{10}\left(\frac{g\beta}{H}\right)^2\frac{1}{r^6}. \tag{2}$$

The Nd ions, having appreciable spin-orbit coupling, interact strongly with the thermally modulated crystalline fields, with a well-known [7] spin phonon relaxation rate T_{1e}^{-1} with the typical value 10^5 sec^{-1} at 20 kG and 1 K. On the other hand, the protons have virtually no thermal relaxation process except via their dipolar coupling with the Nd ions, with a relaxation rate of order σT_{1e}^{-1}, quite negligible compared to T_{1e}^{-1}, but sufficient to establish the thermal equilibrium relative populations in fig. 2, corresponding to a proton polarization $p_0 = \tanh(h\nu_n/2kT)$. Suppose now that forbidden microwave transition W_3 is strongly induced at a rate $W_3 \gg T_{1e}^{-1} \gg T_{1n}^{-1}$. This saturates the transition, i.e., makes the relative populations of states $|--\rangle^0$ and $|++\rangle^0$ equal to, say, unity. The thermal relaxation process T_{1e}^{-1} establishes the populations of the other two states as shown in fig. 2, resulting in a (ideal) dynamically enhanced proton polarization

$$p = \tanh(h\nu_e/2kT). \tag{3}$$

Eq. (3) is just the full value of the Nd spin polarization, which is transferred to the near proton. What then occurs is diffusion of enhanced proton polarization away from the Nd ion to the more distant protons, by mutual proton spin flips between proton neighbors through the $I_{+i}I_{-j}$ terms in the proton-proton dipolar coupling [8]. Thus, ideally, all the protons in hydrogen in the sample eventually acquire the enhanced polarization of eq. (3). If one saturates W_2 instead of W_3, the polarization is equally enhanced but negative, i.e., the protons are held antiparallel to H. This ability to reverse polarization in a constant magnetic field is, of course, a very valuable feature, for it affords the differential counting techniques necessary to measure small spin dependent cross sections. Fig. 3 shows the measured enhancement $E = p_{meas}/p_0$ for $\nu_e = 74$ GHz and $T = 1.5$ K [6]. We found $p_{meas} = 72\%$, compared to the prediction 83% from eq. (3).

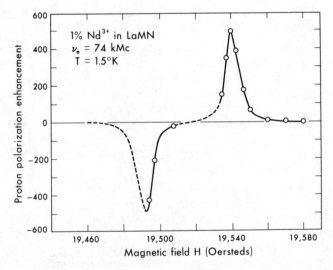

Fig. 3. Observed proton polarization enhancement in Nd:LMN [6].

Although this simple model of dynamic polarization is nearly adequate to explain ideal experimental examples, the exact theory for general substances is quite complex and has not been solved. To consider adequately the very strong H_1 fields used, and to consider properly the spin-spin interactions, one should use a density matrix approach [9, 4], but solutions are untractable except in the high temperature limit exp $(h\nu_e/kT) \approx 1 + h\nu_e/kT$. One should take into account the mechanisms and the finite magnitude of the line width $\Delta\nu_e$ of the microwave resonance, since, if $\Delta\nu_e \geq \nu_n$, the transitions W_1, W_2, and W_3 are not resolved, and the polarization will be reduced. One should consider the details of the proton spin diffusion process and the proton relaxation, both from the intentionally added spins S and from the undesirable extraneous spins S' [10]. The fact that the electron spin-lattice relaxation process may be phonon bottlenecked [7] must be reckoned with [5, 11, 12]. Also, the fact that strong microwave saturation in the wings of the line W_1 leads to electron spin temperature cooling of the S-S dipolar reservoir [9] leads one to another mechanism, the DONKEY effect [13], for nuclear polarization, which may be important in cases where the EPR line is broad. Faced with this array of subtle complex effects, in this brief review we go back to the simple phenomenological model and extend it to include leakage effects and incomplete saturation, obtaining thereby some simple equations that are found to be at least qualitatively correct. Since the typical ratio of proton to electron spins is $(n/N) \approx 10^4$, and each electron spin must relax at a rate T_{1e}^{-1} before the microwave field can flip it again with another proton, the leakage factor

$$f \equiv \frac{n\,T_{1e}}{N\,T_{1n}}, \tag{4}$$

represents the ratio of the bulk proton relaxation rate to the dynamic polarization rate, where T_{1n} is the measured proton relaxation time. In ideal samples like Nd:LMN, it is given by [6, 5]

$$\frac{1}{T_{1n}} \approx \langle \sigma \rangle \frac{\text{sech}^2(h\nu_e/2kT)}{T_{1e}}, \tag{5}$$

where $\langle \sigma \rangle$ is obtained from eq. (2) by taking $r^{-6} \approx r_1^{-3} r_2^{-3}$, where $r_1 \approx$ nearest I,S distance, dependent on crystal structure; and $r_2^{-3} \approx N4\pi/3$, where N is the number of electron spins per unit volume. One can show that $f \ll 1$ from eqs. (4) and (5) for ideal substances. However, for a large number of extraneous spins S', or nuclear relaxation from molecular motion, one finds $f \gg 1$, and p is reduced, as in eq. (8).

To take incomplete saturation into account, one introduces the saturation factor

$$s \equiv 2W_1 T_{1e} = (g\beta/h) H_1^2 T_{1e}/H_{1/2}, \tag{6}$$

where $H_{1/2}$ is the EPR line width, and the value of s for half saturation,

$$s_{1/2} \approx \frac{2Nf}{\langle\sigma\rangle n(1+f)}, \tag{7a}$$

which becomes for $f \ll 1$,

$$s_{1/2} \approx 2 \operatorname{sech}^2(h\nu_e/2kT). \tag{7b}$$

Taking into account both the leakage factor f and incomplete saturation, one replaces eq. (3) by the expression [5]

$$p \approx \frac{1}{1+f}\left(\frac{s}{s+s_{1/2}}\right) \tanh\left(\frac{h\nu_e}{2kT}\right). \tag{8}$$

One can also show that, for $f \ll 1$, the enhanced polarization builds up exponentially at the rate

$$\frac{1}{T_{0n}} \approx \frac{1}{T_{1n}}\left[\frac{s+s_{1/2}}{s_{1/2}}\right]. \tag{9}$$

Eqs. (8) and (9) give the approximate dependence of p and T_{0n} on the microwave power, which is just proportional to s.

Although a large number of hydrogenous substances have been polarized by microwave dynamic polarization, only a few have yielded proton polarizations sufficient for consideration as polarized targets. Table 1 lists the more important results; most of these are obtained at 20 to 25 kG and 1 K using 70 GHz microwaves. Following the use of Ce:LMN in the first thin target [14] and Nd:LMN in the first large target [15], experiments by a number of workers in the last three years have led to the development of new target materials: in particular, frozen solutions of ethanol or butanol containing the free radical porphyrexide, or ethylene glycol, $(CH_2OH)_2$, containing $K_2Cr_2O_7$. At 25 kG and 1 K polarizations in the range of 30 to 50% are obtained, but at temperatures 0.5 K, obtained by ^3He cryostats, polarizations in the range of 67 to 80% are obtained. The fraction of hydrogen by weight is 3.1%, 14%, and 9.7%, respectively, for LMN, alcohol, and ethylene glycol, so that the organic targets are considerably more desirable. Furthermore they are much more resistant to radiation damage: in LMN the polarization drops to half after $2 \times 10^{12}/cm^2$ minimum ionizing beam particles [31], whereas the alcohol and glycol

Table 1. Materials in which significant microwave nuclear polarizations have been achieved (August 1970)

Sample	Nucleus	Polarization (%)	H (kG)	T (K)	T_{on}	% Nucleus by weight	Refs.
Nd:LMN	^{1}H	72	19.5	1.5	5 min	3.1	[6, 15]
Ce:LMN	^{1}H	20	13.3	1.2		3.1	[14]
Ethanol & porphyrexide	^{1}H	35	25	1.05	5 sec	13.5	[16]
Butanol & porphyrexide	^{1}H	40	25	1.0	15 sec	13.5	[17]
"	^{1}H	67	25	0.5	17 min	13.5	[18, 19]
Ethylene glycol & Cr complex	^{1}H	50	25	1.1	25 sec	9.7	[20, 21]
"	^{1}H	80	25	0.5	15 min	9.7	[22]
Glycerol + H_2O + porphyrexide	^{1}H	55	25	0.6		9.9	[23]
Al $NH_4(SO_4)_2 \cdot 12\,H_2O$	^{1}H	51	19	1.0	19 min	6.1	[24]
NH_3 + Cr complex	^{1}H	40	25	1.0		17.6	[25]
Dy:LMN	^{1}H	77	21	1.2		3.1	[26]
Nd:Deuterated LMN	^{2}H	12	17	1.3	2 hrs	6.1	[27]
Deuterated butanol + phorphyrexide	^{2}H	20	25	0.55		24	[28]
Nd:LMN	^{139}La	50	18	1.5		18	[29]
U:CaF_2	^{19}F	60	27	0.7		49	[30]

targets can withstand 5×10^{14} particles/cm^2 before a comparable reduction in polarization [32, 33]. Also, the radiation damage in butanol
targets can be easily thermally annealed [33]. In the most desirable
substances, solid H_2, D_2, or HD, no significant dynamic proton or
deuteron polarizations have yet been observed, because T_{1n} is rather
short owing to molecular rotation and it is difficult to obtain sufficient
concentrations of a fast relaxing electron spin species in these solids.

The technical details of actual target construction and operation
have been adequately described [33–41], and a compilation has been
made of existing targets and the scattering experiments done with
them [42]. Some interesting experiments have also been done by
using a polarized proton target to produce a polarized neutron beam
[43]. Although for thick high energy targets the organic materials are
now to be preferred over Nd:LMN, most thin low energy targets use
Nd:LMN because of its higher thermal conductivity and ease of fabrication into very thin targets. The special difficulties of low energy
targets are discussed in refs. [36], [39], and [40]. Some recent noteworthy developments have been a proton target for use in intense
electron and photon beams [33], a continuous flow ^{3}He cooled target
[48], and an achievement of 20% deuteron polarization in deuterated
butanol [28], probably by the DONKEY effect [13].

Some representative configurations of actual targets are detailed
in figs. 4, 5, and 6.

As for future developments, it is clear that larger polarizations will
invariably be obtained by operating at larger values of H/T; with the
use of superconducting coils and ^{3}He cryostats it is feasible to operate at H/T $\approx$ 150 kG/K. Increasing H/T also gives a marked increase

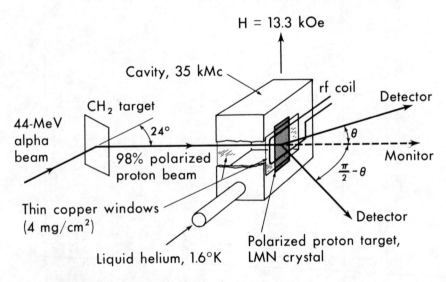

Fig. 4. Thin polarized target details, Saclay [14].

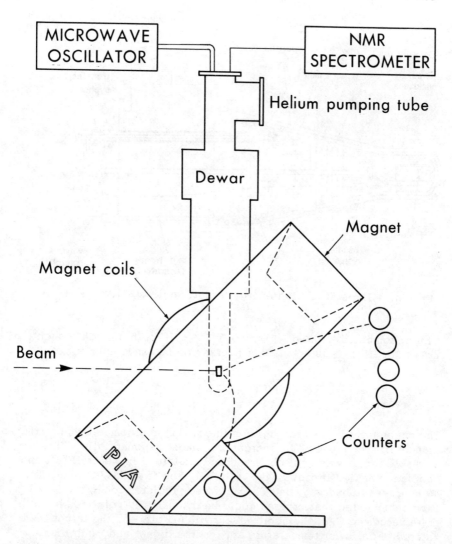

Fig. 5. High energy polarized target, Berkeley [15].

in T_{1n}, as can be seen from eq. (5). For Nd^{3+} or other magnetic species with an odd number of electrons, i.e., those with Kramers' degeneracy, the electron relaxation process is usually given by [7]

$$\frac{1}{T_{1e}} \propto H^5 \coth\left(\frac{g\beta H}{2kT}\right),$$
(10)

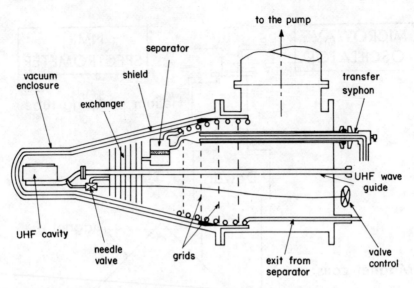

Fig. 6(a). Cryostat of high energy target, Saclay [37].

which, combined with eq. (5) gives $T_{1n}^{-1} \propto H^3 \coth(g\beta H/2kT) \operatorname{sech}^2(g\beta H/2kT)$. For $(H/T) \geq 30$ kG/K, this expression becomes approximately

$$\frac{1}{T_{1n}} \propto H^3 \exp(-g\beta H/kT). \tag{11}$$

The exponentially decreasing term in eq. (11) overcomes the H^3 term, so that as H/T increases, T_{1n} increases, almost exponentially. This remarkable result was first observed and explained in Nd:LMN [6], and has been verified in fig. 7 [44, 45] for H/T up to 40 kG/K, where $T_{1n} = 40$ hours was observed, and $T_{1n} = 10^6$ sec predicted for $H/T = 50$ kG/K. It was suggested that protons could be dynamically polarized at moderate values of H/T, the microwave switched off, and the target then "stored" at a lower temperature and/or in a higher and much less homogeneous field, where it could have better accessibility to beams, counters, etc. Although there are no data yet on T_{1n} for Nd:LMN at $H/T > 40$ kG/K, preliminary measurements on butanol and glycerol targets [19, 23] indicate that T_{1n} follows eq. (11) up to $H/T \approx 40$ kG/K, after which it increases much less rapidly. The reason for the limitation on T_{1n} is not yet clear, but it can be shown that terms in $\mathcal{H}_{I,S}$ of the form S_+I_+, neglected in the derivation of eq. (5), can contribute to the proton relaxation but do not contain the factor $\operatorname{sech}^2(h\nu_e/2kT)$ [46]. Another possibility is that relaxation to the S–S dipole reservoir is occurring [47]. Nevertheless, it still seems likely that accessible targets with very long relaxation times will be developed.

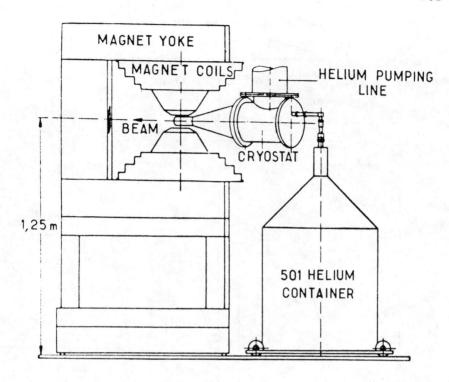

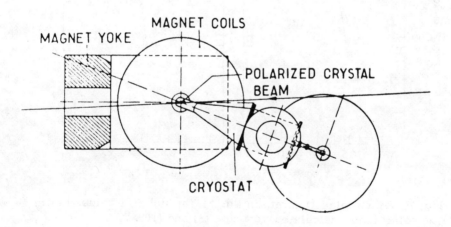

Fig. 6(b). High energy target, Saclay [37].

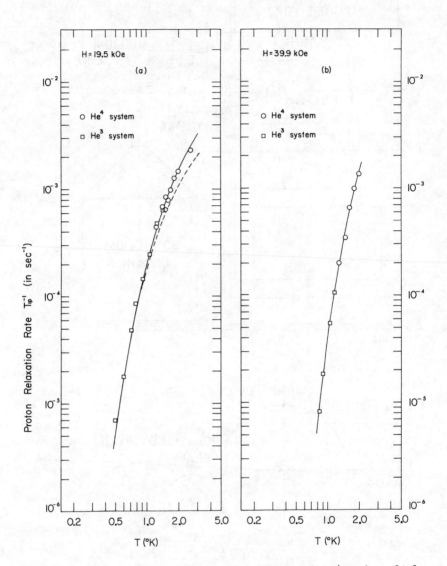

Fig. 7. Proton relaxation rate in Nd:LMN at high H/T values [45].
The solid line is calculated from eqs. (5) and (10).

Saturation of forbidden microwave transitions can also be used to orient dynamically the nuclei of paramagnetic ions or atoms in solids through contact hyperfine (hfs) interaction $A\vec{I}\cdot\vec{S}$ rather than dipole-dipole interaction $\mathcal{H}_{I,S}$. Fig. 2 can be used to explain this case by simply reversing the relative positions of the two lower levels and letting ν_n represent the hyperfine frequency. Actually this case was the first to be investigated experimentally [50], but there is no essential difference between the two cases. The hyperfine case has been adequately reviewed [3], and has been experimentally verified in magnetically dilute crystals at low values of H/T. The nuclei ^{60}Co, ^{54}Mn, ^{52}Mn, ^{122}Sb, and ^{76}As have been oriented in this way, but only in low concentrations. However at large values of H/T, the normally side EPR line width of concentrated paramagnetic crystals becomes extremely narrow [49], and it is possible to saturate selectively the forbidden microwave hyperfine transitions. Experiments are underway at Berkeley to determine the feasibility of dynamic nuclear polarization in samples containing $\sim 50\%$ by weight of the desired nucleus, which would be suitable for targets. The method applies in principle to all the nuclei in the 3d ion group, the 4f lanthanide groups, and the transuranics; in a way it is complementary to the dipolar case, which applies in principle to all the nuclei except these. There is some overlap for elements which have both diamagnetic and paramagnetic oxidation states.

3. OPTICAL PUMPING IN SOLIDS

This subject is related both to paramagnetic resonance and to the optical pumping technique originally applied to gases [51] and later extended to solids [52]. We have outlined some specific schemes for nuclear polarization in solids [53, 54], with these general features: production of an electron spin polarization by optical pumping with, say, circular polarized light; transfer of this polarization to nuclei through any of several phenomena—hfs coupling, selective spin-lattice relaxation, saturation of microwave transitions, or multispin cross relaxation. The justification for considering such schemes is that nuclei not amenable to other methods may be oriented, and that, in principle, sizable polarizations are possible at room temperature, in common with pumping in gases and in contrast to the microwave methods discussed in sect. 2.

Consider a crystal containing a small fraction of paramagnetic ions (or other magnetic species) for which the electronic ground state is represented by

$$\mathcal{H} = g\beta\vec{H}\cdot\vec{J} + A\,\vec{J}\cdot\vec{I}, \tag{12}$$

representing the Zeeman interaction in a field H, and a (smaller) hyper-

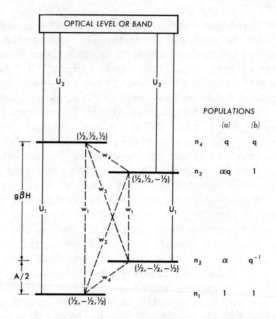

Fig. 8. Schematic level diagram for nuclear
polarization by optical pumping in solids.

fine interaction with the nucleus of the ion. To fix ideas, think of
Tm^{2+} in CaF_2 in which this experiment has been successful, and where
$J = 1/2$ and $I(^{169}Tm) = 1/2$. The energy levels and wave functions (J, J_z,
I_z) are shown in fig. 8 along with an optical level or band to which
we induce transitions by illuminating the crystal with circularly polarized light. We assume that by pumping with right polarized light we
induce the transition probabilities shown, where U_1 is significantly
different from U_2. This comes about because the optical transitions
obey the selection rule $\Delta J_z = +1$, $\Delta I_z = 0$. For example, if the ground
state is $^2S_{1/2}$ and the excited state $^2P_{1/2}$, then $U_1 = 2$, $U_2 = 0$. If the
excited state is $^2P_{3/2}$, then $U_1 = 1$, $U_2 = 3$. However if we pump both
states of the LS multiplet, then $U_1 = U_2$; therefore in solids where the
optical lines or bands may be broad we require a sufficient spin-orbit
coupling to partially resolve the multiplets so that we may selectively
pump out of one of the ground states. For $Tm^{2+}:CaF_2$, one can obtain
$U_1/U_2 \approx 3$ [55].

In fig. 8, $w_1 = T_{1e}^{-1}$ represents the usual spin-lattice relaxation rate
from thermal modulation of the crystalline electric fields; this mechanism also establishes the relaxation rate $w_2 \approx w_1(A/g\beta H)^2$. Relaxation
through modulation of the hyperfine interaction also occurs, but one
usually finds $w_2 \gg w_3$; in fact, this is just what makes possible the
Overhauser effect [56, 57]. If we assume nuclear spin memory ($\Delta I_z = 0$)

in the complete pumping cycle, then ions pumped up from the left side of fig. 8 will decay to the left side, etc., and the overall effect of pumping in competition with w_1 is to establish the relative populations in column (a), where $q \to U_1/U_2$ for strong pumping ($U_1 \gg w_1$) and a is to be determined by the relaxations w_2 and w_3. If indeed $w_2 \gg w_3$, then thermal equilibrium establishes $aq = \exp(-g\beta H/kT)$, and a nuclear polarization

$$p = \frac{q - \exp(-g\beta H/kT)}{q + \exp(-g\beta H/kT)} \,. \tag{13}$$

At room temperature, eq. (12) becomes

$$p = \frac{q-1}{q+1} \to \frac{U_1 - U_2}{U_1 + U_2} \,. \tag{14}$$

Reversing the polarization of the light from right to left reverses the sign of p. If we cannot be sure that $w_2 \gg w_3$, then one knows from dynamic microwave polarization that it is feasible to saturate the forbidden microwave transition $(1/2, 1/2, -1/2) \leftrightarrow (1/2, -1/2, 1/2)$, leading to the populations of column (b) and the polarization given by eq. (14). We see that the saturated polarization is given by optical matrix element ratios and is independent of the temperature, in contrast to the Boltzman factors, as in microwave dynamic polarizations. The chief experimental difficulty is to make $U_1 \gg w_1$ at room temperature. This is easy to do at very low temperatures, however, where one sees that, if $\exp(-g\beta H/kT) \ll 1$, eq. (12) yields a polarization approaching 100%, even if $q = 1$, i.e., for unpolarized light and $U_1 = U_2$. The above discussion has assumed nuclear spin memory, but one can show that simultaneous saturation by optical pumping and forbidden microwave transitions yields eq. (14) even for randomized optical relaxation.

The above scheme has been demonstrated at Berkeley [58] on a crystal of 0.05% $Tm^{2+}:CaF_2$, using the apparatus of fig. 9. It was necessary to work at helium temperatures in order to make the maximum optical pumping rate $U \approx 20 \; \text{sec}^{-1}$, competitive with $w_1 = T_{1e}^{-1}$, which has a comparable rate at 10K due to a Raman process $\propto T^9$. The nuclear polarization of Tm^{169} was measured by an optical technique, by monitoring the magnetic circular dichroism at 4120 Å by a weak beam, while pumping strongly the 4f–5d band at 5400–5800 Å with an Hg arc lamp. At $H = 800G$, $T = 1.6$ K we obtained $p(^{169}Tm) = 18\%$ while also saturating the $J_+ I_-$ forbidden microwave transition; without the microwaves, $p = 9\%$, showing that the assumption $w_2 \gg w_3$ is not entirely valid. Nevertheless this experiment demonstrated for the first time that significant nuclear polarizations can be obtained in solids by optical pumping, although there is a long way to go before useful targets are developed.

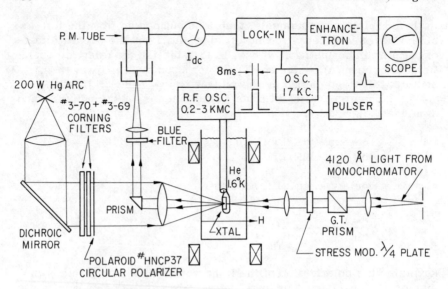

Fig. 9. Experimental arrangement for polarization of ^{169}Tm in Tm^{2+}:CaF$_2$ by optical pumping [58].

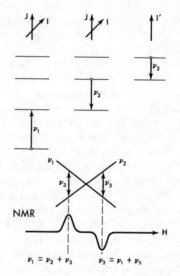

Fig. 10. Polarization of ^{19}F
by three-spin cross relaxation [59].

Fig. 10 shows how we polarized the abundant ^{19}F nuclei of spin I' in the CaF_2 crystal through a three-spin cross relaxation process with a pair of Tm^{2+} ions, each of which is, of course, a 4-level I,J hfs system. At a certain field where $\nu_1 = \nu_2 + \nu_3$, just below a crossing, we may have energy-conserving spin flips in which one ion flips up, the second down, and the ^{19}F flips down, thus enhancing the nuclear polarization. At a slightly higher field where $\nu_2 = \nu_1 + \nu_3$ the polarization will be reversed and enhanced. This behavior is observed [59], but since the ^{169}Tm hfs line width is 50 times larger than the ^{19}F NMR frequency, only a differential effect is observed, with a maximum enhancement of 40.

Small polarizations of the order of $10^{-2}\%$ have been reported [60] for protons in anthracene by optical pumping and spin selective optical decay; and for ^{29}Si in doped silicon using optical pumping and J+I-selective hfs relaxation [61].

4. NUCLEAR SPIN REFRIGERATORS [62]

The nuclear spin refrigerator is a method for cyclically cooling nuclear spins using an anisotropic electron spin system as the working substance. Consider a crystal of yttrium ethylsulfate, $Y(C_2H_5SO_4)_3 \cdot 9H_2O$, in which a few percent of diamagnetic Y^{3+} has been replaced by paramagnetic Yb^{3+} ions, abbreviated Yb:YES. This crystal is placed in a helium dewar at 1.2 K (fig. 11) and in a combination of magnetic fields $H_{d.c.}$ from an external magnet and H_p from a pulsed solenoid, to produce a net field $H(t) \approx 20$ kG, which can be rotated to an angle θ with respect to the crystal c axis. Now the Yb^{3+} ions have a very anisotropic g-factor ($g^2 = g_{\parallel}^2 \cos^2\theta + g_{\perp}^2 \sin^2\theta$, with $g_{\parallel} = 3.4$, $g_{\perp} \approx 0$) as well as an anisotropic relaxation rate ($T_{1e}^{-1} \propto \cos^2\theta \sin^2\theta$). First the field is held at $\theta = 45°$ for a few msec, where the thermal equilibrium populations of the Yb^{3+} Zeeman levels are established as shown in fig. 12(a); the Yb^{3+} ions become almost completely polarized. Then the field is rotated down to $\theta = 90°$, fast compared to T_{1e} but slow compared to the Larmor precession frequency, so that the populations remain unchanged. At 90° the Yb^{3+} g-factor becomes approximately equal to the proton g-factor $g_n = 0.003$, and a mutual spin flip occurs, the Yb flips up and a neighbor proton spin I flips down (fig. 12(c)). The field is rotated back to 45°, etc., and after a sufficient number of cycles this enhanced proton polarization has diffused out to all the protons in the waters of hydration and in the ethyl groups in the entire crystal. Theoretically one expects a nuclear polarization

$$p = \tanh[g(45°)\beta H/2kT], \tag{15}$$

very similar in magnitude to that obtainable in eq. (3) by microwave dynamic polarization, except that no microwaves are needed: one has

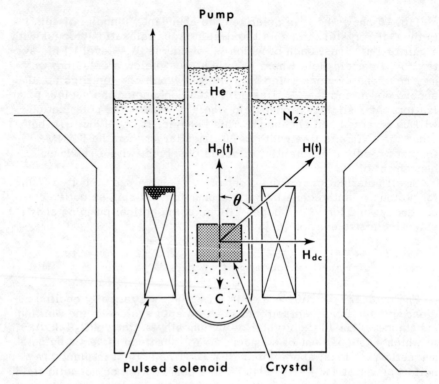

Fig. 11. Pulsed rotating field spin refrigerator [65].

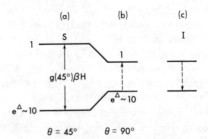

Fig. 12. Energy levels (not to scale)
of Yb^{3+} spins S and proton spins I,
and populations (a) at $\theta = 45°$ and
(b) at $\theta \rightarrow 90°$.

merely to rotate the field, or simpler, to rotate the crystal on a motor
shaft in a fixed field as was first used [63]. We have tried both approaches in Berkeley, with proton polarizations of 18% achieved by
rotating crystals [64], and 35% achieved by pulsed rotating fields as
described above [65]. Since $g_\perp$ is small in a plane, rotated polycrystalline samples of Yb:YES yield almost as large a polarization as
single crystals. In spite of detailed analysis and more recent work
[66], it is not yet clear why the theoretically expected polarizations
of 75% have not been achieved. Yb:YES is 5.3% hydrogen by weight
compared to 3.1% for Nd:LMN. A spin refrigerator target could be
made with excellent geometrical access. Other nuclei polarized in
spin refrigerators include ^{2}H [65], ^{27}Al [67], and ^{54}Mn [68].

5. THERMAL EQUILIBRIUM METHODS

Since the latest review [1], several experimental developments
allow one to reconsider the feasibility of, say, the brute force method
for polarizing nuclei. Dilution ^{3}He-^{4}He refrigerators have become
available, which provide continuous cooling down to $T \approx 0.02$ K.
Superconducting solenoids may produce up to 100 kG; recent advances
in new superconducting alloys may produce 200 kG fields. The discovery [69] that pure solid HD has a very long relaxation time, $T_{1n} >$
10^4 sec, reached after impurity ortho-H_2 molecules had been converted
to para-H_2, lead to the suggestion [70] that bulk samples of HD could
be so polarized to $\approx 50\%$ in 100 kG and 0.02 K and then used as polarized targets in more accessible magnet dewar arrangements with
lower values of H/T, where T_m would still be very long. Although
this has not yet been realized, it is an interesting possibility. Finally,
we note that isentropic compression of liquid to solid ^{3}He at temperatures below 0.318 K will produce cooling [71]. It is predicted that if
the compressional cooling is done in 70 kG from 20×10^{-3} K to $3 \times$
10^{-3} K, this will convert 24% of the liquid to solid ^{3}He, with a nuclear
polarization of 78% in the solid phase.

In this review on solids we have not considered the extensive work
on optical pumping in gases, but merely note that progress is being
made toward compressing and perhaps liquifying polarized gaseous
^{3}He for use as a target [72].

REFERENCES

*Supported in part by the U.S. Atomic Energy Commission.

[1]   For a review, see D. A. Shirley, Ann. Rev. Nucl. Sci. 16 (1966) 89.
[2]   V. L. Sailor et al., Phys. Rev. 127 (1962) 1124.

[3]   C. D. Jeffries, Dynamic Nuclear Orientation (John Wiley & Sons, New York, 1963).

[4]   A. Abragam and M. Borghini, in Progress in Low Temperature Physics, ed. C. J. Gorter (North-Holland Publishing Co., Amsterdam, 1964) vol. 4, p. 384.

[5]   C. D. Jeffries, in Electron Paramagnetic Resonance, vol. 1 (Plenum Press, New York, 1971).

[6]   T. J. Schmugge and C. D. Jeffries, Phys. Rev. 122 (1962) 1781, and 138 (1965) A1785.

[7]   P. L. Scott and C. D. Jeffries, Phys. Rev. 127 (1962) 32.

[8]   N. Bloembergen, Physica 15 (1949) 386.

[9]   B. N. Provorotov, Sov. Phys.—JETP 14 (1962) 1126, and 15 (1962) 611.

[10] H. E. Rorschach, Jr., Physica 30 (1964) 38; G. R. Khutsishvili, Sov. Phys.—Usp. 11 (1969) 802.

[11] M. Borghini, Phys. Rev. Lett. 16 (1966) 318.

[12] T. J. B. Swanenberg, G. M. Vanden Heuvel, and N. J. Poulis, Physica 33 (1967) 707.

[13] M. Borghini, Phys. Lett. 26A (1968) 242.

[14] A. Abragam, M. Borghini, and M. Chapellier, Compt. Rend. 255 (1962) 1343.

[15] O. Chamberlain et al., Phys. Lett. 7 (1963) 293.

[16] M. Borghini, S. Mango, O. Runolfsson, and J. Vermeulen, Proc. Int. Conf. on Polarized Targets and Ion Sources (Saclay, 1966) p. 387.

[17] S. Mango, O. Runolfsson, and M. Borghini, Nucl. Instr. 72 (1969) 45.

[18] D. A. Hill et al., Phys. Rev. Lett. 23 (1969) 460.

[19] D. J. Nicholas, P. H. J. Banks, and D. A. Crag, Rutherford Laboratory preprint RPP/A75 (1970).

[20] V. N. Fedotov, Sov. Phys.—JETP 26 (1968) 1123.

[21] H. Glattli et al., Phys. Lett. 29A (1969) 250.

[22] A. Masaike, H. Glattli, J. Ezratty, and A. Malinovski, Phys. Lett. 30A (1969) 63.

[23] W. G. Williams, D. J. Nichols, P. H. T. Banks, and D. A. Crag, Rutherford Laboratory preprint RPP/A77 (1970).

[24] P. J. Bendt, Phys. Rev. Lett. 25 (1970) 365.

[25] K. Scheffler, to be published in Nucl. Instr.

[26] M. Odehnal, V. I. Loutchikov, and J. Ezratty, J. de Phys. 29 (1968) 941.

[27] V. P. Alfimenkov et al., Phys. Lett. 24B (1969) 151.

[28] M. Borghini and K. Schaeffer, Phys. Lett. A31 (1970) 535; later results to be published.

[29] A. Abragam and M. Chapellier, Phys. Lett. 11 (1964) 207.

[30] M. Chapellier, M. Goldman, V. H. Chau, and A. Abragam, Compt. Rend. (1969).

[31] M. Chapellier, P. Garreta, and J. Kirsch, Second Polarization Symp., p. 126.

[32] J. R. Chen et al., Phys. Rev. Lett. 21 (1968) 1279.
[33] M. Borghini et al., to be published in Nucl. Instr.; UCRL preprint 19724.
[34] G. Shapiro, in Progr. in Nucl. Tech. and Instr., vol. 1, ed. F. J. M. Farley (North-Holland Publishing Co., Amsterdam, 1964) p. 176.
[35] H. Atkinson, Proc. Int. Conf. on Polarized Targets and Ion Sources (Saclay, 1966) p. 41.
[36] D. Garretta, Proc. Int. Conf. on Polarized Targets and Ion Sources (Saclay, 1966) p. 283.
[37] M. Borghini, P. Roubeau, and C. Ryter, Nucl. Instr. 49 (1967) 248, and 49 (1967) 259.
[38] P. Roubeau, Cryogenics 6 (1966) 207.
[39] P. J. Bendt et al., Nucl. Instr. 83 (1970) 201.
[40] A. Masaike et al., Nucl. Instr. 59 (1968) 170.
[41] T. Hasegawa et al., Nucl. Instr. 73 (1969) 349.
[42] M. Borghini, in Methods in Subnuclear Physics, vol. 4 (Gordon and Breach, New York, 1970).
[43] F. L. Shapiro, Proc. Int. Conf. on Polarized Targets and Ion Sources (Saclay, 1966) p. 339.
[44] C. D. Jeffries, Proc. Int. Conf. on Polarized Targets and Ion Sources (Saclay, 1966) p. 95.
[45] T. E. Gunter and C. D. Jeffries, Phys. Rev. 159 (1967) 290.
[46] C. D. Jeffries, to be published.
[47] G. M. Van den Heuvel, C. J. C. Heyning, T. J. B. Swanenberg, and N. J. Poulis, Phys. Lett. 27A (1968) 38.
[48] P. Roubeau et al., to be published in Nucl. Instr.
[49] A. King and C. D. Jeffries, to be published.
[50] C. D. Jeffries, Phys. Rev. 106 (1957) 106, and 117 (1960) 1056.
[51] A. Kastler, J. Phys. Radium 11 (1950) 255.
[52] N. V. Karlov, J. Margerie, and V. Merle D'Aubigne, J. Phys. Radium 24 (1963) 719.
[53] C. D. Jeffries, Proc. Int. Conf. on Hyperfine Structure Detected by Nuclear Radiation, Asilomar, 1967 (North-Holland Publishing Co., Amsterdam, 1968) p. 775.
[54] C. D. Jeffries, Phys. Rev. Lett. 19 (1967) 1221.
[55] C. H. Anderson, H. A. Weakleim, and E. S. Sabisky, Phys. Rev. 143 (1966) 223.
[56] A. Overhauser, Phys. Rev. 89 (1953) 689.
[57] A. Abragam, Phys. Rev. 98 (1955) 1729.
[58] L. F. Mollenauer, W. B. Grant, and C. D. Jeffries, Phys. Rev. Lett. 20 (1968) 488.
[59] W. B. Grant, R. L. Ballard, and L. F. Mollenauer, Technical Report UCB-34P20-T2, University of California, Berkeley.
[60] G. Maier, U. Haeberlen, H. C. Wolf, and K. H. Hausser, Phys. Lett. 25A (1967) 384.
[61] G. Lampel, Phys. Rev. Lett. 20 (1968) 491.

[62] C. D. Jeffries, Cryogenics 3 (1963) 41; A. Abragam, Cryogenics
      3 (1963) 42.
[63] F. N. H. Robinson, Phys. Lett. 4 (1963) 180.
[64] K. H. Langley and C. D. Jeffries, Phys. Rev. Lett. 13 (1963) 808;
      Phys. Rev. 152 (1966) 358.
[65] J. R. McColl and C. D. Jeffries, Phys. Rev. Lett. 16 (1966) 316;
      Phys. Rev. B1 (1970) 2917.
[66] R. L. Ballard, to be published.
[67] W. G. Clark, G. Feher, and M. Weger, Bull. Am. Phys. Soc. 8
      (1963) 463.
[68] J. Lubbers and W. J. Huiskamp, Physica 34 (1967) 193.
[69] W. N. Hardy and J. R. Gaines, Phys. Rev. Lett. 17 (1966) 1278.
[70] A. Honig, Phys. Rev. Lett. 19 (1967) 1009.
[71] R. T. Johnson and J. C. Wheatley, J. Low Temp. Phys. 2 (1970)
      423.
[72] Progress and Status Report for the University of Toronto Polar-
      ized Helium-3 Target, November, 1969 (unpublished).

Polarized Neutrons on Polarized Targets

H. POSTMA, Rijks Universiteit, Groningen, The Netherlands

During the polarization symposia at Basel and Karlsruhe (1960 and 1965) many experiments were described in which either polarized beams or, to a lesser extent, polarized targets have been used. Experiments with both polarized beams and polarized targets have only been touched during these meetings. The reason probably is that such experiments are difficult to perform because of the extremely low temperatures often necessary for polarizing nuclei. Such experiments may therefore have been considered for a long time as curiosities rather than as real possibilities.

Notwithstanding the difficult techniques, a reasonably large number of polarized-neutron polarized-nuclei experiments have been carried out during the past decade. During the Saclay polarization conference in 1966, Shapiro [1] and Schermer [2] reported on transmission experiments with polarized neutrons through targets of polarized nuclei carried out at Dubna and at Brookhaven, respectively. Since then a number of new transmission experiments which are worth reviewing have been carried out.

A number of developments in cryogenics makes one hope that nuclear orientation may be performed better and more easily. Therefore, more interest may evolve in the near future in nuclear physics experiments with polarized targets. The main cryogenic developments that have already shown their great value concern hard superconducting alloys, which make the production of strong magnetic field (100 kG) possible, and the ^{3}He-^{4}He dilution refrigerator [3, 4, 5], which is capable of cooling samples to about 0.01 K continuously. Such refrigerators are now in use in a number of nuclear orientation experiments.

It is of value to consider first the two main ingredients; namely, (a) polarized nuclear targets and (b) polarized neutron beams. Jeffries' review during this symposium of the principles of nuclear orientation makes it considerably easier for me. In most experiments which I want to report thermal equilibrium methods have been applied for polarizing nuclear targets; that is, magnetic hyperfine interaction or direct coupling of the nuclear magnetic moments with a strong external magnetic field (brute force polarization) has been used in samples

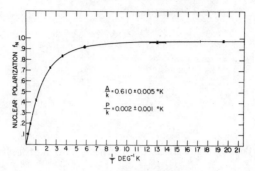

Fig. 1. Nuclear polarization of ^{165}Ho
versus the reciprocal temperature. The
sample was a single crystal of holmium
magnetized along one of the a-axes
(ref. [6]).

in which the nuclear spin systems are supposed to be in thermal
equilibrium with the material as a whole. Temperatures considerably
below 0.1 K are often necessary with a few interesting exceptions,
notably ^{165}Ho as a metal, which is ferromagnetic below 20 K. Fig. 1
shows the nuclear polarization of a Ho single crystal magnetized along
one of the a-axes as a function of the reciprocal temperature. This
polarization has been derived from a transmission experiment with
polarized neutrons [6]. I mention this since polarized Ho has been used,
in the form of single crystals or as polycrystalline metal, in quite a
number of experiments. From fig. 1 it is clear that already at a few
tenths of 1 K sizable degrees of polarization can be achieved for ^{165}Ho.
In many other cases, temperatures of a few hundredths of 1 K or even
lower will be necessary.

A serious drawback for nuclear reaction experiments with targets
at low temperatures is the dramatic decrease of heat conductivity in
the samples themselves and at the boundaries between the samples
and coolants with heat conductors. It is difficult to give exact numbers for allowable heat inputs to samples, since they depend very much
on the target construction, the sample material, the ultimate temperature, and the cooling capacity. A few ergs/s of heat input may already
be too much in several applications. The cooling capacity should not
be a problem nowadays owing to the ^{3}He-^{4}He dilution refrigerator,
which is able to withstand a load of the order of 10^3 erg/s at 0.1 K.
With neutron beams heat dissipation is often not serious. A second
problem for experiments with targets at extremely low temperatures is
related to nuclear relaxation, which may become prohibitively slow.
Thermal decoupling of the nuclear spins with their surroundings may
occur. On the other hand, it is possible to use this decoupling effec-

tively in some of the nuclear orientation schemes, which are then of the non-equilibrium type.

Polarized neutrons can be obtained in a number of ways, depending on the neutron energy region. It is outside the scope of this contribution to discuss them at length; for completeness they will be mentioned briefly. Polarized thermal and subthermal neutrons can be obtained with the aid of total reflection at magnetized cobalt mirrors. Polarizations of about 70—80% are routinely achieved. The neutron energy cannot be varied but at reactors large intensities can be obtained. Diffraction at magnetized single crystals of 94% Co—6% Fe is a very suitable way to get polarized and monochromatic neutron beams with energies easily variable from thermal to several eV. Polarizations very close to 100% are possible when (111) reflection is used, but in practical cases in which one optimizes intensity and polarization simultaneously 90—95% is commonly obtained. The most elaborate precision diffraction spectrometer for polarized neutrons, which is used in conjunction with polarized nuclei, exists at the High Flux Beam Reactor at Brookhaven [7]. It is used in Sailor's group mainly for transmission experiments. Figs. 2 and 3 show vertical and horizontal sections of this equipment. It is suitable for polarized neutrons up to about 20 eV.

A beam of polarized neutrons can also be obtained with the aid of a filter of polarized nuclei which have substantially different cross

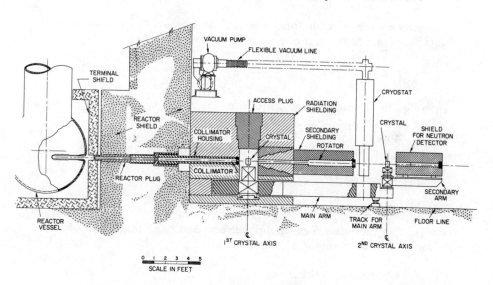

Fig. 2. Vertical section of the diffraction spectrometer for polarized neutrons used in conjunction with polarized nuclei at the HFBR at Brookhaven. (Courtesy of V. L. Sailor.)

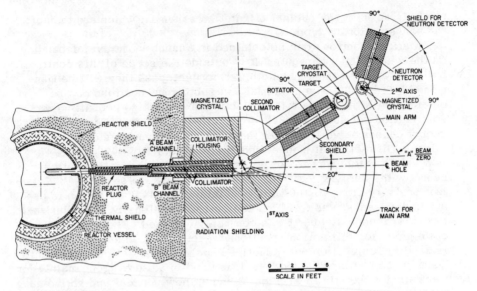

Fig. 3. Horizontal section of the equipment at the HFBR at Brookhaven for polarized neutron—polarized nuclei experiments. A neutron polarization analyzer is also shown. (Courtesy of V. L. Sailor.)

sections $\sigma_\pm$ for neutron spins parallel or antiparallel to the nuclear polarization. If these cross sections are written as

$$\sigma_\pm = \sigma_0 (1 \pm \rho f_N), \tag{1}$$

where σ_0 is the cross section for unpolarized neutrons and $\sigma_p = \sigma_0 \rho f_N$ is the polarization cross section depending on the nuclear polarization f_N and the spin factor ρ, then we find easily that the degree of polarization of a transmitted beam which was originally unpolarized is given by

$$f_n = -\tanh(N\sigma_0 t \rho f_N), \tag{2}$$

where N is the number of nuclei per cm^3 and t the thickness of the filter. The transmission is given by

$$T = \cosh(N\sigma_0 t \rho f_N) \exp(-N\sigma_0 t). \tag{3}$$

There exists an optimum thickness of the filter defined by the maximum value of the figure of merit $f_n^2 T$. This technique for polarizing a neutron beam has been used in Shapiro's group at the pulsed Dubna

reactor [8]. As a polarizing filter a single crystal of lanthanum magnesium double nitrate (LMN) with ^{142}Nd replacing 0.5% of the lanthanum is used. The free protons of this sample are polarized to a
maximum of about 60% using microwave techniques. This interesting
method yields about 60% neutron polarization and $f_n^2T \simeq 0.06$. It is
a useful method for polarizing neutrons in the eV and keV region. The
polarization is constant over a wide energy region up to about 50 keV.
Other isotopes may also be suitable as polarizing filters for neutrons,
e.g., ^{3}He or ^{10}B. Johnson, Symko, and Wheatley [9] have suggested
that a large degree of polarization of ^{3}He may be obtainable by adiabatic compression of ^{3}He in a strong magnetic field starting at about
0.02 K. It is not, however, a continuous process. The neutron cross
section of ^{3}He is related to an $(I - 1/2)^\pi = 0^+$ channel [10, 11], for which
$\rho = -1$ in eq. (1). A disadvantage is, of course, that the cross section
varies approximately as $E^{-1/2}$; however, this is less disturbing when
high degrees of polarization are obtained for ^{3}He. In fig. 4 the neutron
polarization f_n and the figure of merit f_n^2T are plotted versus energy
for LMN with 60% proton polarization (taken from ref. [8]) and for two
filters with solid ^{3}He polarized to 60% and with thicknesses optimized
for 1 and 100 eV respectively.

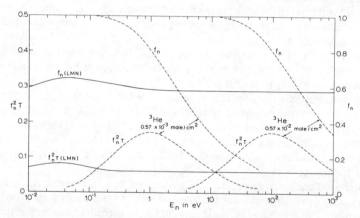

Fig. 4. Neutron polarization, f_n, and the figure of merit
f_n^2T versus neutron energy for LMN with a free-proton
polarization of 60% and for a ^{3}He filter also polarized to
60%. In the latter cases two thicknesses have been
assumed, one optimized for 1 eV and the other for 100 eV.

Polarized fast neutrons can be obtained with the aid of a number of
nuclear reactions using suitably chosen reaction angles and energies.
Such reactions have been reported extensively during this and the fore-
going polarization symposia. Partial moderation of polarized fast neu-

trons has been suggested by Dabbs and Harvey as another method for obtaining polarized neutrons in the intermediate energy regions [12].

All the polarized-neutron polarized-nuclei experiments which I want to report are transmission experiments. They can be divided into two groups according to the neutron energy: (a) those carried out with low-energy neutrons and (b) others with high-energy neutrons. The main difference, apart from experimental techniques, is in the interpretation of the experimental results.

Transmission experiments with polarized low-energy neutrons. Such experiments are carried out in order to determine spins of S-wave resonances in neutron cross sections; they are $J = I \pm 1/2$ if I is the spin of the target nucleus. The cross sections $\sigma_\pm$ for neutrons with spins parallel and antiparallel, respectively, to the direction of nuclear polarization are given by eq. (1). The spin-dependent factor ρ is -1 for $I - 1/2$ resonances and $I/(I+1)$ for $I + 1/2$ resonances. The polarization effect, ϵ, is given by

$$\epsilon = (T_{\uparrow\uparrow} - T_{\uparrow\downarrow})/(T_{\uparrow\uparrow} + T_{\uparrow\downarrow}) = -f_n \tanh(N\sigma_0 t\rho f_N), \tag{4}$$

assuming that depolarization of neutrons does not occur due to spin-flip mechanisms in the target and that the neutron and nuclear polarizations have the same values during the measurement of the parallel and antiparallel transmissions $T_{\uparrow\uparrow}$ and $T_{\uparrow\downarrow}$. The sign of ϵ gives directly the spin of the resonance, provided that the polarization directions are known. Sometimes the sign of the hyperfine interaction is not known a priori; hence the sign of f_N (that is, the nuclear polarization direction) is unknown. Consequently the measurements give in first order a division of the resonances into two classes. By comparing the magnitude of polarization effects it is still possible to make definite spin assignments to the resonances [13]. The polarization method for spin determination is superior to other methods, which are often rather indirect.

Spin assignments to neutron resonances are important for a number of reasons: (a) strength functions and other parameters may be spin-dependent, (b) gamma-ray emission after neutron resonance capture may give conclusive information about spins of low-lying levels if resonance spins are known, and (c) multilevel fits to experimental cross sections of fissile nuclei can be facilitated if resonance spins are known. In this respect ^{235}U is an interesting example. Spins of a series of ^{235}U resonances have been indicated on the basis of various methods, some of them rather indirect. Unfortunately they have given conflicting results for many of the resonances. Spins have been assigned to the three lowest resonances using the polarization method by Schermer et al. at the Brookhaven graphite reactor [14]. It would be of great value to extend these measurements to higher energy resonances in order to obtain reliable spin assignments for this isotope.

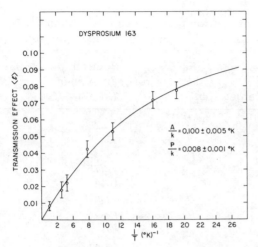

Fig. 5. Polarization effect ε versus the reciprocal temperature for Dy at the 1.71 eV neutron resonance (ref. [15]).

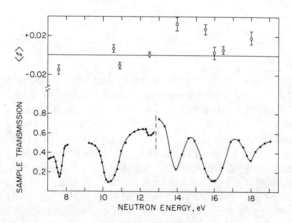

Fig. 6. Transmission and polarization effect ε for Dy versus neutron energy (ref. [15]).

Recently we have used the new polarization facility at the Brookhaven HFBR to assign spins to Dy and Lu resonances. Odd-A Dy isotopes can be polarized with the aid of hyperfine coupling. In fig. 5 the polarization effect ε is shown for neutrons of 1.71 eV as a function of the reciprocal temperature of the Dy sample [15]. From the shape of this

Table 1. Spins of neutron resonances of ^{161}Dy and ^{163}Dy below
20 eV and of ^{175}Lu and ^{176}Lu below 7 eV (refs. [15, 17])

Isotope	Target spin	Neutron resonance energy (in eV)	Resonance spin
^{161}Dy	5/2	2.72	3
		3.69	2
		4.35	2
		7.75	3
		10.40	2
		10.87	3
		14.3	2
		18.5	2
^{163}Dy	5/2	1.71	2
		16.25	3
^{175}Lu	7/2	2.604	4
		4.8	3
		5.22	3
^{176}Lu	7	0.142	13/2
		1.574	15/2
		6.16	15/2

curve the magnetic dipole and electric quadrupole coupling terms of
the hyperfine interaction have been derived. There is agreement of
the magnetic term with nuclear specific heat measurements [16]. It
has been possible to study resonances of Dy up to 18 eV (see fig. 6).
Assigned spins are quoted in table 1. Lu has been similarly studied
[17]. The main difference is in the way of polarizing it with the aid
of direct coupling of its nuclear moments with an external field of
37 kG ("brute force" polarization). Although only small polarizations
can be obtained in this way, it was shown that spin assignments were
possible. In fig. 7 the polarization effects of Lu and In are plotted
versus the reciprocal temperature. The figure clearly shows the linear
relationship between ϵ and T^{-1} expected for low nuclear polarization.
In is used for comparison since the spin of its first resonance at
1.458 eV is known. The transmission and polarization effect of Lu as
a function of energy is shown in fig. 8. Assigned spins of Lu resonances are given in table 1.

Experiments with polarized low-energy neutrons and polarized targets can also be of value for the study of few-nucleon systems; e.g.,
the spin dependence of the D + n and ^{3}He + n reactions have been
studied [18, 11]. In the first example the Dubna group was able to

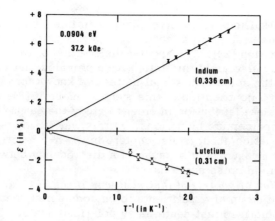

Fig. 7. Polarization effects for Lu and In versus the reciprocal temperature; brute force polarization (ref. [17]).

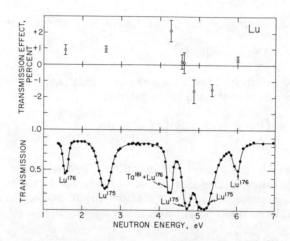

Fig. 8. Transmission and polarization effect of Lu versus neutron energy (ref. [17]).

select the proper set of doublet-quartet scattering amplitudes in a direct way. It may even be of interest to perform this experiment with greater precision in order to study an inconsistency reported in the three-nucleon system [19].

In the case of mixed $I + 1/2$ and $I - 1/2$ capture of slow neutrons, experiments with polarized neutrons on polarized nuclei may be used to derive the $I + 1/2$ and $I - 1/2$ fractions of the primary gamma-ray transitions. It suffices to observe changes in intensity of these transitions when the neutron polarization is reversed with respect to the nuclear polarization [20]. The angular distribution of the primary gamma rays may also yield this information provided that the multipolarities and the spins of the final states are known for these gamma transitions. In a contribution to this symposium Breit [21] suggested the use of angular distribution in order to study the gamma transition following neutron capture by protons. In this way it may be possible to obtain information about triplet capture in the n-p reaction. Breit suggested this measurement because of a discrepancy between the d(γ,n)p and p(n,γ)d cross sections.

Transmission of polarized fast neutrons through polarized targets. A second group of experiments to be considered concerns the scattering of polarized fast neutrons by polarized nuclei. In the analysis of nucleon-nuclear scattering data one often uses the optical-model potential of the Saxon-Woods form. For neutrons this is

$$-(V + iW)\,\rho(r) + 4iW_D \frac{d}{dr}\rho'(r) + \lambdabar_\pi^2 V_{SO} \frac{1}{r}\frac{d}{dr}\rho(r)\,\vec{L}\cdot\vec{\sigma}, \qquad (5)$$

or similar expressions. The radial dependence is often of the form

$$\rho(r) = [1 + \exp(r-R)/a]^{-1}. \qquad (6)$$

In the case of deformed nuclei a non-spherical potential should be used, which can be obtained by substituting

$$R = R_o\left\{1 + \sum_\lambda B_\lambda Y_{\lambda 0}\,(\theta)\right\}$$

where θ refers to the body-fixed coordinate system. The above-given potential contains volume and surface absorption terms and spin-orbit interaction. Feshbach's [22] suggestion was to add to this potential a nuclear-spin nucleon-spin interaction term of the form

$$-V_{SS}\,f(r)\,\vec{\sigma}\cdot\vec{I}, \qquad (7a)$$

or

$$-V_{SS}\,g(r)\,(\vec{I}\cdot\vec{r})\,(\vec{\sigma}\cdot\vec{r}). \qquad (7b)$$

So far only the first form with $f(r) = \rho(r)$ has been considered. There is evidence from the angular distribution of polarization of elastically scattered protons that this spin-spin term is small [23]. The analysis of nuclear reactions involves rather complicated calculations, especially when polarized beams and oriented targets are considered. Several cases and approximations have been studied theoretically in considerable detail. I should like to mention calculations by Davies, Satchler, Drisko, and Bassel [24] and by Davies and Satchler [25] based on DWBA. For deformed nuclei it is more appropriate to use coupled channel (CC) calculations; e.g., those carried out by Barrett [26] and by Tamura [27]. If the particle energy is much higher than the excitation energies the adiabatic approximation of the coupled channel calculations (ACC) can be used; otherwise a non-adiabatic calculation (NACC) must be carried out.

The interest in experiments with polarized fast neutrons on polarized nuclei lies in the fact that they may give direct evidence about the existence of the spin-spin interaction. The simplest method is the transmission experiment in which the spin-spin difference in the total cross section is measured. This involves an integration over the azimuthal angle. The total cross section can then be written as

$$\sigma = \sigma_{un} \left[1 + f_N f_n \left\{ \beta (\vec{n}_N \cdot \vec{n}_n) + \gamma (\vec{n}_N \cdot \vec{k}) (\vec{n}_n \cdot \vec{k}) \right\} \right] \qquad (8)$$

where $\vec{n}_N$ and $\vec{n}_n$ are unit vectors in the polarization directions. Calculations by Davies and Satchler [25] for some typical cases showed that β is the larger parameter. The last term does not occur if there is no spin-orbit coupling.

The first experiment along this line has been carried out by Wagner et al. [28] using [165]Ho polarized at 0.34 K. A change in cross section was observed when the [165]Ho nuclei were oriented, but with unpolarized neutrons. This effect is due to the prolate shape of the nucleus. It is clear from symmetry reasons that this effect is related to nuclear alignment rather than to polarization. It has been observed in all transmission experiments with oriented [165]Ho. Recently it was studied by Marshak and Langsford [29] at Harwell using neutrons up to 135 MeV; an oscillatory behavior was found. Our interest concerns the spin-spin interaction. However spin-spin dependence has not been observed in the neutron cross section of [165]Ho at 340 keV within the experimental limits: $\sigma_{\uparrow\uparrow} - \sigma_{\uparrow\downarrow} = 30 \pm 85$ mb. The total cross section at this energy is 7.94 b. The derived range for the spin-spin interaction strength using coupled channel calculations is $-0.13 < V_{SS} < +0.28$ MeV. This negative result stimulated other groups to repeat this experiment. Kobayashi et al. [30] at Tokyo carried it out at 0.92 MeV. Fisher et al. at Stanford performed the same experiment in a very similar way using polarized neutrons of 7.85 MeV (ref. [31]) and of 0.41 and 0.98 MeV (ref. [32]). Information about all these measurements is collected in table 2. No spin-spin dependence has been detected in any of these measurements. One must be careful when comparing the values of V_{SS} in table 2 since they are sensitive to the way the analysis of the data is carried out.

Table 2. Data concerning the spin-spin interaction in the optical model

Target isotope	T (in K)	E_n (in MeV)	f_N	f_n	σ_{tot} (b)	$\sigma_{\uparrow\uparrow} - \sigma_{\uparrow\downarrow}$ (mb)	V_{SS} (MeV)	Ref.
165Ho *	0.34	0.35	0.15†	0.55†	7.94	+30 ± 85	−0.13 < V_{SS} < 0.28	[28]
"	1.12	0.92	0.29†	0.30†	6.73	−128 ± 66		[30]
						−90 ± 90‡		
"	0.34	7.85	0.55	0.34	4.91	+3 ± 13	−0.34 < V_{SS} < 0.54	[31]
"	0.30	0.41	0.58	0.38		+1.0 ± 8.4 }	$|V_{SS}|$ < 0.09	[32]
"	0.30	0.98	0.58	0.28		7 ± 10 }		[32]
59Co	0.03	7.9	0.43	0.37		+35 ± 80	$|V_{SS}| \leq 1$	[35]
"	0.03	1.1	0.47	0.37		−73 ± 63 }		[36]
"	0.05	1.4	0.30	0.30		−32 ± 42 }		[36]
"	0.04	0.3–1.8	0.35		3 to 5 }	−300 to +200 }	$|V_{SS}| \sim 1$ a good fit is not possible	[37]
"	0.04	3.35	0.35					
"	0.04	7.75	0.35					

*Single crystal; in all other experiments polycrystalline holmium or cobalt are used.
†Polarization perpendicular to the beam, otherwise the polarizations are parallel to the beam direction.
‡If corrected for instrumental asymmetry.

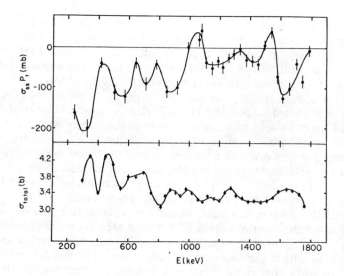

Fig. 9. Spin–spin difference cross section and total cross section for ^{59}Co polarized at 0.04 K measured by Healey et al. [37].

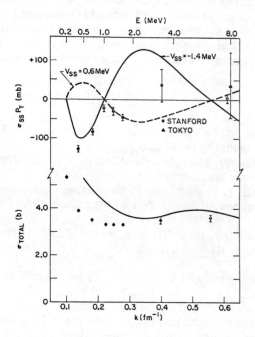

Fig. 10. Analyses of the spin–spin dependent cross section and total cross section between 0.3 and 8 MeV taken from ref. [37].

Recently this type of experiment has been repeated by the Tokyo and Stanford groups using ^{59}Co polarized at about 0.03–0.05 K. The Tokyo group used an adiabatic demagnetization cryostat [33], while at Stanford a dilution refrigerator was built for these experiments [34]. Spin-spin dependence was not detected at 7.90 MeV (ref. [35]) and at 1.1 and 1.4 MeV (ref. [36]) by the Tokyo group. The experiment carried out at Stanford by Healey et al. [37] is a good example of the enormous improvement in polarization experiments over the past few years. They observed a spin-spin dependence in ^{59}Co. In fig. 9 the total cross section and the cross-section difference for parallel and antiparallel polarizations (along the beam direction) are shown for the energy range of 0.3–1.8 MeV. Strong fluctuations exist in both quantities; however, it is difficult to see correlation. Healey et al. tried to fit the total cross section and the spin-spin cross-section difference for the energy region of 0.3 to 8 MeV using DWBA calculations in which spin-orbit coupling is neglected. The results are shown in fig. 10, in which the low-energy points are obtained by averaging over rather wide energy intervals. However, a simultaneous fit could not be obtained. It may be necessary to take into account the interaction of the neutron and $7/2^-$ proton hole using shell-model wave functions. Probably compound nucleus formation and direct reactions cannot be neglected in the calculations of spin-spin interactions.

It would be of great interest to extend the reported experiments with polarized fast neutrons to other polarized nuclei and to differential cross-section measurements in order to obtain a better insight in the way spin-spin interaction must be treated in nuclear reactions.

REFERENCES

[1]   F. L. Shapiro, Proc. Int. Conf. on Polarized Targets and Ion Sources (Saclay, 1966) p. 339.
[2]   R. I. Schermer, Proc. Int. Conf. on Polarized Targets and Ion Sources (Saclay, 1966) p. 357.
[3]   H. E. Hall, J. P. Ford, and K. Thompson, Cryogenics 6 (1966) 80.
[4]   B. S. Neganov, N. Borison, and M. Liburg, Sov. Phys.—JETP 23 (1966) 959.
[5]   J. C. Wheatley, O. E. Vilches, and W. R. Abel, Physics 4 (1968) 1.
[6]   G. Brunhart, H. Postma, and V. L. Sailor, Phys. Rev. B137 (1965) 1485.
[7]   H. L. Foote, Jr., and V. L. Sailor, to be published in Rev. Sci. Instr.
[8]   Yu. Taran and F. L. Shapiro, Sov. Phys.—JETP 17 (1963) 1467; F. L. Shapiro, Proc. Int. Conf. on Nuclear Structure Study with Neutrons, Antwerp, 1965 (North-Holland Publishing Co., Amsterdam) p. 223.

[9] R. T. Johnson, O. G. Symko, and J. C. Wheatley, Phys. Rev. Lett. 23 (1969) 1017.

[10] A. A. Bergman and F. L. Shapiro, Sov. Phys.—JETP 13 (1961) 895.

[11] L. Passell and R. I. Schermer, Phys. Rev. 150 (1966) 146.

[12] J. W. T. Dabbs and J. A. Harvey, Second Polarization Symp., p. 135.

[13] H. Postma, F. J. Shore, and C. A. Reynolds, Physica 30 (1964) 713.

[14] R. I. Schermer et al., Phys. Rev. 167 (1968) 1121.

[15] G. Brunhart et al., Bull. Am. Phys. Soc. 15 (1970) 569.

[16] O. V. Lounasmaa and R. A. Guenther, Phys. Rev. 126 (1962) 1357.

[17] V. L. Sailor, H. Postma, and L. Vanneste, to be published in Phys. Rev.

[18] V. P. Alfimenkov et al., Phys. Lett. 24B (1967) 151.

[19] H. P. Noyes, Proc. Int. Conf. on Three-Body Problem in Nuclear and Particle Physics, Birmingham, 1969 (North-Holland Publishing Co., Amsterdam, 1970) p. 2; J. D. Seagrave, ibid., p. 41; R. Bouchez, private communication.

[20] H. Postma, Ned. T. Natuurk. 34 (1968) 263.

[21] G. Breit, Third Polarization Symp.

[22] H. Feshbach, in Nuclear Spectroscopy, ed. F. Ajzenberg-Selove (Academic Press, 1960) part B, chap. V.A.

[23] L. Rosen, J. E. Brolley, Jr., and L. Stewart, Phys. Rev. 121 (1961) 1423.

[24] K. T. R. Davies, G. R. Satchler, R. M. Drisko, and R. H. Bassel, Nucl. Phys. 44 (1963) 607.

[25] K. T. R. Davies and G. R. Satchler, Nucl. Phys. 53 (1964) 1.

[26] R. C. Barrett, Nucl. Phys. 51 (1964) 27.

[27] T. Tamura, Rev. Mod. Phys. 37 (1965) 679.

[28] R. Wagner, P. D. Miller, T. Tamura, and H. Marshak, Phys. Rev. B139 (1965) 29.

[29] H. Marshak, A. Langsford, C. Y. Wong, and T. Tamura, Phys. Rev. Lett. 20 (1968) 554.

[30] S. Kobayashi et al., Phys. Soc. Japan 22 (1967) 368.

[31] T. R. Fisher et al., Phys. Rev. 157 (1967) 1149.

[32] T. R. Fisher, D. C. Healey, and J. S. McCarthy, Nucl. Phys. A130 (1969) 609.

[33] K. Nagamine, Nucl. Instr. 78 (1970) 285.

[34] T. R. Fisher et al., Rev. Sci. Instr. 41 (1970) 684.

[35] S. Kobayashi et al., Progr. Theoret. Phys. (Kyoto) 40 (1968) 1451.

[36] K. Nagamine, A. Uchida, and S. Kobayashi, Nucl. Phys. A145 (1970) 203.

[37] D. C. Healey, J. S. McCarthy, D. Parks, and T. R. Fisher, Phys. Rev. Lett. 25 (1970) 117.

Summary

EUGENE P. WIGNER, Princeton University, New Jersey, USA

Before starting on this review, let me express my pleasure at having attended a conference together with so many interested people with such a wide range of points of view. For some of us, the wealth of the material presented, the variety of the ideas and experimental arrangements, was the greatest surprise. Let me also mention the usefulness of private conversations, which are part of the attraction of conferences. I am particularly indebted to Drs. Darden and Weidenmüller for illuminating discussions.

The following review of the conference will consist of two parts. The first part will be concerned with questions of central theoretical interest. Not all of these relate directly and exclusively to polarization phenomena; for many of them, the study of the polarization phenomena is only a tool to obtain answers to questions in which one would be interested anyway, even if polarization phenomena did not exist. The second part of the review will recall some of the experimental procedures and phenomena which have been described and which are so striking and ingenious that even a theoretician cannot forget them.

1. POLARIZATION PHENOMENA AND THE COLLISION MATRIX

All of us theoreticians know that the collision is only incompletely described by giving the colliding and the separating particles, together with their momenta, that the complete description of the initial and of the final states includes the dependence on the spins or, more modernly, helicities. However, one is often inclined to disregard this dependence, or at least to postpone its consideration to some indefinite future. It was good, therefore, to learn that a beautiful theory of this dependence has been developed and, even more important, that a wealth of experimental material has been created to be interpreted by that theory.

Since much of the notation used both by theoreticians and by experimentalists is different from that used in other parts of physics, in particular in particle theory, it may be useful if a contact with the other notation is established. This will be restricted to the case in which only two particles collide and the reaction can also yield only two particles, not three or more. In this case, the variable specifying the columns of the collision matrix are, first, the description of the colliding particles and their states of excitation. Next come the momenta of the particles and, finally, their helicities, that is, their angular momenta about their directions of motion. These last variables can assume only discrete values, each $2S + 1$ values for a particle with spin S so that there correspond $(2S_1 + 1)(2S_2 + 1)$ columns to each pair of particles colliding with definite momenta. The rows of the collision matrix are similarly labeled, with the description of the reaction products, their momenta, and their helicities. Hence, to each pair of momenta of each type of reacting particles and reaction products there corresponds a submatrix of the collision matrix with $(2S_1 + 1)(2S_2 + 1)$ columns and $(2S_1' + 1)(2S_2' + 1)$ rows if S_1' and S_2' are the spins of the reaction products. The criterion for time inversion invariance, for instance, implies the symmetry with respect to the interchange of all labels of the rows with all labels of the columns, including the labels of the particles. Naturally, the elements of the collision matrix vanish unless the sum of the momenta of initial and final states are the same, and the collision matrix is also rotationally and Galilei or Lorentz invariant, depending on whether one uses classical or relativistic theories. In the simplest case of zero spins this means that the collision matrix depends only on the scalar product of the moments of the incoming particles—this specifies the energy of the collision— and the scalar product of one of the initial and one of the final momenta—this specifies the momentum transfer. The elements of the collision matrix with rows and columns referring to different total momenta are zero.

The notation of the collision matrix, usual in other parts of physics, has been discussed in such detail because it would appear to me to be useful if a review article were to be written, comparing the notation just described with that used in the studies which we heard discussed. Such a review might well derive the consequences of the various conservation laws, space and time inversions, as they manifest themselves in the two descriptions. It might also contain the observation of Dr. Simonius, who answered at least partially a question which has puzzled many of us for some time: to what extent is the collision matrix truly observable? It may be recalled that Heisenberg's motivation for dealing with the collision matrix was that he wanted to use quantities which are truly observable. In fact, the absolute values of the off-diagonal elements of the collision matrix are reaction cross sections and hence observable. Similarly, the real parts of the diagonal elements are essentially scattering cross sections and also observable. However, one can have serious doubts on the observability of the complex phases of the off-diagonal elements since they relate to relative phases of different reaction products and such phases, for instance the phase of an ^{3}H + ^{1}H pair relative to the phase of an ^{2}H + ^{2}H pair, are apparently unobservable. (I have some reason for saying "apparently.") What Dr. Simonius has taught us is that the relative phase of the ^{3}H + ^{1}H pair with respect to the ^{2}H + ^{2}H pair may be unobservable, but the relative phases of the different polarization states of the ^{3}H + ^{1}H pair, and of the ^{2}H + ^{2}H pair, are observable. This means, more generally, that the relative phases of the elements of the aforementioned submatrix of the collision matrix, with $(2S_1 + 1)(2S_2 + 1)$ columns and $(2S_1' + 1)(2S_2' + 1)$ rows, are observable. The absolute values are, of course, all observable so that the number of unobservable phases in any column of the collision matrix is equal to the different reaction products which the collision can yield—different states of polarization not counting as different products. Similarly, the only phases in a row of the collision matrix which he did not show to be observable are the phases of the elements referring to different colliding pairs, again a change in the state of polarization not constituting a change in the pair in the sense considered. Naturally, it was assumed by Simonius that all the various measurements of polarizations and asymmetries which were discussed can truly be carried out and also that the particles of the colliding pair can be produced in all states of polarization. However, the measurements which we have heard about from the Los Alamos group and from Dr. Postma, give clear justification for these assumptions. The Los Alamos group calls the reactions in question "polarization transfer reactions"—they are the ones assumed to be feasible by Dr. Simonius' arguments.

2. THE DISTORTED WAVE BORN APPROXIMATION

The distorted wave Born approximation (DWBA) played a very important role in the discussions we have heard. For most of us who are not active in this field, this approximation method is a bit foreign. The authors do not always betray how many parameters of their distorted waves were taken from the experiments they are interpreting, and how well these parameters reproduce the elastic scattering cross sections. This is a criticism. A glance at the angular dependence of some of the cross sections, experimentally determined, does on the other hand clearly suggest the validity of a picture at least akin to that used. The oscillations in the angular dependence of the cross sections strongly indicate a diffraction phenomenon. All this has been well known for some time. What is novel is the new measure of the validity of this DWBA method because a new set of experimental data is to be explained with the same theory which was adjusted, in the past, to a smaller number of data. The new data are the results of polarization measurements and of measurements of cross sections of polarized particles.

The preceding statement does contain some exaggeration. Some new parameters unavoidably had to be introduced to calculate the results obtained by polarization measurements. Also, the theory was modified somewhat, but modified in a very natural way, by taking into account the shape dependence of the spin-orbit force. Dr. Satchler called the modified spin-orbit force the "full Thomas form." There is agreement that the DWBA, after these rather minor modifications, reproduces the results of polarization experiments at least qualitatively. In view of the complexity of the experimental results, the strong fluctuations of the polarizations as functions of the directions of the outgoing particles, this is no small success. Nevertheless, the agreement between polarization experiments and DWBA theory is not quantitative and the deviations, at larger deflection angles, are considerable. Gorlov, Satchler, Blair, Graw, Goldfarb, Santos, Haeberli, Grotowski, and Mayer all agree on this. It is also perturbing that it was necessary, in some cases, to assume a greater deformation of the spin-orbit force than is present in the central force. Blair advocates a deformation 1.5 times too large in ^{28}Si. To quote Dr. Goldfarb, "polarization phenomena present the DWBA with a very difficult challenge." I do hope, though, and similar views were voiced by others, that the challenge will be met successfully.

At this point, it would be difficult for me to resist expressing my pleasure at the measurements, reported by Haeberli, for at least one set of reactions of all the cross sections and all the polarizations not involving more than two particles. The reactions in question lead to ^{54}Fe as compound, the compound being obtainable by bringing ^{52}Cr into collision with ^{2}H, or ^{53}Cr with a proton. Altogether, four processes were investigated at Wisconsin and at Saclay—two scattering processes and two reactions. Such complete investigations are more likely to provide convincing evidence for or against a reaction theory than the measurements of only some of the processes, because when more processes—that is, more cross sections and polarizations—are measured, the number of adjustable parameters does not increase but the number of results does. I admit that I have said this before.

3. NUCLEAR FORCES AND SPIN-ORBIT INTERACTION

Dr. Signell's discussion of this subject caused a good deal of apprehension and raised some doubts concerning ideas which were often considered established. He mentioned, in particular, calculations by Landé and Svenne, who, using accepted and in many applications apparently successful potentials, obtained for some nuclei a density about three times too high and a binding energy about 25% too large. This seems to present a very serious problem. Too large a binding energy at too high a density indicates a very serious weakness of the theory. This may be either in the interaction—in the Hamada-Johnston, or Yale, or Tabakin potentials—or in the calculation of the properties of nuclei, using these potentials. This is the Hartree-Fock method, based on Brueckner-Bethe theory, and, considering the success of the potentials in question in explaining various properties of spectra, one will suspect this method in the first place. Needless to say, this view is put forward with much hesitation, and surely this writer cannot propose an alternate for the Brueckner-Bethe theory. It would, however, be wrong to leave this difficulty unmentioned. Possibly we must revert to the point which Breit has reemphasized, that the nucleus contains not only protons and neutrons but also a cloud of pions and other mesons with fluctuating densities so that a more natural description of the nucleus would explicitly refer to these additional particles. This would mean, presumably, the introduction of a Fock space as far as pions and other mesons are concerned.

Altogether, Signell's report on what we do and what we do not understand about the origin of the spin-orbit interaction of nucleons was quite pessimistic. Concerning the description of the nucleon-nucleon scattering experiments, I am a bit disappointed that these are not described, for each angular momentum $J \geq 1$ and each parity, by 2 by 2 matrices rather than by phase shifts. If we do not yet have all the data to do this, it would be better to admit it rather than to use an incomplete description. Dr. Baker made a start in this direction.

4. CHARGE INDEPENDENCE

The problem of obtaining the consequences of the charge independence of nuclear forces—one should say, the near charge independence of nuclear forces—has been with us for a long time. Weidenmüller discussed it in passing, and it was taken up again in McKee's report. We learned that many (d,p) cross sections are similar to the (d,n) cross sections on the same element, but this is not very relevant in the present context unless the element on which the cross section is measured contains an equal number of protons and neutrons. It is relevant, on the other hand, that the ^{3}He(p,p) and the ^{3}H(n,n) cross sections are quite different—surely this statement was meant to imply that they are different even after a correction is applied on account of the Coulomb forces. This is definitely in conflict with what one would have expected on the basis of charge independence. Finally, McKee mentioned that the polarizations resulting from the ^{2}H + ^{2}H → ^{3}He + n and the ^{2}H + ^{2}H → ^{3}H + ^{1}H reactions are quite different, the former being significantly lower. We also heard, from Drs. Perey and Greenlees, that the effective interaction of at least two nuclei with equal proton and neutron contents, of ^{16}O and of ^{40}Ca, is greater for protons than for neutrons—again, after a correction for the Coulomb effect has been applied. This seems to contradict the conclusion which one might arrive at from the validity of the isotopic spin concept, but

Dr. Satchler later called my attention to an additional correction which may restore the agreement between experiment and theory. Nevertheless, these are not good tidings for those of us who would like to see the validity of the isotopic spin quantum number confirmed and who would like to apply this concept in an easy and effortless way. There are so many phenomena, including the mere existence of the analog states, with energies quite accurately given by the charge independence assumption, that the validity of the isotopic spin concept seems to us assured. In a private conversation, Dr. McCullen, of the University of Arizona, told me about another recent confirmation. This refers to the β transition probabilities from the ground state of ^{42}Sc, a nucleus with $J = 0$, $T = 1$ ($T_\zeta = 0$) to two $J = 0$ states of ^{42}Ca. The ground state of this latter nucleus is the $T_\zeta = 1$ member of the isotopic spin triplet of which the normal state of ^{42}Sc is the $T_\zeta = 0$ member. The other $J = 0$ state of ^{42}Ca belongs to another isotopic triplet. Hence, if the isotopic spin is a valid quantum number, the transition from ^{42}Sc to the normal state of ^{42}Ca should be favored, that to the excited state forbidden. In fact, Dr. McCullen told me, it is at least 10,000 times weaker. These β transitions are suited for testing the validity of the isotopic spin concept because, between two $J = 0$ states, only Fermi transitions are possible. If the isotopic spin is a valid quantum number, these should take place only between members of the same T-multiplet—at least if the wave lengths of the electron and of the neutrino which are emitted are large as compared with the size of the nuclei involved. Thus, Dr. McCullen's observation confirms that validity.

5. REMARKABLE AND NOVEL EXPERIMENTAL METHODS AND OBSERVATIONS

Dr. Weidenmüller mentioned Lobashov's measurement of the circular polarization of the $5/2^+ \to 7/2^+$ transition in ^{181}Ta. It amounted to about 10^{-6}. If parity were a truly rigorous concept, the radiation of such a transition would show no circular polarization at all. The degree of circular polarization, around 10^{-6}, is of the order which the weak interaction might well generate.

Kaminsky's method for producing polarized particles is surely the last method a non-professional would have thought of. It will be recalled that he passes a beam of particles through a crystal, its direction parallel to one of the crystal axes. Under such conditions the energy loss is, as Dr. Kaminsky has shown, a good deal smaller than for a beam of particles of random direction. This is the so-called channeling phenomenon. Furthermore, if the crystal is magnetized— Kaminsky used a nickel crystal—the emerging beam of particles is polarized. In fact, it is polarized more strongly than was expected.

Dr. Ebel gave at least a tentative explanation for this phenomenon. He said that the traveling particle picks up an electron from the nickel, interacts with its spin, then loses it again to the crystal. This process is gone through several times. On each occasion, the electron is more likely to be picked up from the filled electron band and can be discharged most easily into the unfilled band so that an angular momentum is transferred to the crystal. This must come, ultimately, from the particle traveling through the crystal, which as a result acquires a polarization. If this explanation is correct, the polarization is related principally to the magnetization of the crystal, and its direction with respect to the direction of the traveling particle can be varied greatly. This would render the method a very flexible one. What is puzzling about Ebel's explanation is that one does not see why the method works only if the direction of the particle coincides with one of the crystal axes or, more generally, is a channeling direction.

Kaminsky's method of producing a polarized beam of particles is not the only experiment making use of the interaction with a solid; Brooks' anthracene crystal provides another one, and it too is a very clever one.

I now come to the method of producing a polarized beam of particles, protons or deuterons, which we believe we understand fully and which seems to be the most efficient one. I am referring to Donnally's Lamb-shift method, which he described so clearly that it is unnecessary to repeat it. Further, as we have learned from Brückman, Glavish, and Walter, the polarization of the protons or deuterons can be transferred by collisions also to other particles. This has a varying effectiveness— generally a rather low one—but Dr. Walter, quoting results obtained by Dr. Simmons and his associates at Los Alamos, mentioned that a fully polarized deuteron beam can produce neutrons with a polarization of 0.9.

Dr. Jeffries discussed other methods of producing polarized beams of particles. Some of these were skillful modifications of older methods, but I found his dynamic polarization method particularly impressive. It also seems very effective. Nevertheless, the problem of producing good beams of particles with arbitrary energy, and also in an arbitrary state of polarization, is far from being completely solved. We can expect to hear more about this question at future polarization conferences.

The present review of our conference is an incomplete one in many respects; its writer can only hope that he will be forgiven by those whose work and results were left unmentioned. Most of the speakers excused themselves for the incompleteness of their discussions, caused by the necessary brevity of their presentations. It is only appropriate that the summary shares this weakness of incompleteness, and the apologies therefor. Even more appropriate is it to thank Dr. H. H. Barschall for his indefatigable help when preparing the final version of this summary. There would be many more mistakes in it had he not stood by.

Concluding Remarks

R. FLEISCHMANN, University of Erlangen-Nürnberg, Germany

At the first symposium in Basel the chief question concerned the production of sources of polarized particles. At the second symposium many results dealing mainly with experimental work were presented and discussed. At the third symposium a large amount of experimental material was shown, and we are glad to see the increasing interest of theoreticians in this field, but a still greater participation on their part would be desirable. In the next years physicists should search for principles which will connect the many pieces of information we have, so that we will gain better insight into the laws governing the structure of nuclei.

Turning now to the organizers of the meeting, may I first express our thanks to Professors Barschall and Haeberli, who had a great deal of work with the symposium. They have done a wonderful job. Even the regrettable event near Sterling Hall could not spoil the conference. During the symposium all participants were a scientific community and we appreciated it. We would be happy if the spirit of harmony would pervade the whole student body of this university, too.

Our thanks are extended further to all those who have contributed in any way to the organization of the meeting. We also want to thank all the speakers who reported on special problems or presented the work of their groups.

Our special gratitude goes to Prof. Wigner, who gave us an excellent summary of the conference. We would be glad if he would include the polarization phenomena in his special field of interest.

We are leaving Madison soon, but we will always have pleasant memories of this city and this symposium.

Formal Description
and Theory

On the Complete Determination of the Scattering Amplitudes from Polarization Measurements

M. SIMONIUS, Seminar für theoretische Physik, ETH, Zürich, Switzerland

What experimental information is necessary and sufficient for a complete determination of the scattering amplitudes for a given process without using a specific model? This question can to a great extent be answered for polarization measurements at fixed dynamical variables (energy and scattering angle). Some of the methods presented here are also applicable if a phase analysis is helpful, i.e., for low enough energies where not more than a few total angular momenta are involved. However, unless explicitly stated otherwise, the results presented apply to the analysis of measurements at fixed dynamical variables without the use of phase analysis.

For definiteness we consider a simple scattering process $a + b \rightarrow c + d$. But the results apply equally well to other processes with the same number of particles with nonzero spin involved (except two-body decay). The process is completely described by N scattering amplitudes $R_\nu = R_{\alpha\beta\gamma\delta}$ (which are functions of energy and scattering angle) where ν and $\alpha\beta\gamma\delta$ are the magnetic quantum numbers of a, b, c, and d, respectively, in some convenient frame of reference. The goal is to determine all R_ν up to one undeterminable phase. A set of measurements which permits one to do this is called complete.

The measurable quantities are of the general form

$$w_\rho = \sum_{\mu\nu} L_\rho^{\mu\nu} R_\mu R_\nu^* \tag{1}$$

where $L_\rho^{\mu\nu}$ are coefficients which depend on the formalism used. In the statistical tensor formalism they are products of Clebsch-Gordan coefficients, one for each particle. For special situations (1) can be reduced to a particularly simple form: using for particle a polarized beams which include all possible tensor ranks one can determine all quantities $H_{\alpha\alpha'} = H_{\alpha'\alpha}$ which depend on the amplitudes according to

$$H_{\alpha\alpha'} = \sum_{\beta\gamma\delta} R_{\alpha\beta\gamma\delta} R_{\alpha'\beta\gamma\delta}^* . \tag{2}$$

To generalize this one divides the set $\{a, b, c, d\}$ into two subsets S_1 and S_2 and denotes by μ and ν all combinations of magnetic quantum numbers of the particles of the set S_1 and S_2, respectively [in (2) $S_1 = \{a\}$ and $S_2 = \{a, b, c\}$]. Measuring all the possible polarization correlations between the particles of S_1 while no polarizations of the particles of S_2 are measured allows one to determine all quantities

$$H_{\nu\nu'} = \sum_{\mu} R_{\nu\mu} R^*_{\nu'\mu} \tag{3}$$

with $R_{\nu\mu} \equiv R_{\alpha\beta\gamma\delta}$ (note the possible reordering of the indices!) or in matrix form

$$H = RR^{\dagger}. \tag{3'}$$

Conversely, knowledge of

$$G = R^{\dagger} R$$

is the maximal information obtainable from measurements where the particles in S_1 are unpolarized. It is the simplicity of these equations which allows one to draw the strongest conclusions.

Necessity condition. If R is a simultaneous solution of eqs. (3) and (4), then $R' = UR$ with unitary U obeying $[U, G] = 0$ is also a solution. Analyzing somewhat closer the rank of the matrices involved one finds that R is determined by H and G up to an overall phase if, and only if, rank $R = 1$. Since in general rank $R > 1$, this implies: If the particles with nonzero spin in the reaction can be divided into two sets S_1 and S_2 such that no correlation is known between the polarizations of a particle of S_1 and one from S_2, then the measurements are not complete. One thus has to perform measurements such that the polarizations of all particles are somehow "connected." A continuous family of solutions exists if this condition is not met.

The Jacobian. A mathematical standard method to find out whether a system of equations completely determines a set of parameters is to study the Jacobian of the system. Apart from giving some principal insight, this method is specially suited for computer treatment and thus provides one with a fairly simple tool to choose, dependent on the experimental facilities at hand, the measurements providing most information on the amplitudes. In addition, this method can also be applied if a phase analysis is performed to find out what angle dependent measurements are necessary to determine all phases up to a given angular momentum.

Because of the simultaneous appearance of R and R^* in eq. (1) one has to use either the real and imaginary parts of R_ν or, equivalently, R_ν and R^*_ν as independent variables which then can be taken as real. The second possibility is less close to physics but mathematically more convenient. Since one overall phase of the N independent amplitudes R is undeterminable, the rank of the Jacobian is (2N-1) for a complete set of measurements and less than that otherwise. In the analysis the real and imaginary parts of w or, equivalently, w_ρ and w_ρ^* must be considered independently (unless the formalism is such that they are real from the outset), i.e., the corresponding equations have to be considered separately in constructing the Jacobian.

Two important features, relevant also for the explicit results given below, are:

(1) The Jacobian is a function of the amplitudes. For its rank to be (2N-1) there has to be a nonzero determinant of order (2N-1). This determinant is a polynomial in the amplitudes and can be zero for special values also if it is not identically zero. It follows that if a set of measurements is complete for some value of the amplitudes, it is complete for almost all values, but it is possible that for special, called singular, values of the amplitudes additional measurements are necessary. This can be decided only in the actual analysis of the data.

(2) The Jacobian method permits to exclude only continuous families of solutions. Even if the rank of the Jacobian is (2N-1), one can have a finite set of different discrete solutions. These can, however, in general be resolved by relatively crude experiments.

To apply the method in practice one can use a computer, proceeding as follows: (1) List the measurements in order of preference (depending on experimental facilities or personal taste) starting with the easiest (or already completed) ones. (2) Choose nonzero random numbers for the amplitudes (for practical reasons they should be of the same order of magnitude). (3) Let the computer find the first (2N-1) linear independent rows in the thus constructed Jacobian matrix. This provides one with a complete set.

Sufficiency conditions. We conclude with some general results obtained by analyzing equations of the form (3) together with their Jacobian. The most important and very comforting result is that one can always find a complete set of measurements containing nothing more difficult than correlations between the polarizations of two particles at a time. Let ①——② signify that the polarization correlations between particle 1 and 2 are measured.

For three particles with nonzero spin j one finds that

①——②
 | is complete unless $j_1 = 1/2$ and $j_2, j_3 > 1/2$. No discrete
 | ambiguities are possible. Thus in ^{3}He(d,p) correlations
③   between d and p together with those between d and ^{3}He

completely determine the amplitudes.

If all four particles have nonzero spin j, then

is complete if $j_1 \geq j_3$ and $j_2 \geq j_4$,

while

 is complete, if not more than one particle has $j > j_1$. In both cases discrete ambiguities are still possible.

In all these statements it is irrelevant which of the four particles are ingoing or outcoming. Photons are equivalent to massive particles with $j = 1/2$.

Clearly the above cases constitute about the simplest way to overcome the earlier discussed necessity conditions.

In an elastic process ⓐ——ⓑ contains the same information as ⓒ——ⓓ because of time reversal invariance. Thus for $j_a \geq j_b$ ⓑ——ⓐ——ⓒ is already complete. Similar statements hold for scattering of identical particles, particle-antiparticle scattering because of CP invariance or proton-neutron scattering because of isospin invariance.

The material covered here is contained in two reports of the author: "Theory of Polarization Measurements I: Formalism and Determination of the Scattering Amplitudes from Measurements" and "Theory of Polarization Measurements II: Mathematical Tools, Bilinear Equations." The necessity statement is also given in M. Simonius, Phys. Rev. Lett. 19 (1967) 279.

DISCUSSION

Wigner:
    What kind of polarization measurements are necessary to determine the density matrices which you assumed to be known?

Simonius:
    For particles with spin $j = 1/2$ one needs the (vector) polarizations in three independent directions. For higher spins one needs the analysis of all possible statistical tensors up to the maximal rank $k = 2j$. The analysis can be done by rotations in the rest frame of the particle by using electromagnetic fields whenever the tensors in question are not identically zero in the beam or analyzing power in the measurement. For my results correlations between two such measurements at the same time are assumed to be known.

Wigner:
    If the reaction can lead to several products, to what degree is

the collision matrix determined? Do I understand correctly that there remains, concerning the measurements you discussed, at least a free phase for each reaction product and each pair of colliding particles which can produce these collision products?

Simonius:

That is right. In fact, there remains even an undetermined phase for each angle and energy.

Raynal:

In the case of N amplitudes and 2N–1 experiments, what is the multiplicity of the discrete solutions? For spin 1/2 or spin 0 elastic scattering, this multiplicity is 2 and the two solutions are possible. For spin 1 on spin 0, there are four solutions. It would be interesting to know which is the minimum number of experiments to be done.

Simonius:

Apart from the case mentioned of not more than three particles with nonzero spin I did not obtain any statement about the number of possible discrete ambiguities. In particular I do not know how many discrete solutions are obtained if one performs exactly the smallest number of measurements allowing the exclusion of continuous families of solutions. However, I never really thought about it, since in my opinion one should always seek a certain overdetermination of the amplitudes anyhow.

A Relationship between Polarization, Analyzing Power, and Asymmetry

K. YAZAKI, University of Tokyo, Japan; M. TANIFUJI, University of
Sydney, Australia

By the invariant-amplitude method [1], those (d,p) reactions are studied
for which the spins of the target and residual nuclei are 1/2 and 0, respectively, and the neutron is captured into an S-state. For such reactions, effects of the proton spin-orbit interaction and the deuteron
spin-orbit interaction have been studied separately [1, 2] in the ratio
P_p/A or P_p/P_d, where P_p is the polarization of the emitted proton, P_d
is the analyzing power for vector-polarized deuterons, and A is the
left-right cross-section asymmetry for a polarized target. In the following, a relationship between P_p, P_d and A will be studied when two
or more spin-orbit forces, e.g., the spin-orbit interactions of the proton
and the deuteron, act at the same time. The transition amplitude is
expanded into the partial-wave invariant amplitude,

$$\langle \nu_p, \vec{k}_p | T | \nu_T, \nu_d, \vec{k}_d \rangle = \sum_{S_1 S} (1 \nu_d \tfrac{1}{2} \nu_T | S_1 \nu_d + \nu_T)$$

$$(S_1 \nu_d + \nu_T \tfrac{1}{2} - \nu_p | Sm)(-)^{1/2 - \nu} p T_m^{S_1 S} (\vec{k}_p, \vec{k}_d) \tag{1}$$

with

$$T_m^{S_1 S} (\vec{k}_p, \vec{k}_d) = (4\pi)^2 \sum_{L_p L_d} \left[Y_{L_p} (\Omega_p) \times Y_{L_d} (\Omega_d) \right]_m^S T_{L_p L_d}^{S_1 S} (E, \cos \theta), \tag{2}$$

where ν's are the z-components of the spins, $\vec{k}$'s are the momenta,
and L's denote the orbital angular momenta. The subscripts d, T, and
p specify the deuteron, the target, and the emitted proton. θ is the
angle between $\vec{k}_d$ and $\vec{k}_p$. When the z-axis is chosen parallel to $\vec{k}_d \times \vec{k}_p$,

$$T_m^{S_1 S} (\vec{k}_p, \vec{k}_d) = (-)^m e^{im\theta} T_m^{S_1 S} (\vec{k}_p, \vec{k}_d), \tag{3}$$

if the following condition is satisfied,

$$T^{S_1S}_{L_pL_d} (E, \cos\theta) = T^{S_1S}_{L_dL_p} (E, \cos\theta). \tag{4}$$

Also, in this reference frame, m = 2, 0, and −2 are selectively allowed by Bohr's theorem [3] and $2P_d - P_p + A$ is proportional to

$$(|T^{\frac{3}{2}2}_{2} (\vec{k}_p, \vec{k}_d)|^2 - |T^{\frac{3}{2}2}_{-2} (\vec{k}_p, \vec{k}_d)|^2).$$

Thus the condition (4) provides

$$A - P_p = -2P_d. \tag{5}$$

Since eq. (4) holds for any combination of spin-orbit forces assumed between the particles and nuclei, the relationship (5) gives a good criterion for the validity of the force assumption when P_p, P_d, and A are measured. An example is the $^3He(d,p)^4He$ reaction, for which the experimental data [4] are described approximately by $A - P_p \sim 2P_d$ and the spin-orbit forces cannot explain the data. Numerical calculations show that $T^{S_1S}_{L_pL_d}$ for $L_p = L_d \pm 2$ improves the theoretical P_p, P_d, and A. This suggests the important contribution of some kind of tensor force. When the spins of the target and residual nuclei are 0 and 1/2, respectively, we get $P_R + P_p = 2P_d$ under the condition (4), where P_R is the polarization of the recoil nucleus.

REFERENCES

[1] M. Tanifuji and K. Yazaki, Prog. Theor. Phys. 40 (1968) 1023.
[2] L. J. B. Goldfarb, Second Polarization Symp.; Van Rij, Nucl. Phys. A102 (1967) 286.
[3] A. Bohr, Nucl. Phys. 10 (1959) 486.
[4] R. I. Brown and W. Haeberli, Phys. Rev. 130 (1963) 1163; S. D. Baker et al., Phys. Rev. Lett. 15 (1965) 115; G. R. Plattner and L. G. Keller, Phys. Lett. 29B (1969) 301.

Angular Correlation Theory
for Reactions with Polarized Beams

G. BERGDOLT, Laboratoire de Physique Nucléaire et Physique des Accélérateurs, Strasbourg-Cronenbourg, France

Polarized ion sources produce a polarization which has cylindrical symmetry. With respect to a general reference frame the polarization of particles of spin S is defined by the two parameters specifying the direction of the polarization axis and $2S + 1$ parameters; the statistical weights $p(m)$ of states of magnetic quantum numbers m relative to the polarization axis or the statistical tensors related to the $p(m)$ by

$$\rho_K = \sum_m \langle SS\, m - m | KO \rangle (-)^{S-m} p(m). \tag{1}$$

For a reaction $a + x \to b + y$ with a polarized beam, unpolarized target, and detection of reaction products with polarization insensitive detectors, the differential cross section can be developed in a series of invariant functions of three directions [1]. Taking the invariant functions defined in ref. [2] the expression is:

$$\frac{d\sigma}{d\omega} = (1/2\pi\hbar) \frac{\rho(E)}{\phi(E)} \frac{1}{\hat{S}_a \hat{S}_x} \sum \rho_{K_0} (i^{K_0} B_{K_1 K_0 K_2}) \hat{K}_2\, P_{K_1 K_0 K_2} (\hat{u}_1, \hat{u}_0, \hat{u}_2) \tag{2}$$

the summation is over all sets of angular momenta $K_1 K_0 K_2$ which can be coupled to 0. $\hat{u}_1$, $\hat{u}_2$ symbolize the two parameters which define the direction of initial and final relative velocity. $\hat{u}_0$ defines the direction of the polarization axis. The transformation properties of the spherical harmonics can be used to show that the $P_{K_1 K_0 K_2}$ functions are independent of the reference frame and depend only on the relative orientation of the 3 directions. $\rho(E)$ is the density of final states of energy E, $\phi(E)$ flux density in the initial state, $\rho(E)$ and $\phi(E)$ depend on the normalization of asymptotic states used in the definition of the S- and the T-matrix.

According to the Wigner-Eckart theorem applied to the scalar operator T, the matrix elements for T are:

$$\langle (L_2 S_2) J_2 m_2 | T | (L_1 S_1) J_1 m_1 \rangle = \delta_{m_2, m_1} \delta_{J_2, J_1} T(q) \tag{3}$$

where q stands for the set of quantum numbers $L_2 S_2 J L_1 S_1$ defining a transition (L is the orbital angular momentum, S the channel spin). In eq. (2) the $B_{K_1 K_0 K_2}$ are quadratic forms of the reduced transition amplitudes $T(q)$:

$$B_{K_1 K_0 K_2} = \sum_{q', q''} Z_g(c_2) Z_g(c_1) T(q') T(q'')^* \tag{4}$$

the summation is over all possible transitions.

The Z_g coefficients are functions of sets of quantum numbers c_1, c_2 belonging respectively and to initial and final state. For the problem at hand those coefficients reduce to

$$Z_g(c_2) = (-)^{K_2 + S_2 - J'} \bar{Z}(L_2' J' L_2'' J''; S_2 K_2) \tag{5}$$

where the $\bar{Z}$ coefficient is defined in ref. [2], and

$$Z_g(c_1) = (-1)^{L_1' + S_1'' + S_a + S_x + K_0} \hat{S}_a \hat{L}_1' \hat{L}_1'' \hat{K}_1 \hat{S}_1' \hat{S}_1'' \hat{K}_0 \hat{J}' \hat{J}'' \begin{pmatrix} L_1' & L_1'' & K_1 \\ 0 & 0 & 0 \end{pmatrix} \begin{Bmatrix} S_a & S_a & K_0 \\ S_1' & S_1'' & S_x \end{Bmatrix} \begin{Bmatrix} L_1' & S_1 & J' \\ L_1'' & S_1'' & J'' \\ K_1 & K_0 & K_2 \end{Bmatrix} \tag{6}$$

where $\hat{S}_a = \sqrt{2S_a + 1}$ etc., the first coefficient in (6) is the symmetrical form of Clebsch-Gordan coefficients; the other coefficients are 6 j and 9 j coefficients.

The K_1 and K_2 arise from the coupling of orbital angular momenta, conservation of parity results in $K_1 + K_2$ being even. It can be shown (1) that this condition and the properties of the recoupling coefficients result in $B_{K_1 K_0 K_2}$ being functions of the real or imaginary part of $T(q') T(q'')^*$ according to K_0 (order of polarization) even or odd. As a corollary, vector polarization has no effect if the reaction is governed by a unique reduced amplitude. In elastic scattering the long range Coulomb interaction gives non-zero transition amplitudes for high orbital angular momenta. The cross section is the sum of Coulomb cross section, nuclear cross section given by a converging expression (2) and interference term

$$\frac{d\sigma_I}{d\omega} = \frac{1}{2\pi\hbar} \frac{\rho(E)}{\phi(E)} 2 \operatorname{Re}(\operatorname{Tr}(\epsilon T_N \rho T_c^\dagger)). \tag{7}$$

The trace can be developed in a series of $P_{L_1 L_2 K_0}$ functions,

$$
\mathrm{Tr}(\epsilon T_N \rho T_c^\dagger) = \frac{(-)^{S_a+S_x}}{4\pi} \hat{S}_a^2 \hat{S}_x T_c^* \left(\sum_{L_1 L_2 K_0} i^{-(L_1+L_2+K_0)} \hat{L}_1^2 \hat{L}_2^2 \hat{K}_0 \right.
$$

$$
\rho_{K_0} P_{L_1 L_2 K_0}(\hat{u}_1, \hat{u}_2, \hat{u}_0) \sum_{S_1 S_2 J} (-)^{S_1+S_2+J} \hat{S}_1 \hat{S}_2 \hat{J}
$$

$$
\left. \begin{Bmatrix} S_1 S_a S_x \\ S_a S_1 K_0 \end{Bmatrix} \begin{Bmatrix} L_1 S_1 J \\ S_2 L_2 K_0 \end{Bmatrix} \exp\left\{ i(\omega_{L_1} + \omega_{L_2}) \right\} \langle L_2 S_2 | T_N(J) | L_1 S_1 \rangle \right)
$$

(8)

where T_c is the point Coulomb amplitude and ω_L a Coulomb phase shift.

REFERENCES

[1] G. Bergdolt, Ann. Phys. (Paris) 10 (1965) 857.
[2] L. C. Biedenharn, Nucl. Spectroscopy, part B, ed. F. Ajzenberg-Selove (Academic Press, 1960).

Formalism for $A(\vec{d},\vec{n},)B$ and $A(\vec{d},\vec{p})B$ Polarization Transfer Reactions

J. L. GAMMEL, P. W. KEATON, Jr., and G. G. OHLSEN, Los Alamos Scientific Laboratory, New Mexico, USA*

If one assumes the conservation of parity, the formal expression for the cross section and nucleon polarization for reactions of the form $A(\vec{d},\vec{n})B$ or $A(\vec{d},\vec{p})B$ may be written in terms of the Goldfarb Cartesian spin-1 operators [1] for the deuteron and in terms of the Pauli operators for the nucleon as follows:

$$I(\theta,\phi) = I_0(\theta)[1 + \frac{3}{2}\langle P_y \rangle P_{\underline{y}}^0 + \frac{2}{3}\langle P_{xz} \rangle P_{\underline{xz}}^0 + \frac{1}{6}\langle P_{xx} - P_{yy} \rangle (P_{xx}^0 - P_{yy}^0) + \frac{1}{2}\langle P_{zz} \rangle P_{zz}^0]$$

$$\langle \sigma_{x'} \rangle I(\theta,\phi) = I_0(\theta)[\frac{3}{2}\langle P_x \rangle P_x^{x'} + \frac{3}{2}\langle P_z \rangle P_{\underline{z}}^{x'} + \frac{2}{3}\langle P_{xy} \rangle P_{\underline{xy}}^{x'} + \frac{2}{3}\langle P_{yz} \rangle P_{yz}^{x'}]$$

$$\langle \sigma_{y'} \rangle I(\theta,\phi) = I_0(\theta)[P_{\underline{0}}^{y'} + \frac{3}{2}\langle P_y \rangle P_y^{y'} + \frac{2}{3}\langle P_{xz} \rangle P_{xz}^{y'} + \frac{1}{6}\langle P_{xx} - P_{yy} \rangle (P_{\underline{xx}}^{y'} - P_{\underline{yy}}^{y'})$$

$$+ \frac{1}{2}\langle P_{zz} \rangle P_{\underline{zz}}^{y'}]$$

$$\langle \sigma_{z'} \rangle I(\theta,\phi) = I_0(\theta)[\frac{3}{2}\langle P_x \rangle P_{\underline{x}}^{z'} + \frac{3}{2}\langle P_z \rangle P_z^{z'} + \frac{2}{3}\langle P_{xy} \rangle P_{xy}^{z'} + \frac{2}{3}\langle P_{yz} \rangle P_{\underline{yz}}^{z'}] \, . \qquad (1)$$

$I_0(\theta)$ is the cross section for an unpolarized beam at the scattering angle θ. The quantities in brackets represent the beam polarization referred to a right-handed (xyz) coordinate system with z along $\vec{k}_{in}$ and y along $\vec{k}_{in} \times \vec{k}_{out}$, while the remaining parameters are polarization transfer coefficients. The symbol zero is used in cases where nothing about the polarization is measured, so that P_y^0 is the ordinary vector analyzing power and $P_0^{y'}$ is the nucleon polarization produced by an unpolarized deuteron beam. From the present point of view, the analyzing tensors are special cases of polarization transfer coefficients. The nucleon polarization is referred to a right-handed (x'y'z') coordinate system where y' is along $\vec{k}_{in} \times \vec{k}_{out}$. The z' axis may be chosen arbitrarily, but is usually most conveniently taken to be the outgoing laboratory nucleon direction. (If z' is the outgoing c.m. nucleon direction, the coefficients are simply related to parameters which describe the inverse reaction [2].)

A beam produced by a polarized ion source may be characterized by a vector polarization p_Z and a tensor polarization p_{ZZ} in a coordinate system with z axis along a quantization axis, $\hat{S}$, together with two angles which specify the quantization axis orientation with respect to the reaction initial (xyz) coordinate system. In terms of the angle between $\hat{S}$ and $\vec{k}_{in}$ (which we call β) and the angle between $\hat{S} \times \vec{k}_{in}$ and the x axis (which we call ϕ), the beam polarization quantities in the xyz coordinate system are

$$\langle P_x \rangle = -p_Z \sin\beta \sin\phi \qquad\qquad \langle P_{xz} \rangle = -\frac{3}{2} p_{ZZ} \sin\beta \cos\beta \sin\phi$$

$$\langle P_y \rangle = p_Z \sin\beta \cos\phi \qquad\qquad \langle P_{yz} \rangle = \frac{3}{2} p_{ZZ} \sin\beta \cos\beta \cos\phi$$

$$\langle P_z \rangle = p_Z \cos\beta \qquad\qquad \frac{1}{2}\langle P_{xx} - P_{yy} \rangle = -\frac{3}{4} p_{ZZ} \sin^2\beta \cos 2\phi$$

$$\langle P_{xy} \rangle = -\frac{3}{2} p_{ZZ} \sin^2\beta \sin\phi \cos\phi \qquad \langle P_{zz} \rangle = \frac{1}{2} p_{ZZ}(3\cos^2\beta - 1).$$

$$(2)$$

If we define "up" scattering to be in the half-plane defined by $\hat{S}$ and $\vec{k}_{in}$, and left scattering to be in the half plane defined by $\hat{S} \times \vec{k}_{in}$ and $\vec{k}_{in}$, then left, right, up, and down directions correspond to the use of $\phi = 0°$, $180°$, $270°$, and $90°$, respectively, in eq. (2).

For clarity, we write a typical observable, $P_{xy}^{x'}$, in terms of the scattering matrix, M:

$$P_{xy}^{x'} = (\mathrm{Tr} M P_{xy} M^\dagger \sigma_{x'})/(\mathrm{Tr} M M^\dagger).$$

The cross section is given in terms of M by $I_0(\theta) = (\mathrm{Tr} M M^\dagger)/3(s_A + 1)$. ($s_A$ is the arbitrary spin for target A.) The subscript always refers to the initial deuteron polarization, and the superscript to the final nucleon polarization.

The beam polarization quantities and polarization transfer coefficients vary between +1 and −1 (vector-type quantities), +1 and −2 (P_{xx}, P_{yy}, or P_{zz} type quantities), or +3/2 and −3/2 (P_{xy}, P_{xz}, P_{yz} or ($P_{xx} - P_{yy}$)/2 type quantities). Each of the coefficients defined is either an even function or an odd function of θ; the odd functions are underlined in eq. (1). It is quite possible to separately measure all of the defined quantities at each scattering angle.

Finally, we rewrite eq. (1) in terms of the spherical tensor spin-1 operators for the incident deuteron:

$$I(\theta,\phi) = I_0(\theta)[1 + 2\mathrm{Im}\langle T_{11}\rangle \underline{\mathrm{Im}T_{11}^0} + \langle T_{20}\rangle T_{20}^0 + 2\mathrm{Re}\langle T_{21}\rangle \underline{\mathrm{Re}T_{21}^0}$$

$$+ 2\mathrm{Re}\langle T_{22}\rangle \mathrm{Re}T_{22}^0]$$

$$\langle \sigma_{x'}\rangle I(\theta,\phi) = I_0(\theta)[\langle T_{10}\rangle \underline{T_{10}^{x'}} + 2\mathrm{Re}\langle T_{11}\rangle \mathrm{Re}T_{11}^{x'} + 2\mathrm{Im}\langle T_{21}\rangle \mathrm{Im}T_{21}^{x'}$$

$$+ 2\mathrm{Im}\langle T_{22}\rangle \mathrm{Im}T_{22}^{x'}]$$

$$\langle \sigma_{y'}\rangle I(\theta,\phi) = I_0(\theta)[P_0^{y'} + 2\mathrm{Im}\langle T_{11}\rangle \mathrm{Im}T_{11}^{y'} + \langle T_{20}\rangle T_{20}^{y'} + 2\mathrm{Re}\langle T_{21}\rangle \mathrm{Re}T_{21}^{y'}$$

$$+ 2\mathrm{Re}\langle T_{22}\rangle \underline{\mathrm{Re}T_{22}^{y'}}]$$

$$\langle \sigma_{z'}\rangle I(\theta,\phi) = I_0(\theta)[\langle T_{10}\rangle T_{10}^{z'} + 2\mathrm{Re}\langle T_{11}\rangle \underline{\mathrm{Re}T_{11}^{z'}} + 2\mathrm{Im}\langle T_{21}\rangle \mathrm{Im}T_{21}^{z'}$$

$$+ 2\mathrm{Im}\langle T_{22}\rangle \mathrm{Im}T_{22}^{z'}]. \tag{3}$$

The functions which are odd in θ are again underlined. The beam quantities are expressed in terms of the initial beam vector polarization (t_{10}), tensor polarization (t_{20}), and the angles, β and ϕ, as follows:

$$\mathrm{Re}\langle T_{11}\rangle = \frac{1}{\sqrt{2}} \sin\beta \sin\phi\, t_{10} \qquad\qquad \mathrm{Im}\langle T_{22}\rangle = -\sqrt{3/8} \sin^2\beta \sin 2\phi\, t_{20}$$

$$\mathrm{Im}\langle T_{11}\rangle = -\frac{1}{\sqrt{2}} \sin\beta \cos\phi\, t_{10} \qquad\qquad \mathrm{Re}\langle T_{21}\rangle = \sqrt{3/2} \sin\beta \cos\beta \sin\phi\, t_{20}$$

$$\langle T_{10}\rangle = \cos\beta\, t_{10} \qquad\qquad \mathrm{Im}\langle T_{21}\rangle = -\sqrt{3/2} \sin\beta \cos\beta \cos\phi\, t_{20}$$

$$\mathrm{Re}\langle T_{22}\rangle = -\sqrt{3/8} \sin^2\beta \cos 2\phi\, t_{20} \qquad \langle T_{20}\rangle = \frac{1}{2}(3\cos^2\beta - 1)\, t_{20}. \tag{4}$$

These matters are treated at greater length in ref. [3].

REFERENCES

*Work performed under the auspices of the U.S. Atomic Energy Commission.

[1]   L. J. B. Goldfarb, Nucl. Phys. 7 (1958) 622.
[2]   G. G. Ohlsen and P. W. Keaton, Jr., Third Polarization Symp.
[3]   J. L. Gammel, P. W. Keaton, Jr., and G. G. Ohlsen, Los Alamos
       Report LA-4492-MS.

Considerations on Polarized Neutron Production by $(\vec{d},\vec{n})$ Polarization Transfer Reactions at Zero Degrees

G. G. OHLSEN, P. W. KEATON, Jr., and J. E. SIMMONS, Los Alamos
Scientific Laboratory, New Mexico, USA*

We consider various aspects of zero-degree polarization transfer for
$(\vec{d},\vec{n})$ reactions with a view toward the practical production of polarized fast neutrons. We use a right-handed coordinate system (xyz)
with the z axis along the incident deuteron direction, $\vec{k}_{in}$, and the y
axis in any direction perpendicular to $\vec{k}_{in}$. The general expressions
for the neutron intensity and polarization [1] at 0° are

$$I(0°) = I_0(0°)[1 + \frac{1}{2}\langle P_{zz}\rangle P_{zz}(0°)$$

$$\langle \sigma_x\rangle I(0°) = I_0(0°)[\frac{3}{2}\langle P_x\rangle P_x^x(0°) + \frac{2}{3}\langle P_{yz}\rangle P_{yz}^x(0°)]$$

$$\langle \sigma_y\rangle I(0°) = I_0(0°)[\frac{3}{2}\langle P_y\rangle P_y^y(0°) + \frac{2}{3}\langle P_{xz}\rangle P_{xz}^y(0°)]$$

$$\langle \sigma_z\rangle I(0°) = I_0(0°)[\frac{3}{2}\langle P_z\rangle P_z^z(0°)]. \tag{1}$$

These relations follow from ref. [1] since all terms odd in the scattering angle, θ, become zero at $\theta = 0$. We have also used the restriction
$P_{xy}^z \equiv 0$, which follows from the fact that the x and y axes may be defined arbitrarily. Other zero-degree conditions, which follow from the
same requirement, are $P_{xx}^0 = P_{yy}^0 = -1/2\, P_{zz}^0$, $P_x^x = P_y^y$, and $P_{xz}^y = -P_{yz}^x$.

In the following, we specialize to cases in which the deuteron
polarization possesses an axis of symmetry, as is obtained from a
polarized ion source, and where this axis is oriented either along the
y axis or the z axis. In these cases, eq. (1) reduces to

$$\langle \sigma_j\rangle = \frac{\frac{3}{2}\langle P_j\rangle P_j^j(0°)}{1 + \frac{1}{2}\langle P_{jj}\rangle P_{jj}^0(0°)}; \quad I(0°) = I_0(0°)[1 + \frac{1}{2}\langle P_{jj}\rangle P_{jj}^0(0°)], \tag{2}$$

where j is either y or z, and where $\langle P_j \rangle$ and $\langle P_{jj} \rangle$ are the deuteron vector and tensor polarization, respectively, with respect to the j axis.

The intensities and polarizations for a pure $m_I = 1$ beam and for a pure vector polarized beam are summarized in table 1. The "figure of merit," $P^2 I$, is always larger for the $m_I = 1$ type of polarization.

Table 1

Beam polarization	I/I_0	Neutron polarization at $0°$
$\langle P_y \rangle = \langle P_{yy} \rangle = 1$	$1 + \frac{1}{2} P_{yy}^0$	$\langle \sigma_y \rangle = \frac{3}{2} P_y^y / (1 + \frac{1}{2} P_{yy}^0)$
$\langle P_y \rangle = \frac{2}{3}, \langle P_{yy} \rangle = 0$	1	$\langle \sigma_y \rangle = P_y^y$
$\langle P_z \rangle = \langle P_{zz} \rangle = 1$	$1 + \frac{1}{2} P_{zz}^0$	$\langle \sigma_z \rangle = \frac{3}{2} P_z^z / (1 + \frac{1}{2} P_{zz}^0)$
$\langle P_z \rangle = \frac{2}{3}, \langle P_{zz} \rangle = 0$	1	$\langle \sigma_z \rangle = P_z^z$

The $T(\vec{d},\vec{n})^4$He reaction has been shown to produce highly polarized neutrons [2] at $0°$. For this case, where the spin structure is relatively simple $(1 + 1/2 \rightarrow 1/2 + 0)$, we can deduce several general properties of the transfer coefficients. First, P_y^y and P_y^0 are connected by an inequality [3],

$$(\frac{3}{2} P_y^y)^2 \le (1 - P_{yy}^0)(1 + 2 P_{yy}^0). \tag{3}$$

This shows, for example, that P_{yy}^0 is always between $-1/2$ and $+1$. Note that the allowed range of P_{yy}^0 is such that the polarized cross section may vary between the extremes $3I_0/4$ and $3I_0/2$, and that if $P_y^y(0°)$ is non-zero, an even smaller range is allowed. For the longitudinal polarization case, P_z^z and P_{zz}^0 obey the relation [3] $3P_z^z(0°)/2 = 1 + P_{zz}^0(0°)/2$, so that $\langle \sigma_z \rangle \equiv 1$ for any energy. In this case, however, the polarized cross section may vary between 0 and $3I_0/2$. In the range $0-15$ MeV [2, 4], $P_{zz}^0(0°) < 0$ so the longitudinal polarized cross section is less than I_0 and the transverse polarized cross section is greater than I_0.

In the general (d,n) case, one would expect a stripping mechanism to dominate at higher energies. Ideal stripping requires $P_y^y(0°) = 2/3$,

$P_z^z(0°) = 2/3$, and $P_{yy}^0 = P_{zz}^0 = 0$. These limits appear to be closely approached in the $D(\vec{d},\vec{n})$ He case [5]. Effects of the deuteron D state are not taken into account in the above limits; we are currently considering this problem.

The cross section for both the $D(d,\vec{n})$ and $T(d,\vec{n})$ reactions at angles where appreciable polarization is obtained with unpolarized projectiles is ~ 5 mb/sr. In contrast, the zero-degree cross section is ~ 100 mb/sr and ~ 20 mb/sr, respectively. In addition, the ability to reverse the neutron spin by controls at the ion source obviates the need for a spin precession solenoid, so that a tighter geometry is permitted for many experiments. Thus, in the transverse neutron polarization case, the present technique will yield a given neutron flux with a polarized deuteron current between 1 and 2 orders of magnitude lower than required in the unpolarized deuteron case. The lower deuteron energy required to deliver neutrons of a given energy would tend to reduce the background, as would the reduced incident flux of deuterons. Thus, the signal-to-noise ratio should also be substantially improved, except where the background comes primarily from breakup in the reaction itself. In addition, the zero-degree geometry often permits, from the background point of view, a more favorable location for the detectors. Thus, it would appear that the polarization transfer technique is a highly competitive method for obtaining polarized neutrons with presently available polarized deuteron beam intensities.

REFERENCES

*Work performed under the auspices of the U.S. Atomic Energy Commission.

[1] J. L. Gammel, P. W. Keaton, Jr., and G. G. Ohlsen, Third Polarization Symp.
[2] W. B. Broste et al., Third Polarization Symp.
[3] G. G. Ohlsen, J. L. Gammel, and P. W. Keaton, Jr., Third Polarization Symp.
[4] G. G. Ohlsen, Phys. Rev. 164 (1967) 1268.
[5] J. E. Simmons, W. B. Broste, G. P. Lawrence, and G. G. Ohlsen, Third Polarization Symp.

Formalism for A($\vec{\text{p}}$,$\vec{\text{p}}$)A and A($\vec{\text{d}}$,$\vec{\text{d}}$)A Polarization Transfer

G. G. OHLSEN and P. W. KEATON, Jr., Los Alamos Scientific
Laboratory, USA[*]

The formalism presented here is valid for the reaction $A + \vec{B} \rightarrow \vec{C} + D$, where B is any polarized spin-one (one-half) particle (photons excluded) in the entrance channel and C is any polarized spin-one (one-half) particle in the exit channel, and where A and D have arbitrary spins. Thus we treat not only elastic and inelastic deuteron and proton scattering, but also such reactions as $^{11}B(\vec{d},p)^{12}B_{g.s.}$ and $^{7}Li(\vec{p},\vec{n})^{7}Be$. However, for simplicity we will refer to these cases hereafter as "deuteron" ("proton") scattering. The structure of the expressions which appear here is the most general allowed by the assumption that parity is conserved.

The form of the observables for deuteron scattering is given in table 1 in terms of the Goldfarb tensor operators [1] for spin-one. $I_0(\theta)$ is the cross section for an unpolarized incident deuteron beam at the scattering angle θ. Initial spin-one polarization of the beam is designated $\langle P_k^i \rangle$ and $\langle P_{k\ell}^i \rangle$ and final spin-one polarization is designated $\langle P_k^f \rangle$ and $\langle P_{k\ell}^f \rangle$. The polarization transfer coefficients are written, for example, $P_{xy}^{y'z'} \equiv P_{xy}^{y'z'}(\theta)$, and are the quantities which are to be measured. We use the coordinate systems described in an accompanying paper [2]. The Goldfarb operators are overcomplete; the operator relation $P_{xx} + P_{yy} + P_{zz} = 0$ can be used to rewrite the terms involving these tensors in various ways (as done, for instance, in ref. [2]).

The form of the observables for proton scattering is given in table 2. We use the Pauli spin-1/2 operators, σ_x, σ_y, and σ_z. The notation is analogous to that which was used for spin-one in table 1. The polarization transfer coefficients are the same as those defined by Wolfenstein [3] with $\sigma_0^{y'}(\theta) = P_2$, $\sigma_y^{y'}(\theta) = D$, $\sigma_x^{x'}(\theta) = R$, $\sigma_z^{x'}(\theta) = A$, $\sigma_x^{z'}(\theta) = R'$, $\sigma_z^{z'}(\theta) = A'$, and are included here for completeness.

Table 1

$$I(\theta,\phi) = I_0[1 + \tfrac{3}{2}\langle P_y^i\rangle P_y^0 + \tfrac{2}{3}\langle P_{xz}^i\rangle P_{xz}^0 + \tfrac{1}{3}\langle P_{xx}^i\rangle P_{xx}^0 + \tfrac{1}{3}\langle P_{yy}^i\rangle P_{yy}^0 + \tfrac{1}{3}\langle P_{zz}^i\rangle P_{zz}^0]$$

$$\langle P_{x'}^f\rangle I = I_0[\tfrac{3}{2}\langle P_x^i\rangle P_x^{x'} + \tfrac{3}{2}\langle P_z^i\rangle P_z^{x'} + \tfrac{2}{3}\langle P_{xy}^i\rangle P_{xy}^{x'} + \tfrac{2}{3}\langle P_{yz}^i\rangle P_{yz}^{x'}]$$

$$\langle P_{y'}^f\rangle I = I_0[P_y^{y'} + \tfrac{3}{2}\langle P_y^i\rangle P_y^{y'} + \tfrac{2}{3}\langle P_{xz}^i\rangle P_{xz}^{y'} + \tfrac{1}{3}\langle P_{xx}^i\rangle P_{xx}^{y'} + \tfrac{1}{3}\langle P_{yy}^i\rangle P_{yy}^{y'} + \tfrac{1}{3}\langle P_{zz}^i\rangle P_{zz}^{y'}]$$

$$\langle P_{z'}^f\rangle I = I_0[\tfrac{3}{2}\langle P_x^i\rangle P_x^{z'} + \tfrac{3}{2}\langle P_z^i\rangle P_z^{z'} + \tfrac{2}{3}\langle P_{xy}^i\rangle P_{xy}^{z'} + \tfrac{2}{3}\langle P_{yz}^i\rangle P_{yz}^{z'}]$$

$$\langle P_{x'y'}^f\rangle I = I_0[\tfrac{3}{2}\langle P_x^i\rangle P_x^{x'y'} + \tfrac{3}{2}\langle P_z^i\rangle P_z^{x'y'} + \tfrac{2}{3}\langle P_{xy}^i\rangle P_{xy}^{x'y'} + \tfrac{2}{3}\langle P_{yz}^i\rangle P_{yz}^{x'y'}]$$

$$\langle P_{x'z'}^f\rangle I = I_0[P_0^{x'z'} + \tfrac{3}{2}\langle P_y^i\rangle P_y^{x'z'} + \tfrac{2}{3}\langle P_{xz}^i\rangle P_{xz}^{x'z'} + \tfrac{1}{3}\langle P_{xx}^i\rangle P_{xx}^{x'z'} + \tfrac{1}{3}\langle P_{yy}^i\rangle P_{yy}^{x'z'} + \tfrac{1}{3}\langle P_{zz}^i\rangle P_{zz}^{x'z'}]$$

$$\langle P_{y'z'}^f\rangle I = I_0[\tfrac{3}{2}\langle P_x^i\rangle P_x^{y'z'} + \tfrac{3}{2}\langle P_z^i\rangle P_z^{y'z'} + \tfrac{2}{3}\langle P_{xy}^i\rangle P_{xy}^{y'z'} + \tfrac{2}{3}\langle P_{yz}^i\rangle P_{yz}^{y'z'}]$$

$$\langle P_{x'x'}^f\rangle I = I_0[P_0^{x'x'} + \tfrac{3}{2}\langle P_y^i\rangle P_y^{x'x'} + \tfrac{2}{3}\langle P_{xz}^i\rangle P_{xz}^{x'x'} + \tfrac{1}{3}\langle P_{xx}^i\rangle P_{xx}^{x'x'} + \tfrac{1}{3}\langle P_{yy}^i\rangle P_{yy}^{x'x'} + \tfrac{1}{3}\langle P_{zz}^i\rangle P_{zz}^{x'x'}]$$

$$\langle P_{y'y'}^f\rangle I = I_0[P_0^{y'y'} + \tfrac{3}{2}\langle P_y^i\rangle P_y^{y'y'} + \tfrac{2}{3}\langle P_{xz}^i\rangle P_{xz}^{y'y'} + \tfrac{1}{3}\langle P_{xx}^i\rangle P_{xx}^{y'y'} + \tfrac{1}{3}\langle P_{yy}^i\rangle P_{yy}^{y'y'} + \tfrac{1}{3}\langle P_{zz}^i\rangle P_{zz}^{y'y'}]$$

$$\langle P_{z'z'}^f\rangle I = I_0[P_0^{z'z'} + \tfrac{3}{2}\langle P_y^i\rangle P_y^{z'z'} + \tfrac{2}{3}\langle P_{xz}^i\rangle P_{xz}^{z'z'} + \tfrac{1}{3}\langle P_{xx}^i\rangle P_{xx}^{z'z'} + \tfrac{1}{3}\langle P_{yy}^i\rangle P_{yy}^{z'z'} + \tfrac{1}{3}\langle P_{zz}^i\rangle P_{zz}^{z'z'}]$$

Table 2

$$I(\theta,\phi) = I_0(\theta)[1 + \langle \sigma_y^i \rangle \underline{\sigma_y^0(\theta)}]$$

$$\langle \sigma_{x'}^f \rangle I = I_0[\langle \sigma_x^i \rangle \sigma_x^{x'}(\theta) + \langle \sigma_z^i \rangle \underline{\sigma_z^{x'}(\theta)}]$$

$$\langle \sigma_{y'}^f \rangle I = I_0[\underline{\sigma_0^{y'}(\theta)} + \langle \sigma_y^i \rangle \sigma_y^{y'}(\theta)]$$

$$\langle \sigma_{z'}^f \rangle I = I_0[\langle \sigma_x^i \rangle \underline{\sigma_x^{z'}(\theta)} + \langle \sigma_z^i \rangle \sigma_z^{z'}(\theta)]$$

Finally, following the line of argument by Csonka and Moravcsik [4], we define the number of times x, y, and z appears in a polarization transfer coefficient as N_x, N_y, and N_z, respectively. We can then classify the coefficients in two useful ways. First, those which have $N_x + N_y$ odd are odd functions of θ. All such terms are underlined in tables 1 and 2. All other transfer coefficients are even in θ. Second, if z' is chosen along the outgoing c.m. direction, those polarization transfer coefficients which have N_x odd reverse sign for the inverse reaction if one assumes time reversal invariance. This leads to, for example,

$$\overline{P_y^0}(\theta) = P_0^{y'}(\theta), \quad \overline{P_{xz}^0}(\theta) = -P_0^{x'z'}(\theta), \text{ and } \overline{P_{xz}^{y'y'}}(\theta) = -P_{yy}^{x'z'}(\theta),$$

where the bars denote observables in the inverse reaction. For the case of proton elastic scattering we arrive at, for example,

$$\sigma_x^{z'}(\theta) = -\sigma_z^{x'}(\theta),$$

which in the p–p scattering case is the well-known relation R' = –A. It should be emphasized that although the formalism of tables 1 and 2 are valid for any choice of the z' axis, statements regarding time reversal invariance hold only for particular coordinate system choices (one of which is cited above) [5].

REFERENCES

*Work performed under the auspices of the U.S. Atomic Energy Commission.

[1] L. J. B. Goldfarb, Nucl. Phys. 7 (1958) 622.
[2] J. L. Gammel, P. W. Keaton, Jr., and Gerald G. Ohlsen, Third Polarization Symp.
[3] L. Wolfenstein, Ann. Rev. Nucl. Sci. 6 (1956) 43.
[4] P. L. Csonka and M. J. Moravcsik, Phys. Rev. 152 (1966) 1310.
[5] P. W. Keaton, Jr., Los Alamos Sci. Lab. Report LA-4373-MS (unpublished).

DISCUSSION

Breit:

Expressions for measurable quantities (the "observables") in terms of the Cartesian-coordinates spin treatment may be found in Volume I of the book on High Energy Physics, edited by E. H. S. Burhop, in a chapter written by R. D. Haracz and myself.

Ohlsen:

We have endeavored to follow the conventions long used in nucleon-nucleon scattering. Our description of polarization transfer where particles of spin-one are involved are the direct generalizations of the Wolfenstein D, R, A, R', A' parameters.

Formalism for Spin-3/2 Beams

P. W. KEATON, Jr., Los Alamos Scientific Laboratory, New Mexico, USA*

Very recently, Holm et al. [1] produced polarized ^{6}Li nuclei using an atomic beam source. Consequently, the polarization of ^{7}Li, which has a nuclear spin of 3/2, would appear to be feasible in the near future. It is the purpose of this paper to discuss the formalism necessary to treat a beam of spin-3/2 particles.

The cross section for the interaction of a particle of spin-j can be written

$$I(\theta) = I_0(\theta) \sum_{k,q} \langle T_{kq} \rangle T_{kq}^*(\theta) , \tag{1}$$

where the spherical tensors [2], T_{kq}, have rank $k \lesssim 2j$ with the restriction $|q| \lesssim k$. The expectation values $\langle T_{kq} \rangle$ refer to the initial beam polarization and the $T_{kq}(\theta)$ are analyzing tensors for the reaction. The spherical tensors are orthogonal in the sense that $\text{Tr}[T_{k'q'}^\dagger, T_{kq}] = (2j+1)\delta_{kk'}\delta_{qq'}$. Furthermore, using the operator property $T_{kq}^\dagger = (-1)^q T_{k-q}$, we see that eq. (1) is an invariant contraction of tensors. $I_0(\theta)$ in eq. (1) is the cross section for an unpolarized beam.

As a coordinate system for the interaction, we choose the z axis ($\hat{k}$) along the incident beam direction, $\vec{k}_{in}$, the y axis ($\hat{n}$) along $\vec{k}_{in} \times \vec{k}_{out}$ (where $\vec{k}_{out}$ is the outgoing particle direction), and the x axis ($\hat{p}$) along $\hat{n} \times \hat{k}$. Under parity inversion $\vec{k}_{in} \rightarrow -\vec{k}_{in}$, and $\vec{k}_{out} \rightarrow -\vec{k}_{out}$, which transforms $\hat{k} \rightarrow -\hat{k}$, $\hat{n} \rightarrow +\hat{n}$, and $\hat{p} \rightarrow -\hat{p}$. Therefore, conservation of parity requires that analyzing tensors be invariant under a 180° rotation of the above defined coordinate system about $\hat{n}$ (y axis). This operation places the restriction $T_{kq}(\theta) = (-1)^k T_{kq}^*(\theta)$. Substituting this relation into eq. (1) yields

$$I(\theta) = I_0(\theta)[1 + 2\text{Im}\langle T_{11} \rangle \text{Im} T_{11}(\theta) + \langle T_{20} \rangle T_{20}(\theta)$$

$$+ 2\text{Re}\langle T_{21} \rangle \text{Re} T_{21}(\theta) + 2\text{Re}\langle T_{22} \rangle \text{Re} T_{22}(\theta) + 2\text{Im}\langle T_{31} \rangle \text{Im} T_{31}(\theta)$$

$$+ 2\text{Im}\langle T_{32} \rangle \text{Im} T_{32}(\theta) + 2\text{Im}\langle T_{33} \rangle \text{Im} T_{33}(\theta)], \tag{2}$$

where all analyzing tensors odd in parity have been set equal to zero in eq. (2).

Due to the cylindrical symmetry of the ion source, the only non-vanishing polarization tensors produced are t_{10}, t_{20}, t_{30}, in a coordinate system with the z axis along the axis of quantization. However, one can precess the spins to a desired orientation in the reaction coordinate system. We define β as the angle between $\hat{k}$ and the spin quantization axis $\vec{J}$. We define ϕ as the angle between $\hat{p}$ and $\vec{J} \times \hat{k}$. Using the fact that spherical tensors have the same rotation properties as spherical harmonics, we can show:

$$\text{Im}\langle T_{11}\rangle = -t_{10}(\sin\beta\cos\phi)/\sqrt{2}$$

$$\langle T_{20}\rangle = t_{20}(3\cos^2\beta - 1)/2$$

$$\text{Re}\langle T_{21}\rangle = t_{20}(\sin\beta\cos\beta\sin\phi)\sqrt{3/2}$$

$$\text{Re}\langle T_{22}\rangle = -t_{20}(\sin^2\beta\cos 2\phi)\sqrt{3/8}$$

$$\text{Im}\langle T_{31}\rangle = -t_{30}(5\cos^2\beta - 1)(\sin\beta\cos\phi)\sqrt{3}/4$$

$$\text{Im}\langle T_{32}\rangle = -t_{30}(\sin^2\beta\cos\beta\sin 2\phi)\sqrt{30}/4$$

$$\text{Im}\langle T_{33}\rangle = t_{30}(\sin^3\beta\cos 3\phi)\sqrt{5}/4 .$$

The spherical tensors in terms of the components of angular momentum for spin-3/2 are

$$T_{00} = \text{unit matrix (4}\times\text{4)}, \quad T_{10} = J_z 2/\sqrt{5}, \quad T_{1\pm 1} = \mp(J_x \pm i J_y)\sqrt{2/5},$$

$$T_{20} = J_z^2 - 5/4, \quad T_{2\pm 1} = \mp[(J_z J_x + J_x J_z) \pm i(J_z J_y + J_y J_z)]/\sqrt{6},$$

$$T_{2\pm 2} = [(J_x^2 - J_y^2) \pm i(J_x J_y + J_y J_x)]/\sqrt{6}, \quad T_{30} = [J_z^3 - (41/20)J_z](2\sqrt{5})/3,$$

$$T_{3\pm 1} = \mp[(J_z J_x J_z - (7/20)J_x) \pm i(J_z J_y J_z - (7/20)J_y)]\sqrt{5}/3,$$

$$T_{3\pm 2} = [(J_x J_z J_x - J_y J_z J_y) \pm i(J_x J_z J_y + J_y J_z J_x)]\sqrt{2/3},$$

$$T_{3\pm 3} = \mp[(J_x^3 - 3J_y J_x J_y - J_x) \mp i(J_y^3 - 3J_x J_y J_x - J_y)]/3.$$

Finally, one might define "Cartesian tensors" for spin 3/2 as $S_k = (2/3) J_k$, $S_{k\ell} = (1/2)(J_k J_\ell + J_\ell J_k) - (5/4)\delta_{k\ell}$, and $S_{k\ell m} = (5/9)[(J_k J_\ell J_m + J_m J_\ell J_k) - ((7/10)\delta_{km} + (34/10)\delta_{m\ell}\delta_{k\ell})J_\ell]$, where $k, \vec{\ell}, m = 1, 2, 3$ (or x, y, z). However, these are overcomplete and $\vec{J} \cdot \vec{J} = j(j+1) = 15/4$ leads to four constraints, namely, $S_{xx} + S_{yy} + S_{zz} = 0$ and $S_{xkx} + S_{yky} + S_{zkz} = 0$ for $k = x, y, z$. Furthermore, $S_{xyz} = S_{yzx} = S_{zxy}$.

It is convenient to use the Cartesian tensors to characterize a polarized beam with an axis of symmetry. Let the fractional population of the nuclear magnetic substate m be written N_{2m}, then the Cartesian tensors become $\langle S_z \rangle = (1/3)[3(N_3 - N_{-3}) + (N_1 - N_{-1})]$, $\langle S_{zz} \rangle = [(N_3 + N_{-3}) - (N_1 + N_{-1})]$ and $\langle S_{zzz} \rangle = (1/3)[(N_3 - N_{-3}) - 3(N_1 - N_{-1})]$, with the constraint $N_3 + N_1 + N_{-1} + N_{-3} = 1$. It can easily be seen that $\langle S_z \rangle$, $\langle S_{zz} \rangle$, and $\langle S_{zzz} \rangle$ are normalized such that each can vary between the limits ± 1.

The author wishes to express appreciation to S. Darden, who brought ref. [1] to his attention, and to D. A. Goldberg and G. G. Olhsen for helpful discussions.

REFERENCES

*Work performed under the auspices of the U.S. Atomic Energy Commission.

[1]   U. Holm et al., Z. Physik 233 (1970) 415.
[2]   W. Lakin, Phys. Rev. 98 (1955) 139.

Two Nucleons

Polarization in the Scattering of
16.2-MeV Neutrons by Protons

R. GARRETT, A. CHISHOLM, J. C. DUDER, D. BROWN, and
H. N. BÜRGISSER, University of Auckland, New Zealand

We report preliminary results of neutron-proton polarization measurements at 16.2 MeV, using as a source of polarized neutrons the
$T(d,n)^4He$ reaction with approximately 6 MeV deuterons. (Further experiments, using the 14-MeV $T(d,n)^4He$ neutrons produced by 150-keV
deuterons from a polarized ion source, are being prepared.) The target was a 0.5 MeV thick self-supporting tritiated titanium disc. Neutrons scattered to the left and right by the protons in a plastic scintillator (S) were detected by two further plastic scintillators (A and B).
The background was reduced by detecting the α-particles from the
$T(d,n)^4He$ reaction and forming fast triple coincidences, αSA and αSB.
The problem caused by the large number of deuterons scattered into
the α-detector was reduced thus. The α-particles were detected in a
plastic scintillator just thick enough to stop them, but thin enough to
let the deuterons pass through and be detected in a second scintillator,
producing a veto pulse that was put in anti-coincidence with the triple
coincidences. To reduce further the random rate, the neutron detectors were shielded from the target, defining slits and beam stop by
paraffin wax and lead. Linear channels were used on the neutron detectors to define their efficiencies accurately and stably. Pulses from
the recoil protons in the neutron scatterer S were fed to a multichannel
analyzer if the requirements of the fast and slow logic had been met.
Typical raw spectra are shown in fig. 1.

We measured left-right asymmetries by rotating the entire assembly
of scatterer and neutron detectors about the neutron beam axis, which
must therefore be accurately known. We found this axis by measuring
neutron beam profiles given by the coincidences between the α-particles
and each of a pair of thin scintillators moved across the beam. In
order to reveal possible misalignments, new profiles were taken and
the apparatus repositioned before each set of data runs. The data from
the four sets taken to date show fluctuations no greater than are to be
expected from statistics. The left-right asymmetries were found by
integrating the parts of the recoil spectra between the arrows (fig. 1).

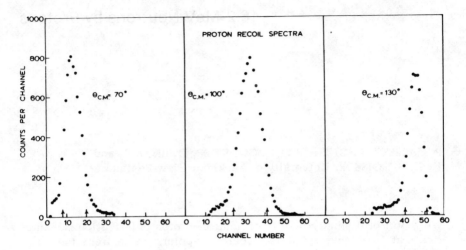

Fig. 1

Our measured n-p polarizations P_2 are shown in fig. 2. The results have been corrected for small finite geometry effects, of less than 0.4%. Our asymmetries were converted to polarizations by assuming the beam polarization P_1 from the $T(d,n)^4He$ reaction at 90° (lab) and for a deuteron energy of 5.6 MeV, to be -43% [1]. We compare our results with the Livermore phase-shift predictions [2].

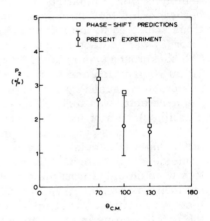

Fig. 2. Angular dependence of n-p polarization at 16.2 MeV.

REFERENCES

[1]   R. B. Perkins and J. E. Simmons, Phys. Rev. 124 (1961) 1153;
      W. Benenson, R. L. Walter, and T. H. May, Phys. Rev. Lett. 8
      (1962) 66; T. H. May, R. L. Walter, and H. H. Barschall, Nucl.
      Phys. 45 (1963) 17.
[2]   R. M. Wright, private communication.

DISCUSSION

Breit:

Am I right in assuming that your comparison was made with the "Livermore ε_1-unconstrained" fit?

Garrett:

Yes.

Breit:

It would, I think, be better to compare with the "Livermore ε_1-constrained" fit or the Yale $(Y-IV)_{npM}$ or $(Y-IV)_{np}$ or $(Y-IV)_{pp+np}$.

Polarization in
Neutron-Proton Scattering at 21.6 MeV

F. D. BROOKS and D. T. L. JONES, University of Cape Town,
South Africa

The polarizations observed [1, 2] and predicted [3] in neutron-proton
scattering at E_n < 30 MeV are of the order of a few percent and are
therefore difficult to measure accurately. In this paper we present
some polarization data obtained by a new method which may have advantages in the study of n-p scattering.

The new method utilizes the directional dependence of scintillation
pulse shapes from an anthracene crystal [4] to determine the left-right
asymmetry of proton recoils within the crystal. The pulse shape is
characterized by the output S of a pulse shape discrimination circuit
[5]. This output S and also the total light output L from a scintillation
depend on the proton direction in the crystal; L and S pass through a
minimum, a maximum, and a shallow saddle as the proton direction is
scanned through the three mutually perpendicular axes (1, 2, and 3;
see fig. 1) of the crystal. The amplitude L is a maximum for the direction which corresponds to a minimum in S and vice-versa.

Two-parameter analyses of (L,S) are shown in fig. 2 for monoenergetic neutrons incident in the 12-plane of the crystal at different
angles θ to axis 1: (a) $\theta = 0°$; (b) $\theta = 60°$; and (c) $\theta = 90°$. The variation
in S for a change of 90° in proton direction is seen by comparing S
values at the points of maximum L (corresponding to forward recoiling
protons) in fig. 2(a, c). Different L values along the proton locus correspond to unique proton energies E_p and recoil angles ϕ related by
$E_p = E_n \cos^2\phi$. At $\theta = 0$ (fig. 2a) the proton locus is relatively narrow

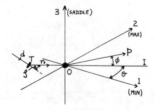

Fig. 1.

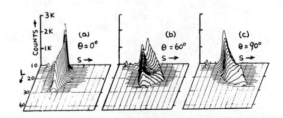

Fig. 2. Two-parameter spectra.

because S is approximately axially symmetric about axis 1. At $\theta = 60°$
recoils to one side or the other of the plane containing axis 3 and the
incident direction OI (see fig. 1) produce lower (if towards axis 1) or
higher (if towards axis 2) S outputs. The (L,S) spectrum for $\theta = 60°$
(fig. 2b) therefore shows two ridges corresponding to left and right
scattering, respectively. The distribution of counts between these
two ridges provides a measure of the left-right asymmetry in n-p
scattering.

The polarization in n-p scattering at 21.6 MeV was studied using
neutrons emitted from the T(d,n) reaction at a laboratory angle $\zeta = 20°$.
The reaction was induced by 5-MeV deuterons from the Van de Graaff
accelerator of the Southern Universities Nuclear Institute, Faure,
South Africa, and the polarization P_{inc} of the neutrons used was taken
to be 0.22 ± 0.05 from the work of Busse et al. [6]. The anthracene
crystal (volume 1.5 cm^3) was aligned with its 12-plane parallel to the
(d,n) reaction plane and pairs of (L,S) spectra were taken with $\zeta = \pm 20°$, with $\theta = \pm 60°$ and with the crystal rotated through 180° about
the incident axis OI (fig. 1). For each pair of spectra the ridge of
lower S (fig. 2b) corresponded to left scattering for one member and
to right scattering for the other member of the pair. Corresponding
members of each pair were summed to form two spectra α and β, and
these two spectra were then compared section by section so as to
determine the left-right asymmetry as a function of L, i.e., as a function of E_p or ϕ. The integral of counts in the sum ($\alpha + \beta$) was first
determined as a function of S for each section, and from this the value
S' of S for which the integral reached half of its total value was determined. Then the integrals of the spectra α and β were determined
first over the range $S < S'$ (giving
$\alpha_<$ and $\beta_<$) and second over the
range $S > S'$ (giving $\alpha_>$ and $\beta_>$).
The left-right asymmetry was
then given by $\epsilon(\phi) = (\alpha_< - \alpha_> + \beta_> - \beta_<)/(\alpha + \beta)$ and the polarization $P_{np}(\phi)$ was calculated
from $\epsilon(\phi) = P_{inc} P_{np}(\phi) \langle \cos \xi \rangle$
where the average value $\langle \cos \xi \rangle = 2/\pi$ was assumed.

The P_{np} values are plotted
against neutron scattering angle
in fig. 3. The polarizations predicted by the phase shift analyses
YLANO and YLAN3M' of Hull et al.
[3] for $E_n = 16$ MeV and 24 MeV
are shown for comparison. The
present data for $E_n = 21.6$ MeV
show a preference for the YLANO
fit, but apart from one datum are
not inconsistent with the YLAN3M'

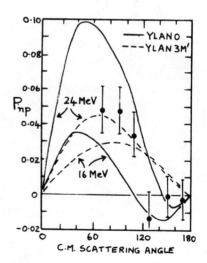

Fig. 3. P_{np} data.

fit. Previous measurements [1, 2] at E_n = 16 MeV and 24 MeV have favored [3] the YLAN3M' fit.

The present technique could clearly be improved by adding a solenoid to rotate the spin of the incident neutrons, and we hope to do this soon. In other respects the technique should compare favorably with the conventional coincidence geometry used to determine P_{np}. Since all scatterings are detected, a range of scattering angles can be studied simultaneously and electronic requirements are simple.

REFERENCES

[1] W. Benenson, R. L. Walter, and T. H. May, Phys. Rev. Lett. 8 (1962) 66.
[2] R. B. Perkins and J. E. Simmons, Phys. Rev. 130 (1963) 272.
[3] M. H. Hull, F. A. McDonald, H. M. Ruppel, and G. Breit, Phys. Rev. Lett. 8 (1962) 68.
[4] K. Tsukada and S. Kikuchi, Nucl. Instr. 17 (1962) 286.
[5] F. D. Brooks, R. W. Pringle, and B. L. Funt, I.R.E. Trans. on Nucl. Sci. NS 7 (1960) 35.
[6] W. Busse et al., Nucl. Phys. A100 (1967) 490.

Polarization in Neutron-Proton Scattering Using 32-MeV Polarized Neutrons from the D-T Reaction

N. RYU, J. SANADA,* H. HASAI, D. C. WORTH, † M. NISHI,
H. HASEGAWA,‡ H. UENO,** M. SEKI, K. IWATANI, Y. NOJIRI,*
and K. KONDO,* University of Hiroshima, Japan

The polarization in n-p scattering was measured by observing the scattering asymmetry when highly polarized 32-MeV neutrons from the D-T reaction were scattered by protons in a scintillator (NE213 or plastic). To determine the neutron beam polarization, n-He scattering in a He gas scintillator (140 atm He; 1 atm Xe) was measured at the same time as the n-p scattering. In both cases the scattered neutron spectra were measured by taking the energy spectra of recoil particles which were coincident with pulses in the lateral detector (similar to the experiments of Benenson et al. [1] and Perkins and Simmons [2]), and also in coincidence with the cyclotron beam pulses.

Energy spectra of He recoils obtained by the above mentioned triple coincidence method are shown in fig. 1. Data points are indicated by circles with error bars. This figure shows clearly the azimuthal asymmetry in the n-He scattering at an angle of 130°, depending on the direction of the neutron polarization. The asymmetry is 0.51 ± 0.04. The polarization is 0.64 ± 0.05 when the polarization in the n-He scattering is assumed to be 80% from the contour maps [3].

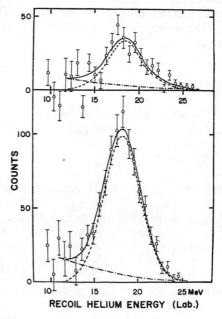

Fig. 1

The scattering asymmetries in n-p scattering were measured at angles of 30°, 40°, and 50° (lab). Since the proton energy spectra con-

tain many background counts of low energy owing to multiple scattering, edge effects and n-C reactions, Monte Carlo calculations under
the same conditions as our ex-
periments were carried out. Our
calculations did not clarify the
background for 30° scattering
in which the recoil proton energy was 8 MeV, though it gave
satisfactory results at 50°. One
of the reasons for this discrepancy may be the neglect of γ-
ray effects in the n-C reactions.
The results are shown in fig. 2,
together with the results of
Langsford et al. [4]. Theoretical curves shown were calculated by Watari [5] using the
phase shift parameters from the
one-boson exchange model [6],
the Hamada-Johnston potential [7] and the corrected Gammel-Christian-
Thaler potential [8]. Since the results have large statistical errors,
unfortunately no clear choice between these theoretical curves can be
made.

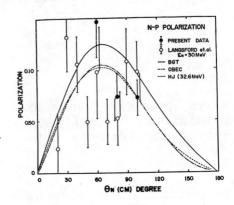

Fig. 2

REFERENCES

   *Toyko University of Education.
   †International Christian University, Mitaka, Tokyo.
   ‡Kyoto University.
   **Tohoku University, Sendai.

[1] W. Benenson, R. L. Walter, and T. H. May, Phys. Rev. Lett. 8
     (1962) 66.
[2] R. B. Perkins and J. E. Simmons, Phys. Rev. 130 (1963) 272.
[3] C. C. Giamati, V. A. Madsen, and R. M. Thaler, Phys. Rev. Lett.
     11 (1963) 163; W. G. Weitkamp and W. Haeberli, Second Polariza-
     tion Symp., p. 484; B. Hoop, Jr., and H. H. Barschall, Nucl. Phys.
     83 (1966) 65.
[4] A. Langsford, P. H. Bowen, G. C. Cox, and G. B. Huxtable, Nucl.
     Phys. 74 (1965) 241.
[5] W. Watari, private communication.
[6] S. Ogawa et al., Prog. Theoret. Phys. Suppl. 39 (1967) 141.
[7] T. Hamada and I. D. Johnston, Nucl. Phys. 34 (1962) 382.
[8] K. A. Brueckner and J. L. Gammel, Phys. Rev. 109 (1958) 1023.

Polarization in Proton-Proton
Scattering between 30 and 50 MeV

J. ARVIEUX, R. DARVES-BLANC, J. L. DURAND, A. FIORE,
N. VAN SEN, and C. PERRIN, Institut des Sciences Nucléaires,
Grenoble, France

We have measured the angular distribution of the polarization in p-p
scattering at 30 MeV, for eight angles, covering the region θ_{cm} = 45° -
90°. Some additional data have been taken at E_p = 42 MeV and 48.8
MeV. The polarized protons produced in a source built by the Saclay
group, are accelerated in the variable energy Grenoble cyclotron onto
a gaseous H_2 target. Four Li-drifted Si detectors are used in a symmetrical arrangement: two detectors are at angles θ_1 and θ_2 on the
left side of the beam and two others are at angles $-\theta_1$ and $-\theta_2$ on the
right. The polarization of the incident protons is continuously monitored using ^{12}C elastic scattering [4, 5]. It varies from 0.55 to 0.68
depending on the energy. The beam intensity is a few nA. The sign
of the polarization is changed every 0.2 s by a system of RF transitions.

Our results are given in table 1. The indicated errors are statistical only, but they take into account statistical uncertainties in the
measurement of the incident polari-
zation. The overall error due to
uncertainties in the analyzing power
of the ^{12}C elastic scattering is
± 5%. The 30-MeV data are shown
in fig. 1. Black points are our
results, the open circle is the
Harwell measurement [3].

Fits are phase-shift analysis
predictions; the full line is that
given by the Livermore group [1]
(X-EDA)$_{pp+np}$, and the dotted line
is the prediction given by the
Yale-Buffalo group [2] (Y-IV)$_{pp+np}$.
Our results seem to be in better
agreement with the former calculations.

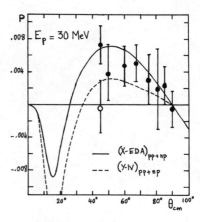

Fig. 1

435

Table 1

Energy in MeV	C.M. angle in degrees	P	ΔP
30	45	0.0073	0.0023
	50	0.0038	0.0034
	60	0.0048	0.0025
	67.5	0.0030	0.0032
	75	0.0038	0.0048
	81	0.0020	0.0048
	85	0.0024	0.0024
	90	−0.0005	0.0022
42	50	+0.0123	0.0065
48.8	54.4	0.0326	0.00026
	80.6	+0.0099	0.0013

REFERENCES

[1]   M. H. MacGregor, R. A. Arndt, and R. M. Wright, Phys. Rev. 182 (1969) 1714.
[2]   R. E. Seamon et al., Phys. Rev. 165 (1968) 1579.
[3]   C. J. Batty, R. S. Gilmore, and G. H. Stafford, Nucl. Phys. 45 (1963) 481.
[4]   R. M. Craig et al., Nucl. Phys. 83 (1966) 493.
[5]   L. N. Blumberg et al., Phys. Rev. 147 (1966) 812.

The 90° Polarization of Photoneutrons from Deuterium at Photon Energies between 5 and 40 MeV

R. NATH, G. W. COLE, Jr., F. W. K. FIRK, and C.-P. WU, Yale
University, New Haven, Connecticut; B. L. BERMAN, Yale University
and Lawrence Radiation Laboratory, Livermore, California, USA

Studies of the photo-disintegration of deuterium involve three aspects: (1) the interaction between the radiation field and the two-nucleon system, (2) the ground-state wave function, and (3) the final-state interaction. In spite of many efforts, a number of basic questions related to all three remain unanswered. One of the least studied topics has been the polarization of photonucleons from deuterium. Although the cross section above about 6 MeV is dominated by El absorption, it is necessary to determine the contribution due to other multipoles. The results of total cross section and angular distribution measurements are not sufficient to determine such admixtures unambiguously. It has long been recognized [1, 2, 3] that any polarization of photo-neutrons at 90° is due to interference between El and opposite parity multipoles. We have therefore undertaken a study of the 90° differential polarization of photoneutrons from the D(γ,n)p reaction in the photon energy range 5 to 40 MeV.

Samples of CD_2 (2 cm thick $\times$ 15 cm diam) were irradiated in the pulsed bremsstrahlung beam ($E_{\gamma max}$ = 54 MeV) obtained from the Yale Electron Accelerator. The polarization was measured using the nano-second time-of-flight spectrometer and the liquid helium polarimeter described in another contribution [6]. In order to determine the contribution from carbon in CD_2, background measurements were made with CH_2 and C targets containing equal number of carbon atoms.

The earlier measurements made at MIT [3], although sparse, are consistent with our results within experimental errors. Our preliminary data (fig. 1) show a more pronounced energy dependence of the 90° polarization than the theoretical predictions of both Partovi [4] and of Feshbach and Loman [5]. Also, in the photon energy range from 5 to 40 MeV, our results indicate a larger polarization than the theoretical values given in refs. [4] and [5].

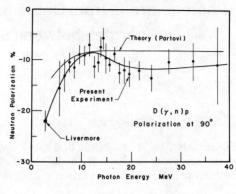

Fig. 1

REFERENCES

[1] R. W. Jewell, W. John, J. E. Sherwood, and D. H. White, Phys.
     Rev. 139 (1965) B71.
[2] J. J. de Swart, W. Czyż, and J. Sawicki, Phys. Rev. Lett. 2 (1959)
     51.
[3] W. Bertozzi et al., Phys. Rev. Lett. 10 (1963) 106.
[4] F. Partovi, Ann. of Phys. 27 (1964) 79.
[5] H. Feshbach and E. Lomon, Advances in High Energy Physics (to
     be published).
[6] G. W. Cole and F. W. K. Firk, Third Polarization Symp.

Proposed Polarized-Target-Beam Test
for $^3S_1$–$^3S_1$ Radiative Transitions
in Thermal n-p Capture

G. BREIT and M. L. RUSTGI, State University of New York at
Buffalo, USA*

The development of techniques for producing polarized proton targets
and polarized neutron beams offers the possibility of setting an upper
limit on the probability of transitions from the $^3S_1$ states in the continuum of the low-energy n + p system to the $^3S_1$ part of the deuteron.
The absence of such transitions has been assumed in traditional treatments of the thermal-energy neutron capture, being an apparent consequence of the orthogonality of two eigenfunctions of a Schroedinger
equation corresponding to two different energies, discounting the
slight effect of the admixture of $^3D_1$ to $^3S_1$ states. In an obvious notation the left side of fig. 1 shows the six Zeeman transitions for M - 1
radiation, the usual spectroscopic designations π and σ referring to
the magnetic rather than the electric vector. The right part similarly
shows the pattern for the spin-flip transition. If the target protons
(p) and beam neutrons (n) are completely polarized with spin directions
parallel and if the quantization axis is taken parallel to that direction,
the only initial state is $^3m_i = 1$ so that the only transitions present
from either $(^3S_1)_i$ or $(^1S_0)_i$ are a and b. If γ emission should be observed at all for this idealized situation its presence would be a proof
of the existence of $^3S_1$–$^3S_1$ transitions. The combined radiation of
Zeeman components a and b has angular-distribution and polarization
properties which make the presence of $^3m_i = 1$ evident even if the beam

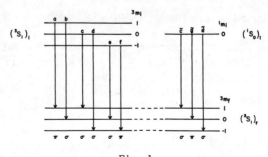

Fig. 1

439

and target polarizations are not complete and the polarizations not mutually parallel [1]. For parallel polarizations the γ-ray intensity distribution is

$$I(\theta) = \mathcal{K}\,[2 + P^n\,P^p \sin^2\theta + |a|^2\,(1 - P^n\,P^p)]$$

where θ is the angle between the direction of observation and the initial spin directions, P^n and P^p are the n and p polarizations and $|a|^2 = {}^1\pi/{}^3\pi$, with ${}^1\pi$ and ${}^3\pi$ representing the individual π-component transition probabilities from ${}^3S_1$ and ${}^1S_0$ sublevels. The constant $\mathcal{K}$ is independent of the other quantities in the formula. If it is *speculatively* supposed that the discrepancy between the d(γ,n)p cross section at threshold and the p(n,γ)d low-E recombination cross section [2] is caused by an inadequacy of traditional theory and that the whole effect is describable by ${}^3S_1$-${}^3S_1$ transitions the ratio $R \equiv I_{max}/I_{min}$ is as in table 1.

Table 1. Values of R corresponding to different $P = (P^n\,P^p)^{1/2}$
for $|a|^2 = 22.2$

P = 0.906	0.954	0.977	1	0.868	0.57735
R = 1.14	1.23	1.32	3/2	1.10	1.01984
$1/(1-P^2)$ = 5.5	11.1	22.2	∞	–	3/2

The value of $|a|^2$ used was adjusted to give 1.09 for the interaction effect, i.e., slightly lower than in ref. [2]. The fifth column has been adjusted to R = 1.10. There are also appreciable effects on the γ-ray polarization. It is clearly impossible to come close to R = 3/2, but with good relative intensity measurements the detection of the effect does not appear hopeless. Estimates indicate only slight effects of the ${}^3D_1$ admixture and there appears to be no obvious conflict with d(γ,n)p data.

REFERENCES

*Work supported by the USAEC (NYO-4022-12).

[1] Brookhaven National Laboratory Colloquium 1963, talk by G. Breit.
[2] N. Austern, Phys. Rev. 92 (1953) 670; N. Austern and E. Rost, Phys. Rev. 117 (1960) 1506.

DISCUSSION

Postma:

    This is a very interesting proposal. The experiment seems possible with current techniques although it will be difficult. Another possible way to determine the triplet part of the n-p capture cross section is to compare experimentally the capture and scattering cross sections using neutrons polarized alternately parallel and antiparallel with respect to the protons. This is sensitive to the triplet part because of the relatively rapid change of a spin-dependent factor. Maybe both experiments should be carried out in order to study the proposed triplet capture.

Three Nucleons

Measurement of the Analyzing Power of Deuterium for 7.89-MeV Neutrons and Phase-Shift Analysis for the n-d Data between 1 and 10 MeV

R. BRÜNING and B. ZEITNITZ, II. Institut für Experimentalphysik, Hamburg, Germany; J. ARVIEUX, Institut des Sciences Nucléaires, Grenoble, France

The elastic scattering of nucleons by deuterons has been the subject of much theoretical [1, 2] and experimental [3, 4] work during the last years. But polarization measurements for n–d scattering are sparse.

Therefore we performed a measurement of the angular distribution of the analyzing power of deuterium for polarized neutrons at a mean incident energy of 7.89 MeV with an energy spread of 220 keV. A target of C_6D_6 (45 cm from the source) was bombarded with polarized neutrons from the $^9Be(\alpha,n_0)^{12}C$ reaction [5]. The α–particles were produced with the 3–MV Van de Graaff generator of the University of Hamburg. A C_6D_6 target served simultaneously as a fast scintillator (4 cm diam., 5 cm high). Pulses from this scintillator gave the start signal for a time-of-flight spectrometer. The stop signals came from a NE 213 scintillator (10 cm diam., 7.5 cm long) positioned at a distance of 105 cm. For both detectors n-γ discrimination was used. The two-dimensional coincidence spectra of the flight times and the deuteron recoil energies were accumulated on a PDP9 computer. The results are shown in fig. 1 together with measurements at 7.80 MeV recently published by Taylor et al. [6].

The solid lines in fig. 1 are results of a phase–shift analysis for nucleon-deuteron scattering performed in the energy range between 1 and 10 MeV using the data given in the compilations of refs. [3, 4]. We used the formalism described in ref. [7]. (Ref. [7] contains a mistake in the sign of the even–order observables so that the results are not valid.) We have checked our modified program at 3 MeV by comparison with the corresponding results of ref. [8]. As starting parameters for the phase-shift analysis we used the unsplit (interpolated) phase shifts (up to $\ell = 6$ at 10 MeV) calculated by Aaron et al. [9]. In the first step the differential elastic and the total inelastic cross sections were fitted by varying the unsplit phase shifts for $\ell \leq 3$. Then the proton and deuteron vector polarization data (at 8 MeV also all neutron polarization data) were included in the fitting procedure in order to obtain the splitting of the quartet-P and D phases. The split quartet-P phases are shown in fig. 2. The solid lines are drawn by hand. The splitting of the quartet-D phases in this energy range is small and shows no unique trend.

We wish to thank Dr. R. G. Seyler for valuable discussions.

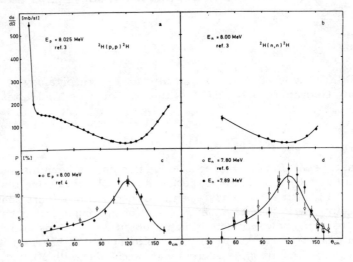

Fig. 1

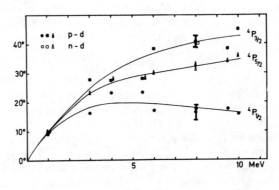

Fig. 2

REFERENCES

[1] R. D. Amado, Ann. Rev. Nucl. Sci. 19 (1969) 61.
[2] L. M. Delves and A. C. Phillips, Rev. Mod. Phys. 41 (1969) 497.
[3] J. D. Seagrave, Proc. Int. Conf. on Three-Body Problem (North-Holland Publishing Co., Amsterdam, 1970) p. 41.

[4] W. Haeberli, Proc. Int. Conf. on Three-Body Problem (North-Holland Publishing Co., Amsterdam, 1970) p. 188.

[5] G. P. Lietz, S. F. Trevino, A. F. Behof, and S. E. Darden, Nucl. Phys. 67 (1965) 193.

[6] J. Taylor et al., Phys. Rev. C1 (1970) 803.

[7] J. Arvieux, Nucl. Phys. A102 (1967) 513.

[8] W. Trächslin, L. Brown, T. B. Clegg, and R. G. Seyler, Phys. Lett. 25B (1967) 585.

[9] R. Aaron, R. D. Amado, and Y. Y. Yam, Phys. Rev. 140 (1965) B1291.

Angular Dependence of Neutron Polarization
in n-d Elastic Scattering at 2.6 MeV

S. JACCARD, J. PIFFARETTI, R. VIENNET, and J. WEBER, Institut de
Physique de l'Université, Neuchâtel, Switzerland[*]

The polarimeter which was used is similar to the one described earlier
[1]. The incident polarized neutron beam was emitted at 20° lab from
the $^{12}C(d,n)^{13}N$ reaction [2]. The amount of multiple scattering was
evaluated by means of a Monte-Carlo code. The P_1 value was measured with a liquid helium analyzer [3] and was found to be, after multiple scattering correction, $P_1 = -0.365 \pm 0.022$, in good agreement
with earlier measurements [2]. The results are summarized in table 1
and in fig. 1, taking for P_1 the mean value $P_1 = -0.40 \pm 0.02$.

Table 1. Corrected asymmetries

	θ_{lab}			
	45°	46°	60°	120°
$\epsilon = P_1 P_2(\theta)$ asymmetry[†]	−0.0039 ±0.0049	−0.0132 ±0.0041	−0.0164 ±0.0037	−0.0048 ±0.0075

[†]The errors contain instrumental and statistical errors.

One can compare these values with the recent results obtained by one
of us (R.V.) in the course of a phenomenological analysis of n-d scattering based on the effective range approximation (ERA) [4,5].

The essential assumptions made so far (but to be relaxed before
completion of this analysis) have been the following: channel spin
and orbital angular momentum are separately conserved, and S, P, and
D states only are taken into account.

The values of polarization given in this paper are not included in
the 118 pieces of data fitted by the ERA analysis. Fig. 1 shows a
comparison between the measured and predicted values calculated
from the best ERA parameters that we have found so far. These ERA
parameters are listed in table 2.

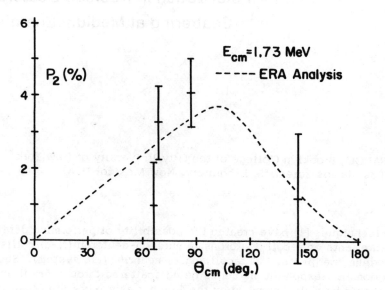

Fig. 1

Table 2

$^{2S+1}\lambda_{2J}$	$^2S_1$	$^4S_3$	$^2P_1$	$^2P_3$	$^4P_1$	$^4P_3$	$^4P_5$	$^2D_3$	$^2D_5$	$^4D_1$	$^4D_3$	$^4D_5$	$^4D_7$
a^{LSJ}	-3.65 E+00	-1.59 E-01	-8.71 E-02	7.17 E-02	3.55 E-03	2.94 E-03	3.85 E-04	-8.43 E-03	-6.78 E-04	2.18 E-03	1.81 E-03	2.79 E-03	2.60 E-03
R^{LSJ}	3.85 E+01	2.30 E+00	-2.36 E+01	-1.58 E-01	9.91 E-01	7.87 E-01	7.96 E-01	1.80 E-01	1.85 E-01	-4.74 E-01	-6.22 E-01	-8.67 E-01	-6.62 E-01

Era: $k^{2L+1} \cotg \delta^{LSJ} = a^{LSJ} + \frac{k^2}{2} R^{LSJ}$ where: k^2 = square of momentum;

δ^{LSJ} = phase shift of the Lth partial wave; $(-a^{LSJ})^{-1}$ = scattering length and

R^{LSJ} = effective range for the Lth partial wave. $\left[a^{LSJ}\right] = fm^{-2L-1}$; $\left[R^{LSJ}\right] = fm^{-2L+1}$.

REFERENCES

*Work supported by the Swiss National Foundation for Scientific Reserach (FNSRS).

[1] J. Piffaretti, J. Weber, and J. Rossel, Helv. Phys. Acta 40 (1967) 805.

[2] J. R. Sawers, F. O. Purser, and R. L. Walter, Phys. Rev. 141 (1966) 825.

[3] J. Piffaretti, J. Rossel, and J. Weber, Second Polarization Symp., p. 152.

[4] R. Viennet and P. L. Huguenin, Helv. Phys. Acta 42 (1969) 562.

[5] R. Viennet, J. Piffaretti, and J. Weber, Helv. Phys. Acta 43 (1970).

Polarization in Nucleon-Deuteron
Scattering at Medium Energies

V. FRANCO, Brooklyn College of the City University of New York[*]
and Los Alamos Scientific Laboratory, New Mexico, USA[†]

New techniques [1] have created the possibility of performing detailed measurements on recoil deuterons in nucleon-deuteron (N-d) collisions at medium energies and at recoil angles near 90° (lab system). Such collisions correspond to nucleons being scattered through small angles and should be well described by the Glauber approximation [2] in which single and double collisions are treated. To describe spin-dependent properties such as polarizations, it is necessary to use the spin-dependent N-N elastic scattering amplitudes in the calculations. Analyses above 800 MeV suffer from a paucity of n-p data. The situation below 800 MeV is more favorable since significant n-p data exist there. Consequently we have calculated cross sections and polarizations for p-d collisions at and below 800 MeV.

Our calculations include single and double scattering effects. To facilitate obtaining a first estimate of the cross sections and polarizations, we assume the deuteron to be a pure S-state. In the double scattering amplitude (only) we neglect those terms which vanish in the forward direction and Coulomb effects and we approximate the N-N amplitudes by Gaussian functions of momentum transfer q (i.e., by $a_i e^{-b_i q^2}$, where a_i and b_i are complex), or by q or q^2 multiplying such functions. Calculations in which these approximations are not made are in progress. In the present analysis we use the Dubna phase shifts [3] for the N-N amplitudes, and the Moravcsik III wave function [4] for the deuteron.

In fig. 1 we show the angular distribution of protons for an unpolarized incident beam. For protons polarized 100% normal to the scattering plane, the intensities differ from that shown by ±33% near 10° and by virtually 0% below 1°. Near 5° $d\sigma/d\Omega$ exhibits a small Coulomb-nuclear destructive interference. The effect of double scattering on $d\sigma/d\Omega$ is negligible below 1°, but increases rapidly with increasing angle above 1°. Near 10° its neglect leads to a cross section that is more than 20% too large.

In fig. 2 we present the vector polarization $\langle T_{11}/i \rangle$ of the deuteron. We show results for unpolarized protons and for protons polarized

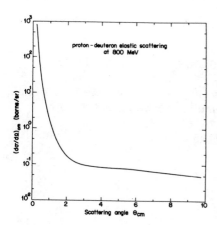

Fig. 1. Angular distribution for scattering of 800 MeV unpolarized protons by deuterium.

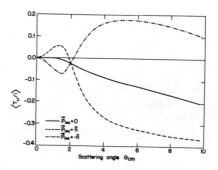

Fig. 2. Angular variation of the vector polarization of the deuteron in 800 MeV p-d scattering. The angle θ_{cm} is the scattering angle of the proton. The solid curve corresponds to unpolarized incident protons. The dashed and dot-dashed curves correspond to incident protons fully polarized normal to the scattering plane.

100% normal to the scattering plane. The direction $\hat{n}$ denotes $\hat{k} \times \hat{k}' / |\hat{k} \times \hat{k}'|$, where $\hat{k}$ and $\hat{k}'$ give the directions of the incident and scattered protons, respectively. For $\theta_{cm} < 2°$, $\langle T_{11}/i \rangle < 0.025$ for unpolarized protons. However, use of polarized protons could yield vector polarizations as large as 0.07 near 1.5°. Furthermore, use of polarized protons reveals structure in $\langle T_{11} \rangle$ for $\theta_{cm} < 2°$. The curves for $\vec{P}_{inc} = \pm\hat{n}$ form the envelope of the curves for arbitrary incident-nucleon polarizations along a direction parallel to $\hat{n}$. To prove this, note that for incident polarizations parallel to $\hat{n}$, $\langle T_{11}/i \rangle$ is a monotonic function of $\vec{P}_{inc} \cdot \hat{n}$ for fixed θ_{cm}. It therefore attains its physically allowable maximum and minimum values at $\vec{P}_{inc} = \pm\hat{n}$.

For $d\sigma/d\Omega$ the relative importance of double scattering increases with increasing θ_{cm}, but for $\langle T_{11}/i \rangle$ no such statement can be made for $\theta_{cm} < 10°$. For $P_{inc} = 0$ double scattering effects change $\langle T_{11}/i \rangle$ by less than 0.002 for $\theta_{cm} \lesssim 10°$, and show no tendency to increase with increasing θ_{cm}. For $\vec{P}_{inc} = \pm\hat{n}$ these effects are significant for $\theta_{cm} \gtrsim 2°$, producing changes in $\langle T_{11}/i \rangle$ that are typically $\sim \pm 0.02$, and showing some tendency to increase with angle for $\theta_{cm} \gtrsim 7°$.

REFERENCES

*Present address.

†Work supported in part by the National Science Foundation, the

City University of New York Faculty Research Program, and the U.S. Atomic Energy Commission.

[1] J. E. Brolley, G. G. Ohlsen, and G. P. Lawrence, Third Polarization Symp.
[2] V. Franco and R. J. Glauber, Phys. Rev. 142 (1966) 1195; V. Franco, Phys. Rev. Lett. 21 (1968) 1360 and references cited therein.
[3] Z. Janout et al., Nucl. Phys. A127 (1969) 449.
[4] M. J. Moravcsik, Nucl. Phys. 7 (1958) 113.

Measurement of the Tensor Moments $\langle iT_{11} \rangle$, $\langle T_{20} \rangle$, $\langle T_{21} \rangle$, $\langle T_{22} \rangle$ in the Elastic Scattering of Polarized Deuterons by Protons

R. E. WHITE, W. GRÜEBLER, R. RISLER, V. KÖNIG, P. A. SCHMELZBACH, and A. RUH, Laboratorium für Kernphysik, ETH, Zürich, Switzerland

The polarized deuteron beam from the ETH tandem accelerator has been used to measure the tensor moments $\langle iT_{11} \rangle$, $\langle T_{20} \rangle$, $\langle T_{21} \rangle$, $\langle T_{22} \rangle$ of deuterons elastically scattered by protons at laboratory energies of 6 and 10 MeV. The experimental arrangement and analysis techniques used are described fully elsewhere [1, 2, 3]. Scattered deuterons and recoil protons from a 1.6 atmos. gas target were detected in surface barrier detectors between laboratory angles of 20° and 65°. The detectors accepted particles over an angular range of ±0.5° lab.

The results are shown in fig. 1; only the statistical errors are shown. Errors from small geometrical asymmetries or incorrect settings of the beam spin axis are estimated to be negligible with the techniques used to evaluate the tensor moments. Observed variations in beam polarization produce changes of less than ±0.003 in the measured moments. The uncertainty introduced by the background correction is estimated to be less than 0.005. This affects most severely the 63.1° and 140° c.m. results, and their reliability was judged from the overall consistency of the data. Measurements at other energies are in progress.

Measurements of the moments $\langle T_{2k} \rangle$ in the 3 to 12 MeV range have been reported at restricted angles by Young et al. [4, 5] and by Extermann [6] who also gives extensive and accurate results for $\langle iT_{11} \rangle$ at 8 and 11 MeV. Preliminary measurements made here at 8 MeV for $\langle iT_{11} \rangle$ agree well with those of Extermann.

These results show that even at very low energies (2 to 3.3 MeV c.m.) spin-dependent forces need to be considered when treating three-nucleon systems. An analysis of these data in terms of possible excited states in ^{3}He is planned.

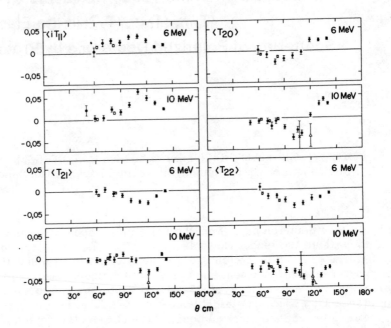

Fig. 1. Tensor moments for d-p elastic scattering. Open
circles show points at which deuterons were recorded.
△ refs. [4, 5]; x ref. [6].

REFERENCES

[1] W. Grüebler, V. König, P. A. Schmelzbach, and P. Marmier, Nucl.
     Phys. A134 (1969) 686.
[2] V. König, W. Grüebler, P. A. Schmelzbach, and P. Marmier, Nucl.
     Phys. A148 (1970) 380.
[3] W. Grüebler, V. König, P. A. Schmelzbach, and P. Marmier, Nucl.
     Phys. A148 (1970) 391.
[4] P. G. Young, M. Ivanovich, and G. G. Ohlsen, Phys. Rev. Lett. 14
     (1965) 831.
[5] P. G. Young and M. Ivanovich, Phys. Lett. 23 (1966) 361.
[6] P. Extermann, Nucl. Phys. A95 (1967) 615.

Measurements of Triple Scattering Parameters in Proton-Deuteron Elastic Scattering at 50 MeV

N. M. STEWART, Wheatstone Laboratory, King's College, London;
W. R. GIBSON, Queen Mary College, University of London, England

Measurements of differential cross sections and polarizations in p-d elastic scattering now exist at several energies up to 160 MeV. The advance of experimental techniques has enabled other observables to be measured, such as tensor polarizations and break-up cross sections, but there is still an absence of triple scattering data except near 140 MeV. This type of measurement is necessary for the determination of a unique set of phase shifts. Trächslin et al. [1] have pointed out that the mixing of S and D states strongly affects the Wolfenstein [2] parameters. Measurements of the parameters D, A, and R have been made at the Rutherford High Energy Laboratory.

The 50-MeV polarized proton beam from the linear accelerator was deflected by a bending magnet through an angle of 47.6° in the horizontal plane. The angle of deflection in the bending magnet was such that an initial transverse (longitudinal) spin was rotated into a longitudinal (transverse) spin due to the anomalous magnetic moment of the proton. The initial polarization was produced by rotating the vertical polarization provided from the ion source in a sextuple magnet or a solenoid.

The polarized protons were scattered from a liquid deuterium target and analyzed with high-pressure helium polarimeters. Scattered protons and recoil deuterons detected in coincidence eliminated unwanted scattering events. Beam position and intensity were monitored by two ion chambers, while the incident beam polarization was measured by carbon polarimeters.

Measurements were made between proton lab scattering angles of 40° and 70° for each of the three parameters. The depolarization parameter D is defined in a completely positive region whereas A is always negative, and R is found to change sign near the center of the angular range investigated.

It is interesting to compare the results reported here with those measured at higher energies at Orsay [3] near 135 MeV and Harvard [4] at 140 MeV. There is a noticeable similarity between the high-energy measurements, which were made in the range 20° to 60° c.m., and the 50 MeV measurements over the angular range of 60° to 100° c.m. This is most striking in the case of the D and R parameters, and to a lesser extent with A. It seems to indicate that the parameters exhibit a small variation with energy from 50 to 140 MeV, at any rate for D and R.

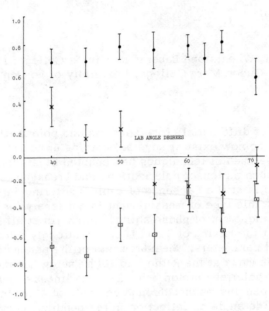

Triple scattering parameters for P–D elastic scattering at 50 MeV

● D Parameter   ✕ R Parameter   ⊡ A Parameter

REFERENCES

[1] W. Trächslin, L. Brown, T. B. Clegg, and R. G. Seyler, Phys. Lett. 25B (1967) 585.

[2] L. Wolfenstein and J. Ashkin, Phys. Rev. 85 (1952) 947; L. Wolfenstein, Ann. Rev. Nucl. Sci. 6 (1956) 43; L. Wolfenstein, Phys. Rev. 101 (1956) 427.

[3] M. Poulet, A. Michalowicz, K. Kuroda, and D. Cronenberger, Nucl. Phys. A99 (1967) 442.

[4] R. A. Hoffman, J. Lefrancois, and E. A. Thorndike, Phys. Rev. 131 (1963) 1671.

Four Nucleons

A DWBA Treatment of the
d + d Reaction at Low Energies

J. C. DUDER, H. F. GLAVISH,[*] and R. RATCLIFF, University of
Auckland, New Zealand

The nucleon polarization produced in the d + d reaction at low energies
($<$ 500 keV) has been a subject of considerable theoretical interest.
The first extensive analysis [1] formulated in terms of deuteron penetrability factors was unsatisfactory in that a strong $^3P \to {}^3F$ transition
was needed [2] to explain the observed nucleon polarization. More
recently [3, 4] attention has been focused on the transitions $^1D \to {}^3D$
and $^3P \to {}^1P$ to explain the polarization data. These transitions are of
particular interest [2] since in first order $^3P \to {}^1P$ is not allowed and
$^1D \to {}^3D$ occurs only if a two-nucleon $\vec{L} \cdot \vec{S}$ force operates. It is possible
to build second order effects into a distorted wave treatment [5, 6] of
the reaction in which case polarization effects can arise from interference between the different J values in $^3P \to {}^3P$ transitions (J = 0, 1, 2).
    The analysis reported here is based on the formulation of ref. [5].
The T-matrix element is written as

$$T_{fi} = \langle \psi_f^- | V_i | \chi_i^+ \rangle$$

where χ_i^+ describes the initial two deuteron state distorted only by the
Coulomb field and $V_i = V_{13} + V_{24} + V_{14} + V_{23}$ is the sum of the two nucleon potentials appropriate to the initial state interaction (1 and 3
refer to protons and 2 and 4 to neutrons). The above form for T_{fi} is
valid even after antisymmetrization is taken into account. In our
approximation ψ_f^- is constructed from Coulomb functions and p-^{3}H (or
n-^{3}He) elastic scattering phase shifts. The integration over the co-
ordinate $\vec{r}_p = \vec{r}_1 - 1/3(\vec{r}_2 + \vec{r}_3 + \vec{r}_4)$ of the outgoing proton is cut off for
$r_p < R$, the p-^{3}H contact radius (3 $\sim$ 4 fm). This approximation is not
as drastic as it might first appear, since there is a large contribution
to T_{fi} from the restricted range of integration due to the large width
($\sim$ 7 fm) of the deuteron. An immediate consequence of this approximation is that V_{13} and V_{14} have negligible contribution in comparison
with V_{24} and V_{23}. Our treatment differs from that of Boersma [6], in this
respect, and in the way the relative motion wave functions are constructed. Physically, we have a cut-off stripping approximation, but

with all the exchange terms included and a formulation that avoids nuclear distortion of the deuteron wave.

The results obtained for some of the T-matrix elements at $E_d = 200$ keV (lab) are shown in table 1. Well-known Gaussian potentials and internal wave functions have been used

$$\text{central:} \quad V_{ij}^C = -\frac{1}{8}(3 - \vec{\tau}_i \cdot \vec{\tau}_j - \vec{\sigma}_i \cdot \vec{\sigma}_j)V_0^C \exp(-r_{ij}^2/r_0^2)$$

$$\text{tensor:} \quad V_{ij}^T = -\frac{1}{4}(1 - \vec{\tau}_i \cdot \vec{\tau}_j) S_{ij} V_0^T (r_{ij}^2/r_0^2) \exp(-r_{ij}^2/r_0^2)$$

$$\Phi_d(12) = (2a^2/\pi)^{3/4} \exp(-a^2 r_{12}^2);$$

$$\Phi_T(234) = (12/\pi^2)^{3/4} \beta^3 \exp\{-\beta^2(r_{23}^2 + r_{24}^2 + r_{34}^2)\}$$

$$r_0 = 1.58 \text{ fm}, \ a = 0.167 \text{ fm}^{-1}, \ \beta = 0.255 \text{ fm}^{-1},$$

$$V_0^C = 59 \text{ MeV}, \ V_0^T = 107 \text{ MeV}.$$

Parameters of the reaction which have been considered are the nucleon polarization P, the ratio C_2/C_0 for the Legendre polynomial coefficients of the unpolarized distribution, and B_3/C_0, where B_3 is the coefficient of $P_1^1(\cos\theta)$ in the angular distribution produced with vector-polarized incident deuterons.

Table 1. d+d reaction matrix elements at $E_d = 200$ keV (lab)

Matrix element			Potentials which	p-^{3}H
J	Real	Imag.	will contribute*	phase shift δ_ℓ^J
$^1S \to {}^1S$ $0(a_0)$	+1.517	-2.910	*central*	-63°
$^5S \to {}^3D$ $2(\gamma_1)$	+0.605	≈ 0	*tensor*, $\vec{L}\cdot\vec{S}$	- 5°
$^3P \to {}^3P$ $0(a_{10})$	-0.399	-0.121	central	+15°
$^3P \to {}^3P$ $1(a_{11})$	+0.138	+0.091	*tensor*	+30°
$^3P \to {}^3P$ $2(a_{12})$	-0.275	-0.182	$\vec{L}\cdot\vec{S}$	+30°

*Potential in italics is the one for which the matrix element has been calculated.

The quintet-singlet transition γ_1 is particularly interesting and in the past has often been neglected. Apart from the small transition ($^1D \to {}^3D$), γ_1 is needed to obtain a non-zero B_3 in first order [4].

$$B_3 = -\frac{9\sqrt{10}}{8} \text{ Im } (\gamma_1 a_{11}^* - \gamma_1 a_{12}^*).$$

With only tensor forces included the a_{1J} are not large enough to produce the observed polarizations for reasonable phase shifts δ_1^J. Computation of the central force contribution to a_{1J} has not yet been made. However, the behavior of the integrations suggests that the central force will be about twice as effective as the tensor force and generate matrix element components of the same sign. Both V_{24} and V_{23} contribute for the central force case, whereas for the tensor force, V_{24} has no effect, since its spin matrix element vanishes identically. To gain some idea of the effectiveness of the $^3P \to {}^3P$ transitions, parameters have been computed for values of a_{1J} a factor of 3 greater than those given in table 1, to take into account the central force. The results are $P(54°) = -0.05$, $C_2/C_0 = +0.19$, $B_3 = +0.09$. The values of all of these coefficients are sensitive to the splitting of the p-wave phase shifts δ_1^J. In particular $P = B_3 = 0$ if there is no splitting. The values chosen for the phase shifts are based on the values obtained [7] for p-^{3}He elastic scattering. It is encouraging to see that the polarizations have the correct signs. While the magnitudes are low by a factor of 2 it is conceivable that including the $\vec{L} \cdot \vec{S}$ force will lead to an improvement.

REFERENCES

*Present address: Stanford University, California.

[1] F. M. Beiduk, J. R. Pruett, and E. J. Konopinski, Phys. Rev. 77 (1950) 622; Phys. Rev. 77 (1950) 628.
[2] R. J. Blin-Stoyle, Proc. Phys. Soc. (London) 65 (1952) 949.
[3] J. R. Rook and L. J. B. Goldfarb, Nucl. Phys. 27 (1961) 79.
[4] D. Fick, Z. Phys. 221 (1969) 451.
[5] H. F. Glavish, Ph.D. Thesis, 1968 (unpublished).
[6] H. J. Boersma, Nucl. Phys. A135 (1969) 609.
[7] L. W. Morrow and W. Haeberli, Nucl. Phys. A126 (1969) 225.

Polarization of Neutrons from the
D(d,n)^{3}He Reaction from 6 to 14 MeV

G. SPALEK, J. TAYLOR, R. A. HARDEKOPF, Th. STAMMBACH, and
R. L. WALTER, Duke University and TUNL, Durham, North Carolina, USA*

On the basis of charge symmetry of nuclear forces one expects that
the polarization of the neutrons from the ^{2}D(d,n)^{3}He reaction should
be nearly the same as the polarization of protons from the ^{2}D(d,p)^{3}H
reaction. The data of Dubbeldam and Walter [1] and Porter and Haeberli [2] strongly indicate that the proton polarization is about 1.4
times greater than the neutron polarization. The present experiment
was undertaken to determine whether the differences in the polarizations are real, taking advantage of advanced coincidence and data
acquisition techniques to improve accuracy.

The experimental arrangement and analysis methods were similar to
those of Taylor et al. [3]. Scattering from a ^{4}He gas scintillator was
employed as a polarization analyzer. The deuterium gas targets were
about 150 keV thick. Recent values for the ^{4}He(n,n)^{4}He elastic scattering phase shifts by Stammbach et al. [4] were used to calculate the
average analyzing power. Results of the experiment are shown in
fig. 1 along with the proton data of Porter and Haeberli. The available
^{2}D(d,n) differential cross sections were used with the present polarization data to generate $P(\theta)\sigma(\theta)$ values which were fitted with an associated Legendre polynomial expansion. The solid curves for $P(45°)$
in fig. 1 were generated from the resulting coefficients and the available cross sections.

A comparison of the experimental polarization excitation function
near 45° c.m. with previous work and with the proton data is also
shown [5-7]. The point from May et al. [7] has been evaluated from
the search of Stammbach and Walter [4]. The solid curves for P (45°)
were obtained from the fits to $P(\theta)\sigma(\theta)$ in this work and in ref. [1].
The proton polarization at 45° c.m. is indeed larger by a factor of 1.15.
It is also evident from the angular distributions that the proton polarization is systematically larger than the neutron polarization at all
angles.

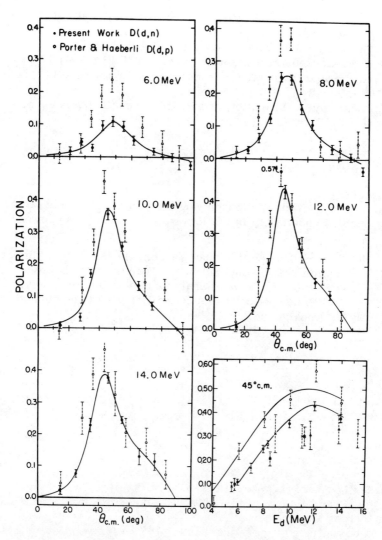

Fig. 1. Polarization angular distributions and polarization excitation functions for the ^{2}D(d,n)^{3}He and ^{2}D(d,p)^{3}H reactions. The earlier neutron values are X from ref. [1], $\triangle$ from ref. [5], ■ from ref. [6], and □ from ref. [7].

REFERENCES

*Work supported by the U.S. Atomic Energy Commission.

[1] P. S. Dubbeldam and R. L. Walter, Nucl. Phys. 28 (1961) 414.

[2] L. E. Porter and W. Haeberli, Phys. Rev. 164 (1967) 164.

[3] J. Taylor et al., Phys. Rev. C1 (1970) 803.

[4] Th. Stammbach and R. L. Walter, to be published.

[5] N. V. Alekseev et al., Soviet Phys.—JETP 18 (1964) 979.

[6] H. Niewodniczanski, J. Szmider, and J. Szymakowski, J. de
     Physique 24 (1963) 871.

[7] T. H. May, R. L. Walter, and H. H. Barschall, Nucl. Phys. 45
     (1963) 17.

DISCUSSION

Simmons:

Do you know of any reason why the proton data could be too high, such as uncertainty in the forward p-^{4}He analyzing power?

Spalek:

No, I do not know of any such reason. I would rather think that the neutron data of Dubbeldam and Walter are probably low because they used a forward analyzing angle and recorded the UP and DOWN number of counts on scalers where one cannot identify the background.

Polarization in ^{2}H(d,n)^{3}He
between 25 and 40 MeV

T. A. CAHILL and R. A. ELDRED, University of California, Davis, USA*

Medium-energy deuteron beams from the Davis 76" isochronous cyclotron have been used in conjunction with a high pressure deuterium gas target to establish a beam of polarized neutrons in the energy region between 23 and 36 MeV. Neutrons at an angle of 29.6° from the deuteron beam are collimated by about 80 cm of copper and 80 cm of iron; the last is placed between the pole pieces of a spin-rotation magnet. Neutrons are scattered from targets at the center of a 2.45-m scattering table and detected by a 5 cm × 15 cm cylindrical NE102 scintillator in a heavily shielded detector assembly on a movable arm. The flux is monitored by two thin plastic scintillators placed behind a CH_2 converter. The coincidence output of these detectors provides the start pulse for a time-to-amplitude converter; the stop pulse is provided by either a beam pick-off unit or the cyclotron r.f. system.

Pending completion of a ^{4}He polarimeter, a carbon polarimeter is being used, and for the purposes of a temporary calibration, it is assumed that the analyzing power for elastic scattering and for inelastic scattering to the 4.43-MeV level of C is the same for protons and neutrons scattered from C at a laboratory angle of 65°. In addition, the relative cross sections for elastic and inelastic scattering are assumed to be the same for neutrons and protons. The C(n,n)C polarization has been measured between 30° and 65° and agreement between these preliminary measurements and the shape of the C(p,p)C curves confirms this choice of angle.

The results show that the ^{2}H(d,n)^{3}He polarization at 29.6° lab declines steadily between 26 and 39 MeV, but less slowly than would be inferred from data below 20 MeV [1]. The polarization is approximately 13% at $E_d = 39$ MeV. Thus, this information, combined with the differential cross sections of Fegley and Cahill [2] indicates that the goodness parameter P^2I is not greatly inferior to that of the ^{3}H(d,n)^{4}He reaction.

REFERENCES

*Work supported in part by the U.S. Atomic Energy Commission.

[1]   N. V. Alekseev et al., Sov. Phys.—JETP 45 (1963) 1416.
[2]   R. W. Fegley and T. A. Cahill, to be published.

DISCUSSION

Simmons:
    Could you clarify the values of the neutron flux from the target?

Cahill:
    The flux at the second target is about 10^6 to 2×10^6 n/sec, but
    the neutrons are spread out over a large area. The highest flux
    is about 3×10^4 n/cm^2/sec at the exit of the collimator. For ex-
    periments which are not limited by target size this is no problem,
    but if target size is important, such as in a measurement of A_{yy},
    the beam is less useful.

Slaus:
    Could you compare the two reactions $D(d,n)^3He$ and $D(d,p)^3H$ in
    the energy region around 30–50 MeV?

Cahill:
    We would very much like to know if anyone is planning to mea-
    sure these parameters in this energy region. If not, we might
    consider doing it ourselves, since the discrepancy seems per-
    sistent.

The Vector Analyzing Power
of the D(d,n)³He Reaction at 12 MeV
Incident Deuteron Energy

C. O. BLYTH, P. B. DUNSCOMBE, J. S. C. McKEE, and C. POPE,
University of Birmingham, England

The data for the reactions D(d,n)³He and D(d,p)³H (see, for example, Seagrave [1]) indicate that at energies above 6 MeV the proton polarizations are roughly 1.4 times the neutron polarizations, contrary to what is expected from charge independence of nuclear forces. Recently also, Bernstein et al. at Los Alamos [2] have measured the vector analyzing power of the D(d,p)T and D(d,n)³He reactions using a polarized beam from a recoil line. Their results are not of sufficient quality to determine consistency with charge symmetry, as only two of the neutron (backward angle) points overlap the proton (forward) angular distribution, and the statistical errors in this region are large. The purpose of the present experiment was therefore to obtain data on the vector analyzing power of the D(d,n)³He reaction at forward angles, at an energy similar to that of the Los Alamos work. These results are shown, in conjunction with the neutron data of Bernstein, in fig. 1.

The 12 MeV polarized deuteron beam from the Radial Ridge Cyclotron [3] was transported to a high-pressure deuterium gas cell previously described [4]. The polarization of the beam was continuously monitored by scattering a little from a thin polythene foil in a small chamber situated directly in the beam line. The scattering chamber was built, and its operation calibrated by Griffith [5]. The average beam polarization was measured as 45% corresponding to a value of 0.40 for $\langle it_{11} \rangle$ and the beam current on target was about 1.5 nA.

Neutrons from the D(d,n)³He reaction were detected in a 7.5-cm diam NE213 scintillator, situated a suitable distance from the gas target for optimum separation of the two- and three-body neutrons. The neutron identification pulse from the pulse shape discriminator unit associated with the 7.5-cm detector,

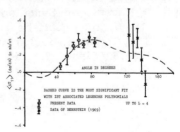

Fig. 1. Graph of $\langle iT_{11} \rangle$ $(d\sigma/d\Omega)$ vs. c.m. angle.

opened a 5 µs linear gate and allowed the TAC output pulse through to
a multichannel analyzer. Asymmetry in the scattering was measured
by switching the polarization of the deuteron beam at the source, the
"up" and "down" events being routed to separate halves of the kick-
sorter memory. The asymmetry was found from the expression:

$$\langle iT_{11} \rangle = \frac{N_{up}/N_{down} - 1}{2(\langle it_{11} \rangle_{up} + N_{up}/N_{down} \langle it_{11} \rangle_{down})} .$$

Data on the vector analyzing power of the reaction $D(d,n)^3He$ and
$D(d,p)T$ are sparse. Huber [6] has presented some beautiful data covering the region 150 keV to 450 keV and the angular dependence is
similar for both reactions. Any differences observed are readily explained in terms of the Coulomb interaction between outgoing particles.
No further experiments have been performed at energies below those
used by Bernstein.

The data of the present experiment are shown in fig. 1; the dotted
line shows the most significant fit to the data. Clearly it is possible
to force a good fit that looks symmetrical around 90° c.m. angle, but
there is no known reason to expect such symmetry. If the Bernstein
data from the $D(d,p)T$ experiments are plotted on the same curve, there
is no evidence for substantial departure from charge independence
over the angular range of the comparison.

REFERENCES

[1]. J. D. Seagrave, Int. Symp. on Few-Body Problems, Light Nuclei
     and Nucleon Reactions, eds. G. Paic and I. Slaus (1969), p. 401.
[2] E. M. Bernstein, G. G. Ohlsen, V. S. Starkovich, and W. G. Simon,
     Nucl. Phys. A126 (1969) 641.
[3] W. B. Powell, Second Polarization Symp., p. 47.
[4] J. Birchall et al., Proc. Int. Conf. Three-Body Problem, eds. J. S. C.
     McKee and P. M. Rolph (1970), p. 423.
[5] J. A. R. Griffith, private communication, and to be published.
[6] P. Huber, Proc. Symp. on Reaction Mechanisms, Laval, Quebec
     (1969), to be published.
[7] S. T. Thornton, Nucl. Phys. A136 (1969) 25; M. D. Goldberg and
     J. M. Le Blanc, Phys. Rev. 119 (1960) 1992.

Preliminary Measurements of Neutron
Polarization from the D($\vec{\text{d}}$,$\vec{\text{n}}$)³He
Reaction Using a Polarized Incident Beam

J. E. SIMMONS, W. B. BROSTE,* G. P. LAWRENCE, and G. G. OHLSEN,
Los Alamos Scientific Laboratory, New Mexico, USA[†]

The D(d,n)³He reaction has long been an important neutron source. At
zero degrees (lab) the differential cross section [1], I(0), is large,
varying from approximately 30 mb/sr at E_d = 1 MeV to 90 mb/sr at E_d =
10 MeV, with a Q value of 3.26 MeV. Much effort has been given to
measurements of neutron polarization from this reaction; a summary of
existing measurements [2] indicates that the polarization at θ = 30°
lab varies from −0.1 at 2 MeV to +0.3 to 12 MeV. Recent measurements
[3] on the outgoing proton polarization show increased values at higher
energies, namely, ≈ 0.5 at 12 MeV. However, the differential cross
section is sharply peaked forward which results in an unfavorable
cross section at the angles for which the neutrons have a significant
polarization. At E_d = 10 MeV [4], for example, the cross section ratio,
I(0)/I(30), has the value 12. Backgrounds would also be serious under
such conditions. The net result is that the d-d reaction has not been
very attractive as a polarized neutron source.

It is clearly of interest to measure the extent to which neutron
polarization may be generated by transfer from an incident polarized
deuteron beam. A preliminary experiment has been performed at the
Los Alamos FN tandem accelerator, as a continuation of comparable
measurements on the T(d,n)⁴He reaction [5]. The beam was accelerated
onto a 3-cm gas target with beam quantization axis pointing in the
vertical direction, here denoted as the y direction. The deuteron beam
was ~ 78% in the m_I = +1 spin state (as measured at the source), that
is, with vector and tensor polarization given by $p_3 = p_{33} = 0.78$ with
deviations of order ±0.01 depending on ion source conditions. Beam
intensities on target varied in the range 10–45 nA.

The y component of outgoing neutron polarization was measured at
0° by means of a liquid helium polarimeter, including a precession
solenoid for which a diagram is given elsewhere [5]. Equations [6]
relating the observables and beam polarization read as follows:

$$\langle \sigma_y \rangle = \frac{3}{2} p_3 \, P_y^y(1)/[1 + \frac{1}{2} p_{33} \, P_{yy}^0(0)] = e/P_2(\theta_2),$$

where $\langle \sigma_y \rangle$ is the measured neutron polarization, e and $P_2(\theta_2)$ are the observed asymmetry and the analyzing power in n-^{4}He scattering, p_3 and p_{33} are the vector and tensor beam polarizations, $P_y^y(0)$ is the polarization transfer parameter, and $P_{yy}^0(0)$ is a polarization efficiency tensor in the reaction .

The preliminary results show large polarization effects. Measurements were made at 0° lab for 6 deuteron energies in the range 4.0 to 13.5 MeV. The raw asymmetries had values near 0.63 with statistical errors of order ±0.015; little if any dependence on energy is discernible. These results may be contrasted with results from the $T(\vec{d},n)^4$He reaction [5], where a spin flip mechanism causes $\langle \sigma_y \rangle$ at 0° to cross zero near 3.9 MeV. It is estimated that multiple scattering and geometry corrections of order 10% will be applicable to these asymmetries, which would result in values of $\langle \sigma_y \rangle$ near 0.70. Auxiliary intensity measurements made during the experiment gave values of $P_{yy}^0(0) \approx 1/4$, roughly independent of energy. These results lead to estimates for $P_y^y(0)$ near 0.66, which would lead to neutrons with $\sim 90\%$ polarization for a 100% polarized ($m_I = 1$) beam of deuterons. The process behaves very much as if the neutron spin was virtually unaffected by the interaction at 0°. This qualitatively is what might be expected from a stripping process. Light as the reaction particles are, good fits to the differential cross section have been obtained [7] by an exchange stripping model.

This preliminary work demonstrates that the $D(\vec{d},\vec{n})^3$He reaction is an excellent source of polarized neutrons at 0° for neutron energies in the range 7 to 16 MeV.

REFERENCES

*Assoc. Western Universities Fellow from the University of Wyoming.
†Work performed under the auspices of the U.S. Atomic Energy Commission.

[1]   J. E. Brolley and J. L. Fowler, Fast Neutron Physics (Interscience Pub., New York, 1960) vol. I, chap. I.C.
[2]   H. Barschall, Second Polarization Symp.
[3]   L. E. Porter and W. Haeberli, Phys. Rev. 164 (1967) 1229.
[4]   S. T. Thornton, Nucl. Phys. A136 (1969) 25.
[5]   W. B. Broste et al., Third Polarization Symp.
[6]   J. L. Gammel, P. W. Keaton, Jr., and G. G. Ohlsen, Third Polarization Symp.
[7]   M. D. Goldberg and J. M. LeBlanc, Phys. Rev. 119 (1960) 1992.

DISCUSSION

Slaus:

Can you say what the polarization of breakup neutrons from the D(d,n)pd reaction is?

Simmons:

The breakup group also shows high polarization of approximately 55% for the whole group.

Polarization Transfer at 0° in the D(d,n)³He and H(d,n)pp Reaction

C. O. BLYTH, P. B. DUNSCOMBE, J. S. C. McKEE, and C. POPE,
University of Birmingham, England

The study of the polarization of nucleons emitted at 0° from reactions initiated by vector polarized deuterons can yield valuable information on the mechanism by which spin is transferred from the incident boson to the outgoing fermion, when the target is unpolarized. One measurement of this kind has been briefly reported by Simmons [1] and concerns the reaction T(d,n)⁴He for which a large polarization of the neutron was found.

As described in another contribution [2], the 12 MeV polarized deuteron beam from the Radial Ridge Cyclotron was transported to a high pressure gas cell, and the polarization of the beam was continuously monitored at a suitable point along the incident beam line. Neutrons emitted at zero degrees from the gas target were passed through a thin beam stop and were scattered in a liquid helium polarimeter [3]. The scattered neutrons were detected in a NE 213 liquid scintillator. A block diagram of the electronics is shown in fig. 1. A "window" can

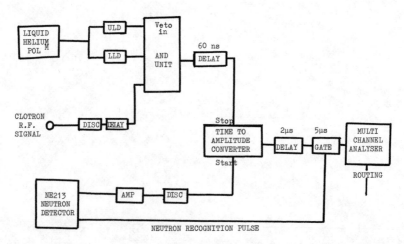

Fig. 1. Block diagram of the neutron detection system.

be set on the recoil alpha scintillation spectrum from the polarimeter to ensure that only those events corresponding to the scattering of neutrons in the direction of the NE 213 detector were selected. Furthermore, only recoils obtained at a time corresponding to the arrival of $D(d,n)^3He$ source neutrons were accepted. The neutron detector was positioned at an angle of $(80 \pm 2)°$ to the neutron beam, and the geometry-corrected analyzing power of helium for this scattering angle was 0.72. Neutrons were routed into the multi-channel analyzer according to the spin "up" or spin "down" designation of the incident deuteron beam.

The polarization of the neutrons emitted at 0° from the reaction $D(d,n)^3He$ was found to be $34 \pm 6\%$. Some time limited data on the polarization of neutrons from the breakup process H(d,n)pp gave a value of $45 \pm 10\%$. These results seem reasonable on first evaluation using a simple model [4], but the correct formulation and handling of tensor components requires much more careful attention.

REFERENCES

[1]   J. E. Simmons et al., Bull. Am. Phys. Soc. 14 (1969) 1230.
[2]   C. O. Blyth, P. B. Dunscombe, J. S. C. McKee, and C. Pope, Third Polarization Symp.
[3]   J. Birchall, M. J. Kenny, J. S. C. McKee, and B. L. Reece, Nucl. Instr. 65 (1968) 117.
[4]   C. O. Blyth, P. B. Dunscombe, J. S. C. McKee, and C. Pope, to be published.

The D($\vec{\text{d}}$,p)T-Reaction below 500 keV

K. JELTSCH, A. JANETT, P. HUBER, and H. R. STRIEBEL, University of Basel, Switzerland

The angular distributions of the protons from the D(d,p)T-reaction have been measured at $E_d = 150$, 230, and 340 keV mean lab energy with polarized (W_{pol}) and unpolarized (W_0) deuterons. From these measurements the components $D_1(\theta)$, $D_{33}(\theta)$, $D_{13}(\theta)$, and $D_{11}(\theta) - D_{22}(\theta)$ were computed and fitted by the adequate set of Legendre polynomials $L_{i,k}(\cos\theta) = P_i^k(\cos\theta)$:

$$W_{pol}(\theta,\phi) = W_0(\theta)\{1 + \frac{3}{2}P_1 D_1(\theta)\sin\phi + \frac{1}{2}P_{33}D_{33}(\theta) + \frac{2}{3}P_{13}D_{13}(\theta)\cos\phi$$

$$+ \frac{1}{6}(P_{11} - P_{22})[D_{11}(\theta) - D_{22}(\theta)]\cos 2\phi\},$$

$$W_0(\theta)D_1(\theta) = \sum_{n\geq 1} a_1^{(n)} L_{n,1}(\cos\theta), \quad W_0(\theta)D_{33}(\theta) = \sum_{n\geq 0} a_{33}^{(n)} L_{n,0}(\cos\theta)$$

$$W_0(\theta)D_{13}(\theta) = \sum_{n\geq 1} a_{13}^{(n)} L_{n,1}(\cos\theta), \quad W_0(\theta)[D_{11} - D_{22}(\theta)] = \sum_{n\geq 2} a_{11-22}^{(n)} L_{n,2}(\cos\theta).$$

$W_0(\theta)$, the angular distribution for unpolarized deuterons, is normalized so that $\sigma_{tot} = 4\pi$. The functions $D_{ik}(\theta)$ and the coefficients of the expansions are shown in the figures. The additional measurements at 460 keV are from ref. [1]. For $D_1(\theta)$ the results are nearly the same for all energies, and therefore only one curve is presented.

Qualitatively the following conclusions may be drawn:

1. The relatively large coefficients $a_{33}^{(1)}$ and $a_{13}^{(1)}$ are exclusively due to the entrance channel spin 2. This means that, in disagreement with former theoretical assumptions [2], quintet states in the entrance channel and spin flips to triplet states in the exit channel must be considered.

2. Only the predominating coefficients $a_{33}^{(2)}$, $a_{13}^{(2)}$ and $a_{11-22}^{(2)}$ may be explained by incoming s-waves alone. All other coefficients presuppose p- or d-waves.

3. Even at low energies too many matrix elements seem to contribute to the reaction, so that a detailed analysis of the reaction from the presented data is impossible.

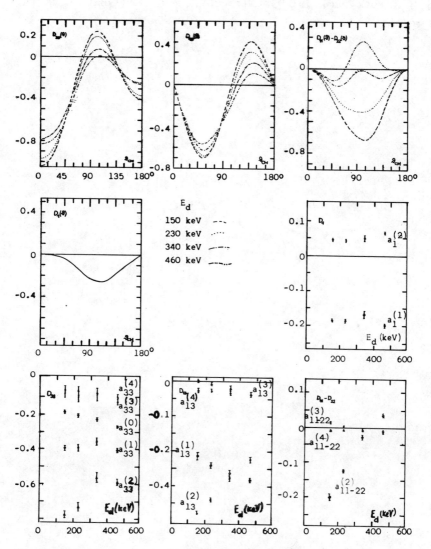

Components of the analyzing power of the D($\vec{d}$,p)T-reaction and coefficients of the expansions by Legendre polynomials.

REFERENCES

[1] Cl. Petitjean, P. Huber, H. Paetz gen. Schieck, and H. R. Striebel, Helv. Phys. Acta 40 (1967) 401.
[2] J. R. Rook and L. J. B. Goldfarb, Nucl. Phys. 27 (1961) 79.

Polarization of 22-MeV Neutrons
Elastically Scattered
from Liquid Tritium and Deuterium

J. D. SEAGRAVE, J. C. HOPKINS, E. C. KERR, R. H. SHERMAN,
A. NIILER, and R. K. WALTER, Los Alamos Scientific Laboratory,
New Mexico, USA*

The asymmetries from T($\vec{n}$,n)T elastic scattering of 22.1-MeV incident
neutrons have been measured for eleven laboratory angles between
40° and 118.5°. Heretofore, information about this interaction has
been limited to one measurement [1] at 1.1 MeV and to predictions
from phase-shift analysis [2] of the cross-section data below 6.0 MeV.
A cryogenic system provided a one-mole cylindrical sample of liquid
tritium.

The Los Alamos vertical Van de Graaff accelerator and Mobley
buncher were used with the T(d,n) reaction at a deuteron energy of
6 MeV. The 22.1-MeV neutrons produced at 29.8° (lab) were (+47.4 ±
1.45)% polarized [3].

Two detectors placed at 2.5 meters from the sample were massively
shielded with copper, polyethylene, and tungsten. In addition, n-γ
discrimination was employed to reduce the gamma-ray background.
Corrections for differences in detector efficiencies were based on
measurements with the detectors interchanged. The data were corrected for various other artificial asymmetries.

The results are shown in fig. 1 and tabulated in table 1. The
^{3}He($\vec{p}$,n)^{3}He polarization data at 21.3 MeV of Tivol [4] are sketched
for comparison. Small corrections were applied for multiple scattering. The solid curve is a phase-shift fit to the polarization data and
the differential cross section [5].

In preparation for the tritium measurements, observations of
D($\vec{n}$,n)D asymmetries were made at the same energy. These results
are consistent with the earlier work of Malanify et al. [6]. Details
of this experiment will be found in the thesis by R. K. Walter [5] and
in two review papers by Seagrave [7, 8]. A paper combining all LASL
work on n-D and n-T cross sections and polarization work is in preparation.

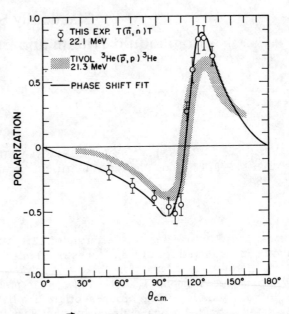

Fig. 1. T($\vec{n}$,n) polarizations at 22.1 MeV.
Comparison is made with the charge-conjugate
reaction studied by Tivol [4], and with a phase-
shift fit to both polarization and differential
cross section [5, 7].

Table 1. T($\vec{n}$,n)T Polarizations $P_2(\theta_2)$ at 22.1 MeV

θ_2 lab (deg)	θ_2 c.m. (deg)	$\cos\theta_2$ c.m.	ϵ	$P_2(\theta_2)$[†]	δP_2[‡] absolute	$(\delta P_2/P_2)$ percent statistics
40	52.4	0.610	−0.095	−0.20	0.058	5.5
55	70.9	0.327	−0.14	−0.30	0.059	5.0
70	88.3	0.029	−0.19	−0.40	0.061	5.2
80	99.2	−0.160	−0.22	−0.47	0.071	8.6
85	104.5	−0.250	−0.24	−0.52	0.080	11.0
90	109.5	−0.334	−0.21	−0.45	0.085	13.0
95	114.5	−0.414	0.13	0.27	0.070	21.0
100	119.2	−0.488	0.24	0.59	0.098	14.0
105	123.8	−0.557	0.35	0.82	0.110	11.0
110 1/4	128.5	−0.623	0.36	0.83	0.096	8.6
118 1/2	135.6	−0.714	0.31	0.69	0.078	6.2

[†]Based on a source polarization of 0.474 ± 0.015
[‡]Also includes ±0.026 uncertainty in artificial asymmetries.

REFERENCES

*Work performed under the auspices of the U.S. Atomic Energy Commission.

[1]   J. D. Seagrave, L. Cranberg, and J. E. Simmons, Phys. Rev. 119
      (1960) 1981.
[2]   T. A. Tombrello, Phys. Rev. 143 (1966) 772.
[3]   W. Broste and J. E. Simmons, private communication.
[4]   W. F. Tivol, Thesis, UCRL-18137 (1968).
[5]   R. K. Walter, Ph.D. Thesis, Brigham Young University (1969);
      Los Alamos Scientific Laboratory Report LA-4334.
[6]   J. J. Malanify, J. E. Simmons, R. B. Perkins, and R. L. Walter,
      Phys. Rev. 146 (1966) 632.
[7]   J. D. Seagrave, Proc. Symp. on Light Nuclei, Few-Body Problems,
      and Nuclear Forces, 1967 (Gordon & Breach, London, 1969),
      vol. II, p. 787.
[8]   J. D. Seagrave, Proc. Int. Conf. on Three-Body Problem, 1969
      (North Holland, Amsterdam, 1970) p. 41.

Nucleon Polarization in p + ³H and n + ³He Elastic Scattering from a Charge-Independent Analysis of ³H(p,p)³H, ³H(p,n)³He, and ³He(n,n)³He

I.Ya. BARIT and V. A. SERGEYEV, P.N. Lebedev Institute of Physics, Moscow, USSR

A charge-independent phase-shift analysis [1] of the differential cross sections of the ³H(p,p)³H, ³H(p,n)³He, and ³He(n,n)³He reactions and of the neutron polarization in the (p,n) reaction was performed for excitation energies of ⁴He corresponding to c.m. energies E in the proton channel between 0 and 3.4 MeV. A contribution of eight states was taken into account, each state a corresponding to certain values of $\ell \leq 2$, $S = 0, 1$, and J. The transitions with a change of ℓ or S and spin-orbit splitting of the triplet D-states were not included in the analysis.

The inverse logarithmic derivatives R_0^a of internal wave functions with isospin $T = 0$ were determined at some fixed energies in the charge independent analysis, the known information about the p+³He scattering being used. The conclusions about the excited states of ⁴He with $T = 0$ can be derived from this analysis without any additional assumptions except the isospin conservation in the region of nuclear interaction. Levels with $J^\pi = 0^+$, 0^-, and 2^- possessing nearly single-particle reduced widths were found at the energies $E_r = 0.5$, 1.3, and 2.3 MeV. The parameters R_0^a in the triplet even states and the singlet P-state were found to be very small, i.e., the appropriate levels were absent. These conclusions are in agreement with the results of a previous analysis [2] and with the theoretical predictions.

The polarizations in p+³H and n+³He scattering calculated with the parameters obtained are in satisfactory agreement with most experimental data [3-6] (see fig. 1). Only the neutron polarization data at $E_n = 2.15$ MeV [6] and $E_n = 3.33$ MeV [7] are not consistent with the results of the charge-independent analysis. More accurate and detailed polarization measurements would be very valuable, especially for n+³He scattering at $E_n < 5$ MeV.

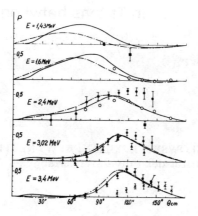

Fig. 1. Polarization in p+³H and
n+³He scattering at five excitation
energies of ⁴He $E_{ex} = 19.8$ MeV +
E. The calculations for protons
and tritons are shown by solid
lines, for neutrons and ³He by
dash-dot lines; the experiments
for protons by ● [3], tritons by ○
[4], neutrons with
$$E_n = 1.1;\ 2.15 \text{ MeV} - ■ \ [6]$$
$$E_n = 3.33 \text{ MeV} - × \ [7]$$
$$E_n = 3 \text{ MeV} - ▲ [5]$$

REFERENCES

[1] I. Ya. Barit and V. A. Sergeyev, Preprint FIAN N 59 (1970).
[2] C. Werntz and W. E. Meyerhof, Nucl. Phys. A121 (1968) 28.
[3] L. Drigo et al., Nuovo Cim. 51B (1967) 43; N. A. Skakun et al.,
     Sov. Phys.—JETP 46 (1964) 120.
[4] L. R. Veeser et al., Nucl. Phys. A140 (1970) 177.
[5] C. E. Hollandsworth et al., Bull. Am. Phys. Soc. 14 (1969) 554.
[6] J. D. Seagrave et al., Phys. Rev. 119 (1960) 1981.
[7] A. F. Behof et al., Nucl. Phys. 84 (1966) 290.

Elastic Scattering of Polarized Protons
on Tritons between 4 and 12 MeV

J. C. FRITZ, R. KANKOWSKY, K. KILIAN, A. W. NEUFERT,[*] and D. FICK,
Physikalisches Institut der Universität Erlangen-Nürnberg, Germany

Among the very light nuclei ^{4}He is the one with the highest symmetry.
This is reflected in the great stability of ^{4}He and the fact that no
bound state except the ground state is known. Therefore it is difficult to investigate experimentally the excited states of ^{4}He. Nevertheless the progress of experimental and theoretical investigations of
^{4}He is clearly demonstrated by the review of Meyerhof and Tombrello
[1]. Especially the one-particle-one-hole shell model calculations
combined with the supermultiplet theory seem to give a good description of the negative parity levels above the p-^{3}H threshold. On the
other hand, there is a lack of experimental information concerning the
elastic p-^{3}H scattering [1]. In the present experiment we therefore
measured the elastic scattering of polarized protons on tritons using
the polarized proton beam of the Erlangen tandem accelerator [2]. In
order to make the phase-shift analysis unique it is an advantage to
use polarized protons. The target consists of a Ti-^{3}H foil on an Al or
Au backing. (Target thickness approximately 25 keV for 5-MeV protons
with 0.5 Ci ^{3}H.) The angular distributions were obtained using up to
16 detectors simultaneously. At forward angles the peaks from the
backscattered ^{3}H could also be observed, and therefore an angular
distribution consists of up to 23 points. The polarization of the protons was determined by elastic P-^{12}C scattering to be $P = 0.7$. Because
for each scattering angle only one detector was used, the analyzing
power was determined in two runs, one with spin up and one with spin
down. From 4 to 12 MeV 42 angular distributions have been measured.
An absolute differential cross section was obtained by normalizing
our relative data to the absolute data of Brolley et al. [3].

In order to reduce this amount of data a Legendre polynomial fit was
tried according to

$$W = (4\pi)^{-1} \sum_k A_k P_k(\cos\theta) \text{ and } WA = (4\pi)^{-1} \sum_k B_k P_{k1}(\cos\theta)$$

W is the differential cross section and A the analyzing power. An ex-

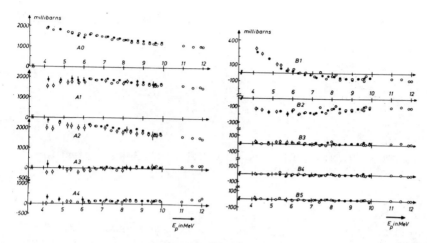

Fig. 1

cellent fit was obtained. The coefficients are shown in fig. 1 (circles: Au backing; black dots: Al backing). Only the coefficients with k = 0, 1, 2 contribute significantly to W (fig. 1). At higher energies there is a small contribution from A_4. The main contributions to AW come from B_1 and B_2. This can be understood by remembering the fact that the excited levels in ^{4}He have the configuration $(1s)^{-1}(1p)$. Around E_p = 5.4 MeV the energy steps of 5 KeV were chosen in order to look for the proposed 2^+-level just above the d-d threshold [4]. (These data points are not displayed in fig. 1.) No obvious structure could be observed. This may be understood if this level has a pure two-particle-two-hole structure. A phase-shift analysis is in progress.

REFERENCES

*Present address: Institut für Kernphysik, Giessen, Germany.

[1]   W. E. Meyerhof and T. A. Tombrello, Nucl. Phys. A109 (1968) 1.
[2]   G. Clausnitzer et al., Nucl. Instr. 80 (1970) 245.
[3]   J. E. Brolley et al., Phys. Rev. 117 (1960) 1307.
[4]   D. Fick, Z. Phys. 221 (1969) 451.

The Scattering of Polarized Protons
by Tritons at 13.6 MeV

J. L. DETCH, Jr., J. H. JETT, and N. JARMIE, Los Alamos Scientific
Laboratory, New Mexico, USA*

The asymmetries in the $T(\vec{p},p)T$ reaction were measured at $13.600 \pm$
0.015 MeV. The polarized protons were supplied by the LASL Lamb-
shift polarized ion source and accelerated by a tandem Van de Graaff.
Previous measurements of proton asymmetries in this energy region
were made by Rosen and Leland [1] at 14.5 MeV.

The experiment was performed with a single detector telescope
using the same apparatus previously used for high accuracy (0.4%
relative accuracy) cross-section measurements. Basically, the equipment consisted of a thin-window (2.3 micrometers Havar) gas scattering cell which was pressurized to 0.25 atmosphere with tritium. The
detector system has a 1° angular acceptance (or a G-factor of $1.06 \times$
10^{-4}). The included angle between left and right detector settings
was known to ±0.03° and the beam was constrained by slits so that it
could not wander more than ±0.06° between left and right measurements.
The beam was integrated in a standard way to an absolute accuracy of
±0.2%. A more detailed description of the apparatus used and error
discussion is contained in ref. [2].

The data were taken at two different times. The majority of the
data were taken during the first run and without an independent measurement of the beam polarization other than by the quench ratio method.
The second data set consisted of 3 data points on the back angle peak
and followed an extensive check of the beam polarization using p-α
scattering as an analyzer. For this last run, the beam polarization
was known to ±1.0% and the two sets of data were consistent.

The resulting data are shown in fig. 1 along with the data of Leland and Rosen [1]. For all but two of the data points, four measurements were made, being two measurements on each side of the chamber with the proton spin up and down. This method eliminates certain
geometrical systematic errors.

The results are listed in table 1 along with the total error, which
includes the error in the knowledge of the beam polarization. The
statistical uncertainty was the predominant contribution to the errors
except for the largest asymmetries.

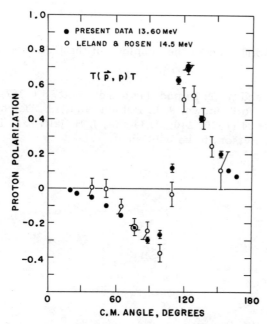

Fig. 1. Asymmetries in T($\vec{p}$,p)T scattering.
The data of Leland and Rosen are from ref. [1].

Table 1. Asymmetries in T($\vec{p}$,p)T scattering
at 13.6 MeV

θ_{Lab}	θ_{CM}	A	δA
15.0	20.0	−0.012	0.004
20.0	26.6	−0.031	0.005
30.0	39.7	−0.051	0.006
40.0	52.5	−0.097	0.008
50.0	65.0	−0.156	0.012
60.0	77.0	−0.224	0.012
70.0	88.5	−0.296	0.015
80.0	99.4	−0.263	0.020
90.0	109.7	0.122	0.025
97.5	116.9	0.626	0.020
105.0	123.9	0.690	0.030
105.0	123.9	0.712	0.018
120.0	136.9	0.405	0.012
140.0	152.5	0.205	0.021
150.0	159.7	0.111	0.013
160.0	166.1	0.075	0.011

REFERENCES

*Work performed under the auspices of the U.S. Atomic Energy Commission.

[1] L. Rosen and W. T. Leland, Phys. Rev. Lett. 8 (1967) 379.
[2] N. Jarmie, R. E. Brown, R. L. Hutson, and J. L. Detch, Jr., Phys. Rev. Lett. 24 (1970) 240; N. Jarmie, J. H. Jett, J. L. Detch, Jr., and R. L. Hutson, to be published.

Asymmetry in the Elastic Scattering of 42-MeV Polarized Protons by ^{3}He and ^{3}H

J. ARVIEUX, R. DARVES-BLANC, J. L. DURAND, A. FIORE,
N. VAN SEN, C. PERRIN, and P. D. LIEN, Institut des Sciences
Nucléaires, Grenoble, France

Some problems arose in the interpretation of the unpolarized cross section of the ^{3}He(p,p)^{3}He and ^{3}H(p,p)^{3}H elastic scattering measured at 156 MeV and analyzed at Orsay [4]. We have done the same experiments at 42 MeV, and we report here the polarization data; the cross section will be published later in a detailed report. We hope also that our p-^{3}He data will help in finding good phases in the analysis of elastic scattering including spin effects. Such an analysis has been performed up to 11.5 MeV by Tombrello [1], who was able to obtain a contour map of the polarization. Between 11.5 and 19.5 MeV there are many numerical solutions fitting the cross section and the polarization, but most of them have been eliminated by a measurement of the spin correlation coefficients A_{xx} and A_{yy} at 19.5 MeV. An unique solution for real phases has been found [2]. At higher energies the only polarization data are the 30- and 50-MeV asymmetries measured by the Harwell group. It seemed interesting to fill the energy gap by measuring the asymmetry at 42 MeV with higher precision.

The polarized proton beam is produced in the polarized source of the Grenoble cyclotron. The intensity was of the order of 5 nA focused onto the target, and the polarization varied from 0.5 to 0.7. For the p + He^3 experiment the target was filled with 99.9% pure ^{3}He gas. Background runs were taken with an empty cell or with a cell filled with air, and ^{3}He data were rejected when the separation between the ^{3}He peak and the background was poor.

For the p + t experiment we have used a solid Ti-t target of 10 μm thickness enriched to 0.6 atom of ^{3}H for 1 atom of Ti. Two solid-state telescopes detect the p and the t in coincidence at conjugate angles. A particle identifier separates the p from the deuterons, the tritons and the ^{3}He-^{4}He group. The pulses coming from a proton and a triton detected in coincidence are summed. This gives a peak having the energy of the incident beam and a background spectrum due to random coincidences between a triton coming from the Ti(p,t) reaction and any inelastic proton. The results are given in fig. 1. The indicated errors are combined statistical errors. When no error bar is given, the error is smaller than the size of the point. The normalization uncertainty is of the order of ± 5%. The He^3 data show asymmetries higher than obtained by the Harwell group [3] at 30 and 50 MeV. The H^3 data show a slightly different shape and smaller negative values.

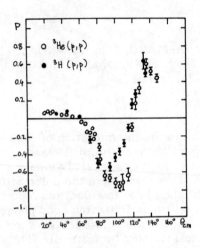

Fig. 1

REFERENCES

[1] T. A. Tombrello, Phys. Rev. 138B (1965) 40.
[2] T. Cahill et al., Bull. Am. Phys. Soc. 12 (1967) 1138.
[3] S. A. Harbison et al., Nucl. Phys. A112 (1968) 137; R. J. Griffiths et al., P.L.A. Progress Report 1968, p. 22.
[4] Ph. Narboni et al., Suppl. J. Phys. (Paris) C2 (1970) 81.

Search for a Narrow State at 23.9 MeV
in ^{4}He in the ^{3}H + p Channel

T. R. DONOGHUE, H. PAETZ gen. SCHIECK, C. E. BUSCH, and
J. A. KEANE, Ohio State University; R. M. PRIOR and K. W. CORRIGAN,
University of Notre Dame, Indiana, USA*

A resonance-like anomaly [1] in the neutron polarization excitation function for the D(d,n)^{3}He reaction near $E_D \approx$ 100 keV suggested the existence of a hitherto unknown state in ^{4}He at 23.9 MeV. Subsequent measurements of the analyzing powers in the D(d,n) and D(d,p) reactions initiated with polarized deuterons show similar anomalies [2] at the same energy. An analysis of the experimental data [3] led to an assignment of $J^\pi = 1^-$ to this state, later revised [4] to $J^\pi = 2^{(+)}$, T = 0 with Γ = 5 keV. Since all of these measurements were made with thick targets (i.e., $\Delta E \gg \Gamma$) for experimental reasons, and since measurements by Behof et al. [5] and Stoppenhagen et al. [6] gave null results, the existence of this state is still in question, and additional experimental investigations are desirable.

We have carried out a search for this level in the ^{3}H + p channel, following the suggestion of Fick [4] that this state might be observable in this channel near 5.45 MeV. Excitation curves of the elastically scattered protons were measured at various angles over the energy region in question using first an unpolarized beam from the Ohio State CN Van de Graaff accelerator and later using the polarized (61%) proton beam from the Notre Dame FN tandem accelerator. Thin Ti-T (1:1) targets with 100 μg/cm^2 titanium (i.e., ~5 keV thick to 5.45-MeV protons) evaporated onto thin Al or Ni backings were used. The unpolarized beam measurements were carried out for 5.33 $\leq E_p \leq$ 5.63 MeV in 2.5 keV (or smaller) energy increments, while the polarized beam measurements were made in the energy interval between 5.4 and 5.5 MeV in 5 keV increments. The respective data are shown in figs. 1 and 2. The dashed curve in fig. 2 is an average value for the analyzing power at each angle. The arrow here indicates the energy where this state is anticipated.

The results of both these investigations are negative. No anomaly appreciably exceeding the statistical fluctuations is observed. Thus, it would appear that there is no state in ^{4}He in this energy range.

Further investigations with better energy resolution in the d + d channel are therefore desirable.

Two of the authors (TRD and HPS) would like to thank Prof. S. E.
Darden for his hospitality while a portion of these measurements was
being made.

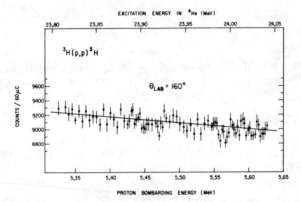

Fig. 1

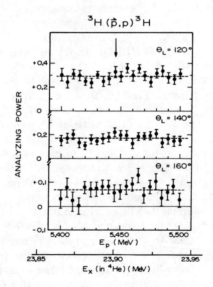

Fig. 2

REFERENCES

*Supported in part by the National Science Foundation.

[1]   H. Haensgen et al., Nucl. Phys. 73 (1965) 417.
[2]   H. W. Franz and D. Fick, Nucl. Phys. A122 (1968) 591.
[3]   D. Fick and H. W. Franz, Phys. Lett. 27B (1968) 541.
[4]   D. Fick, Z. Phys. 221 (1969) 451.
[5]   A. F. Behof et al., Nucl. Phys. A108 (1968) 250.
[6]   W. G. Stoppenhagen and R. W. Findlay, Bull. Am. Phys. Soc. 13 (1968) 873.

DISCUSSION

Hackenbroich:

I think that you used an entrance channel with $s = 1$ whereas the expected resonance should be in a $s = 2$, $\ell = 0$ channel. Only the tensor force can induce a transition which must therefore be small.

Donoghue:

This depends upon the 2^+ assignment made by Fick. Because the recent Kurchatov measurements disagree with the experimental data of Franz and Fick, the assignment to (and the existence of) this state is very questionable.

Calculations Concerning Scattering States
of the Four-, Five-, and Six-Nucleon Systems

H. H. HACKENBROICH, Institut für Theoretische Physik, Universität
zu Köln, Germany

Several calculations have been made concerning the scattering states
of very light nuclei using the refined cluster model technique [1].
Here one starts from a Hamiltonian composed from the internucleon
forces and employs the extended version of the Kohn-Hulthén variational method to calculate the S-matrix elements for reactions which
show two reaction partners in each channel only.

The last calculations used the soft core central potential given by
Eikemeier and this author [2] and the spin-orbit and tensor potentials
of the extended version of this "Tübingen" potential [3]. Calculations
performed in collaboration with Heiss [4] for the $T = 0$ scattering states
of ^{6}Li showed good agreement to experimental α-d phases in the low
energy region. Especially the splitting of the resonances reproduced
the data. At 9 MeV above the ^{3}H-^{3}He threshold we found a broad reso-
nance in the 1^+ state and large α-d $\rightleftarrows$ ^{3}H-^{3}He reaction cross sections.
This state is interpreted as a compound state of structure (^{5}Li*)-n or
(^{5}He*)-p, with the 5-nucleon structure in the $3/2^+$ resonant state.

The calculations for the ^{5}Li and ^{5}He scattering states [5] also gave
good results for the low energy region. The $3/2^+$ resonant state is
explained as a resonance in the $S = 3/2$, $L = 0$, ^{3}He-d and ^{3}H-d channels. This state couples via tensor forces to the $L = 2$ α-p and α-n
channels. The calculated α-p phases agree with the experimental
analysis [6], while the α-n phases differ only slightly from the α-p
phases, contrary to the analysis [7]. At 20 MeV c.m. energy in the
α-p (or α-n) system we get strong coupling between the ^{3}He-d (or
^{3}H-d) $S = 1/2$ channels and the α^*-p (or α^*-n) channels, where α^*
means the second $S = 0$ $J = 0$ state of the α-particle which also shows
up in the elastic phases. Furthermore, at higher energies resonances
in the $S = 3/2$, $L = 2$ (quartet) channels of ^{3}He-d (and ^{3}H-d) are found.

In connection with these reaction calculations we also have investigated the structure of the involved light nuclei. The form factor [8]
of the α-particle ground state is shown to be connected with both
long- and short-range correlations, the d-wave contribution of it is
estimated to be about 10%. A quasi-stationary calculation of the ex-

cited states of the α-particle, using cluster functions of definite spatial symmetry [4], [22], [31] gave as the most important result that a $J^\pi = 2^+$ state with [22] symmetry, $L = 0$, $S = 2$ could appear close to the d–d threshold.

This work was performed with H. Hutzelmeyer and G. John.

REFERENCES

[1] H. H. Hackenbroich and P. Heiss, Z. Phys. 231 (1970) 205;
     P. Heiss and H. H. Hackenbroich, Z. Phys. 235 (1970) 422.
[2] H. Eikemeier and H. H. Hackenbroich, Z. Phys. 195 (1966) 412.
[3] H. Eikemeier and H. H. Hackenbroich, Int. Conf. on Clustering Phenomena in Nuclei (IAEA, Wien, 1969) p. 286.
[4] H. H. Hackenbroich and P. Heiss, Phys. Lett. 32B (1970) 78.
[5] P. Heiss and H. H. Hackenbroich, Z. Phys. 231 (1970) 225;
     P. Heiss and H. H. Hackenbroich, Phys. Lett. 30B (1969) 373;
     H. H. Hackenbroich and P. Heiss, to be published.
[6] W. G. Weitkamp and W. Haeberli, Nucl. Phys. 83 (1966) 46.
[7] B. Hoop and H. H. Barschall, Nucl. Phys. 83 (1966) 65.
[8] H. Hutzelmeyer, to be published in Nucl. Phys.

DISCUSSION

Simmons:

It would be useful to compare ^{3}He(d,p) and T(d,n) polarizations by a charge symmetry operation on your potentials. If possible, this could shed light on certain experimental discrepancies.

Five Nucleons

Elastic Scattering of
Vector Polarized Deuterons on
Tritons between 3.3 and 12.3 MeV

H. TREIBER, K. KILIAN, R. STRAUSS, and D. FICK, Physikalisches
Institut der Universität Erlangen–Nürnberg, Germany

Among the 1p nuclei the spectroscopy of ^{5}He and ^{5}Li is difficult because of the unbound character of these systems. On the other hand,
these systems are of important theoretical interest, since the dimen-
sion of these systems allows a detailed theoretical analysis. Especially the successful particle-hole description of excited states in the
A = 4 system led to similar calculations in the A = 5 system [1]. In addition, cluster model calculations concerning the phase shifts in d-^{3}H
elastic scattering have recently been published [2]. In order to test
these calculations, experiments with polarized particles are very desirable [2].

In the present experiment we therefore measured the elastic scattering of vector polarized deuterons on tritons using the vector polarized
deuteron beam of the Erlangen tandem accelerator [3]. The target consisted of a Ti-^{3}H foil on an Au backing. Angular distributions were
obtained using up to 16 detectors simultaneously. At forward angles
the peaks of the backscattered ^{3}H could also be observed so that an
angular distribution was obtained with up to 21 data. Because only
one detector was used for each scattering angle, the analyzing power
was determined in two runs, one with spin up and one with spin down.
Absolute cross sections were obtained by normalizing our relative
data to the absolute data of Ivanovich et al. [4]. The data were fitted
to a Legendre polynomial series:

$$W = (4\pi)^{-1} \sum_k A_k P_k(\cos\theta) \text{ and } AW = (4\pi)^{-1} \sum_k B_k P_{k1}(\cos\theta)$$

(W = differential cross section and A = analyzing power.) An excellent
fit was obtained. The coefficients are shown in fig. 1. The coefficients with even k exhibit a broad maximum at a deuteron energy of
about 7 MeV. This structure may correspond to reported levels in ^{5}He
at 20 and 22 MeV excitation energies, respectively [5] (arrows). The
striking feature of the coefficients is the resonance-like structure in
the coefficients B_k with odd k (k = 1, 3) at E_d = 5 MeV. The B_k with

497

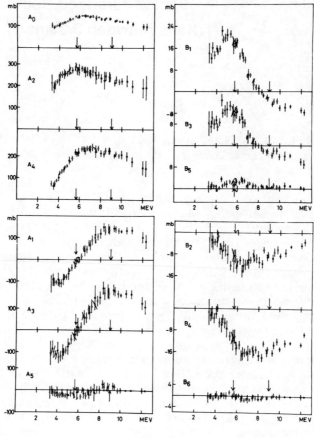

Fig. 1

odd k depend on products of matrix elements with different parity. If
this structure were caused by the matrix elements with positive parity,
it should also be seen in the coefficients A_k and B_k with even k.
Nothing can be seen there. Therefore, the matrix elements with odd
parity should account for this structure. It therefore seems likely that
this structure is connected with an odd parity level in ^{5}He at about
20.5 MeV excitation energy. A further analysis is in progress.

REFERENCES

[1]   K. Ramavataram and S. Ramavataram, Nucl. Phys. A147 (1970) 293.
[2]   P. Heiss and H. H. Hackenbroich, Z. Phys. 231 (1970) 230.
[3]   G. Clausnitzer et al., Nucl. Instr. 80 (1970) 245.
[4]   M. Ivanovich et al., Nucl. Phys. A110 (1968) 441.
[5]   M. P. Baker et al., Contribution 8.12 to Montreal Conf. (1969).

DISCUSSION

Plattner:

Since you have parameterized elastic scattering data in terms of Legendre polynomials, what was the precision with which the Legendre coefficients were determined? Did you leave out forward angle points in order to minimize Coulomb effects?

Treiber:

Yes, we left out the forward angles. We only included data at angles between 33° and 150°.

Plattner:

Did you test whether the coefficients still change if you leave out more forward angle points?

Treiber:

No, we have not done this yet.

Polarization Angular Distributions
from the T(d,n)^{4}He Reaction
at 7.0 and 11.4 MeV

W. B. BROSTE[*] and J. E. SIMMONS, Los Alamos Scientific Laboratory, New Mexico, USA[†]

Angular distributions of the polarization of neutrons from the T(d,n)^{4}He reaction have been measured for incident deuteron energies of 7.0 and 11.4 MeV. By accelerating tritons onto a deuterium target, it was possible to obtain data for T(d,n)^{4}He neutron emission angles which could not be reached when deuterons were accelerated onto tritium.

Neutron polarizations were analyzed by scattering from liquid helium at laboratory angles between 115° and 125°. The measured asymmetries were corrected for the effects of multiple scattering in the helium and finite polarimeter geometry. Polarizations were derived using n-helium analyzing powers calculated from the phase shifts of Hoop and Barschall [1]. Analyzing angles for neutron energies above 20 MeV were chosen where the calculated analyzing powers showed the least discrepancy with recent n-helium polarization measurements.

Figs. 1 and 2 show the present data: Solid circles represent polarizations obtained from the T(d,n)^{4}He reaction, while the triangles are data obtained using the D(t,n)^{4}He reactions at c.m. conditions equivalent to the T(d,n)^{4}He at the lab angles for which the data are plotted. The open circles on the figures are the measurements of Brown and Haeberli [2] on the charge conjugate ^{3}He(d,$\vec{p}$)^{4}He reaction at 8.0 and 12.0 MeV. The two sets agree well, except for the magnitude of the polarization at $\theta = 30°_{lab}$, and small differences in shape between the 11.4 MeV T(d,$\vec{n}$)^{4}He and 12.0 MeV ^{3}He (d,$\vec{p}$)^{4}He data beyond 90° lab. It should be noted that interpolation of the Brown and Haeberli data to 7.0 MeV yields a value for the 30° proton polarization which is consistent with the present value of neutron polarization within experimental errors. However, recent data [3] from this laboratory on the T(d,n)^{4}He polarization at 30° indicate that there are significant differences between the proton and neutron polarization from the two reactions for deuteron energies between 8 and 12 MeV.

Theoretical interpretations of the present results and other data on the T(d,n)^{4}He reaction are being pursued within the framework of a phenomenological reaction matrix analysis. The effect of spin-orbit distortions in entrance and exit channels is also being studied.

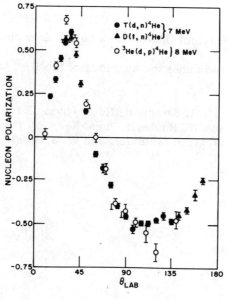

Fig. 1

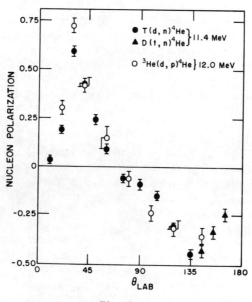

Fig. 2

REFERENCES

*Associated Western Universities Fellow from the University of Wyoming.

†Work performed under the auspices of the U.S. Atomic Energy Commission.

[1]   B. Hoop and H. H. Barschall, Nucl. Phys. 83 (1966) 65.
[2]   R. I. Brown and W. Haeberli, Phys. Rev. 130 (1963) 1163.
[3]   J. E. Simmons, G. S. Mutchler, and W. B. Broste, to be published.

Analyzing Power of the
T($\vec{d},\alpha$)n Reaction at Low Energies

G. G. OHLSEN, J. L. McKIBBEN, and G. P. LAWRENCE, Los Alamos
Scientific Laboratory, New Mexico, USA[*]

For energies less than 100–200 keV, the T($\vec{d}$,a)n reaction is known to
proceed almost totally through S-wave deuteron absorption. If one
assumes S waves only, and further assumes that the polarization of
the incident deuterons possesses an axis of symmetry, the cross sections may be written [1]

$$I = I_0(1 + \frac{1}{2}\langle P_{zz}\rangle P_{zz}^0), \quad P_{zz}^0 = -\frac{g}{2}(3\cos^2\theta_S - 1), \tag{1}$$

where $\langle P_{zz}\rangle$ is the tensor polarization of the deuterons and where θ_S
is the c.m. direction of emission of the observed reaction product,
both with respect to the quantization axis. g is the parameter which
is to be measured; if the reaction proceeds completely through the
J = 3/2+ intermediate state, g = 1. Note that the general analyzing
tensor restriction $-2 < P_{zz}^0 < 1$ requires that $-1 < g < 2$.

In the present paper, we report measurements of the T($\vec{d}$,a)n analyzing power at 51, 76, and 93 keV deuteron energies. Since a thick
tritium–zirconium target was used, the quantities actually determined
were energy averages of g of a certain form [1]. The average values
of g, which we shall call $\bar{g}$, are the quantities directly of interest for
use of the reaction as a calibrated analyzer.

The experiment consisted of the determination of $\bar{g}$p, where p is
the spin state purity of the beam, by a ratio method together with a
determination of p by the quench ratio method. The beam quantization
axis was parallel to the beam direction. Two detectors at 90° and two
detectors at 165° were employed at azimuthal angles such that there
was an unobstructed view of the solid tritium–zirconium target. The
beam was collimated by 5 mm × 5 mm slits 30 cm upstream from the
target.

Ratios between the yields for $m_I = 1$, $m_I = 0$, and $m_I = -1$ beams
were used to determine $\bar{g}$p. (We refer to the states by the m_I value
which characterizes the large magnetic field limit.) This requires an
accurate knowledge of the relative tensor polarization of the three

states (see table 1). Since the field in the negative-ion formation region of the ion source was ~60 G, i.e., strong with respect to the deuterium atom $2S_{1/2}$ hyperfine interaction, these ratios are well known and extremely insensitive to variation of the magnetic field. Experiments with the polarized beam accelerated to tandem energies have verified that the relative polarizations are as expected to an accuracy of ~0.5% or better.

Table 1

State	Tensor polarization
$m_I = 1$	p
$m_I = 0$	-1.966p
$m_I = -1$	0.952p

The quantity $\bar{g}p$ can be determined from the relative yields, Y, observed in a single detector with an $m_I = 1$ beam followed by an $m_I = 0$ beam, or by an $m_I = 0$ beam followed by an $m_I = -1$ beam:

$$\bar{g}p = \frac{4(Y_1 - Y_0)}{(-1.966Y_1 + Y_0)(3\cos^2\theta_s - 1)}$$

or

$$\bar{g}p = \frac{4(Y_0 - Y_{-1})}{(0.952Y_0 - 1.966Y_{-1})(3\cos^2\theta_s - 1)}. \tag{2}$$

The value of $\bar{g}p$ obtained from each of the four detectors and from each of the types of ratios was consistent to within the ±0.007 to ±0.011 statistical error on each determination.

The factor p was measured by a quench ratio method. If we denote the yield and c.m. angle which corresponds to a 165° (90°) detector by Y_{165} and θ_{165} (Y_{90} and θ_{90}), it follows from eq. (1) that the sum Y, where

$$Y = Y_{90} - (\eta_{90}/\eta_{165})[(3\cos^2\theta_{90} - 1)/(3\cos^2\theta_{165} - 1)]Y_{165}, \tag{3}$$

is proportional to the number of deuterons incident on the target during the counting interval, irrespective of their polarization. The factor η_{165}/η_{90} is a relative detector efficiency factor (including kinematic factors); this was determined from the ratio

$$(\eta_{165}/\eta_{90}) = (Y_{165}/Y_{90})[1 - \frac{1}{4}\bar{g}pK_a(3\cos^2\theta_{90} - 1)]/[(1 - \frac{1}{4}\bar{g}pK_a)$$

$$\times (3\cos^2\theta_{165} - 1))], \tag{4}$$

where $K_a = 1$, -1.966, or 0.952 for $m_I = 1$, 0, or -1 beams, respectively, and where the value of $\bar{g}p$ calculated from eq. (2) was used.

The data cycle consisted of five or more sets of six counting periods with the ion source conditions varied in the following sequence: $m_I = 1$ selection, quench, $m_I = 0$ selection, quench, $m_I = -1$ selection, and quench. Timed intervals were used (10 sec, 20 sec, or 40 sec) for each counting period. The source current varied by only a few percent during the acquisition of the data at a given energy, so the summing over five or more data sets is believed to assure that the amount of charge delivered in the various cases does not vary by more than a few tenths of one percent. The beam spin state purity was determined from ratios of the quantity Y with one of the spin states being selected by the spin filter and with the beam quenched, according to the formula

$$p = [Y(\text{filtered}) - Y(\text{quenched})]/Y(\text{filtered}). \tag{5}$$

Note that the data sequence employed measures the quench ratio and $\bar{g}p$ essentially simultaneously.

The final values for $\bar{g}p$, p, and $\bar{g}$ are given in table 2. Also given is $\bar{E}$, the mean energy estimated from yield versus energy data, and the corresponding c.m. angles for the detectors. Statistical errors only are shown on all quantities. We believe systematic effects could not be as large as 0.01. The results agree with preliminary results obtained in a similar experiment at this Laboratory [2], and indicate a value of $\bar{g}$ somewhat larger than was previously believed [3]. Also $\bar{g}$ appears to be independent of energy over the range studied.

Table 2

E_0 (keV)	$\bar{E}$ (keV)	θ_{90} (90° lab)	θ_{165} (165° lab)	$\bar{g}p$	p	$\bar{g}$
51	40	93.8°	165.8°	0.868 ± 0.006	0.905 ± 0.001	0.959 ± 0.006
76	58	94.1°	166.0°	0.856 ± 0.003	0.901 ± 0.001	0.950 ± 0.003
93	67	94.4°	166.1°	0.856 ± 0.004	0.902 ± 0.001	0.949 ± 0.004

REFERENCES

*Work performed under the auspices of the U.S. Atomic Energy Commission.

[1] G. G. Ohlsen, Phys. Rev. 165 (1967) 1268.
[2] G. G. Ohlsen, J. L. McKibben, and G. P. Lawrence, Bull. Am. Phys. Soc. 13 (1968) 1443.
[3] L. Brown, H. A. Christ, and H. Rudin, Nucl. Phys. 79 (1966) 459.

The T($\vec{\text{d}}$,n)^{4}He-Reaction below 1 MeV

H. A. GRUNDER, R. GLEYVOD, J. LIETZ, G. MORGAN, H. RUDIN,
F. SEILER, and A. STRICKER, University of Basel, Switzerland

The analyzing power of the T($\vec{\text{d}}$,n)^{4}He-reaction for vector and tensor
polarization have been measured at 0.2, 0.6, 0.8, and 1.0 MeV, using
a polarized deuteron source installed in the terminal of the Basel 1-
MV accelerator [1]. The beam polarization was measured by the same
reaction below 200 keV, near the 3/2$^+$-resonance. The analyzing
powers $D_i(\theta)$ and $D_{ij}(\theta)$ are shown in the figures, multiplied by
$4\pi \, \sigma_0(\theta)/\sigma_{tot}$. The dashed lines indicate angular distributions for the
s-wave 3/2$^+$-resonance element R_1 alone.

A detailed analysis yields the following results:

1. Below 1 MeV, R_1 is the dominant element.

2. At 200 keV, p-waves are significant enough to require special
precautions, when the reaction is used as an analyzer.

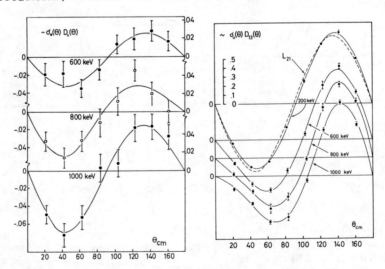

Fig. 1. D_1 and D_{13} components of the analyzing power for
polarized deuterons.

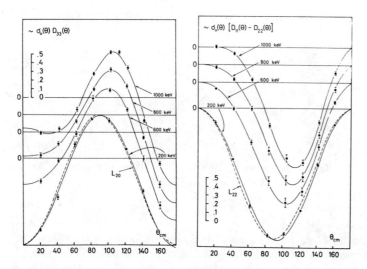

Fig. 2. D_{33} and D_{11}-D_{22} components of the analyzing power for polarized deuterons.

3. At 600 keV d-waves are appreciable, giving rise to the vector analyzing power. The only important d-wave element is the same as R_1 except for $\ell = 2$.

4. Real parts of the interference terms with R_1 or combinations thereof have been determined. A plot is given in another contributed paper [2].

5. The s-wave contribution with $J^\pi = 1/2^+$ cannot be separated from R_1 because the same reaction was used to measure the beam polarization. However, the analysis shows that below 200 keV a reduction in the analyzing power of 5 to 10% is quite possible [3].

REFERENCES

[1] H. A. Grunder et al., Helv. Phys. Acta, to be published.
[2] F. Seiler and E. Baumgartner, Third Polarization Symp.
[3] L. C. McIntyre and W. Haeberli, Nucl. Phys. A91 (1967) 369.

Neutron Polarization in the T(d,n)^{4}He Reaction Using a Polarized Incident Beam

W. B. BROSTE,[*] G. P. LAWRENCE, J. L. McKIBBEN, G. G. OHLSEN, and J. E. SIMMONS, Los Alamos Scientific Laboratory, New Mexico, USA[†]

Measurements have been made of the neutron polarization from the T(d,n)^{4}He reaction with polarized deuterons incident. This type of polarization transfer measurement was the first of its kind at Van de Graaff energies. The experiments were made possible by the recent initial operation of a high performance Lamb-shift polarized ion source on the Los Alamos FN tandem Van de Graaff accelerator, together with the availability of an efficient liquid helium neutron polarimeter.

The polarization of the outgoing neutrons was measured when the incident deuteron beam was 76% in the m_I = +1 state with respect to a quantization axis perpendicular to the horizontal reaction plane. Gammel et al. [1] have provided a formalism for the description of polarization observables in the T(d,n)^{4}He reaction; in their notation, the outgoing neutron polarization for the beam used is given by

$$\langle \sigma_y \rangle = \frac{[P(\theta) + \frac{3}{2}\langle P_y \rangle P_y^y(\theta) + \frac{1}{2}\langle P_{yy} \rangle P_{yy}^y(\theta)]}{[1 + \frac{3}{2}\langle P_y \rangle P_y^0(\theta) + \frac{1}{2}\langle P_{yy} \rangle P_{yy}^0(\theta)]}. \tag{1}$$

The quantities $\langle \sigma_y \rangle$, $P(\theta)$, $P_y^y(\theta)$, $P_{yy}^y(\theta)$, $P_y^0(\theta)$, and $P_{yy}^0(\theta)$ describing the reaction are given in a coordinate system with its y axis along the normal $n = \vec{k}_d \times \vec{k}_n$, where $\vec{k}_d$ and $\vec{k}_n$ are the directions of the deuteron and neutron momenta. For the beam chosen for the experiments $\langle P_y \rangle = \langle P_{yy} \rangle = 0.76$. $P(\theta)$, $P_y^0(\theta)$, and P_{yy}^0 (θ) were measured in separate experiments.

The deuteron beam was accelerated onto a 3-cm long gas target filled with T_2 gas at 5 atm. The neutrons were collimated and had their spins precessed ±90° before scattering from a 5.9 cm × 6.1 cm cylindrical volume of liquid helium either 99 cm or 139 cm from the target. Neutrons scattered at angles of 117° or 125°, depending on their energy, were detected in 5.1 cm × 17.8 cm × 7.6 cm NE102 pro-

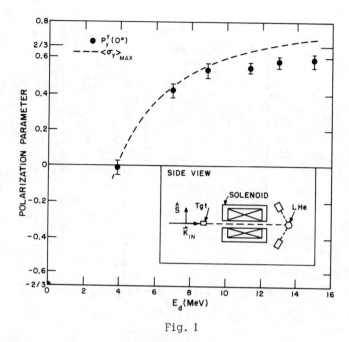

Fig. 1

ton recoil scintillators at a distance of 25.4 cm. The measured asymmetries were corrected for the effects of multiple scattering and finite geometry, then divided by n–helium polarization analyzing powers calculated from the phase shifts of Hoop and Barschall [2] to determine $\langle \sigma_y \rangle$.

For $\theta = 0$, formula (1) simplifies to

$$\langle \sigma_y \rangle = \frac{3}{2} \langle P_y \rangle P_y^y(0) / (1 + \frac{1}{2} \langle P_{yy} \rangle P_{yy}^0(1)). \tag{2}$$

Using independent measurements of P_{yy}^0, the polarization transfer coefficient $P_y^y(0)$ was determined for deuteron energies of 3.9, 7.0, 9.0, 11.4, 13.0, and 15.0 MeV. The results are given in fig. 1. Above 7.0 MeV, $P_y^y(0)$ approaches the value 2/3 as might be expected from a simplified stripping picture for the interaction. As the deuteron energy decreases, the influence of the 107 keV $J^\pi = 3/2^+$ resonance, where $P_y^y(0) = -2/3$, begins to be felt, and $P_y^y(0)$ drops to -0.015 ± 0.028 at $E_d = 3.9$ MeV. The dashed curve on fig. 1 indicates the neutron polarization $\langle \sigma_y \rangle$ which would be obtained for a 100% polarized beam. Actual values of $\langle \sigma_y \rangle$ measured for $\langle P_y \rangle = 0.76$ reached a maximum of 0.565 ± 0.036 for $E_d = 15$ MeV.

The neutron polarization was measured as a function of θ at 7.0 and 11.4 MeV incident deuteron energy. Fig. 2 illustrates the results

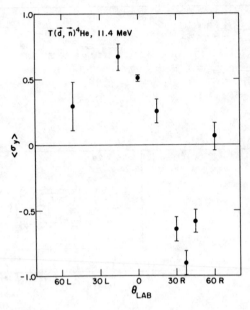

Fig. 2

at 11.4 MeV. Neutron polarization is reckoned positive in the direction along the deuteron spin quantization axis. The data at 7 MeV are very similar to the 11.4 MeV results shown. Although the quantities $P(\theta)$, $P_y^0(\theta)$, and $P_{yy}^0(\theta)$ have been determined, further measurements would be required to obtain values for $P_y^y(\theta)$ and $P_{yy}^y(\theta)$. Nevertheless, the appearance of the very large negative polarization at $\theta = 37\ 1/2°$ right leads one to believe that at this point the reaction is dominated by a spin flip mechanism, in contrast to the almost complete non-spin flip at zero degrees.

REFERENCES

*Assoc. Western Universities Fellow from the University of Wyoming.
†Work performed under the auspices of the U.S. Atomic Energy Commission.

[1] J. L. Gammel, P. W. Keaton, Jr., and G. G. Ohlsen, Third Polarization Symp.
[2] B. Hoop and H. H. Barschall, Nucl. Phys. 83 (1966) 65.

DISCUSSION

McKee:

If we look at this reaction in a simple way, an interesting fact emerges as a result of the requirement that we form a ground state α particle. This is the fact that irrespective of the interaction mechanism the outgoing neutron at 0° must have its spin in the same direction as the incident deuteron, if the incident beam has 100% vector polarization. With such a simple picture in the absence of spin-flip interactions and resonances in the five-body system, 100% transfer of polarization might be expected. Could you comment on this?

Broste:

It is true that in a simple stripping argument one perhaps expects 100% polarization transfer, which means $P_y^y = 2/3$. The data have a trend toward this value as the deuteron energy increases.

Simmons:

There is one condition where the outgoing neutron polarization in the $T(\vec{d},\vec{n})^4He$ reaction is determined completely by angular momentum considerations. At 0° when the incoming deuteron is in the $m = 1$ state with respect to the direction of motion, the outgoing neutron has the same value of longitudinal polarization as the incident beam polarization. This results from the fact that at 0° the Legendre functions that enter the problem have $m = 0$; conservation of spin components then leads to the above conclusion. This point is discussed by Ohlsen et al. elsewhere in the Proceedings.

Cranberg:

How does the polarized neutron source you describe compare with the T(d,n) source using unpolarized projectiles?

Broste:

According to the figure of merit P^2I, the polarization transfer technique will become competitive with polarized beam currents of 0.1 to 0.2 μA. Other considerations make the polarization transfer method attractive even with reduced P^2I.

Quadratic Relations between the $T(\vec{d},\vec{n})^4$He Observables

G. G. OHLSEN, P. W. KEATON, Jr., and J. L. GAMMEL, Los Alamos
Scientific Laboratory, New Mexico, USA*

There are altogether 18 observables which may be measured for the
$T(\vec{d},\vec{n})^4$He polarization transfer reaction, in the case of an unpolarized
tritium target, at each energy and angle (if one assumes conservation
of parity). These are the differential cross section $I_0(\theta)$, four analyzing
tensors $P_y^0(\theta)$, $P_{xz}^0(\theta)$, $1/2[P_{xx}(\theta) - P_{yy}(\theta)]$, and $P_{zz}^0(\theta)$, the polarization
produced by an unpolarized neutron beam, $P_0^{y'}(\theta)$, and twelve "polari-
zation transfer coefficients" as defined in an accompanying paper [1].
In general, the lower subscript indicates the polarization of the incident deuteron and the superscript indicates the neutron polarization
component. Two right-handed coordinate systems with $\vec{k}_{in} \times \vec{k}_{out}$
defining both the y and y' axes are assumed; the xz and x'z' axes are
arbitrary but are usually chosen with z along $\vec{k}_{in}$ and z' along $\vec{k}_{out}$.
Since the scattering matrix, M, which describes the reaction has only
six complex elements, clearly these eighteen experiments are not independent. No linear relations between the observables exist; however, in table 1 twelve *quadratic* relations between the observables
are presented, so that at most six of the experimental observables
are truly independent.

These relations might be expected to serve several useful purposes.
First, each relation is in itself a parity test. Second, these relations
may be applied as consistency checks between sets and types of data
when data of the polarization transfer type become more readily available. (The relations may be regarded as defining the volume in an
eighteen dimensional hyperspace which can be physically "occupied"
by the data sets.) Finally the relations may serve as a guide, in some
cases, as to which of the various experiments that might be performed
are least related to the experiments already completed and, therefore,
contain the greatest amount of new information. Also, the relations
show, at least in principle, that some of the experiments can be
omitted altogether without sacrifice of information.

In table 1, the coefficients which are underlined are odd in the
scattering angle, θ, and therefore vanish at 0° (180°). Using also the

Table 1. Quadratic relations

1. $\left(\frac{3}{2}P_x^{x'}\right)^2 + \left(P_{yz}^{x'}\right)^2 + \left(\frac{3}{2}P_x^{z'}\right)^2 + \left(P_{yz}^{z'}\right)^2 = (1-P_{yy}^0)(1-P_{zz}^0) - (P_0^{y'} - P_{yy}^{y'})(P_0^{y'} - P_{zz}^{y'})$

2. $\left(\frac{3}{2}P_z^{x'}\right)^2 + \left(P_{xy}^{x'}\right)^2 + \left(\frac{3}{2}P_z^{z'}\right)^2 + \left(P_{xy}^{z'}\right)^2 = (1-P_{xx}^0)(1-P_{yy}^0) - (P_0^{y'} - P_{xx}^{y'})(P_0^{y'} - P_{yy}^{y'})$

3. $\left(\frac{3}{2}P_y^0\right)^2 + \left(P_{xz}^0\right)^2 + \left(\frac{3}{2}P_y^{y'}\right)^2 + \left(P_{xz}^{y'}\right)^2 = (1-P_{xx}^0)(1-P_{zz}^0) + (P_0^{y'} - P_{xx}^{y'})(P_0^{y'} - P_{zz}^{y'})$

4. $\frac{3}{2}P_x^{x'} P_{yz}^{z'} - \frac{3}{2}P_x^{z'} P_{yz}^{x'} = \frac{1}{2}(1-P_{yy}^0)(P_0^{y'} - P_{zz}^{y'}) - \frac{1}{2}(1-P_{zz}^0)(P_0^{y'} - P_{yy}^{y'})$

5. $\left(\frac{3}{2}\right)^2 P_y^0 P_y^{y'} + P_{xz}^0 P_{xz}^{y'} = \frac{1}{2}(1-P_{zz}^0)(P_0^{y'} - P_{yy}^{y'}) + \frac{1}{2}(1-P_{xx}^0)(P_0^{y'} - P_{zz}^{y'})$

6. $\frac{3}{2}P_z^{x'} P_{xy}^{z'} - \frac{3}{2}P_z^{z'} P_{xy}^{x'} = -\frac{1}{2}(1-P_{yy}^0)(P_0^{y'} - P_{xx}^{y'}) + \frac{1}{2}(1-P_{xx}^0)(P_0^{y'} - P_{yy}^{y'})$

7. $\frac{3}{2}P_x^{x'} P_{xz}^0 + \frac{3}{2}P_y^0 P_{yz}^{x'} - \left(\frac{3}{2}\right)^2 P_x^{z'} P_y^0 + \frac{3}{2}P_z^{z'} P_{yz}^{y'} = -P_z^{z'}(P_0^{y'} - P_{zz}^{y'}) + \frac{3}{2}P_z^{x'}(1-P_{zz}^0)$

8. $\left(\frac{3}{2}\right)^2 P_x^{x'} P_x^0 - P_{yz}^{x'} P_{yz}^0 + \frac{3}{2}P_z^{z'} P_{xz}^{x'} + \frac{3}{2}P_y^{y'} P_{yz}^{z'} = \frac{3}{2}P_z^{z'}(P_0^{y'} - P_{zz}^{y'}) + P_{xy}^{x'}(1-P_{zz}^0)$

9. $\left(\frac{3}{2}\right)^2 P_x^{x'} P_x^{z'} - P_{yz}^{x'} P_{yz}^{z'} + \left(\frac{3}{2}\right)^2 P_x^0 P_z^{z'} = -P_{xz}^{y'}(P_0^{y'} - P_{yy}^{y'}) + P_{xz}^0(1-P_{yy}^0)$

10. $\frac{3}{2}P_x^{x'} P_{xy}^0 + \frac{3}{2}P_z^{x'} P_{yz}^{x'} + \frac{3}{2}P_z^{z'} P_{xy}^{x'} = -\frac{3}{2}P_y^{y'}(P_0^{y'} - P_{yy}^{y'}) + \frac{3}{2}P_y^0(1-P_{yy}^0)$

11. $\frac{3}{2}P_z^{x'} P_{xz}^0 + \frac{3}{2}P_z^0 P_{xy}^{x'} + \left(\frac{3}{2}\right)^2 P_y^{y'} P_z^{z'} = P_z^{z'}(P_0^{y'} - P_{xx}^{y'}) + \frac{3}{2}P_x^{x'}(1-P_{xx}^0)$

12. $P_{xz}^0 P_{xy}^{x'} - \left(\frac{3}{2}\right)^2 P_y^0 P_z^{x'} + \frac{3}{2}P_y^{y'} P_z^{z'} = \frac{3}{2}P_z^{z'}(P_0^{y'} - P_{xx}^{y'}) - P_{yz}^{x'}(1-P_{xx}^0)$

zero degree relations $P^{z'}_{xy} = 0$, $P^{y'}_{xz} = -P^{x'}_{yz}$, $P^{x'}_{x} = P^{y'}_{y}$, and $P^{0}_{xx} = P^{0}_{yy}$, these twelve equations reduce to only two independent relations at $0°$ $(180°)$:

$$(\frac{3}{2}P^{y'}_{y})^2 + (P^{y'}_{xz})^2 = (1 - P^{0}_{xx})(1 - P^{0}_{zz});$$

$$\frac{3}{2}P^{z'}_{z} = 1 - P^{0}_{xx} \equiv 1 + \frac{1}{2}P^{0}_{zz}.$$

The first relation shows that, at $0°$, it would never be necessary to measure the geometrically complicated quantity $P^{y'}_{xz}$. The second relation can be shown to follow from angular momentum conservation alone; it is interesting in that it implies that the outgoing longitudinal neutron polarization is 100% for a 100% $m_I = 1$ ($\langle P_z \rangle = \langle P_{zz} \rangle = 1$) deuteron beam. This conclusion is independent of all dynamics.

REFERENCE

*Work performed under the auspices of the U.S. Atomic Energy Commission.

[1] J. L. Gammel, P. W. Keaton, Jr., and G. G. Ohlsen, Third Polarization Symp.

Comparison of the Vector Analyzing Power
of the T($\vec{d}$,n) Reaction
with the Mirror Reaction ³He($\vec{d}$,p)

J. C. DAVIS, D. HILSCHER,* and P. A. QUIN, University of Wisconsin, Madison, USA†

Comparison of the polarization of the outgoing nucleon in the mirror reactions T(d,n) and ³He(d,p) [1] indicates that the magnitude of the neutron polarization is systematically smaller than the proton polarization. Similar differences are observed for the D(d,n) and D(d,p) reactions. However, the polarization measurements are sufficiently difficult that experimental effects cannot be disregarded as a possible cause of these differences. Although measurements of the vector analyzing power of mirror reactions do not yield the same information as the polarization experiments, the experiments are considerably easier.

We report measurements of the angular distribution of the vector analyzing power for the T($\vec{d}$,n) reaction at 6, 10, and 11.3 MeV. This work is compared to results for the ³He($\vec{d}$,p) reaction obtained recently at this laboratory [2]. The deuteron-energy spread has been chosen to be the same. The vector analyzing power $\langle iT_{11} \rangle$ was determined from the left-right asymmetry observed when a T_2 gas target was bombarded with the vector-polarized deuteron beam from a polarized ion source. Asymmetry in neutron yield was measured with two 2.5 cm × 2.5 cm scintillators with angular acceptance < 3°. The deuteron-beam polarization was determined from d-α scattering [3] and was monitored during all runs by two large scintillators placed at $\theta_{lab} = 135°$. Pulse-shape gamma discrimination was used for all four neutron detectors.

The results obtained at 6 and 10 MeV are compared in fig. 1 to the ³He($\vec{d}$,p) data of Plattner and Keller [2]. The solid line is a smooth curve through the ³He($\vec{d}$,p) data points. At 11.3-MeV deuteron bombarding energy the c.m. energy in the T(d,n) exit channel equals the c.m. energy in the ³He(d,p) exit channel to 10-MeV deuteron bombarding energy. There is a slight shift toward forward angles in the analyzing power at 11.3 MeV, but no appreciable differences in the magnitude of the analyzing power. The vector analyzing powers of the T($\vec{d}$,n) and ³He($\vec{d}$,p) reactions do not show the large differences observed in the nucleon polarization measurements.

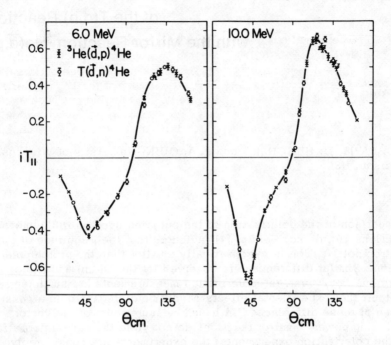

Fig. 1

REFERENCES

*NATO Fellow on leave from Hahn-Meitner Institute, Berlin, Germany.
†Work supported in part by the U.S. Atomic Energy Commission.

[1] H. H. Barschall, Second Polarization Symp., p. 393.
[2] G. R. Plattner and L. G. Keller, Phys. Lett. 29B (1969) 301.
[3] L. G. Keller, Ph.D. Thesis, University of Wisconsin (1969).

DISCUSSION

Keaton:
    We observed a discrepancy between our ^{3}He($\vec{\text{d}}$,p)^{4}He analyzing tensor P_y^0 and Plattner's. This may suggest that the T($\vec{\text{d}}$,n)^{4}He and ^{3}He($\vec{\text{d}}$,p)^{4}He vector analyzing powers are the same, over the total angular range, and that there is an error in Plattner's results for a few back-angle points. In general, it seems that the beam polarization as determined at LASL agrees with that at Wisconsin to less than 2%, perhaps 1%.

J. C. Davis:

We determined our beam polarization exactly as Plattner and Keller did via ^{4}He($\vec{\text{d}}$,d)^{4}He elastic scattering. Thus it is unlikely that there would be a systematic difference between our results and theirs because of an error in the value of the beam polarization. At 10 MeV we do find slightly larger values at back angles in agreement with your results. However, we have too few data points to conclude anything from this difference. In any case, these discrepancies are smaller than the effects we were looking for on the basis of the differences between the polarization of emitted protons and neutrons.

Plattner:

Our d + ^{3}He data were obtained two years ago with the old Lamb-shift source on the Wisconsin tandem accelerator which used to have some fluctuations in the beam polarization because of instability of the Cs-pressure. Thus it is possible that some of the discrepancies between the more recent data and our work are due to such fluctuations which might have gone unnoticed because we did not monitor the beam polarization continuously at that time.

Analysis of the T(d,n)⁴He- and
³He(d,p)⁴He-Reactions below 1 MeV

F. SEILER and E. BAUMGARTNER, University of Basel, Switzerland

Polarization experiments with the T(d,n)⁴He- and ³He(d,p)⁴He-reactions can be described by real efficiencies $C_{q,k,Q,K}^{q',k',0,0}(\theta)$, with beam and target polarizations given by tensor moments $t_{q,k}$ and $t_{Q,K}$, respectively, and primes referring to the outgoing channel [1]. In the general expansion

$$C_{q,k,Q,K}^{q',k',0,0}(\theta) = \sum_L d_{k',k+K}^{(L)}(\theta) \cdot a_{q,k,Q,K}^{q',k'}(L)$$

the d-functions can usually be replaced by Legendre functions when dealing with only one or two polarizations. The coefficients a(L) can be expressed in terms of (ℓ, s, J)-matrix elements R_i

$$a_{q,k,Q,K}^{q',k'}(L) = \sum_{\mu,\nu \geq \mu} a_{q,k,Q,K}^{q',k'}(L,\mu,\nu) \cdot \begin{Bmatrix} \text{Re} \\ \text{Im} \end{Bmatrix} (R_\mu R_\nu^*)$$

$$\text{for } q + Q + q' = \begin{Bmatrix} \text{even} \\ \text{odd} \end{Bmatrix}.$$

The region below 1 MeV is dominated by the $3/2^+$-resonance formed by s-waves (element R_1). Admitting only small additional contributions with $\ell, \ell' \leq 2$ and neglecting the small elements with $J^\pi = 5/2^+$, the calculation is restricted to 9 terms $(R_1 R_i^*)$. Experimental values for $a_{q,k,Q,K}^{q',k'}(L)$ give linear equations in the 16 quantities $M_i = \text{Re}(R_1 R_i^*)/|R_1|^2$ and $N_i = \text{Im}(R_1 R_i^*)/|R_1|^2$, subject to linear relations useful in checking the consistency of the set of elements used [1,2]. On the basis of this model, several "complete sets" of experiments can be laid out, which permit the separation of all M_i and N_i. A representative set is given in table 1, the column ³He(d,p)⁴He reflecting the experimentally more favorable alternative with a polarized target. Information from an incomplete set appears in the plot, taken from the work of Grunder et al. [2] on the T(d,n)⁴He-reaction. The terms M_3 to M_6, involving $\ell = 1$ matrix elements, appear in combinations. M_2 con-

EXPERIMENT involves	T(d,n)^{4}He	^{3}He(d,p)^{4}He
no polarization	$C_{0,0}$	$C_{0,0}$
one polarization	$C_{1,1}$	$C_{1,1}$
	$C_{2,k}$ $_{k=0,1,2}$	$C_{2,k}$
	$C_{0,0}^{1,1}$	$C_{0,0,1,1}$
two polarizations	$C_{1,0}^{1,1}$	$C_{1,0,1,1}$
	$C_{2,0}^{1,1}$	$C_{2,0,1,1}$
"complete set"	$C_{2,\pm2}^{1,1}$ or $C_{2,2}^{1,0}$	$C_{2,2,1,0}$
	$C_{1,1}^{1,1}$ or $C_{1,1}^{1,0}$	$C_{1,1,1,0}$

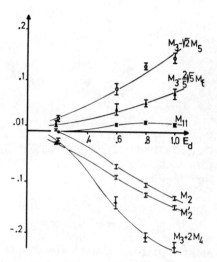

Table 1. Complete set of experiments.

Fig. 1. Matrix element combinations for the T(d,n)^{4}He reaction.

tains the s-wave $J^\pi = 1/2^+$-element and M_{11} the only significant d-wave contribution R_{11}, identical to R_1 except for $\ell = 2$. These two terms are of special interest, since they alter $C_{2,0}(\theta)$ used to determine deuteron polarizations. The numerical influence is difficult to determine in the absence of another analyzer at these energies. A 5% decrease would yield the curve M_2' in the plot, leaving others almost unchanged. The energy dependence of the s-wave parameter M_2' is more plausible than that of M_2 in view of the behavior of the p-wave contributions. On the basis of presently available data [2, 3], the analyzing power is still in doubt by about 5-10%.

REFERENCES

[1] F. Seiler and E. Baumgartner, Nucl. Phys., in print.
[2] H. A. Grunder et al., Helv. Phys. Acta, to be published, and Third Polarization Symp.
[3] P. Huber et al., Third Polarization Symp.; Ch. Leemann et al., Helv. Phys. Acta, to be published, and Third Polarization Symp.; L. C. McIntyre and W. Haeberli, Nucl. Phys. A91 (1967) 369.

A Study of ^{3}He($\vec{d}$,d)^{3}He Scattering
at 10 and 12 MeV

D. C. DODDER, D. D. ARMSTRONG, P. W. KEATON, Jr., G. P. LAWRENCE,
J. L. McKIBBEN, and G. G. OHLSEN, Los Alamos Scientific Laboratory,
New Mexico, USA*

We have measured the vector and tensor analyzing power for d-^{3}He
elastic scattering at 10 and 12 MeV. A "cube" scattering chamber
enabled us to measure the left, right, up, and down yield simultaneously for any polar angle in the range 20° to 160°. An angular resolution of 2° (FWHM) was used. A complete measurement consists of
four counting periods with the chamber rotated around the beam direction by 90° between periods; this procedure removes detector efficiency problems. The deuteron quantization axis was at 54.7° with
respect to the beam direction (in the horizontal plane) for the simultaneous P_y^0, P_{xz}^0, and $(P_{xx}^0 - P_{yy}^0)/2$ measurements and parallel to the
beam direction for the P_{zz}^0 measurements [1]. Only the P_{zz}^0 measurements depend upon accurate current integration. $m_I = 1$ deuterons
($p_Z = p_{ZZ} \approx 0.8$) were used for most of the measurements; $m_I = 0$ deuterons ($p_Z \approx 0$, $p_{ZZ} \approx -1.6$) were used occasionally to check the alignment
of the chamber. False asymmetries were kept small by minimizing the
beam on a properly suppressed collimator which preceded the Faraday
cup. With the present technique P_{xz}^0 is about 2-1/2 times more sensitive to false asymmetries than is P_y^0; for this reason, larger errors
are assumed for P_{xz}^0.

Silicon surface barrier E-ΔE detector telescopes were used to permit mass identification. An on-line computer recorded 512 channel
spectra for one or two masses for each telescope. The computer also
measured the beam polarization (quench method) periodically, rotated
the cube, appropriately summed the spectra, and calculated the values
of the analyzing vector and tensors.

The data obtained are shown in figs. 1 and 2. The P_y^0 data of Plattner and Keller are also shown [2]. All errors shown are the estimated
systematic uncertainties. A preliminary phase-shift fit for the 10-MeV
data is also shown. Since 51 parameters are involved (77 data points)
a meaningful fit can probably be obtained only if data of other types
(^{3}He($\vec{d}$,p)^{4}He, ^{3}He($\vec{d}$,p)^{4}He) is included. Such an analysis is in progress [3].

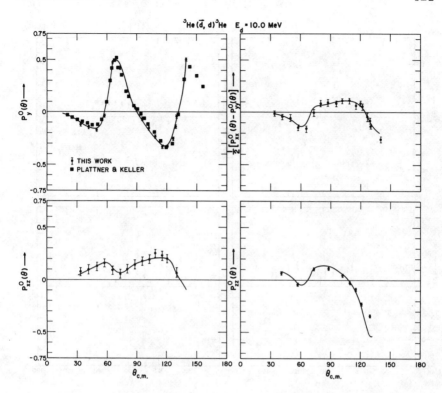

Fig. 1. Vector and tensor analyzing power for $^3\mathrm{He}(\vec{d},d)^3\mathrm{He}$ elastic scattering at 10 MeV.

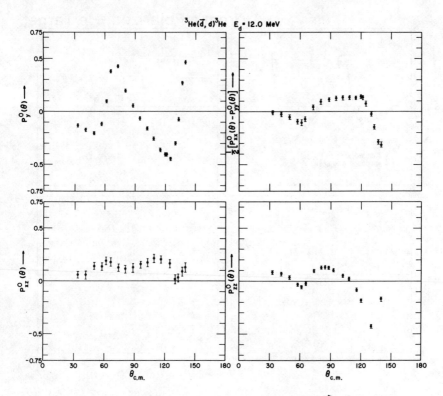

Fig. 2. Vector and tensor analyzing power for ³He($\vec{d}$,d)³He elastic scattering at 12 MeV.

REFERENCES

*Work performed under the auspices of the U.S. Atomic Energy Commission.

[1]   G. P. Lawrence et al., Third Polarization Symp.
[2]   G. R. Plattner and L. G. Keller, Phys. Lett. 29B (1969) 301.
[3]   G. M. Hale, D. C. Dodder, and K. Witte, Third Polarization Symp.

Asymmetries in Deuteron Scattering
from a Polarized ³He Target

B. E. WATT and W. T. LELAND, Los Alamos Scientific Laboratory,
New Mexico, USA[*]

The experiment described herein was performed on the Los Alamos
tandem Van de Graaff accelerator using a polarized ³He gas target
and an unpolarized deuteron beam. Fig. 1 shows schematically the
target cell arrangement. Basically, the cell consists of a pyrex vessel to which appropriate aluminum windows have been attached to
permit passage of the primary beam and to allow scattered or reaction
particles to exit. The primary deuteron beam is collimated to a diameter of 1/8-in. by means of apertures ahead of the cell, passes through
entrance and exit windows of 0.0003-in. aluminum, and is stopped
and monitored with a deep Faraday cup. Scattered deuterons exit
through 0.0005-in. aluminum windows and are detected with silicon
solid state detectors. The angular position of the scattered deuterons
is defined by slits, one near the exit window (not shown) and the
other at the detector. Polarization of the ³He was accomplished by
an optical pumping technique which was developed by Colegrove et
al. [1]. The degree of polarization was calculated from the relation

$$\frac{\Delta I}{I} = \frac{P}{3+P^2} [6 - 2P + 3(1 - P^2)k],$$

where ΔI is the difference in amount of light absorbed by ³He metastables with and without polarization, I is the amount of light absorbed by ³He metastables, P is the nuclear polarization, and k is a
constant which depends on various transition probabilities and the
relative degree of illumination of the F = 1/2 and F = 3/2 levels. Several investigators [1−5] have discussed the relation between P and
the observed optical signals. We have chosen k = 0.3 which agrees
with the measurements made by Hauer and Klinger [3].

A typical point was obtained by positioning the up and down counters
at the appropriate angle and then making two runs with a reversal of
the polarization between runs. The scattering asymmetry was then
calculated from the formula

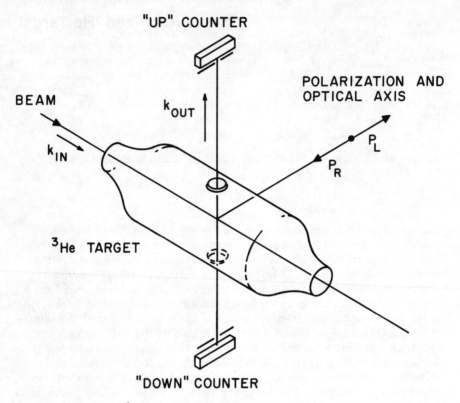

Fig. 1. Polarized ^{3}He target cell.

$$A_S = [(1-R)/(1+R)]\,\frac{1}{2}\left(\frac{1}{P_L} + \frac{1}{P_R}\right),$$

where $R \equiv \sqrt{U_L D_R / D_L U_R}$,

   U_L = "up" detector count with left circular polarization,

   U_R = "up" detector count with right circular polarization,

   D_L = "down" detector count with left circular polarization,

   D_R = "down" detector count with right circular polarization,

   P_L = ^{3}He target polarization, left,

   P_R = ^{3}He target polarization, right.

By combining measurements where only the direction of the polarization is changed we are able to cancel out any asymmetries that are spin independent. Typical target polarizations were in the range 0.17 to 0.20. Tables 1 and 2 tabulate our results for 9.9 MeV and 11.9 MeV deuterons, respectively. Positive asymmetry is reported when a majority of the scattered particles have $\vec{k}_{in} \times \vec{k}_{out}$ parallel to P. Error estimates are based entirely on counting statistics.

Table 1. Deuteron scattering asymmetries for 9.9 MeV

$\theta_{c.m.}$	A_S
78.2	0.139 ± 0.02
83.5	0.087 ± 0.019
88.6	0.004 ± 0.015
94.1	-0.059 ± 0.014
102	-0.121 ± 0.014
109	-0.209 ± 0.018
115	-0.273 ± 0.022
121	-0.242 ± 0.022

Table 2. Deuteron scattering asymmetries for 11.9 MeV

$\theta_{c.m.}$	A_S
78.2	0.285 ± 0.031
	0.29 ± 0.06
83.4	0.079 ± 0.028
89	0.030 ± 0.019
94	-0.030 ± 0.015
100	-0.11 ± 0.026
102	-0.159 ± 0.022
	-0.164 ± 0.038
109	-0.188 ± 0.027
115	-0.282 ± 0.026
121	-0.361 ± 0.031
127	-0.327 ± 0.077

REFERENCES

*Work performed under the auspices of the U.S. Atomic Energy Commission.

[1] F. D. Colegrove, L. D. Schearer, and G. K. Walters, Phys. Rev. 132 (1963) 2561.
[2] R. C. Greenhow, Phys. Rev. 136 (1964) A660.
[3] W. Klinger, Ph.D. Thesis, Friedrich-Alexander-Universität, Erlingen-Nürnberg (1968).
[4] N. Hauer and W. Klinger, Jahresbericht der Arbeitsgruppen am Institut Glückstrasse (Physikalisches Institut Universität, Erlangen-Nürnberg, 1967/68) p. 105.
[5] S. D. Baker, D. H. McSherry, and D. O. Findley, Phys. Rev. 178 (1969) 1616.

Measurement of the Tensor Moments
$\langle iT_{11}\rangle$, $\langle T_{20}\rangle$, $\langle T_{21}\rangle$, and $\langle T_{22}\rangle$
in d-^{3}He Elastic Scattering

V. KÖNIG, W. GRÜEBLER, R. E. WHITE, P. A. SCHMELZBACH, and
P. MARMIER, Laboratorium für Kernphysik, ETH, Zürich, Switzerland

In addition to our polarization measurements on the reaction ^{3}He(d,p)^{4}He,
we have measured polarization in d-^{3}He elastic scattering. The only
previous polarization measurements on the elastic scattering of deu-
terons on ^{3}He is the determination of $\langle iT_{11}\rangle$ by Plattner and Keller [1].
The present work was performed using the polarized deuteron beam
from the ETH tandem accelerator. Four angles were measured simultaneously. The method used in the determination of the polarization
parameters was the same as in our ^{4}He(d,d)^{4}He measurement [2]. This
reaction was also used as an analyzer for the polarized beam. Angular
distributions were measured at c.m. angles between 33.2° and 140° at
energies of 6, 8, and 10 MeV. For the vector parameter, measurements
were also performed at 4 and 5 MeV.

The results are shown in fig. 1. Open circles represent measurements in which the recoil-^{3}He-particles were detected; triangles indicate the values measured by Plattner and Keller. Where no error bars
are shown, the statistical error is smaller than the dots. In general,
the tensor parameters are small, except $\langle T_{20}\rangle$ which can attain large
values at backward angles. For the vector parameter $\langle iT_{11}\rangle$ our measurement shows, especially at backward angles, deviations from the
measurement ref. [1]. The data presented here are preliminary measurements and will be extended in the near future to include lower energies and a larger range of angles. A phase shift analysis will also be
carried out.

REFERENCES

[1] G. R. Plattner and L. G. Keller, Phys. Lett. 29B (1969) 301.
[2] W. Grüebler, V. König, P. A. Schmelzbach, and P. Marmier, Nucl.
     Phys. A134 (1969) 686, Nucl. Phys. A148 (1970) 380, Nucl. Phys.
     A148 (1970) 391.

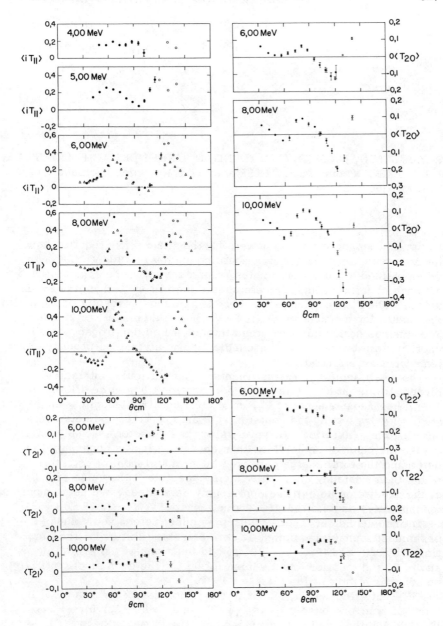

Fig. 1. Tensor moments in d-³He elastic scattering ●●● deuterons, ○○○ recoil-³He-particles, △△△ ref. [1].

A Study of the ^{3}He(d,p)^{4}He Reaction

P. W. KEATON, Jr., D. D. ARMSTRONG, D. C. DODDER, G. P. LAWRENCE,
J. L. McKIBBEN, and G. G. OHLSEN, Los Alamos Scientific Laboratory,
New Mexico, USA*

Vector and tensor analyzing powers for the ^{3}He($\vec{d}$,p)^{4}He reaction have
been measured at a deuteron bombarding energy of 10.0 MeV. For the
data obtained at laboratory angles greater than 20°, the apparatus and
methods of refs. [1] and [2] were used. The slit system for the E-ΔE
detector pairs were arranged for two-degree angular resolutions (FWHM).
The scattering chamber was filled with ^{3}He to a pressure of 200 Torr.
The ^{3}He was periodically recycled through a liquid-nitrogen-cooled
charcoal trap by means of a mechanical pump so that no impurity problems were experienced.

The small angle P^0_{zz} measurements are of particular interest with
respect to the interpretation of 0° polarization transfer coefficients.
Therefore, measurements of P^0_{zz} at 5°, 10°, and 15° (laboratory angle)
were made by setting the deuteron spin axis, β, to 0° and determining
$m_I = 1$ to $m_I = 0$ ratios. However, since the detector assembly intercepted the beam at angles less than about 10°, it was necessary to
determine the total charge delivered by means of monitor detectors
rather than a faraday cup. This is only possible if the value of P^0_{zz}
at the monitor angle has been previously determined. We use a monitor laboratory angle of 45° (53.3° c.m.) where P^0_{zz} is −0.378 ± 0.015.

The present data, together with previously determined values [3, 4]
of the vector analyzing power, P^0_y, are presented in fig. 1. The functions P^0_y, P^0_{xz}, and $(P^0_{xx} - P^0_{yy})/2$ are odd functions of the scattering
angle, θ, and therefore must vanish for $\theta = 0$. The even function, $P^0_{zz}(\theta)$,
approaches a large value (−1.4) at $\theta = 0°$.

The scattering matrix, M, for this reaction can be written [5, 6] in
terms of spin-one base vectors, $\chi^\dagger_1 = (1, 0, 0)$, $\chi^\dagger_0 = (0, 1, 0)$, $\chi^\dagger_{-1} =
(0, 0, 1)$, and the Pauli spin matrices, $\vec{\sigma}$. The M matrix becomes

$$M = A(\vec{\chi}^\dagger \cdot \hat{n}) + B(\vec{\chi}^\dagger \cdot \hat{n})(\vec{\sigma} \cdot \hat{n}) + C(\vec{\chi}^\dagger \cdot \hat{n})(\vec{\sigma} \cdot \hat{p}') + D(\vec{\chi}^\dagger \cdot \hat{p})(\vec{\sigma} \cdot \hat{k}') +$$

$$E(\vec{\chi}^\dagger \cdot \hat{k})(\vec{\sigma} \cdot \hat{p}') + F(\vec{\chi}^\dagger \cdot \hat{k}')(\vec{\sigma} \cdot \hat{k}'),$$

where $\hat{k}(\hat{k}')$ is along $\vec{k}_{in}(\vec{k}_{out})$, $\hat{n}$ is along $\vec{k}_{in} \times \vec{k}_{out}$, and $\hat{p}(\hat{p}')$ is along $\hat{n} \times \hat{k}(\hat{n} \times \hat{k}')$. The spatial components of $\vec{\chi}$ are $\vec{\chi} \cdot \hat{n} = i(\chi_1 + \chi_{-1})/\sqrt{2}$, $\vec{\chi} \cdot \hat{p} = (-\chi_1 + \chi_{-1})/\sqrt{2}$, and $\vec{\chi} \cdot \hat{k} = \chi_0$.

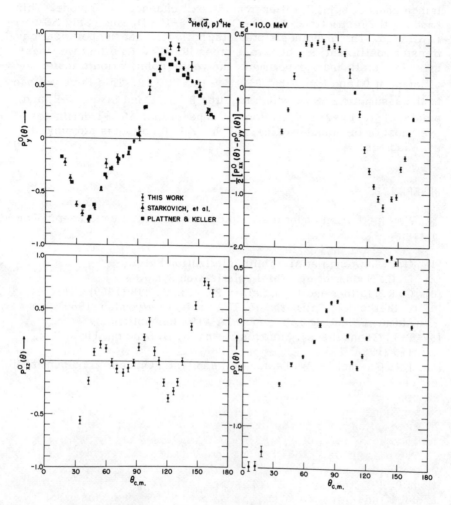

Fig. 1. Measured vector and tensor analyzing power for the $^3\text{He}(\vec{d},p)^4\text{He}$ reaction at 10 MeV.

It has been previously noticed [3] that P_y^0 in the present reaction has an angular dependence similar to the analyzing power observed in $^3\text{He}(d,p)^4\text{He}$ (polarized target) experiments. This condition requires the following relation among the elements of M:

$$4\text{ReAB}^* - 4\text{ImCD}^* - 4\text{ImEF}^* = -4\text{ImCE}^* - 4\text{ImDF}^*.$$

This relation is satisfied non-dynamically by insisting that the entrance channel spin, S, differ from the exit channel spin by one. This imposes the restrictions $F = -(B+C)$ and $E = -iA+D$, since only $\Delta S = -1$ is allowed. This hypothesis predicts, however, that the polarization transfer coefficient, P_y^y, at zero degrees is $-2/3$. As a test we measured, in a preliminary experiment, the vector polarization transfer coefficient at $0°$ and found that $P_y^y(0°) = +0.57 \pm 0.10$. This is very close to the assumption $\Delta S = 0$ which requires $B = C = F$ and $iA = -D = -E$, and predicts $P_y^y(0°) = +2/3$. Therefore, it appears that $\Delta S = -1$ dominates over most of the angular range, but that $\Delta S = 0$ competes strongly at zero degrees.

REFERENCES

*Work performed under the auspices of the U.S. Atomic Energy Commission.

[1]   G. P. Lawrence et al., Third Polarization Symp.
[2]   D. C. Dodder et al., Third Polarization Symp.
[3]   G. R. Plattner and L. G. Keller, Phys. Lett. 29B (1969) 301.
[4]   V. Starkovich, Ph.D. Thesis, University of Wyoming (1969); Los Alamos Scientific Lab. Report LA-4191 (unpublished).
[5]   P. L. Csonka, M. J. Moravcsik, and M. D. Scadron, Phys. Rev. 143 (1966) 775.
[6]   J. L. Gammel, P. W. Keaton, Jr., and G. G. Ohlsen, to be published.

Polarization Phenomena in the
^{3}He(d,p)^{4}He Reaction

N. I. ZAIKA, A. V. MOKHNACH, O. F. NEMETS, P. L. SHMARIN, and
A. M. YASNOGORODSKY, Institute for Nuclear Research, Ukrainian
Academy of Sciences, Kiev, USSR

Studies of the ^{3}He(d,p)^{4}He reaction with polarized deuterons carried
out at our Institute in 1966–67 demonstrated that the reaction is suitable as an analyzer of deuteron vector polarization [1]. To obtain
additional information that could help to explain the reaction we also
measured the proton asymmetries with a polarized deuteron beam [2]
$A_d(\theta) = P_d^{-1} \cdot [\sigma_L(\theta) - \sigma_R(\theta)]/[\sigma_L(\theta) + \sigma_R(\theta)]$ at three deuteron energies
(6, 7.7, and 10 MeV), where the proton polarizations [3] $P(\theta)$ and the
asymmetries with a polarized target [4] $A_T(\theta)$ were known.

Polarized deuterons were obtained by elastic scattering of 13.6-
MeV deuterons by carbon at 25° (lab). The deuteron energy was varied
by foils. The beam polarization parameters measured by different
methods [5] were: $P_d = (2/\sqrt{3}) \langle iT_{11} \rangle = -0.22 \pm 0.02; \langle T_{20} \rangle = -0.02 \pm 0.057;$
$\langle T_{21} \rangle = 0.028 \pm 0.024; \langle T_{22} \rangle = 0.022 \pm 0.043$, i.e., one may consider the
deuterons as only vector polarized.

The results of the $A_d(\theta)$ measurements* are shown in fig. 1 and have
the following characteristics: 1) $A_d(\theta)$ and $A_T(\theta)$ are in phase and
have the same sign; 2) $A_d/A_T = 3/2$, at least near maxima; 3) $A_d(\theta)$
and $P_p(\theta)$ have opposite sign at almost all angles, and their zero-
crossing angles are different; 4) the simple relations between $A_d(\theta)$
and $P_p(\theta)$, resulting from DWBA analysis for $\ell_n = 0$ stripping with
entrance or exit channel spin-orbit distortions ($A_d = (3/2)P_p$ or $A_d = P_p$ respectively [7]), do not hold. In comparison with other $\ell_n = 0$
stripping reactions [1, 8] the most striking feature of the ^{3}He(d,p)^{4}He
reaction is the difference in sign between P_p and A_d.

In his theoretical analysis of A_T and P_p data Tanifuji [9] concluded
that it was necessary to include the p – ^{3}He spin-flip tensor force to
obtain the relation $A_T = -P_p$. Duck [10] also concluded that the presence of large spin-flip amplitudes corresponding to strong tensor
forces is the only way to obtain a deviation from $A_T = -P_p/3$. However, the equality $A_T = -P_p$ may be obtained in the usual DWBA framework for $\ell_n = 0$ stripping with central plus spin-orbit distortion in the
deuteron channel [11]. Therefore it seems that if one is interested in

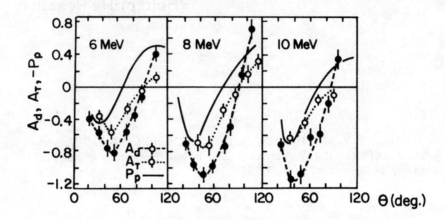

Fig. 1

comparing only A_T and P_p, the conclusion of the p-^{3}He spin-flip
tensor interaction is premature and numerical calculations are desirable. In this connection it is of interest to include $A_d(\theta)$ in the
coupled channels approach, which gives a reasonable fit to the A_T
and P_p data [12]. In recent phenomenological considerations [13] the
conditions were found for which the approximate relations $A_d = -P_p$
and $A_d = -3P_p/2$ hold.

Besides the presence of tensor forces in the d + ^{3}He system, there
may be additional (or alternative) reasons for the described unusual
effects. Both ^{2}H and ^{3}He nuclei are weakly bound, and exchange processes may strongly contribute; therefore the reaction mechanism may
differ from stripping.

Recently we have measured $A_T(\theta)$ at a deuteron energy near 13 MeV
[14]. The ^{3}He target of about 12% polarization at 4.5 Torr was used.
The results are similar to those of Baker et al. [4].

REFERENCES

*Recent asymmetry measurements with a polarized deuteron source
at Wisconsin [6] agree with our data.

[1] Yu. V. Gofman et al., Izv. Akad. Nauk SSSR ser. fiz. 32 (1968) 690.
[2] N. I. Zaika et al., Izv. Akad. Nauk SSSR ser. fiz. 32 (1968) 257.
[3] R. Brown and W. Haeberli, Phys. Rev. 130 (1963) 1163.
[4] S. D. Baker et al., Phys. Rev. Lett. 15 (1965) 115.
[5] N. I. Zaika et al., Yad. Fiz. 7 (1968) 754.

[6] G. R. Plattner and L. G. Keller, Phys. Lett. 29B (1969) 301.
[7] L. J. B. Goldfarb and R. C. Johnson, Nucl. Phys. 18 (1960) 353.
[8] H. H. Cuno et al., Nucl. Phys. A139 (1969) 657.
[9] M. Tanifuji, Phys. Rev. Lett. 15 (1965) 113.
[10] I. Duck, Nucl. Phys. 80 (1966) 617.
[11] V. A. Khangulyan, V. M. Kolybasov, Preprint ITEP No. 714, 1969; M. Tanifuji, K. Yazaki, Progr. Theor. Phys. 40 (1968) 1023.
[12] B. De-Facio et al., Phys. Lett. 25B (1968) 449.
[13] V. A. Khangulyan, Yad. Fiz. 11 (1970) 967.
[14] To be published.

Polarization-Asymmetry Relations
in the Reaction d + ^{3}He→p + ^{4}He

B. DeFACIO, University of Missouri-Columbia; R. K. UMERJEE,
Texas A&M University, College Station, USA*

The combined experiments of Brown and Haeberli [1] and Baker, Roy,
Phillips, and Walters [2] have presented some valuable information concerning the mechanism for the reaction d + ^{3}He →p + ^{4}He. The orthodox
relation P = -A was obtained by several authors [3, 4, 5], but is not a
satisfactory answer. Our approach is that of DeFacio, Umerjee, and
Gammel [6] and is adequate to present corrections to the approximately
correct relationship P = -A.

One source of confusion is the fact that in elastic scattering of
spin 1/2 by spin 0 particles the polarization is equal to the asymmetry.
This is true because of the invariance of the trace of the product of
two square non-singular matrices under commutation, i.e., $\text{Tr}(\hat{A}\cdot\hat{B}) = \text{Tr}(\hat{B}\cdot\hat{A})$.

For the reaction d + ^{3}He →p + ^{4}He the reaction scattering matrix M,
expanded in a complete set of spin states, is 2 × 6 in general because
the d + ^{3}He system is a spin 3/2, 1/2 initial system which scatters to
a final spin 1/2 system. The usual neglect of three-body final states
is plausible for the differential cross sections because of phase space
considerations, but for the scattering matrices is totally inadequate
and our discussion is to be understood to apply to a reduced transition
matrix depleted by an absorptive factor above the n + p + ^{3}He threshold.

In terms of the reaction M matrix the reaction spin polarization vector $\vec{P}(\theta)$ and the reaction asymmetry $\vec{A}(\theta)$ are given by

$$\vec{P}(\theta) = \text{Tr}(MM^\dagger \vec{\sigma}(p))/\text{Tr}(MM^\dagger) \tag{1}$$

and

$$\vec{A}(\theta) = \text{Tr}[\tilde{M}(\vec{\sigma}\,{}^3\text{He})\otimes 1(d)\tilde{M}^\dagger]/\text{Tr}(\tilde{M}\tilde{M}^\dagger). \tag{2}$$

The problem with comparing equations (1) and (2) is that M denotes the
reaction scattering matrix in the channel spin or coupled representation whereas $\tilde{M}$ denotes the scattering matrix in the uncoupled representation. The most general form of the scattering matrix does not
conserve orbital angular momentum because of orbital tensor interaction

terms of the form $(\vec{S}_d \cdot \hat{r})(\vec{\sigma}_{^3He} \cdot \hat{r})$ and does not conserve total spin because of spin-orbit interactions of the second kind $\vec{\ell} \cdot (\vec{S}_d - \vec{S}_{^3He})$. Tanifuji [3], Csonka et al. [4], and Duck [5] implicitly used the orbital tensor interactions but not the total spin breaking interaction. The problem with omission of spin breaking interaction is that the systematics of the full two channel system require these interactions as do the correction terms to the $P = -A$ relation. For example the $^4S_{3/2}(d + {}^3He)$ state is the only state strong enough at low energies to provide the $3/2^+$ state in Li^5 at 18.70 MeV because of the low energy and the angular momentum barrier in the $d + {}^3He$ system. We expand the polarization and asymmetry for the reaction into two parts:

$$\begin{pmatrix} P(\theta) \\ A(\theta) \end{pmatrix} = \begin{pmatrix} P_1(\theta) + P_2(\theta) \\ A_1(\theta) + A_2(\theta) \end{pmatrix} \tag{3}$$

where $P_1 = -A_1$ and P_2, A_2 are matrix elements of the spin orbit interaction of the second kind. In the two-body system the best candidates for contributing to P_2 and A_2 are, (i) the states $^4D_{5/2}(d + {}^3He)$ and $^2D_{5/2}(p + {}^4He)$ suggested in ref. [6] and (ii) the states $^4F_{3/2}(d + {}^3He)$ and $^2F_{3/2}(p + {}^4He)$, with better differential cross-section fits. In each case total spin is broken in the reaction scattering matrix thereby requiring spin orbit interactions of the second kind. Since this is the first reaction for which P and A were measured at the same energy, the simple direct reaction mechanisms must be regarded as *incomplete* models for differential cross-section and spin polarization.

REFERENCES

*This work was supported by the Assistant Professors Research Fund, University of Missouri-Columbia, and by Air Force Research Grant number 69-1817 at Texas A&M University, College Station, Texas.

[1] R. I. Brown and W. Haeberli, Phys. Rev. 130 (1963) 1163.
[2] S. D. Baker, G. Roy, G. C. Phillips, and G. K. Walters, Phys. Rev. Lett. 15 (1965) 115.
[3] M. Tanifuji, Phys. Rev. Lett. 15 (1965) 113.
[4] P. L. Csonka, M. J. Moravcsik, and M. D. Scadron, Phys. Rev. 143, (1966) 1324.
[5] I. Duck, Nucl. Phys. 80 (1966) 617.
[6] B. DeFacio, R. K. Umerjee, and J. L. Gammel, Phys. Lett. 25 (1967) 449.

DISCUSSION

Seiler:

We have made measurements on the $^3\vec{He}(d,p)^4He$ reaction from
2.8 to 10 MeV. A Legendre polynomial analysis of the data compared to one performed on the polarization measurements of
Brown and Haeberli shows that the $L = 1$ and $L = 2$ coefficients are
approximately proportional over the whole energy range, while
the $L = 3$ and 4 coefficients do not seem to show a simple relation.

DeFacio:

Since the uncoupled matrix elements $\widetilde{M}_{16}$ and $\widetilde{M}_{21}$ have azimuthal
dependence $\exp(\pm 2i\phi)$, the final orbital angular momenta which
contribute to my terms $P_2(\theta)$ and $A_2(\theta)$ have values $L' \geq 2$. Furthermore, since these bilinear observables are not symmetric about
$90°$ (c.m.), parities interfere, namely D and F waves in the $p + {}^4He$
channel. This is entirely consistent with Hale's results on this
process. Your observation concerning $L' = 3$ contributions may
well introduce questions on the interpretation of a low energy
parity conservation experiment by Ad'yasevich and Antonenko
which recently appeared in Sov. J. Nucl. Phys. 10 (1970) 278.

Tanifuji:

In the (d,p) reaction on ^3He and the (d,n) reactions on ^3H we found
a simple relationship between the three polarization quantities,
nucleon polarization P, deuteron vector analyzing power P_d, and
asymmetry A using a polarized target (for example, ^3He). The relationship is $A - P = -2P_d$. This relationship generally holds if
there is no tensor interaction between the particles and the nucleus. However, in the experimental data on the $^3He(d,p)$ reaction, one finds the relation $A - P_p \sim 2P_d$. This is direct and clear
evidence of how important tensor forces are in this energy region.
The details of our calculation are presented to this conference.
Are the quantities P_1, A_1, P_2, A_2 related to the force assumption
like the tensor force?

DeFacio:

I agree that tensor forces are important in this energy region. The
transitions contributing to the matrix elements $\widetilde{M}_{16}$ and $\widetilde{M}_{21}$ all
break total spin. One term breaks both L and S. This implies a
product of an orbital tensor and an antisymmetric function of the
spin vectors, i.e., it is a third-rank tensor. The information we
obtain from comparing P and A is that D and F waves are required
for a quantitative description of this reaction between 6 and
10 MeV deuteron energy and that third-rank tensors are definitely
present in this process at these energies.

Vector Analyzing Power
of the ^{3}He(d,p)^{4}He-Reaction
near the 430-keV Resonance

B. FORSSMANN, G. GRAF, H. SCHOBER, and H. P. JOCHIM, Otto-Hahn-Institut, Mainz, Germany

In the course of a general test of the Mainz polarized ion source, which is of the atomic beam type as proposed by Fleischmann and which is installed in the high voltage terminal of the MPI-Mainz cascade generator (0.3 to 1.4 MV), special care was taken to determine the vector analyzing power at the 430 keV-resonance in the ^{3}He(d,p)^{4}He-reaction which was used to test the deuteron performance of the source.

Because of the small effect considerable attention was paid to the problem of beam shift during the measurement, but no such effect could be detected in several independent check experiments. The angular distribution of the vector analyzing power was measured at four energies and five angles. The differential analyzing power was fitted by Legendre-polynomials L_{a1} up to L_{41}. The best fit was found to have $\sqrt{\chi^2/N} = 1.1$ and to include coefficients up to a_{31} (fig. 1).

In these fits the quantity $\sigma(\theta)P^d(\theta)/\sigma(0°)$ was used, resulting in the energy dependence of the coefficients shown in fig. 2. Under the assumption made by Goldfarb and Huq [1] that only p- and d-waves

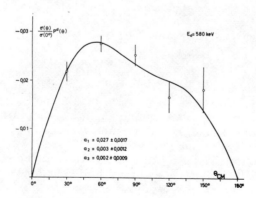

Fig. 1. Differential analyzing power (α-particles).

contribute and only by their interference with the dominant s-wave contribute to the polarization effects, no term proportional to L_{31} should arise. Furthermore, since the term proportional to L_{11} is strongest in all measurements, no d-wave preference occurs, as was assumed because of the positive parity of the 16.65 MeV $3/2^+$ level.

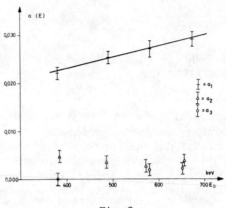

Fig. 2

REFERENCE

[1] L. J. B. Goldfarb and A. Huq, Helv. Phys. Acta 38 (1965) 541;
     A. Huq, Helv. Phys. Acta 39 (1966) 507.

Investigation of the ^{3}He(d,p)^{4}He Reaction
with Vector Polarized Deuterons
between 2 and 13 MeV

W. KLINGER and F. DUSCH, University of Erlangen-Nürnberg, Germany

Several attempts have been made to determine the level structure in the energy range from 16 to 25 MeV above the ground state of ^{5}Li [1, 2]. From the experimental data available at present, however, no definite spin and parity assignment is possible to the one or more levels in this region. More detailed information is necessary and in particular polarization experiments will serve to clarify these questions.

We have measured the angular distribution of the vector analyzing power of the reaction ^{3}He(d,p)^{4}He with a purely vector polarized deuteron beam from $E_{d,Lab} = 2.23$ MeV to 13 MeV in steps of 0.5 MeV. Using Cartesian tensor representation, the differential cross section becomes $\sigma = \sigma_0(1 \pm \frac{3}{2} P A_y)$ for an unpolarized ^{3}He target and a purely vector polarized d-beam whose polarization vector $\vec{P}$ is either parallel (spin up) or antiparallel (spin down) to the scattering normal. σ_0 is the unpolarized differential cross section, and A_y the vector analyzing power. In order to avoid inaccuracies in the beam integration and differences in the solid angle of the detectors, A_y was measured in our experiments by placing detectors symmetrically to the beam and carrying out a run with spin up and spin down at each energy.

A purely vector polarized d-beam of about 500 pA was provided by the polarized ion source of the Erlangen tandem accelerator. A Wien filter and a solenoid enabled us to direct the beam polarization either parallel or antiparallel to the scattering normal. The polarization of the beam was first measured with the reaction ^{12}C(d,p$_0$)^{13}C at $E_{d,Lab} = 10$ MeV [3] and then compared to the analyzing power of the ^{3}He(d,p)^{4}He reaction. In all further experiments, P was determined with the ^{3}He(d,p)^{4}He reaction at 10 MeV. Typically P was 0.52 ± 0.03. A 730 Torr ^{3}He gas target was used and a routing system permitted us to measure with 16 Si-surface barrier detectors simultaneously.

In fig. 1, a few characteristic angular distributions of the vector analyzing power are shown. The error bars show only the statistical accuracy. In the energy range from $E_{d,Lab} \approx 4$ MeV to 13 MeV, A_y is negative at forward and positive at backward angles. The change of sign always occurs near $\theta_{CM} = 90°$. Particularly at high energies, A_y takes large values, up to 0.76 at 13 MeV at $\theta_{CM} = 36.4°$. At energies from 2.23 MeV to 4 MeV, A_y becomes rather small. It is positive in forward directions, changes to negative values and becomes positive again at backward angles.

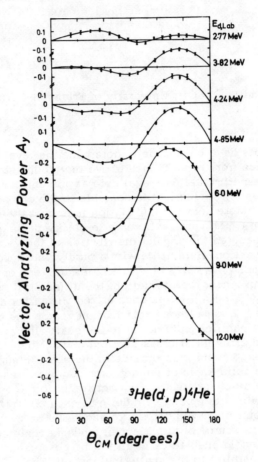

Fig. 1

The angular distributions of the vector analyzing power were fitted by the expansion

$$\sigma_0(\theta,E)A_y(\theta,E) = \sum_{\ell=1}^{6} B_\ell \, P_\ell^1 (\cos\theta).$$

The solid lines in fig. 1 correspond to this least square fit. In fig. 2, the expansion coefficients B_ℓ are shown. There is a pronounced structure in B_2 at $E_{d,\text{Lab}} \approx 7.2$ MeV corresponding to the known state in the ^{5}Li-compound nucleus at about 20.7 MeV. All other expansion coefficients are substantially smaller; B_5 and B_6 in particular are negligible in the energy region investigated. Only B_1 and B_3 show a resonance-like behavior at $E_{d,Lab} \approx 3$ MeV and $E_{d,Lab} \approx 4$ MeV, respectively. This behavior of B_1 and B_3 may possibly indicate further energy states in the ^5Li nucleus which would then have to have the opposite parity to the 20.7-MeV level.

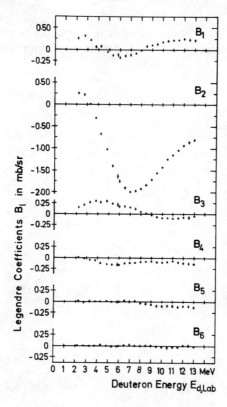

Fig. 2

REFERENCES

[1] T. A. Tombrello, R. J. Spiger, and A. D. Bacher, Phys. Rev. 154 (1967) 935.
[2] H. Schröder and W. Mausberg, Z. Phys. 235 (1970) 234.
[3] H. H. Cuno, G. Clausnitzer, and R. Fleischmann, Nucl. Phys. A139 (1969) 657.

DISCUSSION

Ohlsen:

Does your vector polarized beam have any second-rank polarization component at all? If so, does your experimental technique eliminate the second-rank effects?

Klinger:

There is no second-rank moment in our beam to within the statistical precision of our measurements, i.e., approximately 1%.

Measurement of the Tensor Moments
$\langle iT_{11}\rangle$, $\langle T_{20}\rangle$, $\langle T_{21}\rangle$, and $\langle T_{22}\rangle$
for the Reaction ^{3}He(d,p)^{4}He
between 2.8 and 10 MeV

W. GRÜEBLER, V. KÖNIG, A. RUH, P. A. SCHMELZBACH, and R. E. WHITE,
Laboratorium für Kernphysik, ETH, Zürich, Switzerland

Few-nucleon systems are of considerable interest since they can be analyzed in some detail. In order to check the results of such calculations the properties of the relevant resonance states have to be known. Therefore an investigation of the resonances in ^{5}Li was undertaken by means of the ^{3}He(d,p)^{4}He reaction using the vector and tensor polarized deuteron beam of the ETH tandem accelerator [1]. The tensor moments $\langle iT_{11}\rangle$, $\langle T_{20}\rangle$, $\langle T_{21}\rangle$, and $\langle T_{22}\rangle$ were measured using the experimental arrangement and method which is described elsewhere [2]. The results in the energy range between 2.8 and 10 MeV are shown in fig. 1. The size of the circles corresponds to the statistical error. In general the results for $\langle iT_{11}\rangle$ agree well with those of Plattner [3] in the same energy and angular range.

Characteristic features of the angular distributions are the different behavior of the various tensor moments. While the shape of the angular distributions of $\langle iT_{11}\rangle$ and $\langle T_{20}\rangle$ changes drastically between 2.8 and 6 MeV, the change in $\langle T_{22}\rangle$ is small. A similar observation, but for other tensor moments, can be made between 8 and 10 MeV, i.e., strong variations of $\langle T_{21}\rangle$ and $\langle T_{22}\rangle$ and small variations of $\langle iT_{11}\rangle$ and $\langle T_{20}\rangle$. These variations suggest one or more unknown resonance states in ^{5}Li between 18.2 and 20.0 MeV excitation energy corresponding to an incident deuteron energy between 2.8 and 6 MeV while another state seems to be excited between 8 and 10 MeV deuteron energy. The behavior of the different tensor moments give a rough indication of the orbital angular momenta which are chiefly responsible for forming these states. A detailed analysis in terms of Legendre polynomials is in progress.

The large values of all tensor moments over nearly the whole energy range measured suggest the use of the ^{3}He(d,p)^{4}He reaction as analyser for deuteron vector and tensor polarization. Particularly interesting are the large values of $\langle T_{21}\rangle$ at backward angles since this component is small at all angles and energies for the d-α elastic scattering which is also proposed as deuteron polarization analyser.

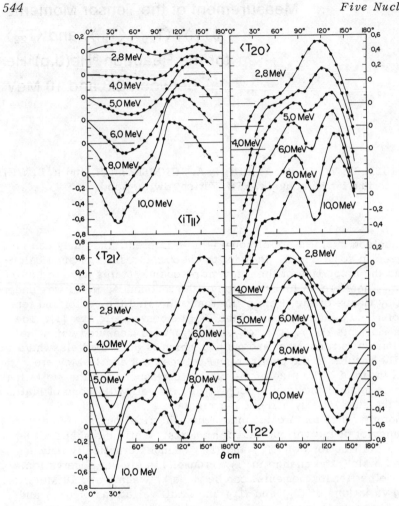

Fig. 1. Tensor moments $\langle iT_{11}\rangle$, $\langle T_{20}\rangle$, $\langle T_{21}\rangle$, and $\langle T_{22}\rangle$
of the ^{3}He(d,p)^{4}He reaction.

REFERENCES

[1] W. Grüebler, V. König, and P. A. Schmelzbach, Third Polarization
     Symp., and to be published in Nucl. Instr.
[2] W. Grüebler, V. König, P. A. Schmelzbach, and P. Marmier, Nucl.
     Phys. A134 (1969) 686, A148 (1970) 380 and A148 (1970) 391.
[3] G. R. Plattner and L. G. Keller, Phys. Lett. 29B (1969) 301.

DISCUSSION

Griffith:

The fact that your data show large values of the analyzing power T_{20} at $0°$ is very interesting. We have measured this same parameter for ^9Be, ^{12}C, ^{28}Si, and ^{29}Si. All of these yield large negative analyzing powers for T_{20} in the (d,p) stripping reactions. It seems very difficult to reproduce this effect in calculations with nuclear models.

The $^3\vec{\text{He}}$(d,p)^{4}He-Reaction below 2.5 MeV

P. HUBER, Ch. LEEMANN, H. MEINER, U. ROHRER, and F. SEILER,
University of Basel, Switzerland

The analyzing power of the $^3\vec{\text{He}}$(d,p)^{4}He-reaction with a polarized ^{3}He target was investigated at six energies below 2.5 MeV. Pyrex target cells, filled with 4 Torr ^{3}He, could be polarized to 18 to 22% for a period of several weeks, depending on age and history [1, 2]. A deuteron beam of about 200 nA was passed through a 3 mm^2 spot on an aluminized glass entrance window and extracted through a similar foil. These 2μm windows are able to withstand atmospheric pressure as well as an energy loss of 100-150 keV by the beam. The analyzing power A(θ) was determined from measurements with opposite target polarization. The related quantity ("efficiency") $C_{0,0,1,1}(\theta) = -\sqrt{2}\,\sigma_0(\theta)$. A(θ), normalized by $\sigma_{tot}/4\pi$, is plotted in fig. 1 with the Legendre polynomial fit. The expansion coefficients are shown in fig. 2 along with some data at higher energies [3]. At energies below 1.5 MeV, $C_{0,0,1,1}(\theta)$ is a function of interference terms involving the s-wave 3/2$^+$-resonance element R_1 and matrix elements with $\ell \geq 1$. The coefficient a_1 determines the interference with p-waves, while a_2 is a

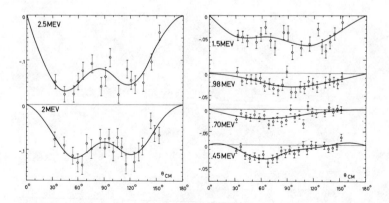

Fig. 1. Efficiencies $(-1/\sqrt{2})C_{0011}\cdot 4\pi/\sigma_t$.

measure for the d-wave amplitudes. Both show considerable enhancement by the resonance. One d-wave element, otherwise identical to
R_1, is mainly responsible for the peak in a_2. Such a contribution will
affect the analyzing power of the reaction for deuteron polarization,
slightly distorting the angular distribution.

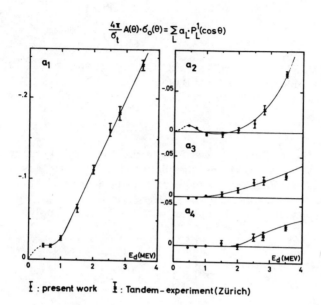

Fig. 2. Legendre polynomial coefficients of
$A(\theta) \cdot \sigma_0(\theta)$.

REFERENCES

[1] Ch. Leemann et al., Helv. Phys. Acta, to be published.
[2] P. Huber, Ch. Leemann, U. Rohrer, and F. Seiler, Helv. Phys.
     Acta 42 (1969) 907.
[3] Ch. Leemann et al., Third Polarization Symp.

DISCUSSION

Bacher:

How does the precision with which you can determine the ^{3}He
target polarization compare with that achieved by the methods
used by other groups?

Huber:

We can measure a $P_{^3\text{He}}$ of 0.2 to 1–2%. It is very constant during
the experiment.

The $^3\vec{\text{He}}$(d,p)^{4}He-Reaction
between 2.8 and 10 MeV

Ch. LEEMANN, H. MEINER, U. ROHRER, J. X. SALADIN, F. SEILER, and P. HUBER, University of Basel; W. GRÜEBLER, V. KÖNIG, and P. MARMIER, Laboratorium für Kernphysik, ETH, Zürich, Switzerland

Data for the analyzing power $A(\theta)$ of the $^3\vec{\text{He}}$(d,p)^{4}He-reaction have been obtained at eight energies between 2.8 and 10 MeV, usually for 16 angles. The target assembly is briefly described in another contribution to this conference [1]. The efficiencies $C_{0,0,1,1}(\theta) = -\sqrt{2}\,A(\theta) \cdot \sigma_0(\theta)$ are given in fig. 1 together with the Legendre polynomial fits.

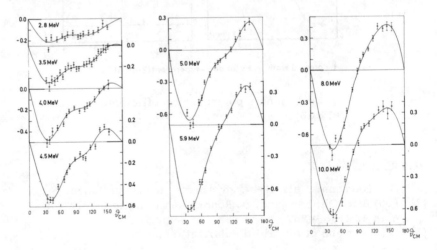

Fig. 1. Efficiencies $(-1/\sqrt{2})\,C_{0,0,1,1}(\theta) \cdot 4\pi/\sigma_{\text{tot}}$.

Agreement with previous measurements [2] is generally good, except for backward angles at 6 MeV. Angular distributions $\sigma_0(\theta)$ for unpolarized particles, supplementing existing data, were measured at the energies of the polarization experiments. The Legendre coefficients in fig. 2 exhibit the known anomalies in the region of 6 and 10 MeV [3]. The expansion coefficients of $C_{0,0,1,1}(\theta)$ in fig. 3 show little effect at 8 to 10 MeV, while the structure at 6 MeV is repeated and the L = 1

548

coefficient shows additional structure between 2 and 4 MeV. The latter is due to the interference of sizable matrix elements of positive and negative parity. An analysis of the measurements, together with polarized beam data [4] and all other experimental information in this energy region, is in progress.

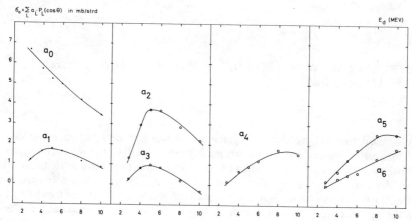

Fig. 2. Legendre polynomial coefficients of $\sigma_0(\theta)$.

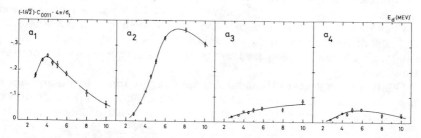

Fig. 3. Legendre polynomial coefficients of $(-1/\sqrt{2})\,C_{0,0,1,1}(\theta)$.

REFERENCES

[1] P. Huber et al., Third Polarization Symp.
[2] S. D. Baker, G. Roy, G. C. Phillips, and G. K. Walters, Phys. Rev. Lett. 15 (1965) 115.
[3] T. A. Tombrello, R. J. Spiger, and A. D. Bacher, Phys. Rev. 154 (1967) 935.
[4] W. Grüebler et al., Third Polarization Symp.

Asymmetries in Protons from the Reaction of Deuterons with a Polarized ³He Target

B. E. WATT and W. T. LELAND, Los Alamos Scientific Laboratory, New Mexico, USA*

We have investigated the asymmetry of the protons produced from ³He(d,p)⁴He reactions when using a polarized ³He target. The general experimental setup and procedures were identical with those described in our paper on deuteron scattering [1], with the exception of the detector system. We used a pair of solid state detectors arranged as a telescope. By using appropriate absorbers, coincidence techniques, and pulse height discrimination, we were able to reduce the background counting rates to a negligible amount.

Tables 1, 2, and 3 present our data along with that previously published by Baker et al. [2] (BRWP). To facilitate direct comparison, we have multiplied the data of BRWP by 1.2 to take into consideration the 20% difference in the target polarizations as calculated by the relation we used as compared to theirs. The listed errors are statistical only.

Plattner and Keller [3] have pointed out the approximate agreement between their measurements of P_d and that of BRWP. Fig. 1 displays data of Plattner and Keller along with ours and that of BRWP multiplied

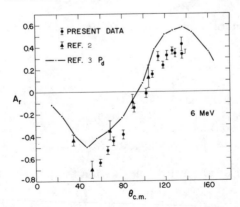

Fig. 1. Proton asymmetry from ³He(d,p)⁴He reaction with polarized ³He.

550

Table 1. Angular asymmetries for reaction
protons at 6 MeV

$\theta_{c.m.}$	Present $A_r(\theta)$	Ref. [2] $1.2\,A_{LR}(\theta)$
34.8°		-0.43 ± 0.05
51.7		-0.70 ± 0.07
59.4	-0.63 ± 0.031	
63.3	-0.525 ± 0.035	
68.3		-0.35 ± 0.08
71.6	-0.44 ± 0.04	
80.6	-0.38 ± 0.04	
89.4		-0.08 ± 0.06
90.6	-0.14 ± 0.04	
102.0	$+0.01 \pm 0.04$	
104.5		$+0.13 \pm 0.1$
108.3	$+0.16 \pm 0.03$	
113.2	$+0.32 \pm 0.04$	
117.8	$+0.24 \pm 0.03$	
121.8	$+0.33 \pm 0.04$	
125.6	$+0.37 \pm 0.03$	
129.2	$+0.35 \pm 0.04$	
135.8	$+0.43 \pm 0.06$	
135.8	$+0.34 \pm 0.04$	

Table 2. Angular asymmetries for reaction
protons at 8 MeV

$\theta_{c.m.}$	Present $A_r(\theta)$	Ref. [2] $1.2\,A_{LR}(\theta)$
41°		-0.83 ± 0.17
52.6		-0.85 ± 0.19
60.3	-0.51 ± 0.04	
69.4		-0.36 ± 0.11
72.7	-0.33 ± 0.04	
81.7	-0.17 ± 0.04	
85.3		-0.1 ± 0.11
91.9	$+0.06 \pm 0.03$	
101		$+0.19 \pm 0.13$
103.3	$+0.29 \pm 0.04$	
114.4	$+0.50 \pm 0.04$	
115.5		$+0.42 \pm 0.17$
122.9	$+0.50 \pm 0.04$	
130.2	$+0.48 \pm 0.04$	
136.7	$+0.50 \pm 0.04$	

Table 3. Angular asymmetries for reaction
protons at 10 MeV

$\theta_{c.m.}$	Present $A_r(\theta)$	Ref. [2] $1.2\,A_{LR}(\theta)$
35.8°		-0.76 ± 0.08
53.3		-0.53 ± 0.07
57.1	-0.39 ± 0.1	
61.1	-0.39 ± 0.1	
70.2		-0.17 ± 0.06
86.3		-0.12 ± 0.1
92.8	$+0.07 \pm 0.08$	
110.4	$+0.03 \pm 0.1$	
115.4	$+0.35 \pm 0.04$	
123.9	$+0.32 \pm 0.04$	
123.9	$+0.41 \pm 0.05$	
137.5	$+0.24 \pm 0.2$	
137.5	$+0.45 \pm 0.04$	

by 1.2. We would agree that $P_d = A_{^3He}$ is a good approximation at
higher energies and forward angles but do not agree that it holds at
lower energies.

REFERENCES

*Work performed under the auspices of the U.S. Atomic Energy Commission.

[1] B. E. Watt and W. T. Leland, Third Polarization Symp.
[2] S. D. Baker, G. Roy, G. C. Phillips, and G. K. Walters, Phys. Rev.
     Lett. 15 (1965) 115.
[3] G. R. Plattner and L. G. Keller, Phys. Lett. 29B (1969) 301.

DISCUSSION

Baker:
    Do you plan to measure $^3\vec{He}(\vec{d},p)^4He$? Do you expect such a mea-
    surement to help in obtaining more accurate values of scattering
    matrix elements; in particular, measurements at 0° might not be
    too difficult.

Ohlsen:
    Measurement of spin correlation effects in the $^3\vec{He}(\vec{d},p)^4He$ reac-
    tion definitely gives new independent information on the reaction
    matrix. Such experiments, both at 0° and elsewhere, are being
    discussed and appear feasible. I hope some will be done at Los
    Alamos in the next year or so.

The $^3\vec{\text{He}}(\vec{\text{d}},\text{p})^4$He-Reaction at 430 keV

Ch. LEEMANN, H. BÜRGISSER, P. HUBER, U. ROHRER, H. PAETZ gen.
SCHIECK, and F. SEILER, University of Basel, Switzerland

For several combinations of beam and target polarization [1] the effi-
ciencies [2] $C_{q,k,Q,K}(\theta)$ of the $^3\vec{\text{He}}(\vec{\text{d}},\text{p})^4$He-reaction have been measured
at 430 keV at the peak of the $3/2^+$ resonance. The solid lines in figs.
1 and 2 are least-square fits. They deviate only slightly from the angular distributions for a single s-wave $J^\pi = 3/2^+$-matrix element R_1.
The sensitivities are normalized by $a_{0,0}(0) = \sigma_{\text{tot}}/4\pi$. An analysis
with a simple model [2] shows that p-waves are present by the occur-
rence of $L = 1$ Legendre coefficients, particularly in $C_{1,1}(\theta)$. D-waves
are also found [1], especially evident in the $L = 2$ term of $C_{1,1}(\theta)$. These
give rise to a small $L = 0$ term in the analyzing power $C_{2,0}(\theta)$.

Those linear relations [2] which involve only interference terms of
R_1 with p- and d-wave elements are satisfied by the data. In the
framework of this model, real parts of these interference terms have
been calculated [3]. In the linear relations involving s- and d-wave
elements, small systematic deviations are found. The most probable
causes are systematic errors in beam and/or target polarization. The
accuracy of our data does not permit the separation of the two effects,
although this would be possible with more precise measurements. The
deviations could be explained, however, by a 4% reduction in the
analyzing power of the T(d,n)⁴He-reaction which was used to measure
deuteron polarization [2].

REFERENCES

[1] P. Huber et al., Third Polarization Symp.
[2] F. Seiler and E. Baumgartner, Third Polarization Symp.
[3] Ch. Leemann et al., Helv. Phys. Acta, to be published.

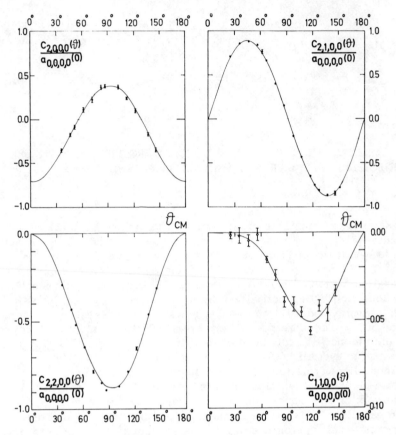

Fig. 1. Efficiencies for deuteron polarization.

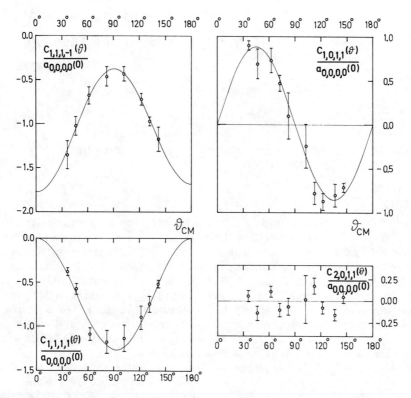

Fig. 2. Efficiencies for deuteron and ³He polarization.

Comparison of the Analyzing Power
for n-⁴He Scattering Calculated
from Various Sets of Phase Shifts

T. C. RHEA, Th. STAMMBACH, and R. L. WALTER, Duke University
and TUNL, Durham, North Carolina, USA*

In most neutron polarization measurements in recent years, scattering
from ⁴He has been employed as the polarization analyzer. Knowledge
of the analyzing power (which is identical to the polarization produced
in scattering unpolarized neutrons) is derived from calculations using
available phase-shift sets. In 1952 Dodder and Gammel [1] presented
a set based on a single level dispersion theory and charge independence of the p-⁴He and n-⁴He systems. Since 1962 much new polarization and cross-section information has become available and with it,
a number of sets of phase shifts based on the new data and (a) the
p-⁴He system [2], (b) the single level dispersion theory [3], (c) an
optical model [4], (d) the effective range parameterization [5], and
(e) the single level dispersion theory plus a weak bias toward the
p-⁴He system [6]. The last set [6] is based on more (recent) data than
the previous sets and thus may be the nearest to reality.

The purpose of this paper is to
present graphically comparisons of
the polarizations computed from the
various sets at many energies and
at those angles where the analyzing
power peaks. See figs. 1 and 2.
From these curves, it is possible
to compare neutron polarization
data which have been reported for
different analyzing angles and/or
for analyzing powers based on different phase shift sets. In many
instances, the accuracy of the past
polarization measurements is such
that a 5% change in analyzing
power will not affect the quoted
result. Thus, there is no adjustment necessary if measurements
have been carried out at analyzing

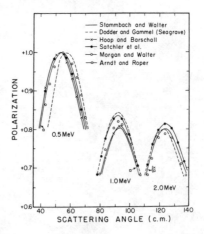

Fig. 1. Polarization for n-⁴He
scattering from 0.5 to 2.0 MeV for
available sets of phase shifts.

556

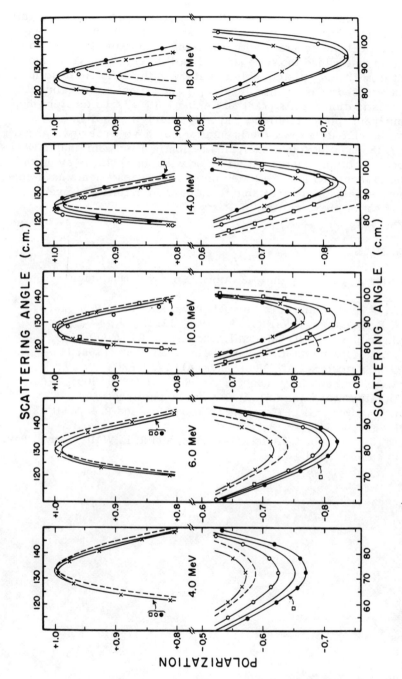

Fig. 2. Polarization for n–⁴He scattering from 4 to 18 MeV for available sets of phase shifts.

angles near the positive peak in the polarization, regardless of which set of phase shifts was utilized. The small differences at low energies reflect the accuracy with which the dominating $P_{3/2}$ resonance has been determined for some time. The differences above 10 MeV result from fitting different sets of data with the different descriptions. The often-used phase shifts of refs. [1] and [2] are not consistent with polarization results [7, 8] at 2.4 and 7.8 MeV. Fortunately though, the analyzing power is affected sizably at only forward angles, a region which only a few experimental groups have employed. Above 4 MeV, the most efficient analyzing region is at forward angles where the cross section is large. It would be wise, however, to avoid this region for energies above 12 MeV in precise measurements until sufficient n-^{4}He data are available to prove the reliability of even the newest set of phase shifts.

REFERENCES

*Work supported by the U.S. Atomic Energy Commission.

[1]   D. C. Dodder and J. L. Gammel, Phys. Rev. 88 (1952) 520;
      J. D. Seagrave, Phys. Rev. 92 (1953) 1222.
[2]   B. Hoop and H. H. Barschall, Nucl. Phys. 83 (1966) 65.
[3]   G. L. Morgan and R. L. Walter, Phys. Rev. 168 (1968) 1114.
[4]   G. R. Satchler et al., Nucl. Phys. A112 (1968) 1.
[5]   R. A. Arndt and L. D. Roper, Phys. Rev. C1 (1970) 903.
[6]   Th. Stammbach and R. L. Walter, to be published.
[7]   J. R. Sawers, G. L. Morgan, L. A. Schaller, and R. L. Walter,
      Phys. Rev. 168 (1968) 1102.
[8]   Th. Stammbach, J. Taylor, G. Spalek, and R. L. Walter, Phys. Rev.,
      to be published.

Phase Shifts from n-^{4}He Elastic Scattering Experiments near 20 MeV

A. NIILER, M. DROSG, J. C. HOPKINS, and J. D. SEAGRAVE, Los Alamos
Scientific Laboratory, New Mexico, USA*

Optical model calculations [1] and a recent energy-dependent phase-
shift analysis [2] of n-^{4}He elastic scattering in the energy range of
15—25 MeV have arrived at sets of phase shifts which at some energies
are in significant disagreement with the set put forward by Hoop and
Barschall [3]. Since the n-^{4}He system is widely used as a neutron
polarization analyzer, a reliable set of phase shifts is essential.

None of the above authors had available the complete set of recent
n-^{4}He polarization data of Broste and Simmons [4] in the energy range
11—30 MeV. Furthermore, all n-^{4}He cross section measurements used
in these analyses were made by the ^{4}He recoil method. By this method,
angular distributions cannot be extended reliably into the forward
angles in this energy region. Thus, rather large uncertainties may
enter into the normalization of the integral of the angular distributions
to total cross sections.

We have measured the n-^{4}He differential cross sections at 17.6,
20.9, and 23.7 MeV with a neutron time-of-flight technique using a
1-mole sample of liquid ^{4}He as a scatterer. The energies of 17.6 and
23.7 MeV were chosen to match Broste and Simmons' polarization
measurements, and 20.9 MeV was chosen to give an intermediate point.
Neutron-proton scattering was also measured to give an absolute normalization of the n-^{4}He points to the well-known n-p cross sections
[5]. Multiple scattering effects were corrected for by the Aldermaston
code MAGGIE [6].

At all three energies, single-energy phase-shift searches were
carried out using the general reaction matrix code EIA2. At 23.7 MeV,
Hoop and Barschall's inelastic parameters were used and kept fixed
in all searches. Starting the phase shifts at several different values,
a unique set was always reached. Maximum L values of 2, 3, 4, and 5
were searched on, and at all energies the minimum weighted variance
was obtained with $L_{max} = 4$. Thus, subsequent searches were made
with $L_{max} = 4$ only. These single-energy sets are shown as the last
entries of table 1. The next step was to obtain a smooth phase-shift
set which would extrapolate smoothly to the optical model and the

Table 1. n-^{4}He phase shifts in degrees

	E_n (MeV)	$S_{1/2}$	$P_{3/2}$	$P_{1/2}$	$D_{5/2}$	$D_{3/2}$	$F_{7/2}$	$F_{5/2}$	$G_{9/2}$	$G_{7/2}$
Ref. [3]	17.6	93	95	56	6	3	1	1		
	20.9	90	92	53	9	6	3	3		
	23.7	86	89	52	12	6	4	4		
Ref. [1]	17.6	89	93	52	7.2	3.2	0.4	0.2		
	20.9	84	88	47	9.4	4.0				
Present Values Smooth Set	17.6	86	95	54	8	7.2	2.1	1.5	0.5	0.2
	20.9	81	92.5	52	11	10	3.3	3	1.0	0.8
	23.7	79.5	91.6	51	17.5	11.2	5.8	5.2	1.5	2.0
Present Values Single Energy	17.6	84	95	52.5	9.1	7.9	2.1	1.8	2.2	0.6
	20.9	83.7	97.7	45.7	8.5	15.7	1.2	4.6	6.5	-0.3
	23.7	80.5	91.6	52.5	17.5	11.2	5.9	5.4	1.4	1.7

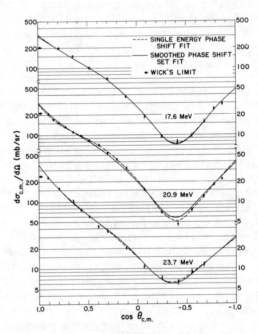

Fig. 1. Present data compared with cross sections from the smooth and single-energy phase-shifts sets.

two-level resonance model [7] values near 10 MeV. This smoothed set is also shown in table 1 along with that of Hoop and Barschall and the optical model set of Satchler et al. [1]. Since both $\sigma(\theta)$ and $P(\theta)$ were available at 17.6 and 23.7 MeV, the phase shifts at these two energies were given more weight than the 20.9-MeV values in obtaining the smoothed set.

The most significant differences between our sets and those of ref. [3] are the smaller value of the $S_{1/2}$ phase, the somewhat larger values for the D- and F-wave phases, and the need for small amounts of G-waves.

Our single-energy set gives excellent fits to the angular distributions at all three energies. Our smoothed set fits equally well at 17.6 and 23.7 but not quite so well at 20.9 MeV. See fig. 1 for these fits. A derivation of the phase shifts from the cross-section data alone (without the constraints of the polarization data) did not give acceptable results for $L_{max} \geq 3$ (at 17.6 MeV) and $L_{max} \geq 4$ (at 23.7 MeV). The usefulness of having both differential cross-section and polarization data available at the same energy cannot be overstated.

The authors are indebted to D. Dodder for use of the code EIA2.

REFERENCES

*Work performed under the auspices of the U.S. Atomic Energy Commission.

[1]   G. R. Satchler et al., Nucl. Phys. A112 (1968) 1.
[2]   R. A. Arndt and L. D. Roper, Phys. Rev. 1C (1970) 903.
[3]   B. Hoop, Jr., and H. H. Barschall, Nucl. Phys. 83 (1966) 65.
[4]   W. B. Broste, J. E. Simmons, and G. S. Mutchler, Bull. Am. Phys. Soc. 14 (1969) 1230, and private communication.
[5]   J. C. Hopkins and G. Breit, to be published.
[6]   J. B. Parker et al., Nucl. Instr. 30 (1964) 77.
[7]   G. L. Morgan and R. L. Walter, Phys. Rev. 168 (1968) 1114.

Neutron-Helium Polarization
at 27.3 and 30.3 MeV

W. B. BROSTE[*] and J. E. SIMMONS, Los Alamos Scientific Laboratory,
New Mexico, USA[†]

As part of an effort to obtain more accurate values for the polarization
analyzing power of n-alpha scattering for neutron energies above 20
MeV, the n-alpha polarization angular distribution has been measured
at this laboratory for six neutron energies between 11.0 and 30.3 MeV.
Data obtained at 11.0, 17.7, 23.7, and 25.7 MeV have been reported
at an earlier meeting [1].

The asymmetry in scattering polarized neutrons from helium was
measured for neutron energies of 27.3 and 30.3 MeV for angles between 50° and 130° (lab). Partially polarized neutrons were obtained
from the $D(t,n)^4$He reaction at a lab angle near 30°. Incident energies
of E_t = 12.0 and 14.9 MeV provided neutron energies of E_n = 27.3 and
30.3 MeV, respectively. The value of the source polarization P_1 was
determined for the 27.3-MeV data in a measurement on the $T(d,n)^4$He
reaction under the same c.m. conditions as were used in the $D(t,n)^4$He
source reaction. The resultant neutron energy is 13.2 MeV; therefore,
P_1 may be determined fairly reliably from the measurement of the asymmetry in n-He scattering at 115° (lab). From this measurement it was
found that $P_1 = 0.53 \pm 0.015$. P_1 for 30.3-MeV data was interpolated
from $T(d,n)^4$He data [2], and had a value $P_1 = 0.46 \pm 0.05$. The asymmetries were corrected for multiple scattering of neutrons in the helium,
and the effects of finite geometry, then divided by P_1 to determine the
n-He polarization function $P(\theta)$.

Figs. 1 and 2 show the final polarization results for the two energies.
Solid curves are the predictions of the Hoop-Barschall phase shift set
[3] at 27.3 and their smooth extrapolation to 30.3 MeV. Fig. 1 also
shows the data of Arifkhanov et al. [4] at a neutron energy of 27.8 ±
0.9 MeV, where their asymmetry data have been converted to polarizations using a value for P_1 obtained in measurements of the $T(d,n)^4$He
30° polarization at this laboratory. Preliminary phase-shift analyses
have indicated that small changes in the Hoop-Barschall phases result in good fits to both cross-section and polarization data at the two
energies.

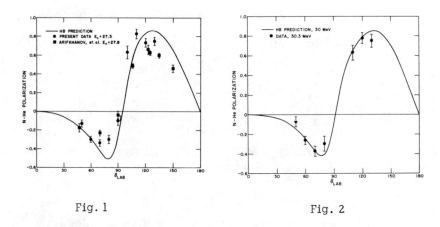

Fig. 1   Fig. 2

REFERENCES

*Associated Western Universities Fellow from the University of Wyoming.

†Work performed under the auspices of the U.S. Atomic Energy Commission.

[1]   W. B. Broste, J. E. Simmons, and G. S. Mutchler, Bull. Am. Phys. Soc. 14 (1969) 1230.
[2]   W. B. Broste and J. E. Simmons, Third Polarization Symp.
[3]   B. Hoop and H. H. Barschall, Nucl. Phys. 83 (1966) 65.
[4]   U. R. Arifkhanov, N. A. Vlasov, V. V. Davydov, and L. N. Samoilov, Sov. J. Nucl. Phys. 2 (1966) 170.

Analysis of the p-⁴He Polarization for
Eₚ=1 to 12 MeV in the Framework
of the Humblet-Rosenfeld S-Matrix Theory

L. KRAUS, I. LINCK, and D. MAGNAC-VALETTE, Laboratoire des
Basses Energies, Strasbourg, France

The differential cross sections [2, 3, 4] for the ^{4}He(p,p)^{4}He elastic
scattering between 0.5 and 12 MeV have been analyzed in the Humblet-
Rosenfeld formalism [1], after the existing data have been supplemented
by measurements between 0.5 and 3 MeV [5].

The experimental cross sections can be satisfactorily described
with two complex resonant terms U_{cc} for the $P_{3/2}$ and $P_{1/2}$ resonances:

$$U_{cc} = e^{2i(\omega_c + \phi_c)} \left\{ 1 - ik_c P_c \frac{X_{cn} e^{i\xi cn}}{E - E_n + i\Gamma_n/2} \right\}$$

with the parameters (a) in table 1. It is seen that the positions of the
^{5}Li $P_{3/2}$ and $P_{1/2}$ levels are very different from those extracted from a
phase-shift interpretation in the R-matrix theory. The validity of our
parameters is tested by calculating the polarization. The agreement
with the measured polarization [6, 7, 8, 9] is satisfactory for E_p up to
4 MeV, less so above this energy [figs. 1, 2, curve (a)]. This discrepancy disappears if cross-section and polarization data are both
fitted and if a constant background Q_{cc} is introduced in U_{cc}, for the
S- and P-waves [figs. 1, 2, curve (b)].

Table 1. Parameters for the $P_{3/2}$ and $P_{1/2}$ levels in ^{5}Li

		[11]	[3]	[4]	Present Results (a)	(b)	[12]	[13]
$P_{3/2}$	E_R	1.965	2.1	2.12	1.6	1.6		
	$\Gamma_{c.m.}$	1.5			1.5	1.4		1.55
$P_{1/2}$	E_R			8.6	4.1	3.8		
	$E_{exc.}^5$Li	5-10	6.5		2.5	2.2	3^{+3}_{-1}	2-5
	$\Gamma_{c.m.}$	3-5			8	9		

All quantities are in MeV and in the c.m. frame.

In this last case, the values (b) given in table 1 are not significantly different from the values (a) for the $P_{3/2}$ ground state; the first excited state $P_{1/2}$ is defined with less accuracy. Nevertheless the $P_{1/2}$-$P_{3/2}$ splitting found here is less than previously reported [11]. The splitting agrees fairly well with the results on some reactions [12, 13] leading to ^{5}Li and is of the same order of magnitude as the ^{5}He $P_{1/2} - P_{3/2}$ splitting [11].

Fig. 3 shows the experimental ^{4}He(p,p)^{4}He phase shifts and the appropriate phase of U_{cc}; $|U|$ is unity within 10% throughout the energy range.

We are indebted to Prof. Goldfarb for stimulating discussions.

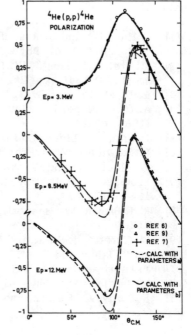

Fig. 1

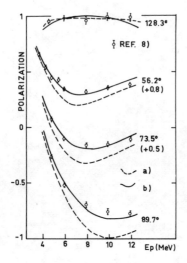

Fig. 2

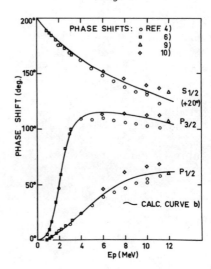

Fig. 3

REFERENCES

[1]   J. Humblet and L. Rosenfeld, Nucl. Phys. 26 (1961) 529;
      J. Humblet, Nucl. Phys. 50 (1964) 1.
[2]   G. Freier, E. Lampi, W. Sleator, and J. H. Williams, Phys. Rev.
      75 (1949) 1345.
[3]   P. D. Miller and G. C. Phillips, Phys. Rev. 112 (1958) 2043.
[4]   A. C. L. Barnard, C. M. Jones, and J. L. Weil, Nucl. Phys. 50
      (1964) 604.
[5]   L. Kraus, I. Linck, and M. Wery, to be published.
[6]   L. Brown and W. Trachslin, Nucl. Phys. A90 (1967) 334; L. Brown,
      W. Haeberli, and W. Trachslin, Nucl. Phys. A90 (1967) 339.
[7]   L. Rosen, J. E. Brolley, M. L. Gursky, and L. Stewart, Phys. Rev.
      124 (1961) 199.
[8]   R. I. Brown, W. Haeberli, and J. X. Saladin, Nucl. Phys. 47 (1963)
      212.
[9]   D. Garetta, J. Sura, and A. Tarrats, Nucl. Phys. A132 (1969) 204.
[10] M. F. Jahns and E. M. Bernstein, Phys. Rev. 162 (1967) 871.
[11] T. Lauritsen and F. Ajzenberg-Selove, Nucl. Phys. 78 (1966) 1.
[12] M. Bernas, Ann. Phys. (Paris) 3 (1968) 213.
[13] J. Cerny, C. Detraz, and R. H. Pehl, Phys. Rev. 152 (1966) 950.

Phase-Shift Representation of
p-⁴He Scattering between 3 and 18 MeV

P. SCHWANDT, Indiana University, Bloomington; T. B. CLEGG,
University of North Carolina, Chapel Hill, USA

Proton scattering from ⁴He has been widely used as a polarization
analyzer for polarized protons. For proton energies within the range
obtainable from tandem accelerators, however, sufficiently precise
measurements of the analyzing power (polarization) have until recently
been available only at a few energies and angles. This lack of adequate polarization data also frustrated attempts to obtain a reliable
energy-dependent set of phase shifts for the p+⁴He system which
would allow one to predict accurately the polarization at any desired
energy and angle. To fill gaps in the existing data, polarization angular distributions were measured for proton energies near 4.6, 6.0,
7.9, 9.9, and 11.9 MeV. The experiment involved measuring the left-
right asymmetry in the scattering from ⁴He of the polarized proton beam
from the Wisconsin tandem accelerator. The primary beam polarization
was calibrated with reference to the precise and absolute double-
scattering measurements of Brown et al. [1] at selected angles. The
average experimental error in the polarization data is less than ± 0.02.

The present polarization measurements and previously available
polarization and cross-section data from 3.0 to 17.5 MeV were used
to determine p+⁴He phase shifts. The aim of the analysis was (1) to
investigate the uniqueness of any "best-fit" phase-shift solution(s),
and (2) to obtain a set of phase shifts which are smooth analytic functions of energy and join smoothly onto existing phase-shift solutions
below 3 MeV and beyond 18 MeV.

A continuum of equally acceptable phase-shift solutions was found
between about 7 and 13 MeV. Very strong correlations were observed
between phase shifts of a given ℓ-value but different $j = \ell \pm 1/2$, i.e.,
the spin-orbit splitting of the phases is considerably better determined
by the data than any individual phase shift. Also observed were somewhat weaker correlations between phase shifts of different ℓ-values,
implying that holding one or more phase shift(s) fixed will significantly
restrict the range of acceptable values for the remaining phase shifts.

As a result of these findings, and because the small d- and f-wave
phases in particular were poorly determined by the data, the magnitude

and energy behavior of these phases were fixed by fitting the available empirical d- and f-wave phase shifts between 20 and 50 MeV with a single-level pole expansion of resonance form and a background term, using the resulting resonance parameters to calculate the phase shifts below 20 MeV. With d- and f-wave phases constrained in this manner, a unique set of optimum s- and p-wave phases could then be determined at each energy. Finally, the energy variation of all phase shifts $\delta_\ell^j(E_p)$ for $\ell = 0, 1, 2, 3$ and for $3 \leq E_p(MeV) \leq 18$ was found to be quite accurately described by a generalized effective-range expansion of the form

$$C_\ell^2(\eta)k^{2\ell+1}\left[\cot\delta_\ell^j(E_p) + 2\eta\, h(\eta)/C_0^2(\eta)\right] = \sum_{n=0}^{N} a_{n\ell j}\, E_p^n,$$

where k is the wave number, η is the Coulomb parameter, and the functions $C_\ell(\eta)$, $h(\eta)$ are defined in ref. [2]. This expansion converges rapidly ($N = 1$ for $\ell = 0, 2, 3$; $N = 2$ for $\ell = 1$), and the effective-range phase shifts obtained by fitting the optimum phase shifts at every energy with the above formula predict p-^{4}He polarizations with an overall uncertainty ΔP about equal to the average experimental error over the energy range 3–18 MeV where ΔP roughly follows the relation $\pm\Delta P \lesssim 0.025 - 0.001\, E_p$.

REFERENCES

[1]   R. I. Brown, W. Haeberli, and J. X. Saladin, Nucl. Phys. 47 (1963) 212.
[2]   M. A. Preston, Physics of Nuclei, Appendix B (Addison-Wesley, 1962).

DISCUSSION

Ohlsen:

Proton-^{4}He scattering at energies between ≈ 6 and ≈ 12 MeV and at an angle near 112° has an analyzing power of nearly 1.0. What is the lowest value you obtained for the analyzing power at this peak for any phase-shift set which fits the data reasonably?

Schwandt:

The predicted polarization P near the maximum is defined to better than ±0.01 around 12 MeV, i.e., P is between 0.99 and 1.00. The polarization maximum is rather insensitive to variations in phase shifts; this was also found to be the case for n-^{4}He scattering by Walter and co-workers.

Polarization and Cross-Section Measurements in p-^{4}He Scattering from 20 to 40 MeV

A. D. BACHER, G. R. PLATTNER, H. E. CONZETT, D. J. CLARK,
H. GRUNDER, and W. F. TIVOL, Lawrence Radiation Laboratory,
Berkeley, California, USA*

The scattering of protons from ^{4}He has been particularly useful as a
proton polarization analyzer, since it produces large polarizations
over a wide range of energies. Also, analyses of resonance effects
corresponding to states in ^{5}Li above the deuteron threshold are simplified if observed in the proton channel because of the single chan-
nel-spin available. In particular, a broad anomaly near an excitation
energy of 20 MeV, which has been seen in the ^{3}He(d,p)^{4}He reaction [1]
and in d-^{3}He elastic scattering [2], has not yet been explained unambiguously. Even though no effect has been seen in p-^{4}He cross-section
excitation functions [3], the ^{3}He(d,p)^{4}He data show that, *if* the anomaly
results from a state in ^{5}Li, the state must have a proton width. Thus,
because the polarization is more sensitive than the cross section
to small changes in a partial-wave amplitude, p-^{4}He polarization
excitation functions could provide information important to the explanation of this anomaly.

We have used the new axially-injected polarized proton beam from
the 88-inch cyclotron to supplement and to improve previous p-^{4}He
elastic scattering data between 20 and 40 MeV. External beams of
80-120 nA with polarization ≈ 0.75 were available. Angular distributions of cross sections and polarizations were measured at 2-MeV
intervals at 20 laboratory angles between 17.5° and 150°. At each
energy the relative uncertainty of the polarization is about ±0.010.
An absolute normalization with an uncertainty of less than 3% at all
energies has been obtained relative to the analyzing power of p-^{4}He
scattering near 14.5 MeV [4]. Absolute cross sections were obtained
by normalizing to the data of ref. [3]. As examples, fig. 1 shows our
results near 24 and 26 MeV. Additional measurements of comparable
precision were made in the region between 20 and 40 MeV to search
for an effect corresponding to the anomaly discussed above.

A contour plot of the experimental proton polarization is shown in
fig. 2. Measurements for the region between 16 and 20 MeV are taken
from ref. [4]. The effect of the 16.65-MeV 3/2$^+$ state of ^{5}Li is clearly

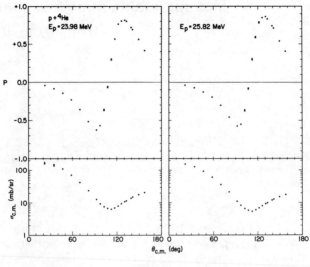

Fig. 1

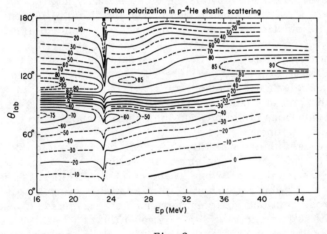

Fig. 2

seen near 23 MeV. Excitation functions were taken across this resonance to provide data for an improved determination of the resonance
parameters [5] and to aid in the continuation of a phase-shift analysis
to the higher energies. With the exception of this narrow resonance
region, the analyzing power near $\theta_{lab} = 120° - 130°$ remains large, and
these measurements provide an accurate proton polarization analyzer
up to 40 MeV. A broad bump between 26 and 32 MeV is apparent in the
contour plot at backward angles. This can be seen more clearly in

fig. 3, which shows an excitation function of the proton polarization
at $\theta_{c.m.} = 102.2°$ (87.5° lab). The
sharp structure near 23 MeV is due
to the 3/2+ level, and the broad
anomaly between 26 and 32 MeV
is seen to correspond in energy
to that seen in the d-^{3}He elastic
scattering cross-section data [2]
at $\theta_{c.m.} = 90°$, which is plotted in
the insert.

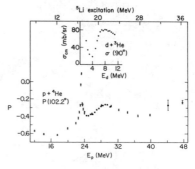

Fig. 3

A phase-shift analysis of these
data is underway, but only qualitative preliminary results based
on the data at 2-MeV intervals
are presently available. Above
30 MeV the phase shifts vary quite smoothly with energy, and the addition of a small g-wave contribution provides significantly better
fits to the data than is possible with analyses limited to s-, p-, d-,
and f-waves. Analysis of the data in the region between 24 and 32 MeV
has not yet progressed to the point where a clear explanation of the
anomaly can be reported here.

REFERENCES

*Work performed under the auspices of the U.S. Atomic Energy
Commission.

[1] L. Stewart, J. E. Brolley, and L. Rosen, Phys. Rev. 119 (1960) 1649.
[2] T. A. Tombrello, R. J. Spiger, and A. D. Bacher, Phys. Rev. 154
(1967) 935.
[3] P. W. Allison and R. Smythe, Nucl. Phys. A121 (1968) 97; S. N.
Bunker et al., Nucl. Phys. A133 (1969) 537.
[4] D. Garreta, J. Sura, and A. Tarrats, Nucl. Phys. A132 (1969) 204;
P. Schwandt, private communication.
[5] W. G. Weitkamp and W. Haeberli, Nucl. Phys. 83 (1966) 46;
P. Darriulat, D. Garreta, A. Tarrats, and J. Testoni, Nucl. Phys.
A108 (1968) 316.

DISCUSSION

Simmons:

McGrath et al. have discussed possible resonances in ^{5}He and/or
^{5}Li at excitations between 22 and 25 MeV. Do any related effects
show in your data?

Conzett:

We see no evidence for such states in our p-^{4}He data, and I don't think assignment of states in that region should be taken very seriously.

A Reactance Matrix Analysis
of the d-³He, p-⁴He System at
10 MeV Deuteron Laboratory Energy

G. M. HALE, D. C. DODDER, and K. WITTE, Los Alamos Scientific
Laboratory, New Mexico, USA*

We are in the process of finding a set of reactance-matrix elements
(presently at a single energy) to represent the scattering data for two-
body reactions in the five-nucleon system. The hope that such an
analysis will produce a single solution is prompted by having such
detailed measurements as polarization transfer coefficients and ana-
lyzing tensors for ^{3}He($\vec{d}$,d)^{3}He [1] and ^{3}He($\vec{d}$,p)^{4}He [2], reported elsewhere in this symposium.

The analysis is done assuming that all states from (p,^{4}He) and
(d,^{3}He) having the same total angular momentum and parity (J^P) are
coupled. In the region where all nuclear potentials vanish, one expects the radial wave function for n states coupled at fixed J^P to be

$$\sim F + GQ_N^{J^P},$$

where F and G are diagonal matrices of the regular and irregular
Coulomb functions for the n states. Q_N is a real, symmetric matrix,
so that the collision matrix,

$$U = e^{i\Omega}(1 + iQ_N)(1 - iQ_N)^{-1}e^{i\Omega},$$

is unitary. Ω is the diagonal matrix of relative Coulomb "phase shifts,"
$\omega_\ell = \sigma_\ell - \sigma_0$. An equivalent description uses the matrices

$$Q = (\sin\Omega + \cos\Omega\, Q_N)(\cos\Omega - \sin\Omega\, Q_N)^{-1}$$

and

$$U = (1 + iQ)(1 - iQ)^{-1}.$$

The search has been carried out with the Los Alamos Energy Independent
Reaction Matrix Code (EIA2) [3], using Q-matrix elements as parameters
to fit simultaneously data from the ^{3}He(d,d)^{3}He, ^{3}He(d,p)^{4}He, and

^{4}He(p,p)^{4}He reactions. Using $\ell_{max} = 4$ in both channels gives 76 parameters to fit the data (232 points), which include differential cross sections, polarizations, analyzing vectors from polarized ^{3}He target experiments, and the above-mentioned measurements for polarized incident deuterons. The best fit to date gives a weighted variance (χ^2 per degree of freedom) of 2.9, whereas ideally one expects a weighted variance of 1. We have found several solutions with higher weighted variances (4—10), but have not searched enough of the parameter space to conclude that our present solution gives the lowest value for the data being used.

One possible use of such a reactance-matrix analysis would be to guide a coupled-channel calculation that uses effective phenomenological potentials to describe the two-body reactions among five nucleons. Effective potentials for ^{4}He(p,p)^{4}He are easy to classify because of the simple structure in spin space. A classification scheme for the reaction potentials has been suggested by J. L. Gammel, based on Moravcsik's expansion for the reaction M-matrix [4]. The idea is to consider the terms of the expansion as momentum-space transforms of all the allowed potential operators. When these terms are transformed to coordinate space, one obtains in general non-local potentials which act, however, like local potentials with the identification of

$$\vec{k}+\vec{k}' \text{ with } \vec{p} = -i\nabla_{\vec{r}}, \ \vec{k}-\vec{k}' \text{ with } \vec{r}, \text{ and } \vec{k} \times \vec{k}' \text{ with } \vec{r} \times \vec{p} = \vec{L}.$$

The coordinate-space operators contain analogs of the familiar spin-orbit, tensor, and spin-spin terms. The matrix elements, which differ somewhat from the familiar case, are cited in table 1. There are obvious generalizations of the scheme that classify the ^{3}He(d,d)^{3}He potentials as well, allowing the five-nucleon reactions to be described by as few as ten effective potentials.

Table 1. Potential matrix elements ($J\ell'S' = 1/2|V|J\ell S$) for analogs of the spin-orbit, tensor, and spin-spin operators. $\vec{\chi}^\dagger$ represents the three deuteron (adjoint) spinors, and $\vec{\sigma}$ the three Pauli spin matrices.

J	ℓ'	$\vec{\chi}^\dagger \cdot \vec{L}$		$3\vec{\chi}^\dagger\cdot\hat{r}\vec{\sigma}\cdot\hat{r} - \vec{\chi}^\dagger\cdot\vec{\sigma}$		$\vec{\chi}^\dagger\cdot\vec{\sigma}$	
		$S=\frac{1}{2}$	$S=\frac{3}{2}$	$S=\frac{1}{2}$	$S=\frac{3}{2}$	$S=\frac{1}{2}$	$S=\frac{3}{2}$
$\ell+\frac{3}{2}$	$\ell+2$				$\sqrt{\frac{\ell+1}{2\ell+3}}$		
$\ell+\frac{1}{2}$		$-\frac{1}{\sqrt{3}}\ell$	$\frac{1}{\sqrt{3}}\sqrt{\ell(2\ell+3)}$	0	$\frac{1}{\sqrt{3}}\sqrt{\frac{\ell}{2\ell+3}}$	$-\sqrt{3}$	0
$\ell-\frac{1}{2}$		$\frac{1}{\sqrt{3}}(\ell+1)$	$\frac{1}{\sqrt{3}}\sqrt{(\ell+1)(2\ell-1)}$	0	$\frac{1}{\sqrt{3}}\sqrt{\frac{\ell+1}{2\ell-1}}$	$-\sqrt{3}$	0
$\ell-\frac{3}{2}$	$\ell-2$				$-\sqrt{\frac{\ell}{2\ell-1}}$		

REFERENCES

*Work performed under the auspices of the U.S. Atomic Energy Commission.

[1]   D. C. Dodder et al., Third Polarization Symp.
[2]   P. W. Keaton, Jr., et al., Third Polarization Symp.
[3]   D. C. Dodder and K. Witte (to be published).
[4]   P. L. Csonka, M. J. Moravcsik, and M. D. Scadron, Phys. Rev. 143 (1966) 775.

DISCUSSION

Rawitscher:

How many parameters entered the determination of the reactance matrix?

Hale:

For the slides shown [ℓ_{max} = 3 in the (d,^{3}He) channel and ℓ_{max} = 4 in the (p,^{4}He) channel], there are 59 independent parameters. For the case of ℓ_{max} = 4 in both channels, there are 76 parameters. Since we use the weighted variance (χ^2 per degree of freedom) as the criterion for determining the fit, the fact that the weighted variance decreased in the latter case indicates that G-waves in the (d,^{3}He) channel are probably necessary at this energy.

Rawitscher:

It would be nice to repeat this analysis for incident proton energies below the (p,d) threshold, and then compare the result with a three-body calculation, similar to that which Shanley might be able to do.

Shanley:

At the energy of your calculation, the deuteron breakup channel is open and, as I understand your calculation, it is not included. Does this mean that its effect is small?

Hale:

The fact that we were able to fit the data assuming two-particle unitarity indicates that the breakup process is not important at this energy.

Hackenbroich:

Our reactance matrix calculations based only on nucleon-nucleon forces show that one cannot get a convergent result if one does not include three-body channels. These channels give rise to a $3/2^-$ resonance.

Hale:

   We also have obtained large $3/2^-$ matrix elements from this
   analysis.

Baker:

   You will be gratified to learn that the ^{3}He analyzing power does
   indeed go negative at small angles. It therefore appears that
   your need for $\ell = 4$ in the $d + {}^3$He channel is substantiated. Why
   did you expect that this would be the case?

Hale:

   I am glad to hear that our curve follows the trend of your small-
   angle measurements. Perhaps I am confusing reactions, but I
   know that the similarity between the vector deuteron and ^{3}He
   analyzing powers has been noted in the ^{3}He$(d,p)^4$He reaction at
   10 MeV. In that case, one can present the non-dynamical argument that the two analyzing vectors are the same if only $\Delta s = 1$
   (quartet to doublet) spin transitions are allowed. There is no
   evidence from our matrix elements that these transitions dominate.

Light Nuclei

Polarization of Neutrons from the ^{3}H(^{3}He,n)^{5}Li Reaction

J. T. KLOPCIC and S. E. DARDEN, University of Notre Dame, Indiana, USA*

Among the neutron-producing reactions involving very light nuclei, the ^{3}H(^{3}He,n) reaction has received relatively little attention to date. The spectrum of neutrons [1] produced in this reaction exhibits a prominent peak corresponding to the ^{3}H(^{3}He,n)^{5}Li reaction (Q = 10.13 MeV), super-imposed on a continuous spectrum of neutrons resulting from breakup processes. We have measured the angular distribution and polarization of neutrons from the reaction ^{3}H(^{3}He,n)^{5}Li for mean ^{3}He bombarding energies of 2.70 and 3.55 MeV. A Ti-^{3}H target having a thickness of 600 keV for 3-MeV ^{3}He ions was used for the measurements. Cross-section measurements were carried out using a hydrogenous liquid scintillator detector employing pulse-shape discrimination. Neutrons from the ^{3}H(^{3}He,n)^{5}Li reaction were separated from breakup neutrons by fitting the scintillator recoil spectra with a superposition of the spectrum produced by a mono-energetic neutron group and that resulting from a continuous neutron spectrum having a statistical energy distribution. Polarization measurements utilized the recoil-coincidence technique with a high pressure ^{4}He gas scintillator. In analyzing the polarization data, the effect of breakup neutrons was taken into account by assuming these neutrons to be unpolarized.

The cross-section angular distributions for ^{3}H(^{3}He,n)^{5}Li exhibit a broad peak at $\theta_{lab} \sim 40°$. At the higher of the two bombarding energies investigated, the cross section on the peak has a value of 2.0 ± 0.2 mb/sr.

Neutron polarizations, corrected for the effect of breakup neutrons as described above, are shown in fig. 1. The uncertainties include statistical uncertainties and uncertainties introduced by the correction for the effect of breakup neutrons. Also shown in fig. 1 is the polarization calculated in DWBA assuming the reaction proceeds by simple two-proton transfer to the ^{3}H nucleus. Although this calculation reproduces qualitatively both the cross-section and polarization data for this reaction, the potentials used do not yield cross sections in agreement with the limited ^{3}H(^{3}He,^{3}He)^{3}H elastic scattering data which are available.

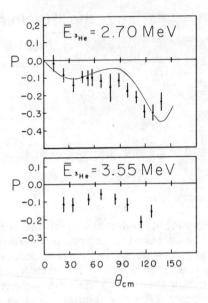

Fig. 1. Polarization of neutrons from the ^{3}H(^{3}He,n)^{5}Li reaction.

REFERENCE

*Work supported in part by the United States Office of Naval Research.

[1] J. Barry, R. Batchelor, and B. Macefield, Proc. Rutherford Jubilee Conf., ed. J. Birks (Heywood and Co., London, 1961) p. 543.

Comparison of Double Scattering and Polarized Ion Source Measurements of ^{4}He(d,d)^{4}He Polarization at 11.5 MeV

V. S. STARKOVICH and G. G. OHLSEN, Los Alamos Scientific Laboratory, New Mexico, USA*

A comparison is made between previously reported d-α double scattering measurements at 11.5 MeV [1] and recent d-α data taken with the Los Alamos Scientific Laboratory (LASL) polarized ion source. The latter measurements were taken at three deuteron energies near 11.5 MeV, and for c.m. angles 108°–126°, in order to permit the calculation of suitable averages for comparison to the low-resolution double scattering measurements. Since the double scattering measurements were regarded as primarily useful in the determination of the vector polarization of a deuteron beam [2], our primary objective was to provide an independent check of the absolute vector polarization of the beam produced by the LASL polarized ion source.

The differential cross section can be written as

$$I(\theta,\phi) = I_0(\theta)[1 + \frac{3}{2}\langle P_y \rangle P_y^0(\theta) + \frac{2}{3}\langle P_{xz} \rangle P_{xz}^0(\theta) + \frac{1}{6}\langle P_{xx} - P_{yy} \rangle$$

$$\times (P_{xx}^0(\theta) - P_{yy}^0(\theta)) + \frac{1}{2}\langle P_{zz} \rangle P_{zz}^0(\theta)], \tag{1}$$

where $I_0(\theta)$ is the cross section for an unpolarized beam. The quantities in brackets represent the beam polarization values and the remaining parameters represent the analyzing tensors for the reaction. A right-handed coordinate system with the z axis along $\vec{k}_{in}$ and the y axis along $\vec{k}_{in} \times \vec{k}_{out}$ is assumed for both the beam and the analyzing tensor quantities.

The analyzing tensors in eq. (1) can be written in terms of the beam polarization which would be produced in the inverse reaction as follows:

$$P_{xz}^0(\theta_{cm}) = -\langle P_{xz}(\theta_{cm}) \rangle, \quad P_{xx}^0(\theta_{cm}) - P_{yy}^0(\theta_{cm}) = \langle P_{xx}(\theta_{cm}) - P_{yy}(\theta_{cm}) \rangle,$$

$$P_{zz}^0(\theta_{cm}) = \langle P_{zz}(\theta_{cm}) \rangle, \quad P_y^0(\theta_{cm}) = \langle P_y(\theta_{cm}) \rangle, \tag{2}$$

where the right-hand quantities in eq. (2) represent the beam polarization produced in the inverse reaction referred to a coordinate system with the outgoing center-of-mass direction, $\vec{k}_{out}$ (c.m.) as the z axis, and $\vec{k}_{in} \times \vec{k}_{out}$ as the y axis.

Expressing both types of quantities in eq. (1) in terms of the "inverse polarizations," we obtain

$$I(\theta,\phi) = I_0(\theta)[1 + (\frac{3}{2}\langle P_y \rangle^{lab} \langle P_y(\theta_{cm}) \rangle - \frac{2}{3}\langle P_{xz} \rangle^{lab} \langle P_{xz}(\theta_{cm}) \rangle)\cos\phi$$

$$+ (\frac{1}{6}\langle P_{xx} - P_{yy} \rangle^{lab} \langle P_{xx}(\theta_{cm}) - P_{yy}(\theta_{cm}) \rangle)\cos 2\phi$$

$$+ \frac{1}{2}\langle P_{zz} \rangle^{lab} \langle P_{zz}(\theta_{cm}) \rangle], \tag{3}$$

where $\phi = 0$ corresponds to scattering to the left. For the first scattering of the double scattering experiment, we used α particles and observed recoil deuterons at 31° (118° c.m.). Thus the "lab" quantities in eq. (3) are related to the "c.m." quantities for $\theta_{cm} = 118°$ by a rotation about the y axis through $\beta = (118° - 31°) \equiv 87°$:

$$\langle P_y \rangle^{lab} = \langle P_y(118°) \rangle$$

$$\langle P_{xx} \rangle^{lab} = \langle P_{xx}(118°) \rangle \cos^2\beta + \langle P_{zz}(118°) \rangle \sin^2\beta - 2\langle P_{xz}(118°) \rangle \sin\beta\cos\beta$$

$$\langle P_{yy} \rangle^{lab} = \langle P_{yy}(118°) \rangle$$

$$\langle P_{zz} \rangle^{lab} = \langle P_{xx}(118°) \rangle \sin^2\beta + \langle P_{zz}(118°) \rangle \cos^2\beta + 2\langle P_{xz}(118°) \rangle \sin\beta\cos\beta$$

$$\langle P_{xz} \rangle^{lab} = \langle P_{xx}(118°) \rangle \sin\beta\cos\beta - \langle P_{zz}(118°) \rangle \sin\beta\cos\beta$$

$$+ \langle P_{xz}(118°) \rangle (\cos^2\beta - \sin^2\beta). \tag{4}$$

Values of the ratio

$$\frac{2(L-R)}{L+R+U+D} = \frac{\frac{3}{2}\langle P_y \rangle^{lab} \langle P_y(\theta_{cm}) \rangle - \frac{2}{3}\langle P_{xz} \rangle^{lab} \langle P_{xz}(\theta_{cm}) \rangle}{1 + \frac{1}{2}\langle P_{zz} \rangle^{lab} \langle P_{zz}(\theta_{cm}) \rangle} \tag{5}$$

obtained in the double scattering experiment at 11.5 MeV and for c.m. angles $\theta = 108° - 126°$ are illustrated in fig. 1 (triangles). In this ex-

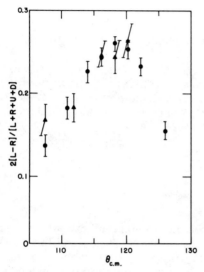

Fig. 1. Comparison of double
scattering (triangles) and ion-
source (circles) measurements.

pression L, R, U, and D represent the mean values of the number of
counts recorded in the left, right, up, and down positions by the four
telescopes. The angular resolution of these measurements was ±4°
(c.m.) and the energy resolution was ±0.5 MeV. The circles represent
values of the ratio [eq. (5)] computed from averages of the analyzing
tensors measured at 11.13, 11.50, and 12.00 MeV with the LASL polarized ion source. The beam polarization was assumed to be correctly
given by the quench ratio measurements [3]. The errors in the values
of the ratio [eq. (5)] derived from the ion-source measurements were
based on an assumed maximum systematic error of ±0.015 for P_y^0.
For the ion-source data, the statistical error is in all cases much
smaller than ±0.015. The errors presented in fig. 1 for the double
scattering measurements are statistical errors only.

If the actual vector and tensor polarization of the ion-source beam
is assumed to differ from the value given by the quench ratio determination by the factor K, a comparison of the value of the ratio [eq.
(5)] obtained from the two methods gives a value K = 0.984 ± 0.028,
where only the statistical error is taken into account for the double
scattering. Thus, this experiment confirms within the error of the
measurement that the ion-source beam polarization is as given by the
quench ratio mentioned.

For additional detail, see Los Alamos Scientific Laboratory Report
LA-4465-MS (unpublished).

REFERENCES

*Work performed under the auspices of the U.S. Atomic Energy Commission.

[1]   V. S. Starkovich, Ph.D. Thesis, University of Wyoming (1969);
      Los Alamos Scientific Laboratory Report LA-4191 (1969).
[2]   G. G. Ohlsen, V. S. Starkovich, W. G. Simon, and E. M. Bernstein,
      Phys. Lett. 28B (1969) 404.
[3]   G. G. Ohlsen et al., Third Polarization Symp.

DISCUSSION

Plattner:
　　There exist large discrepancies between the recent tensor analyzing power data from ETH Zürich and the old McIntyre-Haeberli double scattering measurements of these quantities in d-α scattering. I think there are now three independent recent calibrations of the d-α tensor analyzing power, and such discrepancies should be ironed out. Could you comment on this problem?

Starkovich:
　　We are planning further work on this problem, but have no reason at present to believe our measurements are incorrect. I believe, however, that we should await the $^{16}O(d,\alpha)^{14}N^*$ results that are presently being obtained at several laboratories.

Grüebler:
　　The error in the absolute values of our analyzing power is considered to be ±10%. There are, however, indications from the analyzing power measurements of the $^3He(d,p)^4He$ reaction at 0° that our beam polarization is somewhat higher than we assumed.

Measurement of the Vector and Tensor Analyzing Power of d-⁴He Elastic Scattering at 12 MeV

G. P. LAWRENCE, D. C. DODDER, P. W. KEATON, Jr., D. D. ARMSTRONG, J. L. McKIBBEN, and G. G. OHLSEN, Los Alamos Scientific Laboratory, New Mexico, USA*

The vector and tensor analyzing powers of d-⁴He elastic scattering at 12-MeV deuteron laboratory energy have been measured by means of methods and equipment described in other papers in this symposium [1, 2]. The experimental results together with the phase shift fits to be discussed below are shown in fig. 1. The Cartesian tensors $P_i = S_i$ and $P_{ij} = 3(S_i S_j + S_j S_i)/2 - 2\delta_{ij}$ are used as a basis. Note that the observables must be bounded by $-2 < P_{zz}^0 < 1$, $-3/2 < (P_{xx}^0 - P_{yy}^0)/2 < 3/2$, $-3/2 < P_{xz}^0 < 3/2$, and $-1 < P_y^0 < 1$, and that only a part of the range of possible variation is shown on the graphs. (As is well known, the observables are further related by certain inequalities.)

The tensor analyzing powers are well fit by parameters consistent with those obtained at lower energies by McIntyre and Haeberli [3], while the vector analyzing power is only qualitatively fit. In an initial attempt to obtain an improved fit to the present data, the prediction for the cross section differed violently from the measured cross sections [4] at 11.5 MeV (the closest energy at which data is available). This convinced us that it would be necessary to include cross-section data to obtain a meaningful fit; accordingly, the cross sections of ref. [4] were extrapolated to 12 MeV and included in the analysis. Starting with phase shifts extrapolated from those in ref. [3], a solution was obtained; the parameters are given in table 1. The computer program used in the search was the Los Alamos General Energy Independent Reactance Matrix Analysis Code (E1A2) [5]. This code fits the data with a unitary collision matrix and a set of inelastic parameters which multiply the diagonal elements of the collision matrix. This non-general, non-unitary matrix differs from the non-general, non-unitary matrix used in ref. [3], so the results are not strictly comparable for those states connected with non-diagonal elements of the collision matrix. The 2 × 2 submatrices of the nuclear unitary collision matrix are given in terms of the nuclear bar representation [6].

We have not shown the fit to the extrapolated cross section; it is of quality comparable to the fits of polarization data shown in fig. 1.

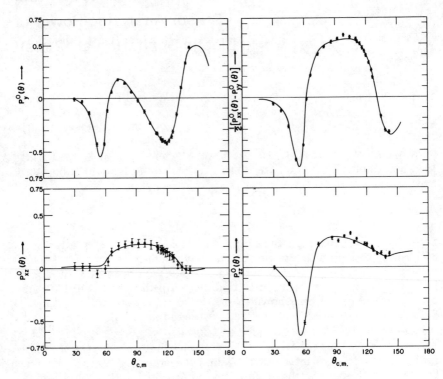

Fig. 1. Analyzing tensors for ^{4}He($\vec{d}$,d)^{4}He elastic scattering at 12 MeV.

Table 1. Nuclear bar phase shifts and inelastic parameters

State	$\bar{\delta}_{LJ}$ or $\bar{\varepsilon}_{L-L'}$ (degrees)	Inelastic parameters
S_1	−134.0	0.95
S_1–D_1	6.0	
D_1	−77.4	0.41
P_2	−3.3	0.64
P_2–F_2	−5.0	
F_2	3.8	0.97
P_1	−14.9	0.93
P_0	−28.1	1.0
D_3	−28.9	0.68
D_3–G_3	0.47	
G_3	4.7	1.0
D_2	−59.8	0.89
F_4	8.2	0.74
F_3	6.8	0.91
G_5	3.8	0.84
G_4	1.7	0.96

The overall weighted variance (χ^2 per degree of freedom) is 1.4. This is subject to change, since the estimated errors for the polarization data were preliminary, and since the errors on the extrapolated cross sections are somewhat crudely estimated.

The solution given here should not be taken as evidence that the presently available data at this energy give a unique set of parameters. Some 15 starts from random sets of parameters have not led to this solution, but have instead led to solutions with larger weighted variance (ranging from 4 to 13). This is not a sufficient search of the space to ensure that other solutions do not exist; the only evidence for that conclusion at present comes from the connection to the lower energy analysis.

REFERENCES

*Work performed under the auspices of the U.S. Atomic Energy Commission.

[1]   G. P. Lawrence et al., Third Polarization Symp.
[2]   D. C. Dodder et al., Third Polarization Symp.
[3]   L. C. McIntyre and W. Haeberli, Nucl. Phys. A91 (1967) 382.
[4]   L. S. Senhouse and T. A. Tombrello, Nucl. Phys. 57 (1964) 624.
[5]   D. C. Dodder and K. Witte, to be published.
[6]   H. P. Stapp, T. J. Ypsilantis, and N. Metropolis, Phys. Rev. 105 (1957) 302.

Measurement of the Tensor Moments
$\langle iT_{11}\rangle$, $\langle T_{20}\rangle$, $\langle T_{21}\rangle$, and $\langle T_{22}\rangle$
in Deuteron-Alpha Elastic Scattering
between 3 and 11.5 MeV

V. KÖNIG, W. GRÜEBLER, P. A. SCHMELZBACH, and P. MARMIER,
Laboratorium für Kernphysik, ETH, Zürich, Switzerland

The polarization parameters $\langle iT_{11}\rangle$, $\langle T_{20}\rangle$, $\langle T_{21}\rangle$, and $\langle T_{22}\rangle$ of deuterons scattered on ^{4}He have been measured using the polarized deuteron beam from the ETH tandem accelerator. In the energy range of 3 to 11.5 MeV eleven angular distributions were measured at c.m. angles between 33.6° and 151°.

For these measurements different directions of the spin axis of the beam are necessary. These could be set by means of a Wien-filter. For the determination of the tensor parameters the sign of the tensor polarization was changed every few seconds and, in addition, for each scattering angle θ two detectors were used, placed at the left and at the right in the scattering plane. Suitable addition of the detector counting rates eliminated the contributions of the unwanted tensor parameters. Thus only a rough determination of the angle of the spin axis is necessary. A more detailed description of the method is given in ref. [1].

The experimental arrangement in the scattering chamber allows the simultaneous measurement at four angles. At forward angles further information is obtained from the recoil-alpha-particles. For the determination of the vector polarization of the beam, the value $\langle iT_{11}\rangle = 0.300$ at $\theta_{LAB} = 45°$ and an energy of 6.85 MeV was taken as the normalization for all measurements; this agrees with the measurement by Trier and Haeberli [2]. At 6.85 MeV the values of $\langle T_{20}\rangle = 0.497 \pm 0.011$ and $\langle T_{22}\rangle = 0.632 \pm 0.012$ at $\theta_{LAB} = 60°$ have been determined simultaneously with a measurement of $\langle iT_{11}\rangle$ at $\theta_{LAB} = 45°$ for the calibration of the tensor polarization.

The measured angular distributions show that the polarization varies slowly over most angles and energies. All parameters, except $\langle T_{21}\rangle$, can attain large values, thus each can be used as an analyzer for vector or tensor polarized beams. Above 6 MeV the shape of the angular distribution of $\langle T_{22}\rangle$ is quite similar to that of $\langle T_{20}\rangle$. Further, the vector parameter $\langle iT_{11}\rangle$ crosses zero at exactly 90°. However, $\langle T_{21}\rangle$ has a maximum in this region. Fig. 1 shows typical angular distributions at three different energies. In general the statistical error of all measurements was 0.01, which is smaller than the dots. The lines between the dots are drawn only for the sake of clarity.

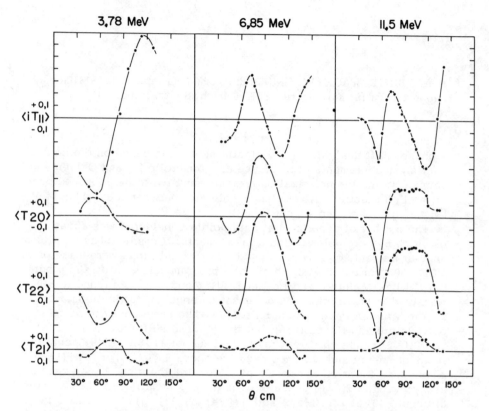

Fig. 1. Tensor moments $\langle iT_{11}\rangle$, $\langle T_{20}\rangle$, $\langle T_{21}\rangle$, and $\langle T_{22}\rangle$ in deuteron–alpha elastic scattering at 3.78, 6.85, and 11.5 MeV.

REFERENCES

[1] W. Grüebler, V. König, P. A. Schmelzbach, and P. Marmier, Nucl. Phys. A134 (1969) 686, Nucl. Phys. A148 (1970) 380, and Nucl. Phys. A148 (1970) 391.
[2] A. Trier and W. Haeberli, Phys. Rev. Lett. 18 (1967) 915.

Phase-Shift Analysis
of d-α Elastic Scattering

P. A. SCHMELZBACH, W. GRÜEBLER, V. KÖNIG, and P. MARMIER,
Laboratorium für Kernphysik, ETH, Zürich, Switzerland

A phase-shift analysis of d-α elastic scattering based on the cross section measurements of Senhouse and Tombrello [1] and Ohlsen and Young [2], and the polarization measurements with the polarized beam of the ETH electrostatic tandem accelerator between 3 and 11.5 MeV [3, 4, 5] has been carried out.

The number of experimental data points considered was about 850 of which 600 are polarization measurements. Complex phase shifts of angular momentum $\ell \leq 3$ were used. The calculations were made on a CDC 6500 computer with the aid of a program derived from the program SPINONE of McIntyre and Haeberli [6]. Good fits were found to all measured cross section and polarization angular distributions.

The S and D-waves phase shifts have the same behavior as those of McIntyre and Haeberli [6], but the P phase shifts are quite different. The P(J = 2) phase shift reaches a maximum of +10° at 6 MeV deuteron incident energy and becomes negative above 9 MeV. For the P(J = 1) phase shift the maximum is +7° at 6 MeV and the sign change occurs at 8 MeV. The P(J = 0) phase shift is always negative and has a value of about −40° at 11.5 MeV. Both F(J = 4) and F(J = 3) phase shifts increase monotonically with energy and reach values of about 10° at 11.5 MeV. The F(J = 2) phase shift has small negative values. We find good agreement between calculated and measured total reaction cross sections at 10 MeV.

Values from smooth curves drawn through the phase shifts found in the search were used to calculate the contour plots shown in fig. 1. The z-axis is given by the direction of the outgoing deuterons in the center-of-mass system. The y-axis is parallel to $\vec{k}_{in} \times \vec{k}_{out}$.

The overall agreement with the measured tensor moments is quite good, except in the 5 MeV region where no good continuation of the eigen-phase shifts as a function of the energy has yet been found. An attempt to increase the quality of the contour plots and to extend this analysis to lower and higher energies will be made.

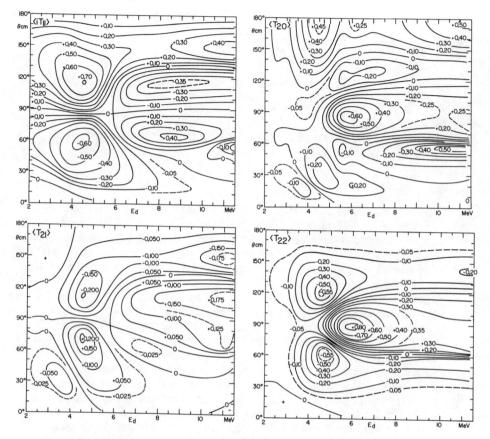

Fig. 1. Contour plots of the polarization tensor moments $\langle T_{qk} \rangle$ in d–α elastic scattering between 2 and 11.5 MeV. The coordinate system used is defined in the text.

REFERENCES

[1] L. S. Senhouse, Jr. and T. A. Tombrello, Nucl. Phys. 57 (1964) 624.
[2] G. C. Ohlsen and P. C. Young, Nucl. Phys. 52 (1964) 134.
[3] W. Grüebler, V. König, P. A. Schmelzbach and P. Marmier, Nucl. Phys. A134 (1969) 686.
[4] V. König, W. Grüebler, P. A. Schmelzbach and P. Marmier, Nucl. Phys. A148 (1970) 380.
[5] W. Grüebler, V. König, P. A. Schmelzbach and P. Marmier, Nucl. Phys. A148 (1970) 391.
[6] L. C. McIntyre and W. Haeberli, Nucl. Phys. 91 (1967) 382.

Polarization in ^{3}He-^{4}He Elastic Scattering at Low Energies

W. R. BOYKIN, D. M. HARDY, and S. D. BAKER, Rice University,
Houston, Texas, USA*

The experimental determination of the phase shifts in ^{3}He-^{4}He elastic
scattering at low energies has depended up to now on differential
cross-section measurements alone. Barnard et al. [1] found that two
sets of phase shifts fit the differential cross section equally well in
the energy range of ^{3}He laboratory energy between 2.5 and about 5.0
MeV (nearly up to the first 7/2$^-$ resonance in ^{7}Be). The principal
difference between these sets is the sign of the splitting of the P-
wave phase shifts. On the resonance, the set for which the $P_{3/2}$
phase is higher than the $P_{1/2}$ phase gave a better fit. This sign of
the splitting appears to maintain itself at higher energies [2]. However, the phase shift search becomes more difficult at higher energies
with the onset of inelastic processes and the necessity to include
more partial waves in the search, and it is difficult to estimate the
precision with which the data specify the various phase shift parameters. It is therefore of interest to examine the scattering process at
energies below the 7/2$^-$ resonance, where the scattering is purely
elastic and the F waves are still relatively unimportant. Using polarized ^{3}He targets prepared by optical pumping, we have measured the
analyzing power P_3 at three c.m. angles: 72°, 87°, and 120°. The upper part of fig. 1 shows P_3 plotted vs the equivalent ^{3}He laboratory
energy $E_{lab}(^3He)$.

The cross-section data of Barnard et al. and the present polariza-
tion data were fitted using $S_{1/2}$, $P_{1/2}$, $P_{3/2}$, $F_{5/2}$, and $F_{7/2}$ phase shifts.
The D-wave phase shifts were set equal to zero because even at higher
energies they do not clearly depart from zero. Only one acceptable
set of phase shifts could be found at each energy. The $P_{3/2}$ phase
shifts are plotted as solid circles in the lowest frame in fig. 1, and
the $P_{1/2}$ phase shifts are plotted as open circles. The error bars are
estimated standard deviations in the phase shifts at each energy. The
lines are single-level fits whose resonance parameters are shown in
table 1. The sign of the splitting of the P-wave phase shifts agrees
with refs. [1] and [2], although the magnitude is somewhat different.
Indeed, there is nothing in table 1 to indicate any sort of anomalous
behavior whatever.

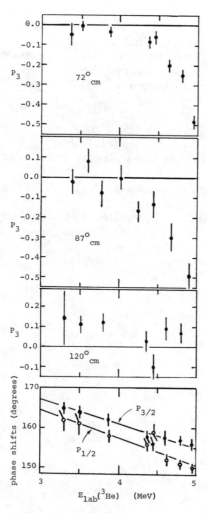

Fig. 1

Table 1

^{7}Be states	J^π	E_{res} (MeV)	γ^2 (MeV)	R (fm)	Ratio to Wigner limit
ground state	$3/2^-$	-1.587	1.237	3.07	0.25
first excited state	$1/2^-$	-1.155	0.762	3.53	0.25

The effects of possible systematic errors in the cross-section data, in the polarization data, and in the choice of the D-wave phase shifts was investigated, and it was found that the P phases can be moved up and down a few degrees but that the error bars and the P-wave splitting are not significantly affected.

REFERENCES

*Supported by the U.S. Atomic Energy Commission.

[1] A. C. L. Barnard, C. M. Jones, and G. C. Phillips, Nucl. Phys. 50 (1964) 629.

[2] R. J. Spiger and T. A. Tombrello, Phys. Rev. 163 (1967) 964.

The $^6Li(\vec{d},\alpha)^4He$-Reaction below 1 MeV

H. RUDIN, P. HUBER, H. P. NÄGELE, R. NEFF, and F. SEILER,
University of Basel, Switzerland

Using the polarized deuteron source of the Basel 1-MV accelerator [1] we have measured the analyzing powers of the $^6Li(\vec{d},\alpha)^4He$-reaction at 0.4, 0.6, 0.8, and 0.96 MeV. Typical deuteron beam characteristics were a polarization of $P_{33} = 0.60$ and a 15-nA beam current on a target spot of 0.4×0.7 cm^2. Fig. 1 shows the measured analyzing powers $D_1(\theta)$, $D_{33}(\theta)$, $D_{13}(\theta)$, and $D_{11}(\theta) - D_{22}(\theta)$ multiplied by the normalized cross section $W_0(\theta)$ for unpolarized particles.

At energies below 1 MeV the s-wave matrix element R_1, forming the 2^+-resonance, dominates and small d-wave contributions R_i have been found [2]. At 400 keV an analysis in terms of $(R_1 R_i^*)$ explains most of the effects observed. Good fits are obtained by including an s-wave 0^+-contribution. At energies above 600 keV the $\ell = 6$ terms in the Legendre polynomial expansion necessitate the introduction of sizable amplitudes from a d-wave 4^+-state. The curves in fig. 1 show the final analysis. For the larger matrix elements R_k terms $|R_k|^2$ have been included in addition to $(R_1 R_k^*)$. Our analysis thus shows contributions from states 2^+, 0^+, and 4^+ in agreement with other authors [3]. Because of our limited energy range no level energies can be assigned.

REFERENCES

[1] H. A. Grunder et al., Helv. Phys. Acta, to be published.
[2] H. Bürgisser et al., Helv. Phys. Acta 40 (1967) 185.
[3] H. Meiner, private communication; R. M. Freeman and G. S. Mani, Proc. Phys. Soc. 85 (1965) 267.

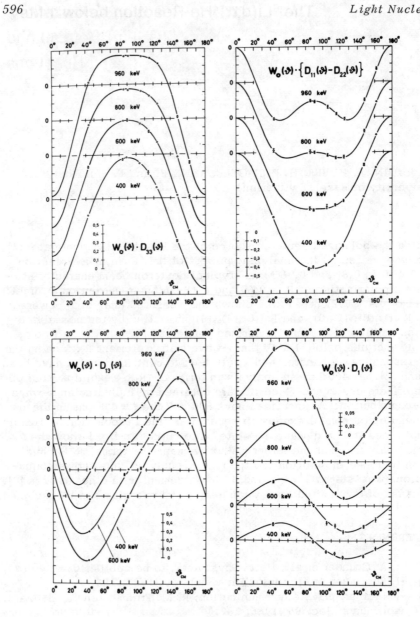

Fig. 1. Components of the analyzing power. The solid
lines show the result of the analysis using 2⁺, 0⁺, and
4⁺-states.

Proton Polarizations
from ^{6}Li(^{3}He,p)^{8}Be(g.s.) and
(f.e.s.) Reactions

D. G. SIMONS, U.S. Naval Ordnance Laboratory, Silver Spring,
Maryland, USA

Proton polarization measurements from the ^{6}Li(^{3}He,p)^{8}Be reaction have
been carried out in the energy range from 1.4 to 2.0 MeV on the 2.5
MV NOL Van de Graaff accelerator. This reaction is observed to have
only two proton groups. Those leading to the ground state (g.s.) and
first excited state (f.e.s.) of ^{8}Be with Q-values of 16.7 and 13.8 MeV,
respectively. Furthermore, excitation curves show a resonance in the
f.e.s. proton group at a ^{3}He energy of 1.6 MeV [1] leading to a compound nucleus of ^{9}B. No similar resonance is observed in the g.s.
proton group.

This reaction therefore offers an interesting opportunity to study the
proton polarization results from a compound nucleus reaction mechanism simultaneously with those which do not appear to result from the
same compound nucleus formation.

Measurements were simultaneously made on the g.s. and f.e.s. proton groups at ^{3}He energies from 1.4 to 2.0 MeV (across the 1.6 MeV
resonance) at lab angles from 0° to 90°. Polarizations were determined
using a carbon polarimeter which has been previously described [2]
with Li drifted Si detectors. The results of the measurements were
quite different for the two proton groups.

The g.s. results are shown in fig. 1. Here we see a large valley of
negative polarization over almost the entire measurement range. There
is no identifying structure at the 1.6 MeV resonance. (The lines connecting the experimental points are merely for identification of points
made at the same beam energy.)

The f.e.s. polarization measurements are shown in fig. 2. Here,
although the magnitude of the polarization is smaller, structure around
the 1.6 MeV resonance appears to be evident as a change in sign of
the polarization.

Future work will consist of an analysis of the f.e.s. results in an
attempt to determine the appropriate spins and parities of compound
states in the ^{9}B nucleus using a resonance analysis method similar
to that used at Ohio State [3].

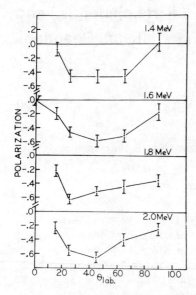

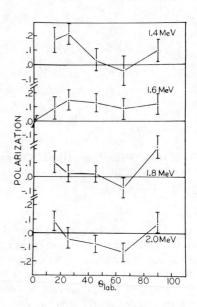

Fig. 1. g.s. proton group.   Fig. 2. f.e.s. proton group.

REFERENCES

[1]   J. P. Schiffer et al., Phys. Rev. 104 (1956) 1064.
[2]   D. G. Simons and R. W. Detenbeck, Phys. Rev. 137 (1965) B1471.
[3]   T. R. Donoghue, private communication.

Remeasurement of the Neutron Polarization from the $^7Li(p,n)^7Be$ Reaction for 3 to 4 MeV Protons

R. A. HARDEKOPF and R. L. WALTER, Duke University and TUNL, Durham, North Carolina; J. M. JOYCE, University of North Carolina at Chapel Hill and TUNL; G. L. MORGAN, North Carolina State University at Raleigh and TUNL, USA*

The $^7Li(p,n)$ reaction is known to be a useful source of low energy polarized neutrons, especially for incident proton energies of 3 to 4 MeV. At 50° (lab), this corresponds to neutron energies of about 1 to 2 MeV. In addition to the large polarization and cross section in this energy range, the use of a solid target and relative ease of target fabrication enhance the usefulness of this reaction.

The most recent measurements of the polarization from the $^7Li(p,n)^7Be$ (ground state) reaction at low energies were made by Andress et al. [1]. The polarization of the $^7Li(p,n)^7Be$ (431 keV) reaction was measured at 3.1, 3.2, and 3.3 MeV by Morgan et al. [2] and found to be near zero. Improvements in experimental methods and computer codes developed for analysis of neutron polarization measurements since that time have led us to remeasure the polarization in this energy range. The goal was to confirm the trend of the polarization found by Andress et al., and to calibrate accurately the reaction as a source of polarized neutrons. A secondary goal was to measure the polarization of the n_1 group and thus extend Morgan's earlier results to higher energies.

For this purpose, measurements were made in 50 keV steps for proton energies from 3 to 4 MeV. The natural Li target was about 30 keV thick to protons at 3.5 MeV. The polarimeter consisted of a high pressure helium scintillator as an analyzer and organic scintillators as side detectors. A spin precession solenoid was used to interchange the role of the side detectors in order to cancel instrumental asymmetries. The data handling program utilized the TUNL on-line computer facility to accumulate simultaneously true-coincidence and random-coincidence spectra as well as to control the spin precession solenoid and monitor experimental quantities.

The combined resolution (FWHM) of the helium cell and the geometry associated with the size of the side detectors was about 20%. This was sufficient to allow separation of the n_1 group from the n_0 group by a search program that fits the coincidence gated recoil spectrum with a sum of gaussians and an unpolarized linear background. This background arose from a number of sources, including room scattered neutrons and gamma radiation. The effects of the linear background and the presence of the n_1 group were studied in detail, since they affect the asymmetry of the n_0 peak. The errors of approximately ±0.014 obtained for the n_0 polarization include the statistical error due to their subtraction, plus a reasonable upper limit on possible error in assignment of the linear background (which averaged about 5% of the peak height for the n_0 group).

The average analyzing power of the polarimeter was calculated at each energy by a computer code that includes finite geometry and double scattering effects. The phase shifts used were those of Stammbach and Walter [3]. Fig. 1 shows the final polarizations obtained compared with the measurements of refs. [1] and [2] corrected by using the Stammbach and Walter phase shifts. The solid line is a least-squares fit to the n_0 data of a cubic polynomial. The results for the n_1 group are consistent with zero polarization.

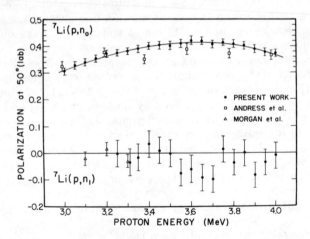

Fig. 1. Polarization of neutrons from the ^{7}Li(p,n) reaction in the region of proton energies from 3 to 4 MeV at an emission angle of 50° (lab).

REFERENCES

*Work supported by U.S. Atomic Energy Commission.

[1] W. D. Andress, Jr., F. O. Purser, J. R. Sawers, and R. L. Walter, Nucl. Phys. 70 (1965) 313.
[2] G. L. Morgan, C. E. Hollandsworth, and R. L. Walter, Second Polarization Symp., p. 523.
[3] Th. Stammbach and R. L. Walter, to be published.

Investigation of the Reactions ^{9}Be(p,d)^{8}Be and ^{9}Be(p,α)^{6}Li with Polarized and Unpolarized Protons of 300 to 1250 keV

R. KECK, H. SCHOBER, and H. P. JOCHIM, Otto-Hahn-Institut, Mainz, Germany

In view of the interest in the ^{10}B nucleus (e.g., isospin impurity of levels; conjectured levels at 6.95, 7.00, and 7.20 MeV; "parity-doublets" at 5 and 7.5 MeV; possible presence of "tensor-spin-orbit" forces) the differential analyzing power and the angular distributions of the reactions ^{9}Be(p,d)^{8}Be and ^{9}Be(p,a)^{6}Li have been measured or re-measured, respectively, in the energy range from 300 to 1250 keV. The data have been fitted by Legendre polynomials in the usual manner according to the least squares method. Figs. 3 and 4 show the energy variation of the various coefficients. The experimental points are connected to guide the eye.

Since so far only four angles (30°, 45°, 90°, and 135°) were included in the polarization measurements, preliminary contour-plots were obtained from interpolation of the fits of the differential analyzing power (figs. 1, 2).

The integrated partial cross sections are shown in fig. 5. Because of the great number of matrix-elements and the variety of possible

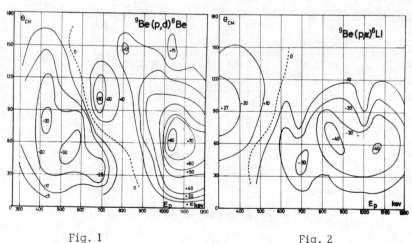

Fig. 1   Fig. 2

direct contributions a satisfactory theoretical analysis does not seem
attainable at the moment. A qualitative inspection of variations of
the coefficients a_1 at 500 keV in the angular distributions and of a_1^1
and a_3^1 at about 700 keV in the polarization data for both reactions
seems to support arguments for the assumption of broad levels at
these energies. Additional structure in the (p,d) polarization at 400
keV is seen. This fact as well as the rather high polarization at the
energy of the well-known 1⁻ level at 6.88 MeV, for which s-wave excitation is assumed, might help in the interpretation of the alleged
high (20%) isospin impurity of this level.

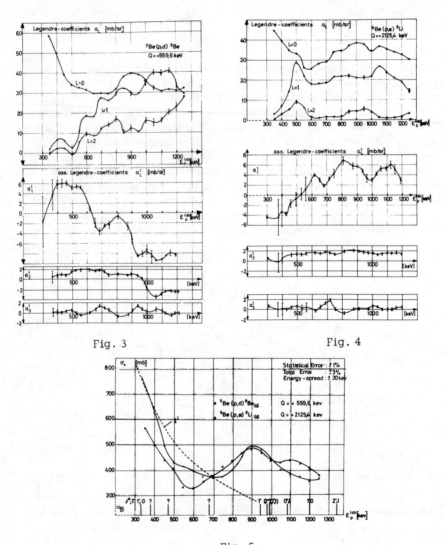

Fig. 3   Fig. 4

Fig. 5

The Polarization of ^{3}He Scattered from
^{9}Be, ^{12}C, and ^{16}O at 18 and 20 MeV

W. S. McEVER, T. B. CLEGG, J. M. JOYCE, and E. J. LUDWIG,
University of North Carolina at Chapel Hill and TUNL; R. L. WALTER,
Duke University and TUNL, Durham, North Carolina, USA*

The present work describes measurements of the polarization of ^{3}He
particles scattered from ^{9}Be, ^{12}C, and ^{16}O in the angular range from
20° to 60° c.m. The analysis of polarization and cross-section angular
distributions can provide valuable information as to the spin dependence and general optical model description of the ^{3}He-nucleus interaction. The polarimeter used in this work has been described elsewhere [1]. Incident ^{3}He particles scatter from the target and enter a
^{4}He gas cell with energies which have been degraded to values between 11.5 and 13 MeV. The analyzing power in this energy range is
approximately 0.70 [1, 2]. The scattered particles were detected using
a pair of matched detector telescopes. Except at the extreme backward
angles the background counts were less than 10% of the peak counts.

Fig. 1 shows the ^{3}He elastic scattering angular distributions at a
mean bombarding energy of 18 MeV for ^{9}Be, ^{12}C, and ^{16}O and at 20 MeV

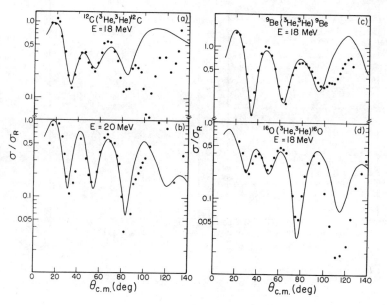

Fig. 1

for ^{12}C. Optical model fits using a standard Woods-Saxon potential
were obtained with the code JIB3 for sets of parameters related by the
discrete and continuous ambiguities and for sets containing volume
or surface absorption.

The polarization data are contained in fig. 2 along with polarization
predictions made with the parameters in table 1. Parameter sets which
are members of the continuous ambiguity yielded predictions of essentially identical cross-section distributions while the polarization calculations differ in the magnitudes of the polarizations predicted for
the same spin-orbit potential. Fig. 2(a) shows that predictions of
larger polarizations result from the shallower real potential wells.
This implies that the spin-orbit potentials required to produce a certain magnitude polarization depend on the depth of the real well. For
parameter sets with the expected real well depths of three times the
nucleon potential, i.e., ≈ 150 MeV, the spin-orbit well depth required
by the polarization data is about 4-5 MeV.

Most sets of optical model potentials related by the discrete ambiguity produced similar predictions for cross-section and polarization
distributions. It has been found, however, that some related parameter
sets which produce similar fits to the cross-section data predict polarizations which are opposite in sign to the measured polarizations.
The dashed curve in fig. 2(c) represents such an example.

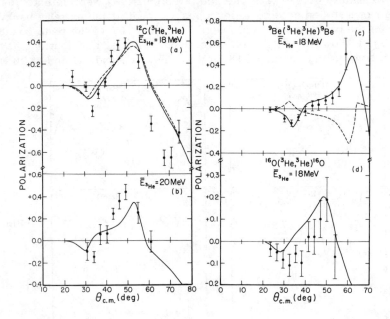

Fig. 2

Table 1

Target	E (MeV)	V (MeV)	r_v (fm)	a_v (fm)	W_s (MeV)	W_D (MeV)	r_w (fm)	a_w (fm)	V_{so} (MeV)
^{12}C (dashed)	18	150.0	1.07	0.69		6.65	1.40	1.02	4.0
^{12}C (solid)	18	130.0	1.20	0.67		7.0	1.35	1.03	4.0
^{12}C	20	150.0	1.07	0.74		13.7	1.50	0.70	5.0
^{9}Be (solid)	18	140.0	1.27	0.63	16.5		1.51	1.04	5.0
^{9}Be (dashed)	18	190.0	1.21	0.69		19.7	1.26	0.75	5.0
^{16}O	18	130.0	1.07	0.79		9.17	1.67	0.72	4.0

$r_{so} = 0.9\, r_v$, $a_{so} = a_v$, $r_c = 1.3$ fm

REFERENCES

*Work supported in part by the U.S. Atomic Energy Commission.

[1] W. S. McEver et al., Phys. Lett. 31B (1970) 560.
[2] D. D. Armstrong, L. L. Catlin, P. W. Keaton, Jr., and L. R. Veeser, Phys. Rev. Lett. 23 (1969) 135.

Neutron Polarization from the
$^{11}B(\alpha,n)^{14}N$ Reaction at Low Energies

C. PENCEA, O. SĂLĂGEAN, A. CIOCĂNEL, and M. MOLEA, Institute for Atomic Physics, Bucharest, Rumania

The reaction $^{11}B(\alpha,n)^{14}N$ in the energy range 3–4 MeV has been studied by Bonner [1], Haddad [2], Calvert [3], and Mani and Dutt [4] who measured the excitation function and angular distributions from 2 to 6 MeV alpha-particle energy (figs. 1, 2 from ref. [4]).

Experimental investigations of nuclear reactions require for completeness polarization measurements, in the case under consideration the polarization of the neutrons leaving ^{14}N in the ground state.

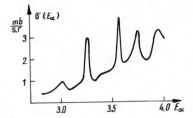

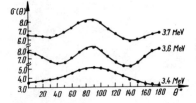

Fig. 1. Excitation function of the $^{11}B(\alpha,n_0)^{14}N$ reaction.

Fig. 2. Angular distribution of the neutrons from $^{11}B(\alpha,n_0)^{14}N$.

Using a He^{4++}-beam (a natural modulation of 3 ns and a current of 5 μA) accelerated on a multiple frequency, the angular distribution of the neutron polarization at four energies (3.4, 3.6, 3.7, 3.8 MeV) situated almost on the resonances and between the resonances were obtained.

The experimental set-up and standard electronics [5] for the time-of-flight technique are shown in fig. 3, together with a typical direct neutron spectrum ($\theta_n = 30°$, $E_\alpha = 3.6$ MeV).

The analyzing power of ^{12}C [6] and measured asymmetries allow the determination of the polarization of the neutrons from the alpha-bombardment of ^{11}B (target thickness 1 mg/cm^2). In fig. 4, the polarizations are fitted by the least-squares method to the expression

$$P(\theta) = \sum_K A_K P_K^1 (\cos \theta) / \sum_k a_k P_k (\cos \theta),$$

where the a_k-coefficients are taken from Mani and Dutt [4].

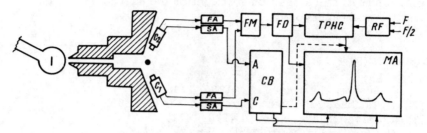

Fig. 3. Experimental arrangement for polarization measurements (C_1, C_2-stilbene detectors, FA-fast preamplifier, FM-fast mixer, FD-fast discriminator, TPHC-time to pulse height converter, CB-command block, RF-radio frequency, MA-multichannel analyzer).

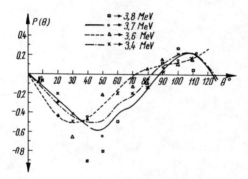

Fig. 4. Angular distribution of the neutron polarization and fitted curves for the ^{11}B(a,n_0)^{14} reaction.

Some relevant aspects in figs. 1, 2, and 4 provide qualitative arguments for a preliminary discussion of the ^{11}B(a,n_0)^{14}N-reaction at low energies. The excitation function shows a resonant structure in this region where one may distinguish five levels ($\Gamma \ll D$) in the excited ^{15}N compound nucleus, with different spins and/or parities. The differential cross sections have fore-aft symmetry and generally change slowly with the energy of the incident particles. The angular distributions of the polarization have maxima in the angular range 30°-50°, in agreement with the supposition that in the reaction small orbital momenta ($\ell \leq 1$) participate and that these maxima are insignificantly shifted ($\pm 10°$) by the variation of the alpha energy. These facts suggest the influence of resonance interferences on the shape and values of the cross sections and polarization.

In conclusion, a direct transition is less probable at low energies, the $^{11}B(a,n_0)^{14}N$ reaction proceeds via a compound mechanism, but only a numerical calculation will confirm the assumption of a predominant compound process.

REFERENCES

[1]   T. Bonner, Phys. Rev. 102 (1956) 1348.
[2]   D. Haddad, Bull. Amer. Phys. Soc. 2 (1957) 309.
[3]   J. Calvert, Nucl. Phys. 31 (1962) 471.
[4]   G. Mani and G. Dutt, Nucl. Phys. 78 (1966) 613.
[5]   A. Ciocănel et al., Rev. Roum. Phys. 15 (1970) 563
[6]   I. Minzatu, Stud. Cerc. Fiz. 17 (1965) 13.

Polarization of 3-7 MeV Neutrons Scattered from 12C

J. L. WEIL and W. GALATI,* University of Kentucky, Lexington, USA†

A polarization map for 3-7 MeV neutrons elastically scattered from
12C is presented in fig. 1. The polarizations were obtained from the
results of a phase-shift analysis of differential cross-section measurements [1, 2]. Polarizations have been calculated using preliminary
phase-shift values read from smooth curves drawn through the measured
phase shifts. Refinements now being made in the normalization of
some of the data may result in small changes in the final values of
the phase shifts and of the polarizations. The regions around 5.4 and
6.3 MeV have been left blank because they are very complex, but the
polarizations are available from the authors.

A comparison is made in figs. 2 and 3 between all the presently
available data in the 3-4 MeV region [3-6] and the polarization calculated from the phase shifts of Wills et al. [7], Meier et al. [8], and
from our new phase-shift analysis [2]. At 135°, the present calculations fit the data slightly better than those of Wills et al. [7] and much

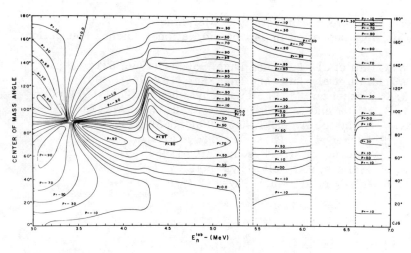

Fig. 1. Polarization calculated from phase shifts.

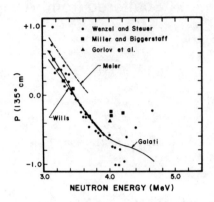

Fig. 2. Comparison with experimental data at 135°.

Fig. 3. Comparison with experimental data at 45°.

better than those of Meier et al. [8] (see fig. 2). At 45°, the present
phase shifts predict the polarization much better than those of either
Wills or Meier (see fig. 3).

The polarizations presented here also agree fairly well with the
angular distribution of Gorlov et al. [5] at 4.0 MeV and that of Kelsey
et al. [9] at 4.38 MeV. Kelsey's [9] three data points at $\theta_{cm} = 54°$ between 6 and 7 MeV are also consistent with the present results. Unfortunately, no other data exist in this energy region against which
the calculated polarizations can be checked.

REFERENCES

*Present address: University of Maryland.
†This work supported in part by the National Science Foundation.

[1] W. Galati et al., Bull. Am. Phys. Soc. 11 (1966) 831.
[2] W. Galati et al., Bull. Am. Phys. Soc. 13 (1968) 1388.
[3] B. E. Wenzel and M. F. Steuer, Phys. Rev. B137 (1965) 80.
[4] T. G. Miller and J. A. Biggerstaff, Nucl. Phys. A124 (1969) 637.
[5] G. V. Gorlov et al., Dokl. Akad. Nauk SSSR 9 (1965) 806.
[6] W. P. Bucher et al., Phys. Rev. 115 (1959) 961.
[7] J. E. Wills, Jr., et al., Phys. Rev. 109 (1958) 891.
[8] R. W. Meier et al., Helv. Phys. Acta 27 (1954) 577.
[9] C. A. Kelsey et al., Nucl. Phys. 68 (1965) 413.

Scattering of 14-MeV Polarized
Neutrons by ^{12}C, ^{16}O, and D

R. SENE, P. DELPIERRE, J. KAHANE, and M. de BILLY de CRESPIN,
Collège de France, Paris, France

We present measurements performed with "Penelope," the polarized-neutron source of the Collège de France [1].

Polarized 14.1-MeV neutrons were scattered by ^{12}C, ^{16}O (H_2O target) [2], and D (D_2O and C_6D_6 targets with deuteron recoil detection for the latter) using an eight-detector time-of-flight spectrometer [9].

The optical-model analysis of ^{12}C and ^{16}O results was performed with MAGALI [3]. The potential used consists of a real central potential of the Woods-Saxon type (V_{Sax}, A_{Sax}, R_{Sax}), a surface-peaked imaginary potential with a derivative Woods-Saxon form factor (W_{sur}, A_{sur}, R_{sur}), and a spin-orbit potential of the Thomas type (V_{ls}, R_{ls}, A_{ls}).

^{12}C *elastic scattering.* The results are presented in fig. 1. Experimental differential cross sections used for the fit are the measurements of Szabo et al. [4]. Two fits are obtained: (a) with $R_{Sax} = R_{sur} = R_{ls}$ and (b) with all the parameters free in the search. The first gives a better χ^2 and a better reaction cross section ($\sigma_R^{exp.} = 600$ mb, $\sigma_R^{theor.} = 478$ mb) than the second, but too small W_{sur}. To take into account the 15.5-MeV resonance seen in the (^{12}C, n) system by Boreli et al. [5], we added a resonance term of the Breit–Wigner form (in the same way as Tamura [6]) to the amplitudes computed with MAGALI. The best fit is obtained with $P_{3/2}$, $E_\lambda = 16.3$ MeV, and $\Gamma_\lambda = 1.2$ MeV.

Inelastic scattering to the 4.43-MeV level. The data are shown in fig. 2. The computation with coupled-channel theory has not yet given acceptable results.

^{16}O. Polarization data are presented in fig. 3. The cross sections used for the fit are those of Beach et al. [7]. Three sets of parameters give a good fit. The first is obtained with $R_{Sax} = R_{sur} = R_{ls}$ and $A_{Sax} = A_{sur} = A_{ls}$. This gives the best value for the reaction cross section ($\sigma_R^{exp.} = 560$ mb, $\sigma_R^{theor.} = 550$ mb). The second fit gives the best polarization. The third gives a different contribution to the backward scattering (there are no data for these angles).

²D. In fig. 4 polarization data are shown compared to the proton data of Faivre et al. [8]. These measurements were the most difficult because both the deuterium analyzing power and the cross section are very small.

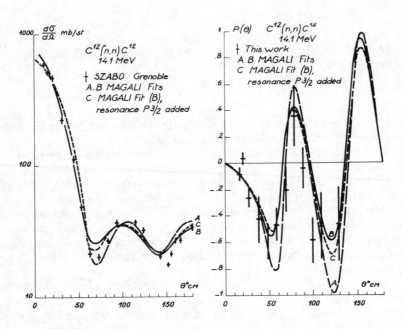

N°	V_{SAX}	R_{SAX}	A_{SAX}	W_{SUR}	R_{SUR}	A_{SUR}	V_{LS}	R_{LS}	A_{LS}	χ^2_σ	χ^2_p	σ_r
A	57.64	1.16	.33	.92	1.16	1.47	5.05	1.16	.12	200	125	478
B	56.07	1.15	.54	11.98	1.49	.13	5.72	1.08	.24	357	76	404
C	same optical potential parameters as B resonance added											

$$\frac{R_\lambda + i\,I_\lambda}{(E - E_\lambda) + i\Gamma_{/2}}$$

R_λ	I_λ	E_λ	Γ_λ	χ^2_σ	χ^2_p
-.246	.31	16.3	1.2	259	100

Fig. 1

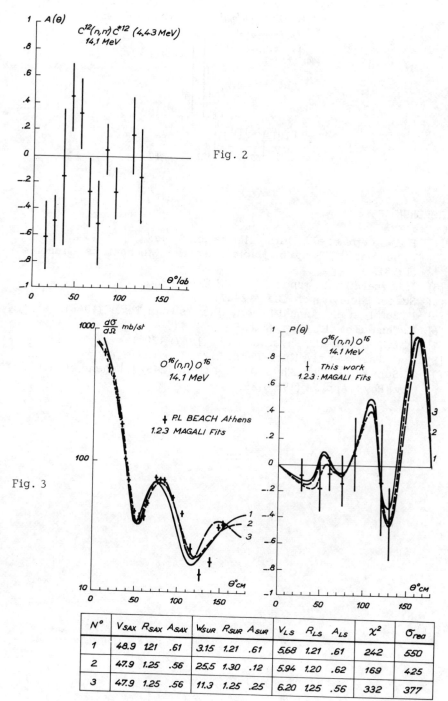

Fig. 2

Fig. 3

N°	V_{SAX}	R_{SAX}	A_{SAX}	W_{SUR}	R_{SUR}	A_{SUR}	V_{LS}	R_{LS}	A_{LS}	χ^2	σ_{rea}
1	48.9	1.21	.61	3.15	1.21	.61	5.68	1.21	.61	242	550
2	47.9	1.25	.56	25.5	1.30	.12	5.94	1.20	.62	169	425
3	47.9	1.25	.56	11.3	1.25	.25	6.20	1.25	.56	332	377

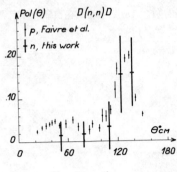

Fig. 4

REFERENCES

[1] P. Delpierre et al., Third Polarization Symp.
[2] R. Sené et al., Symp. on Nuclear Reaction Mechanisms, Grenoble (1970).
[3] J. Raynal, Saclay report CEA DPH T/69-42.
[4] Szabo, Saclay report CEA R 2407.
[5] F. Boreli et al., Annual Report, University of Texas (1966).
[6] T. Tamura et al., Phys. Lett. 8 (1964) 41.
[7] P. L. Beach et al., Phys. Rev. 156 (1967) 1201.
[8] J. C. Faivre et al., Nucl. Phys. A127 (1969) 169.
[9] R. Sené et al., Rev. Phys. Appl. 4 (1969) 245; J. Kahane et al., Rev. Phys. Appl. 4 (1969) 257.

Experiments on the Polarization of 15-MeV Neutrons by Scattering on Light Nuclei

G. MACK, G. HENTSCHEL, C. KLEIN, H. LESIECKI, G. MERTENS,
W. TORNOW, and H. SPIEGELHAUER, University of Tübingen, Germany

The polarization $P(\theta)$ for elastic scattering of neutrons by ^{12}C and the analyzing power $A(\theta)$ for the inelastic channel $^{12}C(n,n')^{12}C(4.44$ MeV) have been determined from 20° to 80° (lab) by measuring the scattering asymmetry of polarized 15.85-MeV neutrons on ^{12}C nuclei in a plastic scintillator. The carbon recoil nuclei were used for time-of-flight measurements. Parts of the results have already been published [1, 2]. With the primary neutron polarization taken as −0.118 which is deduced from n-^{4}He scattering [3] using phase shifts of Arndt and Roper [4], the results are as shown in table 1 and fig. 1.

Table 1

$P(\theta)$ and $A(\theta)$. Errors are statistical. Angular resolution is 2.2°.

θ_{lab}	20°	25°	30°	35°	40°	45°	50°	55°	60°	70°	80°
$P(\theta)$	−11.3	−20.1	−30.9	−52.7	−55.4	−66.7	−78.6	−56.9	−23.7	+6.1	+2.4
$\pm\Delta P$	4.5	4.7	2.4	7.0	3.1	4.0	4.1	4.8	7.1	4.5	5.1
$A(\theta)$	−61	−40	−29	+8	−23	−12	−22	−22	−13	−35	−5
$\pm\Delta A$	70	72	12	56	12	11	9	11	12	13	15

The instrumental asymmetries are less than the statistical errors. Zombeck has found similar results [5] at 14.70 MeV.

For polarization measurements similar to those described for carbon, a high-pressure gas scintillator with N_2 gas has been developed. A cylindrical steel vessel with an inner coating of 1 μm Al, 1 mm MgO and 120 μg/cm^2 p-quaterphenyl was filled with 60 atm N_2 and 23 atm Xe. The pulse height at room temperature was about 1/3 of that of Xe. Neutron time-of-flight spectra show the elastic scattering peak clearly separated from inelastic events.

The recently published n-^{4}He scattering phase shifts (ref. [4]) differ significantly from those of Hoop and Barschall [6]. Absolute measurements by double scattering seem necessary, for which a setup has been built. The measurements will begin soon. It consists of two high pressure He-Xe gas scintillators as first and second scatterers and

615

two liquid scintillators (NE 213) as detectors. Instrumental errors are compensated by rotating the whole assembly around an axis through the first scatterer. The primary neutron energy is 15.0 MeV. Good peak-to-background discrimination has been achieved. A Monte Carlo program simulates the experiment for different sets of phase shifts which will be tested by the measurements.

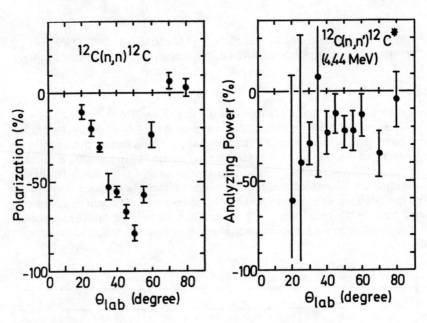

Fig. 1. $P(\theta)$ and $A(\theta)$.

REFERENCES

[1] G. Mertens, Z. Phys. 212 (1968) 347.
[2] G. Mack, Z. Phys. 212 (1968) 365.
[3] G. Hentschel, Z. Phys. 219 (1969) 32.
[4] R. A. Arndt and L. D. Roper, Phys. Rev. C1 (1970) 903.
[5] M. V. Zombeck, Thesis, MIT (1969).
[6] B. Hoop and H. H. Barschall, Nucl. Phys. 83 (1966) 65.

Analyses of Proton-^{12}C Scattering
between 4.8 and 8 MeV

T. TERASAWA, University of Tokyo, Japan; O. MIKOSHIBA, Science
University of Tokyo, Japan; M. TANIFUJI, University of Sydney,
Australia

Scattering of protons by ^{12}C is analyzed using the coupled-channel
method [1] and, assuming the vibrational and rotational models for
^{12}C nuclei, with particular attention to the resonance behavior in the
proton energy range from 4.8 to 8 MeV. Coupling is assumed between
the elastic scattering channel and the inelastic scattering channel to
the first excited state of ^{12}C. The ^{12}C potential is the real Woods-
Saxon type and its Thomas-type derivative for the central and spin-
orbit forces, respectively. Both potentials are deformed. The depth
of the central potential V_c is assumed to be energy-dependent $V_c = V_0 - A_E E_p$. To include effects of nucleonic spin-spin interactions, the
target-spin dependent interaction $V_{\sigma I}(\vec{\sigma} \cdot \vec{I}_c)$ with the Woods-Saxon
radial dependence is phenomenologically introduced [1], where $1/2\,\vec{\sigma}$
and $\vec{I}_c$ are the spin of the proton and that of ^{12}C in the excited state.
The vibrational model is not successful as shown by the disagreement
with the experimental data [2, 3], particularly for the elastic polarization and the inelastic cross section, although the observed $5/2^+$ and
$3/2^+$ resonances in the elastic cross section around $E_p = 5$ MeV are
well reproduced by the spin-spin interaction of considerable strength.
The rotational model is quite successful when large deformations of
negative sign, e.g., $\beta \sim -0.5$, are assumed, for which the magnitude of
$V_{\sigma I}$ is fairly small. Examples of the comparison between the calculations and the experiments are shown in fig. 1, for the excitation curves
of the elastic cross sections and polarizations at 54°, where the rotational model includes the effect of the odd-parity resonances and a
one-level formula with the empirical resonance parameters is used.
The rotational model also gives good fits to the data [2] of angular
distributions for both the differential cross section and the polarization in the elastic scattering.

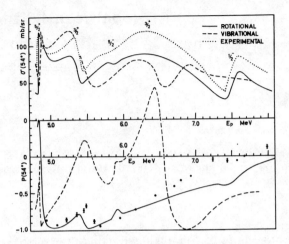

Fig. 1. The potential parameters are V_O = 57.5
MeV, $V_{\sigma I}$ = 0.9 MeV, and A_E = 0 for the vibrational model (β = 0.2) and V_O = 54.1 MeV, $V_{\sigma I}$ =
0.18 MeV, and A_E = 0.73 for the rotational model
(β = -0.5). Other parameters are V_{1s} = 6.5 MeV,
r_O = 1.25 fm, and a = 0.65 fm.

REFERENCES

[1] T. Tamura, Rev. Mod. Phys. 37 (1965) 679.
[2] S. J. Moss and W. Haeberli, Nucl. Phys. 72 (1965) 417.
[3] A. C. L. Barnard et al., Nucl. Phys. 86 (1966) 130.

Proton Polarization in the $^{12}C(^{3}He,p)^{14}N$(g.s.) Reaction at Energies between 2.4 and 3.6 MeV and Angles between 20° and 75°

H. OEHLER, M. I. KRIVOPUSTOV, H. I. VIBIKE, F. ASFOUR, I. V. SIZOV, and G. SCHIRMER, Joint Institute for Nuclear Research, Dubna, USSR

A study of polarization in the $^{12}C(^{3}He,p)^{14}N$(g.s.) reaction was performed on the JINR electrostatic generator using an annular polarimeter whose geometry was suggested by Brinkman in 1960 [1]. Our papers [2,3] present the results of the development of the polarimeter with an annular analyzing target, and the method and the program for the calculation of polarization in such an experiment. The proton polarization was measured at several forward angles for 6 ^{3}He energies ranging from 2.4 to 3.6 MeV. The polarization in the studied regions is negative and has its peak value of $(-87 \pm 5)\%$ at an energy of 2960 keV and $\theta = 70° \pm 3°$.

The polarization data were analyzed by the R-matrix theory of nuclear resonance reactions assuming the formation of isolated levels of the compound nucleus. A modified formula of Simon-Welton [4] for the calculation of differential polarization $d\vec{P}/d\Omega$ was used in a special fitting routine based on the Monte-Carlo method. Fig. 1 shows the energy and angular dependence of $d\vec{P}/d\Omega$ determined from the results of our measurements. A considerable resonance effect at an energy of 2.37 MeV (c.m.) is especially pronounced at small angles. The behavior of $d\vec{P}/d\Omega$ near the resonance is associated with the interference of partial waves due to a small level-spacing in the compound nucleus.

The results of the two-level fitting (the $5/2^{+}$ level at $E_{lab} = 2915$ keV [5] and the $7/2^{-}$ level at 2875 keV) are shown in fig. 1 by the dashed line. From the analysis of the level parameters one can see that a considerable contribution is made by partial waves with large orbital angular momenta ($\ell' = 4$ and 5). At the same time, the nuclear surface penetrability for protons of a given energy with the same ℓ' does not exceed 0.5% of the penetrability for protons with $\ell' = 1$. This contradiction requires introduction of additional terms.

In ref. [6] the authors studied the interference between direct processes and formation of the compound nucleus. We performed a three-level fit (fig. 1, dash-dot line). The parameters which give good agreement between calculated and experimental values of polarization with-

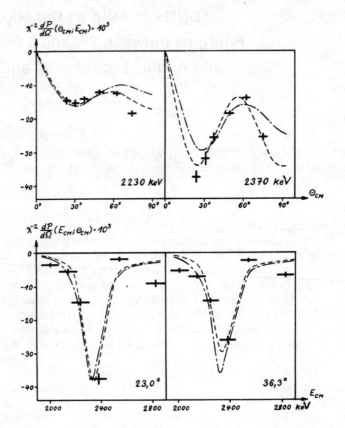

Fig. 1. Comparison between the values of differential polarization calculated by two-level and three-level fitting and the experimental data.

out a significant contribution of large orbital angular momenta are listed in the table.

Consideration of the influence of levels far from the resonance affects the parameters by less than 20%. Apparently the measurement of the polarization at angles $\theta > 90°$ would permit more reliable conclusions to be drawn as to the quantum char-

J^Π	$E_{^3He}$(lab) [keV]	Γ(lab) [keV]	$\Sigma\Gamma_{p_0}$ [keV]
$1/2^+$	2725	110	13.4
$7/2^-$	2870	300	48.4
$5/2^+$	2913	125	49.7

acteristics of the states of the intermediate nucleus ^{15}O and the reaction mechanism. Such measurements are now being performed.

REFERENCES

[1] H. Brinkman, First Polarization Symp., p. 166.
[2] M. I. Krivopustov et al., JINR Preprint P15-3504, Dubna, 1967.
[3] H. Oehler et al., Nucl. Instr. and Meth. 77 (1970) 293.
[4] A. Simon and T. A. Welton, Phys. Rev. 90 (1953) 1036.
[5] Hsin-Min Kuan, T. W. Bonner, and J. R. Risser, Nucl. Phys. 51 (1964) 481.
[6] G. L. Vysotsky and M. A. Chergorian, Izv. Akad. Nauk SSSR, ser. fiz. 34 (1970) 147.

On the Determination of J^π of States in ^{17}O near 8 MeV Excitation Energy via the ^{13}C$(\alpha,n)^{16}$O Reaction

T. R. DONOGHUE, C. E. BUSCH, J. A. KEANE, and H. PAETZ gen.
SCHIECK, Ohio State University, Columbus, USA*

The states of the theoretically important ^{17}O nucleus have been investigated extensively via the ^{13}C$(\alpha,n)^{16}$O reaction because of both experimental and theoretical simplifications. In this reaction, the entrance and exit channel spins are 1/2 and the ground state spins of ^{13}C and ^{16}O are $1/2^-$ and 0^+; hence a state of given J^π is formed (or decays) by a single value of orbital angular momentum ℓ' (or ℓ), where $\ell' = \ell \pm 1$. Although J-values of the contributing states can often be determined, the parity of an isolated state cannot be determined [1] from $\sigma(\theta)$ alone because a change of parity for a given J merely interchanges the values of ℓ' and ℓ, a situation that leaves $\sigma(\theta)$ unchanged. The relative parities of overlapping states, however, can be determined, but an absolute determination of the parity obviously requires additional experimental information.

Recently Schölermann [2] reported neutron polarization measurements for the ^{13}C(α,n) reaction for $1.38 \leq E_\alpha \leq 2.26$ MeV to complement previously measured $\sigma(\theta)$ data [1]. An S-matrix search analysis [2] of the experimental data resulted in a number of J^π assignments to states in ^{17}O for $E_x = 7.4 - 8.3$ MeV. Of particular concern are his J^π assignments of $1/2^+$, $3/2^-$, and $3/2^+$ made to the three principal states at $E_x(^{17}$O$) = 7.97$, 8.07, and 8.197 MeV, in that these parity assignments are opposite to those based on elastic n-^{16}O scattering data [3]. Because of our previous interest [4] in the ^{13}C(α,n) reaction at higher energies, and because the recent measurements [2] used a target thickness of almost 1 MeV (such that $\Delta E \gg \Gamma$), we have re-investigated the $E_\alpha = 2-2.5$ MeV region using 80-110 keV thick targets to clarify these parity assignments. Neutron polarization angular distributions measured at the three resonance energies are shown in fig. 1, together with reconstructed $\sigma(\theta)$ data [1]. An S-matrix search analysis [4] of $\sigma(\theta)$ and $P(\theta)$ was made at each of the three energies to determine which combinations of levels best described the data. It was discovered that an identical description of the data (see fig. 1) could be obtained with either the $1/2^-$, $3/2^+$, $3/2^-$ level sequence [3] or the $1/2^+$, $3/2^-$, $3/2^+$ sequence [2]. This result, unexpected when $P(\theta)$

data are included in the analysis, prompted a theoretical investigation by Seyler [5] who has shown in an accompanying paper that in a reaction with the above cited restrictive properties, an absolute determination of the parities of the states is not possible in an analysis of this type. When a description of $\sigma(\theta)$ and $P(\theta)$ is made in terms of the amplitudes and phases of the contributing states, identical fits result when the parities of all states are reversed if simultaneously the signs of the phases of the SJ^{π} are also reversed. The parity assignments of Schölermann are therefore not unique. A determination of the parities may still be possible, if the phases of all the $S^{J^{\pi}}$ can be determined away from the resonance region, but, as is the case here, this is frequently not possible.

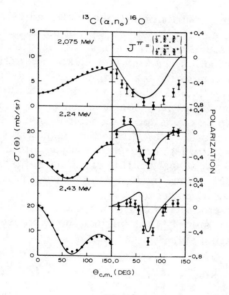

Fig. 1

REFERENCES

*Supported in part by the National Science Foundation.

[1] R. B. Walton, J. D. Clement, and F. Boreli, Phys. Rev. 107 (1957) 1065.
[2] H. Schölermann, Z. Phys. 220 (1969) 211.
[3] C. H. Johnson and J. L. Fowler, Phys. Rev. 162 (1967) 890.
[4] W. L. Baker, C. E. Busch, J. A. Keane, and T. R. Donoghue, Phys. Rev., to be published.
[5] R. G. Seyler, Third Polarization Symp.

On the Possibility of Determining Absolute Parities from an S-Matrix Analysis of Data from Reactions like $^{13}C(\alpha,n)^{16}O$

R. G. SEYLER, Ohio State University, Columbus, USA*

As noted in another contribution [1] the published values for the *absolute* parities of certain ^{17}O states are not all in agreement. We ask whether it is possible to determine absolute parities from such cross-section and polarization *reaction* (not elastic scattering) data. We consider any reaction where the entrance- and exit-channel spin is uniquely equal to $1/2$, and where the intrinsic parities of the two channels are opposite. Parity conservation then requires that the orbital parities also be opposite. We conclude that if the collision- or S-matrix elements $U^{b\pi}_{\ell_2\ell_1}$ are treated as unknowns (as is standardly done in elastic scattering phase-shift analyses) to be determined at each energy by least squares fitting of the data, then only the *relative* parities can be determined by studying the usual cross section, polarization, and analyzing power data.

Following the procedures of Devons and Goldfarb [2] the spin tensors of the exiting spin-$1/2$ particle may be expressed in terms of those of the entering spin-$1/2$ particle,

$$\tau_{k_2\kappa_2} \propto \sum A^{\kappa_2}_{k_2 k}(2) A^{\kappa_1}_{k_1 k}(1)\, (U^{b\pi}_{\ell_2\ell_1} U^{b'\pi'^*}_{\ell'_2\ell'_1} + (-)^{k_1+k_2} U^{b\pi^*}_{\ell_2\ell_1} U^{b'\pi'}_{\ell'_2\ell'_1}) D^k_{\kappa_2\kappa_1}(R_2^{-1}R_1)\tau_{k_1\kappa_1},$$

$$(1)$$

where the summation is over $\ell_1\ell'_1\ell_2\ell'_2 b b' k k_1$ and κ_1, and where

$$A^{\kappa i}_{k_i k}(i) = \hat{\ell}_i \hat{\ell}'_i (-)^{\ell'_i+k_i} \sum_{\bar{k}} \hat{\bar{k}}_i (\ell_i 0, \ell'_i 0|\bar{k}_i 0)(k_i\kappa_i, \bar{k}_i 0|k\kappa_i) X(\tfrac{1}{2}\tfrac{1}{2} k_i; \ell_i \ell'_i \bar{k}_i; b b' k).$$

Here each set of spin tensors is referred to its own z-axis (respective momentum direction) and to a common y-axis (normal to reaction plane).

The above conclusion is already familiar where one has unpolarized cross-section data alone. Here $k_1 = k_2 = \kappa_1 = \kappa_2 = 0$ and thus only the combination $A^0_{0k}(1) A^0_{0k}(2)$ enters (this is more familiar, if written as the product of two $\bar{Z}$ coefficients). Since the entrance and exit channels have opposite orbital parities, whenever $\ell_1 = b \pm 1/2$ we must have $\ell_2 = b \mp 1/2$, the upper signs referring to a spin b resonance state of parity $(-)^{\ell_1}$ times the intrinsic parity of the entrance channel, and the

624

lower signs to a state of opposite parity. Thus changing the parity of *all* the contributing resonance states merely interchanges ℓ_1 with ℓ_2 and ℓ_1' with ℓ_2' which is equivalent to interchanging the arguments of the A_{0k}^0 functions, an operation which obviously leaves their product unchanged. It follows that the unpolarized cross section predicted by any set of states (not necessarily all of the same parity) with elements $U_{\ell_2\ell_1}^{b\pi}$ will be exactly reproduced by a set of opposity parity states if the elements of the latter are obtained from the former by the equation

$$U_{\ell_1\ell_2}^{b,-\pi} = U_{\ell_2\ell_1}^{b,\pi\,*} . \tag{2}$$

Though unnecessary so far, the complex conjugation symbol in eq. (2) becomes necessary when we consider the measurement of either (1) the polarization resulting from an initially unpolarized system, or (2) the analyzing power when the initial system is polarized. Here one finds that $A_{0k}^0(1)A_{1k}^K(2) = -A_{0k}^0(2)A_{1k}^K(1)$, which with eq. (1) implies that the polarization and analyzing power are also insensitive to the absolute parities of the resonance states provided that eq. (2) is satisfied.

For completeness we consider the measurement of the final polarization when the initial system is polarized, $k_1 = k_2 = 1$. Since here A_{1k}^0 need not vanish as before, two measurements are possible (although probably impractical) which do depend on the absolute parities even when eq. (2) is satisfied: (1) the component of the polarization along the final x-axis when the target is polarized along the initial z-axis, and (2) the final z-axis component when the target is polarized along the initial x-axis. However, since this parity dependence is not simple and the experiments seem unlikely, the detailed results will not be quoted here.

If for the data analysis one invokes some model for calculating the S-matrix elements, one may well find that they do not satisfy eq. (2). In this event the two choices of absolute parities would lead to different predictions but any further conclusions would rest on the validity of the model.

REFERENCES

*Supported in part by the National Science Foundation.

[1] T. R. Donoghue, C. E. Busch, J. A. Keane, and H. Paetz gen. Schieck, Third Polarization Symp.
[2] S. Devons and L. B. J. Goldfarb, Handbuch der Physik, vol. 42 (Springer, Berlin, 1957).

Differential Polarizations of Photoneutrons
from the Giant Dipole States of ^{16}O

G. W. COLE, Jr., and F. W. K. FIRK, Yale University, New Haven,
Connecticut, USA

Recent theories of the dipole states in ^{16}O have been concerned with
describing the ground and excited states more realistically and with
estimating the particle decay widths [1]. All such calculations predict that the dipole states are superpositions of 1p-1h states. It is
necessary to test both the predicted admixtures and the purity of the
dipole absorption process [2-4]. We have therefore measured the differential polarization of photoneutrons from ^{16}O at 45° and 90° using
the nanosecond time-of-flight system and liquid He polarimeter associated with the Yale Electron Linac (peak analyzed current: 3A, pulse
width: 5ns, flight path: 25m and repetition rate: 330Hz). Bremsstrahlung, produced in a 0.25 mm thick tungsten foil, irradiated a 7.5 cm
diam cylinder of water. The polarization was measured by right-left
scattering from a liquid He analyzer (7.5 cm diam cylinder). Scintillations caused by recoiling He nuclei were detected in coincidence with
the scattered neutrons (detectors set at ±130°). The results are shown
in fig. 1 (45°, $E_{\gamma max} = 50$ MeV) and fig. 2 (90°, $E_{\gamma max} = 30$ MeV). For
the first time, a resonant behavior of the 45° polarization is observed
[earlier work of Hanser [5] had insufficient resolution]: the 90° polarization is appreciable ($\sim -14\%$) and a smooth function of energy.

[Since completing this
work we have re-measured
the 45° polarization using
a solenoid to precess the
neutron spins. The value
obtained for the polarization at 22.1 MeV (6.1 MeV
neutron energy) is (27 ±
3)%.]

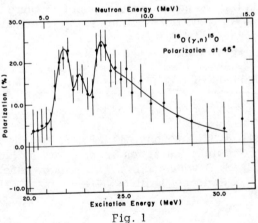

Fig. 1

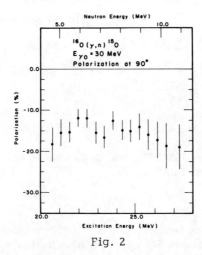

Fig. 2

Assuming pure E1 absorption, typical values of the s- to d-wave amplitudes (a_s/a_d) and phases (Δ_{sd}) deduced from the present work and from the angular distribution data [3] are:

Exc. energy (MeV)	$-A_2/A_0$	a_s/a_d	Δ_{sd}(deg)
21.06	0.263 ± 0.058	0.17 ± 0.03	343
21.29	0.356	0.11	340
21.51	0.548	0.13	352
21.76	0.519	0.17	262
22.00	0.618 ± 0.029	0.20 ± 0.04	240
22.24	0.684	0.25	230
22.48	0.545	0.12	252

Under reasonable assumptions ($a_s/a_d = 0.2$, $\Delta_{sd} = 230°$ and $a_p = 0$), the 90° results give values of $0.05 < a_f/a_d < 0.3$. The lower limit corresponds to only 0.25% by intensity of E2 transitions, thus emphasizing the sensitivity of the method.

Present theories of the giant dipole states in ^{16}O are unable to account for the observed results.

REFERENCES

[1] B. Buck and A. D. Hill, Nucl. Phys. A95 (1967) 271.
[2] E. D. Earle and N. W. Tanner, Nucl. Phys. A95 (1967) 241.
[3] J. E. E. Baglin and M. N. Thompson, Nucl. Phys. A138 (1969) 73.
[4] D. E. Frederick, R. J. Stewart, and R. C. Morrison, Phys. Rev. 186 (1969) 992.
[5] F. Hanser, Thesis, MIT (1967).

Polarization of 8–12 MeV Protons
Elastically Scattered by Oxygen

R. M. PRIOR, K. W. CORRIGAN, and S. E. DARDEN, University of
Notre Dame, Indiana, USA*

The differential cross section for $^{16}O(p,p)^{16}O$ has been measured by
Hardie et al. [1] between 8.5 and 13 MeV. In this energy region there
is pronounced energy-dependent structure in the cross section. Some
of this structure has been interpreted in terms of levels in ^{17}F by
Dangle et al. [2] in an analysis of their $^{16}O(p, \alpha)^{13}N$ and $^{16}O(p,p')^{16}O^*$
data. Phase-shift analyses of $^{16}O(p,p)^{16}O$ have extended only up to
8.5 MeV because the open reaction channels require the phase shifts
to be complex and not enough data have been available to determine
the complex phase shifts. The present data were taken using the
polarized proton beam from the Notre Dame Lamb-shift ion source. The
beam from the FN Tandem accelerator was incident on an oxygen gas
target which was viewed by six solid state detectors. The polarization has been measured for 12 angles at 30 energies between 8.2 and
10 MeV and for fewer angles at 25 energies between 10 and 12 MeV.
Angular distributions at 18 angles have been measured at several energies. The polarization has pronounced energy-dependent structure,
particularly between 8.5 and 9.5 MeV. A phase-shift analysis is now
in progress using the present polarization data and the cross section
data of Hardie et al. Preliminary results indicate possible disagreement with the level assignments of Dangle et al. near 9 MeV.

REFERENCES

*Work supported in part by the National Science Foundation under
Contract No. GP-15560.

[1]   G. Hardie, R. L. Dangle, and L. D. Oppliger, Phys. Rev. 129 (1963)
      353.
[2]   R. L. Dangle, L. D. Oppliger, and G. Hardie, Phys. Rev. 133 (1964)
      B647.

Elastic Scattering
of Nucleons

Experimental Studies of the Polarization
Produced by Elastic Scattering of
Neutrons in the Vicinity of 1 MeV

S. A. COX and E. E. DOWLING WHITING, Argonne National Laboratory,
Illinois, USA*

A systematic study of polarization produced by the elastic scattering
of low energy neutrons for intermediate weight and heavy nuclei has
been carried out over the past three years. Altogether we have studied
29 elements from mass number 48 to 238. The nuclei were selected
to be representative of most of the periodic table. The data are now
extensive enough to indicate some generalizations and systematics
in relation to different regions of the period table. Data for each element were obtained at eight angles using a partially polarized incident
neutron beam emitted at 51° from the Li(p,n) reaction. The incident
neutron energy was always in the energy range from 700 keV to 1.2
MeV, and for purposes of comparison with the optical model, the data
were always averaged over an energy span of at least 200 keV within
that energy range. The data were corrected for effects due to angular
resolution, flux attenuation, multiple scattering, and where necessary,
for inelastically scattered neutrons.

An extensive search of optical model parameters was made for all
29 elements using the Abacus II program. The form of the potential
wells used was a Saxon-Woods real potential, a pure surface imaginary
potential, and a Thomas spin-orbit potential. During the course of
the parameter search it became
clear that the usual formula assumed for the radius (R_1) of the real
potential did not give the best fit
to the data. R_1 was thus allowed
to vary as a free parameter. The
values of R_1 obtained from the resulting fits are given in fig. 1. The
dashed curve follows the usual
formula. The solid curve is the
one which was finally used for
subsequent calculations.

In order to show a systematic
trend with mass number for the

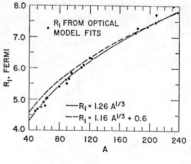

Fig. 1

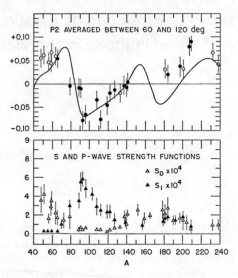

Fig. 2

polarization, we have averaged the values of P2 between 60° and 120°. Three mass characteristic regions are evident in the top part of fig. 2. The solid points are for elements for which we are able to obtain good optical model fits to both the cross section and the polarization data; the open circles are those for which we could not obtain good fits. For A < 80, all the average P2's are positive and all but one were poor fits. For A between 80 and ∼ 140 all the average P2's are negative and all but one were good fits. Above A ∼ 180 we have both good and bad fits, and all the average P2's are positive. There seems to be a correlation between the strength function peaks and the systematic trends that we have found for the polarization. The first mass region, where we could not obtain good fits to the polarization data, is associated with an s-wave peak. The middle region, where we did obtain good fits, is associated with a p-wave peak. The last mass region is one of mixed s- and p-wave strength, and also includes a number of deformed nuclei. A correlation between the quality of the optical model fits and the deformation parameter was considered, but appeared much weaker than was the case with the strength functions.

*This work supported by the U.S. Atomic Energy Commission.

Elastic Scattering of Polarized 4-MeV
Neutrons by Heavy and Medium-Weight Nuclei

G. V. GORLOV, Kurchatov Atomic Energy Institute, Moscow, USSR

During recent years the elastic scattering of polarized 4-MeV neutrons
has been studied over a wide range of scattering angles at the Kurchatov Institute of Atomic Energy by Gorlov, Lebedeva, Morosov, and Zubov.

The neutrons were produced with the D-D reaction (average energy
E_d = 1.2 MeV, polarization of the neutron beam P = 14.8% [1]). Differential cross sections for elastic scattering were measured in the plane
normal to the neutron beam polarization vector to the right and left of
the direction of the incident neutron beam. The total cross sections
were measured as well.

The experimental details will not be discussed here; however, two
conditions required for the success of the work should be pointed out:
(a) a well-collimated beam having sharply defined edges was used [2],
and (b) a stable n-γ separation circuit [3] and stabilizer of the ampli-
fication of the neutron scintillation spectrometer were employed [4].

In the first investigations [5] the elastic scattering of neutrons by
^{59}Co, ^{62}Ni, ^{80}Se, ^{93}Nb, ^{114}Cd, ^{115}In, ^{118}Sn, ^{127}I, Pb, and ^{209}Bi was studied
at scattering angles from 10° to 170°. A diffraction character of the
differential cross sections was observed. The considerable polarization (50% and more) observed for all nuclei is indicative of the important contribution from the spin-orbit interaction in the elastic
scattering.

An optical model potential of the following form [5] was used:

$$V(r) = -V_1 f(r) - iV_2 g(r) + V_3 \left(\frac{\hbar}{\mu c}\right)^2 \frac{1}{r} \frac{df(r)}{dr} (\vec{\sigma}, \vec{\ell});$$

$$f(r) = \left[1 + \exp\left(\frac{r-R}{a}\right)\right]^{-1}; \quad g(r) = \exp\left[-\left(\frac{r-R}{b}\right)^2\right]; \tag{1}$$

$$V_1 = V_0\left(1 - \frac{N-Z}{3A}\right); \quad R = r_0 A^{1/3}.$$

The parameters, which were chosen on the basis of the experimental data on differential cross sections, polarization, and total cross sections, have the following values:

$$V_0 = 50 \text{ MeV}, \qquad V_2 = 8.5 \text{ MeV}, \qquad V_3 \simeq 0.17 \, V_1,$$

$$r_0 = 1.25 \text{ fm}, \qquad a = 0.65 \text{ fm}, \qquad b = 0.98 \text{ fm}.$$

While the agreement with the experimental data on the total cross sections and differential cross sections might be considered good, the agreement with the polarization in many cases proves to be only qualitative (see fig. 1a, b).

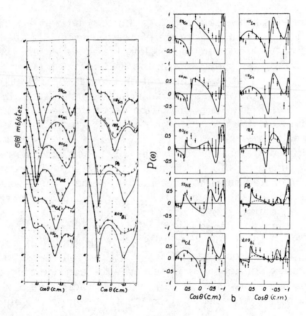

Fig. 1

In further studies [6] the scattering of polarized neutrons by Cu, In, Sn, Pb, Bi, and U at small scattering angles from 2° to 21° was studied. It was observed that for all the nuclei investigated the differential cross sections show an appreciable rise at θ = 2°, and the asymmetry of the scattering increases with decreasing scattering angle in the region of the angles studied (see fig. 2a, b). An analysis of the polarization in the region θ = 2° - 9° has shown that it is in good agreement with the prediction of Schwinger's theory [7] for Coulomb scattering of neutrons at small angles due to the interaction of the neutron magnetic moment with the Coulomb field of the nucleus. The differential cross sections shown in the figure by solid curves were obtained by multi-

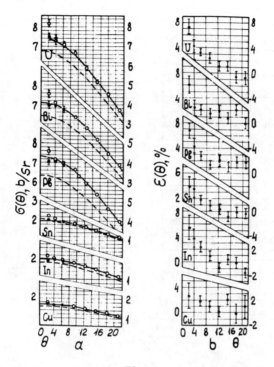

Fig. 2

plying by suitable normalization factors (from 1.05 to 1.13) the differential cross sections calculated with the potential (1) using the
parameters given. The experimentally observed form of the differential cross sections can be well described under the assumption that
only nuclear and Coulomb scattering exist.

In more recent studies the elastic scattering of polarized neutrons
by In, Sn, Pb, Bi, and U is investigated at scattering angles from 146°
to 177° [8]. The differential cross sections and the polarization of the
nuclei investigated are determined (see fig. 3a, b). Peaks in the back
scattering cross sections are found for all the nuclei, except U. The
measured cross sections and polarization differ quantitatively from the
calculated values obtained using the optical potential [1]. The potential (1) predicts a diffraction peak for all the elements investigated,
which does not agree qualitatively with the experimental data on U.
The calculated cross section is twice as large as that observed experimentally. The behavior of the U differential cross section does not
agree either quantitatively or qualitatively with the predictions of several forms of the spherical optical potential. The effect observed may
be due to the non-sphericity of the U nucleus.

In addition to the scattering experiments there have been carried
out measurements of the neutron polarization in the D-D reaction at

incident deuteron energies from 1.2 to 2.7 MeV. The polarizations
obtained are in good agreement with the experimental data shown by
the upper curve in Fig. 4.

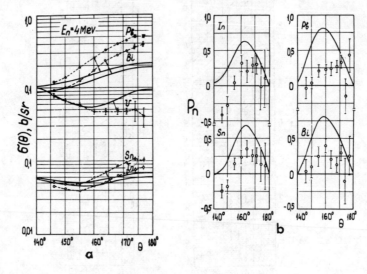

Fig. 3

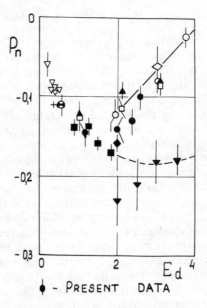

● — PRESENT DATA

Fig. 4

REFERENCES

[1] G. V. Gorlov, N. S. Lebedeva, and V. M. Morozov, Yadernaya Fiz. 4 (1966) 519.
[2] G. V. Gorlov, A. I. Kirilov, and N. S. Lebedeva, Pribori Tekhn. Exp. (USSR) 3 (1966) 27.
[3] V. G. Brovchenko and G. V. Gorlov, Pribori Tekhn. Exp. (USSR) 4 (1961) 49.
[4] V. G. Brovchenko, L. G. Kondratiev, N. S. Lebedeva, and V. M. Morozov. Pribori Tekhn. Exp. (USSR) 1 (1969) 55.
[5] G. V. Gorlov, N. S. Lebedeva, and V. M. Morozov, Yadernaya Fiz. 6 (1967) 910.
[6] G. V. Gorlov, N. S. Lebedeva, and V. M. Morozov, Yadernaya Fiz. 8 (1968) 1086.
[7] J. Schwinger, Phys. Rev. 73 (1948) 407.
[8] V. M. Morozov, G. V. Gorlov, Yu. G. Zubov, and N. S. Lebedeva, Preprint AEI-1962, Moscow (1970).

DISCUSSION

Glashausser:

As you said, your small angle measurements of neutron polarization are a measure of the Mott-Schwinger interaction. Would you indicate how large the effects of this interaction are, and whether calculations which include this term correctly explain your measurements?

Gorlov:

For small angle scattering the data agree with calculations using a nuclear potential plus Mott-Schwinger scattering.

Depolarization in Elastic Scattering
of 1.36-MeV Polarized Neutrons

K. KATORI, T. NAGATA, O. MIKOSHIBA,* and S. KOBAYASHI,†
Tokyo University of Education, Japan

There has been considerable experimental and theoretical interest in the effect of a target-spin dependent interaction on nuclear reactions. One of the direct measurements is the transmission of polarized neutrons through a polarized target (^{165}Ho, ^{59}Co). The results give an upper limit for the spin-spin interaction in the optical potential of 1 MeV. Another method is the measurement of the depolarization of polarized neutrons scattered from a non-zero spin target. In elastic scattering of nucleons from a non-zero spin target, Bohr's theorem allows a spin flip. Then, the depolarization parameter $D(\theta)$ deviates from unity, only when there exists a spin-spin interaction between the projectile and a target nucleus in direct reactions. A few experiments have been performed which showed no detectable deviation of $D(\theta)$ from unity.

The work reported here consists of a triple scattering experiment. An angular dependence of $D(\theta)$ for ^{27}Al ($I = 5/2$), ^{59}Co (7/2), Ni (0), ^{63,65}Cu (3/2), and ^{209}Bi (9/2) has been studied at $E_n = 1.36$ MeV at the laboratory angles $\theta_2 = 30°$, $38°$, $80°$, $120°$, and $150°$. The neutron polarization was measured by a liquid helium, time-of-flight, scintillation polarimeter with a spin precession solenoid. The experimental method and procedure are given elsewhere [1].

The results are shown in figs. 1-4. A departure of $D(\theta)$ from unity for Al, Co, Cu, and Bi was observed particularly at backward angles, but not for Ni. This fact shows that the spin flip of the incident neutrons takes place along the normal to the scattering plane in the elastic scattering process for non-zero spin targets.

From the experimental point of view, it is convenient to choose the z-axis along the normal to the reaction plane. In elastic and inelastic scattering of a spin-one-half projectile, the observables such as the cross section, polarization, and depolarization, respectively, are described in this co-ordinate frame by

$$\sigma(\theta) = \frac{1}{2}(\sigma_{++} + \sigma_{+-} + \sigma_{-+} + \sigma_{--})$$

$$P(\theta) = \frac{\sigma_{++} - \sigma_{--} - \sigma_{+-} + \sigma_{-+}}{\sigma_{++} + \sigma_{+-} + \sigma_{-+} + \sigma_{--}}$$

and $D(\theta) = \dfrac{\sigma_{++} + \sigma_{--} - \sigma_{+-} - \sigma_{-+}}{\sigma_{++} + \sigma_{+-} + \sigma_{-+} + \sigma_{--}}$.

In elastic scattering $\sigma_{+-} = \sigma_{-+}$ because of time reversal invariance. σ_{+-} denotes the partial cross section for initially spin-up and finally spin-down states with respect to the z-axis. The depolarization parameter is related to the spin-flip probability by $D(\theta) = 1 - 2S(\theta)$.

To interpret these experimental results, compound nucleus reaction processes as well as direct reaction processes were considered. The differential cross section $\sigma(\theta)$ is assumed to be described by the in-coherent sum: $\sigma(\theta) = \sigma^{se}(\theta) + \sigma^{ce}(\theta)$. The shape elastic scattering cross section $\sigma^{se}(\theta)$ was calculated with the optical model which includes a spin-spin interaction in the form $-V_{I\sigma}f(r)\ \vec{I}\cdot\vec{\sigma}$. The optical-potential parameters were taken from the work of Perey and Buck. The compound nucleus cross section $\sigma^{ce}(\theta)$ was calculated with the Hauser-Feshbach theory which includes relevant inelastic scatter-ing channels. In figs. 5 and 6 are shown calculated curves of de-polarization D^{se} and partial cross section σ^{se}.

In order to investigate how the spin flip proceeds through the com-

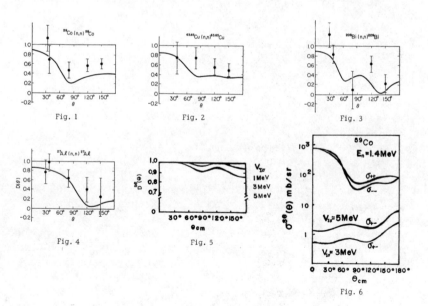

Fig. 1

Fig. 2

Fig. 3

Fig. 4

Fig. 5

Fig. 6

pound nucleus reaction process, the depolarization parameter is represented by the cross section as follows:

$$D(\theta) = \frac{\sigma^{se}_{++} + \sigma^{ce}_{++} + \sigma^{se}_{--} + \sigma^{ce}_{--} - 2(\sigma^{se}_{+-} + \sigma^{ce}_{+-})}{\sigma^{se}_{++} + \sigma^{ce}_{++} + \sigma^{se}_{--} + \sigma^{ce}_{--} + 2(\sigma^{se}_{+-} + \sigma^{ce}_{+-})}$$

$$= \frac{\sigma^{se}(\theta)}{\sigma^{se}(\theta) + \sigma^{ce}(\theta)} D^{se} + \frac{\sigma^{ce}_{++} - \sigma^{ce}_{+-}}{\sigma^{se}(\theta) + \sigma^{ce}(\theta)} \, .$$

In figs. 1-4 full lines show the first term where D^{se} is unity. Calculations of the second term are now in progress. At $E_n = 1.36$ MeV most of the spin-flip proceeds through the compound nucleus process.

REFERENCE

*Permanent address: Science University of Tokyo, Chiba, Japan.
†Permanent address: Kyoto University, Japan.

[1]   K. Katori et al., J. Phys. Soc, Japan 28 (1970) 1116.

Experimental Observation of a Deformation Effect in the Neutron Total Cross Section of ^{59}Co Using a Polarized Target

T. R. FISHER, Lockheed Palo Alto Research Laboratory, California;
D. C. HEALEY, J. S. McCARTHY, and D. PARKS, Stanford University,
California, USA*

Since the suggestion by Visotskii et al. [1] that the scattering of neutrons by oriented, deformed nuclei could be used to determine the sign and magnitude of the nuclear deformation, several experiments of this kind have been performed for the nucleus ^{165}Ho [2]. This nucleus is in the middle of the deformed rare earth region and exhibits a characteristic rotational spectrum. Davies et al. [3] have pointed out that odd vibrational nuclei should show similar, if somewhat smaller effects, due to the polarization of the core by the odd nucleon. This note reports experimental evidence for such a "deformation effect" in the neutron total cross section of ^{59}Co.

The nuclear "deformation effect" is conveniently defined by $\Delta\sigma_{Def} = \sigma_{or} - \sigma_{unor}$ where σ_{or} and σ_{unor} are the total cross sections for the cases of nuclear orientation and no nuclear orientation, respectively. In the present experiment, a nuclear alignment B_2/B_2 (max) = 0.065± 0.012 (equivalently f_2/f_2 (max)) was produced in a 32-g cylinder of polycrystalline Co metal by cooling it to a temperature of 0.040 K in a magnetic field of 8.3 kOe. This target has been described in detail elsewhere [4]. The quantity $\Delta\sigma_{Def}$ was measured by observing the change in the transmitted neutron intensity when the cobalt target was warmed to 1 K, destroying the nuclear alignment. The results of the measurements for neutron energies between 0.3 and 1.8 MeV are shown in fig. 1. The neutron energy spread arising from the thickness of the production target is approximately 80 keV. The data have been corrected for a second order contribution from the spin-spin effect [5] which arises because the target was polarized (B_1/B_1 (max) = 0.35).

The data in fig. 1 should be regarded as preliminary, but it seems established that a sizable deformation effect exists which has the same sign throughout the energy region studied. Efforts are in progress to interpret the data in the light of the DWBA theory of Davies et al. [3]. The sign agrees with a prolate deformation of the core, and the magnitude of the effect is consistent with measurements [6] of the quadrupole moment for ^{59}Co. It would be desirable to repeat this experiment with a single crystal of Co metal in which nuclear alignment but no nuclear polarization was present, and to study the case of alignment perpendicular to the direction of the neutron beam, as well as extending these measurements to higher neutron energies.

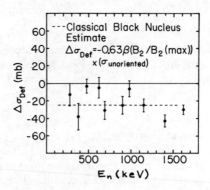

Fig. 1. Deformation effect as a function of neutron energy.

REFERENCES

*Work supported in part by the National Science Foundation, the U.S. Office of Naval Research, and Lockheed Independent Research Funds.

[1] G. L. Vistoskii, E. V. Inopin, and A. A. Kresnin, JETP 9 (1959) 398.
[2] R. Wagner, P. D. Miller, T. Tamura, and H. Marshak, Phys. Rev. 139 (1965) B29; T. R. Fisher et al., Phys. Rev. 157 (1967) 1149; J. S. McCarthy et al., Phys. Rev. Lett. 20 (1968) 502; H. Marshak, A. Langsford, C. Y. Wong, and T. Tamura, Phys. Rev. Lett. 20 (1968) 554.
[3] K. T. R. Davies, G. R. Satchler, R. M. Drisko, and R. H. Bassel, Nucl. Phys. 44 (1963) 607.
[4] T. R. Fisher et al., Rev. Sci. Inst. 41 (1970) 685.
[5] D. C. Healey, J. S. McCarthy, D. Parks, and T. R. Fisher, Phys. Rev. Lett. 25 (1970) 117.
[6] S. C. Fultz et al., Phys. Rev. 128 (1962) 2345.

Proton Polarization Analysis
with a Surface Term in the Real Part
of the Optical Potential

B. C. SINHA, E. J. BURGE,* V. R. W. EDWARDS, King's College, London;
W. H. TAIT, Westham College of Technology, London, England

The real part of the optical potential has been traditionally represented
by a Saxon–Woods form factor on the grounds that the quality of the
fits obtained to proton elastic scattering data are essentially insensitive to the shape chosen for the optical potential. However, a departure from a Saxon–Woods form factor is expected, if one takes into
account effects such as the isospin potential, exchange, target polarization, and nuclear correlation. Extensive analysis has been carried
out on proton elastic scattering data from nuclei ranging from ^{12}C to
^{208}Pb, with a form factor that now includes a Saxon–Woods derivative
term intended to account for the above mentioned effects, so that

$$U_{OPT}(r) = U_R f(r, R_R, a_R) + 4 U_S a_S \frac{d}{dr} f(r, R_S, a_S),$$

where
$$f(r, R_x, a_x) = (1 + \exp[(r - R_x)/a_x])^{-1}$$

A considerable improvement in the fits both for cross–section and
polarization data has been obtained using the present model. The
parameters of the Saxon–Woods derivative term have been found to be
in reasonable agreement with nuclear structure calculations and consistent with other experimental evidence.

One or more of the above–mentioned effects have been found to be
more important than the others, depending on the nucleus. In the case
of ^{12}C and ^{40}Ca, where the isospin term is negligibly small, the presence of the Saxon–Woods derivative term is believed to be due to antisymmetrization and target polarization. A comparison between the two
sets of fits to the polarization data for ^{12}C(p,p) using the present and
the simple optical model are shown in fig. 1.

As can be seen quite clearly, the improvement is striking. For the
case of ^{208}Pb and ^{92,96,98,100}Mo isotopes, the Saxon–Woods derivative
term is largely due to the isospin potential arising from the neutron
excess. The consistent increase in both the radii and the depths of

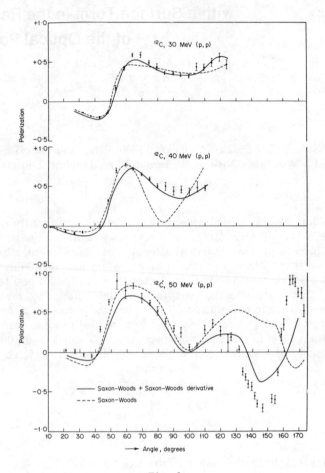

Fig. 1

the Saxon-Woods derivative term found in the case of Mo isotopes
suggests that the excess neutrons are distributed on the nuclear surface. A similar situation arises in the case of ^{208}Pb and also ^{54,56}Fe.
The ratio of the depths of the Saxon-Woods derivative term and the
Saxon-Woods term are in good agreement [1] with nucleon-nucleon
scattering data. The difference between the neutron and proton mean
square radii obtained from this model is in general agreement with nuclear structure predictions and Coulomb displacement energies.

REFERENCE

*Present address: Chelsea College of Technology, London, England.

[1] B. C. Sinha and V. R. W. Edwards, Phys. Lett. 31B (1970) 273.

DISCUSSION

Rawitscher:

The optical potential depends on more than folding the nucleon-nucleon potential into a nuclear matter distribution, because of the presence of two-step direct reaction processes. For example, the proton picks up a neutron to form a deuteron, and in the second part of the two-step process, the proton returns the neutron to the nucleus. Such processes could give rise to effects in the polarization, and might be unimportant in the cross section. Incidentally, this process is not charge independent, since it is easier for a proton to pick up a neutron than for a neutron to pick up a proton at the same incident energy. It would be nice if one would guide the insertion of phenomenological bumps in the optical potential by reaction theoretical considerations.

Liers:

Do you think the fact that you get better fits with a surface term is simply because you have added more parameters to the optical model?

Sinha:

Maybe, but the shape of the potential was our interest. We wanted to compare these shapes with those that could be calculated using various theoretical methods.

Polarization Experiments on Si and Si-Polarimeters

R. BANGERT, B. GONSIOR, P. RÖSNER, M. ROTH, B. STEINMETZ, and
A. STRÖMICH, University of Cologne, Germany

Double scattering techniques can be used to measure the polarization
of elastically scattered protons in order to determine J-values unambiguously. The polarimeters normally use carbon as a second scatterer. Because of the necessary thickness of this second scatterer
the resolution of a carbon polarimeter is typically about 1 MeV. Some
interesting measurements cannot be done with this poor resolution.
The resolution can be greatly improved by a polarimeter using a Si-
ΔE detector as a second scatterer. Such a polarimeter has been constructed and tested by Maddox et al. [1] and some preliminary work
in this field has been done by Dost et al. [2].

Since we are planning measurements of polarization of inelastically
scattered protons, we started experiments with Si-polarimeters. First
we studied the polarization of protons scattered by Si, since only few
results are available in the literature and such results are needed for
estimating the analyzing power of a Si-polarimeter. We used a wafer
of Si 41.5 μm thick as a target and our carbon polarimeters [3] to measure the polarization of the scattered protons. Previous polarization
measurements suggest the use of an angle of 130°, because there the
polarization seems to be largest. The incoming protons had energies
between 7.5 and 11.5 MeV. The results are shown in fig. 1, where the
polarization is given as
a function of the energy.
The eye-guiding line con-
nects the points of the
polarization measure-
ments (P_1) on a Si-target
with carbon polarimeters.
The energy scale gives
the proton energy at the
center of the target. Also
given are data points for
the effective polarization
($\bar{P}_2$) measured with Si as

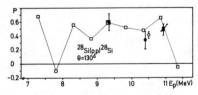

Fig. 1. Polarization of protons
(P_1) on Si and effective polarization $\bar{P}_2$ on the Si-polarimeter.

second scatterer in our polarimeters and with carbon as first target.
Since the angular distribution and the excitation function of the polarization of protons scattered on carbon are well known [4], we can deduce $\bar{P}_2$ from the asymmetry measured after the second scatterer. The
energy scale gives the energy of the protons incident on the Si-scat-
terer in the polarimeter. Fig. 1 also shows one point P_1 from the literature [5]. These results show that the polarization of protons scattered on Si is rather independent of energy between 8.3 and 10.9 MeV.
Since the Si-target as a second scatterer had the same dimensions as
the one we used as a first target measuring P_1 it seems reasonable
that the absolute values of P_1 and $\bar{P}_2$ are about the same. Therefore
$\bar{P}_2$ should also be rather independent of energy in the same energy
region as P_1. The measurements do confirm this. Fig. 2 shows the
pulse height spectra of the two detector
systems in one of the Si-polarimeters.
These systems consist of one Si-ΔE
detector as a second scatterer belonging to both counter telescopes on opposite sides and at equal angles to the
polarimeter axis. The energy loss in the
second scatterer is added to the energy
loss in one of the telescopes depending
on which way the scattered particle trav-
elled. By this method the resolution can
be greatly improved. We obtained an
overall resolution of about 700 keV with
a target thickness of 300 keV at 10 MeV.
The resolution can still be improved
since electronics associated with our
second scatterer was not fast enough.
This was not important in measuring $\bar{P}_2$.

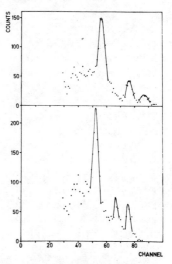

Fig. 2. Spectra obtained
by adding the pulses from
a Si second scatterer and
the two detector telescopes
in the polarimeter.

During these measurements we
could see that the Si-polarimeter works
reliably even with the high currents we
used (1 µA). By using faster electronics
with the second scatterer, we hope to
improve in the near future the energy
resolution appreciably, so that we will be able to try to measure the
polarization of inelastically scattered protons in cases where the use
of beams of polarized protons is restricted.

REFERENCES

[1] D. W. Miller, Second Polarization Symp., p. 429.
[2] M. Dost, S. Fiarman, and H. L. Harney, Jahresbericht des Max-
     Planck-Institutes für Kernphysik, Heidelberg (1968) p. 59.

[3]   R. Bangert, P. Rösner, and M. Roth, Jahresbericht 1968/69 des
      Institutes für Kernphysik der Universität zu Köln.

[4]   S. J. Moss and W. Haeberli, Nucl. Phys. 72 (1965) 417.

[5]   L. Rosen, J. G. Berry, A. S. Goldhaber, and E. H. Auerbach, Ann.
      Phys. (N.Y.) 34 (1965) 34.

Energy Dependence of Cross Sections and Polarizations Measured in Proton Elastic Scattering from ^{48}Ca between 6 and 13 MeV

H. S. LIERS, University of Wisconsin, Madison, USA*

Cross-section and polarization measurements have been made for protons incident on ^{48}Ca in the energy range between 6.0 and 12.75 MeV. The University of Wisconsin tandem accelerator and Lamb-shift polarized ion source were utilized in this experiment. Data were taken at laboratory angles of 60°, 110°, and 140° in energy intervals of 50 keV using Si surface barrier detectors placed at equal angles to the left and right of the incident beam direction. The ratio of the cross sections at a given angle to that at 13° was determined by taking the ratio of yields obtained in detectors at that angle and in monitor counters located at ± 13°.

Results obtained at laboratory angles of 60° and 110° at energies between 7.0 and 12.75 MeV are shown in fig. 1. The most striking feature of these results is the strong fluctuations which take place over the entire energy range. The isobaric analog of the ground state of ^{49}Ca occurs at an incident proton energy of 1.94 MeV [1]. Most of the fluctuations up to an energy of 8.1 MeV can be explained as isobaric analogs of known levels in ^{49}Ca (ref. [2]). It is possible that the structure above 8.1 MeV arises from isobaric analog states which correspond to unbound levels of ^{49}Ca.

In some energy intervals where the cross section does not exhibit large fluctuations the polarization shows large changes, as for instance at 110° near 11.0 MeV. These results lead one to suspect that polarization data taken in this energy region on other targets will show a similar behavior and are, therefore, not suitable for optical model studies.

REFERENCES

*Work supported in part by the U.S. Atomic Energy Commission.

[1] K. W. Jones et al., Phys. Rev. 145 (1966) 894.
[2] E. Kashy, A. Sperduto, H. A. Enge, and W. W. Buechner, Phys. Rev. 135 (1965) B865.

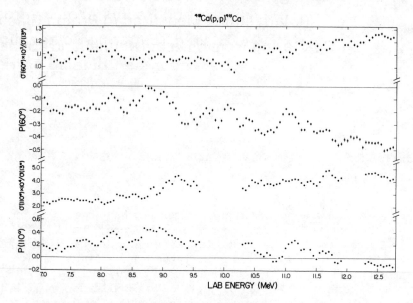

Fig. 1. Energy dependence of polarizations and cross
sections measured in proton elastic scattering from ^{48}Ca
at angles of 60° and 110°. Cross sections are plotted
relative to the cross section at 13°.

DISCUSSION

Rawitscher:

Were there any significant differences between the optical model
parameters for ^{40}Ca and ^{48}Ca?

Liers:

There are no good parameters for ^{40}Ca. There is strong resonant
structure in our energy range.

de Swiniarski:

In your optical model calulations have you done a search on all
parameters?

Liers:

Yes. A number of sets were found which fit the data. They all
show a small diffuseness for the real central potential and a
large one for the imaginary central potential.

Sinha:

Did you find any variation of the real central potential depth with
energy?

Liers:

Yes. The energy dependence is given by $-0.17 \cdot E_p$.

Attempt to Resolve the Optical Model
Spin-Orbit Radius Anomaly from
p + ^{197}Au Elastic Scattering Polarization
Data near the Coulomb Barrier

J. S. ECK, Kansas State University, Manhattan; W. J. THOMPSON, University of North Carolina at Chapel Hill and TUNL, USA

This paper reports an attempt to resolve the SO radius anomaly [1, 2, 3] by observing the onset of polarization as the energy of the incident proton is increased from energies below the Coulomb barrier to energies well above the barrier. The data consisted of p + ^{197}Au elastic scattering cross-section and polarization data in the range from 10-14 MeV [4]. Angular distributions taken with a thin target at 10, 11, and 12 MeV were uncontaminated by any inelastic proton scattering, while an angular distribution at 13.75 and all of the polarization data were taken using a thick target and are contaminated by inelastic scattering from states up to 0.55 MeV in ^{197}Au. A high resolution spectrum taken at 14 MeV at 140° which resolves all inelastic states except the 0.077 MeV state shows that the sum of the contributions from these inelastic states is only about 2% of that of the elastic peak. Calculations carried out at 12 MeV show that the inelastic scattering to the 0.077 MeV state is almost pure Coulomb excitation [5]. It is unlikely that the excited states have a polarization greater than that for the ground state, because of the lower energy (note also that Coulomb excitation gives no polarization). Therefore, the contribution to the measured polarization from the inelastic states is only 1% of the polarization for the elastic scattering, which is less than the statistical error.

A more serious problem is the presence of isobaric analog resonances (IAR). Calculations show that the IAR energy is E = 12.05 ± 0.05 for the g.s. IAR in ^{197}Au(p,p), and the excited states are higher by the excitation energy [6]. However, the IAR polarization is asymmetric about the resonance and would tend to cancel for a thick target [7]. The high g.s. spin of ^{197}Au($3/2^+$) will weaken the resonance polarization effects, since any one IAR is reached from *two* partial waves $\vec{J}_{IAR} = \vec{\ell} + \vec{3/2}$ with the two ℓ values differing by two to conserve parity.

The central-well optical-model parameters were obtained by fitting the angular distributions starting with the parameters suggested by Greenlees et al. [3]. In general these parameters yielded adequate fits except that the diffuseness parameters $a_r = a_I = a_{SO} = 0.464 ± 0.050$.

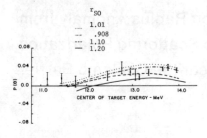

Fig. 1. Polarization at 140° as a function of energy. x's indicate data points. Smooth curves are optical model fits for V_{SO} = 7.2, a_{SO} = 0.474, and r_{SO} as indicated in legend.

Fig. 2. Polarization angular distribution at 13.75 MeV. Smooth curve is optical model fit for V_{SO} = 7.2, a_{SO} = 0.474, and r_{SO} = .908.

The polarization data were then fitted (using V_{SO} = 7.2 ± 1, a_{SO} = 0.464) by varying the values of r_{SO}/r_C from 0.8 to 1.2. The results are shown in fig. 1. The best fit was obtained for r_{SO} = 0.908, r_c = 1.17. The fit to the polarization angular distribution at 13.75 MeV for r_{SO} = 0.908, r_c = 1.17 is shown in fig. 2. Other parameter sets which yielded good fits to the angular distributions were also used to fit the polarization data. In all cases the results were essentially the same with the best fits obtained for r_{SO}/r_C = 0.8-0.9.

It appears clear from this study that the SO force is of shorter range than the central well by about 15-25%. Although strong conclusions concerning the particular derived parameters cannot be drawn due to the unfortunate choice of ^{197}Au as a target nucleus, this technique has been shown to be useful in determining the SO radius.

REFERENCES

[1] D. W. Sprung and P. C. Bhargava, Phys. Rev. 156 (1967) 1185.
[2] L. J. Goldfarb, G. W. Greenlees, and M. B. Hooper, Phys. Rev. 144 (1966) 829.
[3] F. D. Becchetti, Jr., and G. W. Greenlees, Phys. Rev. 182 (1969) 1190; C. M. Perey, F. G. Perey, and J. K. Dickens, Phys. Rev. 175 (1968) 1468.
[4] Polarization data taken by R. Rathmell and W. Haeberli initially reported in Bull. Am. Phys. Soc. 12 (1969) 1230.
[5] K. Alder et al., Rev. Mod. Phys. 28 (1956) 432.
[6] D. D. Long, P. Richard, C. F. Moore, and J. D. Fox, Phys. Rev. 149 (1966) 906.
[7] J. L. Adams, W. J. Thompson, and D. Robson, Nucl. Phys. 89 (1966) 377.

Search for Mott-Schwinger Polarization
in Elastic Proton Scattering

G. ROY, H. SHERIF, and G. A. MOSS, University of Alberta, Edmonton, Canada*

The interaction between the magnetic dipole moment of a particle and the Coulomb field of a nucleus leads to a contribution to the Hamiltonian of the form $\vec{\sigma} \cdot \vec{E} \times \vec{P}$. This may be easily reduced to the form $\vec{\sigma} \cdot \vec{\ell}/r^3$. Such a term can lead to polarization of the particle, and indeed this has been demonstrated for electrons by Mott [1] and for neutrons by Schwinger [2] and Elwyn et al. [3]. We have searched for this polarization (Mott-Schwinger polarization) in elastic proton scattering on Pt at 5° for ~ 5-MeV protons.

Our experimental method was to bombard a thin (~ 200 μgm/cm²) carbon target with 5.04-MeV protons. At angles of 50°, 80°, and 90°, this reaction yields [4] proton polarizations of -0.87, +0.04, and +0.44, respectively. These reaction protons were subsequently scattered from an 8 mg/cm² Pt foil, and the doubly scattered protons were observed by detectors placed at ±5°. No asymmetry beyond the instrumental asymmetry was detected (within statistical error of 0.01) for the three reaction angles. Hence we conclude that the Mott-Schwinger polarization is too small to be measured at these angles.

We have performed some crude calculations on the expected polarization by simply adding a term

$$V = \frac{Ze^2\hbar^2}{2M_p^2c^2}\left(\mu_p - \frac{1}{2}\right)\vec{\sigma}\cdot\vec{\ell} \begin{cases} R_c^{-3} & \text{for } r \leq R_c \\ r^{-3} & \text{for } r > R_c \end{cases}$$

to an optical model program. This yields a crude result because our code is practically limited to ℓ-values of ~ 20, while we may justifiably expect larger ℓ-values to contribute due to this long-range interaction. Therefore we would expect our calculations to underestimate the Mott-Schwinger polarization somewhat. Our calculations indicate extremely small Mott-Schwinger polarization at 5 MeV. We extended our calculations to 155-MeV protons on ^{208}Pb, and again found a very small effect due to the Mott-Schwinger term. We are presently modifying our computer code to yield a more accurate calculation, and are extending the energy range of our calculations to discover if Mott-Schwinger polarization is ever of significance in elastic proton scattering.

REFERENCES

*This work was supported in part by the Atomic Energy Control Board of Canada.

[1]   N. F. Mott, Proc. Roy. Soc. A135 (1932) 429.
[2]   J. Schwinger, Phys. Rev. 73 (1948) 407.
[3]   A. J. Elwyn, J. E. Monahan, R. O. Lane, and F. P. Mooring, Phys. Rev. 142 (1966) 758.
[4]   S. J. Moss and W. Haeberli, Nucl. Phys. 72 (1965) 417.

DISCUSSION

Roman:
I believe that the Mott-Schwinger calculation of Drukarev and Ippolitov for (d,p) reactions contains errors. Under certain circumstances proton polarizations larger than one can be obtained from the formulae mentioned.

Investigation of Threshold Effects
in Isospin Coupled Channels
with Polarized Protons and Deuterons

K. WIENHARD, Institut für Kernphysik der Universität Giessen;
G. GRAW, Physikalisches Institut der Universität Erlangen-Nürnberg,
Germany

Threshold effects were observed in (d,p) reactions on heavy nuclei
at energies where isospin coupled channels are opening [1, 2]. The
excitation function of the (d,p) cross section shows a cusp at backward angles at the (d,n) threshold to the corresponding analog state.
Theoretical interpretations have been published by Coker and Tamura
[3] on the basis of a coupled channel theory and by Zimanyi and Bondorf [4] in the framework of S-matrix theory. The ^{90}Zr(d,p$_0$)^{91}Zr reaction
was investigated with vector-polarized deuterons [5] to provide further
independent experimental information. Fig. 1 shows the excitation
function for the analyzing power at six reaction angles. There is an
easily observable change in behavior at the (d,n) threshold (E$_d$ =
7.05 MeV) which is most obvious at backward angles.

   If the origin of the coupling effect is located in the nucleon channels, then a similar effect is to be expected in elastic proton scattering at the quasielastic (p,n) threshold for the formation of the isobaric analog of the target nucleus. The elastic proton scattering cross
section at this threshold has been measured on a number of nuclei [6,
7]. However, no clear-cut effects could be observed, except in the
elastic scattering from lead isotopes, where a decrease in the differential cross section was observed at the quasielastic (p,n) threshold.
We measured the analyzing power of elastically scattered protons from
^{96}Zr with the polarized proton beam of the Erlangen 12 MeV tandem accelerator [8]. The excitation functions for the analyzing power and the
differential cross section at 160° are shown in fig. 2. The broad fluctuations of the differential cross section are seen also in the analyzing
power. However, the analyzing power exhibits additionally a rapid
change in behavior. The magnitude of the analyzing power changes
abruptly (within 10 keV) from −0.2 to −0.4. This anomaly occurs at an
incident proton energy of 11.7 MeV and is 50 keV broad. This energy
agrees with the energy of the quasielastic (p,n) threshold which is calculated to be at a c.m. proton energy of 11.6 MeV [7].

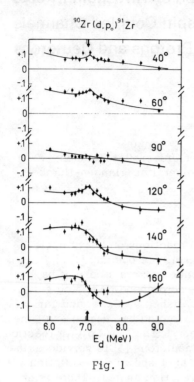

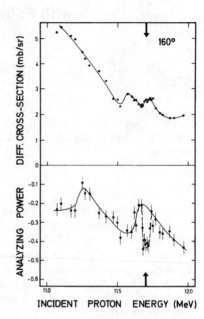

Fig. 1   Fig. 2

REFERENCES

[1] C. F. Moore et al., Phys. Rev. Lett. 17 (1966) 926.
[2] R. Heffner, C. Ling, N. Cue, and P. Richard, Phys. Lett. 26B (1968)
     150; N. Cue and P. Richard, Phys. Rev. 143 (1968) 1310; P. Richard,
     Proc. Conf. on Nuclear Isospin, Asilomar (Academic Press, 1969).
[3] R. Coker and T. Tamura, Phys. Rev. 182 (1969) 1277.
[4] J. Zimanyi and J. P. Bondorf, Nucl. Phys. A146 (1970) 81.
[5] G. Clausnitzer, G. Graw, C. F. Moore, and K. Wienhard, Phys.
     Rev. Lett. 22 (1969) 793.
[6] J. P. Bondorf et al., Nucl. Phys. A101 (1967) 338; E. Friedman,
     A. Ginzburg, A. A. Jaffee, and A. Marinov, Nucl. Phys. A110 (1968)
     300; R. A. Hinrichs, G. W. Phillips, J. G. Cramer, and H. Wieman,
     Phys. Rev. Lett. 22 (1969) 301; N. Stein, Proc. Conf. on Nuclear
     Isospin, Asilomar (Academic Press, 1969).
[7] L. S. Michelman, T. I. Bonner, and J. G. Kulleck, Phys. Lett. 28B
     (1969) 659.
[8] G. Clausnitzer et al., Nucl. Instr. 80 (1970) 245.

Spin-Correlation Parameters in
p-p Scattering and Depolarization
in Nuclear Scattering

P. CATILLON, Centre d'Etudes Nucléaires de Saclay, France

Spin correlation parameters in p-p scattering. The measurement of the spin-correlation parameters A_{xx} and A_{yy} made previously at Saclay [1] has been extended up to 50 MeV by Garreta and Fruneau [2] on the Grenoble cyclotron. These parameters have been measured at two new energies, 37.2 MeV and 46.9 MeV, and, at the same time, at 26.5 MeV in order to deduce the polarization of the beam and of the target from the previous experiments. Therefore the values of these parameters given in table 1 are related to the same normalization point, A_{xx} = -0.984 at 11.4 MeV [3]. The errors do not include the normalization error coming from the uncertainty on the value of A_{xx} at 11.4 MeV.

Table 1

	26.5 MeV	37.2 MeV	46.9 MeV
A_{xx}	-0.912 ± 0.013	-0.874 ± 0.016	-0.850 ± 0.015
A_{yy}	-0.719 ± 0.011	-0.511 ± 0.013	-0.275 ± 0.010

The experimental results for A_{yy} are in good agreement with the predictions and other experiments (cf. fig. 1). Two conclusions can be drawn from the A_{xx} results:

(1) They are in complete disagreement with the Tokyo data [4].

(2) One can see a significant deviation from the values calculated from the X EDA phases [5].

Because of the direct connection between $(1 + A_{xx})$ and the splitting $({}^3P_0 - {}^3P_2)$ it seems that the value of ${}^3P_0$ has to be increased by 1-2° from its expected value. If we use the approximation of Gammel and Thaler [6] for reducing the P-phases in the central, spin-orbit, and tensor parts Δ_C, Δ_{LS}, and Δ_T, this gives a significant change on Δ_{LS} (decrease of $\sim 20\%$) from X EDA in this energy range.

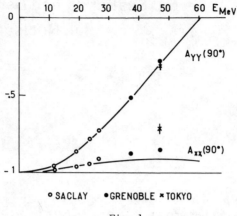

○ SACLAY   ● GRENOBLE × TOKYO

Fig. 1

Depolarization in nuclear elastic scattering. An extensive experimental search to find the effect of a spin-spin interaction in nuclear elastic scattering has given contradictory results. Most of the experiments measure the spin-spin effect in polarized-neutron total cross sections for polarized targets of ^{165}Ho or ^{59}Co [7]. Stamp [8] has shown that this spin-spin interaction has negligible effects on the differential elastic scattering and polarization, but causes the parameter D to diverge from 1 for angles corresponding to the minima in the differential cross section.

We measured this depolarization parameter D in the scattering of polarized protons on ^{9}Be, ^{10}B, and ^{27}Al [9] for angles corresponding to the first minimum of the elastic scattering (fig. 2). The results are given in table 2. The energy is given at the center of the target with half the energy spread due to the target thickness. A test of coherence has been made on ^{28}Si, a target with a spin zero.

Table 2

Target	Spin	Energy (MeV)	θ_{lab} (deg)	D
^{9}Be	3/2	21.4 ± 2.3	58	0.940 ± 0.016
^{10}B	3	19.8 ± 1.3	60	0.926 ± 0.011
^{27}Al	5/2	18.0 ± 0.9	43	0.964 ± 0.021
^{28}Si	0	17.9 ± 1.0	44.5	1.027 ± 0.025

These results indicate a small but real effect. 'A preliminary analysis for ^{27}Al made by Batty [10] gives for a spin-spin potential of the type $V_{ss} f(r) \vec{\sigma} \cdot \vec{I}$ assuming a volume form for $f(r)$ the value

$$V_{ss} = 0.42\ ^{+0.10}_{-0.14}\ \text{MeV}.$$

It is necessary to emphasize that a small contribution from compound elastic scattering could increase the experimental value of V_{ss}. On the other hand, a recent experiment made by the Stanford group [7] indicated that a simple $\vec{I} \cdot \vec{\sigma}$ potential is probably inadequate.

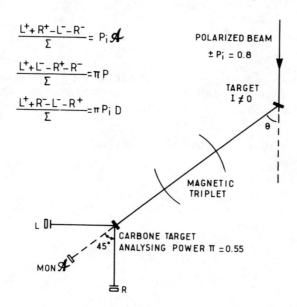

EXPERIMENTAL SET-UP

Fig. 2

REFERENCES

[1] P. Catillon, M. Chapellier, and D. Garreta, Nucl. Phys. B2 (1967) 93.
[2] D. Garreta, N. Nisimura, and M. Fruneau, Phys. Lett. 31B (1970) 363 and to be published.
[3] H. P. Noyes and H. M. Lipinski, Phys. Rev. 162 (1967) 884.
[4] K. Nisimura et al., Progr. Theoret. Phys. (Kyoto) 30 (1963) 719; Phys. Lett. 30B (1969) 612.

[5] M. H. MacGregor et al., Report UCRL 70075; Phys. Rev. 182 (1969) 1714.
[6] J. L. Gammel and R. M. Thaler, Phys. Rev. 107 (1957) 291.
[7] D. C. Healey, J. M. MacCarthy, D. Parks, and T. R. Fisher, Phys. Rev. Lett. 25 (1970) 117; other references can be found in this paper.
[8] A. P. Stamp, Phys. Rev. 153 (1967) 1052.
[9] R. Beurtey, P. Catillon, and P. Schnabel, Compte-rendu d'activité du Département de Physique Nucléaire (1969 and 1970).
[10] C. J. Batty, private communication.

DISCUSSION

Slobodrian:

At a previous conference it was pointed out that spin correlation parameters are relative values. Are they still relative values or was some work done to obtain absolute values?

Catillon:

Still relative. The normalization point is A_{xx} at 11.4 MeV.

Scattering of
Deuterons and Tritons

Investigation of Spin-Orbit and Spin-Spin Interactions in the Optical Model by Vector-Polarized Deuterons

H. WILSCH, G. HEIL, and G. KLIER, Physikalisches Institut der Universität Erlangen-Nürnberg, Germany

The analyzing power $^{*}A(\theta)$ of a range of nuclei for the elastic scattering of vector-polarized deuterons [1] has been measured for $E_d = 9, 10, 11,$ and 12 MeV. Optical model fits to the angular distributions of $A(\theta)$ only, and to $A(\theta)$ and $\sigma(\theta)$ simultaneously yielded quite similar parameter sets for the different nuclei under investigation.[†] Neglecting small individual changes, the optical potential may be represented by

Param.	V	r_o	a_o	W	r_w	a_w	V_{so}	r_{so}	a_{so}
Form	Woods-Saxon			WS–Derivative			Thomas–Fermi		
Value	93 MeV	1.13 fm	0.88 fm	15 MeV	1.58 fm	0.55 fm	7.5 MeV	0.80 fm	0.5 fm
				+			+	+	+

The spin-orbit force is peaked in the interior of the real potential and has relatively small diffuseness. Parameters which show the most pronounced dependence from individual nuclei are marked. Especially neighboring nuclei with different spins I seem to have different spin-orbit parameters. One possible explanation might be a target-spin dependent interaction [2] which has not been incorporated in the optical potential. Preliminary results seem to favor an interaction of the type $\vec{s} \cdot \vec{I}$ ($\vec{s}$: spin of deuteron, $\vec{I}$: spin of target nucleus).

REFERENCES

$^{*}A(\theta) = 2/\sqrt{3} \langle iT_{11}(\theta) \rangle$.

[†] ^{27}Al, ^{28}Si, ^{31}P, ^{32}S, ^{40}Ca, ^{45}Sc, ^{51}V, ^{55}Mn, ^{59}Co. For ^{27}Al, ^{28}Si, ^{40}Ca, ^{52}Cr, ^{56}Fe, ^{58}Ni, ^{60}Ni data from other groups are available.

[1]   G. Clausnitzer et al., Nucl. Instr. and Method. 80 (1970) 245.
[2]   H. Feshbach, Ann. Rev. of Nucl. Sci. 8 (1958) 79.

DISCUSSION

Beurtey:

The difference you observed between the polarizations in the scattering of deuterons from ^{27}Al and ^{28}Si does not prove at all the presence of a spin-spin interaction. Before doing our measurement of the parameter "D" on ^{27}Al for protons, we compared the polarizations in the scattering form ^{26}Mg, ^{27}Al, and ^{28}Si. The absolute value of the polarization in the case of ^{27}Al was much lower than for the two other nuclei ($\approx 40\%$ less). But the spin-flip cross section turned out to be only $\approx 4\%$. There are many other terms explaining such a difference in polarization for a non-zero-spin target.

Raynal:

In the 20-MeV region and around Ni and Ca, I found that the radii of the central part were 1.05 fm, the radii of spin-orbit coupling went down to 0.6 fm, and the diffuseness parameters were of the order of 0.3 or 0.4 fm. The search stopped for the $\vec{L} \cdot \vec{S}$ diffuseness because it was too small. Do you have a real indication that your geometrical parameters cannot be smaller?

Wilsch:

The values of the spin-orbit radius and diffuseness parameters may be even smaller than the mean values I quoted. For some nuclei I found radii of about 0.6 fm and diffuseness parameters of about 0.3 fm.

Fitz:

When comparing asymmetries in deuteron scattering in the region of nuclei around Al, Si, one should also remember the rapidly varying deformation of these nuclei, and one should be careful in interpreting these results in terms of a spin-spin interaction.

Differential Cross Section and Vector Polarization for the Elastic Scattering of 41- to 51-MeV Deuterons on Carbon

W. FETSCHER, K. SATTLER, N. C. SCHMEING, E. SEIBT, R. STAUDT, Ch. WEDDIGEN, and K. WEIGELE, Institut für Experimentelle Kernphysik der Universität und des Kernforschungszentrums Karlsruhe, Germany

The elastic scattering cross section of deuterons on carbon was measured at energies of 41, 46, and 51 MeV, and the polarization was obtained from double scattering experiments. The unusual experimental apparatus is described in another contribution [1]. Double scattering experiments with an unpolarized beam yield an asymmetry B, if there is a tensor polarization, and an asymmetry A which, ignoring tensor quantities, is the product $2 \cdot (iT_{11})_1 \cdot (iT_{11})_2$ of the vector polarization quantities for the scattering from the two targets. Since the asymmetry B showed no significant deviation from zero, the above expression holds. Thus it is possible to extract iT_{11} from a calibration experiment at the energy pairs (51, 46), (46, 41), and (51, 41) MeV. The sign was deduced from calculations. The curves obtained are shown in fig. 1.

We thank G. R. Satchler for fitting the cross section at 51 MeV using the search code "Hunter" (Drisko et al.). The cross sections calculated with a best fit (potential 1 of ref. [2]) are shown in fig. 1, as are predicted polarizations. Using a code of B. A. Robson's we did not find much improvement over the previous fit when we varied the parameters of the spin-orbit potential. Particularly the minimum near 44° could not be fitted. Hence we considered what effects might be relevant at 51 MeV as opposed to lower energies.

Since the deuteron is strongly absorbed, only the surface region is important for the scattering, but at higher energies and for a light target the deuteron sees more of the nucleus before the interior appears black [3]. Thus it is important to take into account the deuteron potential deeper inside the nucleus. In addition, at higher energies particles with higher angular momentum (to about $\ell = 20$) feel the nuclear forces, so that the spin-orbit potential becomes larger. It may no longer be justified to use the Thomas form [4]. Retaining Satchler's central terms, we have tried a volume $\vec{\ell} \cdot \vec{s}$ potential as is usual in structure calculations. This potential is proportional to the central

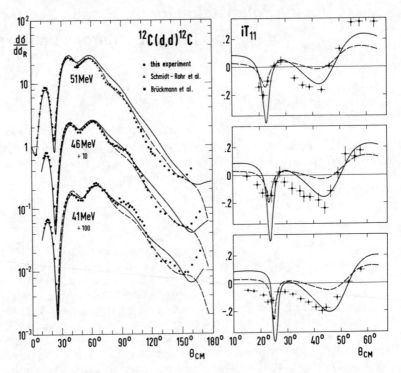

Fig. 1. Differential cross section and polarization at the same
energies. Dashed line: Satchler prediction [2]; solid line:
present calculation.

Woods-Saxon term and has a depth of 5.5 MeV (half that at the nuclear
radius). For $j = \ell + 1$, waves experience more interaction with the nucleus, and perhaps they are absorbed more. We have used a small
surface absorptive $\vec{\ell} \cdot \vec{s}$ potential with the appropriate sign, 0.1 MeV
deep (at the nuclear radius) and proportional to the imaginary central
term. The effect of these changes is: The volume potential indeed
lowers the 44° minimum. The absorptive part lowers the entire polarization to give the fits shown. The minimum in the cross section around
150° is introduced by the small $\vec{\ell} \cdot \vec{s}$ absorption. Both experimental and
calculated minima move toward larger angles with increasing energy.
In view of our failure to obtain fits of the polarization with a Thomas
force, we expect that further experiments could test whether a volume
$\vec{\ell} \cdot \vec{s}$ potential is physically meaningful. Very recent results of a search
on our polarization data by Raynal [5] support the assumption of an
$\vec{\ell} \cdot \vec{s}$ potential with a volume form.

REFERENCES

[1] W. Fetscher et al., Third Polarization Symp.
[2] E. Seibt, Ch. Weddigen, and K. Weigele, Phys. Lett. 27B (1968) 567.
[3] B. Tatischeff and I. Brissaud, to be published.
[4] F. G. Perey, Second Polarization Symp., p. 191.
[5] J. Raynal, private communication.

DISCUSSION

Raynal:

You mentioned three parameters—a depth, a radius, and a diffuseness—but only two of these are meaningful: the diffuseness and
the value of the potential at $r = 0$. The radius parameter can even
have a very large negative value, provided the depth increases
at the same time, without there being a change of the form factor.

Vector Analyzing Power in
the Elastic and Inelastic Scattering
of Deuterons from ^{24}Mg

R. C. BROWN, J. A. R. GRIFFITH, G. HUDSON, O. KARBAN, S. ROMAN, and J. SINGH, University of Birmingham, England

The angular distributions of vector analyzing power for 12.3-MeV deuterons scattered from the ground and 1.37-MeV states of ^{24}Mg have been measured. Though the fit obtained for the elastic data using the optical model was rather poor, the DWBA prediction for the 2^+ state data is in reasonable agreement with experiment.

The experiment was performed on the Radial Ridge cyclotron of Birmingham University using the pure-vector polarized 12.3-MeV deuteron beam. The arrangement of scattering chamber and instrumentation has been described elsewhere [1]. The main difference in technique was that the beam polarization was monitored using a polarimeter, which employed the ^{12}C(d,p)^{13}C (g.s.) reaction at 57° and not the ^{12}C elastic scattering previously described. A target of enriched ^{24}Mg, 4 mg cm^{-2} thick, was used contributing to the overall energy resolution of 200 keV. The (4.122, 4.24 MeV) doublet in Mg could not, therefore, be resolved. The differential cross section for the 1.368-MeV state was obtained from the ratio of the inelastic to elastic counting rates after averaging out the polarization effects. The elastic cross sections were taken from ref. [2].

A preliminary analysis of the elastic data has been made using the optical model. Searches in parameter space were started from the average potentials* of ref. [1], but as can be seen from fig. 1 adequate fits were not obtained to the polarization data. Even so, the parameter set which gave the best fit was used in calculations of the inelastic distributions using distorted waves theory. The code DWUCK was used to perform calculations in the first order theory of the collective model. An equal deformation was assumed for all parameters of the optical potential with the exception of the Coulomb term. The deformation parameter was obtained by normalization of the differential cross section at forward angles and preliminary work suggests a value in reasonable agreement with that obtained from ^{24}Mg(p,p') measurements [3].

As can be seen from the lower part of fig. 1, surprisingly good agreement was obtained from this preliminary analysis, which suggests that the assumption of a collective nature for the 2^+ state is correct. On the other hand, the difficulties encountered in fitting the elastic data could arise either from a compound process, or more likely from the need to use a different formulation of the problem. From the inspection of the elastic and inelastic cross sections it seems clear that the two channels are strongly coupled; therefore a coupled channels calculation should give a more comprehensive treatment of the situation. This should provide more information about the scattering process than a simple distorted waves calculation.

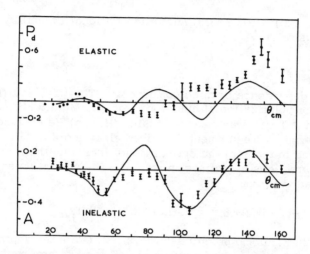

Fig. 1. Angular distributions of vector analyzing power for the scattering of 12.3-MeV deuterons from ^{24}Mg. The upper part shows the elastic scattering, together with an optical model prediction. The lower part shows the data for the 1.37-MeV state and a prediction calculated with the distorted waves theory.

REFERENCES

*The spin-orbit potential strengths given in the tables of ref. [1] are in error. They should be multiplied by a factor $(a_{s.o.}/a_0)$ for any given set.

[1] J. A. R. Griffith et al., Nucl. Phys. A146 (1970) 193.
[2] Max Planck Institute, Heidelberg, Report V13 (1963), unpublished.
[3] A. G. Blair et al., Report UCRL-18927.

Tensor Polarization of Deuterons from Mg(d,d)Mg Elastic Scattering at 7.0 MeV

A. DJALOEIS and J. NURZYNSKI, Australian National University, Canberra, Australia

The angular dependence of three components, $\langle T_{20} \rangle$, $\langle T_{21} \rangle$, and $\langle T_{22} \rangle$ (ref. [1]), of tensor polarization of deuterons from the Mg(d,d)Mg elastic scattering at 7.0 MeV was measured using ^{3}He(d,p)^{4}He analyzer [2], and the experimental equipment described elsewhere [3]. The results were analyzed in terms of the optical model including the Hauser-Feshbach corrections as defined in ref. [3]. The optical potential for deuterons had the form:

$$U = C - Vg - iWf + S\frac{1}{r}\frac{dg}{dr}(\vec{s}\cdot\vec{L}) + Mh\,T_R + Qh\,T_L,$$

where C is the Coulomb potential, V, W, S, M, Q are the depths of various components, g, f, h are functions of r, T_R and T_L are irreducible tensors [4].

In the present analysis g was of Woods-Saxon form, $f \propto |dg/dr|$, and h was a derivative of Woods-Saxon (D), Thomas (T), or Gaussian (G) forms [5]. The results of the experiment and analysis are presented in fig. 1. The main conclusions can be summarized as follows:

1. The type of the potential derived for Al(d,d)Al scattering at low energies [5] can provide an adequate interpretation of $\langle T_{20} \rangle$ and $\langle T_{22} \rangle$ components only, the fit to $\langle T_{21} \rangle$ being far from satisfactory. It has been found necessary to change the tensor potential considerably in order to fit all components of deuteron polarization.

2. The central, the $\vec{s}\cdot\vec{L}$ and the tensor T_R potentials are the only ones required in the interpretation of the present data. An addition of T_L coupling is redundant.

3. All types of the function h produce almost identical curves for $\langle T_{20} \rangle$ and $\langle T_{22} \rangle$ components. The fits to $\langle T_{21} \rangle$ seem to favor, in the χ^2 sense, the derivative of Woods-Saxon form.

4. Set of parameters d-73 produces better overall fit than set d-128.

The $\langle T_{21} \rangle$ component is very sensitive to tensor forces, and measurements of this quantity cannot be neglected in a detailed study of deuteron-nucleus interaction.

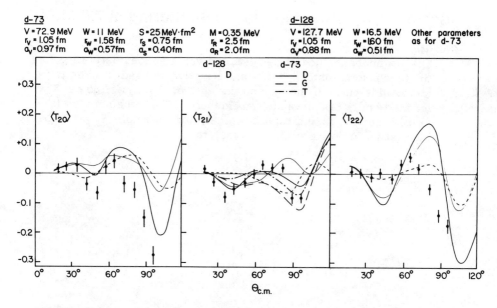

Fig. 1. Tensor polarization of deuterons from Mg(d,d,)Mg scattering at 7.0 MeV. The curves corresponding to two sets of parameters and to various types of form factor h for T_R interaction are displayed. The dotted curves represent the best fit based on the type of the potential derived in ref. [5]. The numerical values for T_R correspond to h = $(dg/dr)4a_R$.

The authors wish to thank N. H. Merrill, H. Cords, and S. Whineray for help in collecting data, and B. A. Robson and P. J. Dallimore for use of their computer codes.

REFERENCES

[1]   W. Lakin, Phys. Rev. 98 (1955) 139.
[2]   L. G. Pondrom and J. W. Daughtry, Phys. Rev. 121 (1961) 1192;
       H. B. Willard, First Polarization Symp., p. 175.
[3]   A. Djaloeis, H. Cords, and J. Nurzyński, submitted for publication
       in Nucl. Phys.
[4]   G. R. Satchler, Nucl. Phys. 21 (1960) 116.
[5]   P. Schwandt and W. Haeberli, Nucl. Phys. A110 (1968) 585.

DISCUSSION

Griffith:

The Watanabe method of derivation gives a second derivative for the tensor form factor. If a first derivative is used, it goes to the outside bump of the second derivative. This was the case for the results of Schwandt and Haeberli. There the nucleus is smaller and the radius must be larger, but not much. It would be interesting to compare the form factors.

Vector Polarization of
Elastically Scattered Deuterons

Yu. V. GOFMAN, N. I. ZAIKA, Yu. V. KIBKALO, E. B. LYOVSHIN, A. V.
MOKHNACH, O. F. NEMETS, P. L. SHMARIN, and A. M. YASNOGORODSKY,
Institute for Nuclear Research, Ukrainian Academy of Sciences, Kiev,
USSR

To obtain information on spin-dependent effects in the deuteron-nucleus
elastic interaction, the polarization of deuterons elastically scattered
by different nuclei was measured in the 15°–18° c.m. angular range, using the 13.6-MeV deuteron cyclotron beam. The particles were registered by one- and double-crystal scintillation and semiconductor spectrometers. The electronics allowed to separate, if necessary, different kinds of particles and to use up to 8 detectors simultaneously.

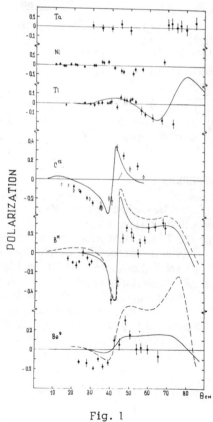

Deuterons scattered on carbon were analyzed by the (d,p) reactions on ^{3}He, ^{12}C, ^{14}N, by the ^{3}He(d,p)^{4}He reaction at the 430 keV resonance, and by double-scattering. The results show that the stripping reactions may be effectively used to determine the deuteron vector polarization and to find all the polarization components for deuterons scattered by carbon at 25° lab ($P_d = 2/\sqrt{3} \cdot \langle iT_{11} \rangle = -0.22 \pm 0.02$, $\langle T_{20} \rangle = -0.02 \pm 0.057$, $\langle T_{21} \rangle = 0.028 \pm 0.024$, $\langle T_{22} \rangle = -0.022 \pm 0.043$) [1].

The analysis of deuterons elastically scattered by Ta and Ni was made by stripping reac-

Fig. 1

tions. In the case of Ti, ^{11}B, ^{9}Be, the polarization was determined
from the elastic scattering asymmetry using the mentioned 25° carbon
scattered 12.6 ± 0.15 MeV polarized deuteron beam with above polarization components [2].

In the angular range measured (fig. 1) the polarization is small or
close to zero for heavy and medium weight nuclei, while for light ones
it is large even at small angles. This may account for the successful
description of the deuteron elastic scattering on heavy and medium
weight nuclei by an optical potential without spin-orbit term and for
the necessity of introducing this term in the case of light nuclei.

Fig. 1 gives also the polarizations calculated from an optical potential with a spin-orbit term, fitted to the experimental differential
cross sections at 13.6 MeV and for ^{9}Be also at 11.8 MeV deuteron energy [3]. The theoretical curve reflects the polarization trend, though
no quantitative agreement is observed in the case of light nuclei.

The optical model parameters used in the calculations are listed
in the table ($r_{so} = r_v$, $a_{so} = a_v$):

	V	W	V_{so}	r_v	a_v	r_w	a_w
Ti	114.2	9.38	9.36	1.005	0.753	1.292	0.881
^{12}C a)	85	8	5	1.18	0.78	1.48	1.9
^{11}B b)	95.9	17.28	7.3	1.043	1.003	1.965	0.374
^{9}Be	149	16.6	14.5	0.817	0.744	1.96	0.52
^{9}Be c)	138.3	9.1	12.54	0.713	1.083	1.92	0.555

a) $r_{so} = 1.7$ fm, $a_{so} = 0.4$ fm;
b) solid curve for $V_{so} = 6$ MeV, $r_{so} = 0.5$ fm, $a_{so} = 0.7$ fm;
c) dotted curve

REFERENCES

[1] Yu. V. Gofman et al., Yad. Fiz. 5 (1967) 718; N. I. Zaika et al.,
     ibid. 7 (1968) 754.
[2] N. I. Zaika et al., Program and Abstracts of Reports to the 20th
     Annual Conference on Nuclear Spectroscopy and Nuclear Structure,
     part 2 (Leningrad, 1970) p. 138.
[3] A. N. Vereshchagin et al., Izv. Akad. Nauk SSSR, ser. fiz. 33 (1969)
     2064; K. O. Terenetsky, Program and Abstracts of Reports to the
     19th Annual Conference on Nuclear Spectroscopy and Nuclear
     Structure, part 2 (Leningrad, 1969) p. 33; P. E. Hodgson, Adv. Phys.
     15 (1966) 329.

Cross-Section and Vector Analyzing Power Measurements for 9-, 11-, and 13-MeV Deuterons Elastically Scattered from ^{27}Al, Si, ^{46}Ti, ^{48}Ti, ^{53}Cr, ^{68}Zn, ^{90}Zr, and ^{120}Sn

J. M. LOHR and W. HAEBERLI, University of Wisconsin, Madison, USA*

Measurements of the cross section and vector analyzing power have been made for deuterons elastically scattered from ^{27}Al, Si, ^{46}Ti, ^{48}Ti, ^{53}Cr, ^{68}Zn, ^{90}Zr, and ^{120}Sn at $E_d = 9$, 11, and 13 MeV for $\theta_{lab} = 30° - 165°$. The purely vector polarized beam produced by the Wisconsin Lamb-shift source was accelerated by the tandem accelerator. With the exception of Si, isotopically enriched targets were used.

Elastically scattered deuterons were separated from protons of the same energy by reducing the bias voltage on the partially depleted Si surface barrier detectors until the deuterons were just stopped in the active region. The beam polarization was monitored with a helium polarimeter located behind the scattering chamber. The calibration of the polarimeter was based on the analyzing powers for d-^{4}He elastic scattering measured by Keller and Haeberli [1].

Fig. 1 shows a representative sample of the analyzing power measurements at $E_d = 13$ MeV. The data resemble the recent measurements [2] from Birmingham near $E_d = 12$ MeV but with the corresponding maxima and minima of $iT_{11}(\theta)$ shifted toward smaller angles and with generally larger values of iT_{11} at the peaks.

A preliminary optical model analysis of the data has been performed. A discussion of the results will be presented.

REFERENCES

*Work supported in part by the U.S. Atomic Energy Commission.

[1] L. G. Keller and W. Haeberli, submitted to Nucl. Phys.
[2] J. A. R. Griffith et al., Nucl. Phys. A146 (1970) 190.

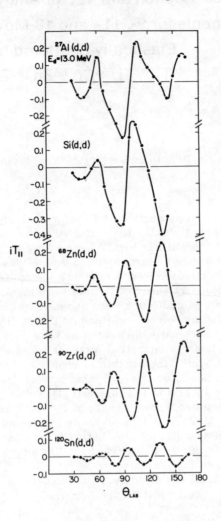

Fig. 1. Representative sample of
the data at 13 MeV. The statistical
errors are smaller than the data
points on the figure. The lines are
to guide the eye.

Elastic Scattering of Polarized Tritons

D. D. ARMSTRONG, P. W. KEATON, Jr., and L. R. VEESER, Los Alamos
Scientific Laboratory, New Mexico, USA*

$^4He(t,\vec{t})^4He$ *reaction.* The measurement at Los Alamos of the polarization of tritons [1] elastically scattered from ^{4}He has established this
reaction as a good polarizer (or analyzer) for tritons [2]. Recently we
have extended the polarization measurements over a sufficient angular
and energy range (see fig. 1) to allow a phase-shift analysis (based
on the work of Spiger and Tombrello [3] and including their cross-
section data) which more accurately describes both the cross-section

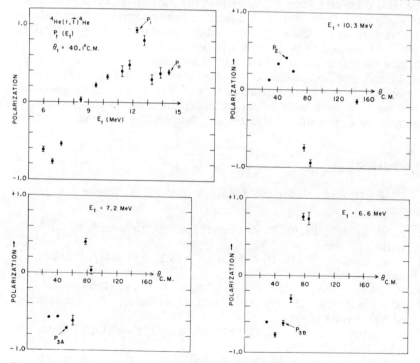

Fig. 1. Polarization of tritons elastically scattered from ^{4}He.

and polarization data. Some information about mass-7 nuclei will result from this analysis. However, the physical interpretation of the analysis is complicated because the decay of the t+^{4}He compound system can occur through many channels.

$H(\vec{t},t)H$ *reaction.* A study of the elastic scattering of polarized tritons from protons gives information about the excited states of ^{4}He. We obtained data for 6.4-9.5-MeV (lab) tritons corresponding to an excitation of the order of 20 MeV in ^{4}He. For this region of excitation only the p+t and n+^{3}He channels need be considered. A two-channel analysis of our data and other relevant data from the literature was done using a computer program which calculates the cross sections and polarizations from the formulas of R-matrix theory [4].

$D(\vec{t},\alpha)n$ *reaction.* Using a polarized ^{3}He target, Baker et al. [5] measured the left-right asymmetry, A, of protons in the ^{3}He(d,p)^{4}He reaction. Theoretical interest was aroused [6] when they pointed out that A $\approx$ -P, where P is the proton polarization in ^{3}He(d,$\vec{p}$)^{4}He. We have taken equivalent data with a polarized ^{3}He beam [7] (D($^3\vec{\text{He}}$, α)H) and also with a polarized triton beam (D($\vec{t},\alpha$)n). Although we have not made any theoretical deductions of our own at this time, it is clear that a theory which assumes or predicts P $\equiv$ -A is in serious error.

Optical model for tritons. Based on the assumption that the optical model potential for a triton is just the sum of the optical model potentials of its constituent nucleons suitably averaged over their internal motion in the triton [8], the polarization of tritons elastically scattered from nuclei is expected to be small. Our measurements (at angles < 60°) give small polarizations for tritons elastically scattered from Ni and ^{54}Fe. However, the measurement of the polarization at back angles is the most definitive test of this model. Such measurements are best done with a beam from a polarized ion source which we expect to be available soon at the Los Alamos Van de Graaff facility.

REFERENCES

*Work performed under the auspices of the U.S. Atomic Energy Commission.

[1] P. W. Keaton, Jr., D. D. Armstrong, and L. R. Veeser, Phys. Rev. Lett. 20 (1968) 1392.
[2] P. W. Keaton, Jr., D. D. Armstrong, and L. R. Veeser, Third Polarization Symp.
[3] R. J. Spiger and T. A. Tombrello, Phys. Rev. 163 (1967) 964.
[4] L. R. Veeser, D. D. Armstrong, and P. W. Keaton, Jr., Nucl. Phys. A140 (1970) 177.
[5] S. D. Baker, G. Roy, G. C. Phillips, and G. K. Walters, Phys. Rev. Lett. 15 (1965) 115.
[6] P. L. Csonka, M. J. Moravcsik, and M. D. Scadron, Phys. Rev. 143 (1966) 775.

[7] D. D. Armstrong, L. L. Catlin, P. W. Keaton, Jr., and L. R. Veeser, Phys. Rev. Lett. 23 (1969) 135.

[8] P. W. Keaton, E. Aufdembrink, and L. R. Veeser, Los Alamos Scientific Laboratory Report LA-4379-MS (1969) (unpublished).

The Absolute Polarization
Determination of Mass-3 Nuclei

P. W. KEATON, Jr., D. D. ARMSTRONG, and L. R. VEESER, Los Alamos
Scientific Laboratory, New Mexico, USA*

The polarizations of tritons [1] and ^{3}He nuclei [2] elastically scattered
from ^{4}He at several energies and angles have been measured using
double-scattering techniques. Absolute polarization values are deduced from the measurements of three asymmetries at appropriate energies and angles.

Nuclei scattered from the primary target are focused 2 m away onto
the secondary target with a three-element magnetic quadrupole lens
system. The lens system and secondary chamber can be rotated to
positive or negative primary angles so that dependence on detector
efficiencies can be eliminated. Nuclei scattered from the secondary
target are detected by E-ΔE silicon-detector pairs and particle identified on an on-line computer code [3]. Typical target thicknesses
were 200 keV for tritons and 800 keV for ^{3}He. Angular spreads were
$\pm 1.5°$, and primary and secondary beam currents were about 1 μA and
1 pA, respectively.

Absolute values (but not signs) for polarizations can be obtained
by measuring left-right asymmetries, A_{ij}, in a set of three double-
scattering experiments such that only three different polarizations
(assuming time-reversal invariance) are involved:

$$A_{12} = P(\theta_1, E_1) P(\theta_2, E_2)$$

$$A_{23} = P(\theta_2, E_2) P(\theta_3, E_3)$$

$$A_{13} = P(\theta_1, E_1) P(\theta_3, E_3),$$

where θ_i and E_i ($i = 1, 2, 3$) are scattering angles and energies, respectively. An aluminum foil was inserted between the first and second
scatterings to slow the nuclei to the correct energy so that A_{13} could
be measured. The signs of the polarization values were determined
by phase shift analyses [4].

Table 1. Absolute polarization values

Reaction	Incident energy (MeV)	Scattering angle		Designation	Polarization
		lab	cm		
$^4\mathrm{He}(\vec{t},t)^4\mathrm{He}$	12.25	23°	40.1°	P_1	0.923 ± 0.027
"	12.25	16.5°	28.9°	P_1'	0.825 ± 0.042
"	10.28	30°	52.1°	$P_2(69)$	0.406 ± 0.013
"	9.5	30°	52.1°	$P_2(68)$	0.370 ± 0.016
"	7.2	30°	52.1°	P_{3A}	-0.724 ± 0.027
"	6.6	30°	52.1°	P_{3B}	-0.615 ± 0.029
"	6.0	30°	52.1°	$P_3(68)$	-0.378 ± 0.020
$^3\mathrm{He}(^4\mathrm{He},^3\vec{\mathrm{He}})^4\mathrm{He}$	13.0	19°	142°	P_1	-0.606 ± 0.033
$^4\mathrm{He}(^3\mathrm{He},^3\vec{\mathrm{He}})^4\mathrm{He}$	13.0	23°	40.1°	P_2	0.639 ± 0.035
"	8.3	23°	40.1°	P_3	-0.604 ± 0.033

The results of these measurements are given in table 1. The energy and angular spreads are presented here so that the values in table 1 can be used as standards by other groups until more precise absolute determinations are made. (See also ref. [5].)

REFERENCES

*Work performed under the auspices of the U.S. Atomic Energy Commission.

[1] P. W. Keaton, Jr., D. D. Armstrong, and L. R. Veeser, Phys. Rev. Lett. 20 (1968) 1392.
[2] D. D. Armstrong, L. L. Catlin, P. W. Keaton, Jr., and L. R. Veeser, Phys. Rev. Lett. 23 (1969) 135.
[3] D. D. Armstrong et al., Nucl. Instr. 70 (1969) 69.
[4] R. J. Spiger and T. A. Tombrello, Phys. Rev. 163 (1967) 964.
[5] D. D. Armstrong, P. W. Keaton, Jr., and L. R. Veeser, Third Polarization Symp.

A General Set of ^{3}He and Triton
Optical-Model Potentials for A $>$ 40, E $<$ 40 MeV

F. D. BECCHETTI, Jr.,* and G. W. GREENLEES, University of
Minnesota, Minneapolis, USA†

A global optical-model analysis [1] of ^{3}He and triton elastic scattering
data has been performed using data taken on targets with mass numbers greater than 40 and using incident ^{3}He and triton energies < 40
MeV. The forms of the potentials used in the analysis are the same
as those in ref. [1]. A listing of the data analyzed [2, 3] is given in
table 1. Since no polarization data were available, the spin-orbit

Table 1. The ^{3}He and triton elastic scattering data
used in the optical-model analysis

---

^{3}He-nucleus $\sigma(\theta)$ data

E (MeV)	Target Nuclei
12	^{50}Ti, ^{58}Ni
19	^{52}Cr, ^{48}Ti, ^{58}Ni, ^{64}Zn, ^{138}Ba
22	^{56,58}Fe, ^{58}Ni
25	^{48}Ti, ^{58,60,62,64}Ni, ^{90,92,94}Zr
29	nat Fe, nat Cu, nat Zn, nat Ag, nat Sn
30	^{60}Ni, ^{90}Zr, ^{114}Cd, ^{116}Sn
33	^{58,60,62,64}Ni
35	^{60}Ni, ^{116}Sn
37.7	^{58}Ni, ^{56}Fe

---

^{3}He-nucleus σ_R data

E(MeV)	Target Nuclei
29	nat Fe, nat Cu, nat Ag

Triton-nucleus $\sigma(\theta)$ data

E(MeV)	Target Nuclei
15	^{52}Cr, ^{62}Ni
20	^{52}Cr, ^{54}Fe, ^{62,64}Ni, ^{90,92,94,96}Zr, ^{116,118,120,122,124}Sn, ^{208}Pb

---

parameters were kept at $V_{SO} = 2.5$ MeV, $r_{SO} = r_R$, $a_{SO} = a_R$, or $V_{SO} = 0$. The best fit to all of the data analyzed, including the reaction cross sections, was obtained with the following parameters [1] for ^{3}He data:

$V_R = 151.9 - 0.17E + 50\xi$ MeV $r_R = 1.20$ fm, $a_R = 0.72$ fm

$W_V = 41.7 - 0.33E + 44\xi$ MeV $r_I = 1.40$ fm, $a_I = 0.88$ fm

triton data:

$V_R = 165.0 - 0.17E - 6.4\xi$ MeV $r_R = 1.20$ fm, $a_R = 0.72$ fm

$W_V = 46.0 - 0.33E - 110\xi$ MeV $r_I = 1.40$ fm, $a_I = 0.84$ fm

and for both particles: $V_{SO} = 2.5$ MeV, $r_{SO} = 1.20$ fm, $a_{SO} = 0.72$ fm, $r_C = 1.30$ fm, and $W_{SF} = 0$, where $\xi = (N-Z)/A$ and E is the incident LAB energy. These parameters yield average χ^2 per point value of 10, 19, and 16 for the ^{3}He differential cross sections, ^{3}He total reaction cross sections, and triton differential cross sections, respectively. The fits obtained without spin-orbit coupling ($V_{SO} = 0$) were comparable to those obtained with $V_{SO} = 2.5$ MeV. The real part of the best fit potentials corresponds to a sum of nucleon-nucleus potentials [1]; the imaginary part, however, does not, so that the parameters determined, and in particular the (N-Z)/A-dependent terms in the expression W_V, must be treated as phenomenological parameters.

REFERENCES

*Present address: The Niels Bohr Institute, Copenhagen, Denmark.
†Work supported by USAEC contract AT (11-1)1265.

[1] F. D. Becchetti, Jr., and G. W. Greenlees, Phys. Rev. 182 (1969) 1190.
[2] J. L. Yntema et al., Phys. Lett. 11 (1964) 302; R. Bock et al., Nucl. Phys. A92 (1967) 539; E. R. Flynn and L. Rosen, Phys. Rev. 153 (1967) 1228; A. G. Blair and D. D. Armstrong, private communication; P. G. Roos et al., Phys. Lett. 24B (1967) 656; D. E. Rundquist et al., Phys. Rev. 168 (1968) 1287; D. J. Baugh et al., Nucl. Phys. A95 (1967) 115; J. W. Luetzelschwab and J. C. Hafele, Phys. Rev. 180 (1969) 1023; R. Balcarcel and J. A. Griffith, Phys. Lett. 26B (1968) 213; R. H. Siemssen et al., Phys. Lett. 18 (1965) 155; E. F. Gibson et al., Phys. Rev. 155 (1967) 1194.
[3] J. C. Hafele et al., Phys. Rev. 155 (1967) 1238; E. R. Flynn et al., Phys. Rev. 182 (1969) 1113.

Inelastic
Nucleon Scattering

Nuclear Structure Effects on Asymmetry
of Proton Inelastic Scattering

J. RAYNAL, Centre d'Etudes Nucléaires de Saclay, France

The asymmetry of polarized protons in inelastic scattering can be described only by taking into account spin-orbit coupling in the interaction. In a macroscopic model, Sherif and Blair [1] have shown that the deformed $\vec{L} \cdot \vec{S}$ coupling must be given by $\vec{\nabla} V(\vec{r}) \wedge (\vec{\nabla}/i) \cdot \vec{\sigma}$. This expression can be written [2]:

$$G_{if}^{\lambda} \left\{ \frac{1}{r} \frac{d}{dr} V(r) \gamma_i + \frac{V_{\lambda}(r)}{r} (\gamma_i - \gamma_f) \frac{d}{dr} + \frac{V_{\lambda}(r)}{2r^2} [\lambda(\lambda+1) - (\gamma_f - \gamma_i)(\gamma_f - \gamma_i - 1)] \right\}, \quad (1)$$

where G_{if}^{λ} is the geometry used for a scalar term between initial and final wave function i and f and the γ's are the eigenvalues of $(\vec{\ell} \cdot \vec{\sigma})$. For $\lambda = 0$, eq. (1) gives back the $\vec{L} \cdot \vec{S}$ coupling of the optical model. In a microscopic model, the nucleon-nucleon $\vec{L} \cdot \vec{S}$ interaction has the same effect. We have introduced it in a DWBA program with antisymmetrization [3] using the helicity formalism [4]. We can use finite range or the following zero range limit [2]

$$\frac{1}{r^2} G_{if}^{\lambda} G_{ph}^{\lambda} \left\{ \left[\lambda(\lambda+1)(a_{if}^{\lambda^2} + a_{hp}^{\lambda^2} - 2) - 1 \right] V_{\lambda}(r) + \right.$$

$$\left[1 - (-)^{\ell_p + \ell_h + \lambda} a_{if}^{\lambda} a_{hp}^{\lambda} \right] \left[-(\gamma_i + \gamma_f + 1) V_{\lambda}(r) + (\gamma_p + \gamma_h - \right.$$

$$\gamma_i - \gamma_f) r \frac{d}{dr} \{ V_{\lambda}(r) \} \right] + \left[1 + (-)^{\ell_p + \ell_h + \lambda} \right] \left[(\gamma_i + \gamma_p + 1) V_{\lambda}(r) + \right.$$

$$\left. \left. (\gamma_h - \gamma_p - \gamma_i + \gamma_f) \left(V_{\lambda}(r) r \frac{d}{dr} + r \phi_h \frac{d}{dr} (\phi_p) \right) \right] \right\} \quad (2)$$

for a particle-hole excitation. Here G_{if}^{λ}, γ_p, and γ_h have the same meaning as in eq. (1), but for bound particles. For a natural parity excitation $a_{jj'}^{\lambda} = (\gamma - \gamma')/\sqrt{\lambda(\lambda+1)}$; for an unnatural one $a_{jj'}^{\lambda} = (\gamma + \gamma' + 2)/\sqrt{\lambda(\lambda+1)}$; $V_{\lambda}(r)$ is the product of particle and hole wave functions.

The interaction (2) is purely $T = 1$. In (pp') inelastic scattering the excitation of proton configurations has twice the effect of neutrons. Experimentally, the asymmetry for nuclei with open shells of protons is larger than for nuclei with open shells of neutrons. The macroscopic limit of the interaction (2) obtained by setting $a_{ph}^{\lambda} = \gamma_h = \gamma_p = \phi_h(d/dr)(\phi_p) = 0$, is not hermitian, but its hermitian part is the expression (1) with a factor 2 before $\lambda(\lambda+1)$. The deformation of the spin-orbit coupling needed to fit the experimental asymmetries is therefore a parameter depending upon the structure of the excited state.

Preliminary calculations [5] with a zero range $\vec{L} \cdot \vec{S}$ interaction showed that the effect is far more important for ^{90}Zr than for ^{62}Ni. The $\vec{L} \cdot \vec{S}$ nucleon-nucleon interaction can be approximated by two Yukawa potentials of range 0.55 and 0.325 fm. The strengths obtained by comparison with the soft core potential of Bressel et al. [6] are 156 and -4400 MeV for protons, 116 and -2600 MeV for neutrons. The zero range limit is -16 MeV, in agreement with optical model values (~ 6.7 MeV for $r_0 = 1.1$ fm). For a given strength, the effect does not change for ranges up to 0.5 fm, but becomes weaker above. Fig. 1 shows the results for ^{90}Zr with the zero range and the finite range interaction. The excited state is described by two quasi-particle excitations for protons and particle-hole excitations for the two shells [6]. The proton configurations are multiplied by 2 to fit the BE2 and the neutron ones by 5.5 to fit the cross section [3]. The experimental data are qualitatively reproduced in the forward direction. The tensor interaction does not contribute. No agreement can be obtained with proton configurations alone. With the macroscopic model, the interaction (1) had to be multiplied by almost 3 to fit the experimental data, as shown in fig. 2.

The first 2^+ of ^{54}Fe [7] is quite well reproduced in the vibrational model ($\beta_2 = 0.14$) with the same factor 3 (fig. 3). Even the cross section [8] is better, as shown in fig. 4. For the second 2^+, the proton and neutron contribution cancel each other and the deformation of $\vec{L} \cdot \vec{S}$ coupling can be neglected (fig. 5). Microscopic calculations give about the same result.

In conclusion, the deformation of $\vec{L} \cdot \vec{S}$ coupling in the macroscopic model, and the deformation of the central potential are quite independent; their ratio is related to the structure of the excited state.

Potentials	R_c	V	R_0	a_0	W_S	R_I	a_I	V_{LS}	R_{LS}	a_{LS}
^{90}Zr	1.25	48.2	1.238	0.618	8.05	1.288	0.638	5.750	1.07	0.526
^{54}Fe	1.25	61.44	1.1	0.75	9.8	1.3	0.55	5.940	1.04	0.55

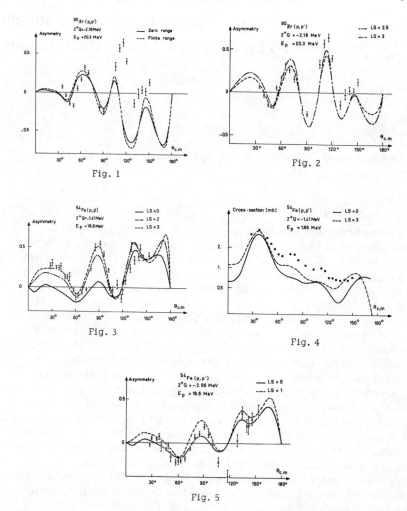

Fig. 1

Fig. 2

Fig. 3

Fig. 4

Fig. 5

REFERENCES

[1] H. Sherif and J. S. Blair, Phys. Lett. 26B (1968) 489; H. Sherif, Nucl. Phys. A131 (1969) 532.

[2] J. Raynal, Proc. Symp. on Nuclear Reaction Mechanisms (Quebec, 1969).

[3] R. Schaeffer, Thesis, Orsay, 1969; Report CEA-R-4000.

[4] J. Raynal, Nucl. Phys. A117 (1968) 101.

[5] C. N. Bressel et al., Nucl. Phys. A124 (1969) 624.

[6] V. Gillet, B. Giraud, and M. Rho, to be published.

[7] C. Glasshauser, R. de Swiniarski, J. Thirion, and A. D. Hill, Phys. Rev. 164 (1967) 1437.

[8] S. F. Eccles, H. F. Lutz, and V. A. Madsen, Phys. Rev. 141 (1966) 1067.

Collective-Model Analysis of Inelastic
Proton Scattering Using Potentials
Derived from Nuclear Matter Distributions

V. HNIZDO, P. D. GREAVES, O. KARBAN, and J. LOWE, University
of Birmingham, England

In a recent reformulation of the optical model, Greenlees et al. [1]
have calculated elastic proton scattering using a potential derived
from folding a nucleon-nucleon force into a nuclear matter distribution.
The present paper describes an extension of this model to describe
inelastic proton scattering, and a comparison with a standard coupled-
channels analysis.

The data [2] consist of cross sections and polarizations for the
elastic scattering of 30.3 MeV protons, and inelastic scattering to
the lowest 2^+ and 3^- states for ^{56}Fe, ^{58}Ni, ^{120}Sn, and ^{208}Pb. For a
standard optical-model analysis the elastic scattering data were first
fitted using an optical-model search program. The inelastic scattering was then calculated using a coupled-channels method, with vibrational wave functions for the nuclear states. The procedure used here
has been described in detail in ref. [2]. The predictions are shown in
the figure (dashed lines).

For the reformulated optical-model analysis, the elastic scattering
data were first fitted with the reformulated optical model. The resulting nuclear matter distributions were then assumed to undergo vibra-
tion, and folding a nucleon-nucleon force into this yields an off-
diagonal coupling potential for a coupled-channels calculation.

The results are shown in the figure (solid lines). The two methods
of analysis predict little difference in the cross sections. However,
for 2^+ states in ^{56}Fe and ^{58}Ni the asymmetry fits are improved at back
angles when the reformulated model is used, although the fit to the
maximum at 120° is worsened slightly. This improvement is presumably
related to the improvement found in a standard coupled-channels analysis [2] when the constraint $r_{so} = r_V$ was imposed in the optical parameters; use of the reformulated model imposes a constraint approximately equivalent to this on the spin-orbit geometry. For the 2^+ state
in ^{120}Sn the standard model gives perhaps a better fit, and for all 3^-
states there is no significant difference between the predictions of the
two models.

690

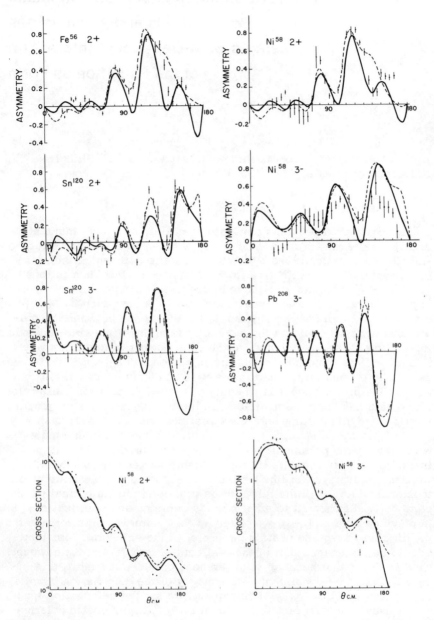

REFERENCES

[1] G. W. Greenlees et al., Phys. Rev. 171 (1968) 1115; Phys. Rev.
     C1 (1970) 1145.
[2] O. Karban et al., Nucl. Phys. A147 (1970) 461.

Further Calculations of Proton Inelastic
Polarizations and Asymmetries
Using a Spin-Dependent Interaction
of the Full Thomas Form

J. S. BLAIR, University of Washington, Seattle, USA; H. SHERIF,
University of Alberta, Edmonton, Canada[*]

Inelastic asymmetries and polarizations of protons whose incident
energies have ranged from 30 to 155 MeV have, with isolated exceptions, been well described by DWBA calculations in which the spin-
dependent part of a collective interaction has the full Thomas form
[1-3]. This model has also proved successful in describing many examples where the incident energy is equal to or less than 20 MeV [4-6]
although, in this latter energy range, the difference between the predictions of this model and those which use a simplified spin-dependent
interaction [7] are diminished; further, the number of instances where
this as well as the other models are in obvious disagreement with
data is here much larger than is the case at higher energies. The purpose of the present paper is to apply our model to additional examples
where data have become available, in the course of which we generalize
the model to situations where the target nucleus has non-zero spin.

$^{208}Pb(p,p')(2.63\ MeV,\ 3^-)$, $E_p = 155\ MeV$. Since previous studies
with high energy protons have been limited to target nuclei with $A \leq 40$,
inelastic polarization data [8] for a sizable nucleus are particularly
welcome. Displayed in fig. 1 are observed inelastic polarizations and
those calculated with the full Thomas interaction using optical parameters derived from a fit to elastic cross sections and polarizations [8].
In previous studies [2-3], we observed that there was a preference for
the dimensionless deformation parameter of the spin-orbit form factor,
β_{so}, to be larger than that of the real central form factor, β_0; however,
we also noted that because of differences of mid-point radii, this inequality might be consistent with equality of the deformation lengths,
$\delta_{so} \equiv \beta_{so} R_{so}$ and $\delta_0 \equiv \beta_0 R_0$. (Throughout, we set the deformation length
of the imaginary form factor, δ_I, equal to δ_0.) The inelastic polariza-
tion of high energy protons scattered into forward angles is particularly
sensitive to the ratio of lengths, (δ_{so}/δ_0). The fits shown in fig. 1
suggest that this ratio equals 0.9 ± 0.2.

We must greet this conclusion with some caution, however,[†] since
the data do fail to show the predicted negative dip at $17.5°$ and the fits
to the inelastic (and even elastic) angular distribution could be much

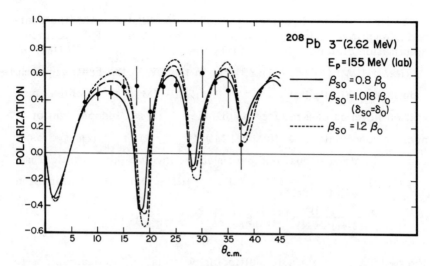

Fig. 1. Polarization of 155-MeV protons exciting 2.62-MeV level of ^{208}Pb.

improved. It thus seems reasonable that there should be some altera-
tions in the optical parameters and we worry to what extent such
changes will affect the inelastic polarization at forward angles. Predictions using optical parameters pertinent to ^{209}Bi (which give an
inferior fit to the ^{208}Pb elastic polarization) yield a poor match to the
inelastic polarization data.

$^{28}Si(p,p')(1.78\ MeV,\ 2^+)$ and $^{40}Ca(p,p')(3.73\ MeV,\ 3^-)$, $E_p = 155\ MeV$.
Re-examination of these cases [9] with a finer grid for the ratio δ_{so}/δ_0
than was used previously [3] indicates a preference for δ_{so} being
somewhat larger than δ_0, in contrast to the conclusion above.

*Inelastic polarizations and asymmetries for target nuclei with
non-zero spin.* The most striking observation that can be made concerning measured [5, 10, 11] inelastic polarizations or asymmetries
for such targets is their strong similarity to those of neighboring zero-
spin targets. This phenomenon finds an easy explanation in many examples through the following rather trivial theorem: If the interaction
Hamiltonian can be described in terms of collective coordinates only
and if the inelastic amplitudes are computed in DWBA with only one
multipolarity of angular momentum transfer being predominant, then
the inelastic relative cross sections, asymmetries, and polarizations
from a nucleus with $J \neq 0$ are the same as those calculated for a hypothetical zero-spin nucleus with some optical and kinematic parameters.
For the weak-coupling model, such a conclusion has already been in-
voked by the Saclay group [5]; the derivation of the theorem above,
however, is not restricted to such a specific model, but draws only on
the Wigner-Eckart theorem and the stated hypotheses. The derivation
is outlined below.

Under the stated hypotheses, nuclear coordinates enter the scattering amplitude linearly through the matrix element of the surface deformation distance, $\langle I'M'|\alpha(\hat{r})|IM\rangle \equiv \langle I'M'|\sum_{\ell m}\xi_{\ell m}Y_\ell^{m*}(\hat{r})|IM\rangle = \sum_{\ell m}(I\ell Mm|I'M')\langle I'||\xi_\ell||I\rangle Y_\ell^{m*}(\hat{r})$. In the DWBA the scattering amplitude is $T_{\mu_f\mu_i}(I'M';IM) = \langle \chi_{\mu_f}^{(-)}|\langle I'M'|\Delta U|IM\rangle|\chi_{\mu_i}^{(+)}\rangle$ which can be written in terms of a reduced amplitude $t_{\mu_f\mu_i}(\ell m)$, independent of nuclear coordinates: $\sum_{\ell m}(I\ell Mm|I'M')\langle I'||\xi_\ell||I\rangle t_{\mu_f\mu_i}(\ell m)$. Cross sections, polarizations, and asymmetries involve the statistical matrix $(2I+1)^{-1}\sum_{M,M'}T_{\bar{\mu}_f\bar{\mu}_i}(I'M';IM)T^*_{\mu_f\mu_i}(I'M';IM)$ which collapses to

$$\sum_\ell \frac{(2I'+1)}{(2I+1)(2\ell+1)}|\langle I'||\xi_\ell||I\rangle|^2 \sum_m t_{\bar{\mu}_f\bar{\mu}_i}(\ell m)t^*_{\mu_f\mu_i}(\ell m).$$

The restriction to a single multipolarity leads, in the expressions for polarization and asymmetry, to cancellation of the factor preceding the sum over m so that these expressions are equivalent to those for a zero-spin target.

$^{11}B(p,p')(2.14\ MeV,\ 1/2^-;\ 4.46\ MeV,\ 5/2^-),\ ^7Li(p,p')(4.63\ MeV, 7/2^-)\ E_p = 155\ MeV$. As an example of the preceding section, we compare in fig. 2 the observed polarization [10] for excitation of the 4.46-MeV level in ^{11}B to that calculated using both the optical parameters found by the Orsay group and those obtained by Haybron and

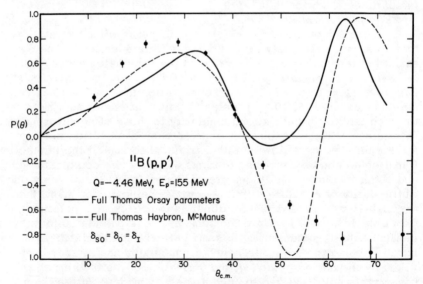

Fig. 2. Polarization of 155-MeV protons exciting 4.46 MeV level of ^{11}B.

McManus [12] for the neighboring nucleus, ^{12}C. A better fit is obtained with the latter parameters but neither fit is good at back angles. Fewer discrepancies are apparent for the excitation of the 2.14 MeV level since here the data points do not extend to such large angles. A fit of comparable quality is also obtained for ^7Li.

$^{11}B(p,p')(2.14 MeV, 1/2^-; 4.46 MeV, 5/2^-) E_p = 30.3 MeV$. We show in fig. 3 several comparisons to the observed asymmetry [11] of inelastically scattered protons which excite the 2.14-MeV level of ^{11}B. The necessity for some form of deformed spin-dependent interaction is apparent from the overall fit while the dip at forward angles gives clear preference to the full Thomas form. The parameters obtained by linearly interpolating those given by Kolata and Galonsky [13] (who fitted the elastic scattering of 26.2- and 40-MeV protons from ^{12}C) yield a tolerable fit to the asymmetry even at back angles. A comparable fit is obtained for excitation of the 4.46-MeV state except that the forward dip is markedly deeper than that predicted. We note that, in addition to favoring the

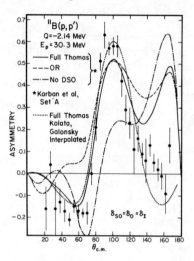

Fig. 3. Asymmetry of 30.3-MeV protons exciting 2.14-MeV level of ^{11}B.

full Thomas form of spin-dependent interaction, the present comparisons provide further support for collective descriptions of very light nuclei.

REFERENCES

*Supported in part by the U.S. Atomic Energy Commission and the Atomic Energy Control Board of Canada.

[1]   H. Sherif and J. S. Blair, Phys. Lett. 26B (1968) 489.
[2]   H. Sherif, Nucl. Phys. A131 (1969) 532.
[3]   H. Sherif and J. S. Blair, Nucl. Phys. A140 (1970) 33.
[4]   H. Sherif and R. DeSwiniarski, Phys. Lett. 28B (1968) 96.
[5]   A. G. Blair et al., Phys. Rev. C1 (1970) 444.
[6]   C. Glasshausser et al., Phys. Rev. 184 (1969) 1217.
[7]   M. P. Fricke et al., Phys. Rev. Lett. 16 (1966) 746.
[8]   A. Willis et al., J. de Phys. 30 (1969) 13.

[9]   A. Willis et al., Nucl. Phys. A112 (1968) 417.

[10] B. Geoffrion et al., Nucl. Phys. A116 (1968) 209.

[11] O. Karban, J. Lowe, P. D. Greaves, and V. Hnizdo, Nucl. Phys. A133 (1969) 255.

[12] R. M. Haybron and H. McManus, Phys. Rev. 140 (1965) B638.

[13] J. J. Kolata and A. Galonsky, Phys. Rev. 182 (1969) 1073.

†*Note added in proof:*

According to the contribution of A. Ingemarsson to this conference, a deep minimum in the inelastic polarization of 185-MeV protons is observed at 16.1°, thus removing the discrepancy mentioned in the text.

Microscopic Analysis of 30.3-MeV Inelastic Proton Scattering by Light Nuclei

P. D. GREAVES, V. HNIZDO, O. KARBAN, and J. LOWE, University of Birmingham, England

This paper presents cross-section and polarization data for ^{13}C(p,p') $(1/2^- \rightarrow 1/2^+$ (3.09 MeV)) and ^{16}O(p,p') $(0^+ \rightarrow 2^-$ (8.88 MeV)), together with a coupled-channels analysis. The data were taken with the polarized proton beam of the Rutherford Laboratory proton linear accelerator.

Diagonal coupling potentials for the coupled-channels calculations were derived from an optical-model analysis of the elastic scattering data. Off-diagonal elements were then calculated using shell-model wave functions, together with a nucleon-nucleon interaction:

$$V = [V_c + V_\sigma \vec{\sigma}_1 \cdot \vec{\sigma}_2 + V_T S_{12} + V_{LS} \vec{L} \cdot \vec{S}] g(r_{12})$$

where $g(r_{12})$ is a Yukawa function of range 1.0 fm. The first three terms were handled exactly, and for $V_{LS} \vec{L} \cdot \vec{S}$, the leading term was taken. Antisymmetrization was neglected. The four coefficients in the effective interaction were used as variable parameters in the analysis. From the rather small cross sections found for each reaction, it was assumed that in each case a small number of shell-model configurations was involved, and that collective effects were not significant.

^{13}C. It was assumed that the transition takes place by the transition of the odd neutron from $1p_{1/2}$ to $2s_{1/2}$ states. The transferred angular momentum J can be either 0 or 1, and predictions for each are shown in the figure. Both cross sections and asymmetries are strongly dependent on J. Also shown in the figure is a prediction for the best effective interaction found; the potential coefficients are given in the table and the range of the Yukawa potential was 1.4 fm in this case. The fit shown corresponds to the J = 1 part of the scattering.

^{16}O. Various one-particle-one-hole and three-particle-three-hole configurations were tried for the 2^- state. The predictions shown here correspond to the configuration $(1p_{1/2}^{-1} 1d_{5/2})_{2^-}$, which gave the best fit. The transition goes via the spin-flip term alone. J is limited to 2, but the transferred orbital angular momentum L can be 1 or 3. As shown in the figure, the asymmetries are sensitive to L, the cross sections are not. The best fit is also shown, and the parameters of the effective interaction are given in the table.

Finally, the table also shows the parameters of the effective interaction found in a similar analysis [1] of $^{11}B(p,p')$ to the $1/2^-$ state at 2.14 MeV. The striking qualitative similarity between the three interactions is surprising in view of the different transitions involved.

Parameters of the Effective Interaction

Nucleus	V_C	V_σ	V_T	V_{LS}
^{11}B	−10.8	+5.3	−23.1	−9.2
^{13}C	−10.0	+5.0	−30.0	−1.3
^{16}O	--	+5.2	−75.0	---

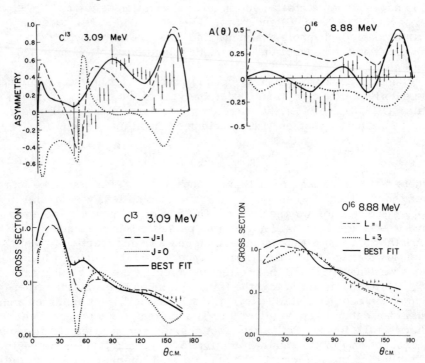

REFERENCE

[1] J. Lowe, Nucl. Phys., to be published.

DISCUSSION

Glashausser:

The 2^- level in ^{16}O is seen rather strongly in the inelastic scattering of alpha particles, indicating that coupled-channels effects are important there. Have you taken these into account?

Lowe:

Only two states, the ground state and the excited state in question, were included in these calculations.

Raynal:

What did you use for the $\vec{L} \cdot \vec{S}$ interaction? Is it equivalent to the phenomenological $\vec{L} \cdot \vec{S}$ term, mentioned by Satchler, or to the full Thomas term? Some approximations can lead to the one, some to the other form.

Lowe:

The $\vec{L} \cdot \vec{S}$ term was approximated by dropping terms of the form $\vec{\sigma}_i \cdot \vec{\ell}_j$ for which $i \neq j$. Not all of the omitted terms can be handled by our program; those which can seemed to have a very small effect in the few cases when we tried them, as would be expected for the non-collective excitations considered here. The force used here is, of course, a nucleon-nucleon force. I don't know to what extent the above approximations are equivalent to using the "Oak-Ridge" spin-orbit force in the nucleon-nucleus case.

Polarization in Octupole Transitions
Induced by 185 MeV Protons

A. INGEMARSSON, Gustaf Werner Institute, Uppsala, Sweden

Calculations on the polarization of medium energy protons inelastically
scattered from even-even nuclei [1, 2] have shown that the use of distorted waves may give rise to a second maximum in the angular distribution. In order to study this effect for different nuclei the polarization has been measured in the excitation of 3^- levels in ^{12}C (9.64 MeV),
in ^{40}Ca (3.74 MeV), in ^{90}Zr (2.75 MeV) and in ^{208}Pb (2.62 MeV) [3, 4].

The experimental arrangement is shown in fig. 1. The energy of the
scattered protons was ana-
lyzed with a magnetic spectrometer and the polarization was measured with a
polarimeter. The analyzing
power of the polarimeter
was determined by a meas-
urement of the asymmetry for
a known polarization [5].
With this equipment the angular resolution was 1.17°
and the total energy resolution apart from energy strag-
gling in the target less than
0.3 MeV (FWHM).

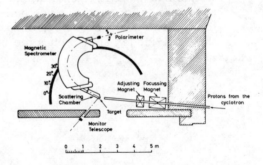

Fig. 1. Schematic view of the experimental area.

The experimental results with statistical errors are shown in fig. 2
together with DWIA calculations (solid lines) and PWIA calculations
(dotted lines) performed by Haybron and McManus [1] for ^{12}C and ^{40}Ca
and by Willis et al. [2] for ^{208}Pb.

For the 9.6 MeV level in ^{12}C the DWIA curve gives quite a good fit
to the experimental values. The angular distribution for the 3.75 MeV
level in ^{40}Ca is not equally well reproduced by the calculations but the
second maximum appears only in the calculation with distorted waves.
No calculation has been made for the 2.75 MeV level in ^{90}Zr but the
shape of the angular distribution is very similar to that obtained for the
3^- level in ^{40}Ca. For the 2.62 MeV level in ^{208}Pb the two minima in the
experimental curve are qualitatively reproduced by the DWIA curve. In

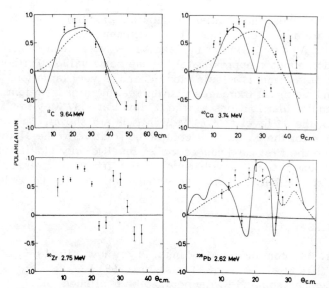

Fig. 2. Polarization results in the inelastic scattering of 185 MeV protons. The dotted and solid lines
are PWIA and DWIA calculations, respectively.

this case, according to ref. [2], the angular distribution is rather sensitive to the optical potential used to generate the distorted waves.
Therefore it seems reasonable that a still better fit may be obtained.

The angular distributions become rather contracted as the atomic
number increases. According to Kerman, McManus and Thaler [6], the
polarization in the inelastic scattering may be expressed in terms of
nucleon-nucleon amplitudes and therefore the angular scale should
have a very weak dependence on A. These measurements indicate, however, that the angular scales transform as $A^{1/3}$. In fig. 3a the polariza-
tion for the 3^- levels in ^{40}Ca and ^{90}Zr is plotted as a function of $\theta_{cm} A^{1/3}$,
and the curves appear to be almost identical.

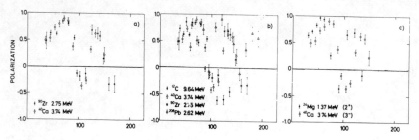

Fig. 3. The polarization plotted as a function of $\theta_{cm} A^{1/3}$.

In fig. 3b, where the angular distributions for the 3^- levels in ^{12}C and ^{208}Pb are also included, the first maximum appears for a smaller value of $\theta_{cm} A^{1/3}$ for a lighter nucleus. However, the first zero for the polarization appears approximately at the same value. Furthermore, the second maximum is missing for ^{12}C and is very large for ^{208}Pb, i.e., the second maximum increases with increasing atomic weight. In fig. 3c the angular distributions for the 2^+ level at 1.37 MeV in ^{24}Mg [5] and for the 3^- level at 3.74 MeV in ^{40}Ca are shown. The angular distribution for the 2^+ level is contracted relative to the 3^- level. However, since the first maximum of the 2^+ level is very similar to the first maximum of the 3^- level in ^{12}C, it is difficult to claim any systematic difference between angular distributions for 2^+ and 3^- levels.

REFERENCES

[1]   R. M. Haybron and H. McManus, Phys. Rev. 140 (1965) B638.
[2]   A. Willis et al., J. Phys. (Paris) 30 (1969) 13.
[3]   A. Ingemarsson and E. Hagberg, to be published.
[4]   A. Ingemarsson and J. Källne, GWI-PH 6/69 (unpublished).
[5]   B. Höistad et al., Nucl. Phys. A119 (1968) 290.
[6]   A. K. Kerman, H. McManus and R. M. Thaler, Ann. Phys. 8 (1959) 551.

DISCUSSION

Sherif:
    The $A^{1/3}$ dependence of the polarization is essentially a plane-wave result; that is why the agreement would be good for light nuclei (say ^{12}C and ^{40}Ca). However, the distortion is more important for heavier nuclei. This explains why you see a big difference between C and Pb.

Elastic and Inelastic Scattering
of Polarized Protons on ^{11}B
in the Giant Resonance Region

E. SALZBORN and G. CLAUSNITZER, University of Giessen, Germany

The elastic and inelastic scattering (Q = -2.13 MeV) of polarized protons from ^{11}B was investigated for proton energies from 3.5 to 10.5
MeV using the polarized beam facility of the Erlangen tandem accelerator [1]. The scattered protons from an isotopically enriched (98%),
self-supporting ^{11}B target (355 $\mu g/cm^2$) were detected simultaneously
by pairs of silicon surface-barrier counters at four laboratory scattering angles between ± 90° and ± 163°. Much care was taken to avoid
carbon contaminations of the target.

The excitation functions of the differential cross section and of
the analyzing power show strong resonance structures both in elastic
and inelastic scattering (figs. 1, 2). The elastic differential cross
section at scattering angles larger than 140° shows a characteristic
decrease just below the maximum of the giant resonance of the compound nucleus ^{12}C. Only a few spin and parity assignments are known
[2] for the large number of levels in the investigated excitation energy
range between 19.2 MeV and 25.5 MeV of the compound nucleus. The
resonance at E_p = 7.73 MeV (E_{ex} = 23.04 MeV) was studied in more detail by measurements of angular distributions of ^{11}B(p,p$_0$)^{11}B, and
^{11}B(p,p')^{11}B (Q = -2.13 MeV) (figs. 3, 4). The drawn lines are guides
to the eye.

All experimental data favor a 2^-, T = 1 assignment for this level
consistent with theoretical predictions of Gillet and Vinh-Mau [3]
from particle-hole calculations for ^{12}C. A phase-shift analysis does
not lead to reliable information (spin assignments) because of the high
spin values involved. A further theoretical interpretation of the excitation functions of both elastic and inelastic scattering is not available at present.

REFERENCES

[1] G. Clausnitzer et al., Nucl. Instr. 80 (1970) 245.
[2] F. Ajzenberg-Selove and T. Lauritsen, Nucl. Phys. A114 (1968) 1.
[3] V. Gillet and N. Vinh-Mau, Nucl. Phys. 54 (1964) 321.

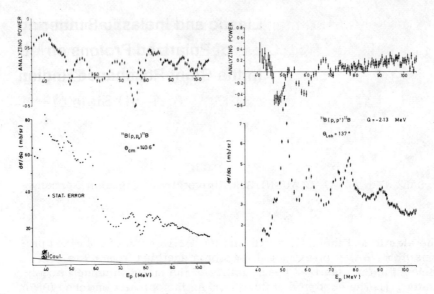

Fig. 1. Differential cross section and analyzing power for protons elastically scattered from ^{11}B.

Fig. 2. Differential cross section and analyzing power for protons inelastically scattered (Q = −2.13 MeV) from ^{11}B.

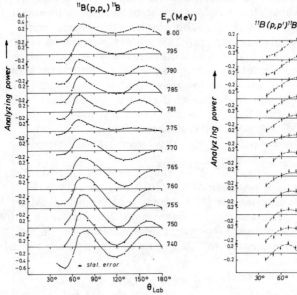

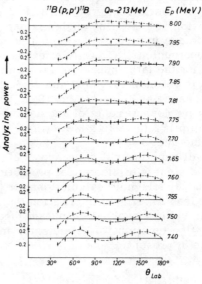

Fig. 3. Angular distributions of the analyzing power for ^{11}B($\vec{p}$,p_0)^{11}B for 7.4 MeV $\leq$ E$_p$ $\leq$ 8.0 MeV

Fig. 4. Angular distributions of the analyzing power for ^{11}B($\vec{p}$,p')^{11}B (Q = −2.13 MeV) for 7.4 MeV $\leq$ E$_p$ $\leq$ 8.0 MeV

Measurement and Microscopic Calculations of the Cross Sections and Asymmetries for the (p,p') Reaction Exciting the 2.31 MeV (0⁺,T=1) State of ¹⁴N

J. L. ESCUDIE, A. TARRATS, and J. RAYNAL, Centre d'Etudes
Nucléaires de Saclay, France

We first obtained experimental evidence that there are no resonant
contributions to the reaction at 24 MeV, the chosen incident energy.
The experiment is difficult because of the small cross section (20 to
30 μb/sr). Thus, it is a problem to separate unambiguously the interesting 2.31-MeV 0⁺ level from the background (this background may
be separated into three parts: an energy independent part from slit
scattering, the low energy tail of the elastic peak, and the peak at
1.78 MeV which is due to inelastic scattering from the detector's Si
of protons elastically scattered by N). We could obtain good results
from 30 to 165°. The analysis of the data, performed by hand, introduces errors which are due to the estimate of the background (up to
15%); in addition, the systematic
normalization errors are about 2%
for the cross sections and 5% for
the asymmetries.

Optical model parameters (fig. 1)
were obtained using the elastic
scattering data. The cancellation
of the Gamow–Teller matrix element describing the β–decay of
¹⁴C must be explained by a tensor
force contribution [1]. The near
proportionality of the GT matrix
element with the one relevant to
the transition we consider, leads
one to expect an important effect
of the tensor force in this case
too [2]. We have used a microscopic DWBA code (DWBA 70 by
Raynal and Schaeffer) in which exchange effects (E) may be included.
The analysis is complicated by
antisymmetrization which involves

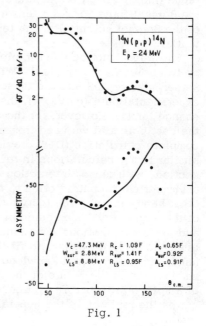

Fig. 1

Table 1

		Force		
Curve	Scalar	$\vec{L}\cdot\vec{S}$	Tensor	Exchange
1	S	zero	OPEP	D+E
2	zero	ZR	zero	–
3	B	FR	B	D+E
4	S	ZR	OPEP	D

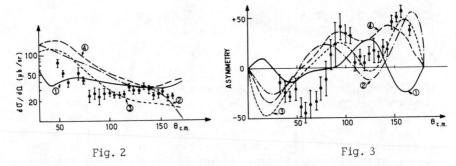

Fig. 2   Fig. 3

terms other than $\vec{\sigma}_1\vec{\sigma}_2\vec{\tau}_1\vec{\tau}_2$ for the calculations (figs. 2, 3). Yukawa
form factors are used. S stands for Serber mixture; B stands for a force
estimated from the Bressel force [3], of which only the long range part
(1.45 fm) is important. For the $\vec{L}\cdot\vec{S}$ interaction, we use either zero
range (ZR) and negative depth, or two finite ranges (FR) estimated from
the Bressel force. For the tensor interaction, we use an approximation
of the Bressel potential or the tensor part of one-pion exchange potential.

The wave function used in plotted calculations describes the states
by two holes in the 1p shell. While direct calculations (D) using a
Serber mixture, $\vec{L}\cdot\vec{S}$ and the tensor force, give reasonable fits to the
experimental data (curves 4), the fit is destroyed by including exchange terms. However, for the calculation 3 (which corresponds to
the best potential we used from a theoretical point of view), only the
complete calculation (D+E) provides a fit. Our calculations (curves 1,
similar to the calculations in ref. [3]), which yield fairly good cross
sections with tensor interaction and without an $\vec{L}\cdot\vec{S}$ contribution, give
asymmetries in contradiction to the experimental evidence. On the
other hand, an $\vec{L}\cdot\vec{S}$ force (curve 2) reproduces qualitatively the shape
of the asymmetry. We could not get good fits for both the asymmetries
and the cross sections simultaneously. However, we confirm that a
tensor force is necessary to fit the cross-section data. On the other
hand, the experimental knowledge of the asymmetries removes important
ambiguities in the nature of the force used. It indicates the necessity
of a spin-orbit term. The introduction of a non-scalar interaction requires the revision of the generally chosen mixture for the scalar potential.

REFERENCES

[1] W. M. Visscher and R. A. Ferrel, Phys. Rev. 107 (1957) 781.
[2] G. M. Crawley et al., Phys. Lett. 32B (1970) 92.
[3] C. N. Bressel, Nucl. Phys. A124 (1969) 624.

DISCUSSION

Blair:

Your data are closer to the measured asymmetries for some strong quadrupole transitions in neighboring even mass targets than to most of the calculations; in particular, the initial negative values, the sharp rise through zero near 80°, and the two succeeding positive maxima are similar to those for excitations of the 4.43 MeV 2^+ level of ^{12}C with protons of this energy observed by the Birmingham group. One would not like to argue that the isospin-changing transition in ^{14}N is collective but the comparison is intriguing.

Cramer:

You said that you measured an excitation function, but you didn't show a slide. Did you see any structure? We measured an excitation function for the same reaction at an energy about 2 MeV lower (22 MeV) and saw some bumps.

Escudié:

We measured an excitation function at six angles from 23.5 to 25 MeV in 300-keV steps. This showed no significant structure. We would, however, like to know your results, which may draw our attention to something we have not seen.

Hexadecapole Deformation and Asymmetry in Inelastic Scattering of 24.5-MeV Polarized Protons from ^{20}Ne and ^{22}Ne

A. D. BACHER, R. de SWINIARSKI,* D. L. HENDRIE, A. LUCCIO,†
G. R. PLATTNER, F. G. RESMINI,† and J. SHERMAN, Lawrence
Radiation Laboratory, Berkeley, California, USA‡

A recent experiment on inelastic scattering with 24.5 MeV protons on ^{20}Ne has shown that a large hexadecapole deformation β_4 was needed to reproduce both the shape and the amplitude of the 2^+, 4^+, 6^+ cross sections of the ground state rotational band in ^{20}Ne [1]. In order to test the sensitivity of the asymmetry to the β_2 and β_4 deformations, we have used the new Berkeley polarized beam to measure the asymmetry of the 0^+ (g.s.), 2^+ (1.63 MeV), 4^+ (4.25 MeV) states in ^{20}Ne with 24.5-MeV polarized protons. Similar measurements have also been obtained for the corresponding states in ^{22}Ne. For the experiment up to 60 nA polarized protons were extracted from the cyclotron with a polarization close to 80%. In fig. 1 are presented the measured asymmetries for ^{20}Ne and ^{22}Ne. It exhibits the similarities between the asymmetries for the 0^+ and 2^+ states in ^{20}Ne and ^{22}Ne and also the large difference between the asymmetries for the 4^+ states in ^{20}Ne and ^{22}Ne. Such a difference may suggest that these two 4^+ states are of different nature. Similar differences have also been found recently for nuclei in the s-d shell [2]. Several coupled-channel, rotational model, calculations (CC) have been performed for the ^{20}Ne data and are shown in fig. 2. The CC code of Hill which includes a deformed spin-orbit term of the Saclay type [2] was used for some of the calculations. As can be seen from this figure, the parameters of ref. [1] essentially fail to reproduce the asymmetries for the 2^+ and 4^+ states in ^{20}Ne (curves 3). Recently Sherif and Blair have obtained significant improvements for asymmetries when using DWBA collective model with the full Thomas form for the deformed spin-orbit interaction [3]. Such a deformed L-S coupling has now been introduced in a coupled-channel program [4]. The CC calculations with full Thomas term using parameters and deformation of ref. [1] are shown in fig. 2 (curve 1). Now the model predicts remarkably well all asymmetries for the 0^+, 2^+, and 4^+ states in ^{20}Ne up to about 120°, while the good agreement already obtained for the corresponding cross sections still remains. On the other hand, this good fit is completely destroyed when β_4 is set equal to zero (curve 2). Such a sensitivity to the β_4 deformation is much less pronounced when the distorted spin-orbit potential is not of the full Thomas form (curve 4).

We conclude that the values of β_2 and β_4 determined from the cross sections are still the best values for these deformations when polarization data are taken into account. Moreover, the asymmetry data emphasizes strongly the need of a β_4 deformation for the ground state rotational band in ^{20}Ne, and the necessity to use the full Thomas form for the distorted spin-orbit potential.

We would like to thank Dr. J. Raynal for performing several calculations mentioned here and Dr. B. G. Harvey and his collaborators at the 88" cyclotron for their enthusiastic support during the experiment.

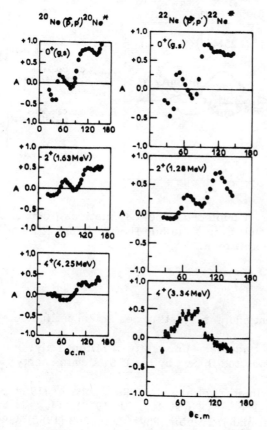

Fig. 1. Measured asymmetries for the ground state rotational band in ^{20}Ne and ^{22}Ne.

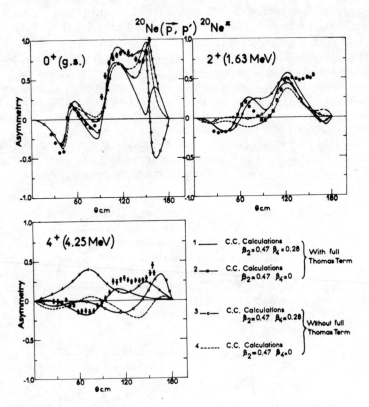

Fig. 2. Coupled-channel calculations (rotational model)
with and without the full Thomas form for the deformed
spin-orbit term.

REFERENCES

*Permanent address: Institut des Sciences Nucléaires de Grenoble,
France.
†Permanent address: University of Milan, Italy.
‡Work supported in part by the U.S. Atomic Energy Commission.

[1]   R. de Swiniarski et al., Phys. Rev. Lett. 23 (1969) 317.
[2]   A. G. Blair et al., Phys. Rev. C1 (1970) 444.
[3]   H. Sherif and J. S. Blair, Phys. Lett. 26B (1968) 489.
[4]   J. Raynal, Proc. Symp. on Nuclear Reaction Mechanisms (Quebec,
       1969).

DISCUSSION

Sherif:

The destruction of the elastic polarization fit when a full Thomas coupled-channel calculation is performed is caused only by the procedure followed, namely, using optical-model parameters determined from a simple elastic fit in a coupled-channel calculation. One really must fit the elastic data using a coupled-channel calculation to determine the parameters. These should be more consistent and should yield better simultaneous fits to both elastic and inelastic data. Therefore the destruction of the elastic fit should not be blamed on the full Thomas term.

Raynal:

Dr. de Swiniarski forgot to say that the $\vec{L} \cdot \vec{S}$ interaction was multiplied by a factor of 2 in these calculations with full Thomas term. The minimum at 100° is not reproduced and the curve remains positive. In the calculations presented last year at Quebec for ^{24}Mg, for which the experimental curves are almost the same, we obtained a negative minimum. Better optical parameters which fit these two experiments equally well can perhaps be found. All the experiments in the 20-MeV regions in the s-d shell must be studied again.

Inelastic Scattering of
Polarized Protons by ^{40}Ca at 24.5 MeV

J. L. ESCUDIE, J. GOSSET, H. KAMITSUBO, and B. MAYER, Centre
d'Etudes Nucléaires de Saclay, France

The inelastic scattering of 20.3-MeV polarized protons from ^{40}Ca has
been studied by Blair et al. at Saclay. In their work, distorted wave
analysis using both the macroscopic and microscopic models in general did not give good fits for the first 3$^-$ and 5$^-$ states. Because
^{40}Ca is one of the most interesting nuclei for nuclear structure study,
we have repeated the asymmetry experiment with better statistics and
better resolution. The beam energy was raised to 24.5 MeV to reduce
the contributions from non-direct reactions.

The experimental results are illustrated in fig. 1 together with the
DWBA macroscopic predictions. The calculated asymmetry is fairly
sensitive to the optical potential parameters. They were determined
by fitting simultaneously elastic scattering cross-section and polarization measurements.

Calculation with the full Thomas term [2] makes the agreement for
the first 3$^-$ state very satisfactory. However, the same calculation
gives a somewhat worse fit for the 5$^-$ state and fails completely for
the second 3$^-$ state. The microscopic calculation at 20.3 MeV [1]
using the wave functions of Gillet and Sanderson shows reasonable
agreement for the asymmetry for the 5$^-$ state at 4.48 MeV, whereas
the agreement is much worse for the 3$^-$ state at 3.73 MeV.

These facts suggest that the 3$^-$ state at 3.73 MeV is quite collective, but the 5$^-$ state at 4.48 MeV is not so collective. It is expected
from the calculation in ref. [1] that the microscopic DWBA analysis
will give a better fit for the asymmetry of the 3$^-$ state at 6.28 MeV.
Calculations with the microscopic model [3] are in progress for these
states.

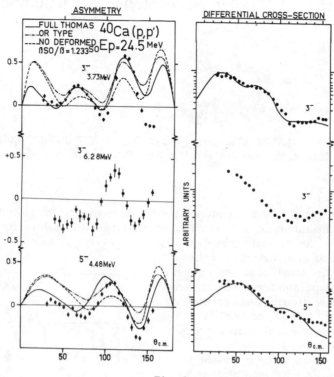

Fig. 1

REFERENCES

[1] A. G. Blair et al., Phys. Rev. 1C (1970) 444.
[2] H. Sherif and J. S. Blair, Phys. Lett. 26B (1968) 489.
[3] R. Schaeffer, Thesis, CEA Report R-4000 (1970).

Asymmetry in the Inelastic Scattering of 15.7-MeV Polarized Protons from ^{63}Cu and ^{65}Cu

M. DOST, C. GLASHAUSSER, C. F. HAYNES, H. PREISSNER, and
H. SCHNEIDER, Rutgers University, New Brunswick, New Jersey, USA*

Legg and Yntema have recently reported [1] evidence for spin dependence in the inelastic scattering of 14- and 16-MeV protons from ^{63}Cu,
^{65}Cu, and ^{67}Zn. In all three nuclei,
the angular distributions for the
lowest $1/2^-$ final state were similar yet consistently different from
the angular distributions for the
lowest $3/2^-$, $5/2^-$, or $7/2^-$ final
states. All these transitions are
presumed to be $\ell = 2$.

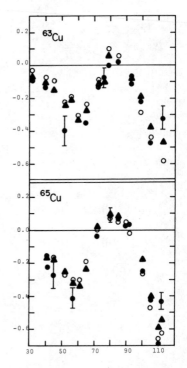

The asymmetry in the inelastic
scattering of polarized protons is
expected to be more sensitive to
spin-dependent effects than the
cross section. Such measurements
were performed with a beam of up
to 2 nA of 15.7-MeV polarized protons from the Rutgers Lamb-shift
polarized ion source. The polarization was usually 50-60%. Four
Si(Li) counters placed 15-20°
apart on the same side of the incident beam detected the scattered
particles.

Asymmetries are shown in fig. 1
for scattering from the states of
^{63}Cu at 0.67-MeV ($1/2^-$), 0.96-MeV
($5/2^-$), and 1.33-MeV ($7/2^-$) and
from the states of ^{65}Cu at 0.77-
MeV ($1/2^-$), 1.11-MeV ($5/2^-$),
and 1.48-MeV ($7/2^-$). A few representative error bars are shown for
the $1/2^-$ results. Errors for the
$5/2^-$ and $7/2^-$ states are generally

Fig. 1. Plot of asymmetry
versus laboratory angle for
^{63}Cu and ^{65}Cu. The data are
shown as follows: $1/2^-$, ●;
$5/2^-$, ○; $7/2^-$, ▲.

714

about half as large as these; errors for ^{65}Cu are generally smaller than for ^{63}Cu. The asymmetries observed for the two nuclei are very similar; there are no large and consistent differences between the $1/2^-$ states and the others. In particular, in the region from about 75° to 100° where the largest differences were observed in the cross sections, the asymmetries are almost identical. There is some evidence that the $1/2^-$ asymmetries at more forward angles are slightly larger than the others.

No simple explanation of both the cross-section and asymmetry results appears possible. Difficulties can be found, e.g., with interpretations based on differences in wave-functions, possible target-spin dependent interactions, and competition from ℓ-values other than two.

REFERENCE

*Supported in part by the National Science Foundation.

[1]   J. C. Legg and J. L. Yntema, Phys. Rev. Lett. 22 (1969) 1005.

DISCUSSION

Raynal:

The Cu isotopes have a closed shell of protons plus an extra proton, and the spin-orbit effects are weaker in these cases. For nuclei with a closed shell of neutrons but an open proton shell, there can be greater effects on the asymmetry attributable to the shell structure.

Polarization of Protons Inelastically Scattered on ^{24}Mg, ^{52}Cr, and ^{58}Ni

A. RUH and P. MARMIER, Laboratorium für Kernphysik, ETH, Zürich, Switzerland

Several authors [1] have obtained very good agreement between the results of (p,p'γ) angular correlation measurements on ^{52}Cr and ^{58}Ni for incident energies between 4 and 7 MeV and the predictions of the statistical CN model. According to this model the polarization of protons inelastically scattered on ^{52}Cr or ^{58}Ni leaving the residual nucleus in its first excited state should vanish if it is averaged over a sufficiently large energy interval [1]. Therefore polarization measurements with relatively poor energy resolution should provide an additional test for the CN model.

In the present work [2] the polarization of protons inelastically scattered on ^{24}Mg, ^{52}Cr and ^{58}Ni leaving the residual nucleus in its first excited state ($J^{\pi} = 2^{+}$) was measured with a carbon polarimeter. The polarization of the elastically scattered protons was also measured. The incident energy was varied between 5.38 and 5.95 MeV and the lab scattering angle between 45° and 120°. In addition, excitation functions were measured for ^{24}Mg(p,p$_1$)^{24}Mg* between 5 and 6 MeV and for ^{52}Cr(p,p$_0$)^{52}Cr and ^{52}Cr(p,p$_1$)^{52}Cr* between 5 and 6.5 MeV. Large values of the polarization were found for protons scattered on ^{24}Mg, since the statistical assumption $\Delta E \gg D$ was not fulfilled and only a small number of states were excited in the compound nucleus. For the same reason the interpretation of (p,p'γ) angular correlation measurements with the CN model did not yield satisfactory agreement [1]. In the case of ^{58}Ni, according to the theoretically expected level density, so many compound levels should be excited that the statistical assumption should be very well fulfilled. In fact, no polarizations having values statistically different from zero were obtained experimentally. For the polarization measurements on ^{52}Cr the statistical assumption should have been especially well satisfied ($\Delta E \approx 400$ keV, $D \approx 0.4$ keV). However, relatively large polarizations were found in the inelastic scattering on ^{52}Cr (see figs. 1 and 2) which could be explained neither by statistical fluctuations nor by a direct interaction contribution. Furthermore, the angular cross-correlation function does not agree with the statistical theory either. This indicates a violation of the assumptions concerning the statistical behavior of the reduced width amplitudes $\gamma_{\mu c}$. These assumptions had been believed to be satisfied because of the good agreement of the CN angular correlation theory with the experimental results. The correlated intermediate structure of the excitation functions can probably be interpreted as an excitation of doorway states. This interpretation would also explain the measured polarizations.

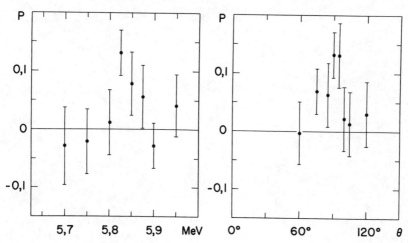

Fig. 1. Polarization for $^{52}Cr(p,p_1)^{52}Cr^*$ ($E^* = 1.434$ MeV) as a function of the lab energy at $\theta_{lab} = 90°$.

Fig. 2. Angular distribution of the polarization for $^{52}Cr(p,p_1)^{52}Cr^*$ ($E^* = 1.434$ MeV) at $E_{lab} = 5.825$ MeV.

REFERENCES

[1] E. Sheldon, Rev. Mod. Phys. 35 (1963) 795 and references therein.
[2] A. Ruh and P. Marmier, to be published in Nucl. Phys.; A. Ruh, Thesis, Juris-Verlag, Zürich, 1970.

Isobaric
Analog Resonances

Polarization in Proton Elastic Scattering
near Isobaric Analog Resonances

G. PISENT, University of Padua, Italy

Let us consider a proton-nucleus scattering process, characterized by a single "doorway" resonance in the channel J, L. If the potential (optical model) scattering matrix is written

$$S_{j\ell} = U_{j\ell} \exp(2i\delta_{j\ell}), \tag{1a}$$

where $\delta_{j\ell}$ is the real nuclear phase shift, the most general Breit-Wigner form of the resonant scattering matrix element is [1]

$$S_{JL} = \exp(2i\delta_{JL})[U_{JL} - i\Gamma' \exp(2i\beta)/(E - E_0 + i\Gamma/2)]. \tag{1b}$$

Since we deal with a single resonance, the "channel" and "level" indices on the resonance parameters E_0, Γ, Γ', β are omitted for brevity.

We can separate the (coherent and spin-flip) scattering amplitudes into a "potential" and "resonant" part as follows:

$$A = A_0 + A_{JL}, \tag{2a}$$

$$B = B_0 + B_{JL}, \tag{2b}$$

where

$$A_0 = f_c + (2ik)^{-1} \Sigma_{j\ell}(j + \tfrac{1}{2}) P_\ell \exp(2i\omega_\ell)[U_{j\ell} \exp(2i\delta_{j\ell}) - 1], \tag{3a}$$

$$B_0 = (2k)^{-1} \Sigma_{j\ell}(-)^{j-\ell-1/2} P_\ell^1 \exp[2i(\omega_\ell + \delta_{j\ell})] U_{j\ell}, \tag{3b}$$

and

$$A_{JL} = -(2k)^{-1}(J + \tfrac{1}{2}) P_L \exp[2i(\omega_L + \delta_{JL} + \beta)]\Gamma'/(E - E_0 + i\Gamma/2), \tag{4a}$$

$$B_{JL} = -(2k)^{-1}(-)^{J-L-1/2} P_L^1 \exp[2i(\omega_L + \delta_{JL} + \beta)]i\Gamma'/(E - E_0 + i\Gamma/2). \tag{4b}$$

We now consider the particular but important case $B_0 = 0$, i.e., the case of zero (or negligible) background polarization.

In this case cross section and polarization assume the following parametric form:

$$\sigma = |A|^2 + |B|^2 = \sigma_0 + \Gamma' \frac{a\sqrt{\sigma_0}\,[(E - E_0)\cos\gamma - (\Gamma/2)\sin\gamma] + (\Gamma'/4)(a^2 + b^2)}{(E - E_0)^2 + (\Gamma/2)^2}$$

(5a)

$$\sigma P = 2\mathrm{Re}(AB^*) = b\Gamma'\sqrt{\sigma_0}\,\frac{(E - E_0)\sin\gamma + (\Gamma/2)\cos\gamma}{(E - E_0)^2 + (\Gamma/2)^2}$$

(5b)

where

$$\sigma_0 = |A_0|^2 = \text{background cross section,}$$

(6a)

$$a = -(J + \tfrac{1}{2})\,P_L/k,$$

(6b)

$$b = (-)^{J-L+1/2}\,P_L^1/k,$$

(6c)

$$\gamma = \arg A_0 - 2(\omega_L + \delta_L + \beta).$$

(6d)

In eqs. (5) the phase γ (in place of β) may be assumed as a free parameter, while a and b are known constants depending only on the spin and parity.

From eqs. (5), the following conclusions may be drawn, some of which are well known:

1. The polarization [cross section] excitation function exhibits a maximum followed by a minimum (when the energy is raised) if the quantity $(-)^{J-L-\frac{1}{2}}\,P_L^1\sin\gamma\,[P_L\cos\gamma]$ is positive, and vice versa. A trivial consequence is that the states $J = L + 1/2$ and $J = L - 1/2$ have opposite signs of polarization.

2. Corresponding to the zeros of P_L we have $a = 0$ and consequently $\sigma = \sigma_0 + (b\Gamma'/2)^2/[(E - E_0)^2 + (\Gamma/2)^2]$. In this case the cross section shows one peak of height $(b\Gamma'/\Gamma)^2$ and half width Γ, over the background σ_0.

3. If the σP excitation function is known near the resonance, at a certain angle (provided P_L nor P_L^1 never vanish), the resonance parameters may be easily calculated from the knowledge of the energies $(E_{1,2})$ and intensities $(y_{1,2} \equiv (\sigma P)_{1,2})$ corresponding to the stationary points of σP. The formulas are

$$\Gamma'/\Gamma = (y_1 - y_2)/2b\sqrt{\sigma_0},$$

(7a)

$$\cos\gamma = (y_1 + y_2)/(y_1 - y_2), \tag{7b}$$

$$\Gamma = \sin\gamma \, (E_1 - E_2), \tag{7c}$$

$$E_0 = \frac{1}{2}(E_1 + E_2 + \Gamma \cot\gamma). \tag{7d}$$

4. Similar formulas may be derived for the analysis of the cross section.

If the nuclear potential scattering is negligible (and this is sometimes a good approximation when dealing with low energy analog resonances), we are able to calculate $\delta_L = -\mathrm{arctg}(F_L/G_L) =$ charged hardsphere phase shift, and consequently A_0 and σ_0 [2]. In this case the resonance phase β can be deduced from γ through eq. (6d).

REFERENCES

[1] G. Terrel et al., Isobaris Spin in Nuclear Physics (Academic Press, 1966) p. 333; J. L. Adams et al., Nucl. Phys. 89 (1966) 337; K. W. McVoy, Ann. Phys. (N.Y.) 54 (1969) 17.
[2] J. M. Blatt and L. C. Biedenharn, Rev. Mod. Phys. 24 (1952) 258.

A Study of Polarization for ^{128}Te(p,p) on Negative Parity Isobaric Analog Resonances

M. ROTH, R. BANGERT, B. GONSIOR, P. RÖSNER, B. STEINMETZ, and A. STRÖMICH, University of Cologne, Germany

In the present paper we investigate spin and parity of four resonances in ^{129}I which correspond to low lying states in ^{129}Te. The excitation energies of these states are respectively 2.09, 2.20, 2.25, and 2.33 MeV. From (d,p) and (d,t) reactions measured by Jolly [1] and (d,p) reactions measured by Moore et al. [2] the energies of the states and the angular momentum could be determined. Foster et al. [3] and Burde et al. [4] studied elastic and inelastic scattering of protons on Te including ^{128}Te. From these measurements the ℓ-values of the states could be identified as $\ell = 3, 3, 1$, and 1, respectively. Foster et al. [3] assume for the $\ell = 3$ states a J-value of $7/2^-$ and assign tentatively a J-value of $3/2^-$ to the $\ell = 1$ states. Burde et al. [4] assume for the $\ell = 3$ states also a J-value of $7/2^-$, but cannot exclude the $5/2^-$ alternative, whereas they assume a J-value of $3/2^-$ for the $\ell = 1$ states.

Since the total angular momentum J of a state can usually not be identified without doubt from proton elastic scattering, we measured the polarization of elastically scattered protons. Such measurements normally determine the J-value unequivocally when the ℓ-value is known.

Protons with energies between 9.8—10.5 MeV from our FN tandem were elastically scattered on Te. The polarization of the scattered protons was measured by double scattering techniques with carbon as a second scatterer. We used two polarimeters of special design mounted symmetrically to the beam axis. In order to determine instrumental asymmetries we measured protons from Rutherford scattering on Au, where the polarization should vanish. We found the instrumental asymmetries to be very small and had only to adjust the polarimeter axis onto the target spot. The Te-targets could withstand a beam current of only about 100 nA. Therefore we constructed a rotating target. With this device we could run our experiments with beam currents up to 2 μA.

The results are presented in fig. 1 together with an excitation curve for inelastic scattering to the 2^+ state in ^{128}Te, also measured at our laboratory [5], to show the position of the resonances. Also given in

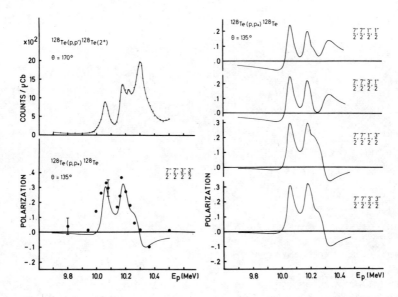

Fig. 1. Measured and calculated proton polarization and measured excitation function for inelastic scattering.

fig. 1 are calculated excitation functions of the polarization assuming four different combinations of J-values for the $\ell = 1$ states. The calculations have been done with the modified optical model code [6] OPTIX 1 using resonance terms of the Breit-Wigner form in the collision matrix. We used the resonance parameters and optical model parameters found by Foster et al. [3], only the resonance energies were taken from our measurements on inelastic scattering [5]. They differ by about 30 keV from those given by Foster et al. [3].

Comparing the theoretical curves with the experimental results we find that the combination of J-values $7/2^-$, $7/2^-$, $3/2^-$, $3/2^-$ fits the data satisfactorily, so that we propose these numbers as the J-values for the states at the excitation energies given in the beginning.

REFERENCES

[1] R. K. Jolly, Phys. Rev. B136 (1964) 683.
[2] W. H. Moore et al., Nucl. Phys. A104 (1967) 327.
[3] J. L. Foster, Jr., P. J. Riley, and C. F. Moore, Phys. Rev. 175 (1968) 1498.
[4] J. Burde et al., Nucl. Phys. A141 (1970) 375.
[5] B. Steinmetz et al., Jahresbericht 1968/69 Institut für Kernphysik der Universität zu Köln, and to be published.
[6] W. J. Thompson and E. Gille, Technical Report OPTIX 1, Tandem Accelerator Laboratory, Florida State University (September 1965).

Transfer Reactions

Deuteron D-State Effect in Tensor Polarization for the ^{9}Be(p,d)^{8}Be Reaction

F. D. SANTOS, Laboratório de Física e Engenharia Nucleares,
Sacavém, Portugal

The second rank tensor components $\langle T_{2q} \rangle$ of the deuteron polarization [1] in the ^{9}Be(p,d)^{8}Be (Q = 0.56 MeV) reaction at proton energies of 2.5 and 3.7 MeV predicted by the distorted wave theory with spin-orbit coupling in the optical potentials and assuming a pure S-state deuteron internal wave function are consistently smaller than the measured values at least by one order of magnitude [2]. Johnson has shown, using plane waves [3], that the inclusion of the D-state of the deuteron gives for $\langle T_{2q} \rangle$ values which are larger than the distorted wave results without the D-state (fig. 1). This result suggests that the D-state effect gives an important contribution to the $\langle T_{2q} \rangle$ observed.

The D-state component of the deuteron wave function was included in the present DWBA calculations using the same approximations and

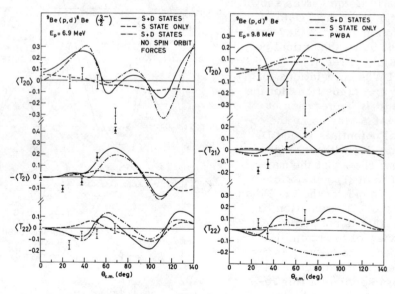

Fig. 1

729

parameters of ref. [5]. The curves shown in fig. 1, together with the experimental results of Ivanovich et al. [4] for E = 6.9 and 9.8 MeV, were calculated using the proton optical potential of ref. [6] for the ^{9}Be(d,p)^{10}Be reaction with $V_s = 0$ and the deuteron optical potential of Satchler [7], respectively, for deuteron energies of 7.8 (U = 89.6 MeV) and 10.2 MeV (U = 78.5 MeV) with spin-orbit coupling. The calculated and experimental results of figs. 1 and 2 are for $(-1)^{\ell}$ $\langle T_{2q} \rangle$ referred to the coordinate system used in refs. [2, 4]. The dot-dash curves in fig. 1 were obtained for $E_p = 6.9$ MeV without the deuteron spin-orbit coupling and for $E_p = 9.8$ MeV from the plane wave formula of ref. [3]. We find that the inclusion of the D-state in the DWBA calculations improves the similarity between theory and experiment at both energies but the agreement is poor particularly for $\langle T_{20} \rangle$. We did not attempt to improve the fit with experiment by variation of the optical potentials; furthermore, tensor forces in the deuteron optical potential should be included in the calculation since they appear to have a non-negligible effect [8] on $\langle T_{2q} \rangle$. The $\langle T_{20} \rangle$ and $\langle T_{21} \rangle$ predicted only with spin-orbit distortion are considerably smaller than the measured values as in the calculations of ref. [2]. However, this discrepancy is not so apparent for $\langle T_{22} \rangle$ at the higher energies. Calculations for

other reactions suggest that $\langle T_{20} \rangle$ and $\langle T_{21} \rangle$ are more dependent on the effect of the D-state and less sensitive to spin-orbit distortion than $\langle T_{22} \rangle$. This can be seen clearly in the curves shown in fig. 2 for the ^{40}Ca(d,p)^{41}Ca reaction at $E_d = 7$ MeV calculated using the optical potentials of ref. [6].

DWBA calculations for two $\ell = 3$ transitions for the ^{56}Fe(p,d)^{55}Fe reaction at $E_p = 18.5$ MeV using the optical potentials Pl and Dl of ref. [9] without spin-orbit terms show that the magnitude of $\langle T_{2q} \rangle$ is noticeably larger for the $7/2^-$ than for the $5/2^-$ transition (fig. 2). This is not surprising in view of the strong j-dependence of the D-state effect in the differential cross section for this reaction [5, 10]: a large

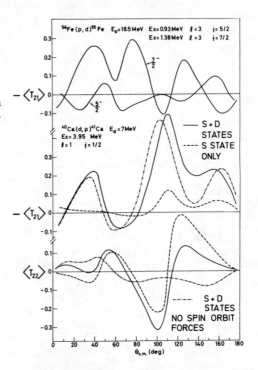

Fig. 2

D-state effect in the differential cross section can be expected to correspond to a sizeable contribution of the D-state to $\langle T_{2q} \rangle$. A recent finite range DWBA calculation [11] with the D-state included shows that its effect of $\langle T_{2q} \rangle$ is larger than that of tensor forces in the deuteron optical potential.

REFERENCES

[1] G. R. Satchler, Nucl. Phys. 55 (1964) 1.
[2] A. J. Froelich and S. E. Darden, Nucl. Phys. A119 (1968) 97.
[3] R. C. Johnson, Nucl. Phys. A90 (1967) 289.
[4] M. Ivanovich et al., Nucl. Phys. A97 (1967) 177.
[5] R. C. Johnson and F. D. Santos, Phys. Rev. Lett. 19 (1967) 364.
[6] T. J. Yule and W. Haeberli, Nucl. Phys. A117 (1968) 1.
[7] P. E. Hodgson, Adv. Phys. 15 (1966) 329.
[8] G. Delic and B. A. Robson, Nucl. Phys. A127 (1969) 234.
[9] C. Glashausser and M. E. Rickey, Phys. Rev. 154 (1967) 1033.
[10] F. D. Santos and R. C. Johnson, to be published.
[11] G. Delic and B. A. Robson, to be published.

Effect of the Deuteron D-State on
DWBA Calculations for $^{16}O(p,d)^{15}O$

G. DELIC, Australian National University, Canberra, Australia

Exact finite range (FR) DWBA calculations including the D-state of
the deuteron [1] were performed for the reaction $^{16}O(p,d)^{15}O$ leading to
the g.s. (Q = -13.44 MeV, j = 1/2) and the 6.18 MeV (Q = -19.62 MeV,
j = 3/2) state for a proton bombarding energy of 30 MeV. In these
cases large D-state effects are expected [2].

The deuteron optical model parameters are those of Chant et al. [3]
and the proton parameters, which included a small volume absorption,
were obtained by using the formula of Becchetti and Greenlees [4] for
"optimum proton-nucleus standard OM parameters." The neutron
bound-state wave functions were obtained by the usual well depth
prescription. Good fits to the cross-section data of Chant et al. [3]
were obtained.

Fig. 1 shows the difference between a zero range (ZR) DWBA calculation (broken curve) and a FR DWBA calculation which includes the
D-state of the deuteron (solid curve). The data for the proton analyzing
powers $P(\theta)$ are those of ref. [3]. The theoretical fits are a significant
improvement over the previous work [3]. The difference in the two
calculations is relatively small for the proton analyzing power $P(\theta)$
and deuteron vector polarization $\langle iT_{11} \rangle$ so that the j-dependent effects
are preserved. However, large differences are evident at forward
angles in the case of the deuteron tensor polarizations $\langle T_{20} \rangle$, $\langle T_{21} \rangle$, and
to a lesser extent $\langle T_{22} \rangle$. This effect is due almost entirely to the inclusion of the deuteron D-state, since the differences between a ZR
and a FR S-state calculation were generally small, particularly for the
deuteron tensor polarizations $\langle T_{2k} \rangle$. Further calculations using elastic
deuteron parameters with a lower radial cut-off at 3 fm have shown
that the effects reported here persist and, if anything, are enhanced.

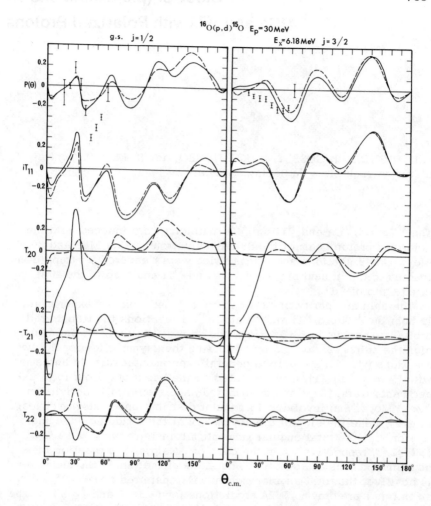

Fig. 1

REFERENCES

[1] G. Delic and B. A. Robson, Third Polarization Symp.
[2] R. C. Johnson, private communication.
[3] N. S. Chant, P. S. Fisher, and D. K. Scott, Nucl. Phys. A99 (1967) 609.
[4] F. D. Becchetti, Jr., and G. W. Greenlees, Phys. Rev. 182 (1969) 1190.

Study of (p,d) Reactions on
^{48}Ca and ^{49}Ti with Polarized Protons

J. L. ESCUDIE, J. GOSSET, H. KAMITSUBO, and B. MAYER, Centre
d'Etudes Nucléaires de Saclay, France

The ^{48}Ca(p,d)^{47}Ca and ^{49}Ti(p,d)^{48}Ti reactions were induced with the
polarized proton beam of the Saclay cyclotron at 22.9 MeV and 20.9
MeV. Cross sections and asymmetries were measured for transitions
leading to the ground states of ^{47}Ca and ^{48}Ti and also to the first excited state of ^{48}Ti.

The spin and parity of ^{47}Ca can be deduced from our experiment.
In both the ^{49}Ti(p,d)^{48}Ti and ^{48}Ca(p,d)^{47}Ca reactions the transferred
orbital momentum is $\ell = 3$. Since the spin of ^{49}Ti is $7/2^-$, one can assign the spin $7/2^-$ to ^{47}Ca by comparing the asymmetries for the two
reactions (fig. 1). Actually a general experimental rule, in agreement
with Newns' rule [1], may be stated on the basis of all our (p,d) experiments (refs. [2, 3] and contributed papers on (p,d) reactions on
^{90}Zr, ^{92}Mo, ^{118}Sm) involving 15 angular distributions of asymmetries:
in every case the sign of the asymmetry at the stripping peak angle
is + or -, if the total angular momentum transferred is $j = \ell + 1/2$ or
$j = \ell - 1/2$, respectively. In the case of ^{48}Ca(p,d)^{47}Ca reaction the
stripping peak is at about 30°, and since at that angle the asymmetry
is positive, the total angular momentum transferred is $j = 7/2^-$.

In fig. 1 are drawn DWBA predictions for $j = 5/2^-$ and $j = 7/2^-$. The
latter gives an asymmetry in much better agreement with the experimental data especially at forward angles. Although the shape of the
cross-section angular distribution is well reproduced, the DWBA results do not give the correct ratio of cross sections at backward and
forward angles. The optical potential parameters (table 1) were cal-

Table 1

Particle	V_0	R_0	a_0	W_D	R_i	a_i	V_{SO}	R_{SO}	a_{SO}	χ_σ^2	χ_p^2	σ_R
p	54.46	1.12	0.75	7.93	1.244	0.7	5	1.03	*0.47*	2.26	1.8	1153
d	92.73	*1.15*	0.81	17.24	*1.34*	*0.68*	6	*1*	0.5	2.5	---	1525

Parameters in italics were kept fixed during the search.

culated to fit both cross sections and polarizations in the proton channel and cross sections in the deuteron channel [4]. The cross sections
calculated in the DWBA are also strongly j-dependent, and the experimental results are closer to the $j = 5/2^-$ curve than to the $j = 7/2^-$ one.
However, a good fit can be obtained with $j = 7/2^-$ with different optical
potential parameters [5]. The deuteron parameters fit scattering data
at 15 MeV.

Experimental results for the ^{49}Ti(p,d)^{48}Ti* reaction leading to the
first excited state of ^{48}Ti are presented in fig. 2. This latter transition involves a mixing of transferred j.

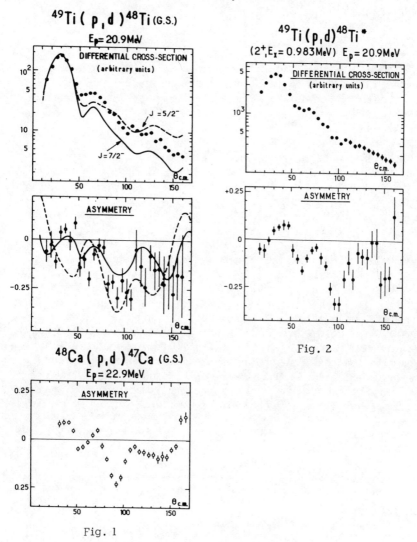

Fig. 1

Fig. 2

REFERENCES

[1] M. C. Newns, Proc. Phys. Soc. (London) A66 (1953) 477.
[2] P. J. Bjorkholm, W. Haeberli, and B. Mayer, Phys. Rev. Lett. 22
     (1969) 955.
[3] J. L. Escudié et al., Phys. Rev. Lett. 23 (1969) 1251.
[4] R. K. Jolly, E. K. Lin, and B. L. Cohen, Phys. Rev. 130 (1963) 2391.
[5] K. Grotowski et al., Report 594/PL, Institute of Nuclear Physics
     (Cracow, 1968).

Study of (p,d) Reactions on
^{90}Zr and ^{92}Mo with Polarized Protons

J. L. ESCUDIE, J. GOSSET, H. KAMITSUBO, and B. MAYER, Centre
d'Etudes Nucléaires de Saclay, France

The improved polarized proton beam of the Saclay variable energy
cyclotron has made possible extensive measurements of asymmetries
(or analyzing powers) in (p,d) reactions. The intensity of the beam
is usually about 5 to 10 nA on target, and the polarization around 80%.

Previous experiments [1, 2] have
shown that the asymmetries for ℓ =
1 transitions exhibit a strong j-
dependence throughout the entire
angular range. The theoretical calculations using the distorted wave
Born approximation (DWBA) gave
reasonable fits for both cross sections and asymmetries. Thus the
DWBA gave good account of the observed j-dependence.

In this paper we report (p,d) reactions on ^{90}Zr at 22.9 MeV and ^{92}Mo
at 24.5 MeV. Both nuclei have filled
neutron shells, so that one expects
to reach levels of similar structure
by the (p,d) reaction. Actually the
angular distributions of cross sections and asymmetries for the first
three levels of ^{89}Zr are very similar
(figs. 1 and 2) to those of the corresponding levels in ^{91}Mo. Therefore,
one can assign spin and parity to
the first three levels of ^{91}Mo: $9/2^+$
for the ground state, $1/2^-$ and $3/2^-$
for the first two levels, in agreement
with ref. [3]. Fig. 2 shows the strong
j-dependence of the asymmetries for
the two excited levels in ^{89}Zr and

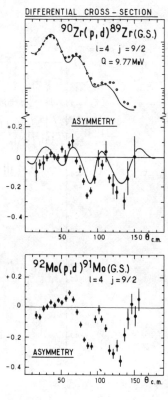

DIFFERENTIAL CROSS - SECTION

^{90}Zr$(p,d)^{89}$Zr(G.S.)
l = 4   j = 9/2
Q = 9.77 MeV

ASYMMETRY

^{92}Mo$(p,d)^{91}$Mo(G.S.)
l = 4   j = 9/2

ASYMMETRY

Fig. 1

^{91}Mo (for these levels the orbital angular momentum transferred is $\ell = 1$).

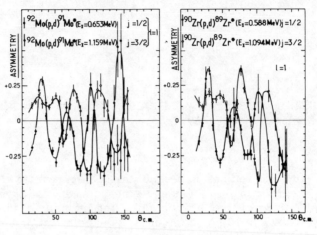

Fig. 2

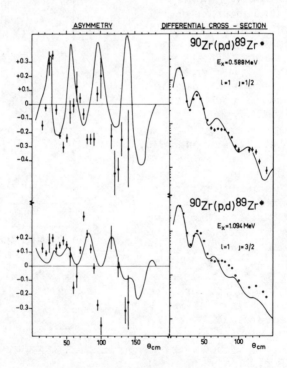

Fig. 3

The DWBA gives good account of the measured cross sections and asymmetries (figs. 1 and 3). The optical potentials used are listed in table 1. The proton potential fits the cross section and polarization for the elastic scattering of protons. The deuteron potential fits the cross section of the elastic scattering of deuterons at 12 MeV [4]. The DWBA calculation includes corrections for finite range and non-locality. These corrections do not affect the asymmetry very much.

Table 1

Particle	V_o	R_o	a_o	W_D	R_i	a_i	V_{SO}	R_{SO}	a_{SO}	x_σ^2	x_p^2	σ_R
p	50.06	1.22	0.63	11.54	1.267	0.54	6.41	*1.15*	*0.5*	0.65	2.5	1234
d	91.85	1.21	0.7	10.9	1.305	0.862	*6*	*1*	*0.4*	0.43	---	1457

The parameters in italics were kept fixed during the search.

REFERENCES

[1] P. J. Bjorkholm, W. Haeberli, and B. Mayer, Phys. Rev. Lett. 22 (1969) 955.
[2] J. L. Escudié et al., Phys. Rev. Lett. 23 (1969) 1251.
[3] G. Bassani and J. Picard, Nucl. Phys. A131 (1969) 653.
[4] F. G. Perey and G. R. Satchler, Nucl. Phys. A97 (1967) 515.

Study of (p,d) Reactions on
^{118}Sn and ^{119}Sn with Polarized Protons

J. L. ESCUDIE, J. GOSSET, H. KAMITSUBO, R. LOMBARD, and
B. MAYER, Centre d'Etudes Nucléaires de Saclay, France

The polarized proton beam of the Saclay cyclotron was used to induce
(p,d) reactions on ^{118}Sn and ^{119}Sn at 24.5 MeV. We measured the cross
sections and asymmetries for transitions to the ground state and the
first five levels of ^{117}Sn, and to the ground state and the first excited
state of ^{118}Sn. In the ^{118}Sn(p,d)^{117}Sn reaction, three transitions occur
with $\ell = 2$; fig. 1 shows the j-dependence of the asymmetry for $\ell = 2$.
The asymmetries for $j = 3/2$ are mainly
negative, while the asymmetries for
$j = 5/2$ are mainly positive, and the two
curves have opposite phase at forward
angles.

A DWBA calculation was performed
with optical potential parameters of
table 1. These parameters were obtained by fitting the elastic scattering
cross sections for both proton and
deuteron channels [1], and the polarization for elastic scattering in the proton channel.

The agreement of the DWBA predictions with the experimental results is
impressive for the $\ell = 0$ transition (fig. 3).
The fit is good for $\ell = 2$ transitions, but
less good for the asymmetries in $\ell = 4$

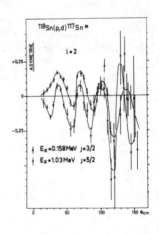

Fig. 1

Table 1

Particle	V_o	R_o	a_o	W_D	R_i	a_i	V_{SO}	R_{SO}	a_{SO}	X_σ^2	X_p^2	σ_R
p.	51.37	1.17	0.763	11.44	1.33	0.622	6.18	1.15	0.552	5.2	13.	1589
d	100.8	*1.15*	*0.81*	15.06	*1.34*	*0.68*	*6.5*	*1.15*	*0.81*	3	---	1570

The parameters in italics were kept fixed during the search.

and $\ell = 5$ transitions although the oscillations are well reproduced (figs. 2 and 3). Better fits can be obtained for either $\ell = 4$ or $\ell = 5$ transitions, but not both, with different optical potential parameters. Actually the asymmetry is more sensitive to the optical potential parameters for large ℓ values than for small ℓ-values.

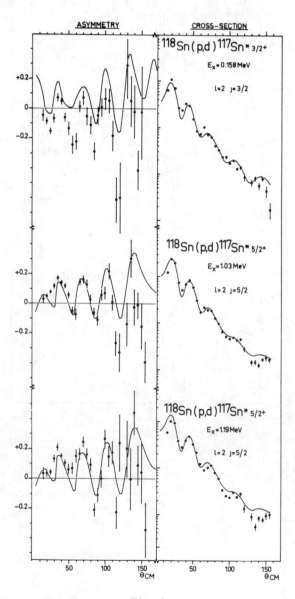

Fig. 2

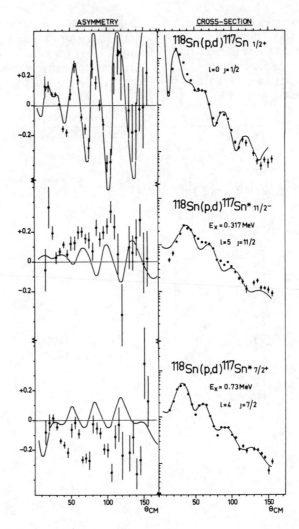

Fig. 3

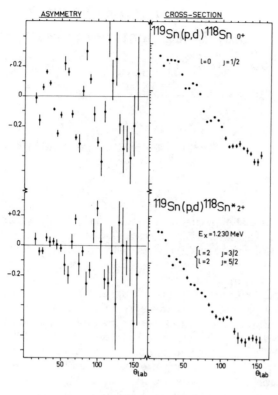

Fig. 4

REFERENCE

[1]   R. K. Jolly et al., Phys. Rev. 130 (1963) 2391.

Effect of Spin-Orbit Terms of the Optical Potentials on the Asymmetry in (p,d) Experiments with Polarized Protons

H. KAMITSUBO and B. MAYER, Centre d'Etudes Nucléaires de Saclay, France

On the basis of our experimental results on ^{118}Sn(p,d)^{117}Sn and ^{119}Sn(p,d)^{118}Sn reactions with polarized protons (contributed paper) the effect of the spin-orbit terms in both proton and deuteron channels is studied. In the case of $\ell = 0$ transfer the asymmetry arises only from those spin-orbit terms.

In fig. 1 the asymmetry in the ^{119}Sn(p,d)^{118}Sn reaction is shown to be very similar to that of protons elastically scattered by ^{119}Sn, mainly above 60°. This correlation between elastic scattering of protons and the (p,d) reaction is well understood in terms of the DWBA. When a spin-orbit term is included in the proton channel only (PSO), the main features of the (p,d) asymmetry are well given by the DWBA (fig. 2). The contribution from the spin-orbit term in the deuteron channel does not change the asymmetry radically. The asymmetry given by a spin-orbit term in the deuteron channel only (DSO) is nearly in phase opposition to the PSO asymmetry [1], and the asymmetry calculated with spin-orbit terms in both channels is roughly equal to the sum of the asymmetries for PSO and DSO [2]. DWBA calculations show that the correlation between proton elastic scattering and (p,d) reactions disappears at lower energy although it remains in the theory of Pearson et al. [3]. Asymmetry measurements at lower energy would therefore provide a good test for the reaction mechanism.

The opposite behavior of the asymmetry for PSO and DSO may be seen in the partial wave scattering amplitudes in fig. 3. The spin-orbit force is more effective at lower partial waves.

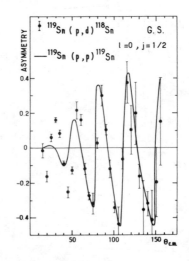

Fig. 1

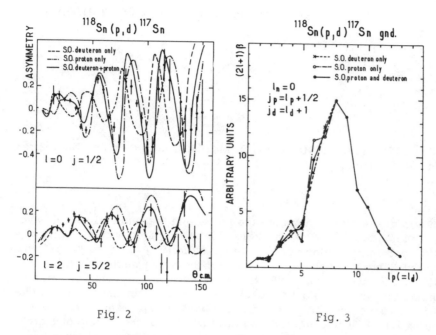

Fig. 2   Fig. 3

In fig. 2 the effects of the spin-orbit term in the proton and deuteron channels are also shown for the case of $\ell_n = 2$ transfer.

REFERENCES

[1] G. R. Satchler, Nucl. Phys. 18 (1960) 110.
[2] L. J. B. Goldfarb and R. G. Seyler, to be published in Nucl. Phys.
[3] C. A. Pearson, J. W. Wilcott, and L. C. McIntyre, Nucl. Phys. A125 (1969) 111.

J-Mixing Determination in
(p,d) Reactions by Asymmetry Measurements

J. GOSSET, H. KAMITSUBO, and B. MAYER, Centre d'Etudes Nucléaires
de Saclay, France

In a (p,d) reaction, when both the target and the residual nuclei have
a spin not equal to zero, the transition takes place with a mixing of
different total angular momenta j ranging from $|J_f - J_i|$ to $J_f + J_i$, where
J_i and J_f are respectively the spins of the target nucleus and the residual nucleus. In the DWBA the cross section, the polarization, and
the asymmetry of the outgoing particle appear as incoherent sums over
different j-values, while interference between different s and ℓ remains [1]. However in (p,d) and (d,p) reactions, the values of ℓ are
restricted to $\ell = j \pm 1/2$; of these two values only one is allowed by
parity conservation. Thus in (p,d) and (d,p) reactions the cross section and the asymmetry can be written as

$$\sigma(\theta) = \sum_j S_{\ell j} \sigma_{\ell j}(\theta) \tag{1}$$

$$\sigma A(\theta) = \sum_j S_{j\ell} \sigma_{\ell j}(\theta) A_{\ell j}(\theta) \tag{2}$$

where the $S_{\ell j}$ are spectroscopic factors, $\sigma_{\ell j}$ and $A_{\ell j}$ cross sections
and asymmetries given by the DWBA. The cross sections $\sigma_{\ell j}$ are very
sensitive to ℓ and can be used to determine the cross-section contribution from different ℓ values. On the other hand, the asymmetries $A_{\ell j}$
are very sensitive to j, so that measurements of the asymmetries permit a determination of cross-section contributions for different j-
values [2].

We have studied several (p,d) reactions in which j-mixing occurs,
by measuring both cross sections and asymmetries. The reaction
^{119}Sn(p,d)^{118}Sn was investigated at 24.5 MeV (see contributed paper).
The target spin is $1/2^+$, and the spin of the first excited state of ^{118}Sn
is 2^+, so that the ℓ transferred has a unique value $\ell = 2$, but there are
two possible values for j in that transition: $j = 3/2$ and $j = 5/2$. Eqs.
(1) and (2) would permit the determination of the spectroscopic factors
$S_{3/2}$ and $S_{5/2}$, if one knows the cross sections $\sigma_{3/2}$ and $\sigma_{5/2}$ and the

asymmetries $A_{3/2}$ and $A_{5/2}$. Instead of calculating them with a DWBA code we have used the experimental results for the reactions $^{118}Sn(p,d)^{117}Sn^*$ (see contributed paper) in which transitions with $\ell = 2$, $j = 3/2$, and $j = 5/2$ were involved (levels at 0.158 MeV and 1.03 MeV in ^{117}Sn). The shapes of the cross section angular distributions for $^{118}Sn(p,d)^{117}Sn^*$ ($E_X = 0.158$ MeV, $J = 3/2^+$) and for $^{118}Sn(p,d)^{117}Sn^*$ ($E_X = 1.03$ MeV, $J = 5/2^+$) are very similar, so that it is reasonable to state $\sigma_{3/2} \approx \sigma_{5/2}$. Then the ratio of the spectroscopic factor is

$$S_{3/2}/S_{5/2} = (A - A_{5/2})/(A_{3/2} - A).$$

The average of $S_{3/2}/S_{5/2}$ over the whole angular range is found to be 1.05 ± 0.18. The contributions of the two orbits $d_{3/2}$ and $d_{5/2}$ are thus nearly equal. This ratio of spectroscopic factors was determined only from experimental values of asymmetries, and depends neither on optical potentials nor on form factors.

The reaction $^{57}Fe(p,d)^{56}Fe^*$ leading to the first excited state in ^{56}Fe was studied at 17.3 MeV. The target and final state spins are $1/2^-$ and 2^+, respectively, so the allowed values of ℓ are 1 and 3. The allowed values of j are 3/2 and 5/2. The spectroscopic factors were calculated with a least χ^2 method, using a DWBA code. In this case the contributions from the two different j-values arise from two different ℓ-values, so that it is possible to determine the mixing from only cross-section measurements. Fig. 1 shows the contributions to the cross section and to the asymmetry, of the $p_{3/2}$ and $f_{5/2}$ orbits. In a similar way, the reaction $^{61}Ni(p,d)^{60}Ni^*$ leading to the first two excited states in ^{60}Ni was studied at 16.6 MeV. The target spin is $3/2^-$, and both final spins are 2^+, so that the orbits involved are: $p_{1/2}$, $p_{3/2}$, $f_{5/2}$, $f_{7/2}$. Figs. 2 and 3 show the fits to cross sections and asymmetries with the determined spectroscopic factors. The contribution of the $p_{1/2}$ and $f_{7/2}$ orbits are found to be negligible, in agreement with the shell model level scheme.

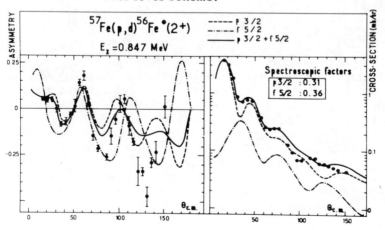

Fig. 1

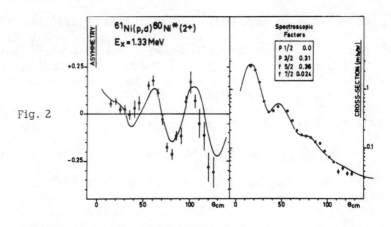

Fig. 2

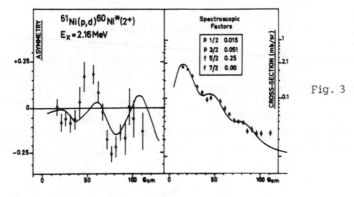

Fig. 3

REFERENCES

[1]   G. R. Satchler, Nucl. Phys. 55 (1964) 1.
[2]   D. C. Kocher and W. Haeberli, Phys. Rev. Lett. 23 (1969) 315.

DISCUSSION

Kocher:

For $^{119}Sn(\vec{p},d)^{118}Sn(2^+)$, you conclude a mixing ratio of $S_{3/2}/S_{5/2} \cong 1$. We have measured $^{117}Sn(\vec{d},p)^{118}Sn$ to the same state and concluded qualitatively that this was pure $3/2^+$, which should provide additional information on the structure of the final state.

In analyzing both $(\vec{p},d)$ and $(\vec{d},p)$ reactions that are a mixture of $\ell = 3$, $j^\pi = 5/2^-$ and $\ell = 1$, $j^\pi = 3/2^-$, $S_{3/2^-}$ and $S_{5/2^-}$ are determined essentially by the cross section only. Since the calibration curves for $3/2^-$ and $5/2^-$ transitions separately are almost identical, no quantitative information on the mixing of the j-values is obtained from a measurement of the analyzing power.

Polarization Produced in the
^{9}Be(d,n) Reactions at 3.0 and 3.5 MeV
and a Comparison to DWBA Calculations

G. SPALEK, R. A. HARDEKOPF, J. TAYLOR, Th. STAMMBACH, and
R. L. WALTER, Duke University and TUNL, Durham, North Carolina,
USA*

A program to study the polarization in (d,n) stripping reactions has
been under way in the 4-MeV laboratory at TUNL. In the 1p shell,
measurements have been reported on ^{6}Li, ^{7}Li, ^{10}B, ^{11}B, ^{12}C, ^{13}C, ^{14}N,
and ^{15}N. In this paper we report the results of a measurement of the
neutron polarization for several groups in the ^{9}Be(d,n)^{10}B reactions.
These reactions are particularly interesting in that four of the lowest
states are produced via $\ell = 1$ stripping with nearly pure 1/2 or 3/2
total angular momentum transfers [1].

The method utilized helium in a high-pressure scintillation cell,
liquid-scintillation side detectors, and a solenoid magnetic field to
precess the neutron spin. An on-line computer recorded the data. The
polarization results at 3.0 and 3.5 MeV are shown in fig. 1 for the neutrons leading to the n_0 ground state and the first four excited states of
^{10}B. The n_2 polarization has been poorly determined because this
group is weak compared to the neighboring groups and unfolding the
peaks introduced considerable statistical uncertainties. Only a small
energy dependence from 3.0 to 3.5 MeV is exhibited.

The ^{9}Be(d,d) scattering cross sections were measured at 3, 4, and
5 MeV to provide optical model parameters for DWBA calculations.
Neutron parameters which were based on those of Watson et al. [2]
and which fit ^{10}B(n,n) cross-section data [3] were employed. Many
sets of "equivalent" deuteron parameters were tried and the sensitivity
of the results to other standard medium-weight-nuclei parameters for
neutrons was also tested. The DWBA results for 3.5 MeV are shown
in fig. 2. For these calculations, a deuteron well with $r_0 = 1.02$ fm,
$V = 120$ MeV, $r_0' = 1.5$ fm, $W' = 8$ MeV and $V_{so} = 9$ MeV was used. Comparison is made to cross-section data of Morrison et al. [4] at 4 MeV
and Ferguson et al. [5] at 3 MeV.

The Q-value dependence and the j-dependence of the DWBA polarization are apparent in fig. 2. The n_1 group is an admixture of $p_{3/2}$ and
$p_{1/2}$ but the n_0, n_2, and n_3 are nearly pure 3/2 transfers and the n_4 is a
pure 1/2 transfer [1], assuming stripping dominates the reactions. Except for the nearly isotropic n_4 group, the comparison between the

3.5 MeV data and the calculations is encouraging, particularly for angles less than 60°. Further adjustment of the deuteron parameters will be carried out in order to study more fully the significance of these results.

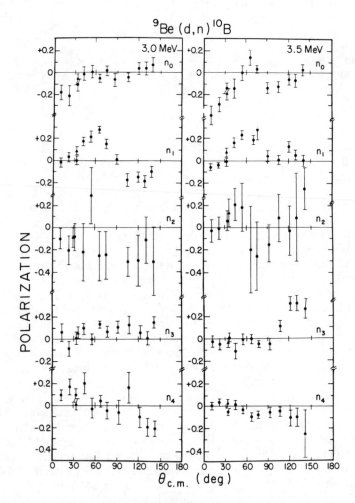

Fig. 1. The polarization of the first five neutron groups from ^{9}Be(d,n) reactions at 3.0 and 3.5 MeV.

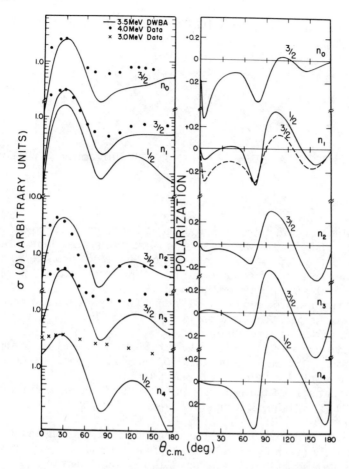

Fig. 2. The DWBA cross section and polarization for the first five neutron groups from ^{9}Be(d,n) reactions at 3.5 MeV. The data are from refs. [4] and [5].

REFERENCES

*Work supported by the U.S. Atomic Energy Commission.

[1] S. Cohen and D. Kurath, Nucl. Phys. 101 (1967) 1.
[2] B. A. Watson, P. P. Singh, and R. E. Segal, Phys. Rev. 182 (1969) 975.
[3] J. C. Hopkins and D. M. Drake, Nucl. Sci. and Eng. 36 (1969) 275.
[4] G. C. Morrison, A. T. G. Evans, and J. E. Evans, Proc. Rutherford Int. Conf. (Manchester, 1961) p. 575.
[5] A. T. G. Ferguson, N. Gale, G. C. Morrison, and R. E. White, Proc. Conf. on Direct Interactions and Nuclear Reaction Mechanisms (Padua, 1962) p. 510.

Vector Analyzing Power of
$\ell_p = 1$ (d,n) Reactions on B, N, and Y

D. HILSCHER,[*] P. A. QUIN, and J. C. DAVIS, University of Wisconsin, Madison, USA[†]

The j-values of transferred neutrons have been shown [1] to determine the vector analyzing power (VAP) of (d,p) stripping reactions. Similar measurements of the VAP of (d,n) reactions have not been performed previously. We have measured the (d,n_0) VAP for reactions on ^{11}B, ^{14}N, and ^{89}Y and the (d,n_1) VAP for ^{11}B.

A polarized source provided a vector polarized deuteron beam with 25-65 nA on target. The neutrons were detected in a symmetric arrangement with angular acceptance of ±1.5°.

We have measured the VAP at 10 and 11.8 MeV for B and N and at 11 MeV for Y. For angles less than 40° there is a considerable energy dependence only for ^{11}B(d,n_0). DWBA calculations had indicated an increase with angle of the (d,p) and (d,n) VAP for $\ell = 1$ reactions for angles < 15°. This prediction agrees qualitatively with the B and N data.

The DWBA calculations at 11.8 MeV shown for the light nuclei use the potentials given by Fitz [2] and Watson [3] and are corrected for non-locality and finite range. For ^{89}Y(d,n_0), which is a unique $j = 1/2$ transition, the potentials given by Bjorkholm [4] and Becchetti [5] were used.

DWBA yields relative spectroscopic factors for different j-values only if the calculations reproduce the results of known j-transfer reactions. As the DWBA calculation does not give agreement with the unique $1p_{3/2}$ transfer in the ^{11}B(d,n_0) reaction, we cannot deduce relative spectroscopic factors for different j-transfers [6] in ^{11}B(d,n_1) and ^{14}N(d,n_0).

For $j = 1/2$ transfer reactions on ^{89}Y calculation and experiment give opposite sign for (d,n) and (d,p) reactions for the VAP at small angles, while for the light nuclei the sign is the same.

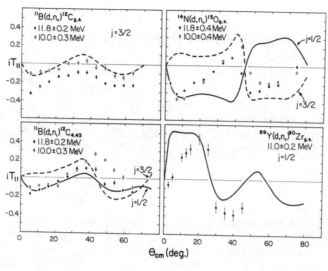

Fig. 1

REFERENCES

*NATO Fellow on leave from Hahn-Meitner Institut, Berlin, Germany.
†Work supported in part by U.S. Atomic Energy Commission.

[1] T. J. Yule and W. Haeberli, Nucl. Phys. A117 (1968) 1.
[2] W. Fitz, R. Jahr, and R. Santo, Nucl. Phys. A101 (1967) 449.
[3] B. A. Watson, P. P. Lingh, and R. E. Segel, Phys. Rev. 182 (1969) 997.
[4] P. J. Bjorkholm and W. Haeberli, to be published.
[5] F. D. Becchetti, Jr., and G. W. Greenlees, Phys. Rev. 182 (1969) 1190.
[6] W. Busse et al., to be published.

Cross-Section and Polarization
Measurements for the ^{11}B(d,n$_0$)
and ^{11}B(d,n$_1$) Reactions from 7 to 12 MeV

J. TAYLOR, Th. STAMMBACH, R. A. HARDEKOPF, G. SPALEK, and
R. L. WALTER, Duke University and TUNL, Durham, North Carolina,
USA*

Neutron polarization angular distributions from the ^{11}B(d,n$_0$) and
^{11}B(d,n$_1$) reactions have been measured at 7.6, 9.6, and 11.7 MeV.
These reactions are expected to demonstrate the j-dependence in
stripping to the 1p shell since the n$_0$ reaction is believed to proceed
through a p$_{3/2}$ transfer and the n$_1$ by a p$_{1/2}$ transfer [1]. Meier et al.
[2] have reported that these reactions below 4 MeV do not follow the
patterns observed also by Reber and Saladin [3] in other $\ell = 1$ transfer
reactions. That is, in the region of the stripping peaks the n$_1$ group
does not show a polarization that is appreciably more negative than
that of the p$_{3/2}$ transfer. The purpose of the present experiment was
to investigate these reactions at higher energies where the momentum
mismatch induced by the high Q-values (13.7 and 9.3 MeV) would be
reduced. The experimental apparatus was quite similar to that used
in ref. [2]. Fig. 1 shows the polarization angular distributions for the
two neutron groups. (The lines do not have any theoretical significance.) The present data differ considerably from the data of Meier
et al. below 4 MeV and that of Busse et al. [4] at 5.5 MeV although a
systematic trend with energy is apparent. However, the polarizations
still do not follow the empirical rule of Reber and Saladin.
 An attempt was made to describe the data using the DWBA formalism.
To do this meaningfully, it was necessary to obtain (d,n) reaction
cross-section distributions and elastic scattering angular distributions.
These data are shown in figs. 2 and 3 respectively. An optical-model
analysis was performed on the elastic data including those of Fitz et
al. [5] at 11.8 MeV. (The forward angle data were weighted more
heavily than the back angle data.) Three kinds of parameter sets were
found which adequately represent the elastic data. These are listed in
table 1 for 9.6 MeV. Data renormalizations from 0.95 to 1.1 were required for the present data and from 0.8 to 1.0 for the data of Fitz et al.
Representative members from each of the sets used are shown in fig. 3.
These sets were then used as input to the distorted wave code DWUCK
of Kunz. Neutron parameters are from Watson et al. [6]. Parameters
from type I and type II sets yielded polarizations in fair agreement

with the ground-state group data for angles less than 90° c.m., but none of the sets consistently predicted both the n_0 and n_1 data. The best agreement with the (d,n) angular distributions was obtained with type I parameters but the spectroscopic factors for the n_0 and n_1 groups were both about 1/5 of their predicted value [1]. Type III gave larger spectroscopic factors but the stripping peak was too pronounced and the predicted polarizations were poor.

Table 1. Optical model sets for 9.6 MeV deuteron energy

Type	V(MeV)	r_0(fm)	a_0(fm)	W'(MeV)	r_w(fm)	a_w(fm)	V_{so}(MeV)	r_{so}(fm)	a_{so}(fm)
I	105.0	1.03	0.70	3.50	1.35	1.40	9.0	0.80	0.80
II	105.0	1.20	0.88	20.0	1.70	0.38	9.0	1.40	1.00
III	115.0	1.65	0.60	35.0	1.80	0.25	9.0	1.30	0.60

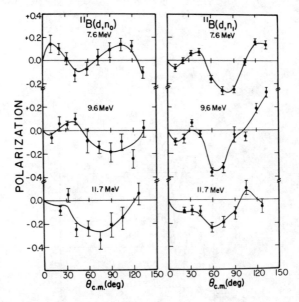

Fig. 1

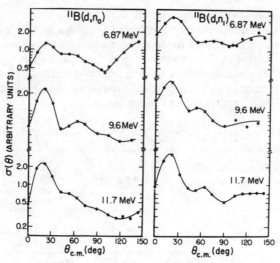

Fig. 2

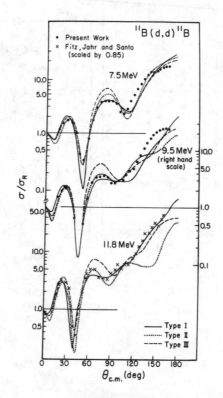

Fig. 3

REFERENCES

*Work supported by the U.S. Atomic Energy Commission.

[1] S. Cohen and D. Kurath, Nucl. Phys. A101 (1967) 1.
[2] M. M. Meier et al., Symp. on Nuclear Reaction Mechanisms and
     Polarization Phenomena (Quebec, 1969), to be published.
[3] L. H. Reber and J. X. Saladin, Phys. Rev. 133 (1964) B1155.
[4] W. Busse et al., to be published.
[5] W. Fitz et al., Nucl. Phys. A101 (1967) 449.
[6] B. A. Watson et al., Phys. Rev. 182 (1969) 977.

Deuteron D-State Effect and J-Dependence
in Vector Polarization for (d,p) Reactions

F. D. SANTOS, Laboratório de Física e Engenharia Nucleares,
Sacavém, Portugal

In the distorted wave theory of (d, p) reactions with spin independent distortion and assuming a pure S-state deuteron internal wave function, the polarization of the outgoing protons $P(\theta)$ for two transitions with the same ℓ, different j and close excitation energies are related by the well known sign rule [1]

$$P(\theta)_{\ell+\frac{1}{2}} \big/ P(\theta)_{\ell-\frac{1}{2}} = -\ell/(\ell+1). \tag{1}$$

The same relation also applies to the deuteron vector analyzing power $P_d(\theta)$, since in this form of the DWBA theory the two quantities are proportional. Measurements of $P_d(\theta)$ in (d,p) reactions for a wide range of target nucleus mass numbers agree at forward angles with this prediction which can be useful to determine the total angular momentum transfer in the reaction [2, 3]. However, the theory is less successful in reproducing proton polarization data [4, 5] and disagreement between experiment and equation (1) is frequent, as for instance in two $\ell = 2$ transitions for the ^{28}Si(d,p)^{29}Si reaction at $E_d = 10.8$ MeV [6]. In the DWBA theory this disagreement can be attributed to a large sensitivity of proton polarization on spin dependent distortion. Reber and Saladin [7], analyzing data for $\ell = 1$ and 2 transitions, noted that for $j = \ell + 1/2$ the proton polarization in the region of the stripping peak is small, while for $j = \ell - 1/2$ it is large and negative.

When the D-state component of the deuteron wave function is included in the DWBA transition matrix, without spin–orbit coupling in the optical potentials, $P(\theta)$ and $P_d(\theta)$ are given by

$$\sigma(\theta)\,P(\theta)_j = \left[\frac{S(\theta)}{2} + 2C(\theta)\right] \frac{(-1)^{\ell+j-\frac{1}{2}}}{2j+1} - I(\theta)_j, \tag{2}$$

$$\sigma(\theta)\,P_d(\theta)_j = [S(\theta) + C(\theta)] \frac{(-1)^{\ell+j-\frac{1}{2}}}{2j+1} + I(\theta)_j, \tag{3}$$

where $S(\theta)$ and $I(\theta)_j$ are respectively the incoherent S- and D-state contributions, $C(\theta)$ the coherent contribution and $\sigma(\theta)$ the proton differential cross section apart from numerical and kinematic factors. No other approximations are assumed in the derivation of equations (2) and (3). The structure of $S(\theta)$ and $I(\theta)_j$ is quite different as regards the contributions from the orbital angular momentum ℓ' transferred from the relative motion in the deuteron channel to the relative motion in the proton channel. In the D-state part of the transition amplitude ℓ' can take the values [8] $\ell - 2$, $\ell - 1$, ℓ and $\ell + 1$ if $j = \ell - 1/2$ and $\ell - 1$, ℓ, $\ell + 1$ and $\ell + 2$ if $j = \ell + 1/2$. The absence of $\ell' = \ell + 2$ ($\ell' = \ell - 2$) contributions for $j = \ell - 1/2$ ($j = \ell + 1/2$) can introduce a j-dependence into D-state effects. All allowed values of ℓ' contribute to $I(\theta)_j$ while in $C(\theta)$ they are restricted by the triangular relation $\Delta(\ell, 1, \ell')$. When the contribution from $\ell' = \ell + 2$ or $\ell' = \ell - 2$ is dominant $I(\theta)_j$ can be larger than $C(\theta)$. Notice that only the incoherent D-state contribution violates equation (1) and $I(\theta)_j$ depends explicitly on the quantum number j. In particular for $\ell = 0$ transitions $C(\theta)$ is zero [9].

The numerical coefficients that multiply $S(\theta)$ and $C(\theta)$ inside the square bracket in equation (2) imply that the effect of the D-state is larger in the proton polarization than in the deuteron vector analyzing power. It is of interest to know to what extent this prediction remains valid, especially at forward angles, with spin-orbit coupling in the optical potentials. The S- and D-state components of the deuteron wave function were included in the present DWBA calculations using the approximations and parameters of ref. [1].

The curves shown in fig. 1 for two $\ell = 1$ transitions in the ^{40}Ca(d,p)^{41}Ca reaction at $E_d = 7$ MeV were calculated using the same optical potentials as in ref. [2] which include spin-orbit terms in both channels. The effect of the D-state at forward angles is smaller in $P_d(\theta)$ than in $P(\theta)$ in agreement with equations (2) and (3). Calculations for the same reactions with the spin-orbit terms set equal to zero show that the D-state effect

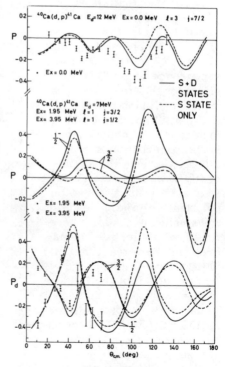

Fig. 1

at forward angles is not very sensitive to spin-orbit coupling. In
these calculations and in those of fig. 1 we find that the D-state effect at angles < 60° decreases $P_d(\theta)$ for both $1/2^-$ and $3/2^-$ transitions which, according to equation (3), indicates that the incoherent
D-state contribution is more important than the coherent contribution.
The agreement with equation (1) and with experiment for $P_d(\theta)$ is not
affected by the inclusion of the D-state. For the proton polarization,
the DWBA curves without the D-state do not agree with equation (1)
at forward angles because of spin dependent distortion.

Measurements of the proton analyzing power for two $\ell = 1$ transitions in the $^{16}O(p,d)^{15}O$ reaction [11] show a pronounced disagreement
with equation (1) and also with the Reber and Saladin rule [6]. This
disagreement may be related to the large reaction Q-value. In DWBA
calculations for these transitions using the adjusted deuteron optical
potential of ref. [11] D-state effects are large, particularly for the $3/2^-$
transition. Generally we find that D-state effects increase with the
reaction Q-value [8]. DWBA calculations for the proton analyzing
power for two $\ell = 3$ transitions
in the $^{55}Fe(p,d)^{56}Fe$ reaction
(fig. 2) at $E_p = 18.5$ MeV using
the optical potentials P1 and
D1 of ref. [12] with spin-orbit
terms show that the D-state
effect is considerably larger
in the $7/2^-$ transition. The j-
dependence of the D-state
effect for this reaction in the
proton analyzing power and
differential cross section [10]
results from a predominance
of the contribution from $\ell' =$
5 over other ℓ' values. Recently it was pointed out that
DWBA predictions for proton
polarization in (d,p) reactions near the Coulomb barrier show a strong j-dependence [13]. For reactions in
the Coulomb field we find
that the effect of the D-state
is small. This can be seen
in the curves shown in fig. 2
for the $^{52}Cr(d,p)^{53}Cr$ reaction
calculated for $E_d = 3$ MeV
assuming a level near $E_x =$
4.2 MeV to be populated by
a $5/2^-$ or a $7/2^-$ transition

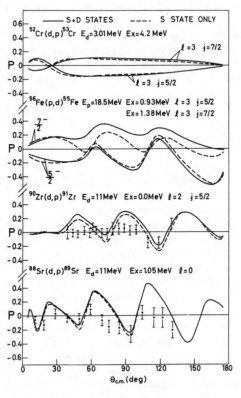

Fig. 2

in order to emphasize j-dependent effects. The optical potentials were taken from refs. [14, 15], and do not include spin-orbit terms.

For $\ell = 0$ transitions we find that D-state effects are in general small. The curves plotted in fig. 2 for the proton polarization in the ^{88}Sr(d,p)^{89}Sr reaction [16], calculated with the optical potentials of ref. [17] show that the inclusion of the D-state does not improve the fit with experiment. In the ^{90}Zr(d,p)^{91}Zr ground state reaction (fig. 2) the D-state effect is larger, but the agreement with the data is not substantially improved using the optical potentials c) and d) of ref. 5. In the calculation for the ^{40}Ca(d,p)^{41}Ca ground state reaction at $E_d = 12$ MeV shown in fig. 1 we used the standard optical potential of ref. [18] and the deuteron optical potential of ref. [19] ($E_d = 11.8$ MeV, set-b), both with spin-orbit terms. The measurements at $E_d = 10.9$ MeV are from ref. [20]. The inclusion of the D-state slightly improves the agreement with experiment at large angles.

In general we find that at forward angles $P(\theta)$ is more sensitive than $P_d(\theta)$ to spin-orbit distortion and the effect of the D-state; this is particularly evident for not too light nuclei and not too high reaction Q-value. Recent calculations using the weakly bound projectile model [21] also show that the corrections due to the inclusion of the D-state are usually larger for $P(\theta)$ than $P_d(\theta)$. The D-state effect on $P(\theta)$ and $P_d(\theta)$ is noticeably more sensitive to spin-orbit distortion at angles $> 90°$ than at forward angles. This may be due to the coherent D-state contributions introduced by spin-orbit forces. The present results indicate that the deuteron D-state should be included in DWBA calculations, particularly for proton polarization and transitions with $\ell > 0$, but the agreement with experiment remains strongly dependent on the choice of spin-orbit forces in the optical potentials.

REFERENCES

[1]   R. Huby et al., Nucl. Phys. 9 (1958) 94.
[2]   T. J. Yule and W. Haeberli, Nucl. Phys. A117 (1968) 1.
[3]   A. M. Baxter et al., Phys. Rev. Lett. 20 (1968) 1114.
[4]   S. A. Hjorth et al., Phys. Rev. 138 (1965) B1425.
[5]   L. S. Michelman et al., Phys. Rev. 180 (1969) 1114.
[6]   W. E. Maddox et al., Phys. Rev. 1C (1970) 476.
[7]   L. H. Reber and J. X. Saladin, Phys. Rev. 133 (1964) B1155.
[8]   F. D. Santos and R. C. Johnson, to be published.
[9]   R. C. Johnson, Nucl. Phys. A90 (1967) 289.
[10] R. C. Johnson and F. D. Santos, Phys. Rev. Lett. 19 (1967) 364.
[11] N. S. Chant et al., Nucl. Phys. A99 (1967) 669.
[12] C. Glashausser and M. E. Rickey, Phys. Rev. 154 (1967) 1033.
[13] J. A. R. Griffith and S. Roman, Phys. Rev. Lett. 24 (1970) 1496.
[14] M. Posner, Phys. Rev. 158 (1967) 1018.
[15] W. R. Smith, Phys. Rev. 137 (1965) B913.

[16] E. J. Ludwig and D. W. Miller, Phys. Rev. 138 (1965) B364.

[17] G. R. Satchler, Oak Ridge Nat. Lab. Report, 1966 (unpublished).

[18] F. D. Becchetti and G. W. Greenless, Phys. Rev. 182 (1969) 1190.

[19] C. M. Perey and F. G. Perey, Phys. Rev. 152 (1966) 923.

[20] S. Kato et al., Nucl. Phys. 64 (1965) 241.

[21] C. A. Pearson et al., Nucl. Phys. A148 (1970) 273.

DISCUSSION

Griffith:

We feel that on the basis of the most naive model of deuteron stripping the polarization transfer should be very sensitive to D-state effects in the reaction. Could detailed calculations be done on this, since measurements of this type are now possible? We report at this conference the transfer of polarization from the incoming polarized deuteron to the outgoing proton in $^{28}Si(d,p)^{29}Si$ (g.s.) at $0°$. The result is 1.0 ± 0.1.

Santos:

I have not done DWBA calculations in order to see the D-state effect on polarization transfer, but they are quite easy to perform with the code used for the present calculations. Although on simple grounds one might expect a noticeable D-state effect, it should also be remarked that the effect increases with the momentum transfer in the reaction and therefore may not be large at $0°$.

Griffith:

Dr. Johnson has suggested that our recent observation of large analyzing power T_{20} at $0°$ in (d,p) reactions could be a signature of D-state effects. Would you agree with this?

Santos:

D-state effects are generally large for T_{2k}, and in particular for T_{20} and T_{21}, as reported in another contribution to this Symposium.

DWBA Calculations for (d,p) Reactions
Including the D-State of the Deuteron

G. DELIĆ and B. A. ROBSON, Australian National University, Canberra,
Australia

Finite-range (FR) DWBA calculations which include exactly both the
deuteron D-state and a T_R tensor interaction in the deuteron-nucleus
potential [1] have been carried out following the treatment of Austern
et al. [2] for the reaction ^{52}Cr(d,p)^{53}Cr leading to the $3/2^-$ ground and
$1/2^-$ 0.57 MeV residual states at 8 MeV deuteron bombarding energy.
The soft-core neutron-proton potential and the corresponding deuteron
wave function of Reid [3] were employed in the calculations. The optical model potential and parameters were taken from earlier analyses
[1, 4] except that a T_R tensor potential M = -4 MeV was used in the
present work.

Fig. 1 shows the results for the $1/2^-$ 0.57 MeV level (similar effects are observed for the $3/2^-$ ground state) referred to a coordinates

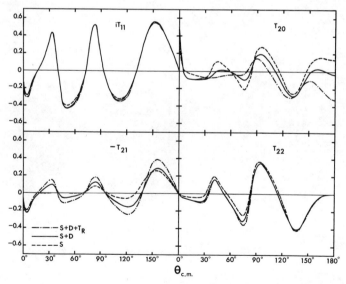

Fig. 1

system with z-axis anti-parallel to the incident deuteron momentum $\vec{k}_d$ and y-axis along $\vec{k}_p \times \vec{k}_d$, where $\vec{k}_p$ is the outgoing proton momentum. The shape (and hence the j-dependent effect) of the vector analyzing power $\langle iT_{11} \rangle$ is insignificantly altered by the inclusion of the D-state in a FR DWBA calculation (cf. curves S and S+D). However, the corresponding tensor analyzing powers, especially $\langle T_{20} \rangle$ and $\langle T_{21} \rangle$ (which is vanishingly small for angles < 90°), are considerably changed. The zero-range predictions lie close to those for the FR S-state calculations (curves S) and for simplicity have been omitted from the diagram.

The inclusion in the D-state FR calculations of a small T_R tensor term (curves S+D+T_R) in the deuteron optical potential caused significant changes in the tensor analyzing powers $\langle T_{20} \rangle$ and $\langle T_{21} \rangle$, but left the corresponding vector analyzing power and tensor analyzing power $\langle T_{22} \rangle$ largely unaffected.

REFERENCES

[1] G. Delic and B. A. Robson, Nucl. Phys. A127 (1969) 234.
[2] N. Austern, R. M. Drisko, E. C. Halbert, and G. R. Satchler, Phys. Rev. 133 (1964) B3.
[3] R. V. Reid, Ann. of Phys. 50 (1968) 411.
[4] G. Delic and B. A. Robson, Nucl. Phys. A134 (1969) 470.

J-Dependence of Vector-Analyzing
Power of (d,p) Reactions

L. D. TOLSMA and B. J. VERHAAR, Eindhoven University of Technology,
The Netherlands

As demonstrated by Yule and Haeberli [1], the experimental vector-analyzing power $P_d(\theta)$ of a (d,p) reaction at forward angles (up to $\sim 60°$) depends strongly on the j_n-value of the transferred neutron and may thus serve to determine j_n, once $\ell_n (\neq 0)$ is known from $\sigma_{unpol}(\theta)$. In a recent paper by Cuno, Clausnitzer and Fleischmann [2] this characteristic dependence on j_n, apparently roughly independent of Q-value and target nucleus, has been summarized in terms of some empirical rules, pertaining to the signs of the quantities C and D introduced below.

In this contribution we present the results of a theoretical investigation into the DWBA prediction for P_d, as it depends on mass number and deuteron energy. Our purpose was to investigate whether the DWBA predictions alone show an independence of mass number and deuteron energy to such an extent that it is possible to formulate sign-rules. In view of the insensitivity of P_d to deuteron and proton spin-orbit interactions at forward angles, suggested by the approximate validity of eq. (9) of ref. [1] and by the DWBA curves in fig. 13 of the same paper, we neglected these interactions. This was of considerable advantage in view of the extensive DWBA investigation intended. We investigated three ℓ_n-values (1, 2, 3), five deuteron lab energies (5, 7.5, 10, 15, 22.5 MeV) and three target nuclei (^{20}Ne, ^{40}Ca, ^{58}Ni). The Q-values chosen were typical for each ℓ_n-value and target nucleus considered. The deuteron and proton optical parameters have been taken from ref. [3], in particular eqs. (9) and (11), and from ref. [4] ($W_D = 3A^{1/3}$ MeV), respectively. In addition, the (^{3}He,d) reaction was studied for the same nuclei at three lab energies (15, 30, 45 MeV).

We selected four features (A to D, see table) to characterize the behavior of $P_d(\theta)$ at forward angles. The signs of these quantities are given in the table for the three lower deuteron energies, since these have been most extensively studied in the above-mentioned papers. Each consecutive combination of three signs corresponds to deuteron energies 5, 7.5 and 10 MeV, in the same order. In all cases $j_n = \ell_n + \frac{1}{2}$; corresponding signs for $j_n = \ell_n - \frac{1}{2}$ follow from eq. (9) of ref. [1].

Reaction	ℓ_n	(A) $dP_d/d\theta\,(0°)$	(B) $P_d(\theta_m)$	(C) $dP_d/d\theta\,(\theta_m)$	(D) $P_d(\theta_m < \theta < 2\theta_m;\ \theta < 90°)$
^{20}Ne(d,p)	1	+ + +	+ + +	+ + +	+ + +
	2	− − −	+ + +	+ + +	+ + +
	3	− + +	+ + +	+ + +	+ + +
^{40}Ca(d,p)	1	+ + +	− + +	− − −	− ? +
	2	− − +	+ + +	− − +	? ? ?
	3	+ − −	+ + +	− ?a) +	? ? ?
^{58}Ni(d,p)	1	+ · + +	+ + +	+ − +	+ ? +
	2	− − −	+ − −	− + +	+ ? −
	3	− − −	?b) − −	?b) + +	?b) ? ?

θ_m = angle of stripping maximum; a) $dP_d/d\theta\,(\theta_m) \approx 0$; b) $\theta_m > 90°$

Conclusions. 1) Criterion B seems most suitable for determining j_n. This is the old Newns' sign-rule which can be explained in terms of a simple classical picture. This rule was not considered by Cuno et al. [2]. 2) Criteria C and D seem to show more irregular results. Question marks in the last column indicate oscillations of P_d through zero in the angular range considered. 3) For $\ell_n = 1$ and 2 criterion A shows an almost regular behavior. In most cases the sign of P_d persists up to 10° at least, so that this criterion may have additional applicability. This sign-rule is in accordance with a simple model [5], incorporating not only the predominance of deuteron absorption relative to proton absorption, but also interference effects not included in the Newns' model. 4) The irregularities tend to increase with mass number, deuteron energy and ℓ_n-value. At even higher deuteron energies no regularity appears to remain with respect to any of the characteristics A to D. For (^{3}He,d) reactions criteria A and B also lend themselves best for the formulation of sign-rules. In this case, however, the respective signs are opposite indicating predominant t-absorption.

REFERENCES

[1] T. J. Yule and W. Haeberli, Nucl. Phys. A117 (1968) 1.
[2] H. H. Cuno, G. Clausnitzer, and R. Fleischmann, Nucl. Phys. A139 (1969) 657.
[3] P. Schwandt and W. Haeberli, Nucl. Phys. A123 (1969) 401.
[4] F. G. Perey, Phys. Rev. 131 (1963) 745.
[5] B. J. Verhaar, Phys. Rev. Lett. 22 (1969) 609.

Recent Polarization Measurements Performed
in Cracow on (d,p) Stripping Reactions

A. BUDZANOWSKI, L. FREINDL, W. KARCZ, B. LAZARSKA, W. ZIPPER,
and K. GROTOWSKI, Cracow Institute of Nuclear Physics, Poland

I would like to present some results of polarization measurements performed at the Cracow Institute of Nuclear Physics.

We can measure the polarization of protons emitted in a nuclear reaction using the apparatus shown in fig. 1. The investigated reaction
takes place in the target located in the center of the reaction chamber
and bombarded by the beam of particles from the Cracow U-120 cyclotron. The polarization of protons from the investigated reaction is
determined by a helium analyzer. The left-right asymmetry of protons
scattered on helium is measured by two semiconductor counter telescopes. In order to decrease the background of neutrons and γ-rays
generated in the reaction chamber, the helium polarization analyzer

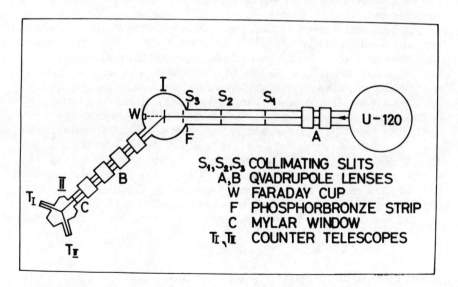

S_1, S_2, S_3 COLLIMATING SLITS
A,B QVADRUPOLE LENSES
W FARADAY CUP
F PHOSPHORBRONZE STRIP
C MYLAR WINDOW
T_I, T_{II} COUNTER TELESCOPES

Fig. 1. Experimental arrangement.

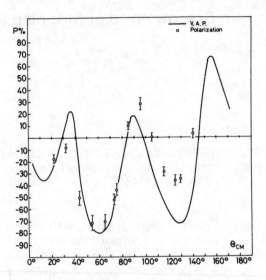

Fig. 2. The experimental results for the
polarization of protons from $^{12}C(d,p)^{13}C$
together with the deuteron vector analyzing
power data of Baxter et al. [1].

is located far from the target at a distance of about 220 cm. A system
of four magnetic quadrupole lenses is placed between the reaction
chamber and the helium analyzer.

The polarization of the outgoing protons from the $^{12}C(d,p)^{13}C$ reaction has been measured using this apparatus. The energy of incident
deuterons was 12.35 MeV, maximum beam intensity available on the
carbon target was 10 μA. The angular dependence of the polarization
was measured in the angular region up to 140° for the ground state,
$\ell = 1$ transition, and up to 95° for the 3.09-MeV state, $\ell = 0$ transition.

The obtained results can be compared with the deuteron vector
analyzing power (VAP) measured by Roman et al. [1] at a deuteron energy very close to ours. (E_d = 12.25 MeV for the $\ell = 1$ transition, and
E_d = 12.0 MeV for the $\ell = 0$ transition.)

For the $\ell = 1$ ground state transition (fig. 2) the polarization follows
the VAP very closely in shape and in the region of angles around 60°
also in amplitude. It was shown by Satchler [2] and Al Jeboori et al. [3]
that, in case of a $\ell \neq 0$ transition, if the distortions are spin independent,

$$P^d(\theta) = 2 P(\theta),$$

where $P^d(\theta)$ is the VAP and $P(\theta)$ denotes the polarization. This relation
is not fulfilled around 60°. This may indicate a contribution of the
spin-dependent interaction.

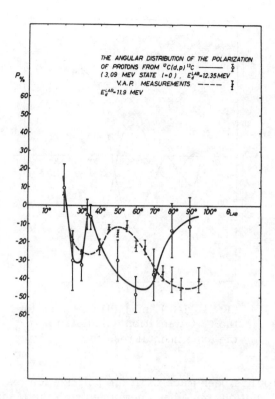

Fig. 3. The experimental results for the
polarization of protons from ^{12}C(d,p)^{13}C
together with vector analyzing power of
Baxter et al. [1].

In the case of $\ell = 0$ stripping (fig. 3) it is possible from the polarization and VAP data to unravel the separate contributions to both these quantities from the spin-orbit parts of the deuteron and proton potentials. According to Goldfarb, Hooper, and Johnson [4,5,6],

$$P(\theta) = J_d(\theta) + J_p(\theta)$$

$$P^d(\theta) = J_d(\theta) + \frac{2}{3} J_p(\theta)$$

where $J_d(\theta)$ and $J_p(\theta)$ are proportional to the deuteron and proton spin-orbit potentials, respectively. Fig. 4 shows $J_d(\theta)$ and $J_p(\theta)$ calculated from the experimental data. The spin-dependent interaction appears to be of equal importance in the deuteron and proton channels.

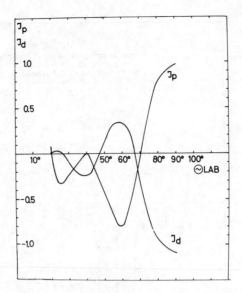

Fig. 4. $J_p(\theta)$ and $J_d(\theta)$ functions for
the $\ell = 0$ transition calculated from
the experimental results.

An attempt has been made to fit the polarization results with the
DWBA. Calculations have been performed with the Macefield spin-
dependent (d,p) program. The optical model potential for the deuteron
channel was obtained from the analysis of the deuteron elastic scattering cross section [7] and polarization [8]. The proton optical model
parameters were taken from Greenlees and Becchetti [9]. The calculated (d,p) cross sections were then compared with the experimental
data of Hamburger [10] taken at $E_d = 12$ MeV. The fits to the differential cross sections are quite good, whereas the DWBA predictions
for the polarization both in $\ell = 1$ and $\ell = 0$ stripping are far from satisfactory. This may mean that the polarization of the outgoing proton
is more sensitive to the details of the interactions in the incoming
and outgoing channels than the cross section.

REFERENCES

[1]   A. M. Baxter, J. A. R. Griffith, S. W. Oh, and S. Roman, Nucl.
      Phys. A112 (1968) 209.
[2]   G. R. Satchler, Nucl. Phys. 6 (1958) 543.
[3]   A. Al Jeboori, M. S. Bokhari, A. Strzałkowski, and B. Hird, Proc.
      Phys. Soc. 75 (1960) 875.

[4] L. J. B. Goldfarb and R. C. Johnson, Nucl. Phys. 18 (1960) 353.

[5] R. C. Johnson, Nucl. Phys. 35 (1962) 644.

[6] M. B. Hooper, Nucl. Phys. 76 (1966) 449.

[7] U. Schmidt-Rohr et al., Max Planck Institute Report V13 (1965).

[8] S. Roman, Proc. Symp. on Nuclear Reaction Mechanisms and Polarization Phenomena (Quebec, 1969), to be published.

[9] G. W. Greenlees and F. D. Becchetti, Minnesota University Annual Report (1967).

[10] W. E. Hamburger, Phys. Rev. 123 (1961) 619.

Differential Cross Section and Polarization in the $^{12}C(d,p)^{13}C$ g.s. Reaction at 51 MeV

W. FETSCHER, K. SATTLER, E. SEIBT, R. STAUDT, and Ch. WEDDIGEN,
Institut für Experimentelle Kernphysik der Universität und des
Kernforschungszentrums Karlsruhe, Germany

At the energies investigated up to now calculations with DWBA have poorly reproduced the cross sections and polarizations in deuteron-induced stripping reactions. We have measured cross sections and polarizations at a higher energy (51 MeV) in the $^{12}C(d,p)^{13}C$ g.s. reaction with unpolarized incident deuterons.

The experiments, which were performed on the Karlsruhe isochronous cyclotron, were unusual in that the same set-up [1] was used to measure the differential cross section and the proton polarization. The apparatus consisted of a first scattering chamber and a movable arm bearing a quadrupole triplet and a second rotating scattering chamber. For cross-section measurements the movable arm was positioned at 0° and the first target was replaced by a circular aperture 1 mm diam. By this technique it was possible to measure down to 3°. In the polarization experiments carbon was used as a polarization analyzer. The analyzing power for protons was taken from ref. [2]. Azimuthal asymmetries up to 0.18 were observed. Spurious asymmetries were measured at angles at which the analyzing power is zero and were found to be of the order of ±0.01.

The differential cross section (fig. 1) was obtained with a relative and absolute error of ±10 and ±20%. We found a pronounced stripping peak of 9 mb/sr at 7° C.M. The first relative minimum is at about 15°. The small value of the integral reaction cross section (1.0 ± 0.2 mb) may be due to the momentum mismatch under our conditions $((kR)_d = 4.6, (kR)_p = 3.5, j_n = 1 - 1/2)$.

The fluctuations in the proton polarization (fig. 2) are correlated with the structure in the differential cross section. Polarizations [3, 4] at 9 and 21 MeV show structure similar to that at 51 MeV. At all energies the polarization tends to be negative for small angles in agreement with Newns' rule.

The DWBA analysis of our data was made with the code "Julie" [5] with the optical model parameter set 3 in ref. [6] for the deuteron channel and parameters from ref. [2] for the proton channel. A fit with a

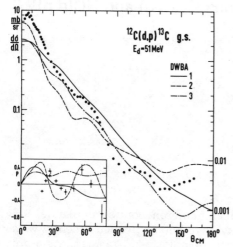

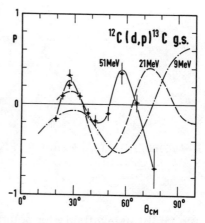

Fig. 1. DWBA fits to the differential cross section and proton polarization with (1) and without (3) spin dependent terms. Fit 2 is like 1 but with a lower cutoff radius of 4 fm.

Fig. 2. Proton polarization in the reaction $^{12}C(d,p)^{13}C$ g.s. at 51 MeV (crosses: this experiment). The curves are experimental results; 9 and 21 MeV are from refs. [3, 4].

neutron wave function as form factor (curves 1 in fig. 1) gives too little structure in the differential cross section. Introduction of a lower cutoff radius of 4 fm (curves 2) creates relative minima in agreement with experiment. Omission of spin-orbit terms (curves 3) in fit 1 results in a more pronounced structure, but the minima are shifted to larger angles. Polarizations could not be reproduced. This kind of DWBA calculation may be inappropriate for describing deuteron-induced stripping reactions at energies near 50 MeV. Considering the good fits obtained by Pearson et al. [7], it seems more promising to apply the WBP model to this energy region.

REFERENCES

[1]   E. Seibt and Ch. Weddigen, to be published in Nucl. Instr.
[2]   R. M. Craig et al., Nucl. Phys. 83 (1966) 493.
[2]   M. S. Bockhari, J. A. Cookson, H. Bird, and B. Wessacul, Proc. Phys. Soc. (London) 72 (1958) 88.
[4]   E. T. Boschitz and J. S. Vincent, Technical Report NASA TR-R-218 (1964).
[5]   R. H. Bassel, R. M. Drisko, and G. R. Satchler, ORNL-3240.
[6]   E. Seibt, K. Weigele, and Ch. Weddigen, Phys. Lett. 27B (1968) 567 and Technical Report KFK 1006.
[7]   C. A. Pearson, D. Rickel, and D. Zissermann, Nucl. Phys. A148 (1970) 273.

Tensor Analyzing Powers of the Reactions ^{12}C(d,d) and ^{12}C(d,p)^{13}C

A. A. DEBENHAM, J. A. R. GRIFFITH, O. KARBAN, S. ROMAN, and
Y. TAKEUCHI, University of Birmingham, England

The facilities available for polarized deuteron work on the Radial Ridge cyclotron of Birmingham University have been extended recently with the successful acceleration of a tensor-polarized beam [1]. As a preliminary experiment we have studied the tensor analyzing powers of the reactions ^{12}C(d,d), ^{12}C(d,p)^{13}C (ground and 3.09-MeV states), and also H(d,d)H. Some large values of analyzing powers have been observed, the most notable being the stripping reaction at small reaction angles.

The experimental arrangement was mainly as described previously for both the beam [1] and for the scattering chamber and instrumentation [2], except that measurements were made in both vertical and horizontal planes. The 12.3-MeV deuteron beam bombarded a target of polyethylene. Four silicon surface barrier detector telescopes were placed in a symmetrical arrangement about the beam axis allowing measurements to be made simultaneously at two reaction angles. The entire scattering chamber was rotated in order to change the angle of azimuth. The detector instrumentation was as previously described except that with the commissioning of a new pulse height analysis facility it was possible to store data for the (d,d) and (d,p) reactions simultaneously.

Following the scattering chamber the beam entered the normal left-right ^{12}C(d,p)^{13}C vector polarimeter to which was added a detector telescope with energy degrader. This was capable of detecting the protons produced at 0°. It was found that the ^{12}C(d,p) reaction at 0° had a great sensitivity to the t_{20} part of the deuteron beam. It proved a useful monitor of the tensor part without interference from vector effects. The main problem in the present work was determining the absolute beam polarization. The normalization is based on an experiment on ^{4}He(d,d) scattering at 9.0 MeV and a comparison with the work of McIntyre and Haeberli [3]. The overall uncertainty in this normalization is about 20%.

In addition to the measurements of angular distributions as described above, separate measurements were made of the sensitivity to t_{20} at 0°

reaction angle of the reactions $^{12}C(d,p)^{13}C$, $^{9}Be(d,p)^{10}Be$, $^{28}Si(d,p)^{29}Si$, and $^{29}Si(d,p)^{30}Si$. All of these showed a significant analyzing power, that for ^{12}C being notably large as can be seen in fig. 1.

Two examples of the angular distributions obtained are shown in figs. 1 and 2. In fig. 1 the sensitivity of the reaction $^{12}C(d,p)^{13}C$ to the t_{20} component of the beam is shown; a preliminary theoretical prediction is also given.

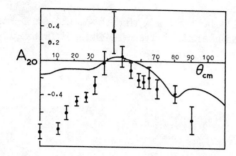

Fig. 1. The analyzing power of the reaction $^{12}C(d,p)^{13}C$ (g.s.) for the t_{20} component. The solid curve is the result of a DWBA calculation with the program DWUCK.

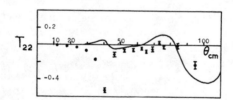

Fig. 2. The component t_{22} produced in the elastic scattering of 12.3-MeV deuterons from ^{12}C. The curve is the result of an optical model prediction.

Neither the result of the distorted waves calculations nor that of the optical model succeeds in giving an adequate description, but work is continuing in this direction.

REFERENCES

[1] Saewoong Oh, Nucl. Instr. 82 (1970) 189.
[2] J. A. R. Griffith et al., Nucl. Phys. A146 (1970) 193.
[3] L. C. McIntyre and W. Haeberli, Nucl. Phys. A91 (1967) 369.

DISCUSSION

Kocher:
  Did you use vector and tensor spin-orbit interactions?

Roman:
  No, vector spin-orbit only.

Johnson:
  Calculations by Santos show that the deuteron D-state can produce large tensor analyzing power when it is included in DWBA
  calculations.

Vector Analyzing Power of the Reactions
^{24}Mg(d,p)^{25}Mg and ^{16}O(d,p)^{17}O

G. HUDSON,* R. C. BROWN, I. GOVIL,† O. KARBAN, S. ROMAN, and
J. A. R. GRIFFITH, University of Birmingham, England

Measurements have been made of the vector analyzing power of (d,p) stripping reactions leading to six separate states of ^{25}Mg and to two states of ^{17}O.

The 12.3-MeV pure-vector polarized deuteron beam of the Birmingham University Radial Ridge cyclotron was used to bombard targets of ^{24}Mg (enriched) and cellulose acetate. Each was placed in a 76 cm scattering chamber, which contained four silicon semiconductor telescopes. Each counter telescope was connected to instrumentation similar to that previously described [1]. Behind the scattering chamber was placed the polarization monitor. This comprised a strip polythene target 12.5 mg cm^{-2} in thickness viewed to the left and right by two silicon detectors set to count the protons from the ^{12}C(d,p)^{13}C (g.s.) reaction. The reaction angle and angle spread were 57°±3°, and under the conditions described this provided a large analyzing power (P_d = -0.75) and a good detection efficiency (about 1 in 10^6) coupled with very little sensitivity to beam alignment. In other respects the technique was as described in ref. [1].

Because of interference from the carbon content of cellulose acetate it was possible to extract the data only for the groups corresponding to (d,p) reactions proceeding to the ground and first excited states of ^{17}O, which are 5+/2 and 1+/2, respectively. In the case of the Mg target, however, it was possible to obtain full angular distributions of vector analyzing power for the ground state and the first five excited states of ^{25}Mg. All of the distributions showed the strong diffraction structure to be expected of direct reactions with the exception of the one relating to the third excited state of ^{25}Mg. This showed little structure and rather low values of analyzing power. Some doubt seems attached to the reaction mechanism prevailing for this state (1.611 MeV). In fig. 1 examples are shown of two cases of stripping reactions proceeding via an orbital angular momentum transfer of zero. The strong similarity between them shows clearly. These cases, one other of the same type and four ℓ_n = 2 transitions are in the process of being analyzed using both the DWBA and the method of Butler [2].

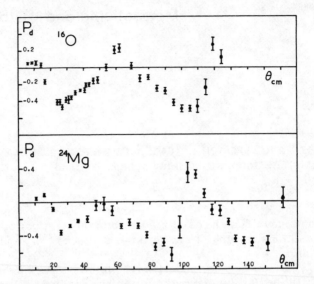

Fig. 1. Two examples of the angular distributions
of vector analyzing power obtained in the study of
$^{16}O(d,p)$ and $^{24}Mg(d,p)$ reactions. The upper curve
is for $^{16}O(d,p)^{17}O$ (0.871-MeV state) and the lower
for $^{24}Mg(d,p)^{25}Mg$ (0.584-MeV state). For both reactions $\ell_n = 0$.

REFERENCES

*Permanent address: Melbourne, Australia.
†Permanent address: Panjab University, Chandigarh, India.

[1]   J. A. R. Griffith et al., Nucl. Phys. A146 (1970) 193.
[2]   S. T. Butler et al., Ann. Phys. (N.Y.) 43 (1967) 282.

Polarization Effects in the
$\ell = 0$ Reaction ^{28}Si(d,p)^{29}Si(g.s.)

R. G. SEYLER, Ohio State University, Columbus, USA; L. J. B. GOLDFARB,
The University, Manchester, England

It has been suggested [1, 2] that separate measurements of the
proton polarization P(p,θ) and the deuteron vector analyzing power
P^d(d,θ) for the same $\ell = 0$ transfer reaction should point to separate effects of spin-dependent distortion (SDD) in the deuteron (D)
channel and the proton (P) channel. Thus if contributions from the
deuteron D-state and non-central V_{np} potentials are neglected, then
according as there is DSDD or PSDD, P(p,θ) is exactly equal to P^d(d,θ)
and 3/2 P^d(d,θ), respectively. Fig. 1 shows for the ground-state
^{28}Si(d,p) reaction, P(p,θ) at 10 MeV [3] and at 10.8 MeV [4] which is
to be compared with P^d(d,θ) as measured at 10 MeV [5] and at 12.3
MeV [6]. The similarity at the different energies, including also 15
MeV [7], attests to the working of a direct mechanism. Although Griffith et al. [6] suggest the equality of P(p,θ) and P^d(d,θ), (pointing thus
to the dominance of DSDD), this is apparent only for θ < 50°. PSDD
dominance is, in fact, suggested for 90° < θ < 120°; otherwise both
channels contribute significantly.

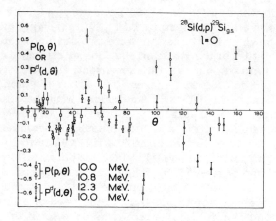

Fig. 1

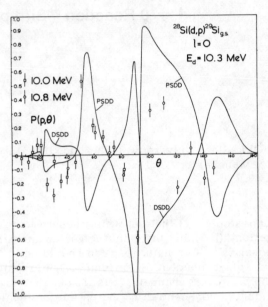

Fig. 2

With the availability of these data, it is tempting to test the DWBA and to unravel the different roles of SDD. First, a test was made of separability which states the equality of results of calculations with the sum of the two results when SDD is acting separately. Such an equality is implicit with first-order treatments of SDD [8], but separability is of greater generality. Calculations were performed using the JULIE code of Drisko with optical-model parameters chosen as in refs. [3, 6] which are consistent with elastic scattering. Separability appears to be generally valid at 10.3 MeV. Thus, $P(p,\theta)$ is found to deviate by no more than ± 0.02 for $\theta < 50°$; but where $P(p,\theta) > 0.25$, separability is less valid. The results for $P(p,\theta)$, assuming either PSDD or DSDD, as shown in fig. 2, demonstrate two striking features: (1) an anticorrelation of the two contributions at all angles and (2) the PSDD contribution is in phase with experiment, i.e., only PSDD reproduces the sharp drop after 20°, the rise to 50° and the drop near 90°, etc. The dominance of PSDD is in accordance with the effects of strong absorption, as emphasized by Hooper [9].

Assuming separability, there is then a need to increase the PSDD contribution or to decrease that of DSDD, if DWBA is to fit experiment. The strengths for the optical-model spin-orbit potentials were allowed to differ from those of refs. [3, 6]. They were taken here to be equal corresponding to a value of 6 MeV, but the deuteron strength could be lowered. The evidence points possibly to the need of a larger value of the absorption potential, as, in fact, was used to fit the data at 15 MeV [9, 10]. The near-equality of $P(p,\theta)$ and $3/2\ P^d(d,\theta)$ which would then follow is, however, not in agreement with the measured values of $P^d(d,\theta)$. This feature, along with the anticorrelation phenomenon, remains unexplained. The latter is, however, not a general feature of all $\ell = 0$ transfers. The reaction $^{118}Sn(p,d)^{117}Sn(g.s.)$ where $E_p = 24.5$ MeV [11] presents a counter-example; but even here the anticorrelation extends over a considerable range of angles.

REFERENCES

[1]   L. J. B. Goldfarb and R. C. Johnson, Nucl. Phys. 18 (1960) 353.
[2]   L. J. B. Goldfarb and R. G. Seyler, Nucl. Phys. A149 (1970) 545.
[3]   W. E. Maddox et al., Phys. Rev. C1 (1970) 476.
[4]   R. W. Bercaw and F. B. Shull, Phys. Rev. 133 (1964) B632.
[5]   H. H. Cuno et al., Nucl. Phys. A139 (1969) 657.
[6]   J. A. R. Griffith et al., Symp. on Nuclear Reaction Mechanisms (Quebec, 1969).
[7]   A. Isoya and M. J. Marrone, Phys. Rev. 128 (1962) 800.
[8]   R. C. Johnson, Nucl. Phys. 35 (1962) 644.
[9]   M. B. Hooper, Nucl. Phys. 76 (1966) 449.
[10] L. J. B. Goldfarb, Second Polarization Symp.
[11] J. L. Escudié et al., Saclay Annual Report (1969), CEA-N-1232.

Polarization Transfer in the Reaction ^{28}Si(d,p)^{29}Si

R. C. BROWN, J. A. R. GRIFFITH, O. KARBAN, and S. ROMAN,
University of Birmingham, England

A measurement has been made of the proton polarization, which results at $0°$ reaction angle when the reaction ^{28}Si(d,p)^{29}Si (g.s. $\ell_n = 0$) is initiated by a vector-polarized deuteron beam. The results agree with the expectations of a simple spectator model.

The 12.3-MeV pure vector-polarized deuteron beam of the Birmingham University Radial Ridge cyclotron was focused onto a natural Si target, 25 mg cm^{-2} thick. This was followed by a Ta foil of thickness chosen to just stop the deuteron beam. The $0°$ protons were only degraded in energy by this and then passed to a proton polarimeter. This comprised a strip target of carbon, of thickness 64 mg cm^{-2}, viewed by two silicon surface barrier detector telescopes placed to the left and right at scattering angles of $45°$, in the lab system. The polarimeter geometry and proton mid-target energy (12.1 MeV) were chosen to coincide closely with those of a known calibration point [1]. A third detector telescope was placed to intercept the protons transmitted through the polarimeter. This served to monitor the proton intensity and energy spectrum and gave an accurate verification that there were no changes of reaction proton intensity with changes in beam polarization parameters. All three telescopes were connected to particle identification circuits, which selected only genuine proton events and assisted in keeping the background counting rate low. The total energy spectra of all three telescopes were stored in split-memory multichannel analyzers. The protons from the first excited state (1.28 MeV) of ^{29}Si contributed only about 3% to the main proton peaks. The experimental technique used the rapid (3 Hz) alternation of beam polarization from spin-up to spin-down as previously described; see ref. [2], for example. Runs were taken with the carbon target removed in order to assess the background counting rate. The beam polarization was monitored in a thin target polarimeter placed in the beam line before the beam switching magnet. It employed the reaction ^{12}C(d,p)^{13}C (g.s.) at $57°$. The beam had $P_d = 0.48$, and 2×10^{10} deuterons/sec were obtained on target. The measurement took one week of machine time.

The result of the measurement was that the ratio of the outgoing proton polarization to the incoming deuteron polarization was 1.0 ± 0.1, where this uncertainty contains all effects and not only statistics. This value would be expected from the simplest model treating the proton as a spectator in the reaction. It may be noted that refinements of this type of measurement should provide a sensitive test of the role played by the deuteron D-state in stripping reactions. A deuteron in the D-state would produce a proton with a polarization opposite to its own.

REFERENCES

[1] R. M. Craig et al., Phys. Lett. 3 (1963) 301.
[2] A. M. Baxter, J. A. R. Griffith, S. W. Oh, and S. Roman, Nucl. Phys. A112 (1968) 209.

Energies of the $2p_{3/2}$, $2p_{1/2}$, $1f_{7/2}$, and $1f_{5/2}$ Single-Particle States in Odd-A Nuclei Obtained from Measurements of the (d,p) Vector Analyzing Power and Cross Section

D. C. KOCHER and W. HAEBERLI, University of Wisconsin, Madison, USA*

The vector analyzing power and absolute cross section were measured for (d,p) reactions on ^{40}Ca, ^{46}Ti, ^{48}Ti, ^{50}Ti, ^{52}Cr, and ^{54}Fe at $E_d = 10$ and 11 MeV for $\theta_{lab} = 15°-85°$. Transitions to states up to $E_x = 5.8$ MeV were observed with an energy resolution of 40-80 keV. Many spin assignments were made from the observed j-dependence of the vector analyzing power. The spectroscopic factors were obtained by normalizing the cross section at the stripping peak to that calculated by the distorted-wave code DWUCK [1].

When the energy distribution of states and the total spectroscopic factor for a given j^π-value indicated that most of the transition strength had been located, the energy of the equivalent single-particle state in the residual nucleus was calculated. The Q-values for the $2p_{3/2}$, $2p_{1/2}$, $1f_{7/2}$, and $1f_{5/2}$ states are shown in fig. 1; dashed lines are the Q-values for the unsplit 2p states. The energies of the states of a given j^π-value are approximately the same in ^{41}Ca and the Ti isotopes. In all cases, the $2p_{3/2}$ state lies lower in energy than the $2p_{1/2}$ state.

The code DWUCK was used to find the real central and spin-orbit depths V_n and V_{so} of the potential well for the odd neutron by adjusting the depths to reproduce the observed Q-values for the $2p_{3/2}$ and $2p_{1/2}$ states. The geometry parameters for the neutron well were $r_0 = 1.2$ fm and $a_0 = 0.65$ fm. The values of V_n and V_{so} in MeV for the target nuclei are (^{40}Ca; 58.2, 6.8), (^{46}Ti; 54.9, 5.5), (^{48}Ti; 52.9. 5.3), (^{50}Ti; 51.6, 5.7), (^{52}Cr; 52.9, 4.4), and (^{54}Fe; 53.9, 4.3). These values are in good agreement with results from analyses of neutron elastic scattering [2]. V_n shows a systematic decrease in magnitude with increasing neutron excess, N–Z, similar to that found in analyses of elastic scattering [2]. The average spin-orbit depth is $V_{so} = 5.2$ MeV.

With the above values of V_n and V_{so}, the predicted Q-values in MeV for the $1f_{7/2}$ states are $Q(^{41}\text{Ca}) = 6.98$, $Q(^{47}\text{Ti}) = 7.17$, and $Q(^{49}\text{Ti}) = 6.66$; for the $1f_{5/2}$ state, $Q(^{55}\text{Fe}) = 5.58$. The predicted Q-values are in reasonable agreement with those shown in fig. 1.

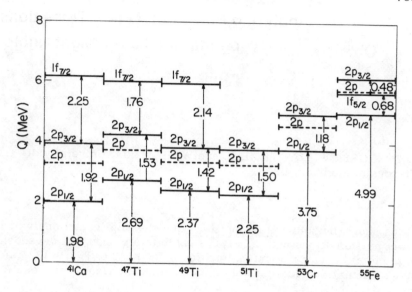

Fig. 1. Observed Q-values for (d,p) reactions populating
$2p_{3/2}$, $2p_{1/2}$, $1f_{7/2}$, and $1f_{5/2}$ neutron single–particle states
(solid lines) and unsplit 2p states (dashed lines) in the
residual nuclei. The neutron binding energy is Q + 2.22 MeV.

REFERENCES

*Work supported in part by the U.S. Atomic Energy Commission.

[1] P. D. Kunz, University of Colorado.
[2] F. D. Becchetti, Jr., and G. W. Greenlees, Phys. Rev. 182 (1969)
     1190.

DISCUSSION

Schiffer:
  Do the strengths of the 2p transitions satisfy the sum rules?

Kocher:
  Yes, within 7% in the worst case.

Johnson:
  Did your analysis indicate any systematic dependence for V_{so}
  (neutron) on N–Z?

Kocher:
  This is not clear from the data.

Roy:
  It is not clear that the radius of the neutron well should vary
  simply as $A^{1/3}$ over this range of nuclei. Unfortunately, any combination of V_{so}, r_n, and a_r can reproduce the p-orbit splitting.

Measurement of the (d,p) Vector Analyzing Power for $\ell_n = 4$ Transitions on Intermediate Weight Nuclei

D. C. KOCHER, R. D. RATHMELL, and W. HAEBERLI, University of Wisconsin, Madison, USA*

It was shown by Yule and Haeberli [1] that the (d,p) vector analyzing power for transitions with $\ell_n = 1, 2, 3$ exhibits a pronounced dependence on the total angular momentum transfer, $j = \ell_n + 1/2$ or $j = \ell_n - 1/2$. DWBA calculations qualitatively reproduce j-dependence for these ℓ_n-values [1].

We report measurements of the vector analyzing power for $\ell_n = 4$ transitions on intermediate weight nuclei. The results for transitions on ^{40}Ca, ^{48}Ti, ^{50}Ti, ^{52}Cr, and ^{90}Zr are shown in fig. 1. The vector analyzing powers for all transitions are small, and the data do not show pronounced oscillations. This behavior differs from results [1] for transitions with $\ell_n \leq 3$, for which j-assignments can readily be made from the observed features of the data at forward angles.

The curves in fig. 1 are the predictions of DWBA calculations [2] employing deuteron and proton optical model parameters obtained from analyses of elastic cross sections and polarizations. For the four nuclei near mass 50, the $j^{\pi} = 9/2^+$ curves agree with the trend of the data at all angles while the $7/2^+$ curves do not. It thus seems that j-assignments can be made by comparing the data with DWBA calculations. The differences in the calculated curves for the same j^{π}-value among ^{48}Ti, ^{50}Ti, and ^{52}Cr result primarily from differences in Q-value for the transitions. For ^{90}Zr, however, neither curve fits the data, and no j-assignment can be made. The spin of the final state in ^{91}Zr is known [3] to be $7/2^+$ from a measurement of the polarization of protons elastically scattered from ^{90}Zr near the isobaric analog resonance in ^{91}Mo.

We conclude that j-assignments for $\ell_n = 4$ transitions on target nuclei near $A = 50$ can be made from a comparison of the data with DWBA calculations in spite of the lack of prominent features in the observed angular distributions. For ^{90}Zr, refinements are needed in the theory to adequately reproduce the data.

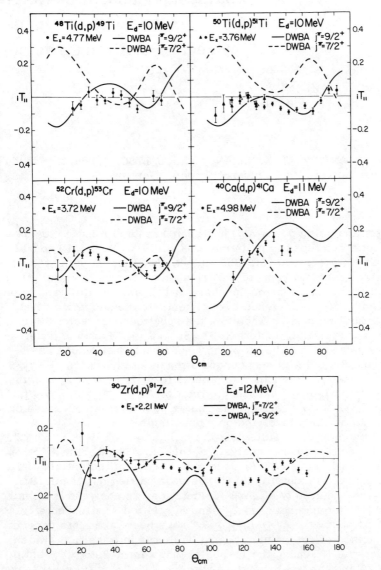

Fig. 1. Measured vector analyzing powers and DWBA predictions for $\ell_n = 4$ (d,p) reactions on ^{40}Ca, ^{48}Ti, ^{50}Ti, ^{52}Cr, and ^{90}Zr.

REFERENCES

*Work supported in part by the U.S. Atomic Energy Commission.

[1] T. J. Yule and W. Haeberli, Nucl. Phys. A117 (1968) 1.
[2] The code DWUCK written by P. D. Kunz, University of Colorado, was used.
[3] J. V. Tyler and W. Haeberli, private communication.

Stripping Reactions on Nuclei
in the 1p-Shell Initiated
by Vector Polarized Deuterons

D. FICK, R. KANKOWSKY, K. KILIAN, and E. SALZBORN,[*] Physikalisches Institut der Universität Erlangen-Nürnberg, Germany

The vector analyzing power $A(\theta)$ has been measured from $15°$ to $75°$ each $5°$ for a number of (d,p)-transitions with $\ell = 1$ at $E_d = 10$ and 12 MeV on the target nuclei ^{6}Li, ^{9}Be, ^{10}B, ^{11}B, ^{13}C, and ^{14}N using the vector polarized deuteron beam of the Erlangen tandem accelerator [1].

For stripping reactions on 1p-nuclei with spin, two different values of $j(1/2$ and $3/2)$ are in general consistent with conservation of angular momentum and parity. Therefore the spectroscopic factor S is a sum over spectroscopic factors S_j $(S = S_{1/2} + S_{3/2})$. From the differential cross section near the stripping peak no information can be obtained about S_j, but $A(\theta)$ may depend strongly on the ratios $p_j = S_j/S$ and can be used to determine the probabilities p_j separately. In the region of the stripping maximum the relation $A = p_{1/2}A_{1/2} + p_{3/2}A_{3/2}$ holds for $A(\theta)$. In this relation the analyzing powers A_j are those obtained for transitions involving only one single j.

Until now only the results near the stripping peak ($\theta = 15°$, $20°$, $25°$) have been used for further analysis of this experiment. The A_j have been calculated using a DWBA code with the optical potentials of Fitz et al. [2] and a 0 fm cutoff. No fit has been carried out at all. Because of the uncertainty of DWBA calculations an analysis is meaningless if the difference between $A_{1/2}$ and $A_{3/2}$ is too small. Therefore only transitions with $|A_{1/2} - A_{3/2}| \geq 0.1$ have been taken into account.

This method of determining p_j can be tested in transitions which have a definite j. Among the 1p-nuclei these are: ^{9}Be(d,p_0), ^{10}B(d,p_0) with $j = 3/2$ and ^{12}C(d,p_0), ^{13}C(d,p_0) with $j = 1/2$. For ^{12}C(d,p_0) it is well known that the optical model approach does not work at our energies [3]. This happens also in ^{13}C(d,p_0), which is shown in fig. 1. There is no similarity between the angular distributions of $A(\theta)$ at 10 and 12 MeV. Therefore this transition cannot be used as a test. Furthermore, all conclusions drawn from experiments with unpolarized deuterons in ^{13}C(d,p_0) are not very reliable [4]. The results for $p_{3/2}$ of the present analysis, which are obtained by averaging $p_{3/2}$ over the results at $15°$, $20°$, $25°$ are shown in table 1 (the two results for each transition correspond to 10 and 12 MeV, respectively). The agreement for the

two transitions with a single j is fair. The ^{10}B(d,p$_0$)-transition seems
to be influenced at 10 MeV by compound nucleus (^{12}C)-contributions.
The transitions with two possible values of j show in general good
agreement with the shell model calculations of Cohen and Kurath [5]
(p$_{3/2}$ theor). For the ^{9}Be(d,p$_1$)-transition Cohen and Kurath [5] predict
two values of p$_{3/2}$ corresponding to two different potentials for the two-
body interaction. The potential parameters leading to p$_{3/2}$ = 0.65 can
be ruled out by our measurements.

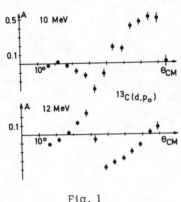

Fig. 1

Table 1

	theor p$_{3/2}$	exp p$_{3/2}$
^{9}Be(d,p$_0$)	1.00	0.48 ± 0.40 0.72 ± 0.18
^{10}B(d,p$_0$)	1.00	0.34 ± 0.14 1.00 ± 0.23
^{6}Li(d,p$_0$)	0.60	0.77 ± 0.05 ---------
^{6}Li(d,p$_1$)	0.96	0.92 ± 0.05 ---------
^{9}Be(d,p$_1$)	$\begin{cases} 0.17 \\ 0.65 \end{cases}$	0.00 ± 0.25 0.04 ± 0.25
^{11}B(d,p$_0$)	0.14	0.30 ± 0.15 ---------
^{14}N(d,p$_0$)	0.02	0.00 ± 0.08 0.27 ± 0.36

REFERENCES

*Present address: Institut für Kernphysik, Giessen, Germany.

[1] G. Clausnitzer et al., Nucl. Instr. 80 (1970) 245.
[2] W. Fitz et al., Nucl. Phys. A101 (1967) 499.
[3] H. Wilsch and G. Clausnitzer, to be published in Nucl. Phys.
[4] J. P. Schiffer et al., Phys. Rev. 164 (1967) 1274.
[5] S. Cohen and D. Kurath, Nucl.Phys. 73 (1965) 1 and A101 (1967) 1.

DISCUSSION

Kocher:

For your quoted $p_{3/2}$ spectroscopic factors, what do the uncertainties signify?

Salzborn:

Only experimental errors.

Kocher:

Do DWBA calculations reproduce the analyzing power for pure $p_{3/2}$-transitions?

Salzborn:

Yes.

Measurements and Analysis of the
Vector Analyzing Power of ^{208}Pb(d,p)^{209}Pb
near the Coulomb Barrier

R. D. RATHMELL and W. HAEBERLI, University of Wisconsin, Madison, USA*

Measurements of the vector analyzing power of ^{208}Pb(d,p)^{209}Pb at 12.3 MeV have recently been reported [1]. It was suggested that the large analyzing power observed for the transition to the $d_{5/2}$ state in ^{209}Pb at $E_x = 1.56$ MeV was caused by a Mott-Schwinger magnetic spin-orbit interaction in the outgoing channel [2]. The present experiment was undertaken to investigate this effect further.

We measured the cross section and vector analyzing power of ^{208}Pb(d,p)^{209}Pb at 12.3 MeV. The results for the transitions leading to the $g_{9/2}$ ground state and $d_{5/2}$ excited state of ^{209}Pb are shown in fig. 1. The present results are generally consistent with the previous measurements [1].

The solid curves in fig. 1 are the result of DWBA calculations [3]. The DWBA curves for $j = 9/2^+$ and $j = 5/2^+$ give a good fit to the vector analyzing power while the predicted analyzing power for $j = 7/2^+$ and $3/2^+$ have the opposite sign. The predicted cross sections for the two possible j-values for each transition are indistinguishable except for spectroscopic factors. Therefore only one curve is shown. Clearly the data can be explained without including a magnetic spin-orbit interaction.

In order to investigate the origin of the vector analyzing power, additional calculations were made. Results of DWBA calculations with no spin-orbit terms in the deuteron and proton potentials (not shown) were nearly identical to the solid curves in fig. 1. Calculations were also made with Coulomb distortions only (CWBA) [4]. The results shown as dashed curves in fig. 1 qualitatively reproduce the vector analyzing power for both transitions except at forward angles. However, the CWBA fit to the cross section is poor.

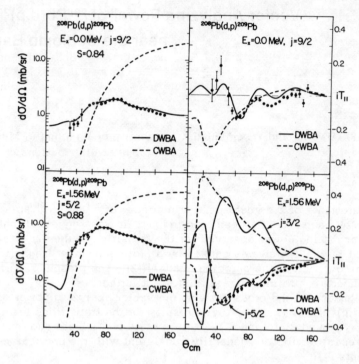

Fig. 1. Cross section and vector analyzing power of
^{208}Pb(d,p)^{209}Pb for $E_x = 0.0$ and 1.56 MeV with $E_d = 12.3$ MeV.

REFERENCES

*Work supported in part by the U.S. Atomic Energy Commission.

[1] S. Roman, Proc. Symp. on Nuclear Reaction Mechanisms and
     Polarization Phenomena (Quebec, 1969), to be published.
[2] G. F. Drukarev and V. T. Ippolitov, Nucl. Phys. A110 (1968) 218.
[3] The code DWUCK written by P. D. Kunz, University of Colorado,
     was used. The deuteron optical parameters were taken from the
     work of G. Muehllehner et al. (Phys. Rev. 159 [1967] 1039); a
     spin-orbit term of $V_{so} = 5.25$ MeV was added to their potential.
     The proton parameters were taken from an analysis of elastic
     cross sections and polarizations for ^{209}Bi at 13 MeV (R. D. Rath-
     mell and W. Haeberli, to be published).
[4] CWBA and DWBA predictions for $\ell_n = 2$ and 4 transitions on ^{209}Bi
     have been reported by J. A. R. Griffith and S. Roman, Phys. Rev.
     Lett. 24 (1970) 1496.

DISCUSSION

Glashausser:

There is an error in the Russian paper on Mott-Schwinger polarization, and I understand a correction will be published in Nuclear Physics soon. Their new calculations show very small polarizations due to the Mott-Schwinger interaction.

Rawitscher:

It is very interesting that the CWBA gives a much better fit to the vector analyzing power than to the angular distributions of the cross section.

Two-Nucleon Transfer Reactions
Induced by Polarized Protons

J. C. HARDY, A. D. BACHER, G. R. PLATTNER, J. A. MACDONALD,
and R. G. SEXTRO, Lawrence Radiation Laboratory, Berkeley, California,
USA*

The two-nucleon transfer reactions, (p,t) and (p,^{3}He), have been previously studied with unpolarized protons not only to determine spins,
parities, and isospins of nuclear levels, but also to investigate wave
functions for the states involved. At forward angles the angular distributions of the differential cross sections for these reactions are
characteristic of the transferred orbital angular momentum L and are
reasonably well reproduced by DWBA calculations. This contribution
summarizes a recent report [1] on the first detailed examination of the
asymmetries produced when these reactions are initiated by polarized
protons; of particular interest was whether these asymmetries are also
characteristic of the quantum numbers of the transferred nucleons.

We have investigated the (p,t) and (p,^{3}He) reactions on ^{16}O and
^{15}N gas targets using a 43.8-MeV polarized proton beam from the
Berkeley 88-inch cyclotron. Sixteen transitions were observed, of
which eleven corresponded to unique sets of transferred quantum numbers. Initially it was hoped that these unique transitions would define
characteristic shapes for angular distributions of the analyzing power
so that the ability of the DWBA calculations to reproduce them could
be tested. However, a very striking result has emerged. Five (p,t)
transitions were observed for which the transferred quantum numbers
are L = 2 and S = 0. The measured analyzing powers from these transitions do *not* have the same angular distribution. They appear instead
to be of two distinct types, one which agrees well with DWBA calculations and one which has virtually opposite phase to the analyzing
power predicted by DWBA. This is particularly surprising in light of
the success we have had in reproducing the shapes of the analyzing-
power angular distributions for transitions with other L-values, and in
fitting cross sections for all the observed transitions. This is not
simply a disagreement with theory; there is a significant discrepancy
observed between transitions in the same nucleus which are characterized by the same transferred quantum numbers. Attempts to understand

these "anomalous" transitions indicate that they cannot be explained by making variations in either the optical-model parameters, the bound-state parameters, or the nuclear wave functions.

The $(p, {}^3He)$ transitions observed may proceed by more than one set of transferred quantum numbers (L, S, and J), and the process is described by the coherent sum of transition amplitudes characterized by the same J but different values of L and S. The extent of the interference between these amplitudes depends upon the strength of the spin-orbit coupling in the entrance and exit channels. The analyzing powers are sensitive to these effects and are being used in their investigation.

REFERENCE

*Work performed under the auspices of the U.S. Atomic Energy Commission.

[1] J. C. Hardy et al., Phys. Rev. Lett. 25 (1970) 298.

Asymmetries in (p,t) and (p,³He) Reactions
Induced by 49.5-MeV Polarized Protons

J. M. NELSON, N. S. CHANT, and P. S. FISHER, Nuclear Physics
Laboratory, Oxford, England

The 49.5-MeV polarized proton beam from the Rutherford Laboratory
linear accelerator has been used in experiments to measure the polarization analyzing powers of the ground state (p,t) reactions on ^{12}C,
^{16}O, and ^{28}Si (L = S = J = 0), and the (p,³He) reactions on ^{16}O leading to
the ground (S = J = 1, L = 0, 2) and first (L = S = J = 0) excited states of
^{14}N. The reaction ^{12}C(p,t) ^{10}C (3.36 MeV, L = J = 2, S = 0) was also
studied. These reactions have been chosen because the transferred
spin-angular momentum, S, is zero (except for the ground state
^{16}O(p,³He)^{14}N reaction where S = 1). The usual spin-independent
distorted-wave theory predicts zero analyzing powers for all S = 0
interactions. The large analyzing powers that are observed demonstrate the necessity of using a spin-dependent theory in the analysis.
The investigation of the dependence of the distorted-wave calculations on spin-orbit effects and on different constructions of the
form factor is the central feature of the analysis. Optical model potentials derived from elastic scattering analyses have been used
throughout, although ³He potentials have had to be used in place of
non-existent triton potentials. Zero-range calculations using harmonic oscillator wave functions for the form factor provided poor fits
to the data in all cases. Variations of the zero-range calculations
which used Saxon-Woods wave functions, or Saxon-Woods functions
expanded in a harmonic oscillator basis, for the form factor were also
unsatisfactory. A common feature of these calculations was the absence of structure in the angular distribution. Considerable improvement could be obtained between zero-range calculations and the data
by substantially increasing the proton absorption strength. This unphysical modification was not considered in any detail. The assumption of a zero-range interaction is a serious weakness in calculations
of this type, and this analysis has shown that the use of a finite-
range interaction significantly improves the theoretical predictions.
A modified version of Bencze and Zimanyi's [1] formulation has been
used.

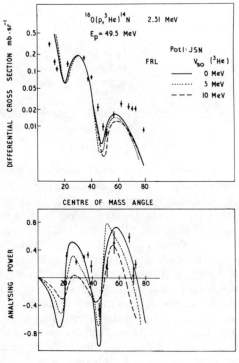

Fig. 1

It was found, in general, that calculations using tri-nucleon potentials that contained a spin-orbit strength less than 5 MeV provided
better fits to the analyzing powers than potentials with a larger spin-
dependent term. The cross sections were fairly insensitive to the
variations in the tri-nucleon spin-orbit strength. An example of the
data and DWBA predictions is shown in fig. 1 for the case of the
$^{16}O(p, ^{3}He)^{14}N$ (2.31 MeV) reaction. The finite-range calculations illustrate the effect of varying the ^{3}He spin-orbit strength. (The proton
spin-orbit potential is 5.48 MeV.)

REFERENCE

[1]   G. Bencze and J. Zimanyi, Nucl. Phys. 81 (1966) 1.

Quadratic Spin-Orbit Effects on
Some $0^+ \rightarrow 0^+$ (^{3}He,t) Reactions

J. RAYNAL, Centre d'Etudes Nucléaires de Saclay, France

The differential cross sections of (^{3}He,t) reactions have been measured for different 0^+ states of the residual nucleus. A typical example
is the ^{3}He(^{40}Ar,^{40}K)t reaction [1] for the analog 0^+ state with Q = 5.88
MeV and the 0^+ state with Q = 3.11 MeV. The analog state exhibits
an L = 0 angular dependence, the other one, an L = 1. Similar effects
have been found also in Fe and Zn.

In the simplest DWBA approach, such an excitation is described by
a scalar form factor A(r); a spin-orbit type form factor $B(r)(\vec{\ell} \cdot \vec{\sigma})$ can be
introduced only via a spin-orbit or a quadratic spin-orbit interaction
between a nucleon of the target and the bombarding particle. The form
factor A(r) gives an L = 0 angular distribution, B(r) an L = 1, if it is
dominant. If we consider the ^{40}Ar nucleus as two neutrons in the $f_{7/2}$
shell and two proton-holes in the $d_{3/2}$ shell, these form factors are
sums of two contributions (creation of a proton in the $f_{7/2}$ shell or
annihilation of a proton-hole in the $d_{3/2}$ shell) which add coherently
for the analog state and subtract for the other 0^+.

The simplest approximation for a nucleon-^{3}He interaction is the
average of the nucleon-nucleon interaction on a ^{3}He wave function.
We are only concerned with the charge exchange part of the potential.
With a Hamada and Johnston [2] type of nucleon-nucleon interaction,
the only contributing parts are, besides the scalar interaction:

1) The spin-orbit coupling $\mathcal{V}_{LS}(r)\{r \wedge (\nabla/i)(\sigma_1 - \sigma_2)\}$ where $r = r_1 - r_2$,
$\nabla = \nabla_1 - (1/3)\nabla_2$, if 1 and 2 are respectively the nucleon and the ^{3}He
coordinates. Its form factor is obtained from the nucleon-nucleon form
factor V_{LS} by

$$\mathcal{V}_{LS} = \int \phi(x,\eta)V_{LS}(|r-x|)\,\phi(x,\eta)(1 - \frac{x}{r}\cos \widehat{xr})d^3x\,d^3\eta, \tag{1}$$

where x is the c.m. coordinate of a nucleon in ^{3}He and η any independent coordinate. The monopole part of this potential is

$$(1 - \frac{r_1}{r_2}\cos \widehat{r_1 r_2})(\ell_1 \sigma_1) - \frac{1}{3}(1 - \frac{r_2}{r_1}\cos \widehat{r_1 r_2})(\ell_2 \sigma_2) \tag{2}$$

The two parts of (2) contribute respectively to A(r) and to B(r). For the anti-analog state the contribution is incoherent for B(r) and coherent for A(r), because $(\ell_1\sigma_1)$ is respectively +3 and -3 for the two components. Using a Yukawa form factor of some range for $\mathcal{V}_{LS}(r)$, one can find a strength such that the angular distribution is L = 1, but the cross section is too small.

2) The charge exchange part of the quadratic LS coupling of Hamada and Johnston which can be written

$$2V_{\sigma\tau}L^2(1-\sigma_1\sigma_2) + (V_\tau - V_{\sigma\tau})\{L^2(\sigma_1\sigma_2) - \frac{1}{2}(L\sigma_1)(L\sigma_2)$$

$$-\frac{1}{2}(L\sigma_2)(L\sigma_1)\}. \tag{3}$$

Averaging on the ^{3}He wave function,

$$L^2 \to \mathfrak{F}_1(r)[r \wedge \nabla]^2 + \mathfrak{F}_2(r)\nabla^2 + \mathfrak{F}_3(r)[r_L \wedge \nabla] + \mathfrak{F}_4(r) \tag{4}$$

$$(L\sigma_1)(L\sigma_2) + (L\sigma_2)(L\sigma_1) \to \mathfrak{F}'_1(r)[(\sigma_1 r \wedge \nabla)(\sigma_2 r \wedge \nabla) + (\sigma_2 r \wedge \nabla)(\sigma_1 r \wedge \nabla)] +$$

$$\mathfrak{F}'_2(r)(\sigma_1 \wedge \nabla)(\sigma_2 \wedge \nabla) + \mathfrak{F}'_3(r)[(\sigma_1 \wedge r)(\sigma_2 \wedge r) +$$

$$(\sigma_2 \wedge \nabla)(\sigma_1 \wedge \nabla)] + \mathfrak{F}'_4(r)S_{12} + \mathfrak{F}'_5(r)(\sigma_1\sigma_2). \tag{5}$$

The L^2 term adds first and second derivatives to A(r). The other terms which do not vanish can be factorized into $F(r)(\ell_1\sigma_1)(\ell_2\sigma_2)$ where

$$F(r) = -\frac{1}{6}[\mathfrak{F}_1(r) - \mathfrak{F}'_1(r)]\sin^2\theta - \frac{1}{3}[\mathfrak{F}_2(r) - \frac{1}{4}\mathfrak{F}'_2(r)]\frac{\cos\theta}{r_1 r_2} +$$

$$\frac{1}{2}\mathfrak{F}'_1(r)\left[1 + \cos^2\theta - \left(\frac{r_1}{r_2} + \frac{r_2}{r_1}\right)\cos\theta\right]. \tag{6}$$

They are coherent for the anti-analog state and can be at the origin of the L = 1 behavior.

We have used the soft core potential of Bressell et al. [3] (with Hamada and Johnston parameters for simplification) and the Irving-Gunn ^{3}He wave function [4]. The results for an optical model with $V_0 = 167$ MeV, $r_0 = 1.16$, $a_0 = 0.715$ and an imaginary part $W_S = 13$ MeV, $r_i = 1.8$, $a_i = 0.872$ are shown in fig. 1. The form factors are very small and of too short range. The cross section is 100 times too small and the maxima too broad. The ratio of cross sections is in agreement with experiment. In contrast to expectation, the quadratic $\vec{L}\cdot\vec{S}$ is not essential except for the L^2 term.

To increase the range, the form factors $\mathcal{V}_{LS}$, $\mathfrak{F}_1$, $\mathfrak{F}'_1$, $\mathfrak{F}_2$, and $\mathfrak{F}'_2$ were replaced by Yukawa potentials of range 1.4 fm and same strength; a Serber force is used for the scalar interaction. The L = 1 shape of

the anti-analog state can be seen in the forward direction. The maxima are too narrow because the quadratic $\vec{L}\cdot\vec{S}$ form factors change sign
at about 6.3 fm. The dotted line is obtained without quadratic $\vec{L}\cdot\vec{S}$
coupling.

If the $L = 1$ shape comes from $(\vec{\ell}\cdot\vec{\sigma})$ form factor (of any origin) the
polarization is greater for the anti-analog state. If measurements
were possible, they would be a good test of this simple mechanism.

I thank R. Schaeffer for having called my attention to a possible
spin-orbit explanation of these phenomena and P. Bonche for checking the calculations.

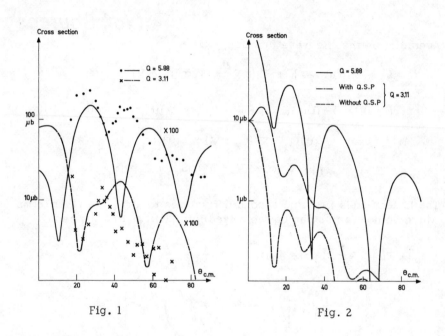

Fig. 1   Fig. 2

REFERENCES

[1] J. J. Wesolowsky et al., Phys. Rev. 172 (1968) 1072.
[2] T. Hamada and I. D. Johnston, Nucl. Phys. 34 (1962) 382.
[3] C. N. Bressel et al., Nucl. Phys. A124 (1969) 624.
[4] J. C. Gunn and J. Irving, Phil. Mag. 42 (1951) 1353.

Design and Calibration
of Ion Sources

Polarization of Channeled Deuterons

M. KAMINSKY, Argonne National Laboratory, Illinois, USA[*]

When energetic ions penetrate a monocrystalline solid, their trajectories can under certain conditions be influenced by the regular arrangement of the lattice atoms. In particular, they may be guided
through the spaces between planes (planar channeling) or along channels formed by parallel rows of atoms along certain crystallographic
directions (axial channeling), and cause "directional effects" in various atomic and nuclear phenomena [1—4].

For the case of energetic light ions (Z = 1 or 2) impinging on metal
monocrystals with energies ranging from 0.1 to 4.0 MeV, earlier work [5]
at this laboratory showed the influence of channeling effects on the
yields of secondary particles, on the energy loss, and on charge-
transfer processes. It was observed that the energy losses and the
secondary-particle yields were smaller in directions in which the
lattice was more transparent. It was also observed that ions trans-
versing and escaping through regions of lower electron density (e.g.,
in the center of an axial channel) have a lower probability of capturing or losing electrons than do those traveling at random in the
crystal.

These findings led to the search for polarization of well-channeled
deuterons. The technique [6] is to pass deuterons accurately along
the [110] direction of a thin monocyrstalline Ni foil magnetized to
saturation in a [111] direction in the plane of the foil. In an appropriate energy range the deuterons capture polarized electrons (e.g.,
from the 3d states in Ni). A portion of the emergent beam which consists of well-channeled polarized deuterium atoms is selected and
passes through a weak uniform magnetic field for a sufficiently long
time that part of the electron polarization is transferred to the deuteron
by hyperfine interaction. If the transition from the strong field inside
the foil to the weak field is adiabatic, a preferential population of
some of the hyperfine states (and thus a polarization of the deuterons)
can be expected.

In our experiments a mass-analyzed, highly-collimated D$^+$ ion
beam with a half angle of 0.01° was incident within 0.1° of a [110]
direction in a Ni(110) foil magnetized to saturation parallel to one of

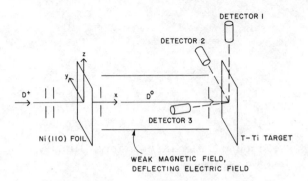

Fig. 1. Schematic diagram of experimental
arrangement.

the [111] axes in the plane of the foil. Two different monocrystalline
Ni(110) foils with thicknesses of 1.24 μm and 2.21 μm were used.
Two polycrystalline nickel foils with thicknesses of 1.14 μm and
1.87 μm were also used to test a method of particle polarization proposed by Zavoiskii [7]. The foils, mounted in the target holder, were
kept in a magnetic field (approximately 160 G) parallel to the original
magnetization direction.

As shown in fig. 1, the atoms emerging from the foil were first
passed through an aperture of half-angle ≈ 0.15° at a scattering angle
of 0° ± 0.1° with respect to the incident beam direction. They then
spent 1.2×10^{-7} s in traversing a homogeneous magnetic field of ~10 G
which can be directed along either the ±z or the ±y axis (i.e., either
parallel or perpendicular to the direction of magnetization of the foil)
by energizing one or the other of a pair of electromagnets. The charged
atoms were electrostatically deflected out of the beam, while the neutral atoms passed through the second collimator system (half angle of
acceptance ≈ 0.15°) and struck either a movable surface-barrier solid-
state detector (to determine the energy spectrum in the emergent beam)
or the T-Ti target. The tensor polarization of the well-channeled
deuterium atoms (now polarized in electron spin and nuclear spin) was
determined by measuring the angular distribution of the α particles
emitted in the $T(d,n)^4He$ reaction. If the incident beam direction was
(within 0.1°) parallel to the [110] axial channel, approximately 93%
of the emergent particles that passed through the collimating system
had suffered a reduced energy loss (with mean energies $\bar{E}_{ch}$) while
most of the remaining 7% had the normal energy loss (with mean energies $\bar{E}_n$).

To avoid detection of this small residue of randomly scattered atoms,
the incident-ion energy was so chosen that $\bar{E}_{ch}$ was in the range 100-
130 keV. Then in the case of the 1.24-μm thick foil, $\bar{E}_n$ was only a few
keV—well below the peak in the yield of the $T(d,n)^4He$ reaction.

The effectiveness of the reaction as a polarization analyzer, which depends on the relationship [8,9] between its cross section $\sigma(\phi)$ and the polarization, is given by

$$\sigma(\phi) = \sigma_0 [1 - \frac{1}{4}(3\cos^2\phi - 1)P_{zz}], \tag{1}$$

where σ_0 is the isotropic unpolarized cross section, P_{zz} is one element of a second-rank tensor (representing the expectation values of the Cartesian spin operator), the z-axis is taken along the axis of polarization, and ϕ is the c.m. angle between the outgoing particle (a or n) and the polarization axis. The tensor polarization P_{zz} is related to the fractional population N_0 of the deuteron $m_1 = 0$ state by

$$P_{zz} = (1 - 3N_0). \tag{2}$$

For the case in which the polarization was in the $\pm z$ direction (fig. 1), solid-state counters 1, 2, and 3 detected the a particles emitted in such c.m. directions that $\phi_1 = 7°$, $\phi_2 = 90°$, and $\phi_3 = 90°$. For an additional test, the polarization axis was reoriented by applying the weak magnetic field in the $\pm y$ direction. For this case, the c.m. angles were $\phi_1 = 90°$, $\phi_2 = 82°$, and $\phi_3 = 7°$. For a determination of the tensor polarization P_{zz} or P_{yy}, it is sufficient to determine the ratio r_{pol} of the differential cross sections σ for two different angles ϕ_μ and ϕ_ν, normalized to the ratio r_{unp} of the cross sections σ_0 for the unpolarized deuterons for the same angles. Since the counting rates n and n_0 are proportional to the cross sections σ and σ_0, respectively, one sees from eq. (1) that P_{zz} (or P_{yy}) can be determined from the ratio

$$R_{\mu,\nu} = \frac{r_{pol}}{r_{unp}} = \left(\frac{\sigma(\phi_\nu)}{\sigma(\phi_\mu)}\right) \bigg/ \left(\frac{\sigma_0(\phi_\nu)}{\sigma_0(\phi_\mu)}\right) = \left(\frac{n(\phi_\nu)}{n(\phi_\mu)}\right)_{pol} \times \left(\frac{n_0(\phi_\mu)}{n_0(\phi_\nu)}\right)_{unp}. \tag{3}$$

The ratio r_{unp} was determined by bombarding the T-Ti target with a collimated, unpolarized primary deuteron beam with mean energies ranging from 100 to 150 keV. The results shown in table 1 reveal a value P_{zz} (or P_{yy}) $\approx -1/3$, corresponding to a fractional occupation $N_0 \approx 4/9$. This value implies an electron polarization of approximately 98%, which seems surprisingly high since the effective magneton number of Ni is only about 0.6.

A possible explanation for the observed large polarization values has been offered recently by Ebel [10]. He suggests a mechanism of pumping electrons from spin-down to spin-up deuterium states (i.e., from states with spins antiparallel to the magnetic field to states with parallel spins). The proposed mechanism is based on the fact that in

Table 1. Values for the tensor polarization P_{zz} or P_{yy} as determined
          for the measured $R_{\mu\nu}$

Polarization direction	Detector angles ϕ_ν	ϕ_μ	$R_{\mu\nu}$	Tensor polarization
‖ to ± z axis	ϕ_1	ϕ_2	+1.260 ± 0.010	P_{zz} = −0.32 ± 0.01
	ϕ_1	ϕ_3	+1.258 ± 0.010	P_{zz} = −0.32 ± 0.01
‖ to ± y axis	ϕ_1	ϕ_3	+0.787 ± 0.011	P_{yy} = −0.32 ± 0.01
	ϕ_2	ϕ_3	+0.852 ± 0.010	P_{yy} = −0.32 ± 0.01

Ni the spin-up states in the 3d band are practically filled while those
with spin-down are not [11]. Therefore, once the captured electron
has spin-up, its chances for loss to the filled spin-up states in the
3d band is very small. Calculations based on this model lead to
electron polarization values which approach those observed in these
experiments.

To test Zavoiskii's proposal for polarizing deuterons, we passed
deuterons through polycrystalline Ni foils magnetized to saturation
and observed P_{zz} = −0.002 ± 0.010 for the 1.14-μm thick foil and P_{zz} =
+0.003 ± 0.010 for the 1.87-μm thick foil. These results indicate no
significant tensor polarization (i.e., $P_{zz} \approx 0$) and support some of the
criticism [12] of the proposal.

The method of polarizing channeled deuterons should work well
also for other particles (e.g., protons, tritons, ^{3}He, etc.). It would be
of interest to try the technique with other ferromagnetic and paramagnetic materials (e.g., Gd), especially to provide additional materials
(e.g., Fe) in order to test Ebel's model, as he suggests [10]. It appears that the method of polarizing channeled particles will permit
the development of a relatively inexpensive source of polarized ions
[13].

REFERENCES

*Work performed under the auspices of the U.S. Atomic Energy
Commission.

[1] R. S. Nelson, The Observation of Atomic Collisions in Crystalline
     Solids (North-Holland Publishing Company, Amsterdam, 1968).

[2]   G. Carter and J. S. Colligon, Ion Bombardment of Solids (American Elsevier Publishing Company, New York, 1968).
[3]   S. Datz, C. Erginsoy, G. Leibfried, and H. O. Lutz, Ann. Rev. Nucl. Sci. 17 (1967) 129.
[4]   M. Kaminsky, Atomic and Ionic Impact Phenomena on Metal Surfaces (Springer-Verlag, Heidelberg/New York, 1965).
[5]   M. Kaminsky, in Recent Advances in Mass Spectrometry, ed. K. Ogata (University of Tokyo Press, Tokyo, in press); J. Vac. Sci. Technol. (in press); Bull. Am. Phys. Soc. 14 (1969) 846; ibid. 13 (1968) 1406; ibid 12 (1967) 635; Adv. Mass Spectrometry 3 (1966) 69; Phys. Rev. 126 (1962) 1267.
[6]   M. Kaminsky, Phys. Rev. Lett. 23 (1969) 819.
[7]   E. K. Zavoiskii, Sov. Phys.—JETP 5 (1957) 378.
[8]   A. Galonsky, H. B. Willard, and T. A. Welton, Phys. Rev. Lett. 2 (1959) 349.
[9]   L. D. B. Goldfarb, Nucl. Phys. 12 (1959) 657.
[10] M. E. Ebel, Phys. Rev. Lett. 24 (1970) 1395.
[11] L. Hodges, H. Ehrenreich, and N. D. Lang, Phys. Rev. 152 (1966) 505.
[12] W. Haeberli, Ann. Rev. Nucl. Sci. 17 (1967) 420; Second Polarization Symp., p. 69.
[13] M. Kaminsky, U.S. patent application, serial # 846788.

DISCUSSION

Ohlsen:

Your result of 0.32 ± 0.01 seems to be too high for 100% polarized electrons when the T(d,α) analyzing power of only ~0.9 is taken into account. The "pumping" effect you described does not affect this remark in any way if I understand correctly.

Kaminsky:

Our result of P_{zz} = -0.32 ± 0.01 implies an electron polarization of 97%—98%. The reported P_{zz}-value has been determined from the measured ratios of the counting rates at two different angles for the polarized and unpolarized beam. The reaction cross section for the $T(d,n)^4He$ reaction was calculated assuming that only s-waves contribute. The pumping mechanism suggested in Dr. Ebel's preliminary treatment leads to electron polarization values between approximately 92%—98%.

Ohlsen:

I am still fascinated by the fact that your data correspond to > 105% electron polarization. As I understand it, there is a large field in the crystal and this must eventually get down to about 10 G. Do you make any special effort to shape the magnetic field transition region, and do you think there is a possibility of some sort of transitions in the interface or transit region which might enhance the expected polarization?

Kaminsky:

As I mentioned before, our value of $P_{zz} = -0.32 \pm 0.01$ implies an electron polarization of 97%—98% and not a value $> 105\%$. The external magnetic field region across the foil (160 G) is indeed shielded on both sides of the foil. The transition region from the strong field region inside the foil to the weak field region (10 G) was so arranged to permit an adiabatic transition. At this time I would certainly not exclude the possibility that an enhancement of the electron polarization may occur near the foil (last layers)-vacuum interface.

Weidenmüller:

Could you explain in some more detail what you mean by the pumping mechanism?

Ebel:

The pumping mechanism, which is envisaged as a possible explanation of the very large ($\sim$98%) electron polarization observed, depends upon the fact that a deuteron passing through the film undergoes a large number of electron pickup and loss processes. The ratio of the number of deuterium atoms with electron spin-up to those with spin-down thus depends both on the ratio of capture cross sections, which is close to unity, and also on the inverse ratio of loss cross sections. Since one component of the 3d band is filled, if the transfer to the 3d band dominates (and there are reasons to believe this, backed up by a crude calculation), then one may expect the probability of loss of a spin-down electron to be much greater than that of a spin-up one.

Clausnitzer:

What is the effect of the tranverse magnetizing field on the deuterons?

Kaminsky:

Negligible.

Clausnitzer:

Concerning the measurement where the weak field was rotated 90°, was the strong magnetizing field rotated at the same time?

Kaminsky:

No.

Clausnitzer:

Did you measure the polarization of non-channeled deuterons?

Kaminsky:

    Such a determination was indeed attempted by off-setting the beam direction by several degrees from the Ni[110] direction, and by increasing the primary deuteron energy so that the energy of the emerging non-channeled beam was ≈ 110 keV. Since in our experiments the collimation of the emergent beam was very high (half angle of acceptance 0.15° at a scattering angle of $0° \pm 0.1°$) and could not be gradually decreased externally, the intensity of the emergent normal beam component was too weak to permit a determination of the degree of polarization in a reasonable amount of time. We plan to do this experiment, however, with a different collimation system. I would like to point out that our experiments with *poly*crystalline Ni foils revealed no significant tensor polarization of the emergent D^0.

Moak:

    Is it possible that some channeled deuterons could emerge (as ions) from the foil but convert to D^0 atoms before being swept out of the beam by the electrostatic field?

Kaminsky:

    Our rest gas pressure, $1-2 \times 10^7$ Torr, was low enough to prevent the formation of any significant amount of D^0 by charge transfer collisions of D^+ with the rest gas.

Huber:

    Can you give some information about the growing of the single-crystal foil?

Kaminsky:

    The foils used in our experiments were produced by strain free spark-cutting slices (approx. 300 mm × 300 mm × 0.10 mm) from a large Ni single-crystal block. These slices were planed by spark-cutting to a thickness of approximately 20−25 µm. Subsequently these slices were "thinned out" by electropolishing to thicknesses varying between 1.2 and 2.5 µm. Our attempts to grow monocrystalline Ni films with the desired orientation epitaxially with a mosaic spread of less than 0.5° have been unsuccessful so far. We have been able to grow 0.8 µm monocrystalline Au films but not monocrystalline Cu or Ni films.

A Source of Polarized Protons and Deuterons for the Terminal of a Pressurized 6-MV Van de Graaff Accelerator

H. PAETZ gen. SCHIECK, C. E. BUSCH, J. A. KEANE, and T. R.
DONOGHUE, Ohio State University, Columbus, USA*

A compact source of polarized protons and deuterons for use in the
terminal of the pressurized 6-MV CN Van de Graaff accelerator is
being constructed at Ohio State University. Such a source seems
highly desirable to fill the gap between the sources working at lower
energies with Cockcroft-Walton accelerators and sources on tandems,
and would therefore cover an energy region only partly explored with
polarized beams. In addition the existing neutron and proton polarimeters would make complicated triple scattering experiments possible.

The basic design of this "conventional" source follows that of the
Basel sources [1] while the low power (4 kW) and small space (1.6 m ×
0.9 m diam) available dictate the design details (fig. 1). After dissociation by a 20 MHz, 250-W rf oscillator the atomic beam is formed by
a multicapillary nozzle (~1000 capillaries with 0.1 mm I.D.). A short
(13.3 cm) sextupole magnet separates the spin states. It uses Alnico
V permanent magnets and 2V Permendur pole tips with a constant separation of 8 mm and a pole tip field of ~9300 G. The beam passes
through a medium and low field transition region with permanent magnets providing fields of 100 ± 10 G and 8.5 ± 4 G over the transition
regions. The transistorized oscillators operate at frequencies of 352
MHz and 8 MHz for deuterium and 12 MHz for hydrogen. The beam is
ionized in the strong (850 G) field of a solenoid. The ions are extracted at 5 kV, focused to a crossover in the center of a rotatable
Wien filter and injected into the accelerator by the main lens. The
Wien filter has an electromagnet and allows the spins to precess into
any desired direction.

The source consists of four square chambers, three of which are
pumped by Ti sublimation pumps which use the chamber walls as pumping areas. With a gas flow of 0.04 Torr ℓ/s pressures of $5 \cdot 10^{-5}$ Torr,
$5 \cdot 10^{-6}$ Torr, and $5 \cdot 10^{-7}$ Torr in the first, second, and third chamber are
obtained. The pumps have Ti supplies which last about 30 days. The
power consumption of the whole source is ~2.5 kW.

Measurements on a test bench yielded an atomic beam intensity of
$7 \cdot 10^{15}$ atoms/cm^2·s at the location of the ionizer. Ions were extracted

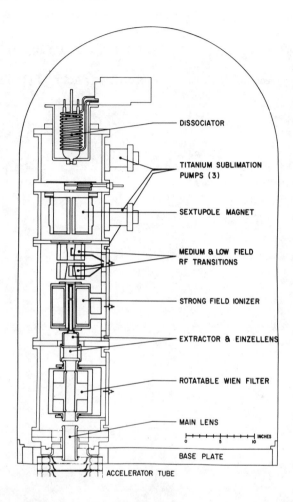

DISSOCIATOR

TITANIUM SUBLIMATION PUMPS (3)

SEXTUPOLE MAGNET

MEDIUM & LOW FIELD RF TRANSITIONS

STRONG FIELD IONIZER

EXTRACTOR & EINZELLENS

ROTATABLE WIEN FILTER

MAIN LENS

INCHES

BASE PLATE

ACCELERATOR TUBE

Fig. 1

and an analyzed deuteron beam of about 40 nA was accelerated to 150 kV to measure the tensor polarization with the $T(d,n)^4He$ reaction. A tensor polarization of −0.43 for the (3) → (5) transition was obtained as well as a partial inversion of the sign of this polarization by subsequent low field transitions at 8 MHz. The results clearly indicate that higher rf power levels of the transistorized oscillators for both transitions are needed.

The source will be mounted on the terminal as soon as a freon cooling system and all remote controls are available.

REFERENCE

*Supported in part by the National Science Foundation.

[1] H. Paetz gen. Schieck et al., Helv. Phys. Acta 40 (1967) 40.

DISCUSSION

Glavish:
   Have you experienced any difficulties with hydrogen instabilities
   in the titanium pumps?

Schieck:
   No.

The University of Minnesota
Polarized Negative Ion Source

C. H. POPPE, V. SHKOLNIK, and D. L. WATSON, University of
Minnesota, Minneapolis, USA*

A polarized negative ion source for use on an MP tandem Van de Graaff
is under construction at the Williams Laboratory of the University of
Minnesota. This source, which is of the atomic-beam type, is designed to produce both polarized $^1H^-$ and $^2H^-$ ions.

The negative ion injector of the MP tandem typically operates at
a potential of -150 kV. The polarized negative ion source will be at
a potential of about -40 kV relative to the injector, so that the energy
of the polarized ion beam presented to the MP accelerator will be the
same as that of the normal unpolarized beam. Consequently, the entire source assembly, at a potential of about -190 kV relative to the
laboratory, is supported on an insulated platform inside the injector
high-voltage cage.

Fig. 1 shows the layout of the polarized source assembly and beam
transport system. The polarized source is at an angle of 45° to the
axis of the accelerator-injector. Source vacuum boxes are supported

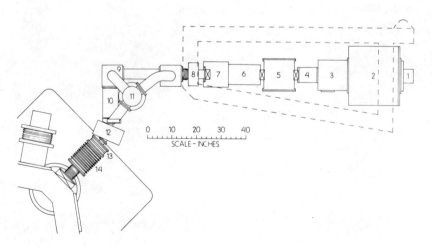

Fig. 1. Layout of the polarized negative ion source.

from an I-beam framework (dotted lines) which sits on the insulated platform. Box 2 houses a three-stage differential pumping column and peelers for formation of the atomic beam from the dissociator (1). Box 3 houses the sextupole magnet and the low-field rf transition unit [1]. High-frequency rf transition units[†] operating at frequencies of 1479 MHz for protons and 332 MHz and 456 MHz for deuterons will be placed at 4. Neglecting an unpolarized background the various rf transition units will make possible polarized proton beams of $P_z = \pm 1$ and polarized deuteron beams of $P_z = \pm 2/3$, $P_{zz} = 0$ or $P_z = \pm 1/3$, $P_{zz} = \pm 1$.

Box 5 houses the strong-field ionizer [1] and the positive ion beam undergoes charge exchange in Na vapor at location 6. A longitudinally polarized negative ion beam of maximum energy 5 keV is then presented to the beam transport system. Pumping manifolds for the charge-exchange canal and the beam transport system are located at positions 7 and 11, respectively.

The negative ion beam is focused by a gridless einzel lens (8) so that, after a 90° deflection by electrostatic mirror (9), it comes to a focus at the center of spin-rotating solenoid (10). After passing through a 45° inflection magnet (12) the beam is focused (13) and accelerated through 40 kV (14) to the injection potential and enters the 0° port of the normal-source box.

REFERENCE

*Work supported by the U.S. Atomic Energy Commission.
†These units were developed by Anac, Ltd. and purchased from Ortec, Inc.

[1]   R. N. Boyd et al., Nucl. Instr. 63 (1968) 210.

Recent Performance of
the Grenoble Polarized Source

J. L. BELMONT, G. BAGIEU, F. RIPOUTEAU, D. BOUTELOUP,
R. V. TRIPIER, and J. ARVIEUX, Institut des Sciences Nucléaires,
Grenoble, France

The Grenoble University variable energy spiral-ridge cyclotron can accelerate protons and α-particles to energies between 20 and 60 MeV, deuterons between 10 and 30 MeV, and ^{3}He-particles between 10 and 80 MeV. A conventional polarized source built by the Saclay group is associated to the cyclotron. The ionization is done in a longitudinal magnetic field ionizer. Its characteristics are given in table 1. The atomic beam is about 10^{16} atoms/s over a spot of 1 cm diam.

Table 1. Characteristics of the strong field
ionizer

Length	200 mm
Magnetic field	800-1200 G
Electron current	80 mA
Electron energy	350 eV
Efficiency	$\sim 10^{-3}$
Emittance	3000 mm × mrad at 12 keV

The polarized source is in a vertical position above the cyclotron. The beam, once ionized, is injected into the cyclotron through a hole drilled in the upper pole piece. The characteristics of the axial injection system are given in table 2. An unpolarized source of the duoplasmatron type can be alternately used with the same injection device. It is possible to shift from one source to the other in a few seconds.

The following currents (in nA) can be obtained for 40-MeV protons: after ionization, 1500–2000; cyclotron center, 600–800; R = 15 cm, 80; extraction radius (R = 86 cm), 45; extracted beam, 30; beam on target, 20. The polarization varies between 0.6 and 0.7. All the values given have been obtained over long periods of time (typically 24–72 hours). Extracted intensity up to 100 nA and polarizations up to 0.8 have been

Table 2. Some of the characteristics of the
         injection device

Voltage at the ionizer	9–16 kV
Hole in the pole piece	
length	1350 mm
diam	71 mm
Diam of the internal tube	16 mm
Number of magnetic lenses	22
Transparency of the over-all system	~40%
Efficiency of the electrostatic mirror	without bunching – 7%
	with bunching – 25%

observed, but they appeared to be not easily reproducible. A bunching device has been installed recently which gives an improvement of a factor of three to four in the internal beam at the radius R = 15 cm.

Some experiments which have already been done using polarized proton beams from 30 to 50 MeV have been reported to this conference [1, 2]. Another experiment in progress is proton elastic and inelastic scattering on medium-weight nuclei with an overall resolution of 50 keV [3].

In the near future we intend to develop polarized deuteron beams. A vector-polarized beam of 6 nA at the target has already been obtained using a low-field transition only. Work is in progress to obtain vector and tensor polarization using two low-field and one high-field rf transition.

The major part of the results reported here are based on internal unpublished reports [4] available on request.

REFERENCES

[1] P. Catillon, Third Polarization Symp.
[2] J. Arvieux et al., Third Polarization Symp.
[3] P. Locard et al., to be published.
[4] J. L. Belmont and G. Bagieu, Report 2/1969, Laboratoire du Cyclotron, Grenoble; F. Ripouteau, D. Bouteloup, and R. V. Tripier, Report 3/1969, Laboratoire du Cyclotron, Grenoble.

The Australian National University
Atomic Beam Polarized Negative Ion Source

H. F. GLAVISH,[*] University of Auckland; B. A. MacKINNON and
I. J. WALKER, ANAC Auckland, New Zealand; S. WHINERAY and
E. W. TITTERTON, Australian National University, Canberra, Australia

A polarized negative ion source has been constructed for the Australian
National University EN tandem accelerator. A schematic diagram is
shown in fig. 1. H or D atoms are formed from molecules in a 20-MHz
discharge. The atoms effuse from the Pyrex discharge tube through a
single canal. Two skimmers of 3.5-mm diam collimate the atomic beam
before it enters the sextupole magnet. The pressure in the discharge
is 1-2 Torr and the gas flow 3 Torr ℓ/s. The sextupole magnet, excited electrically, is 35 cm long, has a uniform taper (0.7 cm diam
entrance and 1.3 cm diam exit aperture) and a maximum pole tip field
of 7.5 kG. Adiabatic passage rf transition units fit externally around
a 2-cm diam $\times$ 22 cm long Pyrex tube located between the sextupole
and ionizer. Transitions $3 \rightarrow 5$ and $2 \rightarrow 6$ are used for D and $2 \rightarrow 4$ for H.

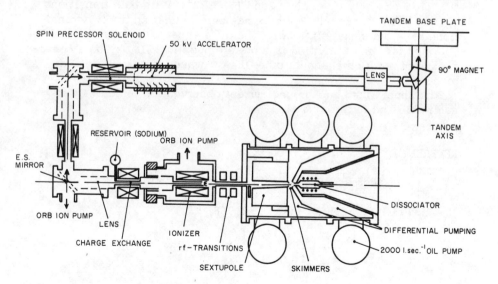

Fig. 1. A.N.U. polarized ion source.

The ionizer is essentially the same as that described by Glavish [1], and its center is 50 cm from the sextupole exit. The atomic flux at the ionizer is $1-2 \times 10^{16}$ atoms/s concentrated in an area < 1 cm diam. The 5 keV output beam is focused into the charge exchange canal (1.6 cm $\times$ 15 cm long) containing sodium vapor generated by heating metallic sodium in an external reservoir. A solenoid around the canal prevents depolarization. The negative ions, after leaving the canal, pass through a sequence of two spin precessors, each consisting of an electrostatic mirror and a solenoid. The combination of the two precessors enable any direction to be obtained for the spin quantization axis relative to the beam direction. An einzel lens just after the charge-exchange canal operates with a sufficient retarding potential to completely stop the transmission of unpolarized H^- (or D^-) ions obtained from H_2^+ (or D_2^+) ions.

A 50-kV accelerator is used to provide the required tandem injection energy. The polarized ions will be injected on to the tandem axis via a 90° magnet just before the tandem base plate. The beam current from the source after the 90° magnet is 100 nA with an emittance area of 6 cm rad $eV^{1/2}$ for the 0% contour and 2 cm rad $eV^{1/2}$ for the 10% contour. Higher currents (200 nA max.) are obtained, if a gridded lens is used between the ionizer and charge-exchange canal. However, sodium accumulation on the grid wires reduces the transmission by a factor of 2 after 40 hours running time. The tandem acceptance is at least 5 cm rad $eV^{1/2}$ at 3 MV terminal potential and reasonable transmission is expected.

Deuteron polarizations of $P_{ZZ} = -0.83$ ($3 \rightarrow 5$) and $P_{ZZ} = +0.74$ ($2 \rightarrow 6$) were measured using the $^3H(d,n)^4He$ reaction and assuming the reaction proceeds only by s-waves and $J^\pi = 3/2^+$. With both transitions operated simultaneously P_{ZZ} becomes zero indicating the two transitions have the same efficiency. The difference in the magnitude of P_{ZZ} obtained for each transition is therefore puzzling. It may be a consequence of reaction channels other than s-waves in the $^3H(d,n)^4He$-reaction.

The unpolarized background current observed for deuterons was negligible and for protons 10-15%.

REFERENCE

*Present address: Stanford University, California

[1] H. F. Glavish, Nucl. Instr. 65 (1968) 1.

Source of Polarized Negative Hydrogen Ions for the ETH Tandem Accelerator

W. GRÜEBLER, V. KÖNIG, and P. A. SCHMELZBACH, Laboratorium für Kernphysik, ETH, Zürich, Switzerland

This paper reports the construction and performance of the polarized ion source, an atomic beam type source with charge exchange in sodium vapor. The source is installed on the ETH tandem accelerator. A schematic diagram of the apparatus is shown in fig. 1. The polarized atomic beam emerges from the quadrupole magnet at the left and enters a region in which first a strong- and then a weak-field r.f. transition may be induced. The strong r.f. field changes the hyperfine component of the deuterium beam from 3 to 5, giving a theoretical tensor polarization of $P_{33} = -1$ after strong-field ionization. The weak-field transition reverses the sign of the tensor polarization, giving $P_{33} = +1$. Using the weak-field transition alone gives the deuteron beam a pure vector

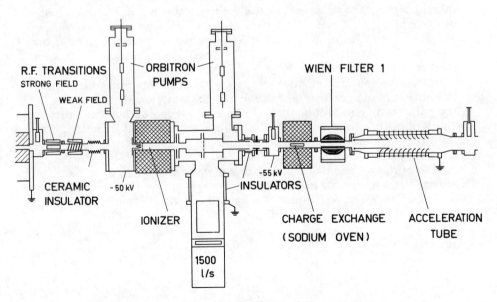

Fig. 1. Schematic diagram of the polarized ion source.

polarization of 2/3. Only the weak-field transition is used for a ^{1}H-atomic beam and a theoretical polarization of P = 1 is obtained.

The atoms are ionized by electron bombardment. The positive ions are accelerated to 5 keV. At this energy charge exchange takes place in sodium vapor in a strong longitudinal magnetic field. Voltages are applied to the various points of the system as indicated in fig. 1. The negative-ion beam emerging from the source has an energy of 60 keV which meets with the beam acceptance requirements of the tandem accelerator.

After ionization, the polarization axis is parallel to the momentum axis of the beam. Since polarization experiments often require certain definite angles between the beam and the polarization axes crossed electric and magnetic fields (Wien-filter) are used to obtain the desired spin orientation. Besides rotation of the spin, the Wien filter has the task of separating the polarized and unpolarized beams. The main part of the unpolarized beam arises from H_2^+ or D_2^+ ions which are produced in the ionizer and later dissociated and transformed to H^- or D^- in the charge exchanger. A second Wien-filter at ground potential (not shown on fig. 1) has larger dimensions such that also the necessary spin angle range for the deuterons can be covered, since the first Wien-filter is too small for this variation. Both Wien-filters can be rotated about the beam axis so that any desired spin orientation can be obtained.

The beam intensity measured in front of the second Wien-filter is 15 nA, decreases to 8 nA after the Wien-filter and to about 2 nA at the entrance of the acceleration tube of the tandem accelerator. The beam intensity focussed on a spot of 0.3 cm diameter in a scattering chamber is typically 1.5 nA. The tensor polarization was measured before and after acceleration to 7 MeV and yields within statistical error the same value, namely $P_{33} = \pm 0.75 \pm 0.01$. The vector polarization P_3 was measured to be 0.50 ± 0.01.

Attractive features of this source are its long term reliability and stability, proved good for more than a year. Only small variations in the polarization and intensity have been observed during this period. The flexibility of the spin rotation system proved to be excellent during the measurements of the vector and the three tensor polarization analyzing powers of deuterons elastically scattered on ^{4}He. An increase of the atomic beam intensity which is at present only 2.6×10^{15} atoms/s and an improvement of the ion beam optics may increase the polarized negative beam intensity by two orders of magnitude.

A Source for Polarized Lithium Ions

H. EBINGHAUS, I. Institut für Experimentalphysik der Universität,
Hamburg, Germany

In Hamburg we have constructed an atomic beam source for polarized
lithium ions [1]. The atomic beam is formed by a laval nozzle in the
front of a steel oven heated to 800-900° C. The hyperfine-structure
components are separated by a magnetic sextupol field. ^{6}Li with nuclear spin I = 1 has 6 HFS components like the deuteron. They are
numbered 1 to 6 in the order of decreasing potential energy. ^{7}Li with
I = 3/2 has 8 components. Separation of the upper three components
from the lower three of ^{6}Li and ionization in a weak field gives a vector polarization $P_3 = 1/3$ and a tensor polarization $P_{33} = -1/3$. High-
frequency transitions in a weak and a strong field and ionization in a
strong field give the following polarizations:

H-F-transitions				Polarization	
strong field	3-5	2-6	weak field	P_3	P_{33}
1	−	−	−	0	0
2	+	−	−	1/3	−1
3	+	−	+	−1/3	1
4	−	−	+	−2/3	0
5	−	+	−	1/3	1
6	−	+	+	−1/3	−1

The polarized beam is ionized by surface-ionization on a heated tungsten tape by the Langmuir-Taylor effect. The lithium atoms strike the
tungsten surface and evaporate after a time of the order of msec as
positive ions. Under suitable conditions the ionization efficiency is
nearly 100%, but it was not certain that surface-ionization conserves
nuclear polarization.

The polarization was measured by the ^{2}H(^{6}Li, ^{4}He)^{4}He-reaction at a
Li energy of 600 keV [2]. For this purpose the Li ions were accelerated
by a Van de Graaff generator with the target and counters in the high
voltage terminal. The analyzing power for the tensor polarization can
be calculated [3] provided that the 22.5 MeV level of the ^{8}Be compound
nucleus has [4] $J^\pi = 2^+$ and the initial orbital angular momentum is zero

at this low energy. Then the differential cross section for a polarized beam is

$$\sigma(\theta, \phi) = \sigma_0(\theta) \{1 + \frac{1}{4} P_{33} (1 - 3 \cos^2 \phi \sin^2 \theta) \}$$

$\sigma_0(\theta)$ is the unpolarized cross section, θ the scattering angle and ϕ the azimutal angle.

We found from the asymmetry measurements:

Transition	P_{33}
3-5	-0.70 ± 0.10
2-6	$+0.68 \pm 0.10$
3-5 and	
weak field	$+0.66 \pm 0.11$

The deviation of about 30% from the ideal tensor polarization $P_{33} = \pm 1$ may be caused partially by surface ionization, but we find that surface ionization conserves at least 70% of the nuclear polarization. The ion beam of 1 - 2 μA was reduced to half if the ionizer was placed in a distance of 100 cm from the high-frequency transitions. Therefore, this source is very suitable for use with a cyclotron, as the neutral beam may be introduced radially and ionized by a simple hot tungsten tape in the center of the cyclotron.

REFERENCES

[1]   H. Ebinghaus, U. Holm, H. V. Klapdor, and H. Neuert, Z. Phys. 199 (1967) 68.
[2]   U. Holm et al., Z. Phys. 233 (1970) 415.
[3]   L. J. B. Goldfarb, Nucl. Phys. 7 (1958) 622.
[4]   H. Bürgisser et al., Helv. Phys. Acta 40 (1967) 185.

DISCUSSION

Keaton:

I find this very exciting. Do you have plans to put ^{7}Li in the source? It would be nice to have a polarized spin-3/2 beam.

Ebinghaus:

We will try ^{7}Li after we have done some measurements with ^{6}Li.

Ohlsen:

Have you thought through the rf-transition scheme that would give suitable flexibility and high polarizations in the ^{7}Li case?

Ebinghaus:

We have tested weak field transitions. For high field transitions you need VHF.

The Karlsruhe Polarized Ion Source

H. BRÜCKMANN,* Institut für Experimentelle Kernphysik des
Kernforschungszentrums und der Universität Karlsruhe, Germany

Lamb-shift-type sources for polarized ions exhibit some special characteristics which make them advantageous for applications at different types of accelerators. The most evident of these features is a
high phase space density of negatively charged polarized ions [1]. The
very small emittance leads to an ion beam with excellent properties
which can be accepted almost completely by any type of accelerator.
Furthermore, the source is rather inexpensive and can easily be
operated.

We have extended our investigations to problems dealing with
the production of positively
charged polarized hydrogen
beams. Such a Lamb-shift-type
ion source C-LASKA was designed and constructed for operation at the Karlsruhe isochronous cyclotron. The design
of the source took into account
that this type of source might
also be operated in producing
negatively charged ions for tandem accelerators. The charge
of the polarized ion beam can
easily be altered by changing
only the gas or vapor used for
the selective ionization process.

The set-up of C-LASKA is
schematically shown in fig. 1.
In comparison to the sources described elsewhere in the literature [2] the Karlsruhe Lamb-shift
source is characterized by the
following special features:

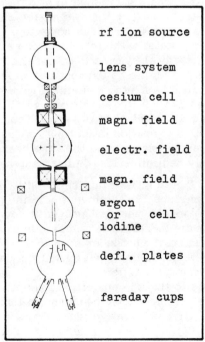

rf ion source

lens system

cesium cell

magn. field

electr. field

magn. field

argon
  or   cell
iodine

defl. plates

faraday cups

Fig. 1. LASKA schematic.

a) As usual the metastable atomic beam is produced by charge exchange in cesium. But the design of our cesium cell was modified in several aspects: a very low cesium consumption was desired; we aimed at a very stable long period operation; a very small contamination by diffusing cesium is desirable; and the cesium should be continuously repurified. These requirements are satisfied by a cell operating in a closed loop cesium distillation circuit. The volume used for the charge-exchange reaction is defined by the spatial overlap of the directed cesium vapor beam with the primary ion beam. To prevent the cesium from diffusing into the surrounding parts of the apparatus, additional baffles were installed.

b) By applying a zero-crossing magnetic field the degree of polarization is increased by a nonadiabatic process [3, 4] which changes the occupation numbers of the hyperfine structure levels. The magnetic field was designed to minimize the length of the source. The distance between the two solenoid magnets is only 23 cm. The shape of the magnetic field is determined by the shape of the iron shielding. The dimensions of the two magnets are 7 cm in length, 4.5 cm i.d., and 18 cm o.d. The process is easily controlled by the magnet current only, and no additional correcting coils are needed.

c) A rotating homogeneous electrical field is provided inside the two magnets mentioned in b). The electric field is needed for two purposes. First, it removes all the charged components from the neutral beam, and second it is necessary for the coupling of the 2S to the $2P_{1/2}$ states. The rotational character of the electrical field assures a rotational symmetric disposal of the charged-beam components and simplifies the adjustment of the beam.

d) The source contains a monitor device. Periodically, the polarized ion beam is electrically deflected into two Faraday cups. This is accomplished by a 2×10^{-4} sec long rectangular electrical pulse with a repetition rate of 50 Hz. This device allows monitoring the beam intensity as well as the beam polarization (by quenching) with only negligible loss in beam intensity.

A prototype of this Lamb-shift source (A-LASKA) has been in operation since 1968 to investigate the physical problems involved with the development of Lamb-shift sources [1, 4, 5, 6, 7, 8]. A very similar source (B-LASKA) is operated at a small Cockcroft-Walton accelerator. This source produces negatively charged polarized ions and the accelerator is designed for tandem operation. Many different charge-exchange reactions were studied systematically. Our main interest was to find the properties required for the production of an intense metastable atomic beam and to ionize the metastables selectively to positive or negative ions.

Fig. 2 shows the characteristic of the cesium charge-exchange reaction. The diagram shown was registered with an X-Y recorder. The temperature of the cesium reservoir is taken to be the abscissa and

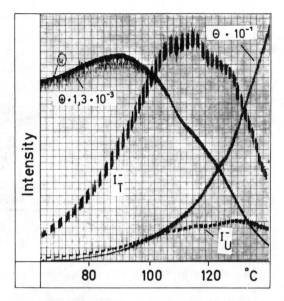

Fig. 2. The absorption and different yields of
cesium measured at 1 keV beam energy.

intensities are plotted on the ordinate. The curve denoted by ⊕ shows
the absorption of the primary positive beam. The curve denoted by ⊖
demonstrates the yield of negative ions which are produced in the
cesium cell. The neutral beam components (metastables and ground
state atoms) were registered by using the charge exchange reaction
in argon. I_T^- denotes the total negative beam intensity produced by
the neutral beam. I_U^- was measured by quenching all the metastables
in the neutral beam. Under optimum conditions this cell is operated
at 150° C. Up to now cesium turned out to be best for the production
of intense metastable beams [5].

The selective ionization was studied in a very similar way. Fig. 3
shows as an example the yields $I_T^\pm$ and $I_U^\pm$ in iodine (positive ions) and
in argon (negative ions) at an energy of 1 keV. The intensities were
measured as a function of the gas density in the charge-exchange cell.
The arbitrary scale of the intensity I is identical for all four measurements. The polarization to be expected can be calculated from these
intensity measurements [8].

The comparison of the argon and iodine reactions shows that iodine
yields an IP^2 which is almost two times greater than the corresponding
value for the argon charge-exchange collision [7]. Knutson has independently investigated the iodine charge-exchange reaction and communicated very similar results for particularly chosen iodine vapor
densities [9].

The nuclear tensor polarization P_{33} of the beam was studied by

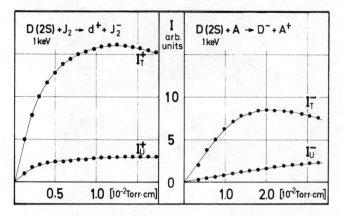

Fig. 3. Yield of the selective ionization by iodine and argon, respectively.

measuring the neutron asymmetry of the $T(d,n)^4$He-reaction. Fig. 4 shows as an example the results obtained for the iodine charge-exchange reaction. P_{33} is almost completely independent of the iodine vapor pressure, and at 1 keV a value of $P_{33} = -0.64$ is obtained which is comparable with the polarizations achieved either for negative ions from Lamb-shift sources or for positive beams from atomic beam sources.

In conclusion, the Lamb-shift-type ion source is very suitable to produce either positively charged or negatively charged polarized beams. Negative beams are reliably obtained with an emittance of less than 0.7 [cm rad $\sqrt{eV}$] and at intensities above 0.1 μA. The source C-LASKA which will produce posi-

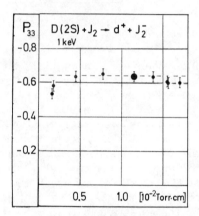

Fig. 4. The experimentally measured tensor polarization P_{33} versus the iodine density of mass.

tive ions for the Karlsruhe isochronous cyclotron will be able to deliver a polarized beam of the order of 1 μA at an emittance of 1 [cm rad $\sqrt{eV}$].

REFERENCES

*Team members: V. Bechtold, D. Finken, L. Friedrich, K. Kamdi, E. Seitz, and P. Ziegler.

[1] H. Brückmann, D. Finken, and L. Friedrich, Z. Phys. 224 (1969) 486.

[2] E.g., W. Haeberli, Ann. Rev. Nucl. Sci. 17 (1967) 406; T. B. Clegg, G. Plattner, L. Keller, and W. Haeberli, Nucl. Instr. 57 (1967) 167.

[3] P. G. Sona, Energia Nucl. 14 (1967) 295.

[4] V. Bechtold, H. Brückmann, D. Finken, and L. Friedrich, Z. Phys. 231 (1970) 98.

[5] V. Bechtold et al., KFK-Report Nr. 1256.

[6] V. Bechtold et al., Third Polarization Symp.

[7] H. Brückmann, D. Finken, and L. Friedrich, Nucl. Instr. (in press).

[8] H. Brückmann, D. Finken, and L. Friedrich, Phys. Lett. 29B (1969) 223.

[9] L. D. Knutson, Phys. Rev., to be published.

DISCUSSION

Cramer:

What are the changes in the optics which will give you the 1 μA positive polarized beam which you mentioned?

Brückmann:

We had not put too much effort into optimizing the design of ion optics between rf-source and the Cs-cell. By redesigning this part we will gain a factor of three.

Performance of the LASL Polarized Source

J. L. McKIBBEN, G. P. LAWRENCE, and G. G. OHLSEN, Los Alamos
Scientific Laboratory, New Mexico, USA*

Our construction of a source of polarized negative ions based on charge
exchange reactions in cesium and argon was initiated in 1965; the
initial design was reported five years ago at Karlsruhe. We have made
many changes from the early design. A description of the source as of
the end of 1968 has been published [1]; this description is still fairly
accurate.

Installation. The source was installed and delivered its first beam
through the tandem accelerator on July 4, 1969, and experiments com-
menced soon after. Experience soon showed that we needed remote
control of a number of variables and this was incorporated during late
1969. During the first half of 1970 a total of 700 hours of operating
time has been accumulated. The installation layout is shown in fig. 1.

The beam is inflected 58° by a special double-focusing magnet;
the choice of angle was dictated by space and focusing considerations.
Rotation of the direction of spin from the axial direction to any required direction is done by means of a crossed-field device which we
call a spin precessor. The precessor can be mechanically rotated
about its axis. The beam is brought to focus at the middle of the precessor by means of an electrostatic quadrupole doublet; a cup can
intercept the beam there for measurement of the source intensity. A
quadrupole singlet which follows the precessor is used to correct for
cylindrical focusing effects of the spin precessor fields. The beam
is brought to a second focus about 1/2 meter ahead of the accelerating
tube by means of the 58° magnet and the einzel lens. A special viewer
near the second focus has proven useful in finding the proper adjustments of focus and of the several steering elements.

Transmission. The emittance of the beam as it leaves the source
is known from the geometry of the source, and by measurement, to be
approximately 1.0 cm rad eV$^{1/2}$. This is substantially smaller than the
acceptance of the tandem Van de Graaff. However, the best transmission we have seen is 65% (from the precessor cup to the target through
a 2.5 × 2.5-mm aperture) for deuterons and 50% for protons.

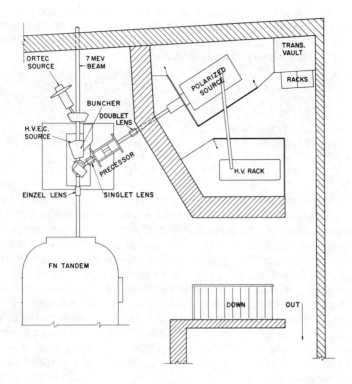

Fig. 1. Installation arrangement for the LASL polarized
ion source.

Polarization and intensity. The quenchable fraction of the beam
varies in the range 86 to 90% for protons and 74 to 80% for deuterons
under normal operating conditions. Better polarization values are obtained by rotating the smallest aperture into position at the focal
point of the first acceleration lens and using about 1/3 the normal
argon flow rates; under these conditions, polarizations of 93% for protons and 86% for deuterons have been measured. The best proton current delivered so far to a useful target is 107 nA; the best deuteron
current so far observed is 208 nA with 74% polarization. Most of our
usage has been with deuterons for which 60 to 100 nA is usually available while the available proton current is usually only about 25% less
than the deuteron current.

Spin states available. Certain spin states require a strong field
(with respect to the hyperfine interaction) in the negative-ion-formation
region, in order that large nuclear polarization can be maintained.
However, too strong a field produces a low intensity beam with poor

emittance. Our practice has been to use 6G for protons and 60G for deuterons; fortunately 60G does not cause a serious loss of intensity for deuterons. The polarizations of ions produced by a pure metastable beam are, for the magnetic fields of interest:

Ion	m_I	P_3	P_{33}	Field (in G)
H⁻	+1/2	1	--	6
H⁻	-1/2	-0.12	--	6
D⁻	+1	1	1	6 (or 60)
D⁻	0	0.012	-1.966	60
D⁻	-1	-0.984	0.952	60

Actual polarizations are obtained by multiplying these numbers by the quenchable percentage value. This is the fraction of the negative ion beam produced through the metastable process, and is measured by quenching the metastable portion. For proton or deuteron vector polarization experiments, the spin is usually precessed to the vertical and the precessor is mechanically rotated to reverse the spin direction. To reverse the vector polarization in the case of deuterons it is often better to turn to the $m_I = -1$ state, since mechanical rotation of the precessor may lose some of the beam; the steering elements can be adjusted to compensate for the beam shift effect, but it takes time. For experiments involving tensor polarization, the spin axis is usually precessed in the horizontal plane and rotation of the counters is used instead of rotation of the precessor.

Stability and intensity. The source runs unattended for many hours at a time. The polarization is quite stable; deviations of less than one percent during a run are common. The temperature of the cesium cell and its reservoir are each thermostatically controlled. The loss of cesium is < 0.1 gram per hour and its cleanup during servicing has been easy. Good current stability does require finding stable operating conditions; in this respect we have been greatly aided by a transverse adjustment on the accel electrode which may be operated while the source is at high voltage. That this source is capable of much greater output we have little doubt, but recently our time has been spent mainly on its installation and utilization.

REFERENCE

*Work performed under the auspices of the U.S. Atomic Energy Commission.

[1] G. P. Lawrence, G. G. Ohlsen, and J. L. McKibben, Phys. Lett. 28B (1969) 594.

DISCUSSION

Catillon:

Are you planning an ion source for the meson factory?

McKibben:

The meson factory people have decided that the polarized proton source should be a negative ion source, since such ions can be easily separated from the positive ions after acceleration. So far they have not asked us for help in building such a source. If they do so, it is clear that a big change in pumps will be required, since they do not allow any diffusion pumps to be connected to their system. The pumping of argon would have to be done cryogenically. Changes in the ion source would no doubt be made. We believe a much higher current can be achieved, but that is only a speculation.

The University of Washington
Lamb-Shift Polarized Ion Source

E. PREIKSCHAT, G. MICHEL, G. W. ROTH, J. G. CRAMER, Jr., and
W. G. WEITKAMP, University of Washington, Seattle, USA*

The Lamb-shift polarized ion source of the University of Washington
utilizes a double-quenching technique as suggested by Sona [1] and
static-quenching fields as already embodied in a number of other
sources [2] (see fig. 1A). A number of source parameters were found
to be critical for an optimized beam output and reliable operation of
the source.

Large fluctuations and nonreproducibility in the metastable beam
intensity were observed when using a straight cesium canal geometry.
Deposits of what is thought to be CsH were formed at the canal walls
reducing the density of neutral Cs in the canal. The operation of the
Cs vapor cell was optimized by installing a reservoir chamber at the
center of the canal, as well as by using a higher temperature difference between the canal and Cs cell and increasing the pumping speed
in that region. This effectively eliminated the formation of deposits.

The distances between the various parts of the duoplasmatron and
the extraction and beam-focusing assembly were found to be less critical than the on-axis alignment. Overall alignment was optimized by
mounting the various components on an optical bench with the duoplasmatron and the magnetic bottle externally adjustable. Apertures 1
and 2 reduce the contamination of the gridded and baffled quenching
plates. The size of each aperture was selected to produce a beam
spread at the argon cell of 5 cm. The divergence of the neutral beam
was easily observable on a fluorescent screen.

The d^+ and d^- components in the beam were analyzed at various
points along the beam line as a function of Cs temperature. Similarly
the neutral component was measured using a secondary electron emission detector. Immediately behind the Cs canal the intensities of the
d^+ and the d^- components cross over at a temperature of 165°C, which
also corresponds to the maximum output of the d^0 beam. For hydrogen
this point occurs at 127°C. Behind the argon cell the crossover point
for deuterium occurs at 145°C while the maximum d^0 output remains
constant at 165°C. Using the beam-handling system shown in fig. 1B
the maximum polarization and metastable beam output occurs for

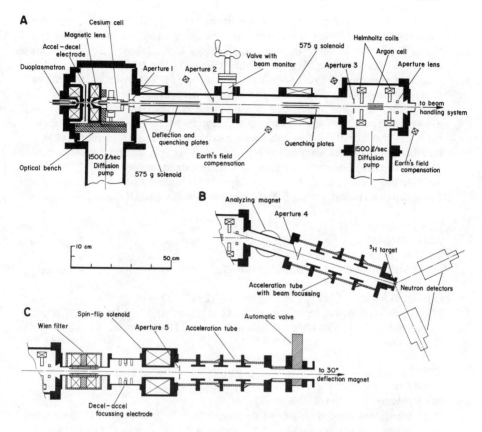

Fig. 1

deuterium at 125°C and for hydrogen at 90°C. This behavior suggests the existence of residual argon and other contaminant gases along the beam path causing premature ionization of the beam and a net reduction in polarization.

After the beam was analyzed and accelerated to 130 kV, the tensor polarization was monitored using the $T(d,n)^4He$ reaction. The largest output obtained with a doubly-quenched deuterium beam was 300 nA with a net polarization of $P_{zz} = -0.763 \pm 0.027$ and an unpolarized background contribution of 25%. Typically, the measured tensor polarization was between 65-70% of the theoretical value. For a singly-quenched beam the polarization was (-0.283 ± 0.020) and (-0.157 ± 0.020) for the first and second quenching assemblies, respectively. Premature ionization of the beam in the high magnetic field region and the presence of transverse magnetic fields off the beam axis are probably the dominant factors reducing the polarization. Efforts are being made to reduce both of these effects.

The transmission of the beam- and polarization-handling system was found to be 100% for the polarized component of the d⁻ beam when the Wien filter was used to precess the nuclear spin by as much as 180°. To achieve this the fields of the filter had to be carefully matched by appropriately shaping the electric field plates and shimming the magnetic field along the beam direction. The filter did produce some defocusing of the beam which made it necessary to install a second aperture lens behind the filter to focus the beam at aperture 5. The acceleration tube has strong focusing properties which can be adjusted by varying the potentials at the various electrodes. The insulating housing for the source is constructed using self-supporting PVC panels and is designed to operate at -150 kV for maximum transmission through the tandem Van de Graaff.

REFERENCES

*Work supported in part by the U.S. Atomic Energy Commission.

[1] P. G. Sona, Energia Nucleare 14 (1967) 295.
[2] T. B. Clegg, G. R. Plattner, L. G. Keller, and W. Haeberli, Nucl. Instr. 57 (1967) 167; G. Michel et al., Nucl. Instr. 78 (1970) 261.

DISCUSSION

McKibben:

I believe that if the hole diameter between the Cs cell and the reservoir were increased to about 0.5 cm, the cell temperature could be lowered to about 82°C where we operate our cell.

Lamb-Shift Polarized Ion Sources for Tandem Accelerators at Wisconsin and Triangle Universities Nuclear Laboratory

T. B. CLEGG and G. A. BISSINGER, University of North Carolina at
Chapel Hill and TUNL; W. HAEBERLI and P. A. QUIN, University of
Wisconsin, Madison, USA*

New Lamb-shift polarized ion sources are operating on tandem accelerators at the University of Wisconsin and at the Triangle Universities
Nuclear Laboratory. Both ion sources use a high intensity duoplasmatron and a symmetric "accel-decel" extraction geometry similar to the
Los Alamos design [1] to produce a 550 eV H^+ or 1100 eV D^+ beam.
This beam is focused by a magnetic lens into a cesium charge exchange
canal. Following the cesium canal are deflection plates to remove
charged components from the beam.

The two ion sources differ in the method to polarize the metastable
component of the beam. At Wisconsin the method of sudden field-
reversal [2] is used to obtain, ideally, complete proton polarization
for hydrogen or pure vector or tensor polarization for deuterium. Seven
coils produce the required static, axial, magnetic field which has a
maximum gradient of 80 G/cm and a gradient in the field-crossover
region of 0.5 G/cm. At TUNL, a "spin filter" [3] is used to retain
metastable atoms in a single hyperfine state while quenching the rest
to the ground state. The selected atoms have complete proton polarization for hydrogen and maximum, pure, tensor polarization for the
$m_I = 0$ component of deuterium. Deuteron beams with $m_I = 1$ have a
mixture of vector and tensor polarization.

The polarized metastable atoms enter an argon charge-exchange
canal and emerge as polarized H^- or D^- ions with spin quantization
axis along the beam. Prior to injection into the accelerator, the negative beam passes through two electrostatic mirrors each producing a
90° deflection and each followed by a spin-precession solenoid. By
suitable choice of the magnetic field in the two solenoids, the spin
quantization axis can be rotated to an arbitrary direction in space.

The polarized sources are operated at -30 to -50 kV with respect
to ground to provide the necessary injection energy for the accelerators.
About 30 kVA of power is required to operate the electrical and vacuum
equipment at high voltage. Cooling is provided by a closed deionized-
water system.

Beams from the Wisconsin ion source were first accelerated in June 1969. At TUNL the first beam was accelerated in July 1970, and initial tests are still in progress. The table below indicates the best performance achieved with both sources.

	Beam	Source current (nA)	Target current (nA)	Polarization Theor. max.	Achieved
Wisconsin	Protons	31	3	$P = 1$	$P = 0.85$
	Vector deuterons	185	65	$P_3 = 2/3$	$P_3 = 0.57$
	Tensor deuterons	70	20	$P_{33} = -1$	$P_{33} = -0.75$
TUNL	Protons	5.7	1	$P = 1$	0.75
	Deuterons $m_z = -1, 0,$ or $+1$	2.8		100%	67%

REFERENCES

*Work supported in part by U.S. Atomic Energy Commission. Work at TUNL supported also by North Carolina Board of Science and Technology and University Research Council at University of North Carolina.

[1] G. P. Lawrence, G. G. Ohlsen, and J. L. McKibben, Phys. Lett. 28B (1969) 594.
[2] T. B. Clegg, G. R. Plattner, and W. Haeberli, Nucl. Instr. 62 (1968) 343.
[3] J. L. McKibben, G. P. Lawrence, and G. G. Ohlsen, Phys. Rev. Lett. 20 (1968) 1180.

DISCUSSION

Donnally:

Is it possible to switch from strong field ionization to weak field ionization in the Wisconsin source without changing the other field coil currents to provide the proper adiabatic transitions in the field reversal region?

Clegg:

Yes. The field-crossover position moves about 2 mm when the field in the argon region is changed from 50 G to 2 G.

A Lamb-Shift Polarized Ion Source

J. SANADA and T. MIKUMO, Tokyo University of Education;
S. KOBAYASHI, Kyoto University; K. KATORI, S. SEKI, Y. TAGISHI,
and Y. NOJIRI, Tokyo University of Education; H. SAKAGUCHI,
Kyoto University, Japan

A Lamb-shift polarized ion source capable of producing pure states of
protons and deuterons was constructed. The polarized beams are obtained by RF-transitions on metastable states of hydrogen or deuterium
atoms [1]. A polarized ion source based on this principle was reported
by the Los Alamos group [2].

The top view of our apparatus is shown in fig. 1. A deuteron beam
from a duoplasmatron is extracted with a potential of about 8 kV, and

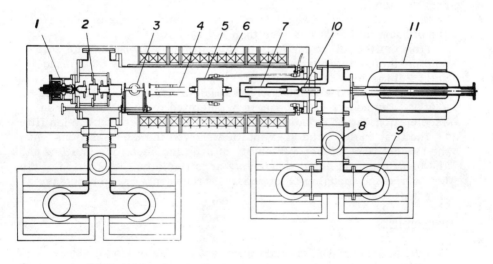

50 cm

Fig. 1. Schematic cross section of the TUE polarized ion source.

1	duoplasmatron	5	RF cavity	9	diffusion pump
2	lens & decelerator	6	solenoid	10	acceleration gap
3	Cs oven	7	Ar gas cell	11	Wien filter
4	deflection plates	8	cold trap		

decelerated to an energy of 1 keV. This beam, after focusing through an electrostatic lens, enters a Cs cell. The beam intensity is a few mA.

A Cs oven-Cs cell system is made of Pyrex glass and produces circulation of Cs vapor. Cs vapor is blown from an oven, 2 cm diam, through a nozzle, 1 cm i.d. and 5 cm long. Cs vapor condensed on the upper wall of the Cs cell, heated to about 50° C with warm water, is trapped and sent back to the Cs oven. With this Cs oven-Cs cell system, Cs vapor does not contaminate the ion source, and causes no discharge in the lens system. The concentration of Cs vapor can be controlled easily. The temperature in the Cs oven is around 200° C.

Deflection plates are 23 cm long and have a 2.5 cm separation. With an electric field of 40 V/cm, more than 99% of the charged particles can be swept out.

Metastable atoms [3] produced in the Cs cell enter an RF-cavity, 18 cm long and 14.2 cm i.d., with entrance and exit holes 1.5 cm diam. An RF oscillator, with a crystal oscillator of 1605 MHz, has a maximum power of 200 mW. The oscillation mode is TM_{010}. The cavity, made of silver plated brass, is of the quadrant sector type. An electric field of a few tens V/cm is produced in the cavity.

Neutral atoms from the cavity enter an Ar cell, 30 cm long and 2 cm i.d., and are ionized selectively [4]. The gas pressure at the center of the Ar cell is 3×10^{-4} Torr, which produces maximum intensity of negative ions.

The vacuum system consists of 4 oil diffusion pumps of 1000 l/s. The vacuum in the tank is less than 3×10^{-5} Torr.

From the Cs cell to the Ar cell, all parts are assembled in a solenoid producing a magnetic field of 575 G. The stability of the magnetic field is less than 1×10^{-4}.

The entire polarized ion source is maintained at the potential of −20 kV. The negative ion beam is accelerated to 20 keV through an acceleration gap. The spin direction is freely oriented by a Wien filter.

The first trials of the RF-transition showed the $\alpha\beta$-transitions. At the present time the beam intensity of negative deuterons is a few nA. The beam transmission depends very sensitively on the position of the ion source and focusing electrodes. Improvements are in progress.

REFERENCES

[1] W. E. Lamb and R. C. Retherford, Phys. Rev. 79 (1950) 549.
[2] J. L. McKibben, G. G. Ohlsen, and G. P. Lawrence, Phys. Letters 28B (1969) 594.
[3] B. L. Donnally et al., Phys. Rev. Lett. 12 (1964) 502.
[4] B. L. Donnally and W. Sawyer, Phys. Rev. Lett. 15 (1965) 439.

The Energy Dependence of Charge Exchange Reactions Applied Successfully to Lamb-Shift-Type Ion Sources

V. BECHTOLD, H. BRÜCKMANN, D. FINKEN, L. FRIEDRICH, K. HAMDI, and E. SEITZ, Institut für Experimentelle Kernphysik des Kernforschungszentrums und der Universität Karlsruhe, Germany

The method which takes advantage of the Lamb-shift and the 2S-2P level crossing to polarize hydrogen ions is very suitable to produce either positively or negatively charged ion beams with a high phase space density. The procedure makes use of specific charge-exchange reactions to ionize the metastable H(2S) atoms selectively. One of the main features that determines the usefulness of such reactions for sources of polarized ions is the energy dependence of the selectivity of these charge exchange processes.

We have systematically investigated various reactions to determine the selectivity for production of positively and negatively charged polarized hydrogen ion beams as a function of beam energy. Some of the measurements with either argon or iodine will be discussed because these two gases seem to be practical in polarized ion sources. The argon reaction was first proposed for the production of negatively charged beams by Donnally et al. [1]. Their measurements seem to exhibit a maximum of the selectivity at 1 keV deuteron beam energy. Positively charged polarized beams were first produced by using hydrogen or helium as charge exchange partners [2]. Later on, halogens were found to be advantageous [3, 4]. Knutson has measured [3] the selectivity and the relative cross section for iodine vapor pressures but without checking the nuclear polarization. Argon and iodine were investigated at the Karlsruhe polarized ion source (A-LASKA) [5, 6, 7, 8] by varying the vapor pressure as well as the primary beam energy.

The selectivity is, in general, a function of the vapor pressure in the charge-exchange cell as was shown in [5]. This effect is due to a superposition of secondary reactions and increases with increasing vapor density. For a reliable comparison of different reactions we have measured the dependence on the vapor density [7, 8] at each beam energy and have compared the selectivities extrapolated to zero vapor pressure. The quantity $d = A/B$ is a measure of the selectivity [8]. The quantity d is defined as the ratio of the two slopes A and B at zero pressure. A is the slope of the I versus p curve when the primary beam

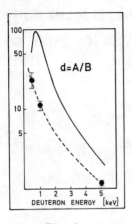

Fig. 1

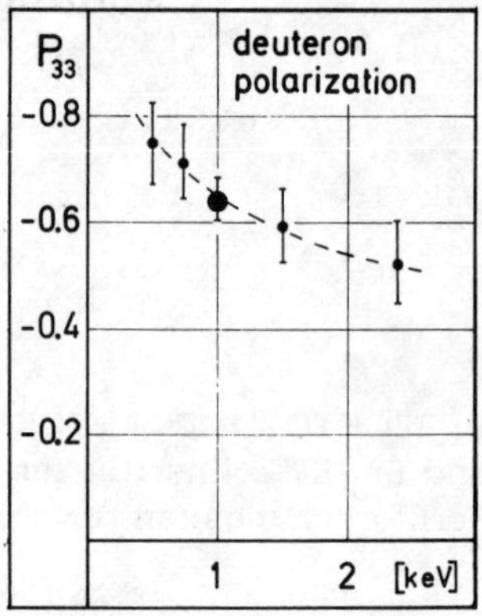

Fig. 2

consists of metastable and ground state atoms. B is the corresponding slope when all the metastable atoms are quenched to the ground
state.

The quantity d is plotted in fig. 1 on a logarithmic scale versus the
beam energy. The dashed line shows that d decreases strongly with
increasing energy. Under somewhat different experimental conditions
earlier measurements have shown the same monotonic decrease of d in
the energy range between 1 and 5 keV (dashed line). These values are
not directly comparable and hence not plotted in fig. 1. For comparison the quantity d was extracted from the measurements of reference
[1] and is shown as a solid curve in fig. 1. The measured selectivity
does not seem to exhibit the peak previously reported at 1 keV.

Lamb-shift type ion sources are also very suitable for the production of positively charged polarized beams [2, 3, 4, 7, 8]. Halogen
charge-exchange partners are at present the most promising candidates [3, 4]. Fig. 2 shows as an example the nuclear tensor polarization P_{33} versus beam energy which was obtained with the same experimental setup by using iodine vapor to produce positively charged ions.
In contrast to argon, the halogens show a selectivity which is almost
independent of the vapor pressure [7]. Consequently one is able to obtain a high nuclear polarization at the maximum yield. P_{33} decreases
monotonically with energy. This behavior seems to be a general feature which can be explained qualitatively with the pseudo-crossing
theory. The monotonic decrease was also observed for all the other
reactions investigated [8].

From the measurements it can be concluded that the optimum energy
to operate a Lamb-shift source depends on the specific geometry used.
but will surely be in the range between a few 100 eV and 1 keV.

REFERENCES

[1] B. L. Donnally and W. Sawyer, Phys. Rev. Lett. 15 (1965) 439.
[2] H. Brückmann, D. Finken, and L. Friedrich, Phys. Lett. 29B (1969) 223.
[3] L. D. Knutson, Phys. Rev., to be published.
[4] H. Brückmann, D. Finken, and L. Friedrich, Nucl. Instr. (in press).
[5] H. Brückmann, D. Finken, and L. Friedrich, Z. Phys. 224 (1969) 486.
[6] V. Bechtold, H. Brückmann, D. Finken, and L. Friedrich, Z. Phys. 231 (1970) 98.
[7] H. Brückmann, Third Polarization Symp.
[8] V. Bechtold et al., KFK-Report Nr. 1256.

Absolute Beam Polarization Determination
for the LASL Lamb-Shift Source
by the Quench Ratio Method

G. G. OHLSEN, G. P. LAWRENCE, P. W. KEATON, Jr., J. L. McKIBBEN,
and D. D. ARMSTRONG, Los Alamos Scientific Laboratory, New
Mexico, USA*

In conjunction with a Lamb-shift polarized ion source, a "nuclear spin
filter" provides an extremely clean way to select single hyperfine
states of either hydrogen or deuterium metastable atoms. The spin
filter action takes place in a ~1600-MHz rf field (~10 V/cm) parallel
to the atomic beam, a (perpendicular) dc electric field (~10 V/cm),
and a (parallel) magnetic field whose strength is adjusted to correspond to resonance for the state desired. The output ion current versus
spin filter magnetic field is schematically shown in fig. 1. The cur-

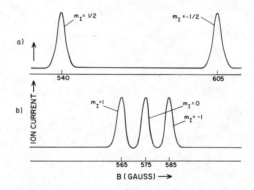

Fig. 1. a) Ion current versus B for H atoms.
b) Ion current versus B for D atoms.

rent which remains when the magnetic field does not correspond to a
resonance arises from the intense ground-state atomic beam which is
also present. For state-1 selection of either hydrogen or deuterium,
the metastable contribution is 100% polarized (P = 1 for hydrogen, P_3 =
P_{33} = 1 for deuterium) while the background current has approximately
zero polarization; thus, the actual beam polarization is approximately
given by

$$p = \frac{I - I_0}{I} \equiv 1 - \frac{1}{Q},\tag{1}$$

where the quenching ratio, Q, is defined by I/I_0, and where p represents either P, P_3, or P_{33}. (If other than state-1 atoms are selected, the strength of the magnetic field in the region of ionization must be taken into account in computing the polarization.)

It is the purpose of the present paper to describe our efforts to evaluate the accuracy of the quench ratio method. We describe two small corrections and list a number of conditions which must be met for the results to be precise. Our conclusion is that the method is capable of $\sim 1/2\%$ absolute accuracy, with about 1% accuracy achieved to date. Most of our tests made use of ^{4}He(p,p)^{4}He scattering at 12 MeV and 112° (lab) where its analyzing power is ~ 1.0.

First, we have noted that the quenched beam is actually slightly polarized (P or $P_3 \approx -0.03$). For typical hydrogen quench ratios of ~ 10, eq. (1) therefore overestimates the polarization by about 0.3%; for a typical deuterium quench ratio of 5, eq. (1) overestimates the vector polarization about 0.6%. The precise mechanism that produces this effect is still under investigation. It appears to be associated with the fact that $m_J = -1/2$ metastable atoms are quenched much earlier than the $m_J = 1/2$ atoms. For more detail, see ref. [1].

Second, the polarized component of the beam has better emittance than the unpolarized component; thus transmission through the accelerator can increase the beam polarization with respect to the value determined by quench ratio measurements made at the source. We have recently determined that this effect can be several percent and depends both on the design of the source output ion-optical systems and on the particular trajectory followed by the beam in the remaining transport elements and the accelerator itself. With one ion source lens design, for example, the effect was never greater than 1% (except for deliberately chosen transport parameters which led to poor transmission); with a recent lens modification, 5% effects (for deuterons) have been observed. This polarization enhancement can be taken into account with the relation $P_T = P_S/[P_S(1-a)+a]$, where P_T and P_S are the polarizations as measured at the target and at the source, respectively, and a is the accelerator transmission ratio for the unpolarized beam component and polarized component; a can be estimated in various ways and need only be crudely known to achieve $\sim 1\%$ accuracy in the derived value of P_T. Alternatively, the quenching ratio can be measured with a Faraday cup situated as close to the target as permits accurate beam current measurements (probably at a point just ahead of the tandem regulating slit jaws). Present data indicate that essentially no further polarization enhancement occurs from this point on, so that quench ratio measurements taken here would yield the true on target beam polarization to within 1% accuracy.

Probably the largest uncertainty arises from the requirement that no conditions in the source change between the filtered and quenched beam current measurements. Experience has shown that the ion source stability is such that at least 0.5–1% reproducibility is obtained. The measurements are usually made under computer control, and require about 7 sec to complete. Other conditions, which appear to be well satisfied, for the method to be absolute are (1) the spin filter parameters must be such that there is no overlap of states; (2) there must be no depolarization of the atomic beam between the spin selection and ionization region; (3) there must be no depolarization of the beam by charge exchange in a poor vacuum region (e.g., in the tandem accelerator terminal); and (4) no contaminant beam (e.g., electrons, heavy ions) may reach the current measuring device. With regard to (4), we have seen as much as 20 nA of unidentified heavy ions in a cup close to the source. A suitable magnetic field, such as the first section of our spin precessor, removes any such component. Finally, good current measurement practices should be used. More detail on these matters is given in ref. [1].

The direction of the spin axis at the target must also be known accurately for some types of experiments. Unwanted bends of 1° or 2° can easily arise if the beam goes through bending magnets off the midplane or if a great deal of steering is used. These effects are discussed from a formal point of view elsewhere [2]. Our experience shows that somewhat better than 1° accuracy for deuterons and 2° accuracy for protons can be obtained with present techniques; better accuracy can be obtained by left–right and up–down polarization monitoring techniques.

REFERENCES

*Work performed under the auspices of the U.S. Atomic Energy Commission.

[1] G. G. Ohlsen, Los Alamos Scientific Laboratory Report LA–4451 (unpublished).
[2] R. R. Stevens, Jr., and G. G. Ohlsen, Third Polarization Symp.

DISCUSSION

Donnally:

A correction needs to be made for the fact that the current I_0 is not the same as the current due to the ground-state atoms before the quenching. Quenching of the remaining hyperfine states increases the number of ground-state atoms. To make this correction, it is necessary to know the fraction of atoms in the 2S state.

Clegg:

Do you understand why your quenched beam is polarized?

Ohlsen:

I believe the effect is associated with the fact that β atoms are quenched earlier than α atoms, and thus fewer are lost by early conversion to negative ions. Thus, more atoms originating from β states are present at the argon region. This simple explanation is not fully consistent with tests we have done, however, and further investigation is planned.

Clausnitzer:

Is there evidence from the production of negative ions of another configuration, $H(1s,2p)$ or $H(2p)^2$, which could be an explanation for a "missing polarization"?

Ohlsen:

I agree that what you mention is a possibility, but we seem not to have any missing polarization to account for. That might be an ultimate limit to the accuracy of the method, however.

Proposal for a Lamb-Shift Polarimeter

J. E. BROLLEY, G. P. LAWRENCE, and G. G. OHLSEN, Los Alamos
Scientific Laboratory, New Mexico, USA*

Spin polarization measurements on low energy charged particles have
not, in general, been technically feasible to date. In this note we
suggest a novel procedure which should prove useful for H, D, (and T)
ions in the energy range 0—500 keV [1]. The particular applications
proposed [2] for the device include the study of spin polarization of
recoil protons and deuterons near 90°, corresponding to very small
angle scattering of 800 MeV protons (LAMPF H- beam) from H_2 and D_2
jet targets presently being developed. Polarization occurring in small
angle (p,p) and (p,d) scattering is not well fitted by predictions calculated from phase shifts. The existing data, however, are meager since
this is a region not readily accessible by conventional techniques;
using the jet targets and the proposed polarimeter it should be possible to considerably augment small angle polarization measurements.

The design aim for the polarimeter is to construct a device which
can measure all polarization components for either protons or deuterons
in the energy range 0—500 keV, with an overall efficiency of 1%. In
fig. 1 we show a schematic diagram of a device which could reasonably be expected to meet this aim. The steps in the process would be
as follows.

Scattered particles are first bent by a wedge magnet; this selects
the energy of the particles to be analyzed and thus the effective scattering angle. A 50.3° bend rotates proton spins 90°, which is a useful
feature; no reasonable bend angle seems better than another in the
deuteron case.

The spin measurement system is expected to work well only in the
range 500 eV to 50 keV, so the incident particles must be decelerated
or accelerated to the desired energy (say 10 keV). If the accepted
energy spread is not too great (< 1%), a crossed field device which
rotates around the beam direction would appear very desirable in that
arbitrary spin orientation could be provided for ions entering the spin
filter.

A large fraction of the beam (~ 10%) is converted to 2S atoms in a
cesium vapor cell. The charge transfer needs to be carried out in a

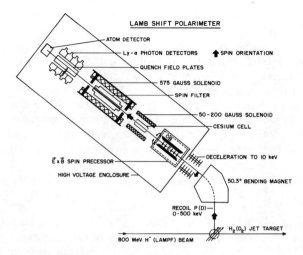

Fig. 1. Schematic drawing of proposed
polarimeter.

"strong" magnetic field ($\sim$ 200 gauss for H, and $\sim$ 50 G for D) in order
to avoid significant depolarization. The 2S atoms now enter a spin
filter which is tuned to pass those in a particular m_I state and quench
the rest. For the case of protons it can be shown [2] that the number
of $m_I = +1/2$ atoms passed is $n_+ \propto (1 + \langle\sigma_z\rangle)$ where $\langle\sigma_z\rangle$ is the polarization component along the polarimeter (z) axis. Similarly if the de-
vice is tuned to transmit $m_I = -1/2$ atoms, the number passed is
$n_- \propto (1 - \langle\sigma_z\rangle)$ so that the polarization component along the z axis is
$\langle\sigma_z\rangle = (n_+ - n_-)/(n_+ + n_-)$, in exact analogy to nuclear scattering methods
(with a 100% efficient analyzing reaction). Other polarization compo-
nents in the incoming beam can be measured with the aid of the spin
orientation device.

For deuterons, with the filter set to pass $m = 1, 0$, and -1 atoms,
respectively, we find: $n_+ \propto [1 + 3/2\langle P_z\rangle + 1/2\langle P_{zz}\rangle]$, $n_0 \propto [1 - \langle P_{zz}\rangle]$,
and $n_- \propto [1 - 3/2\langle P_z\rangle + 1/2\langle P_{zz}\rangle]$, so that

$$\langle P_z\rangle = \frac{2}{3}\left(\frac{n_+ - n_-}{n_+ + n_0 + n_-}\right) \qquad 1 + \frac{1}{2}\langle P_{zz}\rangle = \left(\frac{n_+ + n_- - n_0}{n_+ + n_0 + n_-}\right),$$

again in direct analogy to a nuclear analyzer with 100% vector and/or
tensor analyzing powers. Experimentally, one would cycle between
the spin filter magnetic field values which tune for the various states.
The desired quantities are all ratios so that no absolute efficiencies
are involved, although one must ensure that no efficiency factors change
as a function of magnetic field.

For the applications discussed above, the rate of arrival of particles in the polarimeter will probably be on the order of 10/sec or less. With 10% conversion efficiency to 2S atoms and 25% typical pass efficiency in the spin filter (protons), the arrival rate of the selected 2S atoms in the detection volume is ≤ 0.25/sec. Detection of the metastable atoms must therefore employ single event counting techniques, with particular emphasis on suppression of background events by coincidence and pulse discrimination procedures.

The filtered 2S atoms are passed through a transverse electric field strong enough (500 V/cm) to induce a transition to the ground state with emission of a Lyman-α photon (1216° A). The photons are then detected in coincidence with the associated 10 keV atoms which are incident on a particle detector downstream from the quenching region. The Lyman-α photons can be detected with 10% efficiency by existing image intensifier photomultiplication devices. With a suitable number of such instruments stacked around the quenching region, approximately half the emitted photons can be intercepted, leading to a total photon detection efficiency of 5%, for an overall count rate of ≤ 1 count/min. The associated 10 keV atoms can be detected with nearly 100% efficiency, by a gas gain proportional counter.

Pulses from the photon and atom detectors will be fed to a fast coincidence unit (< 10 nsec time resolution), with the photon pulse suitably delayed to allow for the finite atom transit time. Internally generated noise events will arise mainly from random coincidences between dark counts in the photon detectors and non-selected 10 keV particles incident on the atom detector. With photocathode cooling the dark rate can be reduced to < 20 counts/sec. Coupled with an unwanted atom count rate of ≈ 10/sec, the accidental coincidence rate is $\approx 10^{-4}$/min. It is clear that the polarimeter will be usable even if the externally induced random coincidence rates (due to nuclear interactions in the target area) are three or four orders of magnitude higher, particularly since a large fraction of such events can be eliminated by pulse-height and shape discrimination and anticoincidence techniques.

REFERENCES

*Work performed under the auspices of the U.S. Atomic Energy Commission.

[1] A process several orders of magnitude less efficient has been suggested by G. Clausnitzer and D. Fick, Nucl. Instr. 47 (1967) 171.
[2] See the extended version of this paper in LA-4465-MS (1970) (unpublished).

A Precise Method for the Determination of Secondary Standards for Deuteron Analyzing Tensors

P. W. KEATON, Jr., D. D. ARMSTRONG, G. P. LAWRENCE, J. L. McKIBBEN, and G. G. OHLSEN, Los Alamos Scientific Laboratory, New Mexico, USA[*]

Two-body nuclear reactions in which one of the initial particles has spin-one and the remaining three particles have spin-zero have uniquely determined analyzing tensors [1]. These are

$$P_y^0(\theta) = 0, \quad P_{xz}^0(\theta) = 0, \quad P_{xx}^0(\theta) = 1, \quad P_{yy}^0(\theta) = -2, \quad P_{zz}^0(\theta) = 1,$$

(1)

where we use the right-handed coordinate system with y along $\vec{k}_{in} \times \vec{k}_{out}$ and z along $\vec{k}_{in}$. The general cross section expression is

$$I(\theta,\phi) = I_0(\theta)[1 + \frac{3}{2}\langle P_y \rangle P_y^0 + \frac{2}{3}\langle P_{xz} \rangle P_{xz}^0 + \frac{1}{3}\langle P_{xx} \rangle P_{xx}^0$$

$$+ \frac{1}{3}\langle P_{yy} \rangle P_{yy}^0 + \frac{1}{3}\langle P_{zz} \rangle P_{zz}^0],$$

(2)

where $I_0(\theta)$ is the cross section for an unpolarized beam. Substituting eq. (1) in eq. (2), and using the identity $\langle P_{xx} \rangle + \langle P_{yy} \rangle + \langle P_{zz} \rangle = 0$, we obtain

$$I(\theta,\phi) = I_0(\theta)[1 - \langle P_{yy} \rangle].$$

$\langle P_{yy} \rangle$ may be expressed in terms of the tensor polarization of the beam, p_{ZZ}, with respect to its quantization axis and the angles β, ϕ which describe the orientation of the quantization axis with respect to the xyz coordinate system [2]:

$$\langle P_{yy} \rangle = \frac{1}{2} p_{ZZ} [3 \sin^2 \beta \cos^2 \phi - 1].$$

Thus, if we define the cross section for "left," "right," "up," and "down" as $L = I(\theta, 0°)$, $R = I(\theta, 180°)$, $U = I(\theta, 270°)$, and $D = I(\theta, 90°)$, we have

$$L = R = I_0[1 + \frac{1}{2} P_{ZZ}(1 - 3\sin^2\beta)] \qquad U = D = I_0[1 + \frac{1}{2} P_{ZZ}].$$

The reaction $^{16}O(d,\alpha_1)^{14}N$ (2.31–MeV state) possesses the required spins and has been shown to have a cross section large enough to make its use feasible at deuteron energy 7.1 MeV and laboratory angle 35° [3]. Our purpose here was to find an angle and energy for which the (d,α) reaction has acceptably high yield, which coincides with an angle and energy at which the reaction to be calibrated has a sufficiently large analyzing power to be useful. If the target material for the reaction to be calibrated is gaseous, it may be mixed with the oxygen so that both reactions may be observed simultaneously. The analyzing power being determined may then be deduced from a ratio in which uncertainties or instabilities in the beam polarization and possible misalignments in the detection apparatus have no effect.

We have used this method to determine the analyzing tensors for $^4He(\vec{d},d)^4He$ elastic scattering at 9.80 MeV and at a laboratory scattering angle of 60°. Our results are as follows:

$$\sqrt{3}\, T_{22}(60°) = \frac{1}{2}[P^0_{xx}(60°) - P^0_{yy}(60°)] = +0.553 \pm 0.008$$

$$\sqrt{2}\, T_{20}(60°) = [P^0_{zz}(60°)] = +0.330 \pm 0.005.$$

$\beta = 54.7°$ and left, right, up, and down detectors were used for the $(P^0_{xx} - P^0_{yy})/2$ measurements. With this choice of β, we may write, for any analyzing reaction,

$$\frac{(L + R) - (U + D)}{L + R + U + D} = -\frac{1}{6} P_{ZZ}(P^0_{xx} - P^0_{yy}).$$

Since $P^0_{xx} - P^0_{yy} = +3.0$ for $^{16}O(d,\alpha_1)^{14}N^*$, $P^0_{xx} - P^0_{yy}$ for d-^{4}He scattering can be obtained by a ratio of simultaneously observed asymmetries. For other choice of β the analysis is somewhat more complicated. The quantity P^0_{zz} was obtained by measuring $m_I = 1$ to $m_I = 0$ ratios [2] with $\beta = 0°$.

The principal uncertainty in the present calibration arises from the background contribution to the $^{16}O(d,\alpha_1)^{14}N^*$ yield. The beam polarization calculated from either the $\beta = 54.7°$ data or from the $\beta = 0°$ data is increased by about 1.5% when a reasonable estimate of the background is subtracted from the yields. The quoted errors assume a 70% uncertainty in the background subtraction which is combined quadratically with the statistical error.

A scan of the d-^{4}He analyzing powers versus energy ($\pm$ 200 keV) and angle ($\pm$ 3°) was made in order to establish the accuracy required in reproducing these parameters. We find that neither $(P^0_{xx} - P^0_{yy})/2$ nor P^0_{zz} varies by more than 1.5% if the angle remains within $\pm$ 1° of the nominal value and if the energy remains within $\pm$ 100 keV of the nominal value.

REFERENCES

*Work performed under the auspices of the U.S. Atomic Energy Commission.

[1]   B. A. Jacobsohn and R. M. Ryndin, Nucl. Phys. 24 (1961) 505.
      For a simple derivation based on first principles, see the extended version of the present paper in LA-4465-MS (unpublished).
[2]   G. P. Lawrence et al., Third Polarization Symp.
[3]   R. M. Prior, K. W. Corrigan, and S. E. Darden, Bull. Am. Phys. Soc. 15 (1970) 35.

Figures of Merit for
Polarized Deuteron Beams

P. W. KEATON, Jr., and G. G. OHLSEN, Los Alamos Scientific Laboratory, New Mexico, USA*

The quantity p^2I, where I is the current and p is the polarization, has long been used as a figure of merit for polarized proton beams. We generalize this figure of merit to include the case of vector and tensor polarized deuteron beams.

For a spin-1/2 beam, one measures an asymmetry A, given by $A = (L - R)/(L + R)$, where L and R are the counts recorded in the left detector and in the right detector, respectively. In terms of the beam polarization and the analyzing power of the reaction, $P(\theta)$, we have $A = pP(\theta)$. The statistical error, $\Delta P(\theta)$, associated with such a determination of $P(\theta)$ is

$$\Delta P(\theta) = \frac{\sqrt{1 - A^2}}{p\sqrt{L + R}} \quad \propto \quad \frac{\sqrt{1 - A^2}}{p\sqrt{It}} , \tag{1}$$

since $L + R$ is proportional to It, where t is the data acquisition time. In deriving eq. (1), we have assumed that L and R contain no background counts. In the limit of small A, we find

$$\Delta P(\theta) \propto \frac{1}{p\sqrt{It}} . \tag{2}$$

We define the figure of merit to be inversely proportional to the counting time necessary to produce a given statistical error $\Delta P(\theta)$. Thus we see that the relevant figure of merit is in fact p^2I. Notice, however, that in the case of large $P(\theta)$ this figure underestimates considerably the advantage of large beam polarizations. (This will be true also for the deuteron quantities.)

For deuterons, we consider five asymmetries. These are, with the corresponding observables and (zero-background) statistical error expressions:

$$A_1 = \frac{(L+R)-(U+D)}{L+R+U+D} = \frac{-\frac{1}{4}p_{ZZ}\sin^2\beta[P_{xx}^0 - P_{yy}^0]}{1+\frac{1}{4}p_{ZZ}(3\cos^2\beta - 1)P_{zz}^0} \ ; \quad \Delta A_1 = \sqrt{\frac{1-A_1^2}{L+R+U+D}}$$

$$A_2 = \frac{2(L-R)}{L+R+U+D} = \frac{\frac{3}{2}p_Z\sin\beta\,P_y^0}{1+\frac{1}{4}p_{ZZ}(3\cos^2\beta - 1)P_{zz}^0} \ ; \quad \Delta A_2 = \sqrt{\frac{\frac{4(L+R)}{L+R+U+D}-A_2^2}{L+R+U+D}}$$

$$A_3 = \frac{2(U-D)}{L+R+U+D} = \frac{p_{ZZ}\sin\beta\cos\beta\,P_{xz}^0}{1+\frac{1}{4}p_{ZZ}(3\cos^2\beta - 1)P_{zz}^0} \ ; \quad \Delta A_3 = \sqrt{\frac{\frac{4(U+D)}{L+R+U+D}-A_3^2}{L+R+U+D}}$$

$$A_4 = \frac{L-R}{L+R} = \frac{\frac{3}{2}p_Z\sin\beta\,P_y^0}{1+\frac{1}{2}p_{ZZ}[\sin^2\beta\,P_{yy}^0 + \cos^2\beta\,P_{zz}^0]} \ ; \quad \Delta A_4 = \sqrt{\frac{1-A_4^2}{L+R}}$$

$$A_5 = \frac{U-D}{U+D} = \frac{p_{ZZ}\sin\beta\cos\beta\,P_{xz}^0}{1+\frac{1}{2}p_{ZZ}[\sin^2\beta\,P_{xx}^0 + \cos^2\beta\,P_{zz}^0]} \ ; \quad \Delta A_5 = \sqrt{\frac{1-A_5^2}{U+D}} \qquad (3)$$

where L, R, U, and D are the counts recorded in the left, right, up, and down detectors, respectively, and where p_Z and p_{ZZ} are the incident vector and tensor beam polarizations. (For definitions of β, ϕ, "up," etc., see ref. [1].)

It is clear from eq. (3) that one cannot simply divide out the beam polarization as in the spin-1/2 case. That is, except for the special case of $\cos\beta = 1/\sqrt{3}$ for A_1, A_2, and A_3 [2], the asymmetries are not linearly related to the beam polarization and analyzing tensors. We must first measure separately the denominator factor by means of a ratio of counts with different values of p_{ZZ}. Denoting the counts observed with beam polarization p_{ZZ} by T and with beam polarization p'_{ZZ} by T', we define

$$B = \frac{T-T'}{p_{ZZ}T' - p'_{zz}T} \ ; \quad \Delta B = \frac{(p_{ZZ}-p'_{ZZ})}{(p_{ZZ}T' - p'_{ZZ}T)^2}\sqrt{T^2 T' + T'^2 T} \ ,$$

where B could be $1/4(3\cos^2\beta - 1)P_{zz}^0$, $1/2[\sin^2\beta\,P_{yy}^0 + \cos^2\beta\,P^0]$, or $1/2[\sin^2\beta\,P_{xx}^0 + \cos^2\beta\,P_{zz}^0]$, and T would be $L+R+U+D$, $L+R$, or $U+D$, respectively. In the limit of zero for the analyzing tensors, $T \rightarrow T'$, and

$$\Delta B = \frac{1}{(p_{ZZ} - p'_{ZZ}) \sqrt{T}} \; .$$

The figure of merit relevant for measuring B is, therefore, given by $(\Delta p_{ZZ})^2 I$, where Δp_{ZZ} is the change in p_{ZZ} obtainable between the two modes of operation for the ion source.

Returning to the asymmetries A_1 through A_5, we see that if the denominator factor has already been determined, the figure of merit becomes $p_Z^2 I$ for A_2 and A_4, and $p_{ZZ}^2 I$ for A_1, A_3, and A_5. In summary, deuteron beams are described in terms of three figures of merit: $p_Z^2 I$ for $P_y^0(\theta)$ measurements, $p_{ZZ}^2 I$ for $P_{xx}^0(\theta) - P_{yy}^0(\theta)$ and $P_{xz}^0(\theta)$ measurements, and $(\Delta p_{ZZ})^2 I$ for $P_{zz}^0(\theta)$ measurements.

By way of example, we compare a pure vector polarized beam ($p_Z = 2/3$, $p_{ZZ} = 0$) to a pure $m_I = 1$ beam ($p_Z = p_{ZZ} = 1$). Measurement of A_2 or A_4 to a given accuracy takes 9/4 times as much for the pure vector polarized beam; however, for the $m_I = 1$ beam some additional time must be spent in the evaluation of the denominator expression. The amount of time required depends on the method used—for example, the use of an intense unpolarized beam might reduce the time required markedly. Further, this time should not be charged against the measurement of $P_y^0(\theta)$, since one is measuring two quantities, not one. The ideal figures of merit for three methods are given in table 1. The best mode of operation for each figure of merit is assumed.

Table 1. Ideal figures of merit for three methods of polarizing deuterons with Lamb-shift sources

Method	$p_Z^2 I$	$p_{ZZ}^2 I$	$(\Delta p_{ZZ})^2 I$
spin filter ($p_Z = p_{ZZ} = 1$ or $p_Z = 0$, $p_{ZZ} = -2$)	I	$4I$	$9I$
zero field crossing ($p_Z = 2/3$, $p_{ZZ} = 0$ or $p_Z = 0$, $p_{ZZ} = -1$)	$\frac{4}{9} I$	I	I
adiabatic field reduction ($p_Z = 1/3$, $p_{ZZ} = -1/3$ or unpolarized)	$\frac{1}{9} I$	$\frac{1}{9} I$	$\frac{1}{9} I$

REFERENCES

*Work performed under the auspices of the U.S. Atomic Energy Commission.

[1] J. L. Gammel, P. W. Keaton, Jr., and G. G. Ohlsen, Third Polarization Symp.
[2] G. P. Lawrence et al., Third Polarization Symp.

A Rapid Method for the Measurement
of Analyzing Tensors in
Reactions Induced by Polarized Deuterons

G. P. LAWRENCE, G. G. OHLSEN, J. L. McKIBBEN, P. W. KEATON, Jr., and D. D. ARMSTRONG, Los Alamos Scientific Laboratory, New Mexico, USA[*]

The cross section for a reaction induced by polarized deuterons may be written in the form

$$I(\theta,\phi) = I_0(\theta)\{1 + \frac{3}{2}\langle P_y \rangle P_y^0(\theta) + \frac{2}{3}\langle P_{xz}\rangle P_{xz}^0(\theta) + \frac{1}{6}\langle P_{xx} - P_{yy}\rangle$$

$$\times [P_{xx}^0(\theta) - P_{yy}^0(\theta)] + \frac{1}{2}\langle P_{zz}\rangle P_{zz}^0(\theta)\},$$

where the quantities in brackets represent the beam polarization quantities while $P_y^0(\theta)$, $P_{xz}^0(\theta)$, $P_{xx}^0(\theta)$, $P_{yy}^0(\theta)$, and $P_{zz}^0(\theta)$ represent the analyzing tensors for the reaction. The y axis is along $\vec{k}_{in} \times \vec{k}_{out}$, the z axis is along $\vec{k}_{in}$, and the x axis is chosen to make a right-handed coordinate system.

A beam produced by a polarized ion source is characterized by its vector and tensor polarization with respect to its quantization axis, p_Z and p_{ZZ}, together with the polar and azimuthal angles (β,ϕ), which describe the orientation of the quantization axis with respect to the x,y,z coordinate system described above. The angle ϕ is defined such that scattering to the "left" if the particle has spin "up" corresponds to $\phi = 0$, with positive ϕ corresponding to a rotation of the x axis toward the y axis. With this convention, the beam polarization quantities are

$$\langle P_y \rangle = p_Z \sin \beta \cos \phi \qquad\qquad \langle P_{xx} - P_{yy}\rangle = -\frac{3}{2}p_{ZZ}\sin^2\beta\cos 2\phi$$

$$\langle P_{xz}\rangle = -\frac{3}{2}p_{ZZ}\sin\beta\cos\beta\sin\phi \qquad \langle P_{zz}\rangle = \frac{1}{2}p_{ZZ}(3\cos^2\beta - 1).$$

If we adopt the convention that "up" describes the half-plane containing the quantization axis $(\hat{S})$ and the incident momentum $(\vec{k}_{in})$, then left, right, up, and down will correspond to $\phi = 0$, $180°$, $270°$, and $90°$, respectively. Using the notation L, R, U, and D to denote the counts observed in the left, right, up, and down detectors, we can form the three observable asymmetries

$$A_1 = \frac{2(L-R)}{L+R+U+D} = \left[\frac{3}{2}P_Z \sin\beta\, P_y^0\right] \bigg/ \left[1+\frac{1}{4}P_{ZZ}(3\cos^2\beta-1)P_{zz}^0\right]$$

$$A_2 = \frac{2(U-D)}{L+R+U+D} = \left[P_{ZZ}\sin\beta\cos\beta\, P_{xz}^0\right] \bigg/ \left[1+\frac{1}{4}P_{ZZ}(3\cos^2\beta-1)P_{zz}^0\right]$$

$$A_3 = \frac{(L+R)-(U+D)}{L+R+U+D} = \left[-\frac{1}{4}P_{ZZ}\sin^2\beta(P_{xx}^0-P_{yy}^0)\right] \bigg/ \left[1+\frac{1}{4}P_{ZZ}(3\cos^2\beta-1)P_{zz}^0\right],$$

and the observable ratio

$$R = \frac{T^{(1)}}{T^{(2)}} = \left[1+\frac{1}{4}P_{ZZ}^{(1)}(3\cos^2\beta-1)P_{zz}^0\right] \bigg/ \left[1+\frac{1}{4}P_{ZZ}^{(2)}(3\cos^2\beta-1)P_{zz}^0\right],$$

where the superscripts 1 and 2 refer to two runs taken with different values of the polarization p_{ZZ}, and where $T=L+R+U+D$. The two runs in this case are assumed to correspond to the same integrated current, target thickness, etc. If $\beta = 0$ is chosen, the four detectors may be set at different angles so that P_{zz}^0 at 4 angles may be obtained simultaneously.

The four quantities above suffice for the determination of all four observables. In particular, if β is chosen equal to $54.7°$, we obtain

$$A_1 = \frac{3}{2}P_Z \sin\beta\, P_y^0 \qquad\qquad A_3 = -\frac{1}{4}P_{ZZ}\sin^2\beta\,(P_{xx}^0-P_{yy}^0)$$

$$A_2 = P_{ZZ}\sin\beta\cos\beta\, P_{xz}^0 \qquad\qquad R = 1\,.$$

That is, the ratios determine three of the four tensors in a monitor-free manner exactly analogous to the two-detector scheme long used for accurate spin-1/2 analyzing power measurements. The fourth tensor is determined by choosing $\beta = 0$, in which case,

$$A_1 = A_2 = A_3 = 0 \qquad R = \left[1+\frac{1}{2}p_{ZZ}^{(1)}P_{zz}^0\right] \bigg/ \left[1+\frac{1}{2}p_{ZZ}^{(2)}P_{zz}^0\right];$$

all of the dependence of the measurements on current integration and the like is thus thrown into this term.

At Los Alamos we have applied this technique successfully in a large number of experiments; we use a four-detector reaction chamber which can be rotated azimuthally and set at any of the quadrant angles with 0.1° accuracy by means of a microswitch-controlled two-speed motor. For the ratio measurements we sequentially rotate each of the four counters into each of the four azimuthal positions and use geometric means for L, R, U, D; that is, for example, $L = (L_1 L_2 L_3 L_4)^{1/4}$, where L_1 is the number of counts observed by detector 1 during the run in which it was in the left position, and so on. We use an $\sim 80\%$ $m_I = 1$ beam for the ratio measurements; that is, a beam with $p_Z = p_{ZZ} = \sim 0.8$. Occasionally we use an $\sim 80\%$ $m_I = 0$ beam as a check ($p_Z = 0$, $p_{ZZ} = \sim -1.6$). This ensures, for example, that the P_{xz}^0 measurement is not contaminated by a P_y^0 contribution because of misalignment and that left-right asymmetries vanish when $\langle P_y \rangle = 0$, etc.

For the measurement of P_{zz}^0, no azimuthal rotation of the detectors is used since there is no azimuthal cross-section dependence. For this ratio we use an $m_I = 1$ followed by $m_I = 0$ polarized beam, so that $p_{ZZ}^{(1)} = \sim 0.8$ and $p_{ZZ}^{(2)} = \sim -1.6$.

Application of this technique requires good control and knowledge of the actual direction of the quantization axis, which calls for some care with respect to the effects of bending magnets, electrostatic steerers, etc. We have found reproducibility to one degree or better in β; a ±0.5-degree uncertainty contributes an error of ±0.75% in the observed asymmetry in the least favorable case.

Due to the high rate of data acquisition it has been necessary to automate the described measurement procedure as much as possible. The LASL tandem on-line computer controls the reaction chamber rotation sequence and is interfaced with the polarized ion source to automatically measure beam polarization (one measurements for each ϕ angle change); it accumulates and stores the data at each of the designated angles, and at the end of each sequence calculates the appropriate geometric means, averages beam polarization, and computes the relevant analyzing tensors. In "fully automatic" mode the computer runs through the entire four-fold sequence without operator intervention (once the appropriate gates and indices have been set). With the available beam intensities and the degree of automation noted, it is usually possible to measure one complete angular distribution of the four tensors (5° intervals) in a 24-hour period.

*Work performed under the auspices of the U.S. Atomic Energy Commission.

Transfer Matrix Method for
Calculating Spin Aberrations in
the Transport of Polarized Beams

R. R. STEVENS, Jr., and G. G. OHLSEN, Los Alamos Scientific
Laboratory, New Mexico, USA*

Transfer matrix methods have long been a useful tool for the design
and evaluation of beam transport system. When the beams are polarized, the effect of the transport elements on the spin axis is also of
interest. We here generalize the usual beam transport matrix calculations so as to determine the spin axis orientation in a beam transport system. This development was motivated by experimental observations with the LASL polarized ion source on the tandem Van de Graaff
installation where it has been found that "spin aberrations" induced
by the transport system can be as large as several degrees with im-
proper adjustment of the various transport elements.

In this treatment, we follow closely Penner's [1] development of
first-order beam transport matrices for determining particle trajectories.
We assume a coordinate system with the z axis taken along the beam
line, the y axis vertical, and the x axis taken to make a right-handed
coordinate system. The beam line is a straight line in focusing and
drift elements and a circular arc of radius ρ in a bending element.
The effect of the beam transport elements is characterized by operators
which rotate the spin axis, $\vec{S_0}$, to a new orientation $\vec{S}$. In the Cartesian
representation, we have:

$$\begin{pmatrix} S_x \\ S_y \\ S_z \end{pmatrix} = \begin{pmatrix} a_{11} & a_{12} & a_{13} \\ a_{21} & a_{22} & a_{23} \\ a_{31} & a_{32} & a_{33} \end{pmatrix} \begin{pmatrix} S_{x_0} \\ S_{y_0} \\ S_{z_0} \end{pmatrix}.$$

The rotations induced by these transport elements are determined by
the equation of motion for the spin vector $\vec{S}$ in a magnetic field $\vec{B}$:

$$\frac{d\vec{S}}{dt} = \frac{ge}{2m_p} (\vec{S} \times \vec{B}),$$

where g is the gyromagnetic ratio and m_p is the proton mass. The matrix elements for several beam transport elements of interest are presented in table 1.

Table 1. Polarization transport matrices

Beam transport element	a_{11}	a_{22}	a_{33}	a_{12}	a_{13}	a_{23}
Inclined magnet pole face (entry angle β_1)	1	1	1	$\mp\dfrac{y}{\rho}\left[\dfrac{gM}{2m_p}\right]\dfrac{a}{\lvert a\rvert}$	0	$\dfrac{y}{\rho}\tan\beta_1\left[\mp\dfrac{gM}{2m_p}\right]$
Inclined magnet pole face (exit angle β_2)	1	1	1	$\pm\dfrac{y}{\rho}\left[\dfrac{gM}{2m_p}\right]\dfrac{a}{\lvert a\rvert}$	0	$\dfrac{y}{\rho}\tan\beta_2\left[\mp\dfrac{gM}{2m_p}\right]$
Horizontal bending magnet (bending angle a)	$\cos\delta_y$	1	$\cos\delta_y$	0	$\sin\delta_y$	0
Quadrupole lens (focusing in x-plane)	1	1	1	0	Δ_c	Δ_d
Quadrupole lens (focusing in y-plane)	1	1	1	0	Δ_d	Δ_c
Drift	1	1	1	0	0	0

where $\delta_y = a\left[-1\pm\dfrac{gM}{2m_p}\right]\pm(\theta-\theta_0)\dfrac{gM}{2m_p}$

$$\Delta_c = \pm\left[(\theta-\theta_0)_c\left(\dfrac{gM}{2m_p}\right)\right]$$

$$\Delta_d = \pm\left[(\theta-\theta_0)_d\left(\dfrac{gM}{2m_p}\right)\right]$$

ρ = bending radius.

M = polarized ion mass.

θ_0 = entrance angle to a transport element.

θ = exit angle from a transport element.

$(\theta-\theta_0)_c$ = angular divergence imparted in the converging plane of a quadrupole lens.

$(\theta-\theta_0)_d$ = angular divergence imparted in the diverging plane of a quadrupole lens.

The upper sign is for a positive beam while the lower sign is for a negative beam. For a horizontal bending magnet, a positive bend angle a is assumed to be in the sense that rotates the z axis into the x axis. The off-diagonal elements not listed may be found from the relation $a_{ij} = -a_{ji}$.

In order to use this formalism one must first carry out the usual beam transport calculations to determine the coordinates and slopes at the entrance and exit of each transport element for the particle trajectories of interest. With this information, one can then determine the rotation of the spin axis for these trajectories with the transfer matrices described above. Using this procedure, it has been found that one of the more important sources of spin aberrations is associated with polarized particles entering and leaving a bending magnet off the median plane [2]. This same type of transfer matrix calculation can also be used to determine spin aberrations in the transport of polarized neutron beams.

In conclusion, we find that suitable care is required with the alignment and steering in a beam transport system to obtain undistorted transport of polarized beams.

REFERENCES

*Work performed under the auspices of the U.S. Atomic Energy Commission.

[1]   S. Penner, Rev. Sci. Instr. 32 (1961) 150.
[2]   Additional details are presented in LA-4465-MS (1970) (unpublished).

Remarks on the Measurement of
the Polarization of ^{6}Li-Ions

D. FICK, Physikalisches Institut der Universität Erlangen-Nürnberg,
Germany

Recently Holm et al. [1] reported on an ion source for polarized ^{6}Li
ions, with a very high beam intensity. One of the problems using a
polarized ^{6}Li beam is the determination of the polarization of the ^{6}Li
ions. The purpose of this note is to demonstrate that in the ^{6}Li$(d,\alpha)^4$He-
reaction the analyzing powers A_α for deuterons and ^{6}Li ions are not
independent of each other. The determination of the polarization of
^{6}Li beams can be reduced to the well established determination of the
polarization of deuterons.

The ^{6}Li $(d,\alpha)^4$He-reaction is of spin type $1+1 \rightarrow 0+0$. This type of
reaction has been discussed by Köhler and Fick [2] using a nondynam-
ical description of nuclear reactions (M-matrix). As a result of these
calculations one obtains the identity

$$A^{(d)}_{zz} + A^{(d)}_{xx-yy} = A^{(Li)}_{zz} + A^{(Li)}_{xx-yy} . \tag{1}$$

The A_{ii} are components of an irreducible tensor of rank two. Therefore
eq. (1) can be written as

$$A^{(d)}_{yy} = A^{(Li)}_{yy} . \tag{2}$$

The identities (eqs. 1, 2) are valid for all c.m. energies E and angles
θ, independent of the further reaction mechanism, because these
identities are deduced using the conservation laws of strong interactions only. $A^{(d)}_{yy}$ can be measured for the ^{6}Li$(d,\alpha)^4$He-reaction by usual
techniques. From the only available measurements at low energies
[3] (E = 450 keV) one finds for $A^{(Li)}_{yy}$

$$A^{(d)}_{yy} = A^{(Li)}_{yy} = (-0.52 \pm 0.08) + (0.08 \pm 0.13)P_2(\cos \theta).$$

Until now we did not use the fact that in the ^{6}Li$(d,\alpha)^4$He-reaction the
two products are identical bosons. This implies that the reaction

matrix must be invariant with respect to an interchange of these two bosons. This invariance yields the relation

$$A_a^{(d)} (E, \theta = \pi/2) = A_a^{(Li)} (E, \theta = \pi/2); \quad a = zz, xx-yy \tag{3}$$

for all c.m. energies, independent of the reaction mechanism. Eq. (3) shows that at a c.m. angle of $\theta = \pi/2$ not only eq. (1) but also identities for each of the terms in eq. (1) are valid separately. Using the experimental results [3] at $E = 450$ keV one gets for $A_a^{(Li)}$ ($E = 450$ keV, $\theta = \pi/2$)

$$A_{zz}^{(d)} (\theta = \pi/2) = A_{zz}^{(Li)} (\theta = \pi/2) = 0.45 \pm 0.23$$

$$A_{xx-yy}^{(d)} (\theta = \pi/2) = A_{xx-yy}^{(Li)} (\theta = \pi/2) = -1.5 \pm 0.4.$$

An experimental improvement of these results would be very desirable. The question should be discussed what the conditions for a relation

$$A_a^{(d)} (E, \theta) = A_a^{(Li)} (E, \theta); \quad a = y, zz, xx-yy, xz \tag{4}$$

are for the ^{6}Li(d,α)^{4}He-reaction. In terms of a spin representation of the M-Matrix eq. (4) is equivalent to $\langle 00 | M | 11 \rangle = 0$. This means that if transitions with a channel spin one do not occur, then the analyzing powers for deuterons and ^{6}Li ions are equal. So far nothing is known about this type of transition.

In order to get another view of the condition eq. (4) one can expand the elements of the M-matrix in terms of elements $S_{\ell's'\ell S}^J$ of the S-matrix. Using well-known techniques one gets for $\langle 00 | M | 11 \rangle$

$$\langle 00 | M | 11 \rangle = \frac{-i}{2\sqrt{2} \, k_{in} k_{fin}} \sum_{1} \frac{\hat{\ell}^2}{\sqrt{\ell(\ell+1)}} S_{1011}^{\ell} \cdot P_{\ell 1}(\cos \theta). \tag{5}$$

If S-waves ($\ell = 0$) contribute only, $\langle 00 | M | 11 \rangle = 0$ follows from eq. (5) immediately. This means that at energies where the ^{6}Li(d,α)^{4}He-reaction is affected by S-waves only the identities

$$A_a^{(d)} = A_a^{(Li)}; \quad a = zz, xx-yy \textit{ and } xz$$

are valid for these energies and for all scattering angles. Because the ^{6}Li(d,α)^{4}He-reaction cannot be affected by P-waves, the condition of S-waves only should be satisfied at low energies.

REFERENCES

[1] U. Holm et al., Z. Phys. 233 (1970) 415.
[2] W. E. Köhler and D. Fick, Z. Phys. 215 (1968) 408.
[3] H. Bürgisser et al., Helv. Phys. Acta 40 (1967) 185.

*Neutron Sources and
Polarimeters*

Production of Polarized Neutron Beams from the D(d,n)³He and T(d,n)⁴He Reactions Using the Associated Particle Method

D. G. SCHUSTER and R. L. HAGENGRUBER, University of Wisconsin, Madison, USA*

The reactions $T(p,n)^3He$, $D(d,n)^3He$, and $T(d,n)^4He$ are widely used sources of monoenergetic fast neutrons. However, at high bombarding energies the monoenergetic neutrons may be severely contaminated by neutrons from breakup and other reactions on the target and other materials struck by the incident beam, and by neutrons scattered in the surrounding material. There is no convenient source of neutrons between 8 and 14 MeV because of D-D breakup. When polarized neutrons are produced by using these reactions at angles other than 0°, the background problem becomes worse because the neutron production cross section is smaller and the neutron energy is lower.

In the present work the associated particle method was used with the $D(d,n)^3He$ and $T(d,n)^4He$ reactions to reduce the background and to obtain a collimated monoenergetic beam with fast timing information for time-of-flight applications [1]. By varying the incident energy (up to 14 MeV) and the reaction angle, neutrons with energies from 2 MeV to 26 MeV and with polarizations up to 50% were produced. Previously the method had been used only at low incident energies, typically several hundred keV.

The neutrons were produced in a low-mass chamber which permitted variations of neutron energies, angles, and polarizations over a wide range. Deuterons which passed through the target were stopped in a beam stop 2m beyond the chamber. The targets were deuterated polyethylene and tritiated titanium foils, thinner than the range of the He particles in the foil. The targets were rotated to spread heating by the beam. They could be set at any angle to the beam, and adjusted vertically to expose a fresh target area. The He particles must be detected in the presence of a much larger number of other particles, especially scattered deuterons. A thin totally depleted solid state detector was used, of thickness approximately equal to the range of the He particle in Si, so that pulses from He were much larger than, and well separated from, pulses from deuterons. Complete separation from all other particles was generally neither possible nor necessary. A scattering table with adjustable sample and detector positions and shielding configurations was used for polarization and scattering experiments using the neutron beams. The table rotated on a column supporting the chamber, and could be lowered to swing beneath the beam stop.

The angular range of the He particles accepted was chosen to give a neutron beam with a FWHM of 4° both vertically and horizontally. The neutron intensity was $1770 \ s^{-1} msr^{-1}$ for $D(d,n)^3He$ at $\theta_n = 55°$, $E_d = 10$ MeV, and $110 \ s^{-1} msr^{-1}$ for $T(d,n)^4He$ at $\theta_n = 70°$, $E_d = 13$ MeV, both for $1\mu A$ of deuterons. The coincidence time resolution was 3.5 ns. Polarizations were measured using a triple coincidence between the He particle, the recoil α-particle in a high pressure He gas scintillator, and the scattered neutron. Backgrounds were typically 10% with minimal shielding of the detectors from the uncollimated beam. The measured values of the polarization for $D(d,n)$ neutrons at 11 MeV are consistent with earlier measurements [2, 3], and generally follow the contour plot of Barschall [4]. The value of the polarization at a lab angle of 100° was 0.34 ± 0.04. This value is larger than that shown in a fit to data by Porter and Haeberli [5], but the values measured at other angles are consistent with theirs.

REFERENCES

*Work supported in part by the U.S. Atomic Energy Commission.

[1] D. G. Schuster, Nucl. Instr. 76 (1969) 35.
[2] N. V. Alekseev et al., Sov. Phys.—JETP 18 (1964) 979.
[3] P. S. Dubbeldam and R. L. Walter, Nucl. Phys. 28 (1961) 414.
[4] H. H. Barschall, Second Polarization Symp.
[5] L. E. Porter and W. Haeberli, Phys. Rev. 164 (1967) 1229.

DISCUSSION

Duder:
   A similar associated particle method has been developed at Auckland. It has been used to define a neutron beam at 90° lab from the $T(d,n)^4He$ reaction with 5.7-MeV deuterons. A plastic scintillator detects the α's. It is thick enough to stop them yet thin enough to allow elastically scattered deuterons to pass through. The deuterons are detected by a second scintillator immediately behind the first, but optically shielded from it. A pulse from the deuteron detector is put in anticoincidence with that from the α-detector so that only α-pulses are accepted.

T. G. Miller:
   Do you have a count rate problem in the solid state detector?

Hagengruber:

The count rate does not cause difficulties, since we do not use the linear spectrum for taking data. It is simply used as a monitor. A timing signal is taken from the solid-state detector using a device similar to the commercially available time-pick-off. The timing pulses are quite narrow and therefore no pile-up problems are experienced.

T. G. Miller:

What was the approximate count rate in the solid state detector?

Hagengruber:

The count rate was generally kept below 10^6 counts/sec.

T. G. Miller:

I am surprised that you can obtain count rates of 10^6/sec with a solid state detector.

Simmons:

Have you seen problems with neutron damage to the associated particle detector?

Hagengruber:

No. In two years use the resolution of one 30 μm detector has remained at about 30 keV.

Polarized Neutron Source
Using a Compression Ionizer

P. DELPIERRE, R. SENE, J. KAHANE, and M. de BILLY de CRESPIN,
Collège de France, Paris, France

To obtain 14-MeV polarized neutrons by the $T(d,n)^4He$ reaction we built a polarized deuteron source. The polarized atomic beam of deuterium is produced by the well-known method [1, 2] (see fig. 1).

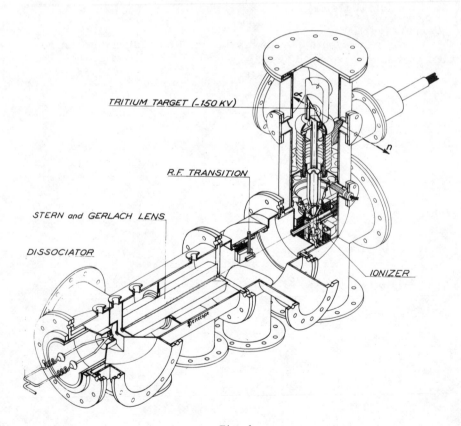

TRITIUM TARGET (-150 KV)

R.F. TRANSITION

STERN and GERLACH LENS

DISSOCIATOR

IONIZER

Fig. 1

R.f. transition. To check the stability of the polarization we first used the (3–5) transition which gives an anisotropy and a good neutron polarization. Then we added the (2–6) transition to obtain the same neutron polarization as with the (3–6) transition but with the possibility of checking the stability of the source by turning off one transition.

Ionizer. The pressure obtained by compression of the beam is too low (10^{-5} Torr.) to produce a plasma with a convenient electronic density. We have to add a neutral buffer gas which enables us to obtain various plasmas (fig. 2). Using the Langmuir double-probe method we have studied plasmas obtained with different buffer gases. The following shows the action of the most important parameters.

decoupling

magnetic field $H_0 \nearrow$ leads to $\begin{cases} \nearrow \text{ decoupling nuclear-electronic spins} \\ \nearrow \text{ ionization efficiency} \\ \nearrow \text{ electron-atom collisions} \end{cases}$

buffer gas

pressure $P_0 \nearrow$ leads to $\begin{cases} \nearrow \text{ absorption of the atomic beam} \\ \searrow \text{ wall-atom collisions} \end{cases}$

r.f. field $\searrow$ leads to $\begin{Bmatrix} \nearrow T_e \\ \searrow n_e \end{Bmatrix}$ which leads to $\begin{cases} \nearrow \text{ ionization efficiency} \\ \searrow \text{ electron-atom collisions} \end{cases}$

The best results are obtained with neon and a decoupling magnetic field of 800 G, an electronic temperature of $T_e = 190\,000°$ K, an electronic density of $n_e = 2 \times 10^8$ ions/cm^3, a deuterium beam of 0.27 μA, an ionization efficiency of 0.4×10^{-3}, a neutron polarization $P_N = 0.14$, a neutron source strength of 6.2×10^6 s^{-1}. These values were obtained during three years of experiments. With the double transition we have obtained under the same conditions $P_N = 0.21$. To measure the instrumental asymmetry in the scattering experiments, we can obtain a high source strength of unpolarized neutrons (up to 2×10^8 s^{-1}) by introducing the deuterium gas directly into the bulb instead of the neon.

Polarimeter. We have built a helium spectrometer (fig. 3) like the one built by Shamu [3]. It is used with a large solid angle (0.1 sr), since it is possible to compute the relation between the vector polarization of the deuterons and the ^{4}He scattering asymmetry in finite geometry [4]. The efficiency is 5×10^{-6}.

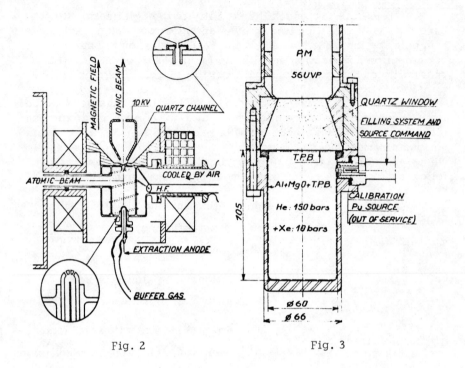

Fig. 2   Fig. 3

REFERENCES

[1]   R. Beurtey, Ph.D. Thesis, CEA Report R-2366 (1964).
[2]   Heyman, Ph.D. Thesis, Paris (1966).
[3]   R. E. Shamu, Nucl. Instr. 14 (1961) 297.
[4]   P. Delpierre et al., Rev. Phys. Appl. 4 (1969) 254.

Measurements of Neutron Polarizations
Resulting from Reactions in Light Nuclei

S. T. THORNTON, R. P. FOGEL, C. L. MORRIS, G. R. NEIL, and
J. R. SMITH, University of Virginia, Charlottesville, USA*

The University of Virginia 5.5-MV Van de Graaff pulsed accelerator
has been used to accelerate protons, deuterons, and helions onto
light nuclei targets in order to measure the polarization of the resulting neutrons. Neutron polarizations have been measured for the following reactions: $T(p,n)^3He$, $D(d,n)^3He$, $^7Li(p,n)^7Be$, $^9Be(d,n)^{10}B$,
$^{12}C(d,n)^{13}N$, $^{12}C(^3He,n)^{14}O$, $^{14}N(d,n)^{15}O$, and $^{16}O(d,n)^{17}F$.

Our neutron polarimeter consists of a liquid-helium dewar with two
large plastic scintillators as side detectors. The neutron time-of-
flight is measured between the target and the helium cell to distinguish
between different energy neutrons. A phototube records scintillations
from the recoil alphas as the neutron scatters in the helium cell.

Extensive measurements have been performed to eliminate sources
of error. In order to check thoroughly our neutron polarimeter and gain
confidence and experience in its use, measurements were made on
seven reactions where existing data were available. These reactions
were:

(1) $T(p,n)^3He$. We made measurements at 33° (lab) for 2.30 and
2.90 MeV. Excellent agreement was obtained with previous results [1].

(2) $D(d,n)^3He$. Angular distributions were measured at 2.0 and 5.0
MeV. At 3 MeV measurements were made for 34.5° and 72° (lab). Good
agreement was obtained with previous results [2].

(3) $^7Li(p,n)^7Be$. Rather extensive measurements were made at 4 MeV
and 50° (lab). The resulting g.s. polarization was 0.32 ± 0.02 which is
slightly inconsistent with the value 0.38 ± 0.02 reported by the Duke
group [3].

(4) $^9Be(d,n)^{10}B$. Measurements at 2.06 MeV and 30° (lab) agree well
with the ground and first three excited states polarization results of
Miller and Biggerstaff [4].

(5) $^{12}C(d,n)^{13}N$. Extensive measurements at 5 MeV were performed
to compare with the results of Morgan et al. [5]. Our polarizations
were generally smaller for the ground state and agreed better with
earlier measurements of Kelsey and Mahajan [6]. The first excited
state polarization measurements agreed reasonably well with those of

871

Morgan et al. [5]. An angular distribution measurement at 5.35 MeV
for the ground state and first excited state agreed well with the results of Donoghue et al. [7]. Measurements of the ground state polarization at 2.8 MeV and 20° and 30° (lab) agreed well with several measurements [8], but is in slight disagreement with the latest Duke result [9].

(6) $^{12}C(^3He,n)^{14}O$. Measurements were made for the ground state
neutrons at 3.76 MeV and 40° (lab), 4.0 MeV and 30° and 75° (lab),
and 4.5 MeV and 27° and 75° (lab). Fair agreement was obtained with
previous results [10].

(7) $^{14}N(d,n)^{15}O$. Measurements at 3.30 MeV and 45° and 85° (lab)
are generally lower, but in reasonable agreement with measurements
at Duke [11].

In addition to the above measurements, new angular distributions
have been obtained at 5.35 MeV for $^{14}N(d,n)^{15}O$ and $^{16}O(d,n)^{17}F$. Ground
state and excited state polarizations have been extracted by using
neutron flight paths up to 3.5 m.

REFERENCES

*Work supported in part by the National Science Foundation.

[1]   R. L. Walter, W. Benenson, P. S. Dubbeldam, and T. H. May,
      Nucl. Phys. 30 (1962) 292; C. A. Kelsey, B. Hoop, Jr., and P.
      Vander Maat, Nucl. Phys. 51 (1964) 395; D. S. Cramer and L.
      Cranberg, Bull. Am. Phys. Soc. 14 (1969) 553.
[2]   F. O. Purser, Jr., J. R. Sawers, Jr., and R. L. Walter, Phys. Rev.
      140 (1965) B870; H. H. Barschall, Second Polarization Symp.
[3]   W. D. Andress, Jr., F. O. Purser, Jr., J. R. Sawers, Jr., and R. L.
      Walter, Nucl. Phys. 70 (1965) 313.
[4]   T. G. Miller and J. A. Biggerstaff, Phys. Rev. C1 (1970) 763.
[5]   G. L. Morgan, R. L. Walter, C. R. Soltesz, and T. R. Donoghue,
      Phys. Rev. 150 (1966) 830.
[6]   C. A. Kelsey and A. S. Mahajan, Nucl. Phys. 71 (1965) 157.
[7]   T. R. Donoghue et al., Phys. Rev. 173 (1968) 952.
[8]   D. I. Garber and E. F. Shrader, Bull. Am. Phys. Soc. 10 (1965)
      510; W. Haeberli and W. W. Rolland, Bull. Am. Phys. Soc. 2
      (1957) 234; J. R. Sawers, Jr., F. O. Purser, Jr., and R. L. Walter,
      Phys. Rev. 141 (1966) 825.
[9]   M. M. Meier, L. A. Schaller, and R. L. Walter, Phys. Rev. 150
      (1966) 821.
[10] L. A. Schaller, R. S. Thomason, N. R. Roberson, and R. L. Walter,
      Bull. Am. Phys. Soc. 12 (1967) 88; C. R. Soltesz et al., Bull. Am.
      Phys. Soc. 12 (1967) 1198.
[11] M. M. Meier, Ph.D. thesis, Duke University (1969).

An Underground Nuclear Explosion
as a Polarized Neutron Source

G. A. KEYWORTH and J. R. LEMLEY, Los Alamos Scientific Laboratory,
New Mexico, USA*

Although an underground nuclear explosion is not the most conventional neutron source, it offers definite advantages, primarily connected with beam intensity. Each burst contains approximately one
mole (10^{24}) of neutrons which are emitted in less than 0.1 μsec. A
moderator is used to increase the neutron flux down to ~20 eV, at
which point the velocities of the moderator and of the neutrons are
equal. Thus, neutrons from 20 eV to several MeV are available and
are resolved by time-of-flight techniques. The experimental techniques for using nuclear explosions as neutron sources have been
amply treated [1] in the literature and will not be discussed here.

The primary purpose of this particular experimental program is to
determine spins of neutron resonances by the straightforward method
of measurements with a polarized target and a polarized neutron beam.
Until fairly recently, however, there was no satisfactory method of
producing a polarized neutron beam from ~50 eV to 100 keV. Shapiro
[2] proposed utilizing the strong spin dependence of the (n,p) interaction to produce such a polarized neutron beam. In this energy range,
the cross section for (n,p) scattering through the singlet state of the
system is about 20 times larger than through the triplet state. Thus,
an unpolarized neutron beam is polarized by passage through a suitably thick target of polarized protons. Shapiro has demonstrated this
technique using neutrons up to 150 eV from a pulsed fast reactor.

The polarized proton target which we have used is a relatively common tool in high energy physics laboratories. It involves dynamic
polarization of the protons in the waters of hydration of crystals of
$La_2Mg_3(NO_3)_{12} \cdot 24 H_2O$, i.e., LMN [3]. The crystal stack (2-cm diam
by 1.8 cm thick) was mounted in a copper microwave cavity which was
located in the center of a superconducting split-pair coil operating at
20.3 kG. The coil was immersed in a liquid helium bath at 4 K in a
cryostat which permitted room temperature access to the coil bore.
The microwave cavity was contained in a separate cryostat whose tail
was inserted into the bore of the superconducting coil. This latter
cryostat contained liquid helium maintained at ~1 K by a 1400 liter/

sec mechanical pump. The cryostats contain thin windows to allow transit of the vertical neutron beam. The coil is located with its bore in the horizontal plane and the neutron beam passes through apertures in the gap.

The axis of the crystals was located perpendicular to the magnetic field direction. Under these conditions, microwaves of frequency 76.6 GHz are required to polarize the free protons. The microwave power was generated by a reflex klystron and transmitted by oversize (8-mm) waveguide to the cavity.

In the summer of 1969, an experiment was performed on the Physics 8 event at the Atomic Energy Commission's Nevada test site to determine the feasibility of polarizing such an intense beam in this manner. Of some concern was the extent of depolarization in the target due to heating from the neutron beam. Comparatively little energy deposit occurs as a result of the initial burst of gamma radiation. Calculations of neutron heating indicated a total temperature rise in the target of 3 K.

One of the features of LMN which makes it an excellent source of polarized protons is its long proton relaxation time (~1 hr at 1 K). This same feature allows us to maintain constant polarization during the 3−4 msec duration between arrival of the highest and lowest energy neutrons. At 4 K, the calculated final temperature, the proton relaxation time is still > 100 sec. This estimate was confirmed on the Physics 8 event.

Measurement of the target polarization is normally made by use of the "Q-meter" nuclear magnetic resonance technique. In this method a small fraction of the protons in the target are caused to flip by an electromagnetic field from a coil near the sample. The coil is part of a tuned circuit whose electrical characteristics are altered by the proton transitions. The most common technique employed involves sweeping the NMR oscillator through the proton line in several minutes. In our case, we wished to monitor the polarization during the ~4 msec duration of the neutron burst. We thus swept the frequency at a rate of 1 kHz and displayed the results of the sweep on an oscilloscope. The oscilloscope was then photographed by a high speed moving-film camera. Subsequent examination of the film showed no change in polarization in the target during the entire film exposure, which ended 1 sec after detonation. Thus we feel confident that this is a practical method of producing polarized neutrons over the energy range 20 eV to 100 keV.

Because this fast sweep NMR system was designed primarily to determine the change in polarization, it was not suited for accurate absolute polarization determination. The thermal equilibrium (TE) signal, i.e., the signal due to brute force polarization before microwave pumping, is compared to the enhanced signal to determine absolute polarization. With the system used in the Physics 8 experiment, the TE signal was nearly unobservable. However, since that experiment, we have acquired a PDP-15 computer with which we average the signal from the fast sweep to accurately determine the TE signal.

As discussed above, the proton polarization was perpendicular to the beam direction in the Physics 8 experiment. This offers the disadvantage that the beam must pass through the coil gap, thereby making it difficult to avoid Majorana depolarization. This problem is serious with the LMN system but far more serious with the actual target system, discussed in another contribution to this conference. In the case of s-wave neutrons, how the polarization vectors are oriented with respect to the beam direction is of no consequence, as there is no transfer of angular momentum. Although in our energy range penetration considerations suggest that s-wave resonances predominate, it is still desirable to preserve the capability of spin determination of p-wave resonances. However, although not tacitly obvious, calculations of resonance formation probabilities reveal that, for targets of reasonably high spin, e.g., 5/2, approximately equal amounts of information regarding the spin configurations of p-wave resonances are determined for the polarization perpendicular or parallel to the beam direction.

We now have under construction a new LMN beam polarization system consisting of a high homogeneity coil (field uniformity ~ 1 G over a 2.5-cm sphere), 10-watt extended interaction oscillator for a microwave power source, and a single cryostat whose 1 K bath is capable of being continuously filled from a 4 K bath. We expect this system to produce neutron beam polarizations of 60—70% with transmissions of 18%. This equipment, with a polarized subthreshold fissionable target, will be used in an experiment on the Physics 9 event in 1971 to measure spins of intermediate structure resonances.

REFERENCES

*Work performed under the auspices of the U.S. Atomic Energy Commission.

[1] B. C. Diven, Proc. Int. Conf. on the Study of Nuclear Structure with Neutrons, Antwerp, 1965 (North-Holland Publishing Co., Amsterdam, 1966) p. 441.

[2] F. L. Shapiro, Proc. Int. Conf. on the Study of Nuclear Structure with Neutrons, Antwerp, 1965 (North-Holland Publishing Co., Amsterdam, 1966) p. 223.

[3] A. Abragham and M. Borghini, Progress in Low Temperature Physics (North-Holland Publishing Co., Amsterdam, 1964) vol. 4, p. 385; C. D. Jeffries, Dynamic Nuclear Orientation (Interscience Publishers, New York, 1963).

DISCUSSION

Postma:

The heat will be mainly generated at the free protons by neutron scattering and not in the bulk of the material. Is the quoted temperature rise of 3 K related to the free protons or to the whole crystal?

Keyworth:

It is related to the entire sample.

Improvements in Performance of a
Liquid Helium Scintillation Counter

J. C. MARTIN, W. B. BROSTE,* and J. E. SIMMONS, Los Alamos
Scientific Laboratory, New Mexico, USA†

Since the liquid helium scintillation polarimeter of Simmons and Perkins [1], many improved LHe scintillators have been constructed [2–8].
For optimum energy resolution in a LHe scintillator, one needs a uniform pulse-height response over the entire volume of the scintillator.
This response depends on the uniformity of the wave shifter and the
light collecting geometry. Recent publications have reported resolutions (FWHM) of between 11% and 30% for alpha sources [2, 3, 5, 7, 8];
and between 9% and 13% for gated He-recoil distributions for elastic
scattered neutrons [4, 6]. Pulse-height uniformity measurements [3,
7, 8] with an alpha source showed variations from 2% to ≈40%. Our
recent work has shown that pulse-height uniformity can be significantly
improved by using a thicker coating of wave shifter on the photomultiplier window of the scintillator cell relative to that on the wall.

In our present scintillation counter the vacuum ultraviolet helium
scintillations are wave shifted by a vacuum evaporated deposit of
p,p'-diphenylstilbene (DPS) on the reflector and window of the scintillation cell. The reflector is a ceramic shell of fired Al_2O_3. The window
is 0.25 cm × 5.0 cm diam sapphire. The active volume of the cell is
150 cm^3 (4.7 moles LHe). The photomultiplier (RCA 7326), housed in
the cryostat vacuum, views the scintillation cell through a hollow pol-
ished aluminum light cone, and is cooled to ≈120 K to minimize the
radiant heat load on the LHe cell. Heretofore, we employed DPS coat-
ings of 100 $\mu g/cm^2$ on the reflector shell wall and 50 $\mu g/cm^2$ on the
window. Under these circumstances our resolution was only about 25%
FWHM for 15 MeV He recoils.

Recent measurements with a test cryostat and a movable α-source
revealed a marked axial non-uniformity in the pulse-height response
for the above DPS prescription. The light output variation along the
cell seemed to indicate a light collection problem in the vicinity of
the window. Consequently we increased the window DPS coating while
holding the reflector coating constant. The result was an improvement in pulse-height uniformity with increasing DPS thickness on the

window. For ≈ 200 $\mu g/cm^2$ on the window the light output variation was reduced to 7 1/2% from an initial value of $\approx 40\%$.

In view of this information, the polarimeter scintillation cell was recoated with ≈ 100 $\mu g/cm^2$ on the reflector and ≈ 200 $\mu g/cm^2$ on the window. Using 4 to 32 MeV neutrons from the $T(p,n)^3He$ and $T(d,n)^4He$ reactions at $0°$, the response of the recoated cells was observed. A NE 102 (5.0 cm $\times$ 18.0 cm $\times$ 7.6 cm) RCA 8575 detector for the recoil neutrons was used in coincidence with the LHe scintillation counter. The geometry of the experiment was: $R_1 = 99.1$ cm, $\theta_1 = 0°$, $\Delta\theta_1 = \pm 1.75°$, $R_1 = 42.2$ cm, $\theta_1 = 137.7°$, and $\Delta\theta_1 = \pm 3.45°$. Both He-recoil singles and coincidence-gated He recoils were observed over the neutron energy range. Fig. 1 shows typical pulse distributions. The left panel shows the singles spectrum with the prominent $n + He \rightarrow d + t$ peak; the right panel shows the coincidence-gated He recoil spectrum. The gated He-recoil resolution (FWHM) ranged from 8% for $E_{He} = 19$ MeV to 22% for $E_{He} = 3$ MeV. The light output as a function of helium-recoil end-point energy exhibits the same non-linear shape as reported by Piffaretti et al. [4]. The resolution for the $n - He$ $(d + t)$ breakup peak varied from 10% for $E_{d+t} = 14$ MeV to 15% for $E_{d+t} = 5$ MeV.

By increasing the DPS thickness on the sapphire window to ≈ 200 $\mu g/cm^2$ we have improved the He-recoil resolution by a factor of three, without appreciably lowering the light output of the scintillator.

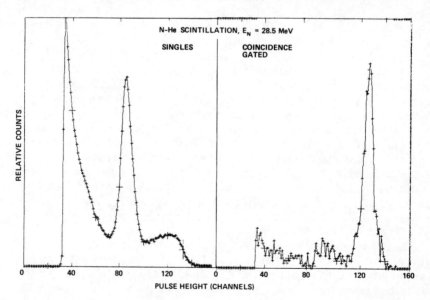

Fig. 1. Singles: $d + t$ breakup peak. Res. (FWHM) = 12%. $E(d + t) =$ 10.8 MeV. Coincidence gated: gated He recoils. Res. (FWHM) = 8%, not corrected for kinematic spread. $E(He) = 16.8$ MeV.

REFERENCES

*Associated Western Universities Fellow from the University of Wyoming.
†Work performed under the auspices of the U.S. Atomic Energy Commission.

[1] J. E. Simmons and R. B. Perkins, Rev. Sci. Instr. 32 (1961) 1173.
[2] T. G. Miller, Nucl. Instr. 40 (1966) 93.
[3] J. R. Kane, R. T. Siegel, and A. Suzuki, Rev. Sci. Instr. 34 (1963) 817.
[4] J. Piffaretti, J. Rossel, and J. Weber, Second Polarization Symp.
[5] S. T. Lam et al., Nucl. Instr. 62 (1968) 1.
[6] J. Birchall et al., Nucl. Instr. 65 (1968) 117.
[7] L. P. Robertson et al., Nucl. Phys. A134 (1969) 545.
[8] D. C. Buckle et al., Nucl. Instr. 77 (1970) 249.

A Simple and Stable
Liquid Helium Neutron Polarimeter

K. KATORI and K. FURUNO, Tokyo University of Education, Japan

We have designed a polarimeter to measure triple scattering parameters in neutron-nucleus scattering at 1.4 MeV. It uses scintillations in liquid helium, a spin precession solenoid, and a time-of-flight neutron spectrometer. The use of scintillations in liquid helium as a timing pulse for the time-of-flight spectrometer reduces backgrounds and defines the path of neutrons.

An illustration of the liquid helium cryostat is shown in fig. 1. Main advantages of this polarimeter are as follows:

1) The photomultiplier is mounted on the side of the cryostat. If the phototube is mounted on the bottom as in previous papers [1, 2], great care must be taken to minimize the absorption of light by ice, frost, or dust dropped into the flask from the upper inlet.

2) The helium flask, the nitrogen heat shield, and the face of the phototube are completely separated. This makes it possible to construct a cryostat that has a simple structure, no vacuum trouble and a low boiling-off rate for both liquid helium and nitrogen.

3) A convex lens is used to collect the scintillation light. The helium flask is made of Pyrex glass in a spherical shape of 7 cm o.d. A thin layer (100 $\mu g/cm^2$) of p, p'-quaterphenyl is evaporated inside the flask in vacuum. White tygon paint is deposited outside the flask except for the surface viewed by the phototube.

A difficult problem is to keep scintillation pulse height stable for a long period. The instability is mainly attributed to the degradation of the p, p'-quaterphenyl. However, no appreciable decrease in the pulse height is observed if the flask is filled with dry helium gas when it is not in use.

The performance was tested by the measurement of the neutron-polarization in the reactions $^{12}C(d,n_0)^{13}N$ and $^9Be(\alpha,n_0)^{12}C$. The signal to background ratio is of the order of ten, and the instrumental asymmetry is less than 1%. Fig. 2 shows a typical neutron time-of-flight spectrum gated by pulses from the liquid helium scintillation counter in the case of $^{12}C(d,n_0)^{13}N$. This polarimeter was successfully used in our triple scattering experiment [3] for about four months.

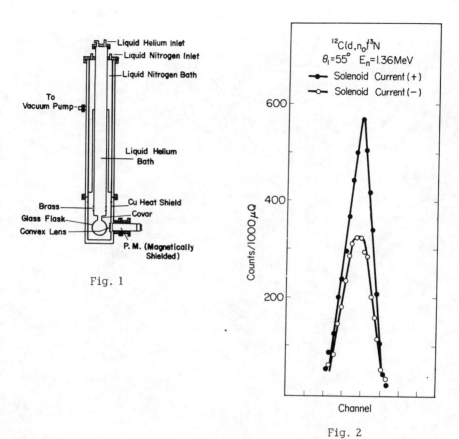

Fig. 1

Fig. 2

REFERENCES

[1]   T. G. Miller, Nucl. Instr. Meth. 40 (1966) 93.
[2]   J. Piffaretti, J. Rossel, and J. Weber, Second Polarization Symp.,
      p. 152.
[3]   K. Katori, T. Nagata, A. Uchida, and S. Kobayashi, J. Phys. Soc.
      Japan 28 (1970) 1116.

Targets

Characteristics of AlK and Al (NH₄) Alum
as New Materials for Polarized Proton Targets

P. J. BENDT, Los Alamos Scientific Laboratory, New Mexico, USA*

Dynamic proton polarization by the "solid effect" has been studied in $AlK(SO_4)_2 \cdot 12H_2O$ and $Al(NH_4)(SO_4)_2 \cdot 12H_2O$, in which $< 0.5\%$ of the diamagnetic Al was replaced with paramagnetic Cr^{3+} ions. Over 50% proton polarization was attained in three alum crystals, at 1 K and ~ 19.5 kG, with limited microwave power ($\sim 1/4$ W) [1]. The alum crystals have higher hydrogen density by weight than lanthanum magnesium nitrate (LMN), by factors of 1.61 for AlK alum and 1.97 for Al(NH₄) alum.

The experiments show it takes three times as much microwave power in alums as in LMN to obtain the same degree of saturation of the forbidden paramagnetic resonance transition. The thermal equilibrium ion polarization is also larger in LMN at any given magnetic field and temperature. At 26 kG, and using a high-power microwave oscillator, these advantages will provide only 5% additional polarization in LMN, while using Al(NH₄) alum will expose approximately twice the number of polarized protons to a beam, for the same charged particle energy loss.

REFERENCE

*Work performed under the auspices of the U.S. Atomic Energy Commission.

[1] P. J. Bendt, Phys. Rev. Lett. 25 (1970) 365.

DISCUSSION

Baker:
How will radiation damage in these crystals compare with that in LMN?

Bendt:

I do not know, because I have had no experience with exposure of the crystals to radiation. Their resistance to radiation damage should be similar to that of LMN.

Jeffries:

The proton line shape should change with increased polarization, contrary to what you reported. Could you comment on that?

Bendt:

The changes must occur at higher polarizations than I reached, i.e., above 50% polarization.

A Proposed Ultra-Low Temperature
Polarized Target for Use with
Single-Burst Neutron Sources

J. R. LEMLEY and G. A. KEYWORTH, Los Alamos Scientific Laboratory,
New Mexico, USA*

An underground nuclear explosion of the type used in physics experiments by the Los Alamos Scientific Laboratory (LASL) [1] produces
approximately 1 mole (10^{24}) of neutrons with a generally continuous
energy spectrum from about 40 eV to several MeV. These neutrons
reach a target at the end of a 250-m flight path over a period of 3 msec;
the neutron energy is determined by a time-of-flight technique. Beam
area at the target is about 3 cm^2. After collimation and transmission
through an LMN spin filter to polarize the beam, about 10^{15} neutrons
reach the target. For a variety of nuclear reactions this high neutron
flux produces events with good statistical frequency over a wide energy range in a very short time. The long flight path increases energy
resolution.

An experiment currently being prepared for the next physics shot
requires both a polarized neutron beam and a polarized target. Polarization of the neutron beam has been discussed in another paper at
this conference [2]. The neutron-induced fission cross sections of
certain nuclei such as ^{237}Np exhibit large subthreshold fission resonances due to coupling of the compound nuclear wave functions with
states in the second potential minimum in the fission barrier. Presumably the coupling is between only states of the same J value.
When a polarized neutron beam impinges upon a polarized ^{237}Np target, the probability of finding certain spin combinations is increased
or decreased depending upon the relative polarizations of beam and
target. Amplitudes of fission resonances with a specific J value will
be increased or decreased relative to their areas when polarization
is zero.

The nature of the neutron beam places several requirements upon
the design of a polarized ^{237}Np target. Beam direction is necessarily
vertical, and the cryostat must be designed to accommodate this.
Since both the LMN target used to polarize the beam and the ^{237}Np
target involve high magnetic fields, Majorana depolarization must be
avoided in the field gradients along the beam path. Relative orientation of target spin, beam spin, and beam momentum must maximize the

effect of the reaction to be studied in the angular distribution of the radiation to be detected. This is not necessarily consistent with minimizing Majorana depolarization when p-wave resonances are to be studied. In order for fission fragments from the ^{237}Np(n,f) reaction to escape from the target and reach counters, the ^{237}Np nuclei must lie within about 5 μm of the target surface but remain within a matrix which will permit polarization of the ^{237}Np nuclei. Target nuclei not close to the surface would be useless for counting purposes, and in addition their fissioning would cause excessive heating, resulting in depolarization of the entire target. It is therefore necessary to use a minimum amount of ^{237}Np and to confine it to the surface of the target. Since the beam contains neutrons in the MeV energy range, no fissionable nuclei other than the ^{237}Np may be present. Many compounds of actinide elements, such as $UO_2Rb(NO_3)_3$, which might be suitable host matrices for dilution and polarization of ^{237}Np are eliminated.

Probably the greatest problem concerned with the ^{237}Np target is dissipation of heat produced by interaction of the beam and target. The major source of heating is the ^{237}Np(n,f) reaction with total fission fragment energy of 170 MeV. Using the neutron flux from a previous experiment and assuming that half of the fission fragment energy is deposited in the target, 10 mg of ^{237}Np will deposit energy at the rate of approximately 10^4 erg/sec.

To prevent immediate warming and depolarization of the ^{237}Np nuclei, we propose to use as a heat sink the heat capacity of a system of nuclear spins in a high magnetic field at ultra-low temperatures. Using a ^{3}He-^{4}He dilution refrigerator, target temperatures of less than 20 mK can be maintained indefinitely before application of the neutron beam. At these temperatures it is possible to polarize the nuclear spins of (nonfissionable) metals such as thallium so that there is a relatively large heat capacity associated with alignment of the spins.

The effectiveness of a large nuclear heat capacity as a heat sink depends on the thermal resistance R_{SL} between the lattice and the nuclear spins. At very low temperatures for a metal in the non-superconducting state, the "lattice" is the thermal bath of the conduction electrons, since lattice vibrations are in their ground states. R_{SL} is related to the nuclear spin-lattice relaxation time T_1 according to $R_{SL}C_S = T_1$, which is the time constant with which the spin temperature approaches that of a lattice maintained at constant temperature. C_S is the nuclear spin heat capacity. R_{SL} may be estimated from Korringa's relation, $T_1T_S = K$ where K is Korringa's constant and T_S is the spin temperature. The Korringa constant is commonly derived from measurements of the Knight shift when T_1 is not measured directly. $R_{SL}C_S = K/T$. If the nuclear heat capacity is given by $C_S = bH^2/T_S^2$ where b is a constant, the heat flow to the nuclei is $\dot{Q} = (T_S - T_L)/R_{SL} = (bH^2/K)(T_L - T_S)/T_S$ where T_L is the lattice (or electron) temperature. For thallium K is small (0.006 sec °K) and the nuclear moment is relatively large (1.6 nm) [3]. For the easily attainable field of 50 kG and 10 cm^3 of Tl, $|(T_S - T_L)/T_S| = 0.1$ when $\dot{Q} = 0.7 \times 10^4$ erg/sec. This ratio is temperature independent as long as the expressions for C_S and T_1 hold. If T_S increases from 15 mK to 100 mK, the heat absorbed is 2.4×10^4 erg.

The problem remains to find an environment in which ^{237}Np can be polarized and which will permit the nuclear spins to be in good thermal contact with the electronic system associated with a nuclear heat sink such as thallium metal. Neptunium-237 has two properties which should facilitate this. Its large magnetic moment (2.8 nm [4]) decreases the size of the effective field required to achieve a given polarization at a given temperature. In a metallic environment the high Z of the nucleus increases the coupling of the nuclear spins and conduction electrons and should decrease spin-lattice relaxation time. There are ^{237}Np compounds and alloys which have hyperfine fields of several MG at the nucleus. At present a metallic matrix seems most desirable, in order to establish sufficient thermal contact with a heat sink via conduction electrons.

An ultra-low temperature system can also probably be adapted for use with pulsed neutron beams produced by linear accelerators, and other nuclei can be polarized for use as targets.

We thank Prof. J. C. Wheatley for valuable discussions about cryogenic aspects of the experiment.

REFERENCES

*Work performed under the auspices of the U.S. Atomic Energy Commission.

[1] B. C. Diven, Proc. Int. Conf. on the Study of Nuclear Structure with Neutrons, Antwerp, 1965 (North-Holland Publishing Co., Amsterdam, 1966) p. 441.
[2] G. A. Keyworth and J. R. Lemley, Third Polarization Symp.
[3] W. R. Abel et al., Physics 1 (1965) 337.
[4] J. A. Stone and W. L. Pillinger, Phys. Rev. 165 (1968) 1319.

A Proposed 0.4 K Polarized
Proton Target Cryostat for LAMPF

F. J. EDESKUTY, C. F. HWANG, and K. D. WILLIAMSON, Jr., Los Alamos
Scientific Laboratory, New Mexico, USA*

A polarized proton target cryostat design is described which is intended for 0.4 K operation in the LAMPF external beams for various
scattering experiments. It has been shown by Hill et al. [1] that proton polarization of the order of 70% can be obtained in butylalcohol
doped with porphyrexide at temperatures near 0.4 K. Previous experiments at CEA [2] and SLAC [3] indicated that an alcohol target is capable of tolerating 10^{10} minimum ionizing particles per sec for periods
of a few hours without losing a substantial part of its initial polarization. When annealed at 77 K at suitable time intervals, an alcohol
target may last as long as 24–36 hours [4]. In consideration of the
expected target life versus reaction yield, the optimized design beam
intensity was chosen as 3×10^9 (300–800 MeV) protons per sec (corresponding to $\sim 10^{10}$ minimum ionizing particles per sec). Under these
circumstances, even with the annealing process, it will be necessary
to replace the target material periodically. The cryostat is designed
to facilitate such target changes as required (daily). In order to avoid
loss of ^{3}He while changing targets, it is not desirable to have ^{3}He in
direct contact with the target material. Therefore to transfer the heat
from the target to the ^{3}He bath, a ^{4}He heat exchanger has been included in the design.

Hill et al. [1] estimated that 30 mW of refrigeration should suffice to
cool a 10-gram target and the microwave cavity. This estimate is consistent with the experience of Masaike [5]. By anchoring the microwave
cavity at 1.0 K it is anticipated that the microwave loading to the ^{3}He
system will be less than 5 mW. For protons of 300–800 MeV traversing
a 2 gm/cm^2 alcohol target, the energy loss is between $0.8–1.2 \times 10^{-9}$
mW-sec per particle or 2–4 mW for design beam intensity. Combining
these two loads a total refrigeration demand of about 10 mW at 0.4 K
in the target material is anticipated.

A schematic diagram of the proposed target is shown in fig. 1. The
precooling of the ^{3}He gas as well as the 1K and 77 K shields is provided by a modified Roubeau [6] cryostat, in which ^{4}He enters by a
tube (13 on fig. 1), is precooled by a heat exchanger (8 on fig. 1),

890

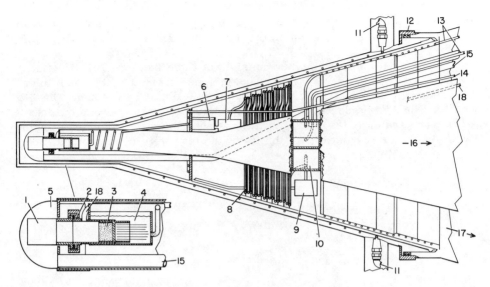

Fig. 1. Schematic diagram of cryostat: (1) target cell, (2) quick disconnect joint, (3) ^{4}He to ^{3}He heat exchanger, (4) ^{3}He evaporator,
(5) microwave cavity, (6)^{4}He evaporator, (7) ^{4}He expansion valve,
(8) ^{4}He heat exchanger, (9) ^{4}He precooling valve, (10) ^{4}He separator,
(11) N_2 supply inlet/outlet, (12) warm O-ring sliding joint, (13) ^{4}He
fill and vent lines for separator, (14) ^{3}He fill line, (15) microwave
guide, (16) ^{3}He pump-out, (17) ^{4}He pump-out, (18) ^{4}He fill line for low
temperature heat exchanger.

and evaporates at near 1 K (6 on fig. 1). The pumped vapors provide
the precooling in the heat exchanger (8 on fig. 1). This Roubeau principle is repeated in the ^{3}He loop by means of inlet tube (14), condenser
on (6), and evaporator (4) (all on fig. 1) with an expected ^{3}He evaporator
temperature of 0.3 K. Heat is transferred from the target to the ^{3}He
evaporator by a closed superfluid ^{4}He heat exchanger. The ^{4}He is in
direct contact with the polarized target material. The chief impedances
in this heat exchanger are three Kapitza resistances and the relatively
poor heat transport capability of the He II in this temperature range.
A separate, preliminary experiment to study these impedances is in
progress.

The features incorporated into the cryostat design to permit rapid
sample changes are a completely removable target-refrigerator system
and a quick disconnect joint between the target and the ^{4}He heat exchanger tube wall. The warm O-ring seal (12 on fig. 1) permits removal
of the entire inner structure of the cryostat and the ^{3}He handling system when one exterior flange (not shown in fig. 1) is unbolted. After
extraction of the double-Roubeau section from the vacuum jacket, the

target cell can be replaced by removing the hemispherical cavity wall
(5 on fig. 1) and by opening the quick disconnect Indium O-ring joint
(2 on fig. 1).

The ^{3}He pumping system consists of three Edwards 9B4 diffusion
booster pumps, each with a pumping speed of 3800 ℓ/sec at an inlet
pressure of $\sim 10^{-3}$ Torr, backed by an Edwards 550 forepump. The ^{3}He
stream can be continuously purified by a cold trap between the forepump and cryostat inlet. With these pumps and the designed pumping
line sizes, we anticipate the following temperature-power input relationships:

Power, mW	10	20	40	80
$T_{evaporator}$, K	0.30	0.32	0.34	0.37
T_{target}, K	0.40	0.42	0.44	0.47
M, $\dfrac{gm\text{-}moles}{hr}$	1.8	3.6	7.2	14.4

REFERENCES

*Work performed under the auspices of the U.S. Atomic Energy Com-
mission.

[1]   D. A. Hill et al., Phys. Rev. Lett. 23 (1969) 460.
[2]   J. R. Chen et al., Phys. Rev. Lett. 21 (1968) 1279.
[3]   T. Powell et al., Phys. Rev. Lett. 24 (1970) 753.
[4]   R. Z. Fuzesy, private communication.
[5]   A. Masaike, private communication.
[6]   M. Borghini, P. Roubeau, and C. Ryter, Nucl. Instr. 49 (1967) 248.

A Frozen Polarized Target

D. J. NICHOLAS, S. F. J. READ, F. M. RUSSELL, and W. G. WILLIAMS,
Rutherford High Energy Laboratory, Chilton, England

A polarized proton target currently under development at the Rutherford
Laboratory is described. The target has the twin objectives of high
polarization and large angular access. In this target the material is
first polarized by the solid-effect and, when a high degree of polarization has been established, the temperature of the target is lowered
until depolarization through spin relaxation becomes very slow. In
this state the polarization is "frozen-in" and the target transferred to
a holding magnetic field of low homogeneity but giving large angular
access.

The polarized target basically is composed of a cylindrical copper
microwave cavity of dimensions 1.5 cm diam × 5 cm long containing
the target material in the form of small spheres of approximately 2.5 mm
diam. The temperature of this cavity can be varied between ~ 1K and
0.3K by means of a ^{3}He refrigerator. Surrounding this cavity, and
mounted one above the other, are two superconducting magnets. These
can be raised and lowered relative to the cavity so that the target can
be subjected to the field of one magnet or the other. The cavity and
target material are cooled by ^{4}He liquid which in turn is refrigerated,
via a heat exchanger, by evaporation of ^{3}He liquid. The principal
reasons for adoption of this arrangement are the high thermal conductivity of ^{4}HeII (0.5K > T > 1.0K), the low cost of ^{4}He, and the considerable experience in its use in targets. In addition, the high heat
flux achievable in ^{4}HeII enables the heat exchanger between the ^{4}He
and ^{3}He liquids to be physically distant from the cavity and target.
This permits the design of the heat exchanger and target to be optimized separately. A schematic drawing of the target in its operating
position, surrounded by a 90° access holding magnet, is shown in
fig. 1. The main ^{3}He/^{4}He heat exchanger is a "mist-type" exchanger
made up of 300 plates of 0.04 mm thick copper perforated by ~ 4900
holes each of 0.045 cm diam. These plates were manufactured by a
photo-etching technique and stacked together to form a cylindrical
heat exchanger (~ 12.7 cm long and 6.0 cm diam) to give maximum

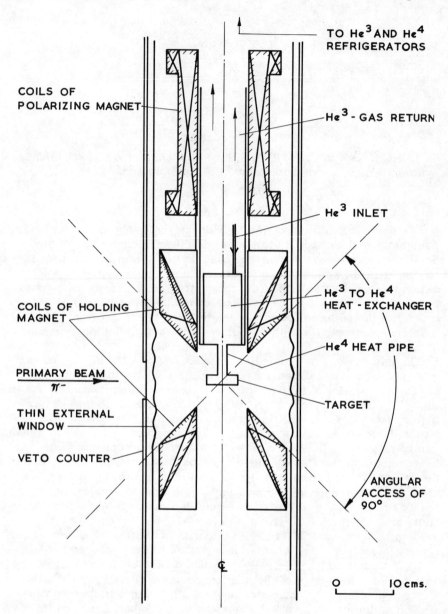

TO He³ AND He⁴
REFRIGERATORS

COILS OF
POLARIZING MAGNET

He³ - GAS RETURN

He³ INLET

COILS OF HOLDING
MAGNET

He³ TO He⁴
HEAT - EXCHANGER

He⁴ HEAT PIPE

PRIMARY BEAM
π^-

TARGET

THIN EXTERNAL
WINDOW

VETO COUNTER

ANGULAR
ACCESS OF
90°

0   10 cms.

Fig. 1. Schematic drawing of the frozen target in operating position.

heat transfer over a range of pressures with minimum pressure drop across the exchanger.

The remainder of the target is basically a continuously operating ^{3}He refrigerator, working in the temperature range 0.9K to 0.3K. The polarization will be measured at the beginning and end of an experimental run by an NMR system which automatically corrects for dispersion errors. Due to the poor homogeneity of the holding field the polarization can only be sampled using a very small coil while the target is in this field.

The validity of the frozen target concept has already been demonstrated on a pilot ^{3}He refrigerator, in particular, establishing polarization at temperature ~ 0.7K and subsequent freezing to 0.3K. Experiments have also been carried out over a range of magnetic fields to simulate the field change during transfer of the target material from the polarizing to the holding magnet.

DISCUSSION

Dabbs:

What are the target materials, fields in the two magnets, and size of the sample spaces?

Nicholas:

As targets, we use only hydrogen-rich materials. The polarizing magnet will be 35 to 40 cm long with a 50 kG field. The holding magnet will be 20 cm long with a 27–30 kG field. The sample can be lowered to 0.9 K.

A New Technique for J-Value Determinations with a Dynamically Polarized Proton Target and Backscattered Neutrons

J. W. T. DABBS, W. W. WALKER,* W. B. DRESS, and P. D. MILLER,
Oak Ridge National Laboratory, Tennessee, USA†

In their paper [1] which suggested the possibility of polarizing inter-mediate energy neutron beams by passage through dynamically polar-ized proton targets, Taran and Shapiro also suggested the idea of using two such targets, the first as polarizer, the second as analyzer for scattered neutrons from an unpolarized nuclear scattering sample. Our modification of this idea is to backscatter s-wave neutrons from the sample and to locate the analyzer crystal (in the present instance $La_2(Nd)Mg_3(NO_3)_{12} \cdot 24H_2O$) in the same magnet, cryostat, and radio-frequency cavity as the polarizer crystal (also LMN).

If we characterize the transmissions of the LMN polarizer and analyzer crystals as T_+ for spin-up neutrons and T_- for spin-down neutrons, the expected effect [2] is given by

$$\Delta = \frac{2}{3(2I + 1)} \frac{(T_+ - T_-)^2}{(T_+ + T_-)} \tag{1}$$

where $\Delta = R_+ - R_-$ and R_+ and R_- are ratios of the analyzed intensity to the unanalyzed (i.e., simply scattered) intensity for $J = I + 1/2$ and $J = I - 1/2$ resonances, respectively, assuming equal solid angles for counting and the same direction of proton polarization in the polarizer and analyzer crystals. It should be noted that normalization of R_+ or R_- to unity will give another factor of $\sim (T_+ + T_-)/2$ in the denominator of eq. (1). For an optimal thickness LMN crystal, values of $T_+ = 0.33$ and $T_- = 0.07$ are expected for 70% proton polarization.

We have examined the possibility for such studies at ORELA (Oak Ridge Electron Linear Accelerator) and have concluded that the tech-nique is feasible for the larger scattering resonances in a fairly large number of nuclides. However, it will not yield results in a reasonable time for the smaller resonances, e.g., in fissionable nuclei. Time-of-flight measurements will serve to separate resonances with resolutions of 0.3-4.5 ns/m up to energies of ~ 50 keV, where the neutron polariza-tion ceases to be independent of energy [1].

Modifications of our polarized proton target apparatus to test these ideas is in progress. We plan to eliminate unwanted spin-orbit effects (right-left asymmetries) by periodic reversal of the proton polarization. The beam polarimeter will consist of a Co-Fe single crystal and a Cu single crystal, both adjusted to measure a sample of the neutron beam in the 2-4 eV region by diffraction. A periodic reversal of left and right Bragg scattering angles will be used to eliminate the spin-orbit effect in this measurement.

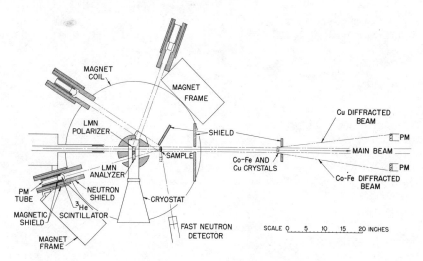

Fig. 1. Schematic drawing of the experimental arrangement.

The polarizer conditions are 18,800 G, 1.0 K, with 4 mm (70 GHz) radiation introduced into the rf cavity to polarize the protons. The main scattered-neutron detectors will be high pressure ^{3}He gas scintillators [3].

An additional scattered-neutron detector at a different angle will probably allow better identification of p-wave resonances. In the case of fissionable nuclei, a fast-neutron detector will be used instead, so that subthreshold fission resonances, e.g., may be located accurately.

REFERENCES

*On sabbatical leave from the University of Alabama 1969-70.
†Research sponsored by the U.S. Atomic Energy Commission under contract with the Union Carbide Corporation.

[1]   Yu. V. Taran and F. L. Shapiro, Sov. Phys.—JETP 17 (1963) 1467.
[2]   We acknowledge a private communication by V. I. Lushchikov and A. Michaudon (1968).
[3]   Kindly loaned by M. M. Hoffman, Los Alamos Scientific Laboratory.

DISCUSSION

Simmons:

Have you measured the neutron beam polarization?

Dabbs:

No, but the LMN target was operated at 45% proton polarization.
We expect 70% polarization when the target is tuned.

Studies with Polarized ³He Targets

S. D. BAKER, Rice University, Houston, Texas, USA

This paper will summarize the work at Rice or in collaboration with Rice using polarized ³He targets since the report [1] at the Karlsruhe conference.

There is no need to describe the optical pumping process by which one obtains polarized ³He nuclei, since it has been well covered at other meetings and in the literature. However, it is worth pointing out that it is now possible to build targets for low-energy scattering experiments on a routine basis which do not require an artisan to construct. The target design is relatively simple. The cell is made primarily of glass except for aluminum fittings, mounted on the ends of flanged glass tube, which hold thin mica or aluminum foils to allow the beam to enter and exit and scattered particles to emerge. Fig. 1 shows the glass work with the entrance window assembly only; the other arms are equipped with somewhat similar fittings designed to accommodate detectors, a beam stop, etc. Collimation slits in the

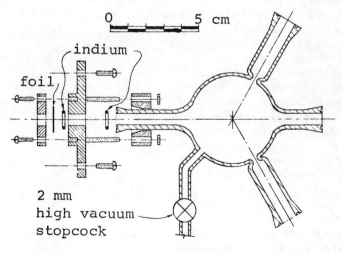

Fig. 1

899

detector arm can be formed by crimping the glass. The glass work is held in a jig during construction to keep the angles of the arms within tolerances. The details of this type of target construction were worked out chiefly by Hardy and Boykin.

The disadvantage of this design is that it requires several cells to be built to cover a range of angles, since it is hard to put the arms much closer together than about 25°. The advantages are ease of cleanup and refill, small size, and dependability. At higher particle energies than we have ordinarily been dealing with it is possible to use all-glass cells. In this case the cleanup procedure goes very quickly if one cleans the glass surfaces with a discharge using a mixture of hydrogen and helium.

Typical target pressure is 5 Torr, and typical polarizations are between 10 and 20%. At present, I do not know of a really dependable way to measure the target polarization by optical means with a possible systematic error of much less than ±15%. This is because of the fact that the light used for pumping will vary from lamp to lamp, and the geometry of the cells may also affect the quality of the transmitted light by internal reflections, birefringence, etc. It is probably safest to calibrate target polarization with a nuclear scattering reaction, particularly ^{3}He-^{4}He, which has recently been studied with double scattering measurements at other laboratories [2, 3] and which gives high analyzing power at certain energies and angles.

Now let us look at some results obtained with polarized ^{3}He targets:

p-^{3}He elastic scattering. In the range from 4 to 11 MeV McSherry and I measured the ^{3}He analyzing power of this reaction at Rice [4]; and in collaboration with Clegg and Plattner, we measured [5] the spin correlation parameter A_{xz} with the Wisconsin polarized proton beam at one energy and two angles. Combining our results with measurements of the proton polarization and differential cross-section data, the search for the phase-shift parameters arrives at a solution in which the parameters are determined with reasonable accuracy with one important exception, which we will come back to in a moment. Fig. 2 shows the S- and P-wave phase shifts in p-^{3}He elastic scattering. The phases are taken to be real, which means breakup processes are ignored. The dots from 4 to 11 MeV show the results of searches at energies for which ^{3}He polarization data were taken at Rice. Error estimates are shown where they have been made. The

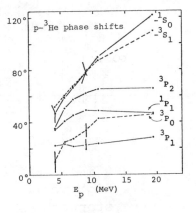

Fig. 2

S-waves look very much like hard sphere scattering, and the P-waves exhibit resonances that were first pointed out by Tombrello [6] and provide experimental evidence for a quartet of T = 1, negative parity levels in the mass-4 system. The higher partial waves contribute very little. An important parameter *not* shown is the amount of mixing between the $^1P_1$ and $^3P_1$ waves. Morrow and Haeberli [7] pointed out that this mixing parameter is very poorly defined by the data, and our search corroborates this. The points at 19.4 MeV are the results of fitting proton- and ^{3}He-polarization data and the spin-correlation parameters A_{yy} and A_{xx} measured at Saclay by Cahill, Catillon, Durand, Garreta, and myself [8]. At this energy the P-wave mixing parameter seems to be better determined, and it indicates very nearly equal mixing. It now appears that further experiments at lower energies, where the resonances actually occur, would be of help in understanding the nature of the states in ^{4}Li.

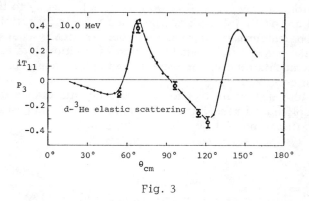

Fig. 3

d-^{3}He elastic scattering and the (d,p) reaction. As an example of our data, figs. 3 and 4 show the comparison of the polarized deuteron data from Wisconsin [9] and our ^{3}He polarization data [10]. In the elastic scattering angular distributions (fig. 3) the value of the ^{3}He analyzing power P_3 (shown as open circles) and the deuteron vector analyzing power iT_{11} (shown as a curve drawn through experimental points) are very nearly identical. A similar statement is true of the (d,p) reaction (fig. 4). Here the deuteron vector analyzing power (shown as a solid line) is normalized somewhat differently, $P_d = (2/3) \sqrt{3}\, iT_{11}$, but the similarity with the ^{3}He analyzing power P_3 (shown as points with error bars) is striking. The dashed line is the negative of

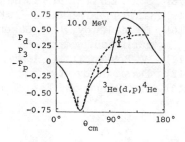

Fig. 4

the proton polarization P_p produced in the (d,p) reaction. The forward
angle P_3 data is old; the back-angle points are new and were picked
up during the elastic scattering measurements. Previous fits to this
sort of data have been fairly crude, and perhaps we can expect new
theoretical treatments to be attempted soon.

   ^3He-^3He *elastic scattering*. The data are skimpy ($60°_{c.m.}$ at E_{lab}
from 4.89 to 9.25 MeV, and 66° at E_{lab} from 9.27 to 17.53 MeV) and so
is the polarization, which is consistent with zero within statistical
error (typically ±0.01 in the lower energy range and ±0.05 in the higher
range). This is consistent with spin-independent fits to the differential cross-section data on this process.

   ^3He-^4He *elastic scattering*. A report [11] on this interaction below
an equivalent ^{3}He laboratory energy of 4.8 MeV is being presented at
this meeting, and the reader is referred to it. At higher energies,
Hardy, Spiger, and I have collaborated with Tombrello and Chen at
Cal Tech to measure the analyzing power up to 13.5 MeV equivalent
^{3}He laboratory energy [12]. This work is complementary to the Los
Alamos [2] and TUNL [3] work and provides a basis for refining the
polarization contour map somewhat without drastically changing the
values of the phase shifts found using cross-section data alone [13].
At the moment this reaction appears to be the best analyzer for a beam
or target of polarized ^{3}He.

   In conclusion, I would like to acknowledge with thanks the contribution of colleagues named above and also the aid and participation
of G. C. Phillips and G. K. Walters in this work.

REFERENCES

   [1]   G. C. Phillips, Second Polarization Symp., p. 113.
   [2]   D. D. Armstrong et al., Phys. Rev. Lett. 23 (1969) 135.
   [3]   W. McEver et al., Phys. Lett. 31B (1970) 560.
   [4]   D. H. McSherry and S. D. Baker, Phys. Rev. C1 (1970) 888.
   [5]   D. H. McSherry et al., Nucl. Phys. A126 (1969) 233.
   [6]   T. A. Tombrello, Phys. Rev. 138 (1965) B40.
   [7]   L. W. Morrow and W. Haeberli, Nucl. Phys. A126 (1969) 225.
   [8]   S. D. Baker et al., to be published.
   [9]   G. R. Plattner and L. G. Keller, Phys. Lett. 29B (1969) 301.
  [10]   D. M. Hardy, S. D. Baker, D. H. McSherry, and R. J. Spiger, to be
         published.
  [11]   W. Boykin, D. M. Hardy, and S. D. Baker, Third Polarization Symp.
  [12]   D. M. Hardy et al., Phys. Lett. 31B (1970) 355.
  [13]   R. J. Spiger and T. A. Tombrello, Phys. Rev. 163 (1967) 964.

An Experiment to Compress Polarized ³He Gas

R. S. TIMSIT, J. M. DANIELS, E. I. DENNIG, A. K. C. KIANG,
P. KIRKBY, and A. D. MAY, University of Toronto, Ontario, Canada

At Toronto we are building a polarized ³He gaseous target. The high density is the important feature of the target. The gas will be at a pressure of one atmosphere compressed from the pressure of 1 Torr used for producing the maximum polarization.

Measurements are described showing that the polarized gas has been successfully compressed to a pressure of 217 Torr and the gas remains polarized. This is part of the preliminary work of investigating the properties of polarized ³He gas leading to the high density target.

The ³He gas is polarized by optical pumping [1]. Polarizations up to 25% are achieved at 1 Torr, which is the pressure for producing optimum polarization. However, once the gas is polarized, it may be compressed to produce a target of high density.

Relaxation of the polarized ³He places strict requirements on the design of the apparatus for compression of polarized ³He gas. This is seen from the following three requirements:

1. The magnetic field must be homogeneous for all sections containing the polarized ³He, for field gradients are a major factor determining the relaxation process. Estimates show that the relaxation time is 1 hour at a pressure of 1 Torr for relative field gradients of 10^{-3} cm^{-1}

2. The helium must be pure. The limit on impurities is approximately 1 part in 10^6.

3. The material containing the ³He must not cause rapid relaxation.

To compress ³He we use a Toepler pump which is made of 1720 glass (relaxation time $\simeq 10^5$ sec) and uses distilled mercury (relaxation time $\simeq 6 \times 10^4$ sec). This pump is controlled externally. The relaxation produced by AC fields generated using this method has been measured and compares well with estimates. The fields have negligible effect on the relaxation time.

Fig. 1 illustrates the arrangement to measure effects when ³He was compressed. The polarizing cell C was a pyrex cell approximately

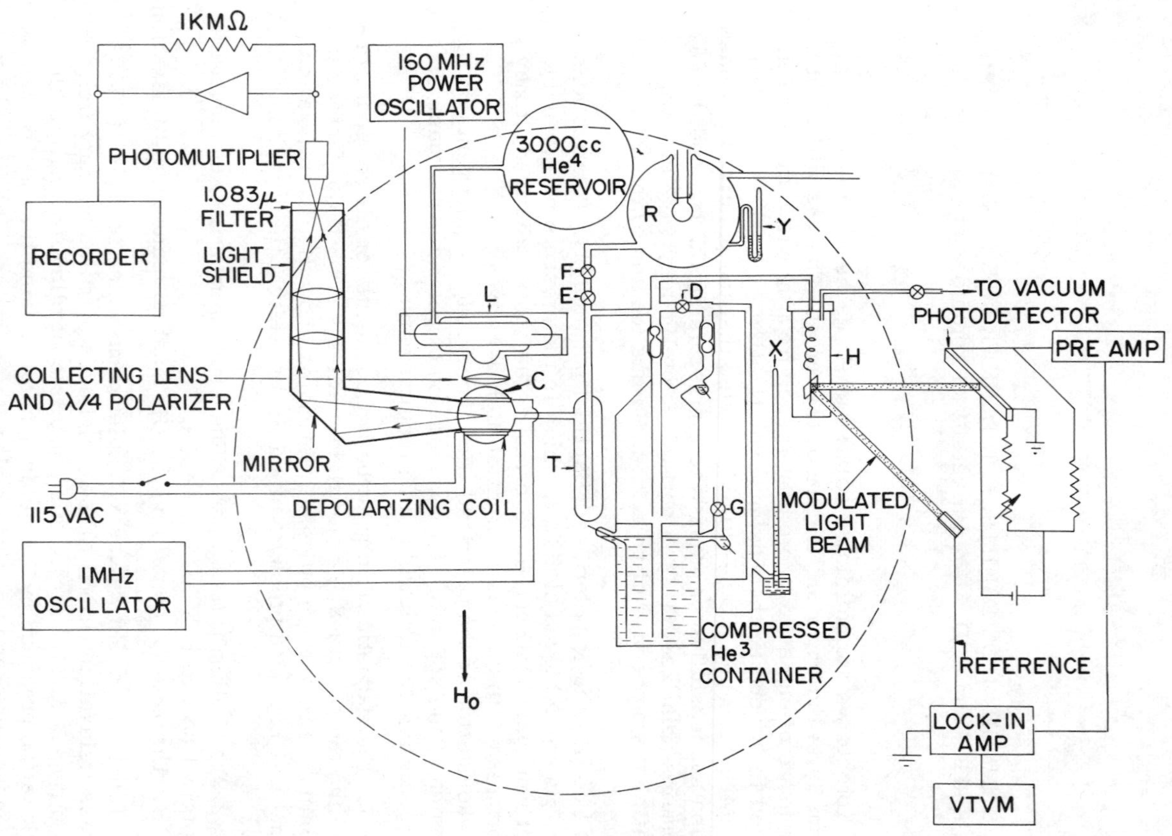
1KMΩ
160 MHz POWER OSCILLATOR
PHOTOMULTIPLIER
3000cc He4 RESERVOIR
1.083μ FILTER
R
Y
RECORDER
LIGHT SHIELD
F
E
L
D
TO VACUUM
PHOTODETECTOR
PRE AMP
X
H
COLLECTING LENS AND λ/4 POLARIZER
C
MIRROR
T
115 VAC
DEPOLARIZING COIL
G
MODULATED LIGHT BEAM
1MHz OSCILLATOR
COMPRESSED He3 CONTAINER
REFERENCE
H0
LOCK-IN AMP
VTVM

Fig. 1

100 cm³ in volume. The high pressure side consisted of a 60 cm³
glass container. The entire system was mounted inside a homogenous
field electromagnet. After the glass was baked out ³He gas was introduced to cell C and polarized. The gas was compressed. A compression cycle consisted of 7 or 8 strokes of the Toepler pump whereby
the pressure in the polarizing cell was reduced in 3 minutes to 0.2 Torr.
More unpolarized gas was then introduced and polarized. The polarization of the optically pumped sample of gas was checked several
times and the compression cycle repeated. After 81 minutes of repeated compression cycles, the gas had been compressed to 90 Torr.
The polarization of the high pressure ³He gas was measured by leak-
ing some of the ³He to a pressure of 1 Torr again. This revealed no
change in the polarization of the sample from the original 4%. The
low value of polarization is attributed to outgasing from the glass
walls.

Compression cycles were continued until the pressure reached
217 Torr. The polarization was found to have dropped to 3% indicating
a slight relaxation of the sample in the 3 hours separating the measurements.

We are in the process of repeating the above measurements incorporating improvements which include measurement of the polarization
of the compressed ³He using an NMR cell.

REFERENCE

[1] F. D. Colegrove, L. D. Schearer, and G. K. Walter, Phys. Rev. 132
     (1963) 2561.

Determination of the Polarization
of a Polarized ³He Gas Target
by the Pulsed NMR Technique

W. KLINGER and N. HAUER, Physikalisches Institut der Universität
Erlangen-Nürnberg, Germany

The degree of polarization of a ³He gas target, polarized by optical
pumping, may be determined by measuring the change in the absorption
of the pumping light in the ³He gas cell [1, 2]. This method requires
an exact knowledge of the spectral line of the pumping light in the
absorption spectrum of the metastable ³He atoms. In addition, one
must consider parameters which are hard to determine accurately,
such as the alignment of the pumping light with the axis of quantization, polarization of the pumping light, reabsorption, and saturation
effects in the pumping region. Since a high degree of ³He polarization
can be obtained only if the discharge for the production of metastable
³He atoms in the pumping cell is very weak, the change in the absorption of the pumping light also becomes very small and a poor signal-
to-noise ratio results.

 Much greater accuracy can be achieved with the NMR-technique.
We are using pulsed NMR for measuring the ³He polarization because
it has several advantages over the method of slow or fast passage
NMR as used in refs. [1, 3], the significantly higher signal-to-noise
ratio being the most important. Also, no assumption on line shape and
relaxation time constants needs to be made as in slow passage NMR,
and the experimental conditions are not so difficult to satisfy as in fast
passage NMR. In the case of pulsed NMR, the degree of polarization
is directly proportional to the maximum amplitude of the free induction
signal following a 90°-rf-pulse and can be easily calibrated by the
NMR signal of a proton-containing sample. In our NMR spectrometer
a polarization of 1% is sufficient to give a signal-to-noise ratio of 1
for a ³He sample of 10 cm³ at 1 Torr pressure. Thus in a sample of
20 cm³ at 5 Torr, polarization above $P = 5\%$ can be determined with an
accuracy greater than $\Delta P/P = 0.02$.

 Fig. 1 shows the experimental set-up. The NMR spectrometer works
at a frequency of 400 kHz, corresponding to a magnetic field of 123.5
Oe for ³He nuclei. The spectrometer is capable of producing single
rf-pulses from 5 to 100 μsec in steps of 5 μsec, as well as Carr-
Purcell [4] series with a Gill-Meiboom correction. Typically, 50 μsec

906

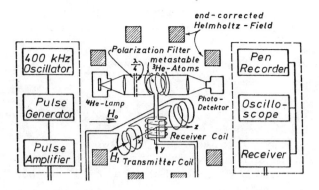

Fig. 1

were used for a 90°-pulse. Fig. 2 shows the decay of an NMR signal following a 90°-pulse. The ³He sample is a dumbbell-shaped container of Pyrex glass. In its upper part polarized ³He gas is produced by optical pumping. The polarization reaches the lower part of the ³He sample by diffusion through a connecting tube and can be measured there by the NMR detection head.

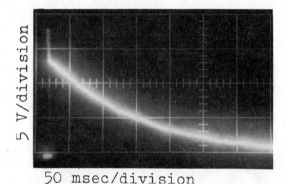

Fig. 2. NMR signal of 1 Torr–40 cm³ ³He sample. $H_0 = 123.5$ Oe, $P = 15\%$.

Since the polarization is completely destroyed every time it is measured by a 90°-rf-pulse, the pulsed NMR technique is not suited for use as a continuous polarization monitor for a polarized ³He gas target. However, this problem can be avoided by connecting the actual target and the cell in the NMR detector by means of a capillary tube. The dimensions of the capillary tube (for instance 5 cm × 0.02 cm) can be chosen in such a way that the time constant which governs the equilibration of the polarization between target and detector caused by diffusion is still smaller than the ³He relaxation time, but becomes much larger than the optical pumping time. Thus every few minutes a polarization measurement can be carried out without destroying the polarization in the target.

REFERENCES

[1] F. D. Colegrove, L. D. Schearer, and G. K. Walters, Phys. Rev. 94 (1963) 630.
[2] R. C. Greenhow, Phys. Rev. 136A (1964) 660.
[3] R. L. Gamblin and Th. R. Carver, Phys. Rev. 138 (1965) 946.
[4] S. Meiboom and D. Gill, Rev. Sci. Instr. 29 (1958) 688.

The Influence of the Absorption of the
Pumping Light on the Degree of Polarization
of an Optically Pumped He Target

H. BRÜCKMANN, W. LÜCK, and F. K. SCHMIDT, Institut für
Experimentelle Kernphysik des Kernforschungszentrums und
der Universität Karlsruhe, Germany

The degree of polarization observed experimentally in an optically
pumped ^{3}He target does not agree with the prediction by Colegrove et
al. [1] regarding its dependence on the intensity of the pumping light.
So far the question of the influence of the pumping light intensity on
the polarization has not been answered and only one theoretical approach [1] to this problem has been published. This approach assumes
a weak absorption. A high absorption of the pumping light by the
metastable atoms might be able to explain the discrepancy between
the theoretical approach and the experiments.

The dependence of the pumping light intensity on the range of the
light in a He gas discharge was investigated experimentally. Light
from a ^{4}He lamp was circularly polarized and sent along the axis
through a cylindrical glass vessel. The vessel contained ^{4}He gas at
a pressure of either 0.05 Torr or 7 Torr. The vessel was 20 cm in
length and 7 cm in diameter. Metastable He atoms were produced by
a weak rf gas discharge. The frequency and the geometry of the electrodes were adjusted carefully in order to obtain a constant density
of metastable atoms along the axis of the cell. With the exception
of the regions close to the entrance and exit windows this condition
was well satisfied. The intensity of the fluorescently scattered light
was taken to be a measure of the local variation of the pumping light
intensity. The fluorescent light emitted at 90° with respect to the
axis of the cell was observed by a movable phototube.

The figure shows the result of such a measurement. The intensity
of the fluorescent light is plotted on a logarithmic scale versus the
distance from the entrance window. The values measured in each of
the two experiments shown are very well represented by straight lines.
The two experiments differed only by the density of the metastable He
atoms. The pressure of the ^{4}He gas was 7 Torr in both cases. The
pure exponential decrease leads to the conclusion that the decay of
only one of the two pumping light components is measured. In ref. [1]
the components are denoted by D_0 and D_3. In a separate measurement
of the polarization of the fluorescent light it was confirmed that only

the D_3 component was present. The mean free path λ of the photons was obtained from the slope of the decay curve. At the lowest discharge level possible the mean free path is $\lambda = 10.8 \pm 0.4$ cm while at a slightly higher discharge level it turns out to be $\lambda = 9.2 \pm 0.4$ cm. The mean free path in ^{3}He gas can easily be evaluated from the ^{4}He data. In this way for ^{3}He a value $\lambda = 13.5 \pm 0.5$ cm was obtained corresponding to the lowest discharge level.

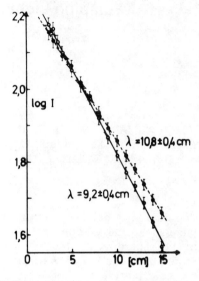

The measurements discussed exhibit such a high photon absorption that the approximation of weak absorption cannot be valid. The mean free path of the photons is comparable to the dimensions of the absorption cell. Therefore reabsorption of fluorescently emitted photons has to be taken into account. This effect decreases the degree of polarization. In addition to the polarized pumping light, a fraction of unpolarized fluorescent light acts in the pumping process. This is equivalent to an additional relaxation mechanism which reduces the maximum obtainable target polarization.

REFERENCE

[1]   F. D. Colegrove, L. D. Schearer, and G. K. Walters, Phys. Rev. 132 (1963) 2561.

DISCUSSION

Baker:
      Have you estimated the error that this effect has on the optical measurement of the degree of polarization?

Brückmann:
      Our estimate is that this rather strong absorption does not affect the optical measurement of the degree of polarization appreciably ($< 5\%$). It limits mainly the maximum achievable polarization.

Participants

R. G. Allas
Naval Research Laboratory
Washington, D.C., USA

B. D. Anderson
Case Western Reserve University
Cleveland, Ohio, USA

J. D. Anderson
Lawrence Radiation Laboratory
Livermore, California, USA

S. Apostolescu
Institute for Atomic Physics
Bucharest, Rumania

D. D. Armstrong
Los Alamos Scientific Laboratory
Los Alamos, New Mexico, USA

J. Arvieux
Institut des Sciences Nucléaires
Grenoble, France

A. D. Bacher
Lawrence Radiation Laboratory
Berkeley, California, USA

S. D. Baker
Rice University
Houston, Texas, USA

F. C. Barker
Michigan State University
East Lansing, Michigan, USA

H. H. Barschall
University of Wisconsin
Madison, Wisconsin, USA

D. J. Baugh
Rutherford High Energy Laboratory
Chilton, Berkshire, England

F. D. Becchetti, Jr.
Niels Bohr Institute
Copenhagen, Denmark

P. J. Bendt
Los Alamos Scientific Laboratory
Los Alamos, New Mexico, USA

R. E. Benenson
State University of New York
Albany, New York, USA

J. A. Benjamin
Brookhaven National Laboratory
Upton, New York, USA

E. M. Bernstein
Western Michigan University
Kalamazoo, Michigan, USA

R. Beurtey
Centre d'Etudes Nucléaires
Saclay, France

G. A. Bissinger
University of North Carolina
Chapel Hill, North Carolina, USA

J. S. Blair
University of Washington
Seattle, Washington, USA

C. H. Blanchard
University of Wisconsin
Madison, Wisconsin, USA

E. Bleuler
Pennsylvania State University
University Park, Pennsylvania,
    USA

R. A. Blue
University of Florida
Gainesville, Florida, USA

H. J. Boersma
Vrije Universiteit
Amsterdam, The Netherlands

R. O. Bondelid
Naval Research Laboratory
Washington, D.C., USA

R. R. Borchers
University of Wisconsin
Madison, Wisconsin, USA

E. T. Boschitz
University of Karlsruhe
Karlsruhe, Germany

G. Breit
State University of New York
Buffalo, New York, USA

F. D. Brooks
University of Cape Town
Cape Town, South Africa

W. B. Broste
Los Alamos Scientific Laboratory
Los Alamos, New Mexico, USA

L. Brown
Carnegie Institution
Washington, D.C., USA

H. Brückmann
University of Karlsruhe
Karlsruhe, Germany

S. N. Bunker
University of Manitoba
Winnipeg, Canada

V. Burke
Case Western Reserve University
Cleveland, Ohio, USA

C. E. Busch
Ohio State University
Columbus, Ohio, USA

W. Busse
Hahn-Meitner-Institut
Berlin, Germany

T. A. Cahill
University of California
Davis, California, USA

P. Catillon
Centre d'Etudes Nucléaires
Saclay, France

R. Ceuleneer
Faculty of Science
Mons, Belgium

J. W. Chien
University of Minnesota
Minneapolis, Minnesota, USA

F. Chmara
High Voltage Engineering
    Corporation
Burlington, Massachusetts, USA

G. Clausnitzer
University of Giessen
Giessen, Germany

B. DeFacio
University of Missouri
Columbia, Missouri, USA

T. B. Clegg
University of North Carolina
Chapel Hill, North Carolina, USA

P. Delpierre
Collège de France
Paris, France

H. E. Conzett
Lawrence Radiation Laboratory
Berkeley, California, USA

R. de Swiniarski
Institut des Sciences Nucléaires
Grenoble, France

S. A. Cox
Argonne National Laboratory
Argonne, Illinois, USA

B. L. Donnally
Lake Forest College
Lake Forest, Illinois, USA

J. G. Cramer, Jr.
University of Washington
Seattle, Washington, USA

T. R. Donoghue
Ohio State University
Columbus, Ohio, USA

L. Cranberg
University of Virginia
Charlottesville, Virginia, USA

M. Drosg
Los Alamos Scientific Laboratory
Los Alamos, New Mexico, USA

C. J. Csikai
Kossuth University
Debrecen, Hungary

J. C. Duder
University of Auckland
Auckland, New Zealand

B. Cujec
Laval University
Quebec, Canada

M. E. Ebel
University of Wisconsin
Madison, Wisconsin, USA

J. W. T. Dabbs
Oak Ridge National Laboratory
Oak Ridge, Tennessee, USA

H. Ebinghaus
Institut für Experimentalphysik
Hamburg, Germany

S. E. Darden
University of Notre Dame
Notre Dame, Indiana, USA

J. S. Eck
Kansas State University
Manhattan, Kansas, USA

J. C. Davis
University of Wisconsin
Madison, Wisconsin, USA

B. Efken
Hahn-Meitner-Institut
Berlin, Germany

R. H. Davis
Florida State University
Talahasse, Florida, USA

D. Ehrlich
University of Munich
Munich, Germany

J. L. Escudié
Centre d'Etudes Nucléaires
Saclay, France

H. Feshbach
Massachusetts Institute of
      Technology
Cambridge, Massachusetts, USA

F. W. K. Firk
Yale University
New Haven, Connecticut, USA

T. R. Fisher
Lockheed Research Laboratory
Palo Alto, California, USA

W. Fitz
Institut für Experimentalphysik
Hamburg, Germany

R. Fleischmann
University of Erlangen–Nürnberg
Erlangen, Germany

W. A. Friedman
Princeton University
Princeton, New Jersey, USA

R. B. Galloway
University of Edinburgh
Edinburgh, Scotland

W. R. Gibson
University of London
London, England

C. Glashausser
Rutgers State University
New Brunswick, New Jersey, USA

H. F. Glavish
University of Auckland
Auckland, New Zealand

L. J. B. Goldfarb
The University
Manchester, England

B. Gonsior
University of Cologne
Cologne, Germany

G. V. Gorlov
Kurchatov Atomic Energy Institute
Moscow, USSR

S. Gorodetzky
Institut de Recherches Nucléaires
Strasbourg-Cronenbourg, France

G. Graw
University of Erlangen–Nürnberg
Erlangen, Germany

B. Greenebaum
University of Wisconsin–Parkside
Kenosha, Wisconsin, USA

G. W. Greenlees
University of Minnesota
Minneapolis, Minnesota, USA

J. A. R. Griffith
University of Birmingham
Birmingham, England

E. E. Gross
Oak Ridge National Laboratory
Oak Ridge, Tennessee, USA

K. Grotowski
Institute for Nuclear Physics
Cracow, Poland

W. Grüebler
Laboratorium für Kernphysik, ETH
Zürich, Switzerland

H. H. Hackenbroich
University of Cologne
Cologne, Germany

W. Haeberli
University of Wisconsin
Madison, Wisconsin, USA

R. L. Hagengruber
University of Wisconsin
Madison, Wisconsin, USA

G. M. Hale
Los Alamos Scientific Laboratory
Los Alamos, New Mexico, USA

R. A. Hardekopf
Duke University
Durham, North Carolina, USA

H. R. Hiddleston
University of Notre Dame
Notre Dame, Indiana, USA

D. Hilscher
University of Wisconsin
Madison, Wisconsin, USA

J. C. Hopkins
Los Alamos Scientific Laboratory
Los Alamos, New Mexico, USA

P. Huber
University of Basel
Basel, Switzerland

A. Ingemarsson
University of Uppsala
Uppsala, Sweden

E. V. Ivash
University of Texas
Austin, Texas, USA

C. D. Jeffries
University of California
Berkeley, California, USA

H. P. Jochim
Max-Planck-Institut
Mainz, Germany

R. C. Johnson
University of Surrey
Guildford, England

M. S. Kaminsky
Argonne National Laboratory
Argonne, Illinois, USA

K. Katori
Tokyo University of Education
Tokyo, Japan

M. Kawai
State University of New York
Stony Brook, New York, USA

J. A. Keane
Ohio State University
Columbus, Ohio, USA

P. W. Keaton, Jr.
Los Alamos Scientific Laboratory
Los Alamos, New Mexico, USA

G. A. Keyworth
Los Alamos Scientific Laboratory
Los Alamos, New Mexico, USA

K. Kilian
University of Erlangen-Nürnberg
Erlangen, Germany

P. Kirkby
Toronto University
Toronto, Canada

W. Klinger
University of Erlangen-Nürnberg
Erlangen, Germany

S. Kobayashi
Kyoto University
Kyoto, Japan

D. C. Kocher
University of Wisconsin
Madison, Wisconsin, USA

K. Kuroda
Faculté des Sciences
Orsay, France

R. O. Lane
Ohio University
Athens, Ohio, USA

A. Langsdorf, Jr.
Argonne National Laboratory
Argonne, Illinois, USA

G. P. Lawrence
Los Alamos Scientific Laboratory
Los Alamos, New Mexico, USA

D. D. Leavitt
University of Minnesota
Minneapolis, Minnesota, USA

W. T. Leland
Los Alamos Scientific Laboratory
Los Alamos, New Mexico, USA

J. R. Lemley
Los Alamos Scientific Laboratory
Los Alamos, New Mexico, USA

J. F. Lemming
Ohio University
Athens, Ohio, USA

H. S. Liers
University of Wisconsin
Madison, Wisconsin, USA

G. P. Lietz
DePaul University
Chicago, Illinois, USA

J. Lilley
University of Minnesota
Minneapolis, Minnesota, USA

J. M. Lohr
University of Wisconsin
Madison, Wisconsin, USA

R. S. Lord
Oak Ridge National Laboratory
Oak Ridge, Tennessee, USA

J. Lowe
University of Birmingham
Birmingham, England

E. J. Ludwig
University of North Carolina
Chapel Hill, North Carolina, USA

M. H. Mac Gregor
Lawrence Radiation Laboratory
Livermore, California, USA

G. Mack
University of Tübingen
Tübingen, Germany

D. Magnac-Valette
Laboratoire des Basses Energies, CRN
Strasbourg, France

G. J. Marmer
Argonne National Laboratory
Argonne, Illinois, USA

B. Mayer
Centre d'Etudes Nucléaires
Saclay, France

J. D. McCullen
University of Arizona
Tucson, Arizona, USA

J. S. C. McKee
University of Birmingham
Birmingham, England

J. L. McKibben
Los Alamos Scientific Laboratory
Los Alamos, New Mexico, USA

D. W. Miller
Indiana University
Bloomington, Indiana, USA

T. G. Miller
Redstone Arsenal
Huntsville, Alabama, USA

C. D. Moak
Oak Ridge National Laboratory
Oak Ridge, Tennessee, USA

J. E. Monahan
Argonne National Laboratory
Argonne, Illinois, USA

F. P. Mooring
Argonne National Laboratory
Argonne, Illinois, USA

G. C. Morrison
Argonne National Laboratory
Argonne, Illinois, USA

N. A. Mulvenon
Oak Ridge National Laboratory
Oak Ridge, Tennessee, USA

R. Nath
Yale University
New Haven, Connecticut, USA

J. M. Nelson
University of Manitoba
Winnipeg, Canada

D. J. Nicholas
Rutherford High Energy Laboratory
Chilton, Berkshire, England

A. Niiler
Los Alamos Scientific Laboratory
Los Alamos, New Mexico, USA

L. C. Northcliffe
Texas A & M University
College Station, Texas, USA

H. Oehler
Joint Institute for Nuclear Research
Dubna, USSR

G. G. Ohlsen
Los Alamos Scientific Laboratory
Los Alamos, New Mexico, USA

M. A. Oothoudt
University of Minnesota
Minneapolis, Minnesota, USA

A. Pascolini
University of Padua
Padua, Italy

N. J. Pattenden
Atomic Energy Research Establishment
Harwell, England

C. A. Pearson
University of Alabama
Birmingham, Alabama, USA

S. Penselin
University of Bonn
Bonn, Germany

F. G. Perey
Oak Ridge National Laboratory
Oak Ridge, Tennessee, USA

G. Pisent
University of Padua
Padua, Italy

G. R. Plattner
University of Basel
Basel, Switzerland

C. H. Poppe
University of Minnesota
Minneapolis, Minnesota, USA

H. Postma
University of Groningen
Groningen, The Netherlands

R. Potenza
University of Catania
Catania, Italy

E. Preikschat
University of Washington
Seattle, Washington, USA

R. M. Prior
University of Notre Dame
Notre Dame, Indiana, USA

L. W. Put
University of Groningen
Groningen, The Netherlands

P. A. Quin
University of Wisconsin
Madison, Wisconsin, USA

R. D. Rathmell
University of Wisconsin
Madison, Wisconsin, USA

G. Rawitscher
University of Connecticut
Storrs, Connecticut, USA

J. Raynal
Centre d'Etudes Nucléaires
Saclay, France

H. T. Richards
University of Wisconsin
Madison, Wisconsin, USA

J. R. Richardson
University of California
Los Angeles, California, USA

E. T. Ritter
U.S. Atomic Energy Commission
Washington, D.C., USA

A. B. Robbins
Rutgers State University
New Brunswick, New Jersey,
     USA

W. S. Rodney
National Science Foundation
Washington, D.C., USA

P. M. Rolph
University of Birmingham
Birmingham, England

S. Roman
University of Birmingham
Birmingham, England

G. W. Roth
University of Washington
Seattle, Washington, USA

M. Roth
University of Cologne
Cologne, Germany

G. Roy
University of Alberta
Edmonton, Canada

H. Rudin
University of Basel
Basel, Switzerland

E. Salzborn
University of Giessen
Giessen, Germany

F. D. Santos
Laboratório de Física e Engenharia
     Nucleares
Sacavém, Portugal

G. R. Satchler
Oak Ridge National Laboratory
Oak Ridge, Tennessee, USA

H. G. Schieck
Ohio State University
Columbus, Ohio, USA

P. Schiemenz
University of Munich
Munich, Germany

D. B. Schlafke
ORTEC
Oak Ridge, Tennessee, USA

E. W. Schmid
University of Tübingen
Tübingen, Germany

P. Schwandt
University of Indiana
Bloomington, Indiana, USA

R. E. Segel
Argonne National Laboratory
Argonne, Illinois, USA

F. Seibel
Los Alamos Scientific Laboratory
Los Alamos, New Mexico, USA

F. Seiler
University of Basel
Basel, Switzerland

J. P. F. Sellschop
University of Witwatersrand
Johannesburg, South Africa

R. Seltz
Centre de Recherches Nucléaires
Strasbourg, France

E. H. Sexton
State University of New York
Albany, New York, USA

R. G. Seyler
Ohio State University
Columbus, Ohio, USA

P. E. Shanley
University of Notre Dame
Notre Dame, Indiana, USA

H. S. Sherif
University of Alberta
Edmonton, Canada

V. Shkolnik
University of Minnesota
Minneapolis, Minnesota, USA

R. H. Siemssen
Argonne National Laboratory
Argonne, Illinois, USA

P. Signell
Michigan State University
East Lansing, Michigan, USA

J. E. Simmons
Los Alamos Scientific Laboratory
Los Alamos, New Mexico, USA

M. Simonius
Seminar für theoretische Physik, ETH
Zürich, Switzerland

D. G. Simons
U.S. Naval Ordnance Laboratory
Silver Spring, Maryland, USA

B. C. Sinha
University of London
London, England

I. V. Sizov
Joint Institute for Nuclear Research
Dubna, USSR

I. Slaus
University of California
Los Angeles, California, USA

R. J. Slobodrian
Laval University
Quebec, Canada

G. V. Solodukhov
Lebedev Institute of Physics
Moscow, USSR

G. Spalek
Duke University
Durham, North Carolina, USA

P. Spilling
Eindhoven University of
    Technology
Eindhoven, The Netherlands

P. J. Van Hall
Eindhoven University of
    Technology
Eindhoven, The Netherlands

V. S. Starkovich
Los Alamos Scientific Laboratory
Los Alamos, New Mexico, USA

L. R. Veeser
Los Alamos Scientific Laboratory
Los Alamos, New Mexico, USA

N. M. Stewart
Kings College
London, England

S. Vigdor
University of Wisconsin
Madison, Wisconsin, USA

H. R. Striebel
University of Basel
Basel, Switzerland

D. von Ehrenstein
Argonne National Laboratory
Argonne, Illinois, USA

M. Tanifuji
Hosei University
Tokyo, Japan

V. G. Vovchenko
Joffe Physical-Technical Institute
Leningrad, USSR

R. F. Taschek
Los Alamos Scientific Laboratory
Los Alamos, New Mexico, USA

R. L. Walter
Duke University
Durham, North Carolina, USA

M. A. Thompson
University of Wisconsin
Madison, Wisconsin, USA

J. Weber
University of Neuchâtel
Neuchâtel, Switzerland

S. T. Thornton
University of Virginia
Charlottesville, Virginia, USA

Ch. Weddigen
University of Karlsruhe
Karlsruhe, Germany

L. D. Tolsma
Eindhoven University of
    Technology
Eindhoven, The Netherlands

H. A. Weidenmüller
University of Heidelberg
Heidelberg, Germany

H. Treiber
University of Erlangen–Nürnberg
Erlangen, Germany

J. L. Weil
University of Kentucky
Lexington, Kentucky, USA

J. L. Underwood
University of Surrey
Guidford, Surrey, England

W. G. Weitkamp
University of Washington
Seattle, Washington, USA

D. Wells
University of Manitoba
Winnipeg, Canada

D. Werren
University of Geneva
Geneva, Switzerland

B. J. Wielinga
Institute for Nuclear Physics
    Research
Amsterdam, The Netherlands

K. Wienhard
University of Giessen
Giessen, Germany

E. P. Wigner
Princeton University
Princeton, New Jersey, USA

H. B. Willard
Case Western Reserve University
Cleveland, Ohio, USA

H. Wilsch
University of Erlangen–Nürnberg
Erlangen, Germany

R. M. Wood
University of Georgia
Athens, Georgia, USA

G. Zannoni
University of Padua
Padua, Italy

B. Zeitnitz
Institut für Experimentalphysik
Hamburg, Germany

E. Zijp
Vrije Universiteit
Amsterdam, The Netherlands

Subject Index

This index follows the same general order as the subject index of the Second Polarization Symp. (p. 531). Page numbers underlined are the first pages of survey papers.

6. Triton and helium-induced reactions

7. γ-ray induced reactions

Author Index